Quantum Machines

École de Physique des Houches

Session XCVI, 4–29 July 2011

Quantum Machines

Measurement and Control of Engineered Quantum Systems

Edited by

Michel Devoret, Benjamin Huard,
Robert Schoelkopf, Leticia F. Cugliandolo

OXFORD
UNIVERSITY PRESS

OXFORD
UNIVERSITY PRESS

Great Clarendon Street, Oxford, OX2 6DP,
United Kingdom

Oxford University Press is a department of the University of Oxford. It furthers the University's objective of excellence in research, scholarship, and education by publishing worldwide. Oxford is a registered trade mark of Oxford University Press in the UK and in certain other countries

First Edition published in 2014
Impression: 1

Published in the United States of America by Oxford University Press
198 Madison Avenue, New York, NY 10016, United States of America

British Library Cataloguing in Publication Data
Data available

Library of Congress Control Number: 2013958193

ISBN 978–0–19–968118–1

École de Physique des Houches

Service inter-universitaire commun
à l'Université Joseph Fourier de Grenoble
et à l'Institut National Polytechnique de Grenoble

Subventionné par l'Université Joseph Fourier de Grenoble,
le Centre National de la Recherche Scientifique,
le Commissariat à l'Énergie Atomique

Directeur:
Leticia F. Cugliandolo, Université Pierre at Marie Curie – Paris VI, France

Directeurs scientifiques de la session XCVI:
Michel Devoret, Department of Applied Physics, Yale University, New Haven, Connecticut, USA
Benjamin Huard, Laboratoire Pierre Aigrain, Ecole Normale Supérieure, Paris, France
Robert Schoelkopf, Department of Applied Physics, Yale University, New Haven, Connecticut, USA
Leticia Cugliandolo, LPTHE, Université Paris VI, Paris, France

Previous sessions

I	1951	Quantum mechanics. Quantum field theory
II	1952	Quantum mechanics. Statistical mechanics. Nuclear physics
III	1953	Quantum mechanics. Solid state physics. Statistical mechanics. Elementary particle physics
IV	1954	Quantum mechanics. Collision theory. Nucleon-nucleon interaction. Quantum electrodynamics
V	1955	Quantum mechanics. Non equilibrium phenomena. Nuclear reactions. Interaction of a nucleus with atomic and molecular fields
VI	1956	Quantum perturbation theory. Low temperature physics. Quantum theory of solids. Ferromagnetism
VII	1957	Scattering theory. Recent developments in field theory. Nuclear and strong interactions. Experiments in high energy physics
VIII	1958	The many body problem
IX	1959	The theory of neutral and ionized gases
X	1960	Elementary particles and dispersion relations
XI	1961	Low temperature physics
XII	1962	Geophysics; the earths environment
XIII	1963	Relativity groups and topology
XIV	1964	Quantum optics and electronics
XV	1965	High energy physics
XVI	1966	High energy astrophysics
XVII	1967	Many body physics
XVIII	1968	Nuclear physics
XIX	1969	Physical problems in biological systems
XX	1970	Statistical mechanics and quantum field theory
XXI	1971	Particle physics
XXII	1972	Plasma physics
XXIII	1972	Black holes
XXIV	1973	Fluids dynamics
XXV	1973	Molecular fluids
XXVI	1974	Atomic and molecular physics and the interstellar matter
XXVII	1975	Frontiers in laser spectroscopy
XXVIII	1975	Methods in field theory
XXIX	1976	Weak and electromagnetic interactions at high energy
XXX	1977	Nuclear physics with heavy ions and mesons
XXXI	1978	Ill condensed matter
XXXII	1979	Membranes and intercellular communication
XXXIII	1979	Physical cosmology

XXXIV	1980	Laser plasma interaction
XXXV	1980	Physics of defects
XXXVI	1981	Chaotic behavior of deterministic systems
XXXVII	1981	Gauge theories in high energy physics
XXXVIII	1982	New trends in atomic physics
XXXIX	1982	Recent advances in field theory and statistical mechanics
XL	1983	Relativity, groups and topology
XLI	1983	Birth and infancy of stars
XLII	1984	Cellular and molecular aspects of developmental biology
XLIII	1984	Critical phenomena, random systems, gauge theories
XLIV	1985	Architecture of fundamental interactions at short distances
XLV	1985	Signal processing
XLVI	1986	Chance and matter
XLVII	1986	Astrophysical fluid dynamics
XLVIII	1988	Liquids at interfaces
XLIX	1988	Fields, strings and critical phenomena
L	1988	Oceanographic and geophysical tomography
LI	1989	Liquids, freezing and glass transition
LII	1989	Chaos and quantum physics
LIII	1990	Fundamental systems in quantum optics
LIV	1990	Supernovae
LV	1991	Particles in the nineties
LVI	1991	Strongly interacting fermions and high Tc superconductivity
LVII	1992	Gravitation and quantizations
LVIII	1992	Progress in picture processing
LIX	1993	Computational fluid dynamics
LX	1993	Cosmology and large scale structure
LXI	1994	Mesoscopic quantum physics
LXII	1994	Fluctuating geometries in statistical mechanics and quantum field theory
LXIII	1995	Quantum fluctuations
LXIV	1995	Quantum symmetries
LXV	1996	From cell to brain
LXVI	1996	Trends in nuclear physics, 100 years later
LXVII	1997	Modeling the earths climate and its variability
LXVIII	1997	Probing the Standard Model of particle interactions
LXIX	1998	Topological aspects of low dimensional systems
LXX	1998	Infrared space astronomy, today and tomorrow
LXXI	1999	The primordial universe
LXXII	1999	Coherent atomic matter waves
LXXIII	2000	Atomic clusters and nanoparticles
LXXIV	2000	New trends in turbulence
LXXV	2001	Physics of bio-molecules and cells
LXXVI	2001	Unity from duality: Gravity, gauge theory and strings

LXXVII	2002	Slow relaxations and nonequilibrium dynamics in condensed matter
LXXVIII	2002	Accretion discs, jets and high energy phenomena in astrophysics
LXXIX	2003	Quantum entanglement and information processing
LXXX	2003	Methods and models in neurophysics
LXXXI	2004	Nanophysics: Coherence and transport
LXXXII	2004	Multiple aspects of DNA and RNA
LXXXIII	2005	Mathematical statistical physics
LXXXIV	2005	Particle physics beyond the Standard Model
LXXXV	2006	Complex systems
LXXXVI	2006	Particle physics and cosmology: the fabric of spacetime
LXXXVII	2007	String theory and the real world: from particle physics to astrophysics
LXXXVIII	2007	Dynamos
LXXXIX	2008	Exact methods in low-dimensional statistical physics and quantum computing
XC	2008	Long-range interacting systems
XCI	2009	Ultracold gases and quantum information
XCII	2009	New trends in the physics and mechanics of biological systems
XCIII	2009	Modern perspectives in lattice QCD: quantum field theory and high performance computing
XCIV	2010	Many-body physics with ultra-cold gases
XCV	2010	Quantum theory from small to large scales
XCVI	2011	Quantum machines: measurement and control of engineered quantum systems

Publishers

- Session VIII: Dunod, Wiley, Methuen
- Sessions IX and X: Herman, Wiley
- Session XI: Gordon and Breach, Presses Universitaires
- Sessions XII–XXV: Gordon and Breach
- Sessions XXVI–LXVIII: North Holland
- Session LXIX–LXXVIII: EDP Sciences, Springer
- Session LXXIX–LXXXVIII: Elsevier
- Session LXXXIX– : Oxford University Press

Preface

What is a quantum machine? Can't we say that devices based on lasers and transistors are quantum machines? After all, physicists advertise the laser and the transistor as the two main spin-offs of the understanding of quantum mechanical phenomena. However, while quantum mechanics must be used to predict the wavelength of a laser and the operation voltage of a transistor, it does not intervene at the level of the signals processed by these active systems. Signals involve macroscopic collective variables such as voltages and currents in a circuit or the amplitude of the oscillating electric field in an electromagnetic cavity resonator. In a true quantum machine, these collective variables that can be controlled and measured and define the state of the machine are themselves quantum operators, just as the position of the electron in a hydrogen atom. Mesoscopic artificial atoms or molecules such as quantum superconducting circuits, quantum dots, or quantum nanomechanical resonators, satisfy the definition of quantum machines. These systems exhibit a few collective dynamical variables, whose fluctuations are well in the quantum regime and whose measurement is essentially limited in precision by the Heisenberg uncertainty principle. Engineered quantum systems derived from quantum optics and based on natural, rather than artificial, atoms can also qualify as quantum machines: examples are trapped ions, single Rydberg atoms in superconducting cavities, and lattices of ultracold atoms.

The goal of this school was to equip students with the basic knowledge needed to investigate and understand all these various novel engineered quantum systems. Because these systems are so much linked to quantum information processing, some of the themes covered by the school could be seen as a continuation of the 2003 Les Houches School on Quantum Entanglement and Information Processing. But the school introduced and discussed also a novel aspect of quantum control, namely quantum feedback. This topic was mainly treated in the course given by Hideo Mabuchi (Chapter 1) who first explained the crucial distinction between measurement-based feedback and autonomous feedback. In the former, a detector performs a measurement of the variable to be controlled, and information passes from the quantum system to a classical system that processes it in order to output a correction signal sent to an actuator acting back on the quantum system. In the latter, the detector-processor-actuator part of the controller is another quantum system coupled to the controlled system and no information needs to pass through classical levels. As an element essential to the detailed understanding of measurement-based feedback, and the design of the filters for the corresponding controller, the course treated the useful formalism of the quantum stochastic master equation. The autonomous aspect of quantum feedback were then illustrated by the discussion of analog, continuous error correction of a qubit by quantum optics elements.

Quantum noise and quantum measurements were another main theme much discussed at the school. Aash Clerk gave a course on this topic (Chapter 2) that covered the basics and the subtle aspects of Caves' theorem on quantum-limited amplifiers. The course introduced on the one hand the susceptibilities of a quantum measurement system and on the other the crucial Nyquist model of dissipation, which is essential to the understanding of the quantum fluctuation-dissipation theorem for a passive system and the constraints on noise for an active system. Great insight on the paradoxical properties of quantum noise can be obtained by comparing, in the calculation of the noise of an active circuit, the point of view of susceptibilities and response functions with the point of view of input-output theory dominated by the formalism of the scattering matrix. In particular, the course provided an explanation of the fact that the optimum working point for a nonmatched amplifier is not the same if one optimizes the gain or the noise.

Quantum information processing with a superconducting quantum circuit was covered in the course given by Steve Girvin (Chapter 3). The lectures carefully showed how both the superconducting gap and long-range Coulomb interactions contribute to the existence of strongly underdamped collective modes, which can be quantized in superconducting circuits. The course also gave an in-depth coverage of quantum harmonic oscillators in all their glory from various perspectives. Superconducting qubits were then introduced, starting with the fundamental theory of the Josephson effect to the presentation of the zoology of existing superconducting qubits. Finally, the opportunities offered for quantum information processing by enclosing a superconducting qubit in a cavity were presented.

Quantum processing systems based on the phase qubit introduced by Steve Girvin were then discussed in the course given by John Martinis. He gave an introduction to a processing architecture developed at UC Santa Barbara in which the tuning capability of the phase qubit is used to shuttle information between different cavities. Such manipulations are also at the root of the construction of an arbitrary superposition of Fock states in a cavity coupled to a qubit, or the cooling of a nanoresonator in its ground state. He then emphasized the importance of the fidelity of quantum logic gates which is the subject of Chapter 4. More on the phase qubit and qubits derived from it could be learned later in the school in the seminar given by Olivier Buisson (Chapter 19).

The subject of circuit QED covered by Steve Girvin in his lectures was further extended in the seminar by Andreas Wallraff (Chapter 15), who treated the analysis of propagating microwave signals by splitters and amplifiers and showed that one can recover enough information to reconstruct the Wigner function of a microwave quantum field emitted by a qubit. The special properties of quantum matter-light interaction in 1D illustrated by a superconducting artificial atom coupled to photons traveling along a microwave transmission line were also covered by a seminar given later by Yasunobu Nakamura (Chapter 18).

Discussions on quantum feedback and measurement back-action were resumed in the seminar by Sasha Korotkov (Chapter 17). This seminar introduced a puzzling property of the monitoring of a qubit by a coherent signal traversing a cavity dispersively coupled to the qubit. If the measurement of the signal is done by a fully efficient phase-sensitive amplifier, no noise is added by the amplifier and the backaction of the amplifier on the qubit induces the Bloch vector of the latter to diffuse along a meridian. When the measurement of the signal is done by a fully efficient phase-preserving amplifier, extra noise of quantum origin is added by the amplifier owing to the presence of the idler port, but the state of the qubit remains pure, the diffusion of the Bloch vector now occurring on the entire Bloch sphere. Sasha's insistence that the added noise of the amplifier does not affect the purity of the qubit after the measurement, and therefore its trackability, was one example of the many great pedagogically illuminating moments of the school.

The subject of persistent Rabi oscillations covered by Sasha Korotkov was treated experimentally in the frequency domain by Daniel Esteve in his course reviewing qubit readout (Chapter 6). Daniel gave food for thought to the students by mentioning the analogy between the gradual rotation of the Bloch vector from north to south pole when no escape to the voltage state is detected in the phase qubit to the no-barking of the hound of the Baskervilles in the Sherlock Holmes story. Daniel's illustration of the quantum Bayesian framework introduced by Sasha was later revisited by Steve Girvin and John Martinis, rendering these difficult concepts easier to grasp.

Quantum error corrections with discrete gates reacting to discrete error syndromes obtained from projective measurement can also be seen as a form of quantum feedback, but this time stabilizing a whole manifold of quantum states rather than a single one. The subject of quantum error correction and fault-tolerant quantum computation was treated by Ike Chuang (Chapter 7). In particular, he presented the various representations of open-system quantum dynamics, insisting on the indistinguishability between different unravellings. This course, which completed the course Ike gave in Les Houches in 2003, contained a useful general introduction to the theory of stabilizers, which is key to the understanding of error correcting codes. Moreover, it explained in full detail the principles behind the size of the overhead needed to make the entire quantum processor fault-tolerant, not just its memory. Ike left a lot of problem as homework and this was particularly appreciated by the students.

The school took place right after the first demonstration of cooling nanomechanical resonators down to their quantum ground state, so we were fortunate to hear lectures on this fast growing area of mesoscopic physics. Jack Harris (Chapter 14) gave an introduction to this field and treated in particular the cooling aspect in a semiclassical manner, from the point of view of retarded friction. The lectures on quantum optomechanics by Florian Marquardt (Chapter 8) and on quantum electromechanics by Konrad Lehnert (Chapter 9) described the recent advances and applications of this field in more depth. Florian showed how quantum optomechanical systems reduce to a linear interaction between the phonons and the fluctuations of the cavity field. He then presented the effect of light on the response and measurement of the mechanical system, drawing an analogy with sideband cooling. He then explained how to prepare exotic states of a mechanical oscillator and eventually confront quantum mechanics with gravity. Konrad presented the basics of quantum electromechanics from the point of view of his recent experiments. In particular, he demonstrated how to treat cooling as autonomous quantum feedback through electrostatic pressure. He also discussed the hybridization of mechanical and electromagnetic degrees of freedom.

Other mesoscopic engineered systems are quantum dots in bidimensional electron gases, which were covered by the course given by Amir Yacoby (Chapter 10). The spin qubits realized in these systems are strongly affected by the surrounding nuclear spins. Amir demonstrated that the nuclear spin bath is not as irreversible as one might think. Echo sequences can efficiently undo the dephasing due to these nuclear spins. Finally, Amir showed that electrostatic $1/f$ noise remains the dominant source of decoherence in these qubits.

Whereas the course of Aash Clerk covered the theoretical aspects of quantum limited amplification, John Clarke (Chapter 12) treated the concrete example of the microwave SQUID, a practical quantum-limited amplifier. The advantage of the SQUID is that it is sufficiently sophisticated to be used in a variety of applications, yet sufficiently simple to lend itself to a detailed analysis of the origin of noise and its processing through the active elements of the device. In particular, the noise of the SQUID can be understood as the mixed-down counterpart of the Johnson noise of the SQUID resistors in the vicinity of the Josephson frequency and its harmonics. In his seminar (Chapter 16), Irfan Siddiqi presented a different kind of quantum-limited amplifier based on parametric amplification and bifurcation of a Josephson junction. He showed how these amplifiers allowed his group to observe the quantum jumps of a superconducting qubit.

Finally, the basics as well as recent results of quantum machines based on "natural atom" systems were covered in courses by Jean-Michel Raimond (Chapter 11), Immanuel Bloch (Chapter 5) and Rainer Blatt (Chapter 13). The recurring theme of the school, quantum feedback, was illustrated in particular by Jean-Michel when he explained the stabilization of a Fock state of Rydberg atom against spontaneous emission in the cavity. The feedback process involves the rapid measurement of atoms passing through the cavity and performing a quantum nondemolition (QND) measurement of its photon number. The information is fed to a processor that in real time computes the right signal to send to the cavity to repump it in the right way. Immanuel showed how to engineer quantum simulators for tunable condensed matter systems out of cold atoms in optical traps. Finally, Rainer showed recent results of the most advanced quantum information processors to date using ion traps. He also showed how to engineer a quantum system to simulate the Klein effect.

In addition to participating in the lectures and seminars, the students gave short talks or presented poster on their research projects, which were often outstanding. A few tutorials and discussions were also organized on related subjects.

Acknowledgments

The summer school and the present volume have been made possible by the financial support of the following institutions, whose contribution is gratefully acknowledged:

- The Université Joseph Fourier.
- The Université Franco-Allemande, whose contribution was obtained with the help of Florian Marquardt. We also thank Otto Iancu, director of the UFA, for giving a seminar during the school.
- The Centre National pour la Recherche Scientifique.

On behalf of all participants, we would like to thank the staff of the school: Murielle and Brigitte for their support in organizing this school and Christian and his team for providing delicious meals in such a warm atmosphere. Thanks also to Anaïs Dreau for making enough delightful kouign-amann for everybody, to Philippe Campagne and Emmanuel Flurin for organizing the pétanque tournament and parties, and to Jérémie Viennot for proofreading. Despite rather bad weather during the last part of the school, hiking trips and the Firemen's dance greatly favored stimulating discussions in the pure tradition of the Les Houches School. In line with this tradition, the organizers must end this preface by reporting that the winners of the pétanque tournament were Mark Girard and Hannah Gordon.

Michel Devoret
Benjamin Huard
Robert Schoelkopf
Leticia F. Cugliandolo

1	S. Barz	16	V. Schmitt	31	C. Laflamme	46	M. Devoret
2	C. Hempel	17	J. Restrepo	32	J. Kerckhoff	47	M. Shi
3	K. Yamanaka	18	K. de Greve	33	A. Rampal	48	B. Huard
4	J. Martinis	19	A. Baksic	34	S. Guilbauld	49	P. Harvey Collard
5	S. Onur	20	S. Aldana	35	G. Zhu	50	C. de Grandi
6	E. Flurin	21	I. Pop	36	T. Weissl	51	C. Eichler
7	P. Campagne-Ibarcq	22	L. Neuhaus	37	G. Kirchmair	52	C. Joshi
8	M. Zwerger	23	M. Cirio	38	A. Dreau	53	J. Grover
9	N. Roch	24	I. Diniz	39	R. Schoelkopf	54	D. Sank
10	C. Macklin	25	H. Sorensen	40	Y. Wang	55	M. Ludwig
11	M. Defeo	26	N. Roeck	41	L. Cugliandolo	56	S. Srinivasan
12	K. Lalumière	27	A. Frisk Kockum	42	Y. Nakamura	57	A. Eltony
13	P. McMahon	28	M. Girard	43	F. Marquardt	58	N. Antler
14	M. Hatridge	29	T. Sugiyama	44	A. Kamal	59	N. Ishida
15	M. Reed	30	F. Qassemi	45	H. Paik		

POLYSHOP

Contents

List of participants

Organizers

DEVORET Michel
Collège de France, Paris, France
HUARD Benjamin
Université Pierre et Marie Curie, Paris, France
SCHOELKOPF Robert
Yale University, New Haven, USA
CUGLIANDOLO Leticia
Université Pierre et Marie Curie, Paris, France

Lecturers

BLATT Rainer
University of Innsbruck, Austria
BLOCH Immanuel
Max Planck Institute, Garching, Germany
BUISSON Olivier
Institut Néel, Grenoble, France
CHUANG Isaac
Massachusetts Institute of Technology, Cambridge, USA
CLARKE John
University of California, Berkeley, USA
CLERK Aashish
McGill University, Montreal, Canada
ESTEVE Daniel
SPEC–CEA Saclay, Gif-sur-Yvette, France
GIRVIN Steven
Yale University, New Haven, USA
HARRIS Jack
Yale University, New Haven, USA
KOROTKOV Alexander
University of California, Riverside, USA
LEHNERT Konrad
JILA, University of Colorado, Boulder, USA

MABUCHI Hideo
Stanford University, USA
MARQUARDT Florian
University of Erlangen, Germany
MARTINIS John
University of California, Santa Barbara, USA
NAKAMURA Yasunobu
The University of Tokyo, Japan
RAIMOND Jean-Michel
LKB–Ecole Normale Supérieure, Paris, France
SIDDIQI Irfan
University of California, Berkeley, USA
WALLRAFF Andreas
ETH Zurich, Switzerland
YACOBY Amir
Harvard University, Cambridge, USA

Participants

ALDANA Samuel
University of Basel, Switzerland
ANTLER Natania
University of California, Berkeley, USA
BAKSIC Alexandre
Université Paris Diderot, Paris, France
BARZ Stefanie
University of Vienna, Austria
CAMPAGNE-IBARCQ Philippe
LPA–Ecole Normale Supérieure, Paris, France
CIRIO Mauro
Marcquarie University, Sydney, Australia
DEFEO Michael
Syracuse University, USA
De GRANDI Claudia
Boston University, USA
De GREVE Kristiaan
Stanford University, USA
DREAU Anaïs
LPQM–Ecole Normale Supérieure, Cachan, France
EICHLER Christopher
ETH Zurich, Switzerland
ELTONY Amira
Massachusetts Institute of Technology, Cambridge, USA

FLURIN Emmanuel
LPA–Ecole Normale Supérieure, Paris, France
FRISK KOCKUM Anton
Chalmers University, Göteborg, Sweden
GIRARD Mark
Freiburg Institute, Germany
GROVER Jeffrey
University of Maryland, Joint Quantum Institute, USA
GUILBAUD Sylvain
LTCI–Telecom Paris Tech, Paris, France
HARVEY-COLLARD Patrick
University of Sherbrooke, Quebec, Canada
HATRIDGE Michael
Yale University, New Haven, USA
HEMPEL Cornelius
IQOQI–Austrian Academy of Sciences, Innsbruck, Austria
ISHIDA Natsuko
University of Tokyo, National Institute of Informatics, Japan
JOSHI Chaitanya
Heriot Watt University, Edinburgh, UK
KAMAL Archana
Yale University, New Haven, USA
KERCKHOFF Joseph
Stanford University, USA
KIRCHMAIER Gerhard
Yale University, New Haven, USA
LAFLAMME Catherine
McGill University, Montreal, Canada
LALUMIERE Kevin
Université de Sherbrooke, Quebec, Canada
LUDWIG Max
University of Erlangen-Nuremberg, Germany
MACKLIN Christopher
University of California, Berkeley, USA
McMAHON Peter
Stanford University, USA
NEUHAUS Leonhard
EPFL, Lausanne, Switzerland
ONUR Sander
Zernike Institute, University of Groningen, Netherlands
PAIK Hanhee
Yale University, New Haven, USA
POP Ioan Mihai
Institut Néel, Grenoble, France

QASSEMI MALOOMEH Farzad
University of Waterloo, Canada
RAMPAL Abhishaik
McMaster University, Hamilton, Canada
REED Matt
Yale University, New Haven, USA
RESTREPO Juan
Université Paris Diderot, Paris, France
ROCH Nicolas
LPA–Ecole Normale Supérieure, Paris, France
ROECK Nadia
University of Innsbruck, Austria
SANK Daniel
University of California, Santa Barbara, USA
SCHMITT Vivien
SPEC-CEA Saclay, Gif-sur-Yvette, France
SHI Molu
Massachusetts Institute of Technology, Cambridge, USA
SORENSEN Heidi Lundgaard
University of Copenhagen, Denmark
SRINIVASAN Srikanth
Princeton University, USA
SUGIYAMA Takanori
University of Tokyo, Japan
TUCHE DE ALMEIDA DINIZ Igor
Institut Néel, Grenoble, France
WANG Yimin
University of Singapore, Singapore
WEISSL Thomas
Institut Néel, Grenoble, France
YAMANAKA Kazuhiro
University of Tokyo, Japan
ZHU Guanyu
Northwestern University, Evanston, USA
ZWERGER Michael
Innsbruck University, Austria

Part I

Lectures

1

Real-time feedback control of quantum optical input-output systems

H. Mabuchi

Edward L. Ginzton Laboratory and Department of Applied Physics
Stanford University Stanford, CA 94305, USA

Quantum Machines: Measurement and Control of Engineered Quantum Systems.
Edited by M. Devoret, B. Huard, R. Schoelkopf, and L. F. Cugliandolo.

Chapter Contents

1.1 Scope and outline of the chapter

Many sophisticated technologies developed in the late twentieth century rely in some critical (though often hidden) way on real-time feedback control, and the use of feedback to achieve robust performance may only become more widespread in the anticipated era of quantum engineering and quantum-enabled technologies. As the interested reader can already find excellent introductions to classical feedback control—such as the textbook (freely available online) by Aström and Murray [1], the first few chapters of which are strongly recommended as background—as well as technical reviews on quantum control—such as those by Doherty *et al.* [2] and Mabuchi and Khaneja [3]—my aim in this chapter will be to provide a sort of interface to help orient the student of modern physics with respect to engineering fundamentals and recent advances in the theory of open quantum systems that are essential for the emerging field of quantum feedback control. A few detailed calculations and derivations are included to help to compensate for the current lack of accessible tutorial material on the use of quantum stochastic differential equations (QSDEs) in modeling coherent feedback. Thus, this chapter is hardly self-contained, but I hope that it will provide a useful resource for readers interested in quantum feedback control that complements what can already be found straightforwardly in the literature.

The following material is divided into two major parts—on coherent-feedback quantum control (Sections 1.2–1.3) and on measurement-based quantum feedback control (Section 1.4). The distinction between these paradigms is discussed in references such as those by James *et al.* [4] and Nurdin *et al.* [5], with a simple illustrative example described by Mabuchi [6]. Section 1.2 introduces a modern framework for modeling coherent feedback systems, based on QSDEs, with simple illustrative examples from cavity quantum electrodynamics (cavity QED) and cavity nonlinear optics [7, 8] that build towards an extended analysis of suppressing spontaneous transitions in dispersive bistability, as considered by Mabuchi [9]. Section 1.3 introduces an important QSDE limit theorem [10] that can be used to facilitate the analysis and simulation of multicomponent coherent-feedback circuits, illustrating its use in a tutorial fashion to derive a simplified model of a cavity QED relay [11]. QSDE modeling techniques and coherent-feedback analysis have not yet gained widespread use in photonics or quantum information research, but explorations such as those described by Kerckhoff *et al.* [12] Mabuchi [13], and Tezak *et al.* [14] suggest that they may have broad utility. Section 1.4 then turns to a systematic development of the concept of a recursive filter (state observer) for measurement-based feedback control, and concludes by drawing explicit connections between state observers and the stochastic Schrödinger equation (SSE) and stochastic master equation (SME), as they are known in quantum optics. Some background on classical recursive filtering seems essential for proper understanding of the SSE and SME (which are central equations for measurement-based quantum feedback control), but can be difficult for a physics student to find in accessible form.

1.2 Input-output models and circuit theory

We begin by introducing a formalism for interconnection of quantum optical input-output models of individual components that can be used to derive master equations for a broad class of photonic circuits. The method works for circuits with feedback, but one must take care to avoid feedback topologies that create "trapped" modes, which cannot be handled straightforwardly. The technical approach is based on a modern formulation of nonrelativistic quantum field theory in the Markov limit, using QSDEs. It is not necessary to know anything about QSDEs if one simply wants to utilize the methodology to derive circuit master equations from the interconnection of given component models, but we will have a quick look "under the hood" at the details of QSDE input-output models in order to establish a very convenient abstraction for nanophotonic circuits that we call the *small volume limit.*

Technical details on the approach can be found in the following original papers [15, 16]:

- J. Gough and M. R. James. The series product and its application to quantum feedforward and feedback networks. *IEEE Trans. Autom. Control*, **54**, 2530 (2009);
- J. Gough and M. R. James, Quantum feedback networks: Hamiltonian formulation. *Commun. Math. Phys.*, **287**, 1109 (2009).

These papers expand upon the following earlier results [17, 18]:

- C. W. Gardiner. Driving a quantum system with the output of another driven quantum system. *Phys. Rev. Lett.*, **70**, 2269 (1993).
- H. J. Carmichael. Quantum trajectory theory for cascaded open quantum systems. *Phys. Rev. Lett.*, **70**, 2273 (1993).

Each input-output component in a photonic circuit is described by a triple (S, L, H), where S is the scattering matrix of the component, L is the coupling vector of the component, and H is the Hamiltonian of the component's internal degrees of freedom. S is required to be a unitary matrix, and its matrix elements can in general be operators on the Hilbert space of the component's internal degrees of freedom (although they are often just complex numbers). The elements of L can also be operators, in which case they describe the way that external fields couple to the component's internal degrees of freedom. H has the usual role of determining the (autonomous) time evolution of the component's internal degrees of freedom. The dimension of L is equal to the number of input-output ports n that the component has (every port must be both an input and an output); the dimension of S is $n \times n$; the Hamiltonian H is scalar. It is important to note that the inclusion of a "scattering matrix" S in this modeling approach does not make it equivalent to "S-parameter" circuit modeling as commonly encountered in microwave engineering. In particular, the formalism we introduce here can be used to build models for time-domain analysis (and recursive filtering) of photonic circuits with nonlinear elements.

Using two composition rules called the series product, denoted $\triangleleft$, and the concatenation product, denoted $\boxplus$, it is possible to combine the individual (S, L, H) models for a number of interconnected components into an overall (S, L, H) model for the whole circuit. Before explaining how the series and concatenation products work, however,

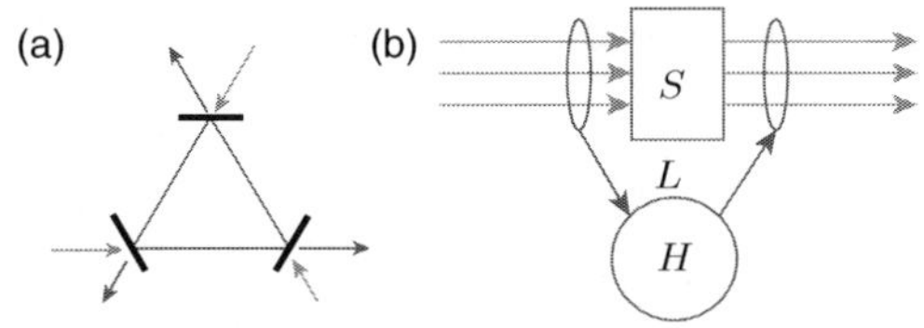

Fig. 1.1 (a) A three-mirror ring resonator representing a generic quantum optical input-output device with input-output ports as shown. (b) Schematic depiction of the parameters S, L and H as they relate to scattering processes and open system dynamics—the matrix S describes any direct scattering among the external fields, L describes the way that the fields couple to the component's internal degrees of freedom, and H generates the autonomous dynamics of the internal degrees of freedom.

let us introduce a few simple component models that we can use for interconnection examples.

A *beamsplitter* is a two-input/output component described by a scattering matrix only—it has no internal dynamic degrees of freedom (see Fig. 1.1). The (S, L, H) triple is thus simply

$$(S, L, H) = \left(\begin{bmatrix} r_{11} & t_{12} \\ t_{21} & r_{22} \end{bmatrix}, \begin{bmatrix} 0 \\ 0 \end{bmatrix}, 0 \right),$$

where since the scattering matrix must be unitary,

$$\begin{bmatrix} r_{11} & t_{12} \\ t_{21} & r_{22} \end{bmatrix} \begin{bmatrix} r_{11}^* & t_{21}^* \\ t_{12}^* & r_{22}^* \end{bmatrix} = \begin{bmatrix} 1 & 0 \\ 0 & 1 \end{bmatrix}.$$

A common example is the 50/50 beamsplitter:

$$\begin{bmatrix} r_{11} & t_{12} \\ t_{21} & r_{22} \end{bmatrix} = \begin{bmatrix} \frac{1}{\sqrt{2}} & -\frac{1}{\sqrt{2}} \\ \frac{1}{\sqrt{2}} & \frac{1}{\sqrt{2}} \end{bmatrix}.$$

Note that from a theoretical standpoint the phase convention we have chosen is arbitrary (for example, the transpose of this matrix could equally well represent a 50/50 beamsplitter), but in modeling an actual experimental setup the phases of the matrix elements will be determined by physical properties of the actual beamsplitting device.

An ideal laser input to a photonic circuit can be represented by a *coherent displacement*, which has the (S, L, H) model

$$(S, L, H) = (1, \alpha, 0),$$

where α is the complex amplitude of the displacement (laser input). The units in these models is such that $|\alpha|^2$ should be a rate (photons per second) corresponding to the optical power divided by the energy per photon.

A propagation phase or *phase shift* can be represented by the (S, L, H) model

$$(S, L, H) = \left(e^{i\varphi}, 0, 0\right).$$

Note that this is really a one-dimensional scattering matrix. Since the scattering matrix is required to be unitary, it is not valid to consider a component such as $(z, 0, 0)$ with $|z| \neq 1$. If we want to represent an attenuator of some kind, we actually have to pass the field of interest through a beamsplitter, with the second input left open, as this is required at the underlying QSDE level in order to preserve fundamental commutation relations.

An empty Fabry-Perot cavity can be represented by an *optical resonator* with two input-output ports (two mirrors),

$$(S, L, H) = \left(\begin{bmatrix} 1 & 0 \\ 0 & 1 \end{bmatrix}, \begin{bmatrix} \sqrt{\kappa_1} a \\ \sqrt{\kappa_2} a \end{bmatrix}, \Delta' a^\dagger a\right).$$

Here Δ' accounts for a detuning between the resonance frequency of the cavity eigenmode that is being considered and the rotating frame of the circuit model, κ_1 and κ_2 are the partial decay rates of the two mirrors, and a is the annihilation operator for the cavity eigenmode. If we want to add a two-level atom, we can write

$$(S, L, H) = \left(\begin{bmatrix} 1 & 0 & 0 \\ 0 & 1 & 0 \\ 0 & 0 & 1 \end{bmatrix}, \begin{bmatrix} \sqrt{\kappa_1} a \\ \sqrt{\kappa_2} a \\ \sqrt{\gamma} \sigma \end{bmatrix}, \Delta' a^\dagger a + \theta \left|e\right\rangle \left\langle e\right| + g\left(a\sigma^\dagger + a^\dagger \sigma\right)\right),$$

where γ is the atomic decay rate, g is the vacuum Rabi parameter [7], and σ is the atomic lowering operator (note that, according to the convention for these circuit models, $\kappa_{1,2}$ and γ are energy decay rates rather than field decay rates; thus, relative to common usage in cavity QED, $\gamma \equiv 2\gamma_\perp$ etc.). Here we have added an extra input-ouput port to represent, in essence, the vacuum electromagnetic fields that interact with the atom and induce spontaneous emission. Note that it would not be realistic to connect the output of this spontaneous emission port to another component because in practice this would require collecting all 4π solid angle of spontaneous emission; on the other hand, we can represent the effect of a laser directly driving the atom (through the "side" of the cavity) by connecting a coherent displacement to the input of this port.

Hopefully, it is obvious how to generalize to ring cavities with three or more mirrors, and to cavities with other internal dynamics. For example, a three-mirror ring cavity with a Kerr nonlinear medium is represented by

$$(S, L, H) = \left(\begin{bmatrix} 1 & 0 & 0 \\ 0 & 1 & 0 \\ 0 & 0 & 1 \end{bmatrix}, \begin{bmatrix} \sqrt{\kappa_1} a \\ \sqrt{\kappa_2} a \\ \sqrt{\kappa_3} a \end{bmatrix}, \Delta a^\dagger a + \chi a^\dagger a^\dagger a a\right).$$

Note that in some cases we may have a ring cavity in mind in terms of the geometry, yet wish to treat some of the mirrors as "high reflectors" that do not have an associated input-output field. In these cases, we can set $\sqrt{\kappa_i} \to 0$ for each such port, and in fact the associated row(s) of the scattering matrix and coupling vector can simply be deleted.

To provide at least one example of a more complex device whose scattering matrix contains operator matrix elements, we make note of the *relay* model

$$(S, L, H) = \left(\begin{bmatrix} \Pi_g & -\Pi_h & 0 & 0 \\ -\Pi_h & \Pi_g & 0 & 0 \\ 0 & 0 & \Pi_h & -\sigma_{gh} \\ 0 & 0 & -\sigma_{hg} & \Pi_g \end{bmatrix}, 0, 0 \right),$$

where the Hilbert space of the internal degrees of freedom is span $\{|g\rangle, |h\rangle\}$ and the operators are $\Pi_g = |g\rangle\langle g|$, $\Pi_h = |h\rangle\langle h|$, $\sigma_{hg} = |h\rangle\langle g|$, $\sigma_{gh} = |g\rangle\langle h|$. The first two input-output ports correspond to fields whose routings are switched by the relay, while the third and fourth ports correspond to the SET and RESET inputs of the relay. We shall see later how to derive such an (S, L, H) as the limit model for a cavity QED system in the small-volume limit [11].

At this point, it is useful to note that we can easily write the master equation for any (S, L, H) model (note that we are taking $\hbar = 1$ here):

$$\dot{\rho} = -i[H, \rho] + \sum_j \left(L_j \rho L_j^\dagger - \frac{1}{2} L_j^\dagger L_j \rho - \frac{1}{2} \rho L_j^\dagger L_j \right),$$

where L_j is the jth component of the coupling vector. Note that the scattering matrix does not enter. It is interesting to note that such master equations can sometimes be written in a form that shuffles terms between the Hamiltonian and Lindblad parts—this will help us to match some models we derive using photonic circuit theory with conventional forms from quantum optics, and can also help in making circuit master equations more amenable to numerical integration.

It is clear from inspection that the (S, L, H) cavity QED model reproduces the usual cavity QED master equation using this approach, although without the cavity driving term:

$$(S, L, H) = \left(\begin{bmatrix} 1 & 0 & 0 \\ 0 & 1 & 0 \\ 0 & 0 & 1 \end{bmatrix}, \begin{bmatrix} \sqrt{\kappa_1} a \\ \sqrt{\kappa_2} a \\ \sqrt{\gamma} \sigma \end{bmatrix}, \Delta' a^\dagger a + \theta |e\rangle\langle e| + g\left(a\sigma^\dagger + a^\dagger \sigma\right) \right),$$

$$\dot{\rho} = -i[H, \rho] + (\kappa_1 + \kappa_2)\left(a\rho a^\dagger - \frac{1}{2} a^\dagger a \rho - \frac{1}{2} \rho a^\dagger a \right) + \gamma \left(\sigma \rho \sigma^\dagger - \frac{1}{2} \sigma^\dagger \sigma \rho - \frac{1}{2} \rho \sigma^\dagger \sigma \right),$$

where we see explicitly the correspondences $\kappa_1 + \kappa_2 \leftrightarrow 2\kappa$ and $\gamma \leftrightarrow 2\gamma_\perp$.

So, what if we want to add a cavity driving term? Here we can illustrate the use of the series and concatenation products. The very simple "circuit" model that we want is

$$D = C \lhd ((1, \alpha, 0) \boxplus (1, 0, 0) \boxplus (1, 0, 0)),$$

where C stands for the undriven cavity QED (S, L, H) model written previously and D will be the (S, L, H) model for the cavity QED model with a coherent driving field (carrying photon rate $|\alpha|^2$) incident on mirror 1.

We have a general rule for the series product of two systems $G_1 = (S_1, L_1, H_1)$ and $G_2 = (S_2, L_2, H_2)$,

$$G_2 \lhd G_1 = \left(S_2S_1, S_2L_1 + L_2, H_1 + H_2 + \mathrm{Im}\left\{L_2^\dagger S_2 L_1\right\}\right),$$

and likewise for the concatenation product,

$$G_2 \boxplus G_1 = \left(\begin{bmatrix} S_1 & 0 \\ 0 & S_2 \end{bmatrix}, \begin{bmatrix} L_1 \\ L_2 \end{bmatrix}, H_1 + H_2\right),$$

where the components of G_1 and G_2 need not commute. Hence we can compute

$$\begin{aligned} D &= \left(I_3, \begin{bmatrix} \sqrt{\kappa_1}a \\ \sqrt{\kappa_2}a \\ \sqrt{\gamma}\sigma \end{bmatrix}, \Delta' a^\dagger a + \theta \left|e\right\rangle\left\langle e\right| + g\left(a\sigma^\dagger + a^\dagger\sigma\right)\right) \lhd \left(I_3, \begin{bmatrix} \alpha \\ 0 \\ 0 \end{bmatrix}, 0\right) \\ &= \left(I_3, \begin{bmatrix} \sqrt{\kappa_1}a + \alpha \\ \sqrt{\kappa_2}a \\ \sqrt{\gamma}\sigma \end{bmatrix}, H_D\right), \end{aligned}$$

where I_3 is the 3×3 identity matrix and

$$\begin{aligned} H_D &= \Delta a^\dagger a + \theta \left|e\right\rangle\left\langle e\right| + g\left(a\sigma^\dagger + a^\dagger\sigma\right) + \mathrm{Im}\left\{\begin{bmatrix} \sqrt{\kappa_1}a^\dagger & \sqrt{\kappa_2}a^\dagger & \sqrt{\gamma}\sigma^\dagger \end{bmatrix}\begin{bmatrix} \alpha \\ 0 \\ 0 \end{bmatrix}\right\} \\ &= \Delta' a^\dagger a + \theta \left|e\right\rangle\left\langle e\right| + g\left(a\sigma^\dagger + a^\dagger\sigma\right) + \mathrm{Im}\left\{\alpha\sqrt{\kappa_1}a^\dagger\right\} \\ &= \Delta' a^\dagger a + \theta \left|e\right\rangle\left\langle e\right| + g\left(a\sigma^\dagger + a^\dagger\sigma\right) + \frac{1}{2i}\sqrt{\kappa_1}\left(\alpha a^\dagger - \alpha^* a\right). \end{aligned}$$

We see that the concatenation product is here used to "stack" three single-field inputs (two of them vacuum) in order to create a G_1 whose dimensions are compatible with those of the cavity QED system G_2. The series product then actually implements driving G_1 "into" G_2 (note that the composition expressions read from right to left).

In order to arrive at a master equation for the driven model, we start by computing the Lindblad terms. Obviously, the terms generated by $L_2 = \sqrt{\kappa_2}a$ and $L_3 = \sqrt{\gamma}\sigma$ will not change, so we just need to compute

$$\begin{aligned} L_1\rho L_1^\dagger &= \left(\sqrt{\kappa_1}a + \alpha\right)\rho\left(\sqrt{\kappa_1}a^\dagger + \alpha^*\right) = \kappa_1 a\rho a^\dagger + \sqrt{\kappa_1}\alpha^* a\rho + \sqrt{\kappa_1}\rho\alpha a^\dagger + |\alpha|^2, \\ L_1^\dagger L_1 &= \left(\sqrt{\kappa_1}a^\dagger + \alpha^*\right)\left(\sqrt{\kappa_1}a + \alpha\right) = \kappa_1 a^\dagger a + \sqrt{\kappa_1}\alpha a^\dagger + \sqrt{\kappa_1}\alpha^* a + |\alpha|^2, \end{aligned}$$

and hence the full Lindblad term for L_1 is

$$\begin{aligned} \{\dot{\rho}\}_{L_1} &= \kappa_1 a\rho a^\dagger + \sqrt{\kappa_1}\alpha^* a\rho + \sqrt{\kappa_1}\rho\alpha a^\dagger + |\alpha|^2 \\ &\quad -\frac{1}{2}\left(\kappa_1 a^\dagger a\rho + \sqrt{\kappa_1}\alpha a^\dagger\rho + \sqrt{\kappa_1}\alpha^* a\rho + |\alpha|^2\rho\right) \\ &\quad -\frac{1}{2}\left(\kappa_1\rho a^\dagger a + \sqrt{\kappa_1}\rho\alpha a^\dagger + \sqrt{\kappa_1}\rho\alpha^* a + |\alpha|^2\rho\right) \end{aligned}$$

$$
\begin{aligned}
&= \kappa_1 \left(a\rho a^\dagger - \frac{1}{2} a^\dagger a \rho - \frac{1}{2} \rho a^\dagger a \right) + \sqrt{\kappa_1} \alpha^* a \rho + \sqrt{\kappa_1} \rho \alpha a^\dagger \\
&\quad - \frac{1}{2} \sqrt{\kappa_1} \alpha a^\dagger \rho - \frac{1}{2} \sqrt{\kappa_1} \alpha^* a \rho - \frac{1}{2} \sqrt{\kappa_1} \rho \alpha a^\dagger - \frac{1}{2} \sqrt{\kappa_1} \rho \alpha^* a \\
&= \kappa_1 \left(a\rho a^\dagger - \frac{1}{2} a^\dagger a \rho - \frac{1}{2} \rho a^\dagger a \right) + \frac{1}{2} \sqrt{\kappa_1} \left(\alpha^* a \rho - \alpha a^\dagger \rho + \rho \alpha a^\dagger - \rho \alpha^* a \right) \\
&= \kappa_1 \left(a\rho a^\dagger - \frac{1}{2} a^\dagger a \rho - \frac{1}{2} \rho a^\dagger a \right) + \frac{1}{2} \sqrt{\kappa_1} \left[\left(\alpha^* a - \alpha a^\dagger \right), \rho \right].
\end{aligned}
$$

Thus, we see that there is a term we could group into the Hamiltonian, leaving us finally with

$$
\begin{aligned}
\dot{\rho} &= -i \left[H_D', \rho \right] \\
&\quad + (\kappa_1 + \kappa_2) \left(a\rho a^\dagger - \frac{1}{2} a^\dagger a \rho - \frac{1}{2} \rho a^\dagger a \right) + \gamma \left(\sigma \rho \sigma^\dagger - \frac{1}{2} \sigma^\dagger \sigma \rho - \frac{1}{2} \rho \sigma^\dagger \sigma \right), \\
H_D' &= \Delta a^\dagger a + \theta \left| e \right\rangle \left\langle e \right| + g \left(a \sigma^\dagger + a^\dagger \sigma \right) + \frac{1}{2i} \sqrt{\kappa_1} \left(\alpha a^\dagger - \alpha^* a \right) - \frac{1}{2i} \sqrt{\kappa_1} \left(\alpha^* a - \alpha a^\dagger \right) \\
&= \Delta a^\dagger a + \theta \left| e \right\rangle \left\langle e \right| + g \left(a \sigma^\dagger + a^\dagger \sigma \right) - \frac{i}{2} \sqrt{\kappa_1} \left(\alpha a^\dagger - \alpha^* a \right) + \frac{i}{2} \sqrt{\kappa_1} \left(\alpha^* a - \alpha a^\dagger \right) \\
&= \Delta a^\dagger a + \theta \left| e \right\rangle \left\langle e \right| + g \left(a \sigma^\dagger + a^\dagger \sigma \right) - i \sqrt{\kappa_1} \left(\alpha a^\dagger - \alpha^* a \right).
\end{aligned}
$$

Hence, we finally see that in comparison with the usual form of the driven-cavity QED master equation, with cavity-driving term

$$
H_{\mathrm{dr}} = iE \left(a^\dagger - a \right),
$$

we have derived the correspondence

$$
E = -\sqrt{\kappa_1} \alpha.
$$

If we consider the master equation with $\Delta = \theta = g = 0$ and initial state $|0, g\rangle$ (reducing to a simple resonantly driven empty cavity),

$$
\begin{aligned}
\frac{d}{dt} \langle a \rangle &= \mathrm{Tr}[a \dot{\rho}] \\
&\rightarrow \sqrt{\kappa_1} \, \mathrm{Tr}\big[a \left[\alpha^* a - \alpha a^\dagger, \rho \right] \big] + (\kappa_1 + \kappa_2) \, \mathrm{Tr}\left[a \left(a\rho a^\dagger - \frac{1}{2} a^\dagger a \rho - \frac{1}{2} \rho a^\dagger a \right) \right] \\
&= \sqrt{\kappa_1} \, \mathrm{Tr}\big[\alpha^* a a \rho - \alpha a a^\dagger \rho - \alpha^* a a \rho + \alpha a^\dagger a \rho \big] + \frac{\kappa_1 + \kappa_2}{2} \mathrm{Tr}\big[a^\dagger a a \rho - a a^\dagger a \rho \big] \\
&= -\alpha \sqrt{\kappa_1} - \frac{\kappa_1 + \kappa_2}{2} \langle a \rangle,
\end{aligned}
$$

and we therefore find that in steady state,

$$
\frac{d}{dt} \langle a \rangle = 0,
$$

$$\langle a \rangle_{\rm ss} = -\frac{2\alpha\sqrt{\kappa_1}}{\kappa_1 + \kappa_2}.$$

Note that the sign in this expression is crucial. From the underlying QSDE formalism, it is possible to derive that for a scenario such as this one with an empty cavity, in which all the optical states should remain simple coherent states, the coherent amplitudes of the output channels should be given by the expectation value of the corresponding components of the coupling vector L (see [15] for a proper statement of the QSDE input-output relations). Hence, in particular, the output state of input-output channels 1 and 2 should be given in steady state by

$$\langle L_1 \rangle_{\rm ss} = \langle \sqrt{\kappa_1} a + \alpha \rangle_{\rm ss} = -\frac{2\alpha\kappa_1}{\kappa_1 + \kappa_2} + \alpha,$$

$$\langle L_2 \rangle_{\rm ss} = \langle \sqrt{\kappa_2} a \rangle_{\rm ss} = -\frac{2\alpha\sqrt{\kappa_1\kappa_2}}{\kappa_1 + \kappa_2}.$$

We find that if we set $\kappa_1 = \kappa_2$, we recover the expected result for a symmetric two-sided cavity that the reflected power drops to zero while the transmission becomes perfect. Even though the output fields will generally be more complicated that coherent states in the full cavity QED setup with strong coupling, the above expressions remain reasonable indications of the optical power flow in many scenarios where the output states remain close to coherent. Hence, for the so-called "bad cavity" regime of cavity QED, a numerical computation based on the master equation with $g = 10$ and $\kappa = 40$ produces the cavity transmission spectrum shown in Fig. 1.2

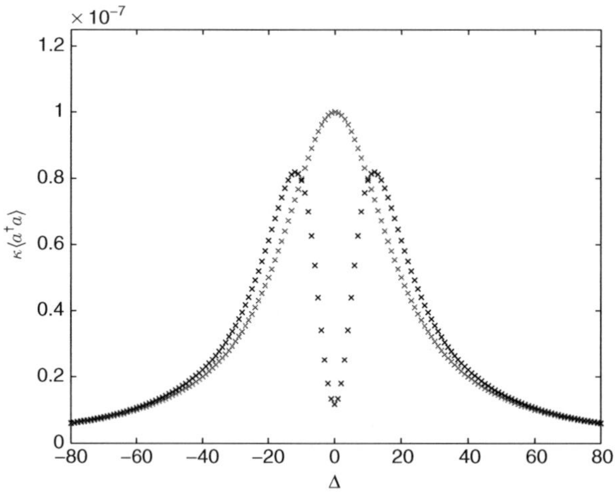

Fig. 1.2 Mean intracavity photon number as a function of probe detuning for a cavity QED model in the "bad cavity" regime.

In this plot, the horizontal axis is the detuning $\Delta' = \theta$ and the vertical axis is $\langle a^\dagger a\rangle$ for the intracavity field (the simple Lorentzian curve is a $g = 0$ reference and the curve that dips around zero detuning is for $g = 10$). On resonance, we see that the intracavity field is very small, and from the above input-output relations we can infer that the incident power is largely reflected (as opposed to dissipated by the atom via spontaneous emission, which would be another reasonable guess). In fact, as this spectrum is calculated for a very weak incident probe field, the reflected and transmitted fields should be quite close to coherent states.

Although we have chosen not to discuss the full QSDE input-output relations here, hopefully it is clear that the photonic circuit modeling tools that we have begun to introduce generally do provide a rigorous basis for analyzing the internal, reflected, and transmitted fields of an (S, L, H) "device" (which could itself be a circuit composed of multiple interconnected components), given a specified incident field amplitude. Note that our very simple calculation above reveals exactly how to relate the parameter E that appeared in our previous master equation to the external amplitude α, as well as how to characterize the partitioning of the total cavity output between reflected and transmitted channels.

To further illustrate the use of series and concatenation products in a more complex scenario, we turn next to an analysis of coherent-feedback suppression of spontaneous switching in ultra-low power dispersive bistability [9]. Reference [9] is freely available online and the reader is referred there for a discussion of the problem setup and numerical results; the following material makes explicit connections between what is presented there and the systematic approach to coherent feedback modeling introduced above.

We first use the series product to derive an open-loop model for the three-port "plant" cavity with a coherent driving field. The plant cavity itself is described by an autonomous dynamical model

$$\begin{aligned}
G_b &= \left(\begin{bmatrix} 1 & 0 & 0 \\ 0 & 1 & 0 \\ 0 & 0 & 1 \end{bmatrix}, \begin{bmatrix} \sqrt{\kappa_{b1}}b \\ \sqrt{\kappa_{b2}}b \\ \sqrt{\kappa_{b3}}b \end{bmatrix}, H_{bu} \right) = G_{b1} \boxplus G_{b2} \boxplus G_{b3}, \\
G_{b1} &= (1, \sqrt{\kappa_{b1}}b, H_{bu}), \quad G_{b2} = (1, \sqrt{\kappa_{b2}}b, 0), \quad G_{b3} = (1, \sqrt{\kappa_{b3}}b, 0), \\
H_{bu} &= \chi_b b^\dagger b^\dagger b b + \Delta_b b^\dagger b.
\end{aligned}$$

In order to include a coherent input field β, we use the series product as before:

$$\begin{aligned}
N &= G_{b1} \boxplus G_{b2} \boxplus (G_{b3} \lhd (1, \beta, 0)) \\
&= G_{b1} \boxplus G_{b2} \boxplus \left(1, \beta + \sqrt{\kappa_{b3}}b, \mathrm{Im}\left\{\sqrt{\kappa_{b3}}b^\dagger \beta\right\}\right) \\
&= \left(\begin{bmatrix} 1 & 0 & 0 \\ 0 & 1 & 0 \\ 0 & 0 & 1 \end{bmatrix}, \begin{bmatrix} \sqrt{\kappa_{b1}}b \\ \sqrt{\kappa_2}b \\ \beta + \sqrt{\kappa_{b3}}b \end{bmatrix}, H_{bu} + \frac{\sqrt{\kappa_{b3}}}{2i}\left(b^\dagger \beta - b\beta^*\right) \right).
\end{aligned}$$

The corresponding open-loop master equation is

$$
\begin{aligned}
\dot{\rho} &= -i\,[H,\rho] + \sum_j \left\{ L_j \rho L_j^\dagger - \frac{1}{2} L_j^\dagger L_j \rho - \frac{1}{2} \rho L_j^\dagger L_j \right\} \\
&\to -i \left[H_{bu} + \frac{\sqrt{\kappa_{b3}}}{2i} \left(b^\dagger \beta - b \beta^* \right), \rho \right] + (\kappa_{b1} + \kappa_{b2}) \left\{ b\rho b^\dagger - \frac{1}{2} b^\dagger b \rho - \frac{1}{2} \rho b^\dagger b \right\} \\
&\quad + (\beta + \sqrt{\kappa_{b3}} b)\, \rho \left(\beta^* + \sqrt{\kappa_{b3}} b^\dagger \right) \\
&\quad - \frac{1}{2} \left(\beta^* + \sqrt{\kappa_{b3}} b^\dagger \right) (\beta + \sqrt{\kappa_{b3}} b)\, \rho - \frac{1}{2} \rho \left(\beta^* + \sqrt{\kappa_{b3}} b^\dagger \right) (\beta + \sqrt{\kappa_{b3}} b) \\
&= -i \left[H_{bu} + \frac{\sqrt{\kappa_{b3}}}{2i} \left(b^\dagger \beta - b \beta^* \right), \rho \right] + (\kappa_{b1} + \kappa_{b2}) \left\{ b\rho b^\dagger - \frac{1}{2} b^\dagger b \rho - \frac{1}{2} \rho b^\dagger b \right\} + |\beta|^2 \rho \\
&\quad + \beta^* \sqrt{\kappa_{b3}} b \rho + \beta \sqrt{\kappa_{b3}} \rho b^\dagger + \kappa_{b3} b \rho b^\dagger - \frac{1}{2} |\beta|^2 \rho - \frac{1}{2} \beta^* \sqrt{\kappa_{b3}} b \rho - \frac{1}{2} \beta \sqrt{\kappa_{b3}} b^\dagger \rho \\
&\quad - \frac{1}{2} \kappa_{b3} b^\dagger b \rho - \frac{1}{2} |\beta|^2 \rho - \frac{1}{2} \beta^* \sqrt{\kappa_{b3}} \rho b - \frac{1}{2} \beta \sqrt{\kappa_{b3}} \rho b^\dagger - \frac{1}{2} \kappa_{b3} \rho b^\dagger b \\
&= -i \left[H_{bu} + \frac{\sqrt{\kappa_{b3}}}{2i} \left(b^\dagger \beta - b \beta^* \right), \rho \right] + (\kappa_{b1} + \kappa_{b2} + \kappa_{b3}) \left\{ b\rho b^\dagger - \frac{1}{2} b^\dagger b \rho - \frac{1}{2} \rho b^\dagger b \right\} \\
&\quad + \frac{1}{2} \beta^* \sqrt{\kappa_{b3}} \left(b\rho - \rho b \right) - \frac{1}{2} \beta \sqrt{\kappa_{b3}} \left(b^\dagger \rho - \rho b^\dagger \right).
\end{aligned}
$$

We note that

$$
\frac{1}{2} \beta^* \sqrt{\kappa_{b3}} \left(b\rho - \rho b \right) - \frac{1}{2} \beta \sqrt{\kappa_{b3}} \left(b^\dagger \rho - \rho b^\dagger \right) = \frac{\sqrt{\kappa_{b3}}}{2} \left[\left(\beta^* b - \beta b^\dagger \right), \rho \right],
$$

and hence, we can pull this remaining term into the Hamiltonian and finally write

$$
\dot{\rho} = -i \left[H_{bu} - i\sqrt{\kappa_{b3}} \left(b^\dagger \beta - b \beta^* \right), \rho \right] + (\kappa_{b1} + \kappa_{b2} + \kappa_{b3}) \left\{ b\rho b^\dagger - \frac{1}{2} b^\dagger b \rho - \frac{1}{2} \rho b^\dagger b \right\}.
$$

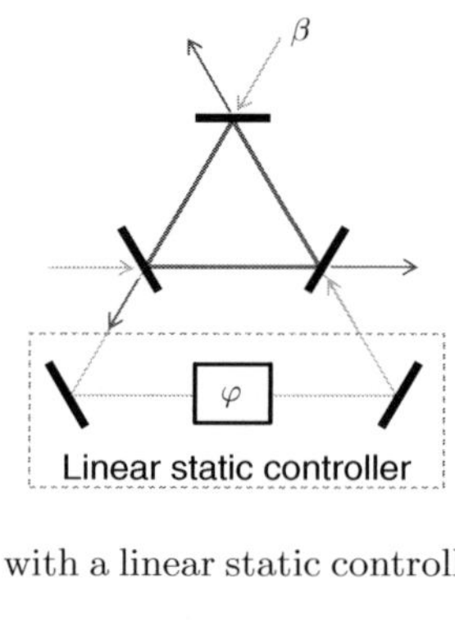

Fig. 1.3 Coherent feedback loop with a linear static controller, corresponding to a simple phase shift.

We thus see clearly that the total cavity decay rate is simply $\kappa_b \equiv \kappa_{b1} + \kappa_{b2} + \kappa_{b3}$, while the effects of the driving term can be absorbed into the system Hamiltonian. The driven-cavity model can thus be written

$$N_d = \left(\begin{bmatrix} 1 & 0 & 0 \\ 0 & 1 & 0 \\ 0 & 0 & 1 \end{bmatrix}, \begin{bmatrix} \sqrt{\kappa_{b1}}b \\ \sqrt{\kappa_{b2}}b \\ \sqrt{\kappa_{b3}}b \end{bmatrix}, H_{bu} - i\sqrt{\kappa_{b3}}\left(b^\dagger\beta - b\beta^*\right) \right).$$

We next consider the effects of linear static coherent feedback, with a simple phase shift (Fig. 1.3). We can write

$$\begin{aligned} N_{LS} &= G_{b1} \lhd \left((e^{i\varphi}, 0, 0) \lhd G_{b2} \right) \boxplus (G_{b3} \lhd (1, \beta, 0)) \\ &= (1, \sqrt{\kappa_{b1}}b, H_{bu}) \lhd \left(e^{i\varphi}, e^{i\varphi}\sqrt{\kappa_{b2}}b, 0 \right) \boxplus (G_{b3} \lhd (1, \beta, 0)) \\ &= \left(e^{i\varphi}, \left(\sqrt{\kappa_{b1}} + e^{i\varphi}\sqrt{\kappa_{b2}}\right) b, H_{bu} + \sin\varphi\sqrt{\kappa_{b1}\kappa_{b2}}b^\dagger b \right) \boxplus (G_{b3} \lhd (1, \beta, 0)) \\ &= \left(\begin{bmatrix} e^{i\varphi} & 0 \\ 0 & 1 \end{bmatrix}, \begin{bmatrix} \left(\sqrt{\kappa_{b1}} + e^{i\varphi}\sqrt{\kappa_{b2}}\right) b \\ \sqrt{\kappa_{b3}}b \end{bmatrix}, \right. \\ &\qquad \left. H_{bu} - i\sqrt{\kappa_{b3}}\left(b^\dagger\beta - b\beta^*\right) + \sin\varphi\sqrt{\kappa_{b1}\kappa_{b2}}b^\dagger b \right), \end{aligned}$$

where we re-use what we have derived above regarding the driving term, yielding the closed-loop master equation

$$\begin{aligned} \dot{\rho} &= -i\left[H_{bu} - i\sqrt{\kappa_{b3}}\left(b^\dagger\beta - b\beta^*\right) + \sin\varphi\sqrt{\kappa_{b1}\kappa_{b2}}b^\dagger b, \rho\right] \\ &\quad + \left(\kappa_{b3} + \left|\sqrt{\kappa_{b1}} + e^{i\varphi}\sqrt{\kappa_{b2}}\right|^2\right) \left\{ b\rho b^\dagger - \frac{1}{2}b^\dagger b\rho - \frac{1}{2}\rho b^\dagger b \right\}. \end{aligned}$$

Hence, the total cavity decay rate is a function of φ, and there is an additional frequency-pulling term in the Hamiltonian. We note that for $\varphi = 0$ we obtain

$$\dot{\rho} \to -i\left[H_{bu} - i\sqrt{\kappa_{b3}}\left(b^\dagger\beta - b\beta^*\right), \rho\right] + \left(\kappa_b + 2\sqrt{\kappa_{b1}\kappa_{b2}}\right) \left\{ b\rho b^\dagger - \frac{1}{2}b^\dagger b\rho - \frac{1}{2}\rho b^\dagger b \right\},$$

while for $\varphi = \pi$ we obtain

$$\dot{\rho} \to -i\left[H_{bu} - i\sqrt{\kappa_{b3}}\left(b^\dagger\beta - b\beta^*\right), \rho\right] + \left(\kappa_b - 2\sqrt{\kappa_{b1}\kappa_{b2}}\right) \left\{ b\rho b^\dagger - \frac{1}{2}b^\dagger b\rho - \frac{1}{2}\rho b^\dagger b \right\}.$$

Hence, in these simple cases, we have either a pure increase or a pure decrease in the cavity decay rate as the only net effect of the feedback. These can be understood as interferometric constructive/destructive interference of the output fields from the κ_{b1} and κ_{b2} cavity mirrors. We infer that since the external driving term (through mirror κ_{b3}) is unaffected, it should be possible to use φ to tune the average intracavity photon number. In particular, if we have a detuned driving field, we should be able to decrease the effective driving strength by decreasing the effective κ_b, and vice versa.

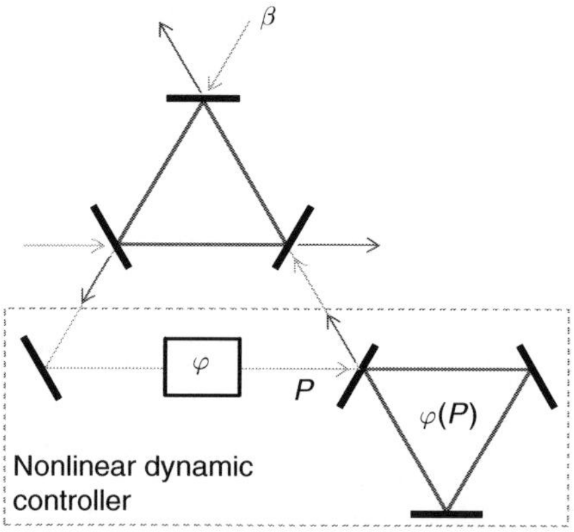

Fig. 1.4 Coherent feedback loop with a dynamic controller based on reflection from a Kerr-nonlinear resonator.

For the nonlinear dynamic controller (Fig. 1.4), we assume two cavities a (controller) and b (plant) with component models

$$G_a = (1, \sqrt{\kappa_a} a\,, H_a),$$
$$G_b = \left(\begin{bmatrix} 1 & 0 & 0 \\ 0 & 1 & 0 \\ 0 & 0 & 1 \end{bmatrix}, \begin{bmatrix} \sqrt{\kappa_{b1}} b \\ \sqrt{\kappa_{b2}} b \\ \sqrt{\kappa_{b3}} b \end{bmatrix}, H_{bu} \right),$$

where

$$H_a = \chi_a a^\dagger a^\dagger a a + \Delta_a a^\dagger a,$$
$$H_{bu} = \chi_b b^\dagger b^\dagger b b + \Delta_b b^\dagger b.$$

We define the partitioning

$$G_b = G_{b1} \boxplus G_{b2} \boxplus G_{b3},$$

where

$$G_{b1} = (1, \sqrt{\kappa_{b1}} b, H_{bu})\,, \quad G_{b2} = (1, \sqrt{\kappa_{b2}} b, 0)\,, \quad G_{b3} = (1, \sqrt{\kappa_{b3}} b, 0)\,.$$

We compute the feedback network as

$$\begin{aligned} N_{ND} &= G_{b1} \lhd \left(G_a \lhd \left((e^{i\varphi_0}, 0, 0) \lhd G_{b2}\right)\right) \boxplus (G_{b3} \lhd (1, \beta, 0)) \\ &= \left((1, \sqrt{\kappa_{b1}} b, H_{bu}) \lhd \left((1, \sqrt{\kappa_a} a\,, H_a) \lhd \left(e^{i\varphi}, e^{i\varphi}\sqrt{\kappa_{b2}} b, 0\right)\right)\right) \\ &\quad \boxplus \left(1, \beta + \sqrt{\kappa_{b3}} b, \mathrm{Im}\left\{\sqrt{\kappa_{b3}} \beta b^\dagger\right\}\right) \\ &= \left((1, \sqrt{\kappa_{b1}} b, H_{bu}) \lhd \left(e^{i\varphi}, \sqrt{\kappa_a} a + e^{i\varphi}\sqrt{\kappa_{b2}} b, H_a + \mathrm{Im}\left\{e^{i\varphi}\sqrt{\kappa_a \kappa_{b2}} a^\dagger b\right\}\right)\right) \\ &\quad \boxplus \left(1, \beta + \sqrt{\kappa_{b3}} b, \mathrm{Im}\left\{\sqrt{\kappa_{b3}} \beta b^\dagger\right\}\right) \end{aligned}$$

$$= \left(e^{i\varphi}, \sqrt{\kappa_a}a + \left(e^{i\varphi}\sqrt{\kappa_{b2}} + \sqrt{\kappa_{b1}}\right)b,\ H_a + H_{bu} + \sin\varphi\sqrt{\kappa_{b1}\kappa_{b2}}b^\dagger b\right.$$
$$\left. + \operatorname{Im}\left\{e^{i\varphi}\sqrt{\kappa_a\kappa_{b2}}a^\dagger b + \sqrt{\kappa_a\kappa_{b1}}ab^\dagger\right\}\right) \boxplus \left(1, \beta + \sqrt{\kappa_{b3}}b, \operatorname{Im}\left\{\sqrt{\kappa_{b3}}\beta b^\dagger\right\}\right)$$
$$= \left(S_{ND}, \begin{bmatrix} \sqrt{\kappa_a}a + \left(e^{i\varphi}\sqrt{\kappa_{b2}} + \sqrt{\kappa_{b1}}\right)b \\ \beta + \sqrt{\kappa_{b3}}b \end{bmatrix}, H_a + H_{bu} + \sin\varphi\sqrt{\kappa_{b1}\kappa_{b2}}b^\dagger b\right.$$
$$\left. + \operatorname{Im}\left\{e^{i\varphi}\sqrt{\kappa_a\kappa_{b2}}a^\dagger b + \sqrt{\kappa_a\kappa_{b1}}ab^\dagger + \sqrt{\kappa_{b3}}\beta b^\dagger\right\}\right).$$

We thus have a total Hamiltonian

$$H = H_a + H_{bu} + \sin\varphi\sqrt{\kappa_{b1}\kappa_{b2}}b^\dagger b + \frac{\sqrt{\kappa_a\kappa_{b2}}}{2i}\left(e^{i\varphi}a^\dagger b - e^{-i\varphi}ab^\dagger\right)$$
$$+ \frac{\sqrt{\kappa_a\kappa_{b1}}}{2i}\left(ab^\dagger - a^\dagger b\right) + \frac{\sqrt{\kappa_{b3}}}{2i}\left(\beta b^\dagger - \beta^* b\right),$$

and (as we did before) we note that the second Lindblad term leads to terms in the master equation,

$$\begin{aligned}
[\dot\rho]_{L_2} &= L_2\rho L_2^\dagger - \frac{1}{2}L_2^\dagger L_2\rho - \frac{1}{2}\rho L_2^\dagger L_2 \\
&= |\beta|^2\rho + \beta\sqrt{\kappa_{b3}}\rho b^\dagger + \beta^*\sqrt{\kappa_{b3}}b\rho + \kappa_{b3}b\rho b^\dagger \\
&\quad -\frac{1}{2}\left(|\beta|^2 + \beta^*\sqrt{\kappa_{b3}}b + \beta\sqrt{\kappa_{b3}}b^\dagger + \kappa_{b3}b^\dagger b\right)\rho \\
&\quad -\frac{1}{2}\rho\left(|\beta|^2 + \beta^*\sqrt{\kappa_{b3}}b + \beta\sqrt{\kappa_{b3}}b^\dagger + \kappa_{b3}b^\dagger b\right) \\
&= \kappa_{b3}\left\{b\rho b^\dagger - \frac{1}{2}b^\dagger b\rho - \frac{1}{2}\rho b^\dagger b\right\} \\
&\quad +\frac{1}{2}\beta\sqrt{\kappa_{b3}}\rho b^\dagger + \frac{1}{2}\beta^*\sqrt{\kappa_{b3}}b\rho - \frac{1}{2}\beta\sqrt{\kappa_{b3}}b^\dagger\rho - \frac{1}{2}\rho\beta^*\sqrt{\kappa_{b3}}b \\
&= \kappa_{b3}\left\{b\rho b^\dagger - \frac{1}{2}b^\dagger b\rho - \frac{1}{2}\rho b^\dagger b\right\} + \frac{\sqrt{\kappa_{b3}}}{2}\left(\beta^* b - \beta b^\dagger\right)\rho + \frac{\sqrt{\kappa_{b3}}}{2}\rho\left(\beta b^\dagger - \beta^* b\right).
\end{aligned}$$

We retain the first term in braces as a modified $L_2 \to \sqrt{\kappa_{b3}}b$ and note that

$$\frac{\sqrt{\kappa_{b3}}}{2}\left(\beta^* b - \beta b^\dagger\right)\rho + \frac{\sqrt{\kappa_{b3}}}{2}\rho\left(\beta b^\dagger - \beta^* b\right) = \left[\frac{\sqrt{\kappa_{b3}}}{2}\left(\beta^* b - \beta b^\dagger\right), \rho\right],$$

meaning that we can combine this with the original Hamiltonian terms to obtain

$$\begin{aligned}
H &\to H_a + H_{bu} + \sin\varphi\sqrt{\kappa_{b1}\kappa_{b2}}b^\dagger b + \frac{\sqrt{\kappa_a\kappa_{b2}}}{2i}\left(e^{i\varphi}a^\dagger b - e^{-i\varphi}ab^\dagger\right) \\
&\quad + \frac{\sqrt{\kappa_a\kappa_{b1}}}{2i}\left(ab^\dagger - a^\dagger b\right) + i\sqrt{\kappa_{b3}}\left(\beta^* b - \beta b^\dagger\right), \\
L_1 &= \sqrt{\kappa_a}a + \left(e^{i\varphi}\sqrt{\kappa_{b2}} + \sqrt{\kappa_{b1}}\right)b, \\
L_2 &= \sqrt{\kappa_{b3}}b.
\end{aligned}$$

1.3 Limit theorem for QSDEs; cavity QED relay model

In this section, we take a very brief look at the use of QSDEs to model open quantum systems, and also at a very useful theorem for QSDEs that allows us to obtain highly simplified component models for certain cavity QED systems in the *small-volume limit.* We illustrate the use of this theorem in the derivation of a simple scattering matrix model for a cavity QED relay, which we will use later in our discussion of continuous quantum error correction.

General references for this section include, on QSDE modeling of quantum optical systems:

- Supplementary Material of J. Kerckhoff *et al.*, Designing quantum memories with embedded control: photonic circuits for autonomous quantum error correction. *Phys. Rev. Lett.*, **105**, 040502 (2010);

on limit theorems for QSDE models:

- L. Bouten, R. van Handel, and A. Silberfarb. Approximation and limit theorems for quantum stochastic models with unbounded coefficients. *J. Funct. Anal.*, **254**, 3123 (2008);

and on the cavity QED relay model (a slightly more complex version than the one we consider here):

- H. Mabuchi. Cavity-QED models of switches for attojoule-scale nanophotonic logic. *Phys. Rev. A*, **80**, 045802 (2009).

1.3.1 Preliminaries: Schrödinger evolution operator, Heisenberg picture, SDEs

The Schrödinger evolution operator $U\left(t_f, t_i\right)$ is defined by

$$|\psi\left(t_f\right)\rangle = U(t_f, t_i)\,|\psi\left(t_i\right)\rangle\,,$$

where the state must satisfy the Schrödinger equation

$$i\hbar\frac{d}{dt}\,|\psi\left(t\right)\rangle = H\,\,|\psi\left(t\right)\rangle\,.$$

Hence, we find the evolution equation

$$\begin{aligned} i\hbar\frac{d}{dt}\left(U\left(t,t_i\right)|\psi\left(t_i\right)\rangle\right) &= HU\left(t,t_i\right)|\psi\left(t_i\right)\rangle\,,\\ i\hbar\left(\frac{d}{dt}U\left(t,t_i\right)\right)|\psi\left(t_i\right)\rangle &= HU(t,t_i)\,|\psi\left(t_i\right)\rangle\,,\\ \frac{d}{dt}U\left(t,t_i\right) &= \frac{1}{i\hbar}HU\left(t,t_i\right), \quad U\left(t_i,t_i\right) = 1. \end{aligned}$$

If the Hamiltonian is constant, this can be solved straightforwardly via the operator exponential,

$$U\left(t,t_i\right) = \exp\left[-iH\left(t-t_i\right)/\hbar\right].$$

We note that any time-dependent operator moment can thus be written

$$\langle O \rangle_t = \langle \psi(t) | \, O \, | \psi(t) \rangle = \langle \psi(t_i) | \, U^\dagger(t, t_i) O U(t, t_i) \, | \psi(t_i) \rangle ,$$

and we note that we can reproduce all such observable quantities in the Heisenberg picture, where the state vector $|\psi(t_i)\rangle$ is considered to be constant (held to its initial condition) while the operators evolve according to

$$O(t) = U^\dagger(t, t_i) O U(t, t_i) .$$

In quantum field theories, it is generally more convenient to work in the Heisenberg picture than the Schrödinger picture.

In the following, we will have a quick look at some quantum input-output models based on QSDEs, which can be thought of as noncommutative generalizations of classical SDEs, which are commonly used in engineering, physics, and mathematical finance [19]. In both the quantum and classical cases, SDEs come in two different flavors: Stratonovich and Itô. When using Itô SDEs, one must be careful to observe the Itô rule, which says (in the classical case) that if x_t obeys the Itô SDE

$$dx_t = A(x_t)\, dt + B(x_t)\, dW_t,$$

then a variable y_t related to x_t via

$$y_t = f(x_t)$$

evolves according to

$$dy_t = \left[A(x_t) \frac{\partial f}{\partial x} + \frac{1}{2} B^2(x_t) \frac{\partial^2 f}{\partial x^2} \right] dt + B(x_t) \frac{\partial f}{\partial x}\, dW_t,$$

where the second-derivative term in the square brackets is known as the Itô correction. We can understand this as corresponding to a type of Taylor expansion in which we keep terms to *second* order and then apply the rules $dW_t^2 = dt$, $dW_t\, dt = dt^2 = 0$. Note that if f is a linear function, the Itô correction vanishes and we recover the prediction of normal calculus. An important advantage of working with Itô SDEs is that if x_t obeys the Itô SDE

$$dx_t = A(x_t)\, dt + B(x_t)\, dW_t,$$

then x_t is uncorrelated with dW_t. This considerably simplifies the computation of statistical moments. For example, consider the linear SDE model

$$dx_t = A\, x_t\, dt + F\, dV_t,$$

with x_t a scalar and $A < 0$ (the Ornstein-Uhlenbeck model). We then have

$$\begin{aligned} d\langle x_t \rangle &= A \langle x_t \rangle\, dt + F \langle dV_t \rangle \\ &= A \langle x_t \rangle\, dt, \\ \langle x_t \rangle &= \langle x_0 \rangle \exp(At) , \end{aligned}$$

and if $y_t = x_t^2$, so that $\langle\, y_t\rangle$ is the variance of x_t,

$$\begin{aligned} dy_t &= \left(2A\,x_t^2 + F^{\,2}\right) dt + 2Fx_t\, dV_t, \\ d\,\langle\, y_t\rangle &= \left(2A\,\langle\, y_t\rangle + F^{\,2}\right) dt + 2F\,\langle x_t\, dV_t\rangle \\ &= \left(2A\,\langle\, y_t\rangle + F^{\,2}\right) dt + 2F\,\langle x_t\rangle\,\langle dV_t\rangle \\ &= \left(2A\,\langle\, y_t\rangle + F^{\,2}\right) dt, \\ \langle\, y_t\rangle &= \exp\left(2At\right)\left[\langle\, y_0\rangle + \int_0^t ds \exp\left(-2As\right) F^{\,2}\right] \\ &= \exp\left(2At\right)\left[\langle\, y_0\rangle + F^{\,2}\int_0^t ds \exp\left(-2As\right)\right]. \end{aligned}$$

If we assume that x_t evolves from a known value x_0 at $t = 0$, then $\langle x_0\rangle = x_0$ and $\langle\, y_0\rangle = x_0^2$, and the mean-square uncertainty in x_t is

$$\begin{aligned} \left\langle x_t^2\right\rangle - \langle x_t\rangle^2 &= \langle\, y_t\rangle - \langle x_t\rangle^2 \\ &= \exp\left(2At\right) F^{\,2}\int_0^t ds \exp\left(-2As\right) \\ &= \exp\left(2At\right) F^{\,2}\left(-\frac{1}{2A}\right)\left[\exp\left(-2At\right) - 1\right] \\ &= -\frac{F^2}{2A}\left[1 - \exp\left(2At\right)\right]. \end{aligned}$$

The mean-square uncertainty thus has a steady-state value as $t \to \infty$ (assuming $A < 0$)

$$\left\langle x_t^2\right\rangle - \langle x_t\rangle^2 \to \frac{F^2}{2\,|A|}.$$

There are quantum generalizations of these "stochastic calculus" methods, which underlie the derivations presented in the following (see e.g. [20]), but these are beyond the scope of this chapter. A key point of similarity is that we will generally need to keep second-order terms in differential expressions and apply quantum generalizations of the rule $dW_t^{\,2} = dt$.

1.3.2 QSDE models of open quantum systems

In the Heisenberg picture, we can consider a unitary evolution operator U_t such that observables evolve according to

$$a_t = U_t\, a_0 U_t^{\,*},$$

which has the differential form

$$da_t = U_t\, a_0\, dU_t^{\,*} + dU_t\, a_0 U_t^{\,*} + dU_t\, a_0\, dU_t^{\,*}.$$

This evolution operator (which we note is the Hermitian conjugate of the usual Schrödinger evolution operator) is assumed to obey a "left" QSDE of the Hudson-Parthasarathy form,

$$dU_t = U_t \left\{ \sum_{i,\,j=1}^{n} \left(N_{i\,j} - \delta_{i\,j}\right) d\Lambda_t^{i\,j} + \sum_{i=1}^{n} M_i \left(dA_t^i\right)^* + \sum_{i=1}^{n} L_i \, dA_t^i + K \, dt \right\},$$

where it is required that

$$K + K^* = -\sum_{i=1}^{n} L_i L_i^*, \quad M_i = -\sum_{j=1}^{n} N_{i\,j} L_j^*, \quad \sum_{j=1}^{n} N_{m\,j} N_{l\,j}^* = \sum_{j=1}^{n} N_{jm}^* N_{jl} = \delta_{ml}.$$

We see from these conditions that the M_i are determined by $N_{i\,j}$ and the L_j, so such a QSDE is fully specified by $\{K, L_i, N_{i\,j}\}$.

With the abstract notation

$$dU_t = U_t \left\{F\right\}, \quad dU_t^* = \left\{F^*\right\} U_t^*,$$

we thus see that

$$\begin{aligned} da_t &= U_t \, a_0 \, dU_t^* + dU_t \, a_0 U_t^* + dU_t \, a_0 \, dU_t^* \\ &= U_t \, a_0 \left\{F^*\right\} U_t^* + U_t \left\{F\right\} a_0 U_t^* + U_t \left\{F\right\} a_0 \left\{F^*\right\} U_t^*. \end{aligned}$$

Following Gardiner's method for deriving the master equation, we can write (averaging over the noise terms)

$$\begin{aligned} d\left\langle a_t\right\rangle &= \left\langle da_t\right\rangle = \left\langle U_t \, a_0 \, dU_t^* + dU_t \, a_0 U_t^* + dU_t \, a_0 \, dU_t^*\right\rangle, \\ \frac{d}{dt}\left\langle a_t\right\rangle &= \operatorname{Tr}\left\{ \left(U_t a_0 K^* U_t^* + U_t K a_0 U_t^* + \sum_{i=1}^{n} U_t L_i a_0 L_i^* U_t^* \right) \rho_0 \right\}, \end{aligned}$$

and combining this with

$$\frac{d}{dt}\left\langle a\right\rangle = \operatorname{Tr}\left\{ a_0 \frac{d\rho_t}{dt} \right\}, \quad \rho_t = U_t^* \rho_0 U_t,$$

we obtain

$$\frac{d\rho_t}{dt} = K^* \rho_t + \rho_t \, K + \sum_{i=1}^{n} L_i^* \rho_t L_i.$$

We should be careful to note that we are here working with "left" QSDEs and that the operators L_i that have been written in the preceding discussion correspond to the Hermitian conjugates of the components of the coupling vector that we used in our discussion of (S, L, H) models. The difference arises because the Gough and James network calculus is conventionally derived from "right" QSDEs for the Schrödinger propagator $dV_t = dU_t^*$, whereas the QSDE limit theorem that we want to discuss has been formulated in terms of left QSDEs for technical reasons having to do with analysis of unbounded operators.

1.3.3 Generic cavity-QED scenario

For a (single-sided) cavity-QED system,

$$\begin{aligned} dU_t^* &= \Big[\sqrt{2\kappa}\,(a\,dA_t^* - a^* dA_t) + \sqrt{\gamma_{\|}}\,(\sigma\,dB_t^* - \sigma^* dB_t) - (i\Delta_c + \kappa)\,a^* a\,dt \\ &\quad - \left(i\Delta_a + \frac{1}{2}\gamma_{\|}\right)\sigma^*\sigma\,dt + g_0\,(a^*\sigma - a\sigma^*)\,dt\Big] U_t^*, \\ dU_t &= U_t\Big[-\sqrt{2\kappa}\,(a\,dA_t^* - a^* dA_t) - \sqrt{\gamma_{\|}}\,(\sigma\,dB_t^* - \sigma^* dB_t) + (i\Delta_c - \kappa)\,a^* a\,dt \\ &\quad + \left(i\Delta_a - \frac{1}{2}\gamma_{\|}\right)\sigma^*\sigma\,dt - g_0\,(a^*\sigma - a\sigma^*)\,dt\Big]. \end{aligned}$$

Comparing this with the generic Hudson-Parthasarathy form, we have

$$\begin{aligned} &K = (i\Delta_c - \kappa)\,a^* a + \left(i\Delta_a - \frac{1}{2}\gamma_{\|}\right)\sigma^*\sigma - g_0\,(a^*\sigma - a\sigma^*), \\ &L_1 = \sqrt{2\kappa}a^*, \quad L_2 = \sqrt{\gamma_{\|}}\sigma^*, \quad M_1 = -\sqrt{2\kappa}a, \quad M_2 = -\sqrt{\gamma_{\|}}\sigma, \quad N_{ij} = \delta_{ij}. \end{aligned}$$

Checking the Hudson-Parthasarathy conditions,

$$\begin{aligned} K + K^* &= (i\Delta_c - \kappa)\,a^* a + \left(i\Delta_a - \frac{1}{2}\gamma_{\|}\right)\sigma^*\sigma - g_0\,(a^*\sigma - a\sigma^*) \\ &\quad - (i\Delta_c + \kappa)\,a^* a - \left(i\Delta_a + \frac{1}{2}\gamma_{\|}\right)\sigma^*\sigma + g_0\,(a^*\sigma - a\sigma^*) \\ &= -2\kappa a^* a - \gamma_{\|}\sigma^*\sigma, \\ -\sum_{i=1}^{n} L_i L_i^* &= -\left(2\kappa a^* a + \gamma_{\|}\sigma^*\sigma\right), \\ -\sum_{j=1}^{n} N_{ij} L_j^* &= -L_i^* = M_i, \end{aligned}$$

and the condition on the N_{ij} is trivial.

Using our result to intuit the master equation, we expect

$$\begin{aligned} K^*\rho_t + \rho_t K &= (-i\Delta_c - \kappa)\,a^* a\,\rho_t + \left(-i\Delta_a - \frac{1}{2}\gamma_{\|}\right)\sigma^*\sigma\,\rho_t + g_0\,(a^*\sigma - a\sigma^*)\,\rho_t \\ &\quad + \rho_t\,(i\Delta_c - \kappa)\,a^* a + \rho_t\left(i\Delta_a - \frac{1}{2}\gamma_{\|}\right)\sigma^*\sigma - \rho_t\,g_0\,(a^*\sigma - a\sigma^*) \\ &= -i\left[\{\Delta_c a^* a + \Delta_a\sigma^*\sigma + ig_0\,(a^*\sigma - a\sigma^*)\}, \rho_t\right] \\ &\quad - \kappa a^* a\rho_t - \kappa\rho_t\,a^* a - \frac{\gamma_{\|}}{2}\sigma^*\sigma\,\rho_t - \frac{\gamma_{\|}}{2}\rho_t\,\sigma^*\sigma, \end{aligned}$$

$$\sum_{i=1}^{n} L_i^*\rho_t L_i = 2\kappa a\rho_t a^* + \gamma_{\|}\sigma\rho_t\sigma^*,$$

and the sum of these two terms gives us the master equation we expect.

1.3.4 Coherent input signal

To work with a coherent-state input, we define the Weyl operator,

$$\psi(f) = W(f)\,\phi,$$

where $f \in L^2(R^+)$ specifies the amplitude and phase of the field at every time $t \in R^+$. All operators should now evolve according to

$$j_t(a) = W(f_t)^* U_t\, a\, U_t^* W(f_t),$$

where f_t denotes the function f truncated at time t. In general, we have

$$dW(f_t) = \left[f(t)\, dA_t^* - f^*(t)\, dA_t - \frac{1}{2} |f(t)|^2\, dt \right] W(f_t),$$

$$dW^*(f_t) = W^*(f_t) \left[f^*(t)\, dA_t - f(t) dA_t^* - \frac{1}{2} |f(t)|^2\, dt \right],$$

and defining

$$U_t^W = W^*(f_t)\, U_t,$$

the Itô rule gives us

$$dU_t^W = dW^*(f_t)\, U_t + W^*(f_t)\, dU_t + dW^*(f_t)\, dU_t.$$

With the general Hudson-Parthasarathy QSDE for dU_t, we thus have

$$\begin{aligned} dU_t^W &= W^*(f_t) \left[f^*(t)\, dA_t^w - f(t)\, dA_t^{w*} - \frac{1}{2} |f(t)|\, dt \right] U_t \\ &\quad + W^*(f_t)\, U_t \left[\sum_{i,j=1}^{n} (N_{ij} - \delta_{ij})\, d\Lambda_t^{ij} + \sum_{i=1}^{n} M_i \left(dA_t^i\right)^* + \sum_{i=1}^{n} L_i\, dA_t^i + K\, dt \right] \\ &\quad + W^*(f_t)\, f^*(t)\, U_t\, M_w\, dt + W^*(f_t)\, U_t\, f^*(t) \sum_{j=1}^{n} (N_{wj} - \delta_{wj})\, dA_t^j \\ &= U_t^W \Bigg\{ \sum_{i,j=1}^{n} (N_{ij} - \delta_{ij})\, d\Lambda_t^{ij} + \sum_{i=1}^{n} [M_i - f(t)\, \delta_{iw}] \left(dA_t^i\right)^* \\ &\quad + \sum_{i=1}^{n} [L_i + f^*(t)\, N_{wi}]\, dA_t^i + \left[K + f^*(t)\, M_w - \frac{1}{2} |f(t)|^2 \right] dt \Bigg\}, \end{aligned}$$

where dA_t^w is the noise term associated with the displacement.

In the case of our cavity QED QSDE with $f = \alpha$ (constant), this would give us (dropping the W superscript on U_t^W)

$$\begin{aligned} dU_t = U_t \Bigg[&\left(\sqrt{2\kappa} a^* + \alpha^*\right) dA_t - \left(\sqrt{2\kappa} a + \alpha\right) dA_t^* - \sqrt{\gamma_\parallel}\, (\sigma\, dB_t^* - \sigma^* dB_t) \\ &- \left(\sqrt{2\kappa} \alpha^* a + \frac{1}{2} |\alpha|^2 \right) dt + (i\Delta_c - \kappa)\, a^* a\, dt + \left(i\Delta_a - \frac{1}{2} \gamma_\parallel \right) \sigma^* \sigma\, dt \\ &- g_0\, (a^* \sigma - a \sigma^*)\, dt \Bigg]. \end{aligned}$$

We confirm that with the new

$$K = -\left(\sqrt{2\kappa}\alpha^* a + \frac{1}{2}|\alpha|^2\right) + (i\Delta_c - \kappa)\, a^* a + \left(i\Delta_a - \frac{1}{2}\gamma_{\|}\right)\sigma^*\sigma - g_0\,(a^*\sigma - a\sigma^*)\,,$$

$$L_1 = \sqrt{2\kappa}a^* + \alpha^*, \quad L_2 = \sqrt{\gamma_{\|}}\sigma^*, \quad M_1 = -\left(\sqrt{2\kappa}a + \alpha\right), \quad M_2 = -\sqrt{\gamma_{\|}}\sigma,$$

and $N_{ij} = \delta_{ij}$, we have

$$\begin{aligned} K + K^* &= -\sqrt{2\kappa}\,(\alpha^* a + \alpha a^*) - 2\kappa a^* a - \gamma_{\|}\sigma^*\sigma - |\alpha|^2\,, \\ -\sum_{i=1}^{n} L_i L_i^* &= -\left(\sqrt{2\kappa}a^* + \alpha^*\right)\left(\sqrt{2\kappa}a + \alpha\right) - \gamma_{\|}\sigma^*\sigma \\ &= -2\kappa a^* a - \sqrt{2\kappa}\,(\alpha a^* + \alpha^* a) - |\alpha|^2 - \gamma_{\|}\sigma^*\sigma, \end{aligned}$$

which still satisfies the Hudson-Parthasarathy conditions. Looking at the implied master equation

$$\begin{aligned} \dot\rho_t &= K^*\rho_t + \rho_t K + \sum_{i=1}^{n} L_i^*\rho_t L_i \\ &\to -i\left[\{\Delta_c a^* a + \Delta_a\sigma^*\sigma + i g_0\,(a^*\sigma - a\sigma^*)\}\,, \rho_t\right] - |\alpha|^2\rho_t - \sqrt{2\kappa}\,(\alpha a^*\rho_t + \alpha^*\rho_t a) \\ &\quad -\kappa\,(\rho_t a^* a + a^* a\rho_t) - \frac{1}{2}\gamma_{\|}\,(\rho_t\sigma^*\sigma + \sigma^*\sigma\rho_t) + \left(\sqrt{2\kappa}a + \alpha\right)\rho_t\left(\sqrt{2\kappa}a^* + \alpha^*\right) \\ &\quad +\gamma_{\|}\sigma\rho_t\sigma^* \\ &= -i\left[\{\Delta_c a^* a + \Delta_a\sigma^*\sigma + i g_0\,(a^*\sigma - a\sigma^*)\}\,, \rho_t\right] + \kappa\,(2\kappa a\rho_t a^* - \rho_t a^* a - a^* a\rho_t) \\ &\quad + \frac{\gamma_{\|}}{2}\,(\sigma\rho_t\sigma^* - \rho_t\sigma^*\sigma - \sigma^*\sigma\rho_t) + \sqrt{2\kappa}\,(\alpha^* a\rho_t - \alpha^*\rho_t a - \alpha a^*\rho_t + \alpha\rho_t a^*) \\ &= -i\left[\left\{\Delta_c a^* a + \Delta_a\sigma^*\sigma + i g_0\,(a^*\sigma - a\sigma^*) - i\sqrt{2\kappa}\,(\alpha a^* - \alpha^* a)\right\}, \rho_t\right] \\ &\quad +\kappa\,(2\kappa a\rho_t a^* - \rho_t a^* a - a^* a\rho_t) + \frac{\gamma_{\|}}{2}\,(\sigma\rho_t\sigma^* - \rho_t\sigma^*\sigma - \sigma^*\sigma\rho_t)\,, \end{aligned}$$

we see that the effective change to the Hamiltonian is indeed the addition of a driving term $-i\sqrt{2\kappa}\,(\alpha a^* - \alpha^* a)$, as we have seen before.

Hopefully, this calculation shows that, in practice, it is much easier to use the $\lhd\ (1, \alpha, 0)$ method! But the Weyl operator transformation just described is in fact the rigorous underpinning of the simplified method.

1.3.5 QSDE limit theorem of Bouten, van Handel, and Silberfarb

Suppose we have a Hudson-Parthasarathy left QSDE whose coefficients contain a scaling factor k :

$$K^{(k)} = k^2 Y + kA + B, \quad L_i^{(k)} = kF_i + G_i, \quad N_{ij}^{(k)} = W_{ij}.$$

Suppose also that there exists a closed subspace $\mathbf{H}_0 \subset \mathbf{H}$ such that the following hold:

(1) $YP_0 = 0$.

(2) There exists $\tilde{Y}$ such that $\tilde{Y}Y = Y\tilde{Y} = P_1$.
(3) $F_j^\dagger P_0 = 0$ for all $1 \leq j \leq n$.
(4) $P_0 \, A \, P_0 = 0$.

Here P_0 is an orthogonal projector onto $\mathbf{H}_0$ and $P_1 = 1 - P_0$ on $\mathbf{H}$. (The full statement of these "structural conditions" is a bit more technical.) Then the theorem states that in the limit $k \to \infty$, the scaled QSDE limits to a QSDE on $\mathbf{H}_0$ with coefficients

$$K = P_0 \left(B - A\tilde{Y}A \right) P_0, \quad L_i = P_0 \left(G_i - A\tilde{Y}F_i \right) P_0,$$

$$M_i = -\sum_{j=1}^{n} P_0 W_{ij} \left(G_j^\dagger - F_j^\dagger \tilde{Y} A \right) P_0, \quad N_{ij} = \sum_{l=1}^{n} P_0 W_{il} \left(F_l^\dagger \tilde{Y} F_j + \delta_{lj} \right) P_0.$$

This limit holds in the sense that

$$\lim_{k\to\infty} \sup_{0 \leq t \leq T} \left\| U_t^{(k)*} \psi - U_t^* \psi \right\| = 0 \quad \forall \psi \in \mathbf{H}_0 \otimes \mathbf{F}, \quad T < \infty,$$

where $U_t^{(k)}$ is the solution of the Hudson-Parthasarathy equation (the Heisenberg-picture propagator) in the original model with scaling parameter k, and U_t is the solution in the limit QSDE. Thus, $U_t^* \psi$ is the solution of the Schrödinger equation on the joint state space of the system and its input-output fields, and the limit stated here amounts to strong uniform convergence on compact time intervals. The proof is rather technical and relies on a version of the Trotter-Kato theorem for QSDEs [10].

1.3.6 The "small-volume" limit

We are interested in using the theorem just described to obtain component model abstractions in limits that are relevant to nanophotonic device physics. One such limit is the small-volume limit, in which g and κ both naturally become large. In particular, we may be interested in scenarios in which $g, \kappa \to \infty$ such that g/κ stays fixed, or perhaps g^2/κ stays fixed. Generally, g^2 scales inversely with the volume of a resonator, and for a Fabry-Perot–type cavity, one can note that κ will tend to increase as the cavity volume decreases simply because the circulating light will bounce off of the end mirrors more often. For monolithic or photonic-crystal defect resonators, κ also increases as the volume decreases because of fundamental limits on the confinement of light within structures of subwavelength dimension. The other parameters appearing in these models, which retain their original values in the limit model, are things such as atomic decay rates and input field powers. As briefly discussed by Bouter *et al.* [10], the regime with $(g \sim \kappa) \gg \gamma$ seems accessible with current systems of interest for nanophotonic cavity QED.

1.3.7 Four-state, three-cavity model for the SR flip-flop switch

We now attempt to derive an SR flip-flop switch model using a four-level atom coupled to three cavity modes. The QSDE terms are as follows:

$$
\begin{aligned}
K =& - k_1^2\kappa_a a^* a - k_1^2 g_a \left(a^*\sigma_{ge} - a\sigma_{ge}^*\right) - k_2^2\kappa_b b^* b - k_2^2\kappa_c c^* c \\
& - k_2 g_b \left(b^*\sigma_{gr} - b\sigma_{gr}^*\right) - k_2 g_b \left(b^*\sigma_{he} - b\sigma_{he}^*\right) - k_2 g_c \left(c^*\sigma_{hr} - c\sigma_{hr}^*\right), \\
L_1 =& k_1\sqrt{\kappa_a}a^*, \quad L_2 = k_1\sqrt{\kappa_a}a^*, \quad L_3 = k_2\sqrt{2\kappa_b}b^*, \quad L_4 = k_2\sqrt{2\kappa_c}c^*,
\end{aligned}
$$

and $N_{ij} = \delta_{ij}$. The kind of situation we have in mind here involves a multilevel "atom" with ground states $\{|h\rangle, |g\rangle\}$ and excited states $\{|e\rangle, |r\rangle\}$. Cavity mode a connects $|g\rangle \leftrightarrow |e\rangle$ only, cavity mode b connects $|g\rangle \leftrightarrow |r\rangle$ (SET) and $|h\rangle \leftrightarrow |e\rangle$, and cavity mode c connects $|h\rangle \leftrightarrow |r\rangle$ (RESET). (Note that this QSDE corresponds to Fig. 1.5 in this chapter, but Bouter *et al.* [10] assume a different atomic level structure that can be used when the control field and routed field for the switch/relay have very different frequencies.)

We will first take the limit $k_1 \to \infty$ and then $k_2 \to \infty$. For the first elimination,

$$
\begin{aligned}
K &= k^2 Y + kA + B, \quad L_i = kF_i + G_i, \quad N_{ij} = W_{ij}, \\
Y &= -\kappa_a a^* a - g_a \left(a^*\sigma_{ge} - a\sigma_{ge}^*\right), \quad A = 0, \\
B &= -k_2^2\kappa_b b^* b - k_2^2\kappa_c c^* c - k_2 g_b \left(b^*\sigma_{gr} - b\sigma_{gr}^*\right) - k_2 g_b \left(b^*\sigma_{he} - b\sigma_{he}^*\right) \\
&\quad -k_2 g_c \left(c^*\sigma_{hr} - c\sigma_{hr}^*\right),
\end{aligned}
$$

$$
F = \begin{pmatrix} \sqrt{\kappa_a}a^* \\ \sqrt{\kappa_a}a^* \\ 0 \\ 0 \end{pmatrix}, \quad G = \begin{pmatrix} 0 \\ 0 \\ k_2\sqrt{2\kappa_b}b^* \\ k_2\sqrt{2\kappa_c}c^* \end{pmatrix}, \quad W_{ij} = \delta_{ij}.
$$

We choose $H_0 = \text{span}\{|g\,0_a\,n_b\,n_c\rangle, |h\,0_a\,n_b\,n_c\rangle, |r\,0_a\,n_b\,n_c\rangle\}$, which clearly lies within the kernel of Y. We define $\tilde{Y}$ by its action on an orthonormal basis of states. We begin with basis states of the form $\{|g\,0_a\,n_b\,n_c\rangle, |h\,0_a\,n_b\,n_c\rangle, |r\,0_a\,n_b\,n_c\rangle\}$, for which

$$
\begin{aligned}
Y\ \{|g\,0_a\,n_b\,n_c\rangle, |h\,0_a\,n_b\,n_c\rangle, |r\,0_a\,n_b\,n_c\rangle\} &= 0, \\
\tilde{Y}\ \{|g\,0_a\,n_b\,n_c\rangle, |h\,0_a\,n_b\,n_c\rangle, |r\,0_a\,n_b\,n_c\rangle\} &= 0.
\end{aligned}
$$

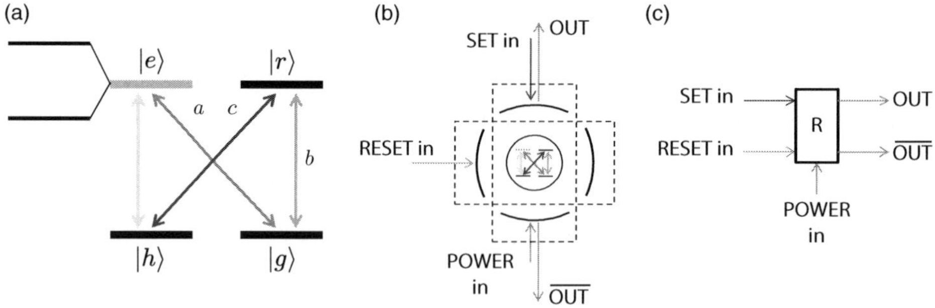

Fig. 1.5 (a) Level diagram for the "atomic" system coupled to cavity modes in the relay construction. (b) Arrangement of resonator modes and correspondence of inputs/outputs to logical signals. (c) Component diagram for the relay, showing logical signals only.

This ensures that $Y\tilde{Y}P_0 = \tilde{Y}YP_0 = 0$. We next note that, for $n_a \geq 1$,

$$\begin{aligned}
Y\,|g\,n_a\,n_b\,n_c\rangle &= -\kappa_a n_a\,|g\,n_a\,n_b\,n_c\rangle + g_a\sqrt{n_a}\,|e\,(n_a-1)_a\,n_b\,n_c\rangle\,,\\
Y\,|e\,(n-1)_a\,n_b\,n_c\rangle &= -\kappa_a\,(n_a-1)\,|e\,(n_a-1)_a\,n_b\,n_c\rangle - g_a\sqrt{n_a}\,|g\,n_a\,n_b\,n_c\rangle\,,\\
Y\,|h\,n_a\,n_b\,n_c\rangle &= -\kappa_a n_a\,|h\,n_a\,n_b\,n_c\rangle\,,\\
Y\,|r\,n_a\,n_b\,n_c\rangle &= -\kappa_a n_a\,|r\,n_a\,n_b\,n_c\rangle\,.
\end{aligned}$$

We can thus define (still for $n_a \geq 1$)

$$\begin{aligned}
\tilde{Y}\,|r\,n_a\,n_b\,n_c\rangle &= -\frac{1}{\kappa_a n_a}\,|r\,n_a\,n_b\,n_c\rangle\,,\\
\tilde{Y}\,|h\,n_a\,n_b\,n_c\rangle &= -\frac{1}{\kappa_a n_a}\,|h\,n_a\,n_b\,n_c\rangle\,,\\
\tilde{Y}\,|e\,(n-1)_a\,n_b\,n_c\rangle &= \frac{g_a\sqrt{n_a}}{\kappa_a^2 n_a\,(n_a-1) - g_a^2 n_a}\,|g\,n_a\,n_b\,n_c\rangle\\
&\quad -\frac{\kappa_a n_a}{\kappa_a^2 n_a\,(n_a-1) - g_a^2 n_a}\,|e\,(n_a-1)_a\,n_b\,n_c\rangle\,,\\
\tilde{Y}\,|g\,n_a\,n_b\,n_c\rangle &= -\frac{\kappa_a\,(n_a-1)}{\kappa_a^2 n_a\,(n_a-1) - g_a^2 n_a}\,|g\,n_a\,n_b\,n_c\rangle\\
&\quad +\frac{g_a\sqrt{n_a}}{\kappa_a^2 n_a\,(n_a-1) - g_a^2 n_a}\,|e\,(n_a-1)_a\,n_b\,n_c\rangle\,,
\end{aligned}$$

and thus satisfy $Y\tilde{Y} = \tilde{Y}Y = P_1$. Checking the remaining structural conditions, we require

$$F_i^* P_0 = 0,$$

which is evident by inspection, and

$$P_0 A P_0 = 0,$$

which is likewise trivial. Moving on to compute the limit coefficients

$$\begin{aligned}
&\breve{K} = P_0\left(B - A\tilde{Y}A\right)P_0, \quad \breve{L}_i = P_0\left(G_i - A\tilde{Y}F_i\right)P_0,\\
&\breve{M}_i = -\sum_{j=1}^{n} P_0 W_{ij}\left(G_j^* - F_j^*\tilde{Y}A\right)P_0, \quad \breve{N}_{ij} = \sum_{l=1}^{n} P_0 W_{il}\left(F_l^*\tilde{Y}F_j + \delta_{lj}\right)P_0,
\end{aligned}$$

we first obtain

$$\begin{aligned}
P_0 B P_0 &= \left[-k_2^2\kappa_b b^* b - k_2^2\kappa_c c^* c - k_2 g_b\left(b^*\sigma_{gr} - b\sigma_{gr}^*\right) - k_2 g_b\left(b^*\sigma_{he} - b\sigma_{he}^*\right)\right.\\
&\quad \left.-k_2 g_c\left(c^*\sigma_{hr} - c\sigma_{hr}^*\right)\right]P_0\\
&= -k_2^2\kappa_b b^* b - k_2^2\kappa_c c^* c - k_2 g_b\left(b^*\sigma_{gr} - b\sigma_{gr}^*\right) - k_2 g_b P_0\left(b^*\sigma_{he} - b\sigma_{he}^*\right)P_0\\
&\quad -k_2 g_c\left(c^*\sigma_{hr} - c\sigma_{hr}^*\right)\\
&= -k_2^2\kappa_b b^* b - k_2^2\kappa_c c^* c - k_2 g_b\left(b^*\sigma_{gr} - b\sigma_{gr}^*\right) - k_2 g_c\left(c^*\sigma_{hr} - c\sigma_{hr}^*\right),\\
\breve{K} &= -k_2^2\kappa_b b^* b - k_2^2\kappa_c c^* c - k_2 g_b\left(b^*\sigma_{gr} - b\sigma_{gr}^*\right) - k_2 g_c\left(c^*\sigma_{hr} - c\sigma_{hr}^*\right).
\end{aligned}$$

Next

$$\breve{L}_1 = \breve{L}_2 = 0, \quad \breve{L}_3 = k_2\sqrt{2\kappa_b}b^*, \quad \breve{L}_4 = k_2\sqrt{2\kappa_c}c^*,$$

and

$$\breve{M}_1 = \breve{M}_2 = 0, \quad \breve{M}_3 = -k_2\sqrt{2\kappa_b}b, \quad \breve{M}_3 = -k_2\sqrt{2\kappa_c}c.$$

To compute the scattering matrix, we first note

$$\begin{aligned}
a^*P_0 &= |\, g\, 1_a\rangle\, \langle g\, 0_a|\otimes 1_b \otimes 1_c + |\, h\, 1_a\rangle\, \langle h\, 0_a|\otimes 1_b\otimes 1_c \\
&\quad + |\, r\, 1_a\rangle\, \langle r\, 0_a|\otimes 1_b\otimes 1_c, \\
\tilde{Y}a^*P_0 &= -\frac{1}{g_a}\, |\, e\, 0_a\rangle\, \langle g\, 0_a|\otimes 1_b\otimes 1_c - \frac{1}{\kappa_a}\, |\, h\, 1_a\rangle\, \langle h\, 0_a|\otimes 1_b\otimes 1_c \\
&\quad -\frac{1}{\kappa_a}\, |\, r\, 1_a\rangle\, \langle r\, 0_a|\otimes 1_b\otimes 1_c, \\
a\tilde{Y}a^*P_0 &= -\frac{1}{\kappa_a}\, |\, h\, 0_a\rangle\, \langle h\, 0_a|\otimes 1_b\otimes 1_c - \frac{1}{\kappa_a}\, |\, r\, 0_a\rangle\, \langle r\, 0_a|\otimes 1_b\otimes 1_c, \\
\kappa_a P_0 a\tilde{Y}a^*P_0 &= -\left\{\Pi_{h0_a} + \Pi_r 0_a\right\}\otimes 1_b\otimes 1_c.
\end{aligned}$$

$$\breve{N}_{11} = P_0\left(F_1^*\tilde{Y}F_1 + 1\right)P_0 = 1 + \kappa_a P_0 a\tilde{Y}a^*P_0 = \Pi_g,$$

$$\breve{N}_{12} = P_0F_1^*\tilde{Y}F_2P_0 = \kappa_a P_0 a\tilde{Y}a^*P_0 = -\Pi_{hr},$$

$$\breve{N}_{13} = P_0\left(F_1^*\tilde{Y}F_3\right)P_0 = 0, \quad \breve{N}_{14} = P_0\left(F_1^*\tilde{Y}F_4\right)P_0 = 0,$$

$$\breve{N}_{21} = P_0\left(F_2^*\tilde{Y}F_1\right)P_0 = \kappa_a P_0 a\tilde{Y}a^*P_0 = -\Pi_{hr},$$

$$\breve{N}_{22} = P_0\left(F_2^*\tilde{Y}F_2 + 1\right)P_0 = 1 + \kappa_a P_0 a\tilde{Y}a^*P_0 = \Pi_g,$$

$$\breve{N}_{23} = P_0\left(F_2^*\tilde{Y}F_3\right)P_0 = 0, \quad \breve{N}_{24} = P_0\left(F_2^*\tilde{Y}F_4\right)P_0 = 0,$$

$$\breve{N}_{31} = P_0\left(F_3^*\tilde{Y}F_1\right)P_0 = 0, \quad \breve{N}_{32} = P_0\left(F_3^*\tilde{Y}F_2 + 1\right)P_0 = 0,$$

$$\breve{N}_{33} = P_0\left(F_3^*\tilde{Y}F_3 + 1\right)P_0 = 1, \quad \breve{N}_{34} = P_0\left(F_4^*\tilde{Y}F_4\right)P_0 = 0,$$

$$\breve{N}_{41} = P_0\left(F_4^*\tilde{Y}F_1\right)P_0 = 0, \quad \breve{N}_{42} = P_0\left(F_4^*\tilde{Y}F_2\right)P_0 = 0,$$

$$\breve{N}_{43} = P_0\left(F_4^*\tilde{Y}F_3\right)P_0 = 0, \quad \breve{N}_{44} = P_0\left(F_4^*\tilde{Y}F_4 + 1\right)P_0 = 1.$$

Collecting together the results, we have

$$K = -k_2^2\kappa_b b^*b - k_2^2\kappa_c c^*c - k_2 g_b\left(b^*\sigma_{gr} - b\sigma_{gr}^*\right) - k_2 g_c\left(c^*\sigma_{hr} - c\sigma_{hr}^*\right),$$

$$L = \begin{pmatrix} 0 \\ 0 \\ k_2\sqrt{2\kappa_b}b^* \\ k_2\sqrt{2\kappa_c}c^* \end{pmatrix}, \quad M = \begin{pmatrix} 0 \\ 0 \\ -k_2\sqrt{2\kappa_b}b \\ -k_2\sqrt{2\kappa_c}c \end{pmatrix}, \quad N = \begin{pmatrix} \Pi_g & -\Pi_{hr} & 0 & 0 \\ -\Pi_{hr} & \Pi_g & 0 & 0 \\ 0 & 0 & 1 & 0 \\ 0 & 0 & 0 & 1 \end{pmatrix}.$$

Checking the Hudson-Parthasarathy conditions,

$$K + K^* = -2k_2^2\kappa_b b^* b - 2k_2^2\kappa_c c^* c,$$

$$-\sum_{i=1}^{4} L_i L_i^* = -k_2^2 2\kappa_b b^* b - k_2^2 2\kappa_c c^* c,$$

and

$$-NL^* = \begin{pmatrix} -\Pi_g & \Pi_{hr} & 0 & 0 \\ \Pi_{hr} & -\Pi_g & 0 & 0 \\ 0 & 0 & -1 & 0 \\ 0 & 0 & 0 & -1 \end{pmatrix} \begin{pmatrix} 0 \\ 0 \\ k_2\sqrt{2\kappa_b}b \\ k_2\sqrt{2\kappa_c}c \end{pmatrix} = \begin{pmatrix} 0 \\ 0 \\ -k_2\sqrt{2\kappa_b}b \\ -k_2\sqrt{2\kappa_c}c \end{pmatrix} = M,$$

$$NN = \begin{pmatrix} \Pi_g & -\Pi_{hr} & 0 & 0 \\ -\Pi_{hr} & \Pi_g & 0 & 0 \\ 0 & 0 & 1 & 0 \\ 0 & 0 & 0 & 1 \end{pmatrix} \begin{pmatrix} \Pi_g & -\Pi_{hr} & 0 & 0 \\ -\Pi_{hr} & \Pi_g & 0 & 0 \\ 0 & 0 & 1 & 0 \\ 0 & 0 & 0 & 1 \end{pmatrix} = \begin{pmatrix} 1 & 0 & 0 & 0 \\ 0 & 1 & 0 & 0 \\ 0 & 0 & 1 & 0 \\ 0 & 0 & 0 & 1 \end{pmatrix},$$

so everything looks fine after the first elimination.

For the second elimination, we have

$$Y = -\kappa_b b^* b - \kappa_c c^* c, \quad A = -g_b\left(b^*\sigma_{gr} - b\sigma_{gr}^*\right) - g_c\left(c^*\sigma_{hr} - c\sigma_{hr}^*\right), \quad B = 0,$$

$$F = \begin{pmatrix} 0 \\ 0 \\ \sqrt{2\kappa_b}b^* \\ \sqrt{2\kappa_c}c^* \end{pmatrix}, \quad G = \begin{pmatrix} 0 \\ 0 \\ 0 \\ 0 \end{pmatrix}, \quad W = \begin{pmatrix} \Pi_g & -\Pi_{hr} & 0 & 0 \\ -\Pi_{hr} & \Pi_g & 0 & 0 \\ 0 & 0 & 1 & 0 \\ 0 & 0 & 0 & 1 \end{pmatrix}.$$

We choose $H_0 = \mathrm{span}\left\{|\,g\,0_b\,0_c\rangle, |\,h\,0_b\,0_c\rangle, |\,r\,0_b\,0_c\rangle\right\}$, which clearly lies in the kernel of Y. We define

$$\tilde{Y}\;|\,g\,0_b\,0_c\rangle = \tilde{Y}\;|\,h\,0_b\,0_c\rangle = \tilde{Y}\;|\,r\,0_b\,0_c\rangle = 0,$$

and

$$\tilde{Y}\;|\,g\,n_b\,n_c\rangle = -\frac{1}{\kappa_b n_b + \kappa_c n_c}\,|\,g\,n_b\,n_c\rangle, \quad n_b + n_c \geq 1,$$

$$\tilde{Y}\;|\,h\,n_b\,n_c\rangle = -\frac{1}{\kappa_b n_b + \kappa_c n_c}\,|\,h\,n_b\,n_c\rangle, \quad n_b + n_c \geq 1,$$

$$\tilde{Y}\;|\,r\,n_b\,n_c\rangle = -\frac{1}{\kappa_b n_b + \kappa_c n_c}\,|\,r\,n_b\,n_c\rangle, \quad n_b + n_c \geq 1,$$

which achieves the desired condition $Y\tilde{Y} = \tilde{Y}Y = P_1$. Checking the remaining structural conditions,

$$F_1^* P_0 = F_2^* P_0 = 0, \quad F_3^* P_0 = \sqrt{2\kappa_b}bP_0 = 0, \quad F_4^* P_0 = \sqrt{2\kappa_c}cP_0 = 0,$$

and

$$
\begin{aligned}
AP_0 &= \left[-g_b\left(b^*\sigma_{gr} - b\sigma^*_{gr}\right) - g_c\left(c^*\sigma_{hr} - c\sigma^*_{hr}\right)\right]\left(\Pi_g + \Pi_h + \Pi_r\right) \otimes |\,0_b 0_c\rangle\,\langle 0_b 0_c| \\
&= -g_b\sigma_{gr} \otimes |\,1_b 0_c\rangle\,\langle 0_b 0_c| - g_c\sigma_{hr} \otimes |\,0_b 1_c\rangle\,\langle 0_b 0_c|\,, \\
P_0 A P_0 &= 0.
\end{aligned}
$$

Hence, we can move on to computing the limit coefficients,

$$
\begin{aligned}
&\breve{K} = P_0\left(B - A\tilde{Y}A\right)P_0, \quad \breve{L}_i = P_0\left(G_i - A\tilde{Y}F_i\right)P_0, \\
&\breve{M}_i = -\sum_{j=1}^{n} P_0 W_{ij}\left(G_j^* - F_j^*\tilde{Y}A\right)P_0, \quad \breve{N}_{ij} = \sum_{l=1}^{n} P_0 W_{il}\left(F_l^*\tilde{Y}F_j + \delta_{lj}\right)P_0.
\end{aligned}
$$

Starting with the first, we have

$$
\begin{aligned}
AP_0 &= -g_b\sigma_{gr} \otimes |\,1_b 0_c\rangle\,\langle 0_b 0_c| - g_c\sigma_{hr} \otimes |\,0_b 1_c\rangle\,\langle 0_b 0_c|\,, \\
\tilde{Y}AP_0 &= \frac{g_b}{\kappa_b}\sigma_{gr} \otimes |\,1_b 0_c\rangle\,\langle 0_b 0_c| + \frac{g_c}{\kappa_c}\sigma_{hr} \otimes |\,0_b 1_c\rangle\,\langle 0_b 0_c|\,, \\
A\tilde{Y}AP_0 &= \frac{g_b^2}{\kappa_b}\Pi_r\,|\,0_b 0_c\rangle\,\langle 0_b 0_c| + \frac{g_c^2}{\kappa_c}\Pi_r\,|\,0_b 0_c\rangle\,\langle 0_b 0_c|\,, \\
\breve{K} &= -P_0 A\tilde{Y}AP_0 = -\left(\frac{g_b^2}{\kappa_b} + \frac{g_c^2}{\kappa_c}\right)\Pi_r.
\end{aligned}
$$

Next, $\breve{L}_1 = \breve{L}_2 = 0$,

$$
\begin{aligned}
F_3 P_0 &= \sqrt{2\kappa_b}\left(\Pi_g + \Pi_h + \Pi_r\right) \otimes |\,1_b 0_c\rangle\,\langle 0_b 0_c|\,, \\
\tilde{Y}F_3 P_0 &= -\sqrt{\frac{2}{\kappa_b}}\left(\Pi_g + \Pi_h + \Pi_r\right) \otimes |\,1_b 0_c\rangle\,\langle 0_b 0_c|\,, \\
A\tilde{Y}F_3 P_0 &= -\sqrt{\frac{2}{\kappa_b}}\Big(g_b\sigma_{rg} \otimes |\,0_b 0_c\rangle\,\langle 0_b 0_c| - \sqrt{2}g_b\sigma_{gr} \otimes |\,2_b 0_c\rangle\,\langle 0_b 0_c| \\
&\quad - g_c\sigma_{hr} \otimes |\,1_b 1_c\rangle\,\langle 0_b 0_c|\Big), \\
\breve{L}_3 &= -P_0 A\tilde{Y}F_3 P_0 = \sqrt{\frac{2g_b^2}{\kappa_b}}\sigma_{rg},
\end{aligned}
$$

$$
\begin{aligned}
F_4 P_0 &= \sqrt{2\kappa_c}\left(\Pi_g + \Pi_h + \Pi_r\right) \otimes |\,0_b 1_c\rangle\,\langle 0_b 0_c|\,, \\
\tilde{Y}F_4 P_0 &= -\sqrt{\frac{2}{\kappa_c}}\left(\Pi_g + \Pi_h + \Pi_r\right) \otimes |\,0_b 1_c\rangle\,\langle 0_b 0_c|\,, \\
A\tilde{Y}F_4 P_0 &= -\sqrt{\frac{2}{\kappa_c}}\Big(g_c\sigma_{rh} \otimes |\,0_b 0_c\rangle\,\langle 0_b 0_c| - g_b\sigma_{gr} \otimes |\,1_b 1_c\rangle\,\langle 0_b 0_c| \\
&\quad - \sqrt{2}g_c\sigma_{hr} \otimes |\,0_b 2_c\rangle\,\langle 0_b 0_c|\Big), \\
\breve{L}_4 &= -P_0 A\tilde{Y}F_4 P_0 = \sqrt{\frac{2g_c^2}{\kappa_c}}\sigma_{rh}.
\end{aligned}
$$

Next,

$$\breve{M}_1 = P_0\Pi_g F_1^*\tilde{Y}AP_0 - P_0\Pi_{hr}F_2^*\tilde{Y}AP_0 = 0,$$

$$\breve{M}_2 = -P_0\Pi_{hr}F_1^*\tilde{Y}AP_0 + P_0\Pi_g F_2^*\tilde{Y}AP_0 = 0,$$

$$\begin{aligned}\breve{M}_3 &= -P_0\left(-F_3^*\tilde{Y}A\right)P_0\\ &= P_0\sqrt{2\kappa_b}b\left(\frac{g_b}{\kappa_b}\sigma_{gr}\otimes|1_b0_c\rangle\langle 0_b0_c| + \frac{g_c}{\kappa_c}\sigma_{hr}\otimes|0_b1_c\rangle\langle 0_b0_c|\right) = \sqrt{\frac{2g_b^2}{\kappa_b}}\sigma_{gr},\end{aligned}$$

$$\begin{aligned}\breve{M}_4 &= -P_0\left(-F_4^*\tilde{Y}A\right)P_0\\ &= P_0\sqrt{2\kappa_c}c\left(\frac{g_b}{\kappa_b}\sigma_{gr}\otimes|1_b0_c\rangle\langle 0_b0_c| + \frac{g_c}{\kappa_c}\sigma_{hr}\otimes|0_b1_c\rangle\langle 0_b0_c|\right) = \sqrt{\frac{2g_c^2}{\kappa_c}}\sigma_{hr}.\end{aligned}$$

Finally

$$\breve{N}_{11} = P_0W_{11}\left(F_1^*\tilde{Y}F_1 + \delta_{11}\right)P_0 + P_0W_{12}\left(F_2^*\tilde{Y}F_1 + \delta_{21}\right)P_0 = P_0W_{11}P_0 = \Pi_g,$$

$$\breve{N}_{12} = P_0W_{11}\left(F_1^*\tilde{Y}F_2 + \delta_{12}\right)P_0 + P_0W_{12}\left(F_2^*\tilde{Y}F_2 + \delta_{22}\right)P_0 = P_0W_{12}P_0 = -\Pi_{hr},$$

$$\breve{N}_{21} = P_0W_{21}\left(F_1^*\tilde{Y}F_1 + \delta_{11}\right)P_0 + P_0W_{22}\left(F_2^*\tilde{Y}F_1 + \delta_{21}\right)P_0 = P_0W_{21}P_0 = -\Pi_{hr},$$

$$\breve{N}_{22} = P_0W_{21}\left(F_1^*\tilde{Y}F_2 + \delta_{12}\right)P_0 + P_0W_{22}\left(F_2^*\tilde{Y}F_2 + \delta_{22}\right)P_0 = P_0W_{22}P_0 = \Pi_g,$$

$$\breve{N}_{33} = P_0W_{33}\left(F_3^*\tilde{Y}F_3 + \delta_{33}\right)P_0 = 1 - 2\left(\Pi_g + \Pi_h + \Pi_r\right)\otimes|0_b0_c\rangle\langle 0_b0_c| = -1,$$

$$\breve{N}_{34} = P_0W_{33}\left(F_3^*\tilde{Y}F_4 + \delta_{34}\right)P_0 = P_0 2\sqrt{\kappa_b\kappa_c}b\left(\Pi_g + \Pi_h + \Pi_r\right)\otimes|0_b1_c\rangle\langle 0_b0_c| = 0,$$

$$\breve{N}_{43} = P_0W_{44}\left(F_4^*\tilde{Y}F_3 + \delta_{43}\right)P_0 = P_0 2\sqrt{\kappa_b\kappa_c}c\left(\Pi_g + \Pi_h + \Pi_r\right)\otimes|1_b0_c\rangle\langle 0_b0_c| = 0,$$

$$\breve{N}_{44} = P_0W_{44}\left(F_4^*\tilde{Y}F_4 + \delta_{44}\right)P_0 = 1 - 2\left(\Pi_g + \Pi_h + \Pi_r\right)\otimes|0_b0_c\rangle\langle 0_b0_c| = -1.$$

Collecting the results and setting

$$\frac{2g_b^2}{\kappa_b} = \frac{2g_c^2}{\kappa_c} \equiv \gamma, \quad K = -\gamma\Pi_r,$$

we have

$$L = \begin{pmatrix} 0 \\ 0 \\ \sqrt{\gamma}\sigma_{rg} \\ \sqrt{\gamma}\sigma_{rh} \end{pmatrix}, \quad M = \begin{pmatrix} 0 \\ 0 \\ \sqrt{\gamma}\sigma_{gr} \\ \sqrt{\gamma}\sigma_{hr} \end{pmatrix}, \quad N = \begin{pmatrix} \Pi_g & -\Pi_{hr} & 0 & 0 \\ -\Pi_{hr} & \Pi_g & 0 & 0 \\ 0 & 0 & -1 & 0 \\ 0 & 0 & 0 & -1 \end{pmatrix}.$$

Now, we are finally ready to take the limit $\gamma \to \infty$. We set

$$Y = -\gamma\Pi_r, \quad A = B = 0,$$

$$F = \begin{pmatrix} 0 \\ 0 \\ \sqrt{\gamma}\sigma_{rg} \\ \sqrt{\gamma}\sigma_{rh} \end{pmatrix}, \quad G = \begin{pmatrix} 0 \\ 0 \\ 0 \\ 0 \end{pmatrix}, \quad W = \begin{pmatrix} \Pi_g & -\Pi_{hr} & 0 & 0 \\ -\Pi_{hr} & \Pi_g & 0 & 0 \\ 0 & 0 & -1 & 0 \\ 0 & 0 & 0 & -1 \end{pmatrix}.$$

Then, $H_0 = \operatorname{span}\{|g\rangle, |h\rangle\}$, and we set

$$\tilde{Y}\,|g\rangle = \tilde{Y}\,|h\rangle = 0, \quad \tilde{Y}\,|r\rangle = -\gamma^{-1}|r\rangle.$$

Then,

$$\breve{K} = 0,$$

$$\breve{L}_1 = \breve{L}_2 = \breve{L}_3 = \breve{L}_4 = 0,$$

$$\breve{M}_1 = \breve{M}_2 = \breve{M}_3 = \breve{M}_4 = 0,$$

$$\breve{N}_{11} = P_0 W_{11}\left(F_1^*\tilde{Y}F_1 + \delta_{11}\right)P_0 + P_0 W_{12}\left(F_2^*\tilde{Y}F_1 + \delta_{21}\right)P_0 = \Pi_g,$$

$$\breve{N}_{12} = P_0 W_{11}\left(F_1^*\tilde{Y}F_2 + \delta_{12}\right)P_0 + P_0 W_{12}\left(F_2^*\tilde{Y}F_2 + \delta_{22}\right)P_0 = -\Pi_h,$$

$$\breve{N}_{21} = P_0 W_{21}\left(F_1^*\tilde{Y}F_1 + \delta_{11}\right)P_0 + P_0 W_{22}\left(F_2^*\tilde{Y}F_1 + \delta_{21}\right)P_0 = -\Pi_h,$$

$$\breve{N}_{22} = P_0 W_{21}\left(F_1^*\tilde{Y}F_2 + \delta_{12}\right)P_0 + P_0 W_{22}\left(F_2^*\tilde{Y}F_2 + \delta_{22}\right)P_0 = \Pi_g,$$

$$\breve{N}_{33} = P_0 W_{33}\left(F_3^*\tilde{Y}F_3 + \delta_{33}\right)P_0 = 1 - P_0 F_3^*\tilde{Y}F_3 P_0 = 1 - \Pi_g = \Pi_h,$$

$$\breve{N}_{34} = P_0 W_{33}\left(F_3^*\tilde{Y}F_4 + \delta_{34}\right)P_0 = -P_0 F_3^*\tilde{Y}F_4 P_0 = -\sigma_{gr}\sigma_{rh} = -\sigma_{gh},$$

$$\breve{N}_{43} = P_0 W_{44}\left(F_4^*\tilde{Y}F_3 + \delta_{43}\right)P_0 = -P_0 F_4^*\tilde{Y}F_3 P_0 = -\sigma_{hr}\sigma_{rg} = -\sigma_{hg},$$

$$\breve{N}_{44} = P_0 W_{44}\left(F_4^*\tilde{Y}F_4 + \delta_{44}\right)P_0 = 1 - P_0 F_4^*\tilde{Y}F_4 P_0 = 1 - \Pi_h = \Pi_g.$$

Collecting the results together,

$$K = 0, \quad L = 0, \quad M = 0, \quad N = \begin{pmatrix} \Pi_g & -\Pi_h & 0 & 0 \\ -\Pi_h & \Pi_g & 0 & 0 \\ 0 & 0 & \Pi_h & -\sigma_{gh} \\ 0 & 0 & -\sigma_{hg} & \Pi_g \end{pmatrix}.$$

We should keep in mind that these are Hudson-Parthasarathy coefficients for a left QSDE model; in general, we convert to the (S_r, L_r, H_r) triple for a right QSDE model via

$$S_r = N^{\dagger}, \quad L_r = L^{\sharp}, \quad H_r = \mathrm{Im}\{K\},$$

where for a vector or matrix of operators or complex numbers $A = [a_{ij}]$, we define $A^{\#} \equiv [a^*_{ij}]$, where a^*_{ij} is the complex conjugate of a number or the Hermitian conjugate of an operator, and $A^{\dagger} \equiv [a^*_{ji}] = (A^{\#})^T$.

1.3.8 Model with coherent signals

Finally, we add displacements to modes 1, 3, and 4 to obtain a model with known, coherent signals. With the general rules

$$M_i \to M_i - f_w \delta_{iw}, \quad L_i \to L_i + f^*_w N_{wi}, \quad K \to K + f^*_w M_w - \frac{1}{2}|f_w|^2,$$

and $f_1 = \beta$, $f_3 = \alpha_s$, $f_4 = \alpha_r$,

$$K = -\frac{1}{2}|\beta|^2 - \frac{1}{2}|\alpha_s|^2 - \frac{1}{2}|\alpha_r|^2,$$

$$L = \begin{pmatrix} \beta^*\Pi_g \\ -\beta^*\Pi_h \\ \alpha^*_s\Pi_h - \alpha^*_r\sigma_{hg} \\ -\alpha^*_s\sigma_{gh} + \alpha^*_r\Pi_g \end{pmatrix}, \quad M = \begin{pmatrix} -\beta \\ 0 \\ -\alpha_s \\ -\alpha_r \end{pmatrix}, \quad N = \begin{pmatrix} \Pi_g & -\Pi_h & 0 & 0 \\ -\Pi_h & \Pi_g & 0 & 0 \\ 0 & 0 & \Pi_h & -\sigma_{gh} \\ 0 & 0 & -\sigma_{hg} & \Pi_g \end{pmatrix}.$$

The master equation then reads

$$\begin{aligned}
\dot{\rho}_t = &-|\beta|^2\rho_t - |\alpha_s|^2\rho_t - |\alpha_r|^2\rho_t + |\beta|^2(\Pi_g\rho_t\Pi_g + \Pi_h\rho_t\Pi_h) \\
&+ (\alpha_s\Pi_h - \alpha_r\sigma_{gh})\rho_t(\alpha^*_s\Pi_h - \alpha^*_r\sigma^*_{gh}) + (\alpha_s\sigma_{hg} - \alpha_r\Pi_g)\rho_t(\alpha^*_s\sigma^*_{hg} - \alpha^*_r\Pi_g) \\
= &-|\beta|^2(\Pi_g\rho_t\Pi_h + \Pi_h\rho_t\Pi_g) - |\alpha_s|^2\rho_t - |\alpha_r|^2\rho_t \\
&+ (\alpha_s\Pi_h - \alpha_r\sigma_{gh})\rho_t(\alpha^*_s\Pi_h - \alpha^*_r\sigma^*_{gh}) + (\alpha_s\sigma_{hg} - \alpha_r\Pi_g)\rho_t(\alpha^*_s\sigma^*_{hg} - \alpha^*_r\Pi_g) \\
= &\left(|\beta|^2 + \frac{|\alpha_s|^2}{2} + \frac{|\alpha_r|^2}{2}\right)(Z\rho_t Z - \rho_t) \\
&+ \frac{|\alpha_s|^2}{2}\left(2\sigma_{hg}\rho_t\sigma^*_{hg} - \sigma^*_{hg}\sigma_{hg}\rho_t - \rho_t\sigma^*_{hg}\sigma_{hg}\right) \\
&+ \frac{|\alpha_r|^2}{2}\left(2\sigma_{gh}\rho_t\sigma^*_{gh} - \sigma^*_{gh}\sigma_{gh}\rho_t - \rho_t\sigma^*_{gh}\sigma_{gh}\right) \\
&- (\alpha^*_s\alpha_r\sigma_{gh})\rho_t\Pi_h - \Pi_h\rho_t(\alpha^*_s\alpha_r\sigma_{gh})^* - (\alpha_s\alpha^*_r\sigma_{hg})\rho_t\Pi_g - \Pi_g\rho_t(\alpha_s\alpha^*_r\sigma_{hg})^*,
\end{aligned}$$

where $Z \equiv \Pi_g - \Pi_h$.

1.3.9 Pre-limit master equation with displacements and spontaneous emission

For use in numerical simulations, we apply displacements to the pre-limit model as well, obtaining

$$
\begin{aligned}
K &= -k_1^2\kappa_a a^* a - k_2^2\kappa_b b^* b - k_2^2\kappa_c c^* c - k_1^2 g_a\left(a^*\sigma_{ge} - a\sigma_{ge}^*\right) - k_2 g_b\left(b^*\sigma_{gr} - b\sigma_{gr}^*\right)\\
&\quad - k_2 g_b\left(b^*\sigma_{he} - b\sigma_{he}^*\right) - k_2 g_c\left(c^*\sigma_{hr} - c\sigma_{hr}^*\right)\\
&\quad -\beta^* k_1\sqrt{\kappa_a}a - \alpha_s^* k_2\sqrt{2\kappa_b}b - \alpha_r^* k_3\sqrt{2\kappa_c}c - \frac{1}{2}|\beta|^2 - \frac{1}{2}|\alpha_s|^2 - \frac{1}{2}|\alpha_r|^2,\\
L_1 &= k_1\sqrt{\kappa_a}a^* + \beta^*, \quad M_1 = -k_1\sqrt{\kappa_a}a - \beta,\\
L_2 &= k_1\sqrt{\kappa_a}a^*, \quad M_2 = -k_1\sqrt{\kappa_a}a,\\
L_3 &= k_2\sqrt{2\kappa_b}b^* + \alpha_s^*, \quad M_3 = -k_2\sqrt{2\kappa_b}b - \alpha_s,\\
L_4 &= k_2\sqrt{2\kappa_c}c^* + \alpha_r^*, \quad M_4 = -k_2\sqrt{2\kappa_c}c - \alpha_r, \quad N_{ij} = \delta_{ij}.
\end{aligned}
$$

The corresponding master equation is then

$$
\begin{aligned}
\dot{\rho}_t &= -i\left[H', \rho_t\right] + \left(-\beta k_1\sqrt{\kappa_a}a^* - \alpha_s k_2\sqrt{2\kappa_b}b^* - \alpha_r k_3\sqrt{2\kappa_c}c^*\right)\rho_t\\
&\quad + \rho_t\left(-\beta^* k_1\sqrt{\kappa_a}a - \alpha_s^* k_2\sqrt{2\kappa_b}b - \alpha_r^* k_3\sqrt{2\kappa_c}c\right) - |\beta|^2\rho_t - |\alpha_s|^2\rho_t - |\alpha_r|^2\rho_t\\
&\quad + k_1^2\kappa_a\left(-a^* a\rho_t - \rho_t a^* a\right) + k_2^2\kappa_b\left(-b^* b\rho_t - \rho_t b^* b\right) + k_2^2\kappa_c\left(-c^* c\rho_t - \rho_t c^* c\right)\\
&\quad + \left(k_1\sqrt{\kappa_a}a + \beta\right)\rho_t\left(k_1\sqrt{\kappa_a}a^* + \beta^*\right) + \left(k_1\sqrt{\kappa_a}a\right)\rho_t\left(k_1\sqrt{\kappa_a}a^*\right)\\
&\quad + \left(k_2\sqrt{2\kappa_b}b + \alpha_s\right)\rho_t\left(k_2\sqrt{2\kappa_b}b^* + \alpha_s^*\right) + \left(k_2\sqrt{2\kappa_c}c + \alpha_r\right)\rho_t\left(k_2\sqrt{2\kappa_c}c^* + \alpha_r^*\right),\\
H' &= ik_1^2 g_a\left(a^*\sigma_{ge} - a\sigma_{ge}^*\right) + ik_2 g_b\left(b^*\sigma_{gr} - b\sigma_{gr}^*\right) + ik_2 g_b\left(b^*\sigma_{he} - b\sigma_{he}^*\right)\\
&\quad + ik_2 g_c\left(c^*\sigma_{hr} - c\sigma_{hr}^*\right).
\end{aligned}
$$

This can all be simplified to

$$
\begin{aligned}
\dot{\rho}_t &= -i\left[H, \rho_t\right] + k_1^2\kappa_a\left(2a\rho_t a^* - a^* a\rho_t - \rho_t a^* a\right)\\
&\quad + k_2^2\kappa_b\left(2b\rho_t b^* - b^* b\rho_t - \rho_t b^* b\right) + k_2^2\kappa_c\left(2c\rho_t c^* - c^* c\rho_t - \rho_t c^* c\right),\\
H &= H' + +ik_1\sqrt{\kappa_a}\left(\beta^* a - \beta a^*\right) + ik_2\sqrt{2\kappa_b}\left(\alpha_s^* b - \alpha_s b^*\right) + ik_2\sqrt{2\kappa_c}\left(\alpha_r^* c - \alpha_r c^*\right).
\end{aligned}
$$

Finally, we add atomic spontaneous emission terms of all four kinds:

$$
\begin{aligned}
\dot{\rho}_t &= -i\left[H, \rho_t\right] + \gamma_{ge}\left(2\sigma_{ge}\rho_t\sigma_{ge}^* - \sigma_{ge}^*\sigma_{ge}\rho_t - \rho_t\sigma_{ge}^*\sigma_{ge}\right)\\
&\quad + k_1^2\kappa_a\left(2a\rho_t a^* - a^* a\rho_t - \rho_t a^* a\right) + \gamma_{hr}\left(2\sigma_{hr}\rho_t\sigma_{hr}^* - \sigma_{hr}^*\sigma_{hr}\rho_t - \rho_t\sigma_{hr}^*\sigma_{hr}\right)\\
&\quad + k_2^2\kappa_b\left(2b\rho_t b^* - b^* b\rho_t - \rho_t b^* b\right) + \gamma_{gr}\left(2\sigma_{gr}\rho_t\sigma_{gr}^* - \sigma_{gr}^*\sigma_{gr}\rho_t - \rho_t\sigma_{gr}^*\sigma_{gr}\right)\\
&\quad + k_2^2\kappa_c\left(2c\rho_t c^* - c^* c\rho_t - \rho_t c^* c\right) + \gamma_{he}\left(2\sigma_{he}\rho_t\sigma_{he}^* - \sigma_{he}^*\sigma_{he}\rho_t - \rho_t\sigma_{he}^*\sigma_{he}\right).
\end{aligned}
$$

From the adiabatic eliminations studied above, we know that we would like to have k_1, k_2, g_b^2/κ_b and g_c^2/κ_c all large. We presumably also need $k_1 \gg k_2$ in order to justify the order in which we performed the adiabatic eliminations. We thus propose the following master equation for simulation:

$$
\begin{aligned}
\dot{\rho}_t &= -i\left[H,\rho_t\right] + \left(2\sigma_{ge}\rho_t\sigma_{ge}^* - \sigma_{ge}^*\sigma_{ge}\rho_t - \rho_t\sigma_{ge}^*\sigma_{ge}\right)\\
&\quad + k_1^2\left(2a\rho_t a^* - a^*a\rho_t - \rho_t a^*a\right) + \left(2\sigma_{hr}\rho_t\sigma_{hr}^* - \sigma_{hr}^*\sigma_{hr}\rho_t - \rho_t\sigma_{hr}^*\sigma_{hr}\right)\\
&\quad + k_2^2\left(2b\rho_t b^* - b^*b\rho_t - \rho_t b^*b\right) + \left(2\sigma_{gr}\rho_t\sigma_{gr}^* - \sigma_{gr}^*\sigma_{gr}\rho_t - \rho_t\sigma_{gr}^*\sigma_{gr}\right)\\
&\quad + k_2^2\left(2c\rho_t c^* - c^*c\rho_t - \rho_t c^*c\right) + \left(2\sigma_{he}\rho_t\sigma_{he}^* - \sigma_{he}^*\sigma_{he}\rho_t - \rho_t\sigma_{he}^*\sigma_{he}\right),\\
H &= ik_1^2\left(a^*\sigma_{ge} - a\sigma_{ge}^*\right) + ik_2 g\left(b^*\sigma_{gr} - b\sigma_{gr}^*\right) + ik_2 g\left(b^*\sigma_{he} - b\sigma_{he}^*\right)\\
&\quad + ikg\left(c^*\sigma_{hr} - c\sigma_{hr}^*\right) + ik_1\left(\beta^* a - \beta a^*\right) + ik_2\sqrt{2}\left(\alpha_s^* b - \alpha_s b^*\right)\\
&\quad + ik_2\sqrt{2}\left(\alpha_r^* c - \alpha_r c^*\right),
\end{aligned}
$$

to be studied in the regime $k_1 \gg k_2 \gg 1$. It may also be of interest to vary the ratio g, which could be assigned a default value of 1.

Computing finally the output modes, we use the general expression [15]

$$
(d\mathbf{A}_t)_{\text{out}} = \mathbf{S}\, d\mathbf{A}_t + \mathbf{L}\, dt,
$$

where the boldface symbols represent matrices or vectors. Since we are particularly interested in modes 1 and 2 and we have $\mathbf{S} = \mathbf{I}$,

$$
\begin{aligned}
L_1 &= k_1\sqrt{\kappa_a}a^* + \beta^*, \quad M_1 = -k_1\sqrt{\kappa_a}a - \beta,\\
L_2 &= k_1\sqrt{\kappa_a}a^*, \quad M_2 = -k_1\sqrt{\kappa_a}a,
\end{aligned}
$$

we find

$$
\begin{aligned}
\left(dA_t^1\right)_{\text{out}} &= \left(k_1\sqrt{\kappa_a}a^* + \beta^*\right)dt + dA_t^1,\\
\left(dA_t^2\right)_{\text{out}} &= k_1\sqrt{\kappa_a}a^* dt + dA_t^2.
\end{aligned}
$$

With the idea that the average intracavity photon number in the empty (no-atom) cavity would be

$$
\langle a^*a\rangle \to \frac{\left|k_1\sqrt{\kappa_a}\beta\right|^2}{k_1^4\kappa_a^2} = \frac{|\beta|^2}{k_1^2\kappa_a},
$$

we would expect the output fields out each mirror to have magnitude $k_1\sqrt{\kappa_a}\sqrt{\langle a^*a\rangle} \to \beta$, so this should all work provided the phases do. Explicitly, with the parameter definitions that we have proposed, we should check

$$
\left\langle \left(dA_t^1\right)_{\text{out}}\right\rangle = \left(k_1\left\langle a^*\right\rangle + \beta^*\right)dt, \quad \left\langle \left(dA_t^2\right)_{\text{out}}\right\rangle = k_1\left\langle a^*\right\rangle dt,
$$

as α_s and α_r are alternatively activated. We would expect that when α_s is on, the cavity should go transparent and we would have $\left\langle (dA_t^1)_{\text{out}}\right\rangle = 0$, $\left\langle (dA_t^2)_{\text{out}}\right\rangle = -\beta^* dt$, while if α_r is on, the cavity should go reflective and $\left\langle (dA_t^1)_{\text{out}}\right\rangle = \beta^* dt$, $\left\langle (dA_t^2)_{\text{out}}\right\rangle = 0$. Comparing this with the adiabatically eliminated model with

$$
L = \begin{pmatrix} \beta^*\Pi_g \\ -\beta^*\Pi_h \\ \alpha_s^*\Pi_h - \alpha_r^*\sigma_{hg} \\ -\alpha_s^*\sigma_{gh} + \alpha_r^*\Pi_g \end{pmatrix}, \quad N = \begin{pmatrix} \Pi_g & -\Pi_h & 0 & 0 \\ -\Pi_h & \Pi_g & 0 & 0 \\ 0 & 0 & \Pi_h & -\sigma_{gh} \\ 0 & 0 & -\sigma_{hg} & \Pi_g \end{pmatrix},
$$

we would have

$$\begin{aligned}(d\mathbf{A}_t)_{\rm out} &= \mathbf{S}\,d\mathbf{A}_t + \mathbf{L}\,dt,\\ \left(dA_t^1\right)_{\rm out} &= \Pi_g dA_t^1 - \Pi_h dA_t^2 + \beta^*\Pi_g\,dt,\\ \left(dA_t^2\right)_{\rm out} &= -\Pi_h dA_t^1 + \Pi_g dA_t^2 - \beta^*\Pi_h\,dt.\end{aligned}$$

If $\alpha_s = \alpha_r = 0$ and $\rho = \Pi_h$ (cavity transparent), we would have

$$\left\langle \left(dA_t^1\right)_{\rm out}\right\rangle = 0, \quad \left\langle \left(dA_t^2\right)_{\rm out}\right\rangle = -\beta^*\,dt,$$

while if $\rho = \Pi_g$ (cavity reflective), we would have

$$\left\langle \left(dA_t^1\right)_{\rm out}\right\rangle = \beta^*\,dt, \quad \left\langle \left(dA_t^2\right)_{\rm out}\right\rangle = 0.$$

So that seems to agree fully.

A direct comparison of an analogous limit model with numerical results from integrating the corresponding pre-limit master equation, for a slightly more complex cavity QED relay model, can be found in [11].

1.4 State observers and recursive filters in classical feedback control theory

We turn now to a tutorial introduction to state observers for linear dynamical input-output systems. The concept of an observer as an "online" signal-processing component within a real-time feedback loop is fundamental in modern control theory, but seems hardly to be known in the general physics community. For this reason, the discussion here will focus on the elementary example of second-order systems (damped harmonic oscillators), with or without noise driving the oscillator motion and corrupting the measurement signal, and builds basic concepts of observability and optimal observer gain. It is hoped that exposure to these control-theoretic ideas will make enable an interested student to understand properly the role that the stochastic Schrödinger and stochastic master equation play (as optimal recursive filters) in measurement-based quantum feedback control.

1.4.1 State-feedback control example: second-order system

Consider the driven second-order system

$$\ddot{q} = -2\zeta\omega_0\dot{q} - \omega_0^2 q + u, \quad x_1 \equiv q, \quad x_2 \equiv \dot{q},$$

$$\frac{d}{dt}\begin{bmatrix} x_1 \\ x_2 \end{bmatrix} = \begin{bmatrix} 0 & 1 \\ -\omega_0^2 & -2\zeta\omega_0 \end{bmatrix}\begin{bmatrix} x_1 \\ x_2 \end{bmatrix} + \begin{bmatrix} 0 \\ u \end{bmatrix}.$$

Here, u could represent an external applied force (in a mechanical mass-spring-damper system) or voltage (in the LCR circuit realization). Anticipating conventional control-theoretic notation that we will introduce below, let us also write this as

$$\frac{d}{dt}\vec{x} = A\,\vec{x} + Bu, \quad A = \begin{bmatrix} 0 & 1 \\ -\omega_0^2 & -2\zeta\omega_0 \end{bmatrix}, \quad B = \begin{bmatrix} 0 \\ 1 \end{bmatrix}.$$

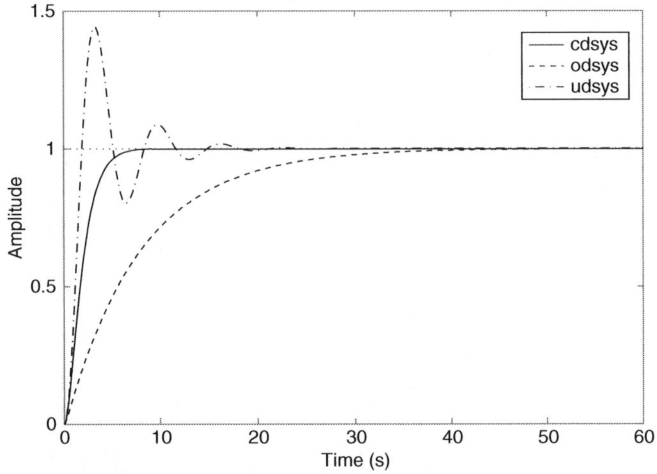

Fig. 1.6 Step responses for a damped harmonic oscillator: udsys, underdamped response; cdsys, critically damped response; odsys, overdamped response.

Recall that the damping ratio $\zeta \geq 0$ sets important properties of the step (transient) response, determining whether the response is overdamped ($\zeta > 1$), critically damped ($\zeta = 1$) or underdamped.

We can write a general solution (Fig. 1.6) for the initial-value problem for such a system via the matrix exponential

$$\begin{bmatrix} x_1(t) \\ x_2(t) \end{bmatrix} = \exp(At) \begin{bmatrix} x_1(0) \\ x_2(0) \end{bmatrix},$$

where methods for computing $\exp(At)$ can be found in elementary linear algebra textbooks:

$$\zeta > 1 \rightarrow \frac{e^{-\gamma t}}{\sqrt{\zeta^2 - 1}} \begin{bmatrix} \sqrt{\zeta^2 - 1}\cosh(\delta t) + \zeta \sinh(\delta t) & \frac{1}{\omega_0}\sinh(\delta t) \\ -\omega_0 \sinh(\delta t) & \sqrt{\zeta^2 - 1}\cosh(\delta t) - \zeta \sinh(\delta t) \end{bmatrix},$$

$$\zeta = 1 \rightarrow \begin{bmatrix} e^{-\omega_0 t} + t\omega_0 e^{-\omega_0 t} & te^{-\omega_0 t} \\ -\omega_0 e^{-\omega_0 t} + \omega_0 \left(e^{-\omega_0 t} - t\omega_0 e^{-\omega_0 t}\right) & e^{-\omega_0 t} - t\omega_0 e^{-\omega_0 t} \end{bmatrix},$$

$$\zeta < 1 \rightarrow \frac{e^{-\gamma t}}{\sqrt{1 - \zeta^2}} \begin{bmatrix} \zeta \sin(\nu t) + \sqrt{1 - \zeta^2}\cos(\nu t) & \frac{1}{\omega_0}\sin(\nu t) \\ -\omega_0 \sin(\nu t) & -\zeta \sin(\nu t) + \sqrt{1 - \zeta^2}\cos(\nu t) \end{bmatrix}.$$

Let us briefly consider the linear state-feedback control scenario

$$u = -K\,\vec{x},$$

where (since we have taken u to be scalar)

$$K = \begin{bmatrix} k_1 & k_2 \end{bmatrix}.$$

With this feedback law, we have

$$\frac{d}{dt}\vec{x} = A\vec{x} + Bu = (A - BK)\,\vec{x},$$

where

$$\begin{aligned} A - BK \to & \begin{bmatrix} 0 & 1 \\ -\omega_0^2 & -2\zeta\omega_0 \end{bmatrix} - \begin{bmatrix} 0 \\ 1 \end{bmatrix} \begin{bmatrix} k_1 & k_2 \end{bmatrix} \\ &= \begin{bmatrix} 0 & 1 \\ -\omega_0^2 & -2\zeta\omega_0 \end{bmatrix} - \begin{bmatrix} 0 & 0 \\ k_1 & k_2 \end{bmatrix} \\ &= \begin{bmatrix} 0 & 1 \\ -\omega_0^2 - k_1 & -2\zeta\omega_0 - k_2 \end{bmatrix}. \end{aligned}$$

Clearly, we can view this as a modified second-order system. If we define

$$\omega_0' = \sqrt{\omega_0^2 + k_1},$$

then apparently we have

$$\begin{aligned} \zeta'\omega_0' &= \zeta\omega_0 + \frac{k_2}{2}, \\ \zeta' &= \frac{\zeta\omega_0 + k_2/2}{\sqrt{\omega_0^2 + k_1}}, \end{aligned}$$

and it follows that we can actually set ω_0' to any value we like through choice of the feedback parameter k_1, and once this is set (assuming $\omega_0' \neq 0$), we can adjust ζ' as desired through choice of k_2. Hence, if we do not like the transient behavior of our original open-loop system (without feedback), we can in principle modify it however we like by use of *state feedback* as described above. We program the desired behavior by setting the feedback gain K.

Note that if accurate, instantaneous measurements of the components of $\vec{x}$ are available as electronic signals, it is straightforward to produce the feedback signal u with an op-amp circuit.

The fact that we have full power to reprogram the oscillation frequency and damping ratio in this state-feedback scenario can be made obvious if we look at the dynamics with feedback, transformed back to a single second-order ordinary differential equation:

$$\begin{gathered} \frac{d}{dt}\begin{bmatrix} q \\ \dot{q} \end{bmatrix} = \begin{bmatrix} 0 & 1 \\ -\omega_0^2 - k_1 & -2\zeta\omega_0 - k_2 \end{bmatrix} \begin{bmatrix} q \\ \dot{q} \end{bmatrix}, \\ 0 = \ddot{q} + (2\zeta\omega_0 + k_2)\,\dot{q} + \left(\omega_0^2 + k_1\right) q. \end{gathered}$$

Hence, k_1 is associated with a modification of the harmonic restoring force while k_2 directly and independently modifies the velocity damping.

Just for fun, let us calculate the eigenvalues of the feedback-modified A matrix:

$$\begin{aligned} 0 &= \det(A - BK - I\lambda) \to \begin{bmatrix} -\lambda & 1 \\ -\omega_0^2 - k_1 & -2\zeta\omega_0 - k_2 - \lambda \end{bmatrix}, \\ 0 &= \lambda(2\zeta\omega_0 + k_2 + \lambda) + \omega_0^2 + k_1 \\ &= \lambda^2 + (2\zeta\omega_0 + k_2)\lambda + \omega_0^2 + k_1. \end{aligned}$$

Considering the solutions given by the quadratic equation,

$$\lambda = \frac{1}{2}\left(-2\zeta\omega_0 - k_2 \pm \sqrt{(2\zeta\omega_0 + k_2)^2 - 4\omega_0^2 - 4k_1}\right),$$

we see that we can achieve any desired pair of eigenvalues $\{\lambda_+, \lambda_-\}$ by setting

$$\begin{aligned} \lambda_+ + \lambda_- &= -2\zeta\omega_0 - k_2, \\ k_2 &= -2\zeta\omega_0 - \lambda_+ - \lambda_-, \end{aligned}$$

and

$$\begin{aligned} \lambda_+ - \lambda_- &= \sqrt{(2\zeta\omega_0 + k_2)^2 - 4\omega_0^2 - 4k_1}, \\ k_1 &= \frac{1}{4}(2\zeta\omega_0 + k_2)^2 - \omega_0^2 - \frac{1}{4}(\lambda_+ - \lambda_-)^2 \\ &= \frac{1}{4}(\lambda_+ + \lambda_-)^2 - \omega_0^2 - \frac{1}{4}(\lambda_+ - \lambda_-)^2 \\ &= \lambda_+\lambda_- - \omega_0^2. \end{aligned}$$

Hence, we find that we can arbitrarily program the eigenvalues of the closed-loop dynamics matrix, a design method called "pole placement."

Before turning to some generalities, we note that the second-order system with the control structure we are considering can be pushed into arbitrary states by appropriately driving u. Suppose we are starting at the origin at $t = 0$ and want to reach $\begin{bmatrix} q & \dot{q} \end{bmatrix}'$ at a target time T. If we use the input u and allow our input signals to be unbounded, we can in principle take the following approach. First, give the system a momentum impulse at time $t = 0$ that will cause the system to pass through position q at time T. To convince ourselves that this is really possible, we could look at the matrix exponential solution and show that we can always solve for $\dot{q}_0$ such that

$$\begin{bmatrix} q \\ \cdot \end{bmatrix} = \exp(AT)\begin{bmatrix} 0 \\ \dot{q}_0 \end{bmatrix}.$$

In the overdamped case, for example, this leads to

$$\dot{q}_0 = \frac{\omega_0\sqrt{\zeta^2 - 1}}{e^{-\omega_0\zeta T}\sinh\left(\omega_0 T\sqrt{\zeta^2 - 1}\right)} q.$$

Then, at time T, apply a second momentum impulse that "corrects" the final velocity $\dot{q}$ to its desired value. Note that, even with this rather singular strategy, in which we rely on our ability to apply controls so strong that they overwhelm most of the system's natural dynamics (harmonic restoring force and velocity damping), we still rely on the inherent integrator structure

$$\frac{d}{dt}q = \dot{q}$$

that allows us to use an input to $\dot{q}$ only to affect the position q.

1.4.2 Reachability rank condition

We say that a linear system

$$\dot{x} = Ax + Bu$$

is *reachable* if, for any initial state $x(0) = x_0$, desired final state x_f, and "target time" T, it is possible to find a control input $u(t)$, $t \in [0, T]$ that steers the system to reach $x(T) = x_f$. There is a theorem (see [1], pp. 167–170) that says that a system is reachable if its *reachability matrix*

$$W_r \equiv \begin{bmatrix} B & AB & A^2B & \cdots & A^{n-1}B \end{bmatrix}$$

has full rank (here $x \in \mathbb{R}^n$). If a system is reachable, it is furthermore possible to solve the *pole-placement* problem, in which we want to design a state-feedback law

$$u = -Kx$$

such that we can pick any eigenvalues we want for the controlled dynamics

$$\begin{aligned} \dot{x} &= Ax + Bu \\ &= (A - BK)\,x. \end{aligned}$$

Here A and B are given, and we must find a K to achieve the desired eigenvalues for $A - BK$. There is a very convenient Matlab Control Toolbox function, place, that can be used to accomplish this numerically.

In the case of our second-order system with

$$A = \begin{bmatrix} 0 & 1 \\ -\omega_0^2 & -2\zeta\omega_0 \end{bmatrix}, \quad B = \begin{bmatrix} 0 \\ 1 \end{bmatrix},$$

we have

$$AB = \begin{bmatrix} 0 & 1 \\ -\omega_0^2 & -2\zeta\omega_0 \end{bmatrix} \begin{bmatrix} 0 \\ 1 \end{bmatrix} = \begin{bmatrix} 1 \\ -2\zeta\omega_0 \end{bmatrix}, \quad W_r = \begin{bmatrix} 0 & 1 \\ 1 & -2\zeta\omega_0 \end{bmatrix},$$

and clearly W_r has full rank as $\det(W_r) = -1$.

Before moving on, let us look at a simple example (from A&M, Ex. 5.3) of a system that is *not* reachable:

$$\frac{d}{dt}\begin{bmatrix} x_1 \\ x_2 \end{bmatrix} = -\begin{bmatrix} 1 & 0 \\ 0 & 1 \end{bmatrix}\begin{bmatrix} x_1 \\ x_2 \end{bmatrix} + \begin{bmatrix} 1 \\ 1 \end{bmatrix} u.$$

Here we can easily compute

$$W_r = \begin{bmatrix} 1 & -1 \\ 1 & -1 \end{bmatrix},$$

which clearly has determinant zero. This can be understood by noting the complete "symmetry" of the way that u modifies the evolution of x_1 and x_2. For example, if $x_1(0) = x_2(0)$, there is no way to use u to achieve $x_1(T) \neq x_2(T)$ at any later time.

1.4.3 State feedback versus output feedback

Note that in our discussion of stabilization and pole placement so far, we have assumed that it makes sense to design a control law of the form

$$u = -Kx.$$

This is called a "state-feedback" law since, in order to determine the control input $u(t)$ at time t, we generally need to have full knowledge of the state $x(t)$. In practice, this is often not possible, and thus we usually specify the available output signals when defining a control design problem:

$$\dot{x} = Ax + Bu,$$
$$y = Cx.$$

Here, the output signal $y(t)$, which can in principle be a vector of any dimension, represents the information about the evolving system state that is made available to the controller via sensors. An "output-feedback" law must take the form

$$u(t) = f\ [y(\tau \leq t)],$$

where, in general, we can allow $u(t)$ to depend on the entire history of $y(\tau)$ with $\tau \leq t$ (this is discussed in more detail later in this chapter and elsewhere in this volume (chapters 7, 11 and 17)). Output feedback is a natural setting for practical applications. For example, if we are talking about cruise control for an automobile, x may represent a complex set of variables having to do with the internal state of the engine, wheels, and chassis, while y is only a readout from the speedometer. Hopefully, it will seem natural that it is usually prohibitively difficult to install a sensor to monitor every coordinate of the system's state space, and also that it will often be unnecessary to do so (cruise control electronics can function quite well with just the car's speed).

One simple example of a system in which full state knowledge is clearly not necessary is (asymptotic) stabilization of a simple harmonic oscillator. If the natural dynamics of the plant is

$$m\ddot{x} = -kx,$$

and our actuation mechanism is to apply forces directly on the mass, then the control system looks like

$$\frac{d}{dt}\begin{bmatrix} x_1 \\ x_2 \end{bmatrix} = \begin{bmatrix} 0 & 1 \\ \frac{-k}{m} & 0 \end{bmatrix}\begin{bmatrix} x_1 \\ x_2 \end{bmatrix} + \begin{bmatrix} 0 \\ 1 \end{bmatrix} u$$

(where x_1 is now the position and x_2 the velocity). We can clearly make the equilibrium point at the origin asymptotically stable via the linear state-feedback control law

$$u = -bx_2 = \begin{bmatrix} 0 & -b \end{bmatrix}\begin{bmatrix} x_1 \\ x_2 \end{bmatrix},$$

which makes the overall equation of motion

$$\ddot{x}_1 = -\frac{k}{m}x_1 - b\dot{x}_1,$$

which we recognize as a damped harmonic oscillator. Thus, it is clear that the controller only needs to know the velocity of the oscillator in order to implement a successful feedback strategy. So, even if we go to a SISO (single-input single-output) formulation of this problem,

$$\dot{x} = Ax + Bu,$$
$$y = Cx,$$

we are obviously fine for any C of the form ($\alpha \neq 0$)

$$C = \begin{bmatrix} 0 & \alpha \end{bmatrix},$$

since $x_2 = y/\alpha$ and we can implement an output-feedback law of the form

$$u = -bx_2 = -\frac{b}{\alpha}y.$$

Clearly, if C is a square matrix and y has the same dimension as x, output feedback is essentially equivalent to state feedback if C is invertible. As a generalization of what we did for the simple harmonic oscillator, we could just design a state-feedback controller K, set

$$\hat{x} = C^{-1}y,$$

and apply feedback

$$u = -K\hat{x} = -KC^{-1}y.$$

However, this is a special case and not the sort of convenience we want to count on!

1.4.4 State estimation

At this point, it might seem as if we might need completely new theorems about reachability and pole placement for output-feedback laws, when $u(t)$ is only allowed to depend on $y(\tau < t)$. However, it turns out that we can build naturally on our previous results by appealing to a *separation method.* The basic idea is that we will try to construct a procedure for processing the data $y(\tau < t)$ to obtain an estimate $\hat{x}(t)$ of the true system state $x(t)$, and then apply a feedback law $u = -K\,\hat{x}$ based on this estimate. This can be possible even when C is not invertible (not even square). The controller thus assumes the structure of a dynamical system itself, with $y(t)$ as its input, $u(t)$ as its output and $\hat{x}(t)$ as its internal state. There are various ways of designing "state estimators" to extract $\hat{x}(t)$ from $y(\tau < t)$, of which we will discuss two, and there is also a convenient procedure for determining whether or not y contains enough information to make full state reconstruction possible in principle. The latter test looks a lot like the test for reachability, for reasons that are not accidental.

Let us start by thinking about the simple harmonic oscillator again. We noted that in order to asymptotically stabilize the equilibrium point at the origin, it would be most convenient to have an output signal that told us directly about its velocity x_2. However, the reader may have already realized that, in a scenario with $(\alpha \neq 0)$,

$$C = \begin{bmatrix} \alpha & 0 \end{bmatrix},$$
$$y = \alpha x_1,$$

it should be simple to obtain a good estimate of x_2 via

$$\hat{x}_2 = \frac{d}{dt}\alpha^{-1}y.$$

This is certainly a valid procedure for estimating x_2, although in practice one should be wary of taking derivatives of measured data since that tends to accentuate high-frequency noise.

In a similar spirit, we note that for any dynamical system

$$\dot{x} = Ax + Bu,$$
$$y = Cx,$$

if we hold u at zero, we can make use of the general relations

$$\begin{aligned}
\dot{y} &= C\dot{x} = CAx,\\
\ddot{y} &= C\ddot{x} = C\frac{d}{dt}\dot{x} = C\frac{d}{dt}Ax = CA\dot{x} = CA^2x,\\
&\vdots\\
\frac{d^n}{dt^n}y &= CA^n x.
\end{aligned}$$

If we look at how this applies to our modified simple harmonic oscillator example with

$$C = \begin{bmatrix} \alpha & 0 \end{bmatrix}, \quad A = \begin{bmatrix} 0 & 1 \\ -\dfrac{k}{m} & 0 \end{bmatrix},$$
$$y = C\begin{bmatrix} x_1 \\ x_2 \end{bmatrix} = \alpha x_1,$$

we have

$$CA = \begin{bmatrix} \alpha & 0 \end{bmatrix} \begin{bmatrix} 0 & 1 \\ -\dfrac{k}{m} & 0 \end{bmatrix} = \begin{bmatrix} 0 & \alpha \end{bmatrix},$$

$$\dot{y} = CA \begin{bmatrix} x_1 \\ x_2 \end{bmatrix} = \alpha x_2,$$

and we start to get a sense for how the natural dynamics A can move information about state-space variables into the "support" of C. Hopefully, it should thus seem reasonable that, in order for a system to be *observable*, we require that the observability matrix

$$W_o = \begin{bmatrix} C \\ CA \\ CA^2 \\ \cdot \\ \cdot \\ \cdot \\ CA^{n-1} \end{bmatrix}$$

have full rank. Informally, if a system is observable, then we are guaranteed that we can design a procedure (such as the derivatives scheme above) to extract a faithful estimate $\hat{x}$ from y. However, it will generally be necessary to monitor y for some time (and with good accuracy) before the estimation error

$$\tilde{x}(t) \equiv x(t) - \hat{x}(t)$$

can be made small. In the derivatives scheme, for instance, we cannot estimate high-order derivatives of $y(t)$ until we see enough of it to get an accurate determination of its slope, curvature, etc.

1.4.5 State observer with innovations

A more common (and more robust) method for estimating x from y is to construct a state observer that applies corrections to an initial guess $\hat{x}$ until $C\hat{x}$ becomes an accurate predictor of y.

Suppose that at some arbitrary point in time t we have an estimate $\hat{x}(t)$. How should we update this estimate to generate estimates of the state $x(t')$ with $t' > t$? Most simply, we could integrate

$$\frac{d}{dt}\hat{x} = A\hat{x} + Bu,$$

assuming we know A and B for the plant. It is generally assumed that we know u, since this signal is under our control! Then we notice that the estimation error $\tilde{x}$ evolves as

$$\begin{aligned} \frac{d}{dt}\tilde{x} &= \frac{d}{dt}(x - \hat{x}) \\ &= (Ax + Bu) - (A\hat{x} + Bu) \\ &= A(x - \hat{x}) \\ &= A\tilde{x}. \end{aligned}$$

Hence, this strategy has the nice feature that, if A is stable,

$$\lim_{t \to \infty} \tilde{x} = 0,$$

meaning that our estimate will eventually converge to the true system state. Note that this works even if B and u are nonzero.

What if we are not so lucky as to have sufficiently stable natural dynamics A? As already mentioned, a good strategy is to try to apply corrections to $\hat{x}$ at every time step, in proportion to the so-called innovation,

$$w \equiv y - C\hat{x}.$$

Here $y - C\hat{x}$ is the error we make in predicting $y(t)$ on the basis of $\hat{x}(t)$. Clearly, when $\tilde{x}$ is small, so is w. A "Luenberger state observer" can thus be constructed as

$$\frac{d}{dt}\hat{x} = A\hat{x} + Bu + L(y - C\hat{x}),$$

where L is a "gain" matrix that is left to our design. This observer equation results in

$$\begin{aligned}
\frac{d}{dt}\tilde{x} &= \dot{x} - \frac{d}{dt}\hat{x} \\
&= (Ax - Bu) - [A\hat{x} + Bu + L(y - C\hat{x})] \\
&= A(x - \hat{x}) - L(y - C\hat{x}) \\
&= A(x - \hat{x}) - LC(x - \hat{x}) \\
&= (A - LC)\tilde{x}.
\end{aligned}$$

Hence, we see that our design task should be to choose L, given A and C, such that $A - LC$ has nice stable eigenvalues.

This should immediately be reminiscent of the pole-placement problem in state feedback, in which we wanted to choose K, given A and B, such that $A - BK$ had desired eigenvalues. Indeed, one can map between the two problems by noting that the transpose of a matrix M^T has the same eigenvalues as M. Thus, we can view our observer design problem as being the choice of L^T such that

$$(A - LC)^T = A^T - C^T L^T$$

has nice stable eigenvalues, and this now has precisely the same structure as before. Indeed, there is a complete "duality" between state feedback and observer design, with correspondences

$$A \leftrightarrow A^T, \quad B \leftrightarrow C^T, \quad K \leftrightarrow L^T, \quad W_r \leftrightarrow W_o^T.$$

Hence, it should be clear, for example, how the Matlab function place that we have mentioned previously can be used also for observer design. And as long as the observability matrix has full rank, we are guaranteed to be able to find an L such that $A - LC$ has arbitrary desired eigenvalues.

1.4.6 Pole placement with output feedback

As discussed in Section 7.3 of [1], the following theorem holds (here we simplify to the $r = 0$ case):

Theorem *For a system*

$$\dot{x} = Ax + Bu,$$
$$y = Cx,$$

the controller described by

$$u = -K\hat{x},$$
$$\frac{d}{dt}\hat{x} = A\hat{x} + Bu + L\left(y - C\hat{x}\right)$$

gives a closed-loop system with the characteristic polynomial

$$\det\left(sI - A + BK\right)\det\left(sI - A + LC\right).$$

This polynomial can be assigned arbitrary roots if the system is observable and reachable.

The overall setup is summarized in Fig. 1.7.

Next, we need to examine how this sort of strategy performs in the presence of noise.

1.4.7 State observer examples

We begin by illustrating some of the concepts that we have introduced, on state observers for linear systems without noise.

Consider, as our plant, a simple harmonic oscillator:

$$\frac{d}{dt}\begin{bmatrix} x_1 \\ x_2 \end{bmatrix} = \begin{bmatrix} 0 & 1 \\ -\omega_0^2 & 0 \end{bmatrix}\begin{bmatrix} x_1 \\ x_2 \end{bmatrix} + \begin{bmatrix} 0 \\ 1 \end{bmatrix} u,$$
$$\frac{d}{dt}\vec{x} = A\vec{x} + Bu,$$

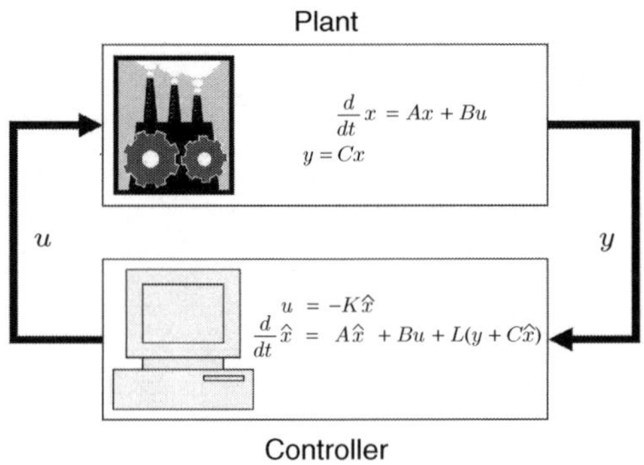

Fig. 1.7 Schematic diagram of observer-based feedback control.

where as usual $x_1 \leftrightarrow q$, $x_2 \leftrightarrow \dot{q}$. The dynamics as given fix A and B, and let us consider a general output signal related linearly to the state:

$$y = C\vec{x} = \begin{bmatrix} C_1 & C_2 \end{bmatrix} \begin{bmatrix} x_1 \\ x_2 \end{bmatrix}.$$

We have seen that the observability criterion is that we have full rank for the matrix

$$W_o = \begin{bmatrix} C \\ CA \end{bmatrix},$$

and since

$$CA = \begin{bmatrix} C_1 & C_2 \end{bmatrix} \begin{bmatrix} 0 & 1 \\ -\omega_0^2 & 0 \end{bmatrix} = \begin{bmatrix} -\omega_0^2 C_2 & C_1 \end{bmatrix},$$

we have

$$\det W_0 = \det \begin{bmatrix} C_1 & C_2 \\ -\omega_0^2 C_2 & C_1 \end{bmatrix} = C_1^2 + \omega_0^2 C_2^2.$$

Hence, as long as $\omega_0 \neq 0$, the system is observable as long as C is nonzero.

For general C, our linear (Luenberger) observer structure is

$$\frac{d}{dt}\hat{x} = A\hat{x} + Bu + L\,(y - C\hat{x}),$$

which induces the following dynamics for the estimation error:

$$\frac{d}{dt}\tilde{x} = (A - LC)\,\tilde{x}.$$

We thus want to design L to make the eigenvalues of $A - LC$ have negative real part. We noted that we could do this by using Matlab's pole-placement routine, place, with

$$A \leftrightarrow A^T, \quad B \leftrightarrow C^T, \quad K \leftrightarrow L^T.$$

We try the following examples, setting $\omega_0 = 1$ and designing for eigenvalues $\{-1, -2\}$:

$$C = \begin{bmatrix} 1 & 0 \end{bmatrix} : L = \begin{bmatrix} 3 \\ 1 \end{bmatrix}, \quad L\,(y - C\hat{x}) = \begin{bmatrix} 3 \\ 1 \end{bmatrix} (x_1 - \hat{x}_1),$$

$$C = \begin{bmatrix} 0 & 1 \end{bmatrix} : L = \begin{bmatrix} -1 \\ 3 \end{bmatrix}, \quad L\,(y - C\hat{x}) = \begin{bmatrix} -1 \\ 3 \end{bmatrix} (x_2 - \hat{x}_2),$$

$$C = \begin{bmatrix} 1 & 1 \end{bmatrix} : L = \begin{bmatrix} 1 \\ 2 \end{bmatrix}, \quad L\,(y - C\hat{x}) = \begin{bmatrix} 1 \\ 2 \end{bmatrix} (x_1 - \hat{x}_1 + x_2 - \hat{x}_2).$$

To examine how the observers work, we perform some numerical integrations. Setting $u = 0$ and

$$\vec{x}(0) = \begin{bmatrix} 1 \\ 1 \end{bmatrix},$$

we have for the plant evolution

$$\vec{x}(t) = \exp(At)\,\vec{x}(0) \mapsto \begin{bmatrix} \cos t & \sin t \\ -\sin t & \cos t \end{bmatrix} \begin{bmatrix} 1 \\ 1 \end{bmatrix} = \begin{bmatrix} \cos t + \sin t \\ \cos t - \sin t \end{bmatrix}.$$

For the observer, we assume no knowledge of the initial state, and thus set

$$\hat{x}(0) = \begin{bmatrix} 0 \\ 0 \end{bmatrix}.$$

The dynamics of the state estimate is

$$\begin{aligned} \frac{d}{dt}\hat{x} &= \begin{bmatrix} 0 & 1 \\ -\omega_0^2 & 0 \end{bmatrix} \begin{bmatrix} \hat{x}_1 \\ \hat{x}_2 \end{bmatrix} + Ly - LC\hat{x} \\ &= \left(\begin{bmatrix} 0 & 1 \\ -\omega_0^2 & 0 \end{bmatrix} - LC \right) \begin{bmatrix} \hat{x}_1 \\ \hat{x}_2 \end{bmatrix} + Ly. \end{aligned}$$

For the purposes of this example, we could integrate this analytically, treating y as a driving term. However, in the spirit of recursive state estimation, we instead integrate numerically (Matlab example).

```
Ttot=10; Nsteps=5000;
  t=linspace(0,Ttot,Nsteps); dt=t(2)-t(1);
  x=[cos(t)+sin(t);cos(t)-sin(t)];
  xhat=zeros(2,Nsteps);
  for ii=2:Nsteps,
   y=C*x(:,ii);
   xhat(:,ii) = xhat(:,ii-1) + dt*(A*xhat(:,ii-1)+L*(y-C*xhat(:,ii-1)));
  end;
```

In Figs. 1.8, 1.9, and 1.10, we plot the results for $C = [1\ 0]$, $C = [0\ 1]$, and $C = [1\ 1]$, respectively, with (a) x_1 (solid) and $\hat{x}_1$ (dashed) and (b) x_2 (solid) and $\hat{x}_2$ (dashed).

Could we have guessed the forms of these observers?

For $C = \begin{bmatrix} 1 & 0 \end{bmatrix}$, note that we can construct an "intuitive" observer by setting $\hat{x}_1 = y$, $\hat{x}_2 = \dot{y}$, which should work well as long as there is negligible measurement noise.

For $C = \begin{bmatrix} 0 & 1 \end{bmatrix}$, we obviously have $\hat{x}_2 = y$, and we can guess

$$\hat{x}_1 = \int_0^t ds\, y(s),$$

but it is not entirely clear how we should choose $\hat{x}_1(0)$. Note that the Luenberger observer does something more complicated, as it integrates

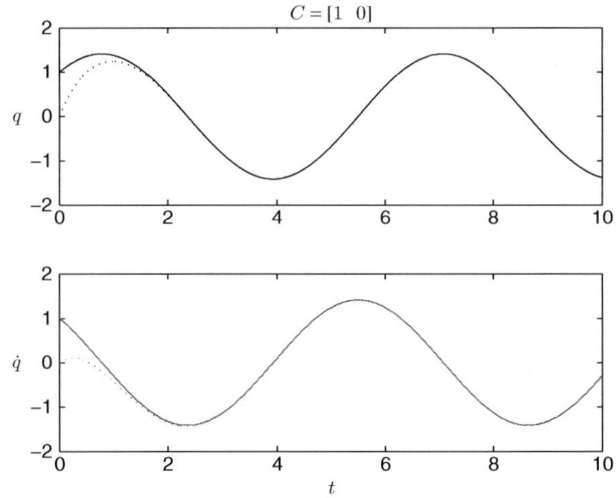

Fig. 1.8

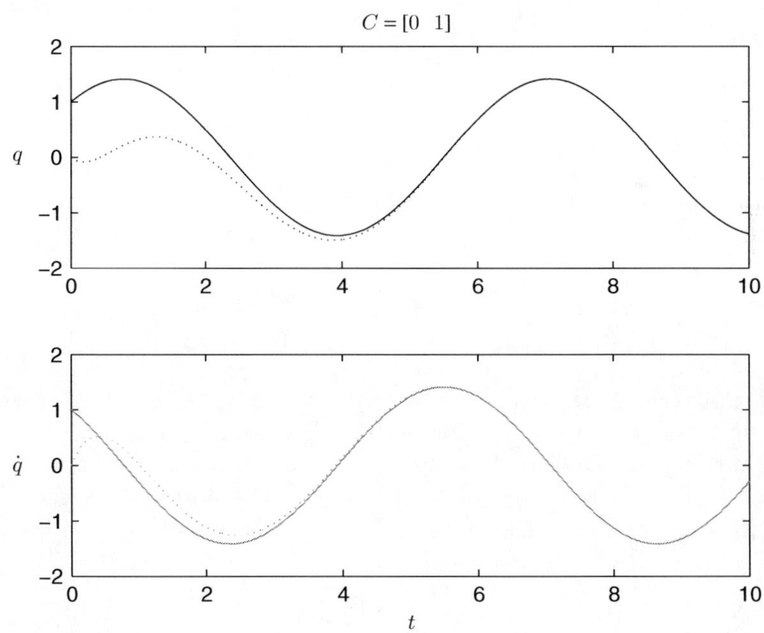

Fig. 1.9

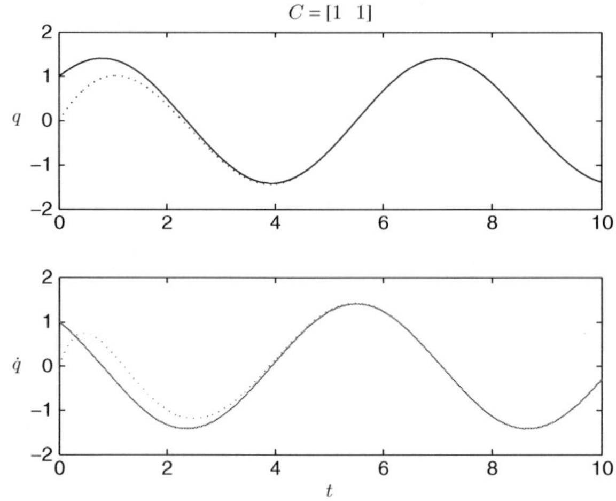

Fig. 1.10

$$\frac{d}{dt}\hat{x} = \left(\begin{bmatrix} 0 & 1 \\ -1 & 0 \end{bmatrix} - LC\right)\begin{bmatrix} \hat{x}_1 \\ \hat{x}_2 \end{bmatrix} + Ly$$

$$\equiv A'\hat{x} + Ly, \ A' = \begin{bmatrix} 0 & 2 \\ -1 & -3 \end{bmatrix} = \begin{bmatrix} 1 & 1 \\ -1 & -\frac{1}{2} \end{bmatrix}\begin{bmatrix} -2 & 0 \\ 0 & -1 \end{bmatrix}\begin{bmatrix} -1 & -2 \\ 2 & 2 \end{bmatrix},$$

$$\exp(A't) = \begin{bmatrix} 1 & 1 \\ -1 & -\frac{1}{2} \end{bmatrix}\begin{bmatrix} e^{-2t} & 0 \\ 0 & e^{-t} \end{bmatrix}\begin{bmatrix} -1 & -2 \\ 2 & 2 \end{bmatrix}$$

$$= \begin{bmatrix} 2e^{-t} - e^{-2t} & 2e^{-t} - 2e^{-2t} \\ e^{-2t} - e^{-t} & 2e^{-2t} - e^{-t} \end{bmatrix},$$

$$\hat{x}(t) = \exp(A't)\left\{\hat{x}(0) + \int_0^t ds \exp(-A's)\, Ly\right\}$$

$$= \begin{bmatrix} 2e^{-t} - e^{-2t} & 2e^{-t} - 2e^{-2t} \\ e^{-2t} - e^{-t} & 2e^{-2t} - e^{-t} \end{bmatrix}\left\{\hat{x}(0) + \int_0^t ds \begin{bmatrix} 4e^{-s} - 5e^{-2s} \\ 5e^{-2s} - 2e^{-s} \end{bmatrix} y\right\},$$

with $\hat{x}(0)$ arbitrary.

There does not seem to be an obvious intuitive strategy for $C = \begin{bmatrix} 1 & 1 \end{bmatrix}$.

Before turning to consider noisy observation scenarios, we take a brief look at the behavior of the Luenberger observer for $C = \begin{bmatrix} 1 & 0 \end{bmatrix}$ and varying gain. Recall that when we designed for eigenvalues $\{-1, -2\}$, we obtained

$$L \rightarrow \begin{bmatrix} 3 \\ 1 \end{bmatrix}.$$

What happens if we try simply multiplying this gain by 10? As shown in Fig. 1.11, the convergence of $\hat{x}_2$ actually gets worse.

If, instead, we design for eigenvalues $\{-10, -20\}$, we get

$$L \rightarrow \begin{bmatrix} 30 \\ 199 \end{bmatrix},$$

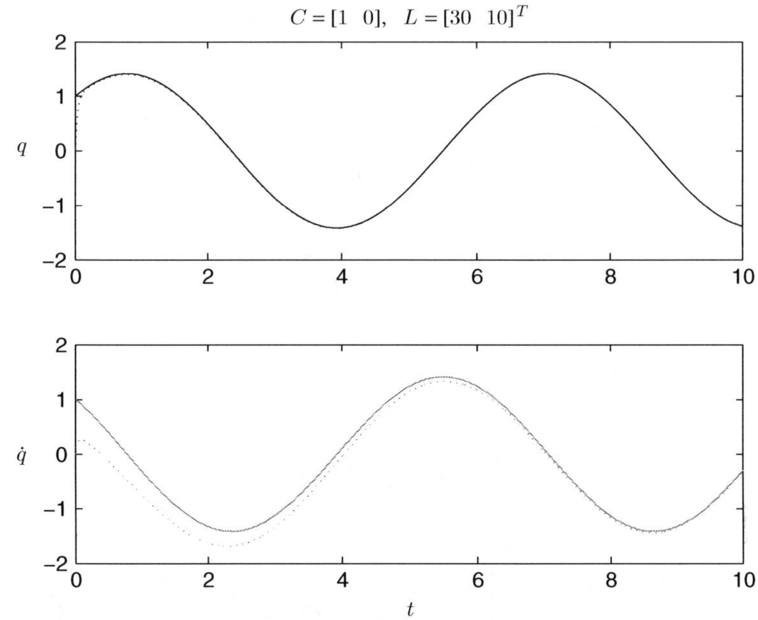

Fig. 1.11

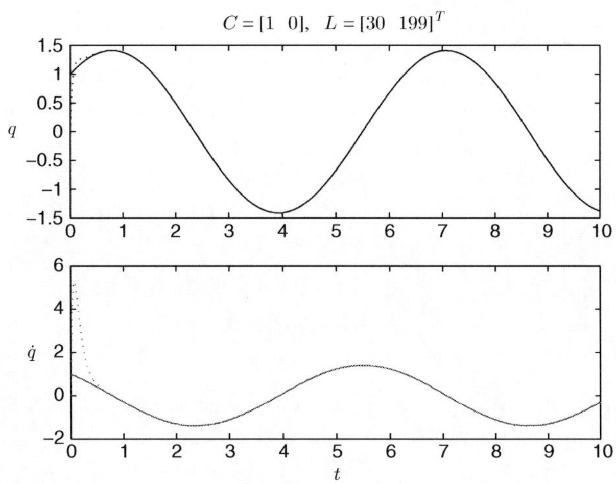

Fig. 1.12

and, as shown in Fig. 1.12, the performance is transiently bad, but does indeed settle quite quickly.

1.4.8 Stochastic models (notation)

We will normally write linear stochastic control models in the form

$$\begin{aligned} dx_t &= A\,x_t\,dt + B\,u_t\,dt + F\,dV_t, \\ dy_t &= C\,x_t\,dt + G\,dW_t. \end{aligned}$$

Here, the subscripts serve to remind us of things that depend on time, and the vector nature of x and/or y is implicit. The (Itô) stochastic increments dV_t and dW_t satisfy

$$\begin{aligned} \langle dV_t \rangle &= \langle dW_t \rangle = 0, \\ dV_t^{\,2} &= dW_t^{\,2} = dt, \\ dV_t\,dt &= dW_t\,dt = 0, \end{aligned}$$

and, for $s \neq t$, we have $\langle dV_s\,dV_t \rangle = \langle dW_s\,dW_t \rangle = 0$. We can informally think of dV_t and dW_t as Gaussian white noises with zero mean and variance dt. It is conventional to refer to dV_t as process noise and to dW_t as measurement noise or observation noise.

It is important to be aware of the fact that SDEs of the type we have written are, rigorously speaking, a sort of shorthand notation for stochastic integrals. There is an important distinction between Itô and Stratonovich stochastic integrals, and therefore between Itô and Stratonovich SDEs. In control theory, one normally works with Itô SDEs, as they are much more natural for recursive filtering, and in any case there is a straightforward recipe for converting a model between Itô and Stratonovich forms. For Itô SDEs, one must in general be careful to observe the Itô rule and associated Itô calculus, as briefly introduced at the beginning of Section 1.3.

In numerical simulations, we can simply update x_t according to

$$\begin{aligned} x_{t+dt} &= x_t + A\,x_t\,dt + B\,u_t\,dt + F\,dV_t, \\ dy_t &= C\,x_t\,dt + G\,dW_t, \end{aligned}$$

where dV_t and dW_t are independent normal random variables with variance dt. In Matlab, if `dt` is a variable with some assigned numerical value,

```
dVt=sqrt(dt)*randn(1); dWt=sqrt(dt)*randn(1);
```

This simple procedure is known as the Itô-Euler stochastic integration routine, which is easy to implement but has the disadvantage that it only converges to order $(dt)^{1/2}$. Higher-order integrators can be found in various computer packages (including SDE toolboxes for Matlab), and are described in textbooks [21].

1.4.9 State observers: performance with noise

First, we consider the case of process noise only. Returning to our simple harmonic oscillator with $C = \begin{bmatrix} 1 & 0 \end{bmatrix}$, we can add a noisy force acting on the oscillator by setting

$$F = \begin{bmatrix} 0 \\ 0.3 \end{bmatrix}.$$

Simulating both the plant and the response of the Luenberger observer, which we now write as

$$\hat{x}_{t+dt} = \hat{x}_t + (A - LC)\,\hat{x}_t\,dt + B\,u_t\,dt + L\,dy_t,$$

we obtain the performance depicted in Fig. 1.13. Here, we used the L computed for target eigenvalues $\{-1, -2\}$. It is clear that the observer still tracks the state, but with degraded performance due to the process noise. If we try turning up the observer gain by using values computed for target eigenvalues $\{-10, -20\}$, we do much better, as shown in Fig. 1.14.

Next, we set F to zero but $G = 0.001$, with the results shown in Fig. 1.15. Here, part (c) shows the performance of a naive velocity estimator, $\hat{x}_2 = \dot{y}$ (there is of course quite a bit of aliasing in the plot). The Luenberger observer does much better, as expected.

In order to bother the Luenberger observer, we turn G all the way up to 0.1, as shown in Fig. 1.16.

Now, if we try to get greedy with this much observation noise, by turning up L to the values that would achieve eigenvalues $\{-10, -20\}$ in the noiseless system, we see that the estimation of velocity becomes very poor (Fig. 1.17). So, apparently too

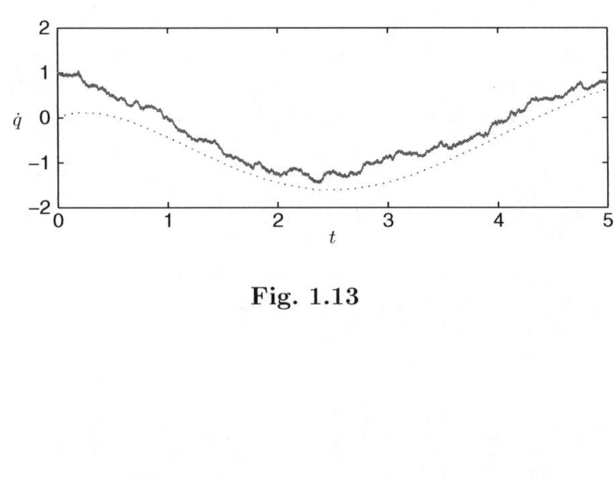

Fig. 1.13

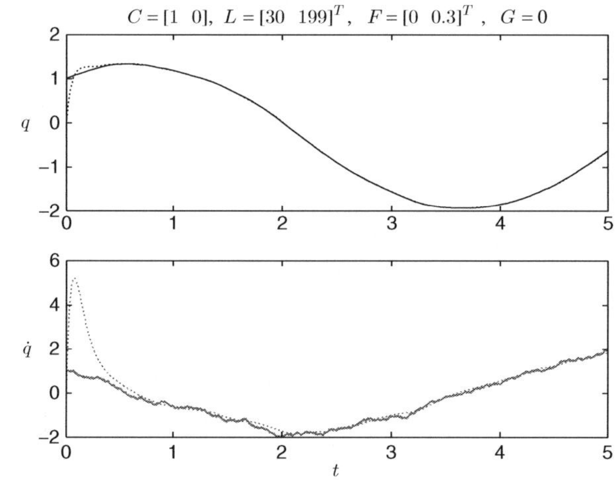

Fig. 1.14

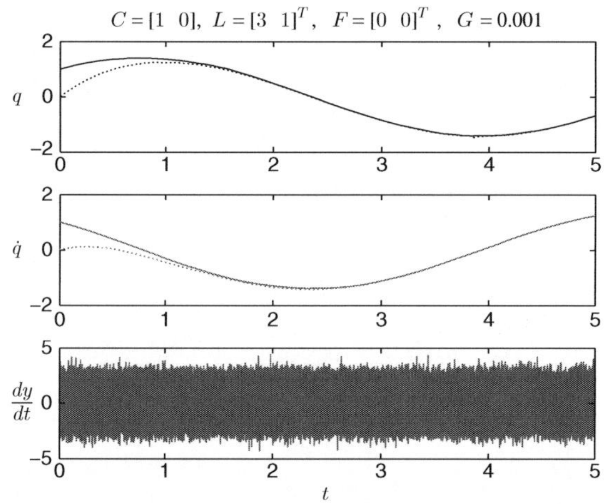

Fig. 1.15

much observer gain is a bad thing, when there is noise. Is there an optimal value of the Luenberger gain? This would seem to be an especially important question when there is both process noise and measurement noise.

1.4.10 The Kalman-Bucy filter

To answer this sort of question, we first have to state how we judge the observer's performance quantitatively. It is most common to adopt a minimum-least-squares

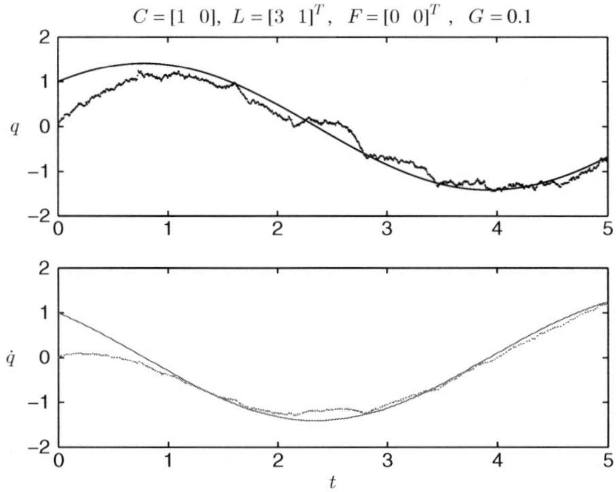

Fig. 1.16

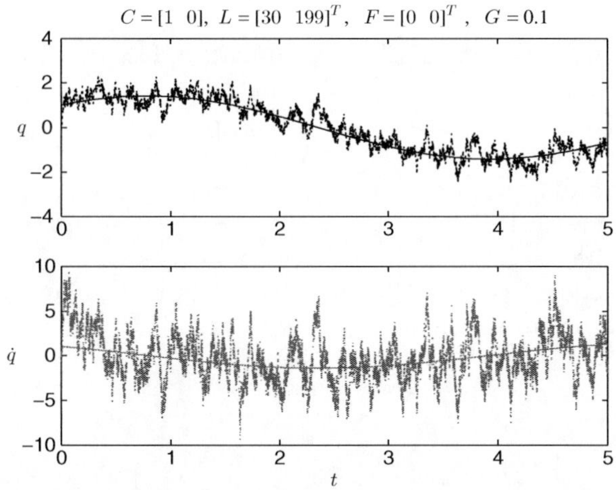

Fig. 1.17

framework, in which our objective is to design the estimator (method of generating $\hat{x}_t$ from knowledge of y_s with $s \leq t$) that achieves the lowest possible value of $\langle (x_t - \hat{x}_t)\,(x_t - \hat{x}_t)^T \rangle$. As discussed in Section 7.4 of [1], for the plant and observation model

$$\begin{aligned} dx_t &= A\,x_t\,dt + B\,u_t\,dt + F\,dV_t, \\ dy_t &= C\,x_t\,dt + G\,dW_t, \end{aligned}$$

we have the following theorem:

Theorem [Kalman-Bucy, 1961] *The optimal estimator has the form of a linear observer*

$$d\hat{x}_t = (A\,\hat{x}_t + B\,u_t)\,dt + L_t\,(dy_t - C\,\hat{x}_t dt)\,, \quad \hat{x}_0 = \langle x_0 \rangle\,,$$

where $L_t = P_t\,C^T\left[GG^T\right]^{-1}$ *and* $P_t = \langle (x_t - \hat{x}_t)\,(x_t - \hat{x}_t)^T \rangle$ *is the (symmetric and positive-definite) estimation error covariance matrix that satisfies the following matrix Riccati equation:*

$$\frac{d}{dt}P_t = FF^T + AP_t + P_t\,A^T - P_t\,C^T\left[GG^T\right]^{-1}C\,P_t, \quad P_0 = \left\langle x_0 x_0^T \right\rangle.$$

It is important to note that the Kalman filter provides both a point estimate of the evolving system state and a computation of the estimation error covariance matrix—it provides its best guess *and* a numerical uncertainty. When the system is stationary and if P_t converges, the observer gain settles to a constant:

$$L = PC^T\left[GG^T\right]^{-1}, \quad FF^T + AP + P\,A^T - P\,C^T\left[GG^T\right]^{-1}C\,P = 0.$$

The second equation is called the *algebraic Riccati equation*, and may be solved using `Matlab's lqe` function.

We see that the essence of Kalman filtering is an optimal choice of the observer gain, which may be time-dependent in a way that reflects our evolving degree of confidence in our state estimate. The general structure is to apply high observer gain when we have large uncertainty, and to reduce it when our uncertainty approaches a limiting value set by the process and measurement noises.

As an example, let us compute the steady-state Kalman gain for our simple harmonic oscillator example with

$$F = \begin{bmatrix} 0 \\ 0.3 \end{bmatrix}, \quad G = 0.1.$$

This results in

$$L \approx \begin{bmatrix} 2.08 \\ 2.16 \end{bmatrix},$$

and a simulation looks as shown in Fig. 1.18.

It is interesting to note that, as a consequence of its least-squares optimality, the Kalman-Bucy filter achieves what is known as *whitening* of the innovations process $dy_t - C\,\hat{x}_t\,dt$. That is, if $\hat{x}_t$ is propagated by the Kalman-Bucy filter, then $dy_t - C\,\hat{x}_t\,dt$ becomes a completely random signal (Gaussian white noise); roughly, we can think that $\hat{x}_t$ becomes good enough that subtracting $C\,\hat{x}_t\,dt$ from dy_t removes all the information from the observed signal. The notions of least-squares optimal state estimation, the innovations process, and whitening all carry over to nonlinear scenarios.

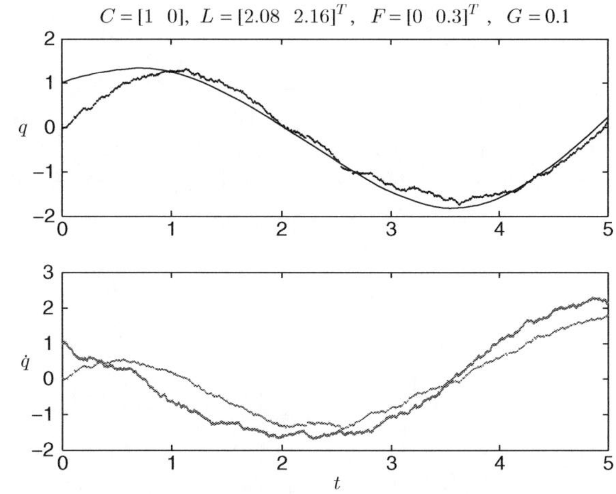

Fig. 1.18

1.4.11 Quantum filtering equations

As the careful reader is now ready to learn from articles such as those by Doherty *et al.* [2] and [20], it is possible to derive optimal filtering equations for open quantum systems whose plant models are given by QSDEs of the form introduced in Section 1.3. Least-squares optimal quantum filters are also known in the quantum optics context as stochastic Schrödinger or stochastic Master equations (SSEs or SMEs), depending on whether the recursive state estimate is propagated as a conditional state vector (possible only when all output channels are observed with perfect fidelity) or conditional density operator. Hence, SSEs/SMEs have a key role in measurement-based quantum feedback control, analogous to the state observers introduced above in the simple context of linear output-feedback control.

Measurement-based quantum feedback control can present some daunting technical challenges associated with the typically short timescales of coherent evolution in experimentally accessible systems. SSEs/SMEs typically have rather high variable count (corresponding to the dimension of the system Hilbert space, or its square) and are thus difficult to integrate numerically in real time if it is necessary to apply feedback ("close the loop") on microsecond or shorter timescales. This is especially challenging in sampled-data systems incorporating digital signal processors. As a result, there is ample reason to explore the application of filter reduction methods from classical nonlinear stochastic control theory to SSE/SME models, as in [22]. For the physicist, it is interesting to consider how such control-theoretic model reduction tools relate to more traditional approaches to nonequilibrium statistical mechanics.

Finally, it seems worth noting that while the general form of the SSE and SME can be derived by considering a continuous limit of standard methods in quantum measurement theory [23], from the stochastic control perspective it can be said that these particular forms of recursive quantum filter are not unique, in that they are optimal only relative to the particular choice of a mean-square error criterion. Other formulations of the recursive filtering problem are possible, such as "risk-sensitive" filtering, and can be applied to quantum scenarios leading to modified versions of the usual SSE/SME [24]. It is interesting to consider the implications of this—and indeed of all, the formal parallels between classical filtering theory and "quantum trajectory theory" (as the SSE/SME methodology is known in quantum optics)—for our interpretation of the quantum state and state "collapse" in standard quantum measurement theory.

References

[1] Aström, K. J. and Murray, R. M. (2008). *Feedback Systems: An Introduction for Scientists and Engineers.* Princeton University Press, Princeton, NJ. Available online at <http://www.cds.caltech.edu/ murray/amwiki/>.
[2] Doherty, A. C., Habib, S., Jacobs, K., Mabuchi, H., and Tan, S. M. (2000). Quantum feedback control and classical control theory. *Phys. Rev. A*, **62**, 012105.
[3] Mabuchi, H. and Khaneja, N. (2005). Principles and applications of control in quantum systems. *Int. J. Robust Nonlinear Control*, **15**, 647.
[4] James, M. R., Nurdin, H. I., and Petersen, I. R. (2008). H^{∞} control of linear quantum systems. *IEEE Trans. Autom. Control*, **53**, 1787.
[5] Nurdin, H. I., James, M. R., and Petersen, I. R. (2009). Coherent quantum LQG control. *Automatica*, **45**, 1837.
[6] Mabuchi, H. (2008). Coherent-feedback quantum control with a dynamic compensator. *Phys. Rev. A*, **78**, 032323.
[7] Carmichael, H. J. (2008). *Statistical Methods in Quantum Optics 2: Non-Classical Fields.* Springer-Verlag, Berlin.
[8] Kimble, H. J. (1998). Strong interactions of single atoms and photons in cavity QED. *Physica Scripta*, **T76**, 127.
[9] Mabuchi, H. (2011). Coherent-feedback control strategy to suppress spontaneous switching in ultralow power optical bistability. *Appl. Phys. Lett.*, **98**, 193109.
[10] Bouten, L., Van Handel, R., and Silberfarb, A. (2008). Approximation and limit theorems for quantum stochastic models with unbounded coefficients. *J. Funct. Anal.*, **254**, 3123.
[11] Mabuchi, H. (2009). Cavity-QED models of switches for attojoule-scale nanophotonic logic. *Phys. Rev. A*, **80**, 045802.
[12] Kerckhoff, J., Nurdin, H. I., Pavlichin, D. S., and Mabuchi, H. (2010). Designing quantum memories with embedded control: Photonic circuits for autonomous quantum error correction. *Phys. Rev. Lett.*, **105**, 040502.
[13] Mabuchi, H. (2011). Nonlinear interferometry approach to photonic sequential logic. *Appl. Phys. Lett.*, **99**, 153103.
[14] Tezak, N., Niederberger, A., Pavlichin, D. S., Sarma, G., and Mabuchi, H. (2012). Specification of photonic circuits using quantum hardware description language. *Phil. Trans. R. Soc. Lond. A*, **370**, 5270.

[15] Gough, J., and James, M. R. (2009). The Series product and its application to quantum feedforward and feedback networks. *IEEE Trans. Autom. Control*, **54**, 2530.
[16] Gough, J. and James, M. R. (2009). Quantum feedback networks: Hamiltonian formulation. *Communi. Math. Phys.*, **287**, 1109.
[17] Gardiner, C. W. (1993). Driving a quantum system with the output of another driven quantum system. *Phys. Rev. Lett.*, **70**, 2269.
[18] Carmichael, H. J. (1993). Quantum trajectory theory for cascaded open quantum systems. *Phys. Rev. Lett.*, **70**, 2273.
[19] Øksendal, B. K. (2002). *Stochastic Differential Equations: An Introduction with Applications.* Springer-Verlag, Berlin.
[20] Bouten, L., Van Handel, R., and James, M. R. (2007). An introduction to quantum filtering. *SIAM J. Control Optim.*, **46**, 2199.
[21] Kloeden, P. E., Platen, E., and Schurz, H. (2003). *Numerical Solution of SDE Through Computer Experiments.* Springer-Verlag, Berlin.
[22] Nielsen, A. E. B., Hopkins, A. S., and Mabuchi, H. (2009). Quantum filter reduction for measurement-feedback control via unsupervised manifold learning. *New J. Phys.*, **11**, 105043.
[23] Caves, C. M. and Milburn, G. J. (1987). Quantum-mechanical model for continuous position measurements. *Phys. Rev. A*, **36**, 5543.
[24] James, M. R. (2004). Risk-sensitive optimal control of quantum systems. *Phys. Rev. A*, **69**, 032108.

2

Quantum noise and quantum measurement

A. CLERK

Department of Physics, McGill University,
Montreal, Quebec, Canada H3A 2T8

Quantum Machines: Measurement and Control of Engineered Quantum Systems.
Edited by M. Devoret, B. Huard, R. Schoelkopf, and L. F. Cugliandolo.

Chapter Contents

2.1 Introduction

The fact that quantum mechanics can place restrictions on our ability to make measurements is something we all encounter in our first quantum mechanics class. One is typically presented with the example of the Heisenberg microscope [1], where the position of a particle is measured by scattering light off it. The smaller the wavelength of light used, the better the precision of the measurement. However, decreasing the wavelength also increases the magnitude of backaction disturbance of the particle's momentum by the measurement (i.e., the momentum kick delivered to the particle by the scattering event). One finds that the imprecision of the measurement $\Delta x_{\rm imp}$ and the backaction momentum disturbance $\Delta p_{\rm BA}$ of the particle are constrained in a way suggestive of the usual Heisenberg uncertainty principle

$$\Delta x_{\rm imp} \Delta p_{\rm BA} \gtrsim h. \tag{2.1}$$

As instructive as this example is, it hardly provides a systematic way for formulating precise quantum limits on measurement in more general settings. Recent progress in the general area of engineered quantum systems has rekindled interest in such fundamental limits on measurement and amplification. One would like to have a precise formulation of what these constraints are, and, moreover, an understanding of how one can achieve these ultimate limits in realistic and experimentally-relevant setups. In these lectures, I will describe an extremely useful and powerful method for doing this—the so-called "quantum noise" approach. The focus will be on weak, continuous measurements, where the measured system is only weakly coupled to the detector, and information is obtained only gradually in time. Even on a purely classical level, such measurements are limited by the presence of detector noise. Quantum mechanically, one finds that there are fundamental quantum limits on the noise properties of any system capable of acting as a detector or amplifier. These constraints then directly yield quantum limits on various measurement tasks.

For the most part, these notes draw extensively from material presented in our recent review article [2], but do not attempt to be as exhaustive as that work. I have slightly reworked the discussion of several key points to hopefully add clarity. I also include in these notes a few topics not found in the review article. These include a discussion of the quantum shot noise of a quantum point contact (QPC) (Section 2.2.3), a formulation of the quantum limit on quantum nondemolition (QND) qubit detection beyond weak coupling (Section 2.3.5), and a heuristic discussion of how a QPC can miss that limit due to correlated backaction-imprecision noise (Section 2.3.3).

2.2 Quantum noise spectral densities: some essential features

In this chapter, we give a compact (and no doubt highly incomplete) review of some basic properties of spectral densities describing quantum noise.

2.2.1 Classical noise basics

Consider a classical random signal $I(t)$. The signal is characterized by zero mean $\langle I(t) \rangle = 0$ and autocorrelation function

$$G_{II}(t,t') = \langle I(t)I(t')\rangle. \tag{2.2}$$

The autocorrelation function is analogous to a covariance matrix: for $t = t'$, it tells us the variance of the fluctuations of $I(t)$, whereas for $t \neq t'$, it tells us if and how fluctuations of $I(t)$ are correlated with those at $I(t')$. Some crucial concepts regarding noise are as follows:

- *Stationary noise.* The statistical properties of the fluctuations are time-translation invariant, and hence $G_{II}(t,t') = G_{II}(t-t')$.
- *Gaussian fluctuations.* The noise is fully characterized by its autocorrelation function; there are no higher-order cumulants.
- *Correlation time.* This timescale τ_c governs the decay of $G_{II}(t)$: $I(t)$ and $I(t')$ are uncorrelated (i.e., $G_{II}(t-t') \to 0$) when $|t-t'| \gg \tau_c$.

For stationary noise, it is often most useful to think about the fluctuations in the frequency domain. In the same way that $I(t)$ is a Gaussian random variable with zero mean, so is its Fourier transform, which we define as

$$I_T[\omega] = \frac{1}{\sqrt{T}} \int_{-T/2}^{+T/2} dt\, e^{i\omega t} I(t), \tag{2.3}$$

where T is the sampling time. In the limit $T \gg \tau_c$ the integral is a sum of a large number $N \approx T/\tau_c$ of random uncorrelated terms. We can think of the value of the integral as the endpoint of a random walk in the complex plane starting at the origin. Because the distance traveled will scale with $\sqrt{T}$, our choice of normalization makes the statistical properties of $I[\omega]$ independent of the sampling time T (for sufficiently large T). Notice that $I_T[\omega]$ has the peculiar units of $[I]\sqrt{\text{seconds}}$, which is usually denoted $[I]/\sqrt{\text{Hz}}$.

The spectral density of the noise (or power spectrum) $\mathcal{S}_{II}[\omega]$ answers the question "how big is the noise at frequency ω?" It is simply the variance of $I_T(\omega)$ in the large-time limit:

$$\mathcal{S}_{II}[\omega] \equiv \lim_{T\to\infty} \langle |I_T[\omega]|^2\rangle = \lim_{T\to\infty} \langle I_T[\omega] I_T[-\omega]\rangle. \tag{2.4}$$

A reasonably straightforward manipulation (known as the Wiener-Khinchin theorem) tells us that the spectral density is equal to the Fourier transform of the autocorrelation function:

$$\mathcal{S}_{II}[\omega] = \int_{-\infty}^{+\infty} dt\, e^{i\omega t} G_{II}(t). \tag{2.5}$$

We stress that Eq. (2.4) provides a simple intuitive understanding of what a spectral density represents, whereas, in theoretical calculations, one almost always starts with the expression in Eq. (2.5). We also stress that since the autocorrelation function $G_{II}(t)$ is real, $\mathcal{S}_{II}[\omega] = \mathcal{S}_{II}[-\omega]$. This is of course in keeping with Eq. (2.3), which tells us that negative- and positive-frequency components of the noise are related by complex conjugation, and hence necessarily have the same magnitude.

As a simple example, consider a simple harmonic oscillator of mass M and frequency Ω. The oscillator is maintained in equilibrium with a large heat bath at temperature T via some infinitesimal coupling that we will ignore in considering the dynamics. The solution of Hamilton's equations of motion gives

$$x(t) = x(0)\cos(\Omega t) + p(0)\frac{1}{M\Omega}\sin(\Omega t), \tag{2.6}$$

where $x(0)$ and $p(0)$ are the (random) values of the position and momentum at time $t = 0$. It follows that the position autocorrelation function is

$$\begin{aligned} G_{xx}(t) &= \langle x(t)x(0)\rangle \\ &= \langle x(0)x(0)\rangle\cos(\Omega t) + \langle p(0)x(0)\rangle\frac{1}{M\Omega}\sin(\Omega t). \end{aligned} \tag{2.7}$$

Classically, in equilibrium, there are no correlations between position and momentum. Hence, the second term vanishes. Using the equipartition theorem $\frac{1}{2}M\Omega^2\langle x^2\rangle = \frac{1}{2}k_\mathrm{B}T$, we arrive at

$$G_{xx}(t) = \frac{k_\mathrm{B}T}{M\Omega^2}\cos(\Omega t), \tag{2.8}$$

which leads to the spectral density

$$S_{xx}[\omega] = \pi\frac{k_\mathrm{B}T}{M\Omega^2}\left[\delta(\omega - \Omega) + \delta(\omega + \Omega)\right], \tag{2.9}$$

which is indeed symmetric in frequency.

2.2.2 Quantum noise spectral densities

Definition In formulating quantum noise, one turns from a noisy classical signal $I(t)$ to a Heisenberg-picture Hermitian operator $\hat{I}(t)$. Similar to our noisy classical signal, one needs to think about measurements of $\hat{I}(t)$ statistically. One can thus introduce a quantum noise spectral density that completely mimics the classical definition, for example,

$$S_{xx}[\omega] = \int_{-\infty}^{+\infty} dt\, e^{i\omega t}\langle\hat{x}(t)\hat{x}(0)\rangle. \tag{2.10}$$

We have simply inserted the quantum autocorrelation function in the classical definition. The expectation value is the quantum statistical average with respect to the noisy system's density matrix; we assume that this is time-independent, which then also gives us an autocorrelation function which is time-translational invariant.

What makes quantum noise so quantum? There are at least three answers to this question that we will explore in turn:

- *Zero-point motion.* While a classical system at zero temperature has no noise (see Eq. (2.9)), quantum mechanically there are still fluctuations, that is, $S_{xx}[\omega]$ need not be zero.

- *Frequency asymmetry.* Quantum mechanically, $\hat{x}(t)$ and $\hat{x}(t')$ need not commute when $t \neq t'$. As a result, the autocorrelation function $\langle \hat{x}(t)\hat{x}(t')\rangle$ can be complex, and $S_{xx}[\omega]$ need not equal $S_{xx}[-\omega]$. This of course can never happen for a classical noise spectral density.
- *Heisenberg constraints.* For any system that can act as a detector or amplifier, there are fundamental quantum constraints that bound its noise. These constraints have their origin in the uncertainty principle, and have no classical counterpart.

Let's start our discussion with the second of these points, and make things concrete by again considering the example of a harmonic oscillator in thermal equilibrium. We again assume that the oscillator is maintained in equilibrium with a large heat bath at temperature T via some infinitesimal coupling that we will ignore in considering the dynamics. The solutions of the Heisenberg equations of motion are the same as for the classical case but with the initial position x and momentum p replaced by the corresponding quantum operators. It follows that the position autocorrelation function is

$$\begin{aligned} G_{xx}(t) &= \langle \hat{x}(t)\hat{x}(0)\rangle \\ &= \langle \hat{x}(0)\hat{x}(0)\rangle \cos(\Omega t) + \langle \hat{p}(0)\hat{x}(0)\rangle \frac{1}{M\Omega}\sin(\Omega t). \end{aligned} \tag{2.11}$$

Unlike the classical case, the second term on the right-hand side cannot be zero: that would violate the commutation relation $[\hat{x}(0), \hat{p}(0)] = i\hbar$. Writing $\hat{x}$ and $\hat{p}$ in terms of ladder operators, one finds that $\langle \hat{x}(0)\hat{p}(0)\rangle = i\hbar/2$: it is purely imaginary. One shouldn't be too troubled by this, as $\hat{x}(0)\hat{p}(0)$ is not Hermitian and hence is not an observable quantity. Evaluating $\langle \hat{x}(0)^2\rangle$ in a similar manner, we find

$$G_{xx}(t) = x_{\mathrm{ZPF}}^2 \left\{ n_{\mathrm{B}}(\hbar\Omega)e^{+i\Omega t} + [n_{\mathrm{B}}(\hbar\Omega)+1]e^{-i\Omega t}\right\}, \tag{2.12}$$

where $x_{\mathrm{ZPF}}^2 \equiv \hbar/2M\Omega$ is the root-mean-square zero-point uncertainty of x in the quantum ground state, and n_{B} is the Bose-Einstein occupation factor. Fourier transforming then yields a spectral density that is asymmetric in frequency:

$$S_{xx}[\omega] = 2\pi x_{\mathrm{ZPF}}^2 \left\{ n_{\mathrm{B}}(\hbar\Omega)\delta(\omega+\Omega) + [n_{\mathrm{B}}(\hbar\Omega)+1]\delta(\omega-\Omega)\right\}. \tag{2.13}$$

Note that in the high-temperature limit $k_{\mathrm{B}}T \gg \hbar\Omega$, we have $n_{\mathrm{B}}(\hbar\Omega) \sim n_{\mathrm{B}}(\hbar\Omega)+1 \sim k_{\mathrm{B}}T/\hbar\Omega$. Thus, in this limit, $S_{xx}[\omega]$ becomes symmetric in frequency, as expected classically, and coincides with the classical expression for the position spectral density (see Eq. (2.9)).

The Bose-Einstein factors suggest a way to understand the frequency asymmetry of Eq. (2.13): the positive-frequency part of the spectral density has to do with stimulated emission of energy *into* the oscillator and the negative-frequency part has to do with emission of energy *by* the oscillator. That is, the positive-frequency part of the spectral density is a measure of the ability of the oscillator to *absorb* energy, while the negative-frequency part is a measure of the ability of the oscillator to *emit* energy.

The above interpretation is of course not restricted to harmonic oscillators or thermal states. Consider now the quantum noise associated with a general observable $\hat{I}(t)$, and let $\hat{\rho}$ be the system's density matrix. We will also let $|j\rangle$ denote the system's energy eigenstates (eigenenergy E_j), where for simplicity we take j to be a discrete index. Letting $\hat{U}(t)$ denote the time evolution operator, the quantum noise spectral density $S_{II}[\omega]$ can be written

$$S_{II}[\omega] \equiv \int_{-\infty}^{+\infty} dt\, e^{i\omega t} \langle \hat{I}(t)\hat{I}(0)\rangle = \int_{-\infty}^{+\infty} dt\, e^{i\omega t}\, \mathrm{Tr}\left[\hat{\rho}\hat{U}^\dagger(t)\hat{I}(0)\hat{U}(t)\hat{I}(0)\right]$$
$$= 2\pi\hbar \sum_{i,j} \langle i|\hat{\rho}|i\rangle \left|\langle j|\hat{I}|i\rangle\right|^2 \delta(\hbar\omega - (E_j - E_i)). \tag{2.14}$$

We have used the fact that since $\hat{\rho}$ is time-independent, it must be diagonal in the basis of energy eigenstates. The expression (2.14) is a standard Lehman representation for a quantum correlation function. It is nothing more than a sum of Fermi Golden Rule transition rates (from the state $|i\rangle$ to $|j\rangle$), mediated by the operator $\hat{I}$. More explicitly, if we coupled $\hat{I}$ to a qubit such that

$$H_{\mathrm{qb}} = \frac{\hbar\Omega}{2}\hat{\sigma}_z + A\hat{I}\hat{\sigma}_x, \tag{2.15}$$

and then treated the second term via Fermi's Golden Rule, the qubit excitation rate Γ_+ and relaxation rate Γ_- would be simply given by

$$\Gamma_\pm = \frac{A^2}{\hbar^2} S_{II}[\mp\Omega]. \tag{2.16}$$

The origin of the frequency asymmetry is exactly the same as in our oscillator example. For $\omega > 0$, the spectral density describes transitions where the noise source absorbs energy, whereas for $\omega < 0$, it describes transitions where energy is emitted by the noise source. The noise at positive and negative frequencies need not be equal, as, in general, absorption will occur at a higher rate than emission. In the extreme limit of zero temperature, where the system is in its ground state, we see that there is still noise at positive frequencies, as the noise source can still absorb energy.

Finally, we note that in thermal equilibrium, $\langle i|\hat{\rho}|i\rangle \propto \exp(-E_i/k_{\mathrm{B}}T)$, which necessarily relates the positive- and negative-frequency noise via

$$S_{II}[\omega]/S_{II}[-\omega] = \exp(\hbar\omega/k_{\mathrm{B}}T), \tag{2.17}$$

where T is the temperature of the noise source. On a physical level, this ensures that the transition rates in Eq. (2.16) obey detailed balance, and thus cause the qubit to also relax to an equilibrium state at the same temperature as the noise source

Classical interpretation Our analysis so far seems to indicate that, apart from the formal similarity in their definitions (see Eqs. (2.5) and (2.10)), classical and quantum noise spectral densities have very little in common. The classical noise spectral density tells us the "size" of the noise at a particular frequency, whereas, in contrast, the quantum noise spectral density tells us the magnitude of Golden Rule transition rates for emission or absorption events. The frequency asymmetry of a quantum noise spectral density also gives the appearance that it contains more information than one has classically. One is tempted to conclude that referring to $S_{II}[\omega]$ as noise is an abuse of terminology.

To overcome these apprehensions and to gain a deeper understanding of quantum noise spectral densities, it is useful to think of our noisy quantity as a force $\hat{F}$, and see what happens when we weakly couple this force to a harmonic oscillator, that is,

$$\hat{H}_{\text{int}} = -\hat{x}\hat{F}. \tag{2.18}$$

Classically, including this force in Newton's equation yields a Langevin equation:

$$M\ddot{x} = -M\Omega^2 x - M\gamma_{\text{cl}}\dot{x} + F_{\text{cl}}(t). \tag{2.19}$$

In addition to the noisy force, we have included a damping term (rate γ_{cl}). This will prevent the oscillator from being infinitely heated by the noise source—we can think of it as describing the *average* value of the force exerted on the oscillator by the noise source, which is now playing the role of a dissipative bath. If this bath is in thermal equilibrium at temperature T, we also expect the oscillator to equilibrate to the same temperature. This implies that the heating effect of $F_{\text{cl}}(t)$ must be precisely balanced by the energy-loss effect of the damping force. More explicitly, one can use Eq. (2.19) to derive an equation for the average energy of the oscillator $\langle E\rangle$. As we are assuming a weak coupling between the bath and the oscillator, we can take $\gamma_{\text{cl}} \ll \Omega$, and hence find

$$\frac{d}{dt}\langle E\rangle = -\gamma_{\text{cl}}\langle E\rangle + \frac{S_{FF}[\Omega]}{2M}. \tag{2.20}$$

Insisting that the stationary $\langle E\rangle$ obey equipartition then leads directly to the classical fluctuation-dissipation relation:

$$S_{FF}[\omega] = 2M\gamma_{\text{cl}}k_{\text{B}}T. \tag{2.21}$$

Let us now look at our problem quantum mechanically. Writing $\hat{x}$ in terms of ladder operators, we see that $\hat{H}_{\text{int}}$ will cause transitions between different oscillator Fock states. Treating $\hat{H}_{\text{int}}$ in perturbation theory, we thus derive Fermi Golden Rule transition rates $\Gamma_{n\pm1,n}$ for transitions from the n to the $n\pm1$ Fock state:

$$\Gamma_{n+1,n} = (n+1)\frac{x_{\text{ZPF}}^2}{\hbar^2}S_{FF}[-\Omega] \equiv (n+1)\Gamma_{\uparrow}, \tag{2.22}$$

$$\Gamma_{n-1,n} = (n)\frac{x_{\text{ZPF}}^2}{\hbar^2}S_{FF}[\Omega] \equiv n\Gamma_{\downarrow}. \tag{2.23}$$

We could then write a simple master equation for the probability $p_n(t)$ that the oscillator is in the nth Fock state:

$$\frac{d}{dt}p_n = [n\Gamma_{\uparrow}p_{n-1} + (n+1)\Gamma_{\downarrow}p_{n+1}] - [n\Gamma_{\downarrow} + (n+1)\Gamma_{\uparrow}]\,p_n. \tag{2.24}$$

At this stage, the connection between the classical and quantum pictures is still murky. To connect them one uses the quantum equation for p_n to derive an equation for the average oscillator energy $\langle E \rangle$. One obtains

$$\frac{d}{dt}\langle E \rangle = -\gamma \langle E \rangle + \frac{\bar{S}_{FF}[\Omega]}{2M}, \tag{2.25}$$

where

$$\gamma = \frac{x_{\mathrm{ZPF}}^2}{\hbar^2}\left(S_{FF}[\Omega] - S_{FF}[-\Omega]\right), \tag{2.26}$$

$$\bar{S}_{FF}[\Omega] = \frac{S_{FF}[\Omega] + S_{FF}[-\Omega]}{2}. \tag{2.27}$$

We see that the quantum equation for the average energy, Eq. (2.25), has an identical form to the classical equation, Eq. (2.20), which gives us a simple way to connect our quantum noise spectral density to quantities in the classical theory:

- The symmetrized quantum noise spectral density $\bar{S}_{FF}[\Omega]$ defined in Eq. (2.27) has the same role as the classical noise spectral density $\mathcal{S}_{FF}[\Omega]$: it heats the oscillator the same way a classical stochastic force would.
- The asymmetric-in-frequency part of the quantum noise spectral density $S_{FF}[\Omega]$ is directly related to the damping rate γ in the classical theory. The asymmetry between absorption and emission events leads to a net energy flow between the oscillator and the noise source, analogous to what one obtains from a classical viscous damping force.

We thus see that there is a direct connection to a classical noise spectral density, and moreover the "extra information" in the asymmetry of a quantum noise spectral density also corresponds to a seemingly distinct classical quantity, a damping rate. This latter connection is not so surprising. The asymmetry of the quantum noise is a direct consequence of (here) $[\hat{F}(t), \hat{F}(t')] \neq 0$. However, this same noncommutation causes the average value of $\langle \hat{F} \rangle$ to change in response to $\hat{x}(t)$ via the interaction Hamiltonian of Eq. (2.18). Using standard quantum linear response (i.e., first-order time-dependent perturbation theory; see, e.g., Chapter 6 of [3]), one finds

$$\delta \langle \hat{F}(t) \rangle = \int_{-\infty}^{\infty} dt' \, \chi_{FF}(t-t') \langle \hat{x}(t') \rangle, \tag{2.28}$$

where the force-force susceptibility is given by the Kubo formula:

$$\chi_{FF}(t) \equiv \frac{-i}{\hbar}\theta(t)\langle [\hat{F}(t), \hat{F}(0)] \rangle. \tag{2.29}$$

From the classical Langevin equation (2.19), we see that the part of $\langle \hat{F}(t) \rangle$ that is in phase with $\dot{x}$ is the damping force. This leads to the definition

$$\gamma = \frac{1}{M\Omega}\left(-\mathrm{Im}\, \chi_{FF}[\Omega]\right). \tag{2.30}$$

An explicit calculation shows that this above definition is identical to Eq. (2.26), which expresses γ in terms of the noise asymmetry. Note that in the language of many-body Green functions, $-\text{Im}\,\chi_{FF}$ is referred to as a spectral function, whereas the symmetrized noise $\bar{S}_{FF}[\omega]$ is known (up to a constant) as the "Keldysh" Green function.

Quantum fluctuation-dissipation theorem and notion of effective temperature We saw previously (see Eq. (2.17)) that in thermal equilibrium, the negative- and positive-frequency noise are related by a Boltzmann factor. Using the definitions in Eqs. (2.26) and (2.27), it is then straightforward to derive the quantum version of the fluctuation-dissipation theorem:

$$\bar{S}_{FF}[\Omega] = M\gamma[\Omega]\,\hbar\Omega \coth\left(\frac{\hbar\Omega}{2k_{\text{B}}T}\right) = M\gamma[\Omega]\,\hbar\Omega\,(1 + 2n_B[\Omega])\,. \tag{2.31}$$

For $k_{\text{B}}T \gg \hbar\Omega$, this reproduces the classical result of Eq. (2.21), whereas in the opposite limit, it describes zero-point noise. In our example of an oscillator coupled to a noise source, we saw that the zero-point noise corresponds to transitions where the noise source absorbs energy—at zero temperature, emission events are not possible.

What happens if our noise source is *not* in thermal equilibrium? In that case, one could simply use Eq. (2.31) to define an effective temperature $T_{\text{eff}}[\Omega]$ from the ratio of the symmetrized noise and damping. Re-writing things in terms of the quantum noise spectral density, one finds

$$k_{\text{B}}T_{\text{eff}}[\Omega] \equiv \hbar\Omega \left[\ln\left(\frac{S_{FF}[\Omega]}{S_{FF}[-\Omega]}\right)\right]^{-1}. \tag{2.32}$$

The effective temperature at a given frequency Ω characterizes the asymmetry between absorption and emission rates of energy $\hbar\Omega$; a large temperature indicates that these rates are almost equal, whereas a small temperature indicates that emission by the noise source is greatly suppressed compared with absorption by the source. Away from thermal equilibrium, there is no guarantee that the ratio on the right-hand side of Eq. (2.32) will be frequency-independent, and hence T_{eff} will generally have a frequency dependence.

The effective temperature of a nonequilibrium system is indeed physically meaningful, especially in the case where one only probes the noise at a single, well-defined frequency. For example, in our oscillator system, only the noise at $\pm\Omega$ is important, as we are considering the limit of a very high quality factor. It is straightforward to show that the stationary energy distribution of the oscillator found by solving Eq. (2.24) is a thermal distribution evaluated at the temperature $T_{\text{eff}}[\Omega]$ defined in Eq. (2.32). The notion of an effective temperature plays an important role in so-called backaction cooling techniques [4–7]; it also has a significant role in the quantum theory of linear amplification, as we shall soon see.

2.2.3 Brief example: current noise of a quantum point contact

We have seen that we can view quantum noise either in terms of transition rates between energy eigenstates of the noise source or in analogy to a classical noise process. An example that illustrates this dichotomy very nicely is the current noise of a quantum point contact. We sketch this result quickly in what follows; more details can be found in [8].

A quantum point contact (QPC) is a quantum electronic conductor consisting of a narrow constriction in a two-dimensional electron gas (2DEG)—typically formed using gate electrodes placed above the 2DEG. The constriction is narrow enough that the transverse momentum of electrons is quantized; we will focus on the single-channel case, where only a single transverse mode is occupied. Such QPCs can be used as extremely sensitive charge sensors [9], and are routinely used as detectors of quantum dot qubits [10, 11] and even mechanical oscillators [12]. The basic idea of the detection is that the input signal (a charge) changes the potential of the QPC, and hence changes its conductance. By monitoring the QPC current, one can thus learn about the signal.

Given its use as a detector, one is naturally very interested in the current noise of a quantum point contact. To a good approximation, one can view this system as a simple one-dimensional scattering problem, with the QPC constriction acting as a scattering potential separating two ideal one-dimensional wires (the "leads"). One can easily find the single-particle scattering states $\phi_\alpha(\varepsilon, z)$, which describe the scattering of an electron incident from the lead α (in a plane-wave state) having energy ε ($\alpha = L, R$). Turning to a many-particle description, each single-particle scattering state is now described by a fermionic annihilation operator $\hat{c}_\alpha(\varepsilon)$. It is straightforward to write the electronic field operator in terms of these states, and then use it to write the current operator (say in the left lead) with the result

$$\hat{I}(t) = \int d\varepsilon_1 \int d\varepsilon_2 \sum_{\alpha,\beta=L,R} A_{\alpha\beta}(\varepsilon_1, \varepsilon_2) e^{-i(\varepsilon_2-\varepsilon_1)t}\, \hat{c}^\dagger_\alpha(\varepsilon_1)\hat{c}_\beta(\varepsilon_2). \tag{2.33}$$

The matrix A here is determined by the scattering matrix of the system (an explicit form is given in, e.g., [8]). For finite temperature and voltage, the QPC "leads" are taken to be in equilibrium, but at different chemical potentials, with $\mu_L - \mu_R = eV$ giving the applied bias voltage. Using Eq. (2.14), we see that the current noise at frequency ω naturally corresponds to transitions between scattering states whose energy differs by $\hbar\omega$. The probability for a given scattering state incident from lead α to be occupied is just given by a Fermi function evaluated at the chemical potential μ_α. At zero temperature, the Fermi function becomes a step function, and all states are either empty, or are occupied with probability one. In this case, the only possible transitions at $\omega = 0$ involve states in the energy window $\mu_R < \varepsilon < \mu_L$, with the relevant transitions being the removal of a left scattering state, and the creation of a right scattering state at the same energy.

We thus obtain a picture of the current noise that appears to be explicitly quantum mechanical, involving transitions between energy eigenstates of the detector (here, just scattering states). Remarkably, a simple classical interpretation in terms of a classical noise process is also possible. Note first that at zero temperature (and ignoring any energy dependence of the scattering), the average current through the QPC obtained from Eq. (2.33) is simply given by the Landauer-Büttiker result

$$\frac{1}{e}\langle \hat{I} \rangle = \frac{eV}{h}\mathcal{T}, \tag{2.34}$$

where $0 \le \mathcal{T} \le 1$ is the transmission probability of the QPC. One also finds from Eq. (2.33) that the zero-frequency current noise is given by

$$\frac{1}{e^2}\bar{S}_{II}[0] = \frac{eV}{h}\mathcal{T}(1-\mathcal{T}). \tag{2.35}$$

These results correspond to a simple classical noise process, where electrons are launched one at a time toward the scattering potential at a rate eV/h. In each event, the electron is either transmitted (probability $\mathcal{T}$) or reflected (probability $1-\mathcal{T}$). We thus have a binomial process, similar to flipping an unevenly weighted coin. The average current just corresponds to the average number of electrons that are transmitted, while the noise corresponds to the variance of the binomial process. If we imagine integrating $\hat{I}(t)$ up from $t=0$ (i.e., counting electrons), then, in the long-time limit, we have a variance

$$\frac{1}{e^2}\left\langle\left\langle\left[\int_0^t dt'\,\hat{I}(t')\right]^2\right\rangle\right\rangle = \frac{eVt}{h}\mathcal{T}(1-\mathcal{T}) \equiv N_{\mathrm{A}}\mathcal{T}(1-\mathcal{T}). \tag{2.36}$$

Here, we can view N_{A} as the number of attempts at getting an electron through the QPC. We thus have a concrete example showing the two complementary ways that one can look at quantum noise in general (i.e., in terms of transitions between eigenstates, versus in analogy to a classical noise process).

2.2.4 Heisenberg inequality on detector quantum noise

Generic two-port linear-response detector Having discussed two of the ways in which quantum noise spectral densities differ from their classical counterparts (zero-point noise, frequency asymmetry), we now turn to the third distinguishing feature: there are purely quantum constraints on the noise properties of any system capable of acting as a detector or amplifier. We will be interested in the generic two-port detector sketched in Fig. 2.1. The detector has an input port characterized by an operator

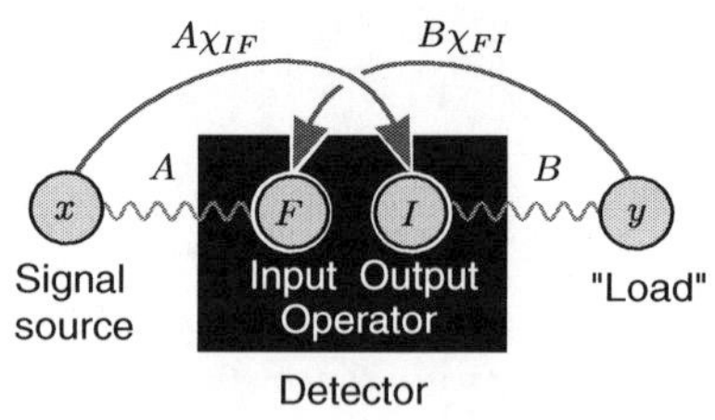

Fig. 2.1 Schematic of a generic linear-response detector.

$\hat{F}$: this is the detector quantity that couples to the system we wish to measure. Similarly, the output port is characterized by an operator $\hat{I}$: this is the detector quantity that we will readout to learn about the system coupled to the input. For example, for a QPC detector coupled to a double dot, the state of the qubit σ_z changes the potential of electrons in the QPC. The operator $\hat{F}$ will thus involve the charge-density operator of the QPC electrons. In contrast, the quantity that is actually measured is the QPC current; hence, $\hat{I}$ will be the QPC current operator.

We will be interested almost exclusively in detector-signal couplings weak enough that one can use linear-response theory to describe how $\hat{I}$ changes in response to the signal. For example, if we couple an input signal $\hat{z}$ to our detector via an interaction Hamiltonian

$$\hat{H}_{\text{int}} = \hat{z} \cdot \hat{F}, \tag{2.37}$$

linear-response theory tells us that the change in the detector output will be given by

$$\delta\langle \hat{I}(t)\rangle = \int_{-\infty}^{\infty} dt' \, \chi_{IF}(t - t')\langle \hat{z}(t')\rangle, \tag{2.38}$$

$$\chi_{IF}(t) = -\frac{i}{\hbar}\theta(t)\langle[\hat{I}(t), \hat{F}(0)]\rangle. \tag{2.39}$$

This is completely analogous to the way we discussed damping; see Eq. (2.30). As is standard in linear-response theory, the expectation values above are all with respect to the state of the system (signal plus detector) at zero coupling (i.e., $\hat{H}_{\text{int}} = 0$). Also, without loss of generality, we will assume that both $\langle\hat{I}\rangle$ and $\langle\hat{F}\rangle$ are zero in the absence of any coupling to the input signal.

Even on a classical level, any noise in the input and output ports will limit our ability to make measurements with the detector. Quantum mechanically, we have seen that it is the symmetrized quantum spectral densities that have a role analogous to classical noise spectral densities. We will thus be interested in the quantities $\bar{S}_{II}[\omega]$ and $\bar{S}_{FF}[\omega]$. Given our interest in weak detector-signal couplings, it will be sufficient to characterize the detector noise at zero coupling to the detector (though we will go beyond this assumption in our discussion of qubit detection).

In addition to $\bar{S}_{II}, \bar{S}_{FF}$, we will also have to contend with the fact that the noise in $\hat{I}$ and $\hat{F}$ may be correlated. Classically, we would describe such correlations via a correlation spectral density $\mathcal{S}_{IF}[\omega]$:

$$\mathcal{S}_{IF}[\omega] \equiv \lim_{T\to\infty} \langle I_T[\omega] \left(F_T[\omega]\right)^* \rangle = \int_{-\infty}^{\infty} dt \, \langle I(t)F(0)\rangle e^{i\omega t}, \tag{2.40}$$

where the Fourier transforms $I_T[\omega]$ and $F_T[\omega]$ are defined analogously to Eq. (2.3). Not surprisingly, such classical correlations correspond to a symmetrized quantum noise spectral density

$$\bar{S}_{IF}[\omega] \equiv \frac{1}{2}\int_{-\infty}^{\infty} dt \, \langle\{\hat{I}(t), \hat{F}(0)\}\rangle e^{i\omega t}. \tag{2.41}$$

Note that the classical correlation density $\mathcal{S}_{IF}[\omega]$ is generally complex, and is only guaranteed to be real at $\omega = 0$; the same is true of $\bar{S}_{IF}[\omega]$.

Finally, we normally are only concerned about how large the output noise is compared with the magnitude of the "amplified" input signal at the output (i.e., Eq. (2.38)). It is thus common to think of the output noise at a given frequency $\delta I_T[\omega]$ as an equivalent fluctuation of the signal $\delta z_{\rm imp}[\omega] \equiv \delta I_T[\omega]/\chi_{IF}[\omega]$. We thus define the imprecision noise spectral density and imprecision-backaction correlation density as

$$\bar{S}_{zz}[\omega] \equiv \frac{\bar{S}_{II}[\omega]}{|\chi_{IF}[\omega]|^2}, \quad \bar{S}_{zF}[\omega] \equiv \frac{\bar{S}_{IF}[\omega]}{\chi_{IF}[\omega]}. \tag{2.42}$$

Motivation and derivation of noise constraint We can now ask what sort of constraints exist on the detector noise. In almost all relevant cases, our detector will be some sort of driven quantum system, and hence will not be in thermal equilibrium. As a result, any meaningful constraint should not rely on having a thermal equilibrium state. Classically, all we can say is that the correlations in the noise cannot be bigger than the noise itself. This constraint takes the form of a a Schwartz inequality, yielding

$$\mathcal{S}_{zz}[\omega]\mathcal{S}_{FF}[\omega] \geq |\mathcal{S}_{zF}[\omega]|^2. \tag{2.43}$$

Equality here implies a perfect correlation, that is, $I_T[\omega] \propto F_T[\omega]$.

Quantum mechanically, additional constraints will emerge. Heuristically, this can be expected by making an analogy to the example of the Heisenberg microscope. In that example, one finds that there is a tradeoff between the imprecision of the measurement (i.e., the position resolution) and the backaction of the measurement (i.e., the momentum kick delivered to the particle). In our detector, noise in $\hat{I}$ will correspond to the imprecision of the measurement (i.e., the bigger this noise, the harder it will be to resolve the signal described by Eq. (2.38)). Similarly, noise in $\hat{F}$ is the backaction: as we already saw, by virtue of the detector-signal coupling, $\hat{F}$ acts as a noisy force on the measured quantity $\hat{z}$. Making the analogy to Eq. (2.1) for the Heisenberg microscope, we thus might expect a bound on the product of $\bar{S}_{zz}\bar{S}_{FF}$.

Alternatively, we see from Eq. (2.39) that for our detector to have any response at all, $\hat{I}(t)$ and $\hat{F}(t')$ cannot commute for all times. Quantum mechanically, we know that uncertainty relations apply any time we have noncommuting observables; here things are somewhat different, as the noncommutation is between Heisenberg-picture operators at different times. Nonetheless, we can still use the standard derivation of an uncertainty relation to obtain a useful constraint. Recall that for two non-commuting observables $\hat{A}$ and $\hat{B}$, the full Heisenberg inequality is (see, e.g., [13])

$$(\Delta A)^2(\Delta B)^2 \geq \frac{1}{4}\langle\{\hat{A},\hat{B}\}\rangle^2 + \frac{1}{4}\left|\langle[\hat{A},\hat{B}]\rangle\right|^2. \tag{2.44}$$

Here, we have assumed $\langle\hat{A}\rangle = \langle\hat{B}\rangle = 0$. We now take $\hat{A}$ and $\hat{B}$ to be cosine transforms of $\hat{I}$ and $\hat{F}$, respectively, over a finite time interval T:

$$\hat{A} \equiv \sqrt{\frac{2}{T}}\int_{-T/2}^{T/2} dt\, \cos(\omega t + \delta)\, \hat{I}(t), \quad \hat{B} \equiv \sqrt{\frac{2}{T}}\int_{-T/2}^{T/2} dt\, \cos(\omega t)\, \hat{F}(t). \tag{2.45}$$

Note that we have phase-shifted the transform of $\hat{I}$ relative to that of $\hat{F}$ by a phase δ. In the limit $T \to \infty$, we find

$$\bar{S}_{zz}[\omega]\bar{S}_{FF}[\omega] \geq \left[\mathrm{Re}\left(e^{i\delta}\bar{S}_{zF}[\omega]\right)\right]^2 + \frac{\hbar^2}{4}\left[\mathrm{Re}\; e^{i\delta}\left(1 - \frac{(\chi_{FI}[\omega])^*}{\chi_{IF}[\omega]}\right)\right]^2. \tag{2.46}$$

We have introduced here a new susceptibility $\chi_{FI}[\omega]$, which describes the reverse response coefficient or reverse gain of our detector. This is the response coefficient relevant if we used our detector in reverse: couple the input signal $\hat{z}$ to $\hat{I}$, and see how $\langle\hat{F}\rangle$ changes. A linear-response relation analogous to Eq. (2.38) would then apply, with $F \leftrightarrow I$ everywhere. We define the ratio of the detector response coefficients to be

$$r[\omega] = \frac{(\chi_{FI}[\omega])^*}{\chi_{IF}[\omega]}. \tag{2.47}$$

If we now maximize the right-hand side of Eq. (2.46) over all values of δ, we are left with the optimal bound

$$\bar{S}_{zz}[\omega]\bar{S}_{FF}[\omega] - \left|\bar{S}_{zF}[\omega]\right|^2 \geq \frac{\hbar^2}{4}\left|1 - r[\omega]\right|^2\left(1 + \Delta\left[\frac{2\bar{S}_{zF}[\omega]}{\hbar(1 - r[\omega])}\right]\right), \tag{2.48}$$

where

$$\Delta[y] = \frac{\left|1 + y^2\right| - \left(1 + |y|^2\right)}{2}, \tag{2.49}$$

Note that for any complex number y, $1 + \Delta[y] > 0$. Related noise constraints on linear-response detectors are presented in [14, 15].

We see that applying the uncertainty principle to our detector has given us a rigorous constraint on the detector's noise that is stronger than the simple classical bound of Eq. (2.43) on its correlations. This extra quantum constraint vanishes if our detector has completely symmetric response coefficients, that is, $r[\omega] = 1$. For simplicity, consider first the $\omega \to 0$ limit, where all noise spectral densities and susceptibilities are real, and hence the term involving $\Delta[y]$ vanishes. For a nonsymmetric detector, the extra quantum term on the right-hand side of Eq. (2.46) then implies the following:

- The product of the imprecision noise $\bar{S}_{zz}$ and backaction noise $\bar{S}_{FF}$ cannot be zero. The magnitude of both kinds of fluctuations must be nonzero.
- Moreover, these fluctuations cannot be perfectly correlated with one another: we cannot have $\left(\bar{S}_{zF}\right)^2 = \bar{S}_{zz}\bar{S}_{FF}$.

The presence of these extra quantum constraints on noise will lead to fundamental quantum limits on various things we might try to do with our detector; this will be the focus of the remainder of this chapter.

Comments on the reverse gain of a detector Before moving on, it is worth commenting further on the reverse gain χ_{FI}: both its meaning and its role in the quantum noise inequality have been the subject of some confusion. Several points are worth noting:

- If our detector is in thermal equilibrium and in a time-reversal symmetric state, then the relationship between χ_{IF} and χ_{FI} is constrained by Onsager reciprocity relations (see, e.g., [16] for an elementary discussion). One has $\chi_{IF}[\omega] = \pm\chi_{FI}[\omega]^*$, where the + (−) sign corresponds to the case where both $\hat{F}$ and $\hat{I}$ have the same (opposite) parity under time reversal. For example, in a QPC detector, $\hat{F}$ is a charge and hence even under time reversal, whereas $\hat{I}$ is a current and odd under time reversal; one thus has $\chi_{IF}[\omega] = -\chi_{FI}[\omega]^*$. It follows that in thermal equilibrium, if one has forward response, then one must necessarily also have reverse response.
- In general, it is highly undesirable to have nonzero reverse gain. To make a measurement of the output operator $\hat{I}$, we must necessarily couple to it in some manner. If $\chi_{FI} \neq 0$, the noise associated with this coupling could in turn lead to additional backaction noise in the operator $\hat{F}$, above and beyond the intrinsic fluctuations described by $\bar{S}_{FF}$. This is clearly something to be avoided. Thus, the ideal situation is to have $\chi_{FI} = 0$, implying a high asymmetry between the input and output of the detector, and requiring the detector to be in a state far from thermodynamic equilibrium.

Ideal quantum noise We can now define in a general and sensible manner what it means for a detector to possess "ideal" quantum noise at a frequency ω: we require that the detector optimize the fundamental quantum noise inequality; that is, Eq. (2.48) holds as an equality. We will see that having such ideal quantum noise properties is a prerequisite for achieving various quantum limits on measurement. It also places tight constraints on the property of our detector. One can show having ideal quantum noise necessarily implies the following [2]:

- In a certain restricted sense, the operators $\hat{I}$ and $\hat{F}$ must be proportional to one another. More formally, $\bar{S}_{II}[\omega]$ and $\bar{S}_{FF}[\omega]$ can be written as sums of transitions between detector energy eigenstates $|i\rangle$ and $|f\rangle$ whose energy differs by $\pm\hbar\omega$ (see Eq. (2.14)). To have quantum ideal noise, one needs that for each such contributing transitions, the ratio of the matrix elements $\langle f|\hat{I}|i\rangle/\langle f|\hat{F}|i\rangle$ is the same.
- As long as $|r[\omega]| \neq 1$, the detector *cannot* be in a thermal equilibrium state.
- For a general nonequilibrium system, the effective temperature defined using the asymmetry of the quantum noise spectral density $S_{II}[\omega]$ (see Eq. (2.32)) need not be the same as that defined using $S_{FF}[\omega]$. However, having ideal quantum-noise necessarily implies that both these effective temperatures are the same. Paradoxically, while such a detector cannot be in equilibrium, its effective temperature is in some sense more universal than that of an arbitrary nonequilibrium system.

At this stage, the true meaning of the quantum noise inequality may seem quite opaque. It may also seem that there is no hope of understanding in general what one needs to do to achieve ideal quantum noise. However, by considering concrete examples, we will gain insights into both these issues.

2.3 Quantum limit on QND qubit detection

2.3.1 Measurement rate and dephasing rate

Armed now with a basic understanding of quantum noise and the Heisenberg bounds that constrain it, we can finally consider doing something useful with our generic linear-response detector. To that end, we consider a qubit whose Hamiltonian is

$$\hat{H}_{\text{qb}} = \frac{\hbar\Omega}{2}\hat{\sigma}_z. \tag{2.50}$$

Suppose we want to measure whether the qubit is in its ground or excited state. We start by coupling its $\hat{\sigma}_z$ operator to the input of our detector:

$$\hat{H}_{\text{int}} = A\hat{\sigma}_z\hat{F}. \tag{2.51}$$

By virtue of Eq. (2.38), the two different qubit eigenstates $|\uparrow\rangle$, $|\downarrow\rangle$ will lead to two different average values of the detector $\langle\hat{I}\rangle$; thus, by looking at the detector output, we can measure the value of σ_z.

Note crucially that $[\hat{H}_{\text{int}}, \hat{H}_{\text{qb}}] = 0$. As such, $\langle\hat{\sigma}_z\rangle$ is a constant of the motion even when the qubit is coupled to the detector: if the qubit starts in an energy eigenstate, it will remain in that state. Detection schemes where the coupling commutes with the system Hamiltonian are known as being "quantum nondemolition" (QND). On a practical level, this can be extremely useful, as one can leave the measurement on for a long time (or make multiple measurements) to improve precision without worrying about the measurement process altering the value of the measured observable.

Because of the intrinsic noise in the output of our detector (described by $\bar{S}_{II}$), it will take some time before we can tell whether the qubit is up or down. We only gradually obtain information about the qubit state, and can rigorously define a rate to characterize this process, the so-called measurement rate. Imagine we turn the measurement on at $t = 0$, and start to integrate up the output $I(t)$ of our detector:

$$\hat{m}(t) = \int_0^t dt'\,\hat{I}(t'). \tag{2.52}$$

The probability distribution of the integrated output $\hat{m}(t)$ will depend on the state of the qubit; for long times, we may approximate the distribution corresponding to each qubit state as being Gaussian. Noting that we have chosen $\hat{I}$ so that its expectation value vanishes at zero coupling, the average value of $\langle\hat{m}(t)\rangle$ corresponding to each qubit state is

$$\langle\hat{m}(t)\rangle_\uparrow = A\chi_{IF}[0]t, \quad \langle\hat{m}(t)\rangle_\downarrow = -A\chi_{IF}[0]t. \tag{2.53}$$

Note that we are assuming integration times t much longer than any internal detector timescale, and thus only the zero-frequency response coefficient χ_{IF} appears. Taking the long-time limit here is consistent with our assumption of a weak detector-qubit coupling: it will take a long time before we get information on the qubit state.

Next, let's consider the uncertainty in the quantity m as described by its variance $\langle\langle m^2\rangle\rangle \equiv \langle\hat{m}^2\rangle - \langle\hat{m}\rangle^2$. For weak coupling, we can ignore the fact that the variance will have a small dependence on the qubit state, as this will only lead to higher-order-in-A corrections to our expression for the measurement rate. We thus have

$$\langle \hat{m}^2(t)\rangle \equiv \int_0^t dt_1 \int_0^t dt_2 \, \langle \hat{I}(t_1)\hat{I}(t_2)\rangle \rightarrow \bar{S}_{II}[0]t. \tag{2.54}$$

We have again taken the limit where t is much larger than the correlation time of the detector noise, and hence the variance is completely determined by the zero-frequency output noise.

We can now define the measurement rate by how quickly the resolving power of the measurement grows:[1]

$$\frac{1}{4}\frac{[\langle \hat{m}(t)\rangle_\uparrow - \langle \hat{m}(t)\rangle_\downarrow]^2}{\langle\langle \hat{m}^2(t)\rangle\rangle_\uparrow + \langle\langle \hat{m}^2(t)\rangle\rangle_\downarrow} \equiv \Gamma_{\text{meas}} t. \tag{2.55}$$

This yields

$$\Gamma_{\text{meas}} = \frac{A^2 \left(\chi_{IF}\right)^2}{2\bar{S}_{II}}. \tag{2.56}$$

We can think of $1/\Gamma_{\text{meas}}$ as a measurement time, that is, the amount of time we have to wait before we can reliably determine whether the qubit is up or down (above the intrinsic noise in the detector output).

Having characterized the imprecision of the measurement, we now turn to its backaction. At first glance, one might think the fact that we have a QND setup implies the complete absence of measurement backaction. This is not true. We are making a measurement of σ_z, and hence there must be a backaction disturbance of the conjugate quantities σ_x, σ_y. More explicitly, if we start the qubit out in a superposition of energy eigenstates, then the phase information of this superposition will be lost gradually in time owing to the backaction of the measurement.

To describe this backaction effect, note first that we can incorporate the coupling into the qubit's Hamiltonian as

$$\hat{H}_{\text{qb}} + \hat{H}_{\text{int}} = \left(\frac{\hbar\Omega}{2} + A\hat{F}\right)\hat{\sigma}_z. \tag{2.57}$$

Thus, from the qubit's point of view, the coupling to the detector means that its splitting frequency has a randomly fluctuating part described by $\Delta\Omega = 2A\hat{F}/\hbar$. This effective frequency fluctuation will cause a diffusion of the qubit's phase in the long-time limit according to

$$\left\langle e^{-i\varphi}\right\rangle = \left\langle e^{-i\int_0^t d\tau\, \Delta\Omega(\tau)}\right\rangle. \tag{2.58}$$

For weak coupling, the dephasing rate is slow, and thus we are interested in long times t. In this limit, the integral is a sum of a large number of statistically independent

[1] The strange looking factor of $1/4$ here is purely chosen for convenience, as it will let us formulate the quantum limit in a way that involves no numerical prefactors. Interestingly enough, the prefactor can be rigorously justified if one uses the accessible information (a standard information-theoretic measure) to quantify the difference between the output distributions; the measurement rate as defined is precisely the rate of growth of the accessible information [17].

terms, and thus we can take the accumulated phase to be Gaussian-distributed. Using the cumulant expansion, we then obtain

$$\begin{aligned}\langle e^{-i\varphi}\rangle &= \exp\left(-\frac{1}{2}\left\langle\left[\int_0^t d\tau\,\Delta\Omega(\tau)\right]^2\right\rangle\right)\\ &= \exp\left(-\frac{2A^2}{\hbar^2}\bar{S}_{FF}[0]t\right) \equiv \exp\left(-\Gamma_\varphi t\right). \end{aligned} \tag{2.59}$$

We have again taken the long-time limit, which means that the only the zero-frequency backaction noise spectral density enters. Equation (2.59) yields the dephasing rate

$$\Gamma_\varphi = \frac{2A^2}{\hbar^2}\bar{S}_{FF}[0]. \tag{2.60}$$

2.3.2 Efficiency ratio

On a completely heuristic level, we can easily argue that the measurement and dephasing rates of our setup should be related. Imagine a simple case where at $t = 0$ the qubit is in a superposition state and the detector is in some pure state $|D_0\rangle$:

$$|\psi(0)\rangle = \frac{1}{\sqrt{2}}\left(|\uparrow\rangle + e^{i\varphi_0}|\downarrow\rangle\right)\otimes|D_0\rangle. \tag{2.61}$$

At some later time, owing to the qubit-detector interaction, the qubit can become entangled with the detector, and we have

$$|\psi(t)\rangle = \frac{1}{\sqrt{2}}\left(|\uparrow\rangle\otimes|D_\uparrow(t)\rangle + e^{i\varphi_0}|\downarrow\rangle\otimes|D_\downarrow(t)\rangle\right), \tag{2.62}$$

where the two detector states are not necessarily equal: $|D_\uparrow(t)\rangle \neq |D_\downarrow(t)\rangle$.

To see if the qubit has dephased or not, consider an off-diagonal element of its reduced density matrix:

$$\rho_{\downarrow\uparrow}(t) \equiv \langle\downarrow|\hat{\rho}(t)|\uparrow\rangle = \frac{e^{i\varphi_0}}{2}\langle D_\uparrow(t)|D_\downarrow(t)\rangle. \tag{2.63}$$

At $t = 0$, $|D_\uparrow(0)\rangle = |D_\downarrow(0)\rangle = |D_0\rangle$, and the off-diagonal density matrix element $\rho_{\downarrow\uparrow}(0)$ just tells us the initial qubit phase. As t increases from 0, $|D_\uparrow(t)\rangle$ and $|D_\downarrow(t)\rangle$ will in general be different, causing the magnitude of $\rho_{\downarrow\uparrow}(t)$ to decay with time. Comparing against Eq. (2.59), we would thus associate backaction dephasing with the fact that the two detector states $|D_\uparrow(t)\rangle$, $|D_\downarrow(t)\rangle$ have an overlap of magnitude less than one.

In contrast, the measurement rate in Eq. (2.56) does not directly involve the overlap of the two detector states $|D_\uparrow(t)\rangle$, $|D_\downarrow(t)\rangle$. Rather, it involves how different the distributions of $\hat{m}$ are in these two states. Clearly, if the two states $|D_\uparrow\rangle$, $|D_\downarrow\rangle$ can be distinguished by looking at m, then they must have an overlap < 1: *measurement implies dephasing.* The converse is not true: the two detector states could be orthogonal because the qubit has become entangled with extraneous detector degrees of freedom, without these states yielding different distributions of m. Hence, *dephasing does not imply measurement.*

Putting this together, one roughly expects that the measurement rate should be bounded by the dephasing rate. We can test this expectation by using the linear-response expressions derived in Section 2.2.4, and making use of the quantum noise inequality of Eq. (2.48). If we assume the ideal case of zero reverse gain in our detector, we find

$$\eta \equiv \frac{\Gamma_{\text{meas}}}{\Gamma_{\varphi}} = \frac{\hbar^2/4}{\bar{S}_{zz}\bar{S}_{FF}} \leq 1. \tag{2.64}$$

Thus, in the absence of any detector reverse gain, we obtain the expected result: the dephasing rate must be at least as large as the measurement rate. This is the quantum limit on QND qubit detection [17–21]. This derivation does more than prove the bound, it also indicates what we need to do to reach it. We need both:

(1) A detector with quantum ideal noise at zero frequency; that is, it must saturate the inequality of Eq. (2.48) at $\omega = 0$.
(2) There must be *no* backaction-imprecision noise correlations at zero frequency: $\bar{S}_{zF}$ must be zero

Violating the quantum limit with reverse gain? Despite the intuitive reasonableness of the above quantum limit on QND qubit detection, there would seem to be a troubling loophole in the case where our detector has a nonzero reverse gain χ_{FI}. In this case, the right-hand side of our so-called "quantum limit" is now $(1-r[0])^2$ (where $r[0]$ is the ratio of the reverse to forward gains at zero, frequency; see Eq. (2.47)), and can be made arbitrarily small by having our detector's forward and reverse responses be symmetric. One is tempted to conclude that there is in fact no quantum limit on QND qubit detection. This is of course an invalid inference: as discussed, $\chi_{FI} \neq 0$ implies that we must necessarily consider the effects of extra noise injected into the detector's output port when one measures $\hat{I}$, as the reverse gain will bring this noise back to the qubit, causing extra dephasing. The result is that one can do no better than $\eta = 1$. To see this explicitly, consider the extreme case $\chi_{IF} = \chi_{FI}$ and $\bar{S}_{II} = \bar{S}_{FF} = 0$, and suppose we use a second detector to read out the output $\hat{I}$ of the first detector. This second detector has input and output operators $\hat{F}_2$, $\hat{I}_2$; we also take it to have a vanishing reverse gain, so that we do not have to also worry about how its output is read out. Coupling the detectors linearly in the standard way (i.e., $H_{\text{int},2} = \hat{I}\hat{F}_2$), the overall gain of the two detectors in series is $\chi_{I_2F_2}\chi_{IF}$, while the backaction driving the qubit dephasing is described by the spectral density $(\chi_{FI})^2 S_{F_2F_2}$. Using the fact that our second detector must itself satisfy the quantum noise inequality, we have

$$\left[(\chi_{FI})^2\,\bar{S}_{F_2F_2}\right]\bar{S}_{I_2I_2} \geq \frac{\hbar^2}{4}\left(\chi_{I_2F_2}\chi_{IF}\right)^2. \tag{2.65}$$

Thus, the overall chain of detectors satisfies the usual, zero-reverse-gain quantum noise inequality, implying that we will still have $\eta \leq 1$.

2.3.3 Example: QPC detector

Let's consider again the single-channel QPC detector of Section 2.2.3, and imagine that we connect it to a single-electron double quantum dot. The single electron can be in either the left or the right dot; these will correspond to the σ_z eigenstates of the effective qubit formed by the dot. If we assume that interdot tunnelling has been switched off, then these two states are also energy eigenstates of the qubit. Finally, these two states will lead to to different electrostatic potentials for the QPC electrons: we thus have a coupling of the form given in Eq. (2.37), where the input $\hat{F}$ operator is actually a charge in the QPC.

A rigorous treatment of the measurement properties of a QPC detector is given in [17, 22, 23]. Here, we provide a more heuristic treatment that brings out the main aspects of the physics, and also helps motivate the crucial connection between the quantum limit on QND qubit detection, the quantum noise constraint of Eq. (2.48), and the principle of "no wasted information."

Recall first our results for the current and current noise of a QPC detector (see Eqs. (2.34) and (2.35)); both depend on $\mathcal{T}$, the probability of electron transmission through the QPC. Including the coupling to the qubit, each qubit eigenstate will correspond to two different effective QPC potentials, and hence two different transmission coefficients $\mathcal{T}_\uparrow, \mathcal{T}_\downarrow$:

$$\mathcal{T}_\uparrow \equiv \mathcal{T}_0 + \Delta\mathcal{T}, \quad \mathcal{T}_\downarrow \equiv \mathcal{T}_0 - \Delta\mathcal{T}. \tag{2.66}$$

Using Eqs. (2.34) and (2.35) for $\langle \hat{I} \rangle$ and $\bar{S}_{II}$, Eq. (2.56) for the measurement rate immediately yields

$$\Gamma_{\text{meas}} = \frac{1}{2}\frac{(\Delta\mathcal{T})^2}{\mathcal{T}_0(1-\mathcal{T}_0)}\frac{eV}{h}. \tag{2.67}$$

Turning to the dephasing rate, the backaction charge fluctuations $\bar{S}_{FF}$ can be calculated using scattering theory [17]. We instead take a more heuristic approach that yields the correct answer and provides us the general insight we are after. Let's describe the transmitted charge m through the QPC with an approximate wavefunction; further, let's ignore the discreteness of charge, and treat m as being continuous. Thus, if the qubit was initially in state α, the transmitted charge through the QPC at time t might reasonably be described by a wavefunction

$$|\psi_{\text{QPC},\alpha}(t)\rangle = \int dm\, \phi_\alpha(m)|m\rangle. \tag{2.68}$$

We pick $\phi_\alpha(m)$ to yield the expected (Gaussian) probability distribution of m,

$$|\phi_\alpha(m)|^2 = \frac{1}{\sqrt{2\pi\sigma^2}}\exp\left[-\frac{(m-\bar{m}_\alpha)^2}{2\sigma^2}\right], \tag{2.69}$$

where the mean and variance match what we already calculated:

$$\bar{m}_\alpha = \frac{eVt}{h}\left(\mathcal{T}_0 \pm \Delta\mathcal{T}\right), \quad \sigma = \frac{1}{e^2}\bar{S}_{II}t = \frac{eVt}{h}\mathcal{T}_0\left(1-\mathcal{T}_0\right). \tag{2.70}$$

We still have to worry about the phase of our phenomenological QPC wavefunction. Recall that our description of the QPC is based on a simple scattering picture. An incident electron is either transmitted with an amplitude $\sqrt{\mathcal{T}}e^{i\theta_t}$ or reflected with an amplitude $\sqrt{1-\mathcal{T}}e^{i\theta_r}$. Here, θ_t and θ_r are phases in the scattering matrix. If we use the state where all electrons are reflected as our phase reference, we see that each transmission event is associated with a net phase shift

$$\exp\left[i(\theta_t - \theta_r)\right] \equiv \exp\left(i\theta\right). \tag{2.71}$$

Further, in the same way that the two states of the qubit can change the transmission probability $\mathcal{T}$, they could also cause the value of this phase difference to change. We thus write

$$\theta_{\uparrow} = \theta_0 + \Delta\theta, \quad \theta_{\downarrow} = \theta_0 - \Delta\theta. \tag{2.72}$$

Based on this picture, it is reasonable to write our final heuristic wavefunction in the form

$$\phi_\alpha(m) = |\phi_\alpha(m)|e^{im\theta_\alpha}. \tag{2.73}$$

We are now in a position to calculate the backaction dephasing rate of the qubit. Using Eq. (2.63), we see that this is just determined by the overlap of the two QPC states:

$$\exp(-\Gamma_\varphi t) \equiv |\langle\psi_{\mathrm{QPC},\uparrow}|\psi_{\mathrm{QPC},\downarrow}\rangle|\,. \tag{2.74}$$

We can easily calculate the required overlap, as it amounts to a simple Gaussian integration. We find

$$\Gamma_\varphi = \Gamma_{\mathrm{meas}} + 2\left(\Delta\theta\right)^2 \mathcal{T}_0\left(1-\mathcal{T}_0\right)\frac{eV}{h} \equiv \Gamma_{\mathrm{meas}} + \Gamma_{\mathrm{meas},\theta}. \tag{2.75}$$

where Γ_{meas} is just the measurement rate given in Eq. (2.67). We note that this expressions matches exactly what is found from a rigorous calculation of the backaction noise $\bar{S}_{FF}$ [17, 22]. We see that if $\Delta\theta \neq 0$, the QPC misses the quantum limit on QND qubit detection: the dephasing rate is larger than the measurement rate. The extra term in the dephasing rate can be directly interpreted as the measurement rate of an experiment where one tried to determine the qubit state by interfering transmitted and reflected beams of electrons. This "phase" contribution to the dephasing rate has even been measured in experiments using quantum Hall edge states [24].

Several comments are in order:

- We see that a failure to reach the QND quantum limit corresponds to the existence of "wasted information": there are other quantities besides $\hat{I}$ that one could measure to learn about the state of the qubit. The corollary is that to reach the quantum limit, there should be no wasted information: there should be no other degrees of freedom in the detector that could provide more information on the qubit state besides that available in $\hat{I}$. This idea is of course more general than just this example, and provides a powerful way of assessing whether a given system will reach the quantum limit.

- The same reasoning applies to the noise properties of the detector: a failure to optimize the quantum noise inequality of Eq. (2.48) is in general associated with "wasted information" in the detector. Many more examples of this are given in [2].

As a final comment, we note that reaching the quantum limit on QND qubit detection not only requires having a detector with "ideal" quantum noise, but, in addition, there must be no backaction-imprecision noise correlations. Such correlated backaction noise is always in excess of the absolute minimum value of $\bar{S}_{FF}$ required by Eq. (2.48). As we will see, in non-QND measurements, one can make use of these correlations, and in some cases one even requires their presence to reach the quantum limit on the measurement. In the QND case, however, backaction is irrelevant to what shows up in the output of the detector, and hence one cannot make use of any backaction-imprecision correlation. As such, the correlated backaction also represents a kind of wasted information.

It is interesting to note that in the QPC example, the "phase" contribution to the backaction noise is in fact perfectly correlated with the imprecision current noise. Consider the simple case where $\Delta T = 0$, and there is only "phase information" on the state of the qubit. If the qubit is initially in a superposition state with a phase ϕ_0 (see Eq. (2.61)), then, at time t, it follows from Eqs. (2.63) and (2.73) that its off-diagonal density matrix element will be given by

$$\langle e^{-i\varphi} \rangle = e^{-i\varphi_0} \int dm \, |\phi(m)|^2 \, e^{-2im\Delta\theta} = \exp[-i\phi_0] \int dm \, p(m) e^{-2im\Delta\theta}. \quad (2.76)$$

Thus, if the value of m were definite, the qubit would pick up a deterministic phase shift $2\Delta\theta m$; however, as m fluctuates, one gets a random phase shift and hence dephasing. Crucially though, this random phase shift (i.e., backaction noise) is correlated with the fluctuations of m (i.e., the imprecision current noise). In the simple limit $\Delta T \to 0$ considered here, the QPC saturates the quantum limit of Eq. (2.48), but with zero gain: $\chi_{IF} = 0$.

2.3.4 Significance of the quantum limit on QND qubit detection

At this stage, one might legitimately wonder why anyone would care about this quantum limit on qubit measurement. If our only goal is to determine whether the qubit is up or down, why should we care about whether the qubit is dephased as slowly as is allowed by quantum mechanics? This would seem to have no bearing on our ability to make a measurement.

The full answer to this question involves the world of conditional measurement: what happens to the qubit in a *single* run of the experiment? More concretely, in a given run of the experiment, one obtains a specific, noisy time trace of $I(t)$. Given this time trace, what can one say about the qubit? Such knowledge is of course crucial if one wishes to use the measurement record in a feedback protocol to control the qubit state. The QND qubit quantum limit plays a crucial role here: if the detector reaches the quantum limit, then, in a *particular* run of the experiment, there is no measurement-induced qubit dephasing (see, e.g., [25]). Rather, the qubit's phase undergoes a seemingly random evolution that is in fact correlated with the noise in the detector output, $I(t)$. This phase evolution only looks like dephasing when one does not have access to the measurement record. These fascinating ideas will be treated in chapters 1, 7, 11 and 17 of this book.

2.3.5 QND quantum limit beyond linear response

What if the qubit-detector coupling A is not sufficiently small to allow the neglect of higher-order contributions to the dephasing and measurement rates? We can still formulate a quantum limit on the backaction dephasing rate, by saying that it is bounded below by its value in the most ideal case. The most ideal case corresponds to the situation in Eq. (2.62), where each qubit state leads to a different detector pure state. Further, the most ideal situation is where the overlap of these two states is *completely* determined by the probability distribution of the integrated detector output m (i.e., like our heuristic QPC discussion in the case where $\Delta\theta = 0$). If we further assume the long-time limit and take the distributions of m corresponding to each qubit state to be Gaussian, we end up with the quantum limit

$$\Gamma_\varphi \geq \Gamma_{\varphi,\text{info}}, \tag{2.77}$$

where the minimum dephasing rate required by the information gain of the measurement, $\Gamma_{\varphi,\text{info}}$, is given by

$$\Gamma_{\varphi,\text{info}} \equiv - \lim_{t\to\infty} \ln\left[\int dm \sqrt{p_\uparrow(m) p_\downarrow(m)}\right] = \frac{1}{4}\frac{(\langle \hat{I}\rangle_\uparrow - \langle \hat{I}\rangle_\downarrow)^2}{\bar{S}_{II,\uparrow} + \bar{S}_{II,\uparrow}}. \tag{2.78}$$

In this expression, we have allowed for the fact that detector output noise could be different in the two qubit states. As usual, taking the long-time limit implies that only the zero-frequency noise correlators enter this definition.

2.4 Quantum limit on linear amplification: the op-amp mode

2.4.1 Weak continuous position detection

We now turn to a more general situation, where we use our detector to amplify some time-dependent signal that is coupled to the input. For concreteness, we start with the case of continuous position detection, where the input signal is the position $\hat{x}$ of a simple harmonic oscillator of frequency Ω and mass M. The coupling Hamiltonian is thus

$$\hat{H}_{\text{int}} = A\hat{x} \cdot \hat{F}, \tag{2.79}$$

and the output $\langle \hat{I}(t)\rangle$ will respond linearly to $\langle \hat{x}(t)\rangle$. Similar to the case of qubit detection, because of the intrinsic noise in the detector output (i.e., $\bar{S}_{II}$), it will take us some time before we can resolve the signal due to the oscillator. We will focus on weak couplings, such that we only learn about the oscillator's motion on a timescale long compared with its period. As such, the goal is *not* to measure the instantaneous value of $x(t)$, but rather the slow quadrature amplitudes $X(t)$, $Y(t)$ defined via

$$\hat{x}(t) = \hat{X}(t)\cos(\Omega t) + \hat{Y}(t)\sin(\Omega t). \tag{2.80}$$

As we have already seen in great detail, the fluctuations of the input operator $\hat{F}$ correspond to a noisy backaction force that will both heat and damp the oscillator. Unlike the qubit measurement discussed in Section 2.3, this backaction will impair our ability to measure, as the measurement is not QND: $[\hat{H}_{\text{int}}, \hat{H}_{\text{osc}}] \neq 0$. The noise in the oscillator's momentum caused by $\hat{F}$ will translate into extra position fluctuations at later times, and hence extra noise in the output of the detector. As we will see, this will place a fundamental limit on how well we can continuously monitor position. Alternatively, note that the two quadrature operators $\hat{X}$ and $\hat{Y}$ are canonically conjugate. As such, we are attempting to simultaneously measure two noncommuting observables, and hence a quantum limit is expected (i.e., we cannot know about both $\hat{X}$ and $\hat{Y}$ to arbitrary precision).

For reasons that will become clear in later sections, we will term the amplifier operation mode used here (and consequent quantum limit) the "op-amp" mode. In this mode of operation, the detector is so weakly coupled to the signal source (i.e., the oscillator) that it has almost no effect on the total oscillator damping. This is similar to an ideal voltage op-amp, where the input impedance is extremely large, and thus there is no appreciable change in the impedance of the voltage source producing the input signal. This mode of operation will be contrasted to the scattering mode of operation, where the amplifier-detector coupling is no longer weak in the above sense.

Defining the quantum limit To begin, let's treat the detector output in the presence of the oscillator as a classically noisy quantity; we will also ignore the frequency-dependence of the detector response coefficient χ_{IF} to keep things simple. Letting $x(t)$ denote the signal we are trying to measure (i.e., the position of the oscillator in the absence of any corrupting backaction effect), we then have

$$I_{\text{tot}}(t) = A\chi_{IF}\left[x(t) + \delta x_{\text{BA}}(t)\right] + \delta I(t) \equiv A\chi_{IF}\left[x(t) + \delta x_{\text{add}}(t)\right], \tag{2.81}$$

where

$$\delta x_{\text{add}}(t) = \delta x_{\text{BA}}(t) + \delta x_{\text{imp}}(t) \equiv \delta x_{\text{BA}}(t) + \frac{\delta I(t)}{A\chi_{IF}} \tag{2.82}$$

describes the total added noise of the measurement, viewed as an equivalent position fluctuation. We see that there are two distinct contributions:

- The intrinsic output fluctuations in $\hat{I}$, $\delta I(t)$, which, when referred back to the oscillator, give us the imprecision noise $\delta x_{\text{imp}}(t)$. Making the coupling A (or response χ_{IF}) larger reduces the magnitude of $\delta x_{\text{imp}}(t)$.
- Backaction fluctuations: noise in $\hat{F}$ drives extra position fluctuations of the resonator δx_{BA}. On a classical level, we could describe these with a Langevin equation similar to Eq. (2.19), which would give

$$\delta x_{\text{BA}}[\omega] = A\chi_{xx}[\omega]\delta F[\omega], \tag{2.83}$$

where $\chi_{xx}[\omega]$ is the oscillator's force susceptibility and is given by

$$M\chi_{xx}[\omega] = \left(\omega^2 - \Omega_M^2 + i\omega\gamma_0\right)^{-1}. \tag{2.84}$$

This contribution to the added noise scales as A, and hence gets *worse* the larger one makes A.

To optimize our measurement, we would of course like to make $\delta x_{\rm add}(t)$ as small as possible. In the absence of backaction noise, we could make the added noise arbitrarily small by just increasing the coupling strength A. However, because of backaction, the best we can do is to tune A to balance the contributions from backaction and imprecision; we will be left with something nonzero.

To state the quantum limit on position detection, we first define the measured position $x_{\rm meas}(t)$ as simply the total detector output $I_{\rm tot}(t)$ referred back to the oscillator:

$$x_{\rm meas}(t) = I_{\rm tot}(t)/A\chi_{IF}. \tag{2.85}$$

If there were *no* added noise and, further, if the oscillator were in thermal equilibrium at temperature T, the spectral density describing the fluctuations $\delta x_{\rm meas}(t)$ would simply be the equilibrium fluctuations of the oscillator, as given by the fluctuation-dissipation theorem:

$$\bar{S}_{xx}^{\rm meas}[\omega] = \bar{S}_{xx}^{\rm eq}[\omega, T] = \hbar \coth\left(\frac{\hbar\omega}{2k_{\rm B}T}\right)(-{\rm Im}\ \chi_{xx}[\omega]) \tag{2.86}$$

$$= \frac{x_{\rm ZPF}^2(1+2n_B)}{2}\sum_{\sigma=\pm}\frac{\gamma_0}{(\omega-\sigma\Omega)^2+(\gamma_0/2)^2}. \tag{2.87}$$

Here, γ_0 is the intrinsic damping rate of the oscillator, which we have assumed to be $\ll \Omega$.

Including the added noise, and for the moment ignoring the possibility of any additional oscillator damping due to the coupling to the detector, the above result becomes

$$\bar{S}_{xx}^{\rm meas}[\omega] = \bar{S}_{xx}^{\rm eq}[\omega, T] + \bar{S}_{xx}^{\rm add}[\omega] \tag{2.88}$$

where the last term is the spectral density of the added noise (both backaction and imprecision noise).

We can now, finally, state the standard quantum limit on continuous position detection (which is equivalent to that on linear, phase-preserving amplification): at each frequency ω, we must have

$$\bar{S}_{xx}^{\rm add}[\omega] \geq \bar{S}_{xx}^{\rm eq}[\omega, T=0]. \tag{2.89}$$

The spectral density of the added noise cannot be made arbitrarily small: at each frequency, it must be at least as large as the corresponding zero-point noise.

We now refine the above result to include the presence of backaction damping of the oscillator (at a rate $\gamma_{\rm BA}$) due to the coupling to the detector. Such damping is described by the asymmetry of the detector's $S_{FF}[\omega]$ quantum noise spectrum, as in Eq. (2.26). Including nonzero backaction damping, the added noise is defined as

$$\bar{S}_{xx}^{\rm meas}[\omega] = \frac{\gamma_0}{\gamma_{\rm BA}+\gamma_0}\bar{S}_{xx}^{\rm eq}[\omega, T] + \bar{S}_{xx}^{\rm add}[\omega], \tag{2.90}$$

where the susceptibility χ_{xx} now involves the total damping of the oscillator, that is,

$$M\chi_{xx}[\omega] = \left[\omega^2 - \Omega^2 + i\omega(\gamma_0 + \gamma_{\mathrm{BA}})\right]^{-1}. \tag{2.91}$$

With this definition, the quantum limit on the added noise is unchanged from the limit stated in Eq. (2.89).

2.4.2 A possible correlation-based loophole?

Our heuristic formulation of the quantum limit naturally leads to a possible concern. Even though quantum mechanics may require a position measurement to have a backaction (as position and momentum are conjugate quantities), couldn't this backaction noise be perfectly anticorrelated with the imprecision noise? If this were the case, the added noise $\delta x(t)$ (which is the sum of the two contributions; see Eq. (2.82)) could be made to vanish.

One might hope that this sort of loophole would be explicitly forbidden by the quantum noise inequality of Eq. (2.48). However, this is not the case. Even in the ideal case of zero reverse gain, one can achieve a situation where backaction and imprecision are perfectly correlated at a given nonzero frequency ω. One needs the following:

- The correlator $\bar{S}_{IF}[\omega]$ should be purely imaginary; this implies that the part of $F(t)$ that is correlated with $I(t)$ is $90°$ out of phase. Note that $\bar{S}_{IF}[\omega]$ can only be imaginary at nonzero frequencies.
- The magnitude of $\bar{S}_{IF}[\omega]$ should be larger than $\hbar/2$.

Under these circumstances, one can verify that there is no additional quantum constrain on the noise beyond what exists classically, and hence the perfect correlation condition of $\bar{S}_{FF}[\omega]\bar{S}_{II}[\omega] = |\bar{S}_{IF}[\omega]|^2$ is allowable. The $\pi/2$ phase of the backaction-imprecision correlations are precisely what is needed to make $\delta x_{\mathrm{add}}[\omega]$ vanish at the oscillator resonance, $\omega = \Omega$.

As might be expected, this seeming loophole is *not* a route toward amplification free from any quantum constraints. The problem is that we have not been sufficiently careful to specify what we want our detector to do, namely, the condition that the detector *amplifies* the motion of the oscillator—the signal should be "bigger" at the output than it is at the input. It is only when we insist on amplification that there are quantum constraints on added noise; a passive transducer need not add any noise. On a heuristic level, one could view amplification as an effective expansion of the phase space of the oscillator. Such a pure expansion is of course forbidden by Liouville's theorem, which tells us that volume in phase space is conserved. The way out is to introduce additional degrees of freedom, such that for these degrees of freedom phase space contracts. Quantum mechanically, such degrees of freedom necessarily have noise associated with them (at the very least, zero-point noise); this then is the source of the limit on added noise. We will see that this heuristic argument can be converted into a rigorous formulation of the quantum limit (albeit of a different sort) in Section 2.5.

More concretely, we need to define what we mean by amplification in our linear response detector. We can then rigorously insist that our detector amplifies. The result will be additional constraints beyond the quantum noise inequality of Eq. (2.48) that make the perfect correlation described above impossible.

2.4.3 Power gain

To be able to say that our detector truly amplifies the motion of the oscillator, it is not sufficient to simply say the response function χ_{IF} must be large (note that χ_{IF} is not dimensionless!). Instead, true amplification requires that the *power* delivered by the detector to a following amplifier be much larger than the power drawn by the detector at its input—that is, the detector must have a dimensionless power gain $G_P[\omega]$ much larger than one. If the power gain were not large, we would need to worry about the next stage in the amplification of our signal, and how much noise would be added in that process. Having a large power gain means that, by the time our signal reaches the following amplifier, it is so large that the added noise of this following amplifier is unimportant

To make the above more precise, we start with the ideal case of no reverse gain, $\chi_{FI} = 0$. We will define the power gain $G_P[\omega]$ of our generic position detector in a way that is analogous to the power gain of a voltage amplifier. Imagine we drive the oscillator we are trying to measure (whose position is x) with a force $2F_D \cos \omega t$; this will cause the output of our detector $\langle \hat{I}(t) \rangle$ to also oscillate at frequency ω. To optimally detect this signal in the detector output, we further couple the detector output I to a second oscillator with natural frequency ω, mass M, and position y: there is a new coupling term in our Hamiltonian, $H'_{\text{int}} = B\hat{I} \cdot \hat{y}$, where B is a coupling strength. The oscillations in $\langle I(t) \rangle$ will now act as a driving force on the auxiliary oscillator y (see Fig. 2.2). We can consider the auxiliary oscillator y as a "load" we are trying to drive with the output of our detector.

To find the power gain, we need to consider both P_{out}, the power supplied to the output oscillator y from the detector, and P_{in}, the power fed into the input of the amplifier. Consider first P_{in}. This is simply the time-averaged power dissipation of the input oscillator x caused by the backaction damping $\gamma_{\text{BA}}[\omega]$. Using a bar to denote a time average, we have

$$P_{\text{in}} \equiv M\gamma_{\text{BA}}[\omega]\overline{\dot{x}^2} = M\gamma_{\text{BA}}[\omega]\omega^2|\chi_{xx}[\omega]|^2 F_D^2. \tag{2.92}$$

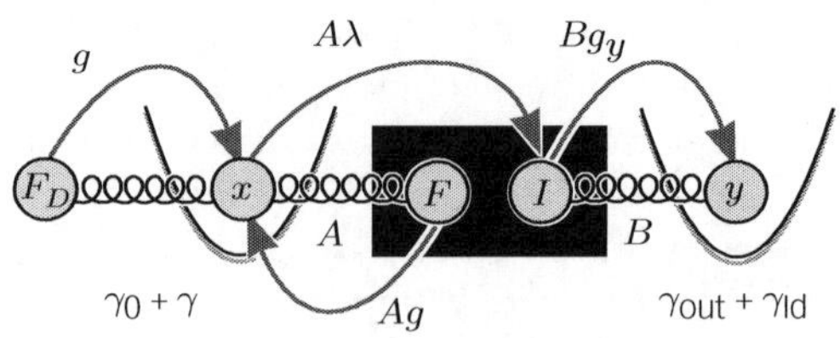

Fig. 2.2 Schematic of a generic linear-response position detector, where an auxiliary oscillator y is driven by the detector output.

Note that the oscillator susceptibility $\chi_{xx}[\omega]$ includes the effects of γ_{BA}; see Eq. (2.91).

Next, we need to consider the power supplied to the "load" oscillator y at the detector output. This oscillator will have some intrinsic, detector-independent damping γ_{ld}, as well as a backaction damping γ_{out}. In the same way that the backaction damping γ_{BA} of the input oscillator x is determined by the quantum noise in $\hat{F}$ (see Eq. (2.26)), the backaction damping of the load oscillator y is determined by the quantum noise in the output operator $\hat{I}$:

$$\begin{aligned}\gamma_{\mathrm{out}}[\omega] &= \frac{B^2}{M\omega}\left(-\operatorname{Im}\chi_{II}[\omega]\right)\\ &= \frac{B^2}{M\hbar\omega}\left(\frac{S_{II}[\omega]-S_{II}[-\omega]}{2}\right),\end{aligned} \tag{2.93}$$

where χ_{II} is the linear-response susceptibility, which determines how $\langle\hat{I}\rangle$ responds to a perturbation coupling to $\hat{I}$:

$$\chi_{II}[\omega] = -\frac{i}{\hbar}\int_0^\infty dt\langle[\hat{I}(t),\hat{I}(0)]\rangle e^{i\omega t}. \tag{2.94}$$

As the oscillator y is being driven on resonance, the relation between y and I is given by $y[\omega] = \chi_{yy}[\omega]I[\omega]$, with $\chi_{yy}[\omega] = -i\left(\omega M\gamma_{\mathrm{out}}[\omega]\right)^{-1}$. From conservation of energy, we have that the *net* power flow into the output oscillator from the detector is equal to the power dissipated out of the oscillator through the intrinsic damping γ_{ld}. We thus have

$$\begin{aligned}P_{\mathrm{out}} &\equiv M\gamma_{\mathrm{ld}}\overline{\dot{y}^2}\\ &= M\gamma_{\mathrm{ld}}\omega^2|\chi_{yy}[\omega]|^2|BA\chi_{IF}\chi_{xx}[\omega]F_D|^2\\ &= \frac{1}{M}\frac{\gamma_{\mathrm{ld}}}{\left(\gamma_{\mathrm{ld}}+\gamma_{\mathrm{out}}[\omega]\right)^2}|BA\chi_{IF}\chi_{xx}[\omega]F_D|^2.\end{aligned} \tag{2.95}$$

Using the above definitions, we find that the ratio between P_{out} and P_{in} is independent of γ_0, but depends on γ_{ld}:

$$\frac{P_{\mathrm{out}}}{P_{\mathrm{in}}} = \frac{1}{M^2\omega^2}\frac{A^2B^2|\chi_{IF}[\omega]|^2}{\gamma_{\mathrm{out}}[\omega]\gamma_{\mathrm{BA}}[\omega]}\frac{\gamma_{\mathrm{ld}}/\gamma_{\mathrm{out}}[\omega]}{\left(1+\gamma_{\mathrm{ld}}/\gamma_{\mathrm{out}}[\omega]\right)^2}. \tag{2.96}$$

We now define the detector power gain $G_P[\omega]$ as the value of this ratio maximized over the choice of γ_{ld}. The maximum occurs for $\gamma_{\mathrm{ld}} = \gamma_{\mathrm{out}}[\omega]$ (i.e., the load oscillator is "matched" to the output of the detector), resulting in

$$\begin{aligned}G_P[\omega] &\equiv \max\left[\frac{P_{\mathrm{out}}}{P_{\mathrm{in}}}\right]\\ &= \frac{1}{4M^2\omega^2}\frac{A^2B^2|\chi_{IF}|^2}{\gamma_{\mathrm{out}}\gamma_{\mathrm{BA}}}\\ &= \frac{|\chi_{IF}[\omega]|^2}{4\operatorname{Im}\chi_{FF}[\omega]\operatorname{Im}\chi_{II}[\omega]}\end{aligned} \tag{2.97}$$

In the last line, we have used the relation between the damping rates $\gamma_{\mathrm{BA}}[\omega]$ and $\gamma_{\mathrm{out}}[\omega]$ and the linear-response susceptibilities $\chi_{FF}[\omega]$ and $\chi_{II}[\omega]$; see Eq. (2.30). We thus find that the power gain is a simple dimensionless ratio formed by the three different response coefficients characterizing the detector, and is independent of the coupling constants A and B. As we will see, it is completely analogous to the power gain of a voltage amplifier, which is also determined by three parameters: the voltage gain, the input impedance, and the output impedance.

Finally, we note that these results can be generalized to include a nonzero detector reverse gain, χ_{FI}, [2]. We saw previously in Section 2.4.2 that if $\chi_{FI} = \chi_{IF}^*$, then there is no additional quantum constraint on the noise beyond what exists classically. In this case of a perfectly symmetric detector, one can show that the power gain is at most equal to one: true amplification is never possible in this case.

2.4.4 Simplifications for a detector with ideal quantum noise and large power gain

Requiring both the quantum noise inequality in Eq. (2.48) to be saturated at frequency ω as well as a large power gain (i.e., $G_P[\omega] \gg 1$) leads to some important additional constraints on the detector, as derived in Appendix I of [2]:

- $(2/\hbar)\,\mathrm{Im}\ \bar{S}_{zF}[\omega]$ is small like $1/\sqrt{G_P[\omega]}$. Hence, the possibility of having a perfect backaction-imprecision noise correlations as discussed in Section 2.4.2 is excluded.
- The detector's effective temperature must be much larger than $\hbar\omega$; one finds

$$k_{\mathrm{B}} T_{\mathrm{eff}}[\omega] \sim \sqrt{G_P[\omega]}\hbar\omega. \tag{2.98}$$

  Conversely, it is the largeness of the detector's effective temperature that allows it to have a large power gain.

2.4.5 Derivation of the quantum limit

We now turn to a rigorous proof of the quantum limit on the added noise given in Eq. (2.89). From the classical-looking Eq. (2.82), we expect that the symmetrized quantum noise spectral density describing the added noise will be given by

$$\bar{S}_{xx,\mathrm{add}}[\omega] = \frac{\bar{S}_{II}}{|\chi_{IF}|^2 A^2} + A^2 \left|\chi_{xx}\right|^2 \bar{S}_{FF} + \frac{2\,\mathrm{Re}\big[\chi_{IF}^* \left(\chi_{xx}\right)^* \bar{S}_{IF}\big]}{|\chi_{IF}|^2} \tag{2.99}$$

$$= \frac{\bar{S}_{zz}}{A^2} + A^2 \left|\chi_{xx}\right|^2 \bar{S}_{FF} + 2\,\mathrm{Re}\big[\left(\chi_{xx}\right)^* \bar{S}_{zF}\big]. \tag{2.100}$$

In the second line, we have introduced the imprecision noise $\bar{S}_{zz}$ and imprecision backaction correlation $\bar{S}_{zF}$ as in Eq. (2.42). We have also omitted writing the explicit frequency dependence of the gain χ_{IF}, susceptibility χ_{xx}, and noise correlators; they should all be evaluated at the frequency ω. Finally, the oscillator susceptibility χ_{xx} here is given by Eq. (2.91), and includes the effects of backaction damping. While we have motivated this equation from a seemingly classical noise description, the full quantum theory also yields the same result: one simply calculates the detector output noise perturbatively in the coupling to the oscillator [26].

The first step in determining the limit on the added noise is to consider its dependence on the coupling strength A. If we ignore for a moment the detector-dependent damping of the oscillator, there will be an optimal value of the coupling strength A that corresponds to a tradeoff between imprecision noise and backaction (i.e., the first and second terms in Eq. (2.99)). We would thus expect $\bar{S}_{xx,\text{add}}[\omega]$ to attain a minimum value at an optimal choice of coupling $A = A_{\text{opt}}$ where both these terms make equal contributions. Defining $\phi[\omega] = \arg \chi_{xx}[\omega]$, we thus have the bound

$$\bar{S}_{xx,\text{add}}[\omega] \geq 2|\chi_{xx}[\omega]| \left[\sqrt{\bar{S}_{zz}\bar{S}_{FF}} + \text{Re}\left(e^{-i\phi[\omega]}\bar{S}_{zF}\right)\right], \tag{2.101}$$

where the minimum value at frequency ω is achieved when

$$A^2_{\text{opt}} = \sqrt{\frac{\bar{S}_{zz}[\omega]}{|\chi_{xx}[\omega]|^2 \bar{S}_{FF}[\omega]}}. \tag{2.102}$$

Using the inequality $X^2 + Y^2 \geq 2|XY|$, we see that this value serves as a lower bound on $\bar{S}_{xx,\text{add}}$ even in the presence of detector-dependent damping. In the case where the detector-dependent damping is negligible, the right-hand side of Eq. (2.101) is independent of A, and thus Eq. (2.102) can be satisfied by simply tuning the coupling strength A; in the more general case where there is detector-dependent damping, the right-hand side is also a function of A (through the response function $\chi_{xx}[\omega]$), and it may no longer be possible to achieve Eq. (2.102) by simply tuning A.

While Eq. (2.101) is certainly a bound on the added displacement noise $\bar{S}_{xx,\text{add}}[\omega]$, it does not in itself represent the quantum limit. Reaching the quantum limit requires more than simply balancing the detector backaction and intrinsic output noises (i.e., the first two terms in Eq. (2.99)); *one also needs a detector with "quantum-ideal" noise properties, that is a detector that optimizes Eq. (2.48).* Using the quantum noise constraint of Eq. (2.48) to further bound $\bar{S}_{xx,\text{add}}[\omega]$, we obtain

$$\bar{S}_{xx,\text{add}}[\omega] \geq \hbar\,|\chi_{xx}[\omega]| \left[\sqrt{\left(1 + \Delta\left[\frac{\bar{S}_{zF}}{\hbar/2}\right]\right) + \left|\frac{\bar{S}_{zF}}{\hbar/2}\right|^2} + \frac{\text{Re}\left(e^{-i\phi[\omega]}\bar{S}_{zF}\right)}{\hbar/2}\right], \tag{2.103}$$

where the function $\Delta[z]$ is defined in Eq. (2.49). The minimum value of $\bar{S}_{xx,\text{add}}[\omega]$ in Eq. (2.103) is now achieved when one has *both* an optimal coupling (i.e., Eq. (2.102)) *and* a quantum limited detector, that is, one that satisfies Eq. (2.48) as an equality.

Next, we consider the relevant case where our detector is a good amplifier and has a power gain $G_P[\omega] \gg 1$ over the width of the oscillator resonance. As we have discussed, this implies that the ratio $\bar{S}_{zF}$ is purely real, up to small $1/G_P$ corrections (see Appendix I of [2] for more details). This in turn implies that $\Delta[2\bar{S}_{zF}/\hbar] = 0$; we thus have

$$\bar{S}_{xx,\text{add}}[\omega] \geq \hbar|\chi_{xx}[\omega]| \left[\sqrt{1 + \left(\frac{\bar{S}_{zF}}{\hbar/2}\right)^2} + \cos\left(\phi[\omega]\right)\frac{\bar{S}_{zF}}{\hbar/2}\right]. \tag{2.104}$$

Finally, as there is no further constraint on $\bar{S}_{zF}$ (beyond the fact that it is real), we can minimize the expression over its value. The minimum $\bar{S}_{xx,\mathrm{add}}[\omega]$ is achieved for a detector whose cross-correlator satisfies

$$\bar{S}_{zF}[\omega]\Big|_{\mathrm{optimal}} = -\frac{\hbar}{2}\cot(\phi[\omega]), \tag{2.105}$$

with the minimum value of the added noise being given precisely by

$$\bar{S}_{xx,\mathrm{add}}[\omega]\Big|_{\mathrm{min}} = \hbar|\mathrm{Im}\ \chi_{xx}[\omega]| = \lim_{T\to 0}\bar{S}_{xx,\mathrm{eq}}[\omega,T], \tag{2.106}$$

in agreement with Eq. (2.89). Thus, in the limit of a large power gain, we have that *at each frequency, the minimum displacement noise added by the detector is precisely equal to the noise arising from a zero-temperature bath.* This conclusion is irrespective of the strength of the intrinsic (detector-independent) oscillator damping.

We have thus derived the amplifier quantum limit (in the context of position detection) for a two-port amplifier used in the "op-amp" mode of operation. Moreover, our derivation shows that to reach the quantum limit on the added displacement noise $\bar{S}_{xx,\mathrm{add}}[\omega]$ with a large power gain, one needs:

(1) A detector with quantum-limited noise properties, that is, one that optimizes the inequality of Eq. (2.48). Similar to our discussion of QND qubit detection, optimizing this inequality corresponds to the heuristic requirement of "no wasted information."
(2) A coupling A that satisfies Eq. (2.102).
(3) A detector cross-correlator $\bar{S}_{IF}$ that satisfies Eq. (2.105).

It is worth stressing that Eq. (2.105) implies that it will *not* in general be possible to achieve the quantum limit simultaneously at all frequencies. On resonance, this condition tells us that $\bar{S}_{zF}[\omega]$ should be zero. In contrast, far from resonance, it implies that one needs strong correlations, $\bar{S}_{zF} \gg \hbar/2$. There are systems in which one is indeed interested in minimizing the added noise far from resonance. For example, in interferometers used for gravitational wave detection, the test masses used are almost in the free-mass limit, and thus one is interested in frequencies much larger than the resonance frequency of the test mass. A way to achieve such large imprecision-backaction correlations using a nonlinear cavity detector was discussed recently in [27].

If one focuses on optimizing $\bar{S}_{xx,\mathrm{add}}[\omega]$ at resonance (i.e., $\omega = \Omega$), and if one is using a quantum-limited detector with a large power gain ($k_{\mathrm{B}}T_{\mathrm{eff}} \gg \hbar\Omega$), the remaining condition on the coupling A, Eq. (2.102), may be written

$$\frac{\gamma_{\mathrm{BA}}[A_{\mathrm{opt}}]}{\gamma_0 + \gamma_{\mathrm{BA}}[A_{\mathrm{opt}}]} = \frac{\hbar\Omega}{4k_{\mathrm{B}}T_{\mathrm{eff}}}. \tag{2.107}$$

As $\gamma_{\mathrm{BA}}[A] \propto A^2$ is the detector-dependent damping of the oscillator, we thus have that *to achieve the quantum-limited value of* $\bar{S}_{xx,add}[\Omega]$ *with a large power gain, one needs the intrinsic damping of the oscillator to be much larger than the detector-dependent damping*. The detector-dependent damping must be small enough to compensate the large effective temperature of the detector; if the bath temperature satisfies $\hbar\Omega/k_B \ll T_{\text{bath}} \ll T_{\text{eff}}$, then Eq. (2.107) implies that, at the quantum limit, the temperature of the oscillator will be given by

$$T_{\text{osc}} \equiv \frac{\gamma_{\text{BA}} T_{\text{eff}} + \gamma_0 T_{\text{bath}}}{\gamma_{\text{BA}} + \gamma_0} \to \frac{\hbar\Omega}{4k_B} + T_{\text{bath}}. \tag{2.108}$$

Thus, at the quantum limit and for large T_{eff}, the detector raises the oscillator's temperature by $\hbar\Omega/4k_B$.[2] As expected, this additional heating is only *half* the zero-point energy; in contrast, the quantum-limited value of $\bar{S}_{xx,\text{add}}[\omega]$ corresponds to the full zero-point result, as it also includes the contribution of the intrinsic output noise of the detector.

Finally, we return to Eq. (2.103); this is the constraint on the added noise $\bar{S}_{xx,\text{add}}[\omega]$ *before* we assumed our detector to have a large power gain, and consequently a large T_{eff}. Note crucially that *if* we did not require a large power gain, then there need not be *any* added noise. Without the assumption of a large power gain, the ratio $\bar{S}_{IF}/\chi_{IF}$ can be made imaginary with a large magnitude. In this limit, $1 + \Delta[2\bar{S}_{IF}/\chi_{IF}] \to 0$: the quantum constraint on the amplifier noises (i.e., the right-hand side of Eq. (2.48)) vanishes. One can then easily use Eq. (2.103) to show that the added noise $\bar{S}_{xx,\text{add}}[\omega]$ can be zero.

2.4.6 Noise temperature

There is an alternate but roughly equivalent way of phrasing the amplifier quantum limit we have been discussing, where the crucial quantity to be bounded is the so-called "noise temperature" $T_N[\omega]$. For a given frequency ω, this is defined via the equation

$$\bar{S}_{xx}^{\text{eq}}[\omega, T + T_N[\omega]] \equiv \bar{S}_{xx}^{\text{eq}}[\omega, T] + \bar{S}_{xx}^{\text{add}}[\omega], \tag{2.109}$$

in the limit where $T \gg \hbar\omega$ (which ensures that the definition of T_N is independent of T).[3] In words, the added noise position fluctuations at a frequency ω can be viewed as an effective heating of the oscillator from a temperature T to a temperature $T + T_N[\omega]$. One finds

$$\frac{2k_B T_N[\omega]}{\omega} = \frac{\bar{S}_{xx}^{\text{add}}[\omega]}{-\text{Im}\ \chi_{xx}[\omega]} = \frac{1}{-\text{Im}\ \chi_{xx}[\omega]}\left[\frac{\bar{S}_{zz}}{A^2} + A^2 |\chi_{xx}|^2 \bar{S}_{FF} + 2\,\text{Re}\left(\chi_{xx}^* \bar{S}_{zF}\right)\right]. \tag{2.110}$$

[2] If, in contrast, our oscillator was initially at zero temperature (i.e., $T_{\text{bath}} = 0$), one finds that the effect of the backaction (at the quantum limit and for $G_P \gg 1$) is to heat the oscillator to a temperature $\hbar\Omega/(k_B \ln 5) \simeq 0.62\hbar\Omega/k_B$.

[3] Other, less common, definitions of the noise temperature also exist in the literature. For example, Caves [28] defines the noise temperature using Eq. (2.109), but takes the initial temperature T to be zero. With this definition, the quantum limit takes the somewhat awkward form $k_B T_N[\omega] \geq \hbar\omega/\ln 3$.

Our bound on the added noise spectral density (in the high-power-gain limit) immediately implies that, for large power gain,

$$k_B T_N[\omega] \geq \hbar\omega/2. \tag{2.111}$$

In complete analogy to the quantum limit on the added noise, reaching the quantum limit on the noise temperature first requires one to balance the contributions from imprecision and backaction by, for example, tuning the value of the coupling A. For the noise temperature, note that the imprecision contribution scales as $1/|\chi_{xx}|$, whereas the backaction contribution scales as $|\chi_{xx}|$. Thus, for the noise temperature, one could keep A fixed, and balance backaction and imprecision noise contributions by tuning the value of the oscillator susceptibility $|\chi_{xx}[\omega]|$ (e.g., imagine one could tune the value of Ω_M). In a similar fashion, as opposed to tuning the value of the detector cross-correlator $\bar{S}_{zF}$, one could tune the phase of χ_{xx}. While this way of thinking about the optimization may seem quite unnatural in the context of a position detector, it is completely natural if we now think of our detector as a voltage amplifier. As we will see, in this case the role of χ_{xx} is played by Z_{src}, the source impedance of the system producing the signal that is to be amplified.

2.4.7 Quantum limit on an "op-amp" style voltage amplifier

We now use our general linear-response machinery to tackle the quantum limit on the noise temperature of a voltage amplifier, again used in the "op-amp" mode of operation. For the most part, this just involves a simple relabelling of various quantities in our position detector; this is done in Table 2.1. The input signal to the detector is now a voltage $v_{\mathrm{in}}(t)$, and the output quantity that is read out is also a voltage: the output operator $\hat{I}$ thus now becomes an operator $\hat{V}_{\mathrm{out}}$.

Figure 2.3 shows a standard schematic description of a voltage amplifier. The input voltage to be amplified, $v_{\mathrm{in}}(t)$, is produced by a circuit that has a Thevenin-equivalent impedance Z_{s}, the source impedance. The amplifier itself has an input impedance Z_{in} and an output impedance Z_{out}, as well as a voltage gain coefficient λ_V: assuming no current is drawn at the output (i.e., $Z_{\mathrm{load}} \to \infty$ in Fig. 2.3), the output voltage $V_{\mathrm{out}}(t)$ is simply λ_V times the voltage across the input terminals of the amplifier. We focus here on the simple case of no reverse gain.

The added noise of the amplifier is represented in a standard way by two noise sources placed at the amplifier input. There is both a voltage noise source $\tilde{V}(t)$ in series with the input voltage source and a current noise source $\tilde{I}(t)$ in parallel with the input voltage source (Fig. 2.3). The voltage noise produces a fluctuating voltage $\tilde{V}(t)$ (spectral density $\bar{S}_{\tilde{V}\tilde{V}}[\omega]$) which simply adds to the signal voltage at the amplifier input, and is amplified at the output; as such, it is completely analogous to the imprecision noise $\bar{S}_{zz} = \bar{S}_{II}/|\chi_{IF}|^2$ of our linear-response detector. In contrast, the current noise source of the voltage amplifier represents backaction: this fluctuating current (spectral density $\bar{S}_{\tilde{I}\tilde{I}}[\omega]$) flows back across the parallel combination of the source impedance and amplifier input impedance, producing an additional fluctuating voltage at its input. The current noise is thus analogous to the backaction noise $\bar{S}_{FF}$ of our generic linear-response detector.

Table 2.1 Correspondence between position detector and voltage amplified

QUANTITY	POSITION DETECTOR	VOLTAGE AMPLIFIER
Input signal	$\hat{x}(t)$	$\hat{v}_{\text{in}}(t)$
Output operator	$\hat{I}(t)$	$\hat{V}_{\text{out}}(t)$
Backaction operator (i.e., quantity that couples to input signal)	$\hat{F}[\omega]$	$\hat{Q}[\omega] = \hat{\tilde{I}}[\omega]/(-i\omega)$
Linear gain coefficient	$\chi_{IF}[\omega]$	$\lambda_V[\omega] = \chi_{V_{\text{out}}Q}[\omega]$
Imprecision noise operator	$z[\omega] = I[\omega]/\chi_{IF}[\omega]$	$\tilde{V}[\omega] = \hat{V}_{\text{out}}[\omega]/\lambda_V[\omega]$
Imprecision noise	$\bar{S}_{zz}[\omega] = \bar{S}_{II}[\omega]/\lvert\chi_{IF}[\omega]\rvert^2$	$\bar{S}_{\tilde{V}\tilde{V}}[\omega]$
Backaction noise	$\bar{S}_{FF}[\omega]$	$\bar{S}_{\tilde{I}\tilde{I}}[\omega]/\omega^2$
Backaction-imprecision correlator	$\bar{S}_{zF}[\omega]$	$\bar{S}_{\tilde{V}\tilde{I}}[\omega]/(i\omega)$
Detector-induced input dissipation	$-\text{Im}\ \chi_{FF}[\omega] = M\omega\gamma_{\text{BA}}[\omega]$	$-\text{Im}\ \chi_{QQ}[\omega] = \frac{1}{\omega}\text{Re}\frac{1}{Z_{\text{in}}[\omega]}$
Detector-induced output dissipation	$-\text{Im}\ \chi_{II}[\omega] = M\omega\gamma_y[\omega]$	$-\text{Im}\ \chi_{V_{\text{out}}V_{\text{out}}}[\omega] = \omega\,\text{Re}\,Z_{\text{out}}[\omega]$
Power gain G_P	$\frac{\lvert\chi_{IF}[\omega]\rvert^2/4}{\text{Im}\ \chi_{FF}[\omega]\ \text{Im}\ \chi_{II}[\omega]}$	$\frac{\lvert\lambda_V\rvert^2/4}{\text{Re}(Z_{\text{out}})\ \text{Re}(1/Z_{\text{in}})}$

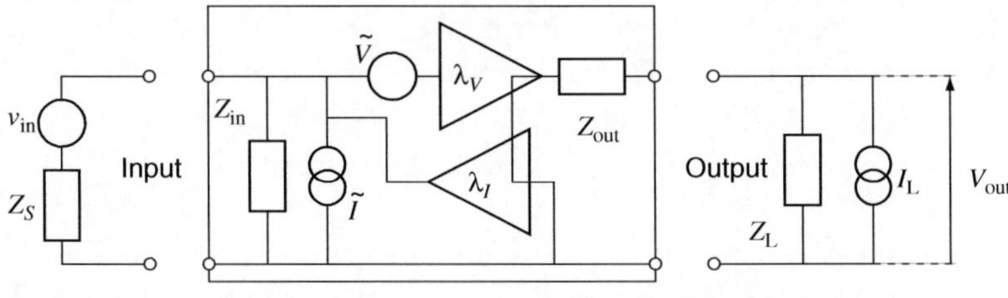

Fig. 2.3 Schematic of a linear voltage amplifier, including a reverse gain λ_I. $\tilde{V}$ and $\tilde{I}$ represent the standard voltage and current noises of the amplifier, as discussed further in the text. The case with reverse gain is discussed in Section 2.5.2.

Putting the above together, and treating for the moment the output voltage as a classically noisy variable, we have

$$v_{\mathrm{out,tot}}(t) = \lambda_V \left\{ \frac{Z_{\mathrm{in}}}{Z_{\mathrm{in}} + Z_{\mathrm{s}}} [v_{\mathrm{in}}(t) + \tilde{V}(t)] - \frac{Z_{\mathrm{s}} Z_{\mathrm{in}}}{Z_{\mathrm{s}} + Z_{\mathrm{in}}} \tilde{I}(t) \right\}$$

$$\simeq \lambda_V [v_{\mathrm{in}}(t) + \tilde{V}(t) - Z_{\mathrm{s}} \tilde{I}(t)]. \tag{2.112}$$

In the second line, we have taken the usual limit of an ideal voltage amplifier that has an infinite input impedance (i.e., the amplifier draws zero current). We are left with an equation that is completely analogous to the corresponding equation for a position detector, Eq. (2.81). For simplicity, we have ignored any frequency dependences of λ_V, Z_{s}, and Z_{in}; these are easily restored.

If we refer the output voltage fluctuations back to the input (via λ_V) to obtain the measured voltage fluctuations, we find they are described by a spectral density

$$\bar{S}_{VV,\mathrm{tot}}[\omega] = \bar{S}_{v_{\mathrm{in}} v_{\mathrm{in}}}[\omega] + \bar{S}_{VV,\mathrm{add}}[\omega]. \tag{2.113}$$

As in our discussion of position detection, we have used the fact that quantum expressions for symmetrized noise densities will match what is expected from the classical theory. Here $\bar{S}_{v_{\mathrm{in}} v_{\mathrm{in}}}[\omega]$ is the spectral density of the voltage fluctuations of the input signal $v_{\mathrm{in}}(t)$; assuming the signal source is in equilibrium at temperature T, they will be given by

$$\bar{S}_{v_{\mathrm{in}} v_{\mathrm{in}}}[\omega] = \mathrm{Re}\,(Z_s[\omega]) \hbar\omega \coth(\hbar\omega/2k_{\mathrm{B}}T), \tag{2.114}$$

consistent with the fluctuation dissipation theorem. In contrast, $\bar{S}_{VV,\mathrm{add}}$ is the total amplifier added noise (referred to the input), and has both a backaction and an imprecision noise contribution:

$$\bar{S}_{VV,\mathrm{add}}[\omega] = \bar{S}_{\tilde{V}\tilde{V}} + |Z_{\mathrm{s}}|^2 \bar{S}_{\tilde{I}\tilde{I}} - 2\,\mathrm{Re}\big(Z_{\mathrm{s}}^* \bar{S}_{\tilde{V}\tilde{I}}\big). \tag{2.115}$$

For clarity, we have dropped the frequency index for the spectral densities appearing on the right-hand side of this equation.

Finally, we can define the noise temperature exactly as in Eq.(2.110) by viewing the added noise at frequency ω as being due to an effective heating of the source from temperature $T \gg \hbar\omega/k_{\mathrm{B}}$ to $T + T_{\mathrm{N}}[\omega]$. Writing $Z_{\mathrm{s}} = |Z_{\mathrm{s}}| e^{i\phi}$, we find

$$2k_{\mathrm{B}} T_{\mathrm{N}} = \frac{1}{\cos\phi} \left[\frac{\bar{S}_{\tilde{V}\tilde{V}}}{|Z_{\mathrm{s}}|} + |Z_{\mathrm{s}}| \bar{S}_{\tilde{I}\tilde{I}} - 2\mathrm{Re}\big(e^{-i\phi} \bar{S}_{\tilde{V}\tilde{I}}\big) \right] \tag{2.116}$$

We can now apply the same line of reasoning as we used for the T_N of a position detector, and thus, in the large-power-gain limit, the quantum limit of Eq. (2.111) applies to $T_N[\omega]$. Achieving the quantum limit requires optimizing the quantum noise constraint of Eq. (2.48), which here takes the form

$$\bar{S}_{\tilde{V}\tilde{V}}[\omega] \bar{S}_{\tilde{I}\tilde{I}}[\omega] - \big(\mathrm{Im}\ \bar{S}_{\tilde{V}\tilde{I}}\big)^2 \geq \left(\frac{\hbar\omega}{2}\right)^2. \tag{2.117}$$

In addition, one also needs to balance the contribution of backaction and imprecision (see Eq. (2.101)), as well as properly taking advantage of any backaction-imprecision correlations (see Eq. (2.105)). Unlike the oscillator case, this is most naturally done by tuning both the magnitude and phase of the source impedance Z_s. If one simply minimizes $T_N[\omega]$ with respect to Z_s, one finds a completely classical minimum bound on T_N,

$$k_B T_N \geq \sqrt{\bar{S}_{\tilde{V}\tilde{V}}\bar{S}_{\tilde{I}\tilde{I}} - \left(\text{Im } \bar{S}_{\tilde{V}\tilde{I}}\right)^2} - \text{Re } \bar{S}_{\tilde{V}\tilde{I}}, \tag{2.118}$$

where the minimum is achieved for an optimal source impedance that satisfies

$$|Z_s[\omega]|_{\text{opt}} = \sqrt{\frac{\bar{S}_{\tilde{V}\tilde{V}}[\omega]}{\bar{S}_{\tilde{I}\tilde{I}}[\omega]}} \equiv Z_N \tag{2.119}$$

$$\sin\phi[\omega]\Big|_{\text{opt}} = -\frac{\text{Im } \bar{S}_{\tilde{V}\tilde{I}}[\omega]}{\sqrt{\bar{S}_{\tilde{V}\tilde{V}}[\omega]\bar{S}_{\tilde{I}\tilde{I}}[\omega]}} \tag{2.120}$$

Recall that in the position detector, achieving the quantum limit required a coupling A so weak that the detector-induced damping γ_{BA} was much less than the intrinsic oscillator damping (see Eq. (2.107)). Similarly, in the present case, one can show [2] that for a large-power-gain amplifier with ideal quantum noise, the noise impedance Z_N satisfies

$$\left|\frac{Z_N[\omega]}{\text{Re } Z_{\text{in}}[\omega]}\right| = \frac{1}{2\sqrt{G_P[\omega]}} \ll 1. \tag{2.121}$$

It follows that $|Z_N| \ll |Z_{\text{in}}|$ in the large-power-gain, large-effective-temperature regime of interest, thus justifying the form of Eq. (2.115).

Role of noise cross-correlations Before leaving the topic of a linear voltage amplifier, we pause to note the role of cross-correlations in current and voltage noise in reaching the quantum limit. First, note from Eq. (2.120) that in both the classical and quantum treatments, the noise impedance Z_N of the amplifier will have a reactive part (i.e., Im $Z_N \neq 0$) if there are out-of-phase correlations between the amplifier's current and voltage noises (i.e., if Im $S_{VI} \neq 0$). Thus, if such correlations exist, it will not be possible to minimize the noise temperature (and hence reach the quantum limit), if one uses a purely real source impedance Z_s.

More significantly, note that the final classical expression for the noise temperature T_N explicitly involves the real part of the S_{VI} correlator (see Eq. (2.118)). In contrast, we have argued that in the quantum case, Re $\bar{S}_{VI}$ *must* be zero if one wishes to reach the quantum limit while having a large power gain (see Section 2.4.4 and Appendix I of [2]); as such, this quantity does not appear in the final expression for the minimal T_N. It also follows that *to reach the quantum limit while having a large power gain, an amplifier cannot have significant in-phase correlations between its current and voltage noise.*

This last statement can be given a heuristic explanation. If there are out-of-phase correlations between current and voltage noise, we can easily make use of these by appropriately choosing our source impedance. However, if there are in-phase correlations between current and voltage noise, we cannot use these simply by tuning the source impedance. We *could*, however, have used them by implementing feedback in our amplifier. The fact that we have not done this means that these correlations represent a kind of missing information; as a result, we must necessarily miss the quantum limit. In Section 2.5.2.2, we explicitly give an example of a voltage amplifier that misses the quantum limit owing to the presence of in-phase current and voltage fluctuations.

2.5 Quantum limit on a linear amplifier: scattering mode

We now consider an alternate mode of amplifier operation, where the input signal is the amplitude of a wave incident on the amplifier and the output signal is the amplitude of a wave leaving the amplifier. Unlike the op-amp mode considered above, backaction is irrelevant here: the assumption is that the input signal entering the amplifier is completely insensitive to any fluctuations emanating from the amplifier. To achieve this, one typically has to impedance-match the signal source to the amplifier input so that there are no reflections. This is very different from the op-amp mode, where the amplifier input impedance is much larger than the source impedance. This condition ensures that in the op-amp mode, the coupling to the amplifier only weakly increases the dissipation of the signal source. In contrast, the impedance matching implies that the coupling to the amplifier will have a more pronounced impact on the signal source dissipation.

While the quantum limit on the added noise of an amplifier in this scattering mode of operation has the same form as that in the op-amp mode, we will see that they are not equivalent: an amplifier can be quantum-limited in one mode, but fail to reach the quantum limit in the other mode.

2.5.1 Caves-Haus formulation of the scattering-mode quantum limit

The derivation of the amplifier quantum limit in the scattering mode of operation is in many ways better known that the op-amp quantum limit presented above, and is simpler to present. This derivation is originally due to Haus and Mullen [29], and was both clarified and extended by Caves [28]; the amplifier quantum limit was also motivated in a slightly different manner by Heffner [30].[4]

The starting assumption of this derivation is that both the input and output ports of the amplifier can be described by sets of bosonic modes. If we focus on a narrow bandwidth centered on frequency ω, we can describe a classical signal $E(t)$ in terms of a complex number a defining the amplitude and phase of the signal (or equivalently the two quadrature amplitudes) [29, 31].

$$E(t) \propto i(ae^{-i\omega t} - a^* e^{+i\omega t}). \tag{2.122}$$

In the quantum case, the two signal quadratures of $E(t)$ (i.e., the real and imaginary parts of $a(t)$) cannot be measured simultaneously because they are canonically

[4] Note that Caves [28] provides a thorough discussion of why the derivation of the amplifier quantum limit given by Heffner [30] is not rigorously correct

conjugate; this is in complete analogy to a harmonic oscillator (see Eq. (2.80)). As a result, a, a^* must be elevated to the status of photon ladder operators: $a\rangle\hat{a}, a^*\rangle\hat{a}^\dagger$.

Consider the simplest case, where there is only a single mode at both the input and output, with corresponding operators $\hat{a}$ and $\hat{b}$. It follows that the input signal into the amplifier is described by the expectation value $\langle\hat{a}\rangle$, while the output signal is described by $\langle\hat{b}\rangle$. Correspondingly, the symmetrized noise in both these quantities is described by

$$(\Delta a)^2 \equiv \frac{1}{2}\left\langle\{\hat{a}, \hat{a}^\dagger\}\right\rangle - |\langle\hat{a}\rangle|^2, \tag{2.123}$$

with an analogous definition for $(\Delta b)^2$.

To derive a quantum limit on the added noise of the amplifier, one uses two simple facts. First, both the input and the output operators must satisfy the usual commutation relations:

$$\left[\hat{a}, \hat{a}^\dagger\right] = 1, \quad [\hat{b}, \hat{b}^\dagger] = 1 \tag{2.124}$$

Second, the linearity of the amplifier and the fact that it is phase-preserving (i.e., both signal quadratures are amplified the same way) imply a simple relation between the output operator $\hat{b}$ and the input operator $\hat{a}$:

$$\hat{b} = \sqrt{G}\hat{a}, \quad \hat{b}^\dagger = \sqrt{G}\hat{a}^\dagger, \tag{2.125}$$

where G is the dimensionless *"photon number gain"* of the amplifier. It is immediately clear, however, that this expression cannot possibly be correct as written because it violates the fundamental bosonic commutation relation $[\hat{b}, \hat{b}^\dagger] = 1$. We are therefore forced to write

$$\hat{b} = \sqrt{G}\hat{a} + \hat{\mathcal{F}}, \quad \hat{b}^\dagger = \sqrt{G}\hat{a}^\dagger + \hat{\mathcal{F}}^\dagger, \tag{2.126}$$

where $\hat{\mathcal{F}}$ is an operator representing additional noise added by the amplifier. Based on the discussion in the previous subsection (Sec. 2.4), we can anticipate what $\hat{\mathcal{F}}$ represents: it is noise associated with the additional degrees of freedom that must invariably be present in a phase-preserving amplifier.

As $\hat{\mathcal{F}}$ represents noise, it has a vanishing expectation value; in addition, one also assumes that this noise is uncorrelated with the input signal, implying $[\hat{\mathcal{F}}, \hat{a}] = [\hat{\mathcal{F}}, \hat{a}^\dagger] = 0$ and $\langle\hat{\mathcal{F}}\hat{a}\rangle = \langle\hat{\mathcal{F}}\hat{a}^\dagger\rangle = 0$. Insisting that $[\hat{b}, \hat{b}^\dagger] = 1$ thus yields

$$[\hat{\mathcal{F}}, \hat{\mathcal{F}}^\dagger] = 1 - G. \tag{2.127}$$

The question now becomes: how small can we make the noise described by $\hat{\mathcal{F}}$? Using Eqs. (2.126), the noise at the amplifier output Δb is given by

$$\begin{aligned}(\Delta b)^2 &= G\,(\Delta a)^2 + \frac{1}{2}\langle\{\hat{\mathcal{F}}, \hat{\mathcal{F}}^\dagger\}\rangle \\ &\geq G\,(\Delta a)^2 + \frac{1}{2}|\langle[\hat{\mathcal{F}}, \hat{\mathcal{F}}^\dagger]\rangle| \\ &\geq G\,(\Delta a)^2 + \frac{|G-1|}{2}.\end{aligned} \tag{2.128}$$

We have used here a standard inequality to bound the expectation of $\{\hat{\mathcal{F}}, \hat{\mathcal{F}}^\dagger\}$. The first term here is simply the amplified noise of the input, while the second term represents the noise added by the amplifier. Note that if there is no amplification (i.e., $G = 1$), there need not be any added noise. However, in the more relevant case of large amplification ($G \gg 1$), the added noise cannot vanish. It is useful to express the noise at the output as an equivalent noise at ("referred to") the input by simply dividing out the photon gain G. Taking the large-G limit, we have

$$\frac{(\Delta b)^2}{G} \geq (\Delta a)^2 + \frac{1}{2}. \tag{2.129}$$

Thus, we have a very simple demonstration that *an amplifier with a large photon gain must add at least half a quantum of noise to the input signal.* Equivalently, the minimum value of the added noise is simply equal to the zero-point noise associated with the input mode; the total output noise (referred to the input) is at least twice the zero point input noise. Note the complete analogy to the quantum limit we found for the added noise of an op-amp amplifier (see Eqs. (2.89) and (2.111)).

As already discussed, the added-noise operator $\hat{\mathcal{F}}$ is associated with additional degrees of freedom (beyond input and output modes) necessary for phase-preserving amplification. To see this more concretely, note that every linear amplifier is inevitably a *nonlinear* system consisting of an energy source and a "spigot" controlled by the input signal, which redirects the energy source partly to the output channel and partly to some other channel(s). Hence there are inevitably other degrees of freedom involved in the amplification process beyond the input and output channels. An explicit example is the quantum parametric amplifier (see, e.g., Section V.C of [2]).

To see explicitly the role of the additional degrees of freedom, note first that, for $G > 1$, the right-hand side of Eq. (2.127) is negative. Hence, the simplest possible form for the added noise is:

$$\hat{\mathcal{F}} = \sqrt{G-1}\hat{d}^\dagger, \quad \hat{\mathcal{F}}^\dagger = \sqrt{G-1}\hat{d}, \tag{2.130}$$

where $\hat{d}$ and $\hat{d}^\dagger$ represent a single additional mode of the system. This is the minimum number of additional degrees of freedom that must inevitably be involved in the amplification process. Note that, for this case, the inequality in Eq. (2.128) is satisfied as an equality, and the added noise takes on its minimum possible value. If, instead, we had, say, two additional modes (coupled inequivalently),

$$\hat{\mathcal{F}} = \sqrt{G-1}(\cosh\theta\hat{d}_1^\dagger + \sinh\theta\hat{d}_2), \tag{2.131}$$

it is straightforward to show that the added noise is inevitably larger than the minimum. This again can be interpreted in terms of wasted information, as the extra degrees of freedom are not being monitored as part of the measurement process and so information is being lost.

2.5.2 Bosonic scattering description of a two-port amplifier

To highlight the differences between the op-amp and scattering modes of amplifier operation (and the quantum limit in each case), we will now consider a generic device that can be used in both modes. We will see explicitly that if we construct this device to be ideal as scattering amplifier, then it cannot be used as a quantum-limited op-amp-style amplifier. The basic reason is simple. Being ideal in the scattering mode means having an amplifier that is ideally suited for use with a signal source that is impedance-matched to the amplifier input. In contrast, being ideal in the op-amp mode requires a signal impedance that is much smaller than the source impedance.

2.5.2.1 Scattering versus op-amp representations

In the bosonic scattering approach, a generic linear amplifier is modeled as a set of coupled bosonic modes. To make matters concrete, we will consider the specific case of a voltage amplifier with distinct input and output ports, where each port is a semi-infinite transmission line (see Fig. 2.4). We start by recalling that a quantum transmission line can be described as a set of noninteracting bosonic modes (see Appendix D of [2] for a quick review). Denoting the input transmission line with an a and the output transmission line with a b, the current and voltage operators in these lines may be written as

$$\hat{V}_q(t) = \int_0^{\infty} \frac{d\omega}{2\pi} \left(\hat{V}_q[\omega] e^{-i\omega t} + \text{h.c.} \right), \tag{2.132a}$$

$$\hat{I}_q(t) = \sigma_q \int_0^{\infty} \frac{d\omega}{2\pi} \left(\hat{I}_q[\omega] e^{-i\omega t} + \text{h.c.} \right), \tag{2.132b}$$

with

$$\hat{V}_q[\omega] = \sqrt{\frac{\hbar\omega}{2}} Z_q \left(\hat{q}_{\text{in}}[\omega] + \hat{q}_{\text{out}}[\omega] \right), \tag{2.133a}$$

$$\hat{I}_q[\omega] = \sqrt{\frac{\hbar\omega}{2Z_q}} \left(\hat{q}_{\text{in}}[\omega] - \hat{q}_{\text{out}}[\omega] \right). \tag{2.133b}$$

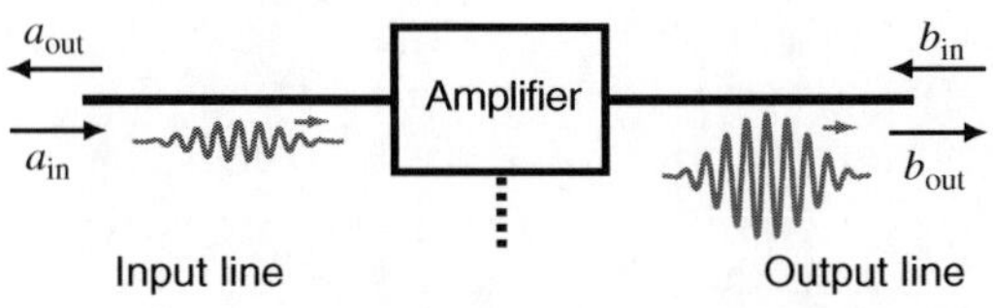

Fig. 2.4 Schematic of a two-port bosonic amplifier. Both the input and outputs of the amplifier are attached to transmission lines. The incoming and outgoing wave amplitudes in the input (output) transmission line are labeled $\hat{a}_{\text{in}}, \hat{a}_{\text{out}}$ ($\hat{b}_{\text{in}}, \hat{b}_{\text{out}}$), respectively. The voltages at the ends of the two lines ($\hat{V}_a, \hat{V}_b$) are linear combinations of incoming and outgoing wave amplitudes.

Here, q can be equal to a or b, and we have $\sigma_a = 1, \sigma_b = -1$. The operators $\hat{a}_{\rm in}[\omega], \hat{a}_{\rm out}[\omega]$ are bosonic annihilation operators; $\hat{a}_{\rm in}[\omega]$ describes an incoming wave in the input transmission line (i.e., incident on the amplifier) having frequency ω, while $\hat{a}_{\rm out}[\omega]$ describes an outgoing wave with frequency ω. The operators $\hat{b}_{\rm in}[\omega]$ and $\hat{b}_{\rm out}[\omega]$ describe analogous waves in the output transmission line. We can think of $\hat{V}_a$ as the input voltage to our amplifier and $\hat{V}_b$ as the output voltage. Similarly, $\hat{I}_a$ is the current drawn by the amplifier at the input and $\hat{I}_b$ the current drawn at the output of the amplifier. Finally, Z_a (Z_b) is the characteristic impedance of the input (output) transmission line.

As we have seen, amplification invariably requires additional degrees of freedom. Thus, to amplify a signal at a particular frequency ω, there will be $2N$ bosonic modes involved, where the integer N is necessarily larger than 2. Four of these modes are simply the frequency-ω modes in the input and output lines (i.e., $\hat{a}_{\rm in}[\omega]$, $\hat{a}_{\rm out}[\omega]$, $\hat{b}_{\rm in}[\omega]$, and $\hat{b}_{\rm out}[\omega]$). The remaining $2(N-2)$ modes describe auxiliary degrees of freedom involved in the amplification process; these additional modes could correspond to frequencies different from the signal frequency ω. The auxiliary modes can also be divided into incoming and outgoing modes. It is thus convenient to represent them as additional transmission lines attached to the amplifier; these additional lines could be semi-infinite, or could be terminated by active elements.

Scattering representation In general, our generic two-port bosonic amplifier will be described by an $N \times N$ scattering matrix that determines the relation between the outgoing-mode operators and incoming-mode operators. The form of this matrix is constrained by the requirement that the output modes obey the usual canonical bosonic commutation relations. It is convenient to express the scattering matrix in a form that *only* involves the input and output lines explicitly:

$$\begin{pmatrix} \hat{a}_{\rm out}[\omega] \\ \hat{b}_{\rm out}[\omega] \end{pmatrix} = \begin{pmatrix} s_{11}[\omega] & s_{12}[\omega] \\ s_{21}[\omega] & s_{22}[\omega] \end{pmatrix} \begin{pmatrix} \hat{a}_{\rm in}[\omega] \\ \hat{b}_{\rm in}[\omega] \end{pmatrix} + \begin{pmatrix} \hat{\mathcal{F}}_a[\omega] \\ \hat{\mathcal{F}}_b[\omega] \end{pmatrix}. \tag{2.134}$$

Here $\hat{\mathcal{F}}_a[\omega]$ and $\hat{\mathcal{F}}_b[\omega]$ are each an unspecified linear combination of the incoming auxiliary modes introduced above. They thus describe noise in the outgoing modes of the input and output transmission lines that arises from the auxiliary modes involved in the amplification process. Similar to Eq. (2.126) in our discussion of a one-port amplifier, the $\hat{\mathcal{F}}$ operators also ensure that canonical commutation relations are preserved.

In the quantum optics literature, one typically views Eq. (2.134) as the defining equation of the amplifier; we will call this the scattering representation of our amplifier. The representation is best suited to understanding the amplifier when used in the scattering mode of operation. In this mode, the experimentalist ensures that $\langle \hat{a}_{\rm in}[\omega] \rangle$ is precisely equal to the signal to be amplified, *irrespective* of what is coming out of the amplifier. Similarly, the output signal from the amplifier is the amplitude of the outgoing wave in the output line, $\langle \hat{b}_{\rm out}[\omega] \rangle$. If we focus on $\hat{b}_{\rm out}$, we have precisely the same situation as described in Section 2.5.1, where we presented the Haus-Caves derivation of the quantum limit (see Eq. (2.126)). It thus follows that in the scattering mode of operation, the matrix element $s_{21}[\omega]$ represents the gain of our amplifier at frequency ω, $|s_{21}[\omega]|^2$ the corresponding "photon number gain," and $\hat{\mathcal{F}}_b$ the added noise operator of the amplifier. The operator $\hat{\mathcal{F}}_a$ represents the backaction noise in the scattering mode of operation; this backaction has no effect on the added noise of the amplifier in the scattering mode.

Op-amp representation In the op-amp amplifier mode of operation, the input and output signals are not simply incoming/outgoing wave amplitudes; thus, the scattering representation is not an optimal description of our amplifier. The system we are describing here is a voltage amplifier: thus, in the op-amp mode, the experimentalist would ensure that the voltage at the end of the input line ($\hat{V}_a$) is equal to the signal to be amplified, and would read out the voltage at the end of the output transmission line ($\hat{V}_b$) as the output of the amplifier. From Eq. (2.132a), we see that this implies that the amplitude of the wave going into the amplifier, a_{in}, will depend on the amplitude of the wave exiting the amplifier, a_{out}.

Thus, if we want to use our amplifier as a voltage amplifier, we would like to find a description that is better tailored to our needs than the scattering representation of Eq. (2.134). This can be found by simply re-expressing the scattering matrix relation of Eq. (2.134) in terms of voltages and currents. The result will be what we term the "op-amp" representation of our amplifier, a representation that is standard in the discussion of classical amplifiers (see, e.g., [32]). In this representation, one views $\hat{V}_a$ and $\hat{I}_b$ as inputs to the amplifier: $\hat{V}_a$ is set by whatever we connect to the amplifier input, while $\hat{I}_b$ is set by whatever we connect to the amplifier output. In contrast, the outputs of our amplifier are the voltage in the output line, $\hat{V}_b$, and the current drawn by the amplifier at the input, $\hat{I}_a$.

Using Eqs. (2.132a,b), and suppressing frequency labels for clarity, Eq. (2.134) may be written explicitly in terms of the voltages and current in the input ($\hat{V}_a, \hat{I}_a$) and output ($\hat{V}_b, \hat{I}_b$) transmission lines:

$$\begin{pmatrix} \hat{V}_b \\ \hat{I}_a \end{pmatrix} = \begin{pmatrix} \lambda_V & -Z_{\text{out}} \\ 1/Z_{\text{in}} & \lambda_I' \end{pmatrix} \begin{pmatrix} \hat{V}_a \\ \hat{I}_b \end{pmatrix} + \begin{pmatrix} \lambda_V \hat{\tilde{V}} \\ \hat{\tilde{I}} \end{pmatrix}. \tag{2.135}$$

The coefficients in the matrix in Eq. (2.135) are completely determined by the scattering matrix of Eq. (2.134) (see Eqs. (2.138)); moreover, they are familiar from the discussion of a voltage amplifier in Section 2.4.7. $\lambda_V[\omega]$ is the voltage gain of the amplifier, $\lambda_I'[\omega]$ is the reverse current gain of the amplifier, Z_{out} is the output impedance, and Z_{in} is the input impedance. The last term on the right-hand side of Eq. (2.135) describes the two familiar kinds of amplifier noise. $\hat{\tilde{V}}$ is the usual voltage noise of the amplifier (referred back to the amplifier input), while $\hat{\tilde{I}}$ is the usual current noise of the amplifier. Recall that in this standard description of a voltage amplifier (see Section 2.4.7), $\tilde{I}$ represents the backaction of the amplifier: the system producing the input signal responds to these current fluctuations, resulting in an additional fluctuation in the input signal going into the amplifier. Similarly, $\lambda_V \tilde{V}$ represents the intrinsic output noise of the amplifier: this contribution to the total output noise does not depend on properties of the input signal. Note that we are using a sign convention where a positive $\langle \hat{I}_a \rangle$ indicates a current flowing *into* the amplifier at its input, while a positive $\langle \hat{I}_b \rangle$ indicates a current flowing *out of* the amplifier at its output. Also note that the operators $\hat{V}_a$ and $\hat{I}_b$ on the right-hand side of Eq. (2.135) will have noise; this noise is entirely due to the systems attached to the input and output of the amplifier, and, as such, should not be included in what we call the added noise of the amplifier.

Additional important properties of our amplifier follow immediately from quantities in the op-amp representation. As discussed in Section 2.4.3, the most important measure of gain in our amplifier is the dimensionless power gain. This is the ratio between power dissipated at the output to that dissipated at the input, taking the output current I_B to be V_B/Z_{out}:

$$G_P \equiv \frac{(\lambda_V)^2}{4} \frac{Z_{\text{in}}}{Z_{\text{out}}} \left(1 + \frac{\lambda_V \lambda_I'}{2} \frac{Z_{\text{in}}}{Z_{\text{out}}} \right)^{-1}. \tag{2.136}$$

Another important quantity is the loaded input impedance: what is the input impedance of the amplifier in the presence of a load attached to the output? In the presence of reverse current gain $\lambda_I' \neq 0$, the input impedance will depend on the output load. Taking the load impedance to be Z_{load}, some simple algebra yields

$$\frac{1}{Z_{\text{in,loaded}}} = \frac{1}{Z_{\text{in}}} + \frac{\lambda_I' \lambda_V}{Z_{\text{load}} + Z_{\text{out}}}. \tag{2.137}$$

It is of course undesirable to have an input impedance that depends on the load. Thus, we see yet again that it is undesirable to have appreciable reverse gain in our amplifier.

Converting between representations Some straightforward algebra now lets us express the op-amp parameters appearing in Eq. (2.135) in terms of the scattering matrix appearing in Eq. (2.134):

$$\lambda_V = 2\sqrt{\frac{Z_b}{Z_a}} \frac{s_{21}}{D}, \tag{2.138a}$$

$$\lambda_I' = 2\sqrt{\frac{Z_b}{Z_a}} \frac{s_{12}}{D}, \tag{2.138b}$$

$$Z_{\text{out}} = Z_b \frac{(1 + s_{11})(1 + s_{22}) - s_{12}s_{21}}{D}, \tag{2.138c}$$

$$\frac{1}{Z_{\text{in}}} = \frac{1}{Z_a} \frac{(1 - s_{11})(1 - s_{22}) - s_{12}s_{21}}{D}. \tag{2.138d}$$

All quantities are evaluated at the same frequency ω, and D is defined as

$$D = (1 + s_{11})(1 - s_{22}) + s_{12}s_{21}. \tag{2.139}$$

Further, the voltage and current noises in the op-amp representation are simple linear combinations of the noises $\hat{\mathcal{F}}_a$ and $\hat{\mathcal{F}}_b$ appearing in the scattering representation:

$$\begin{pmatrix} \hat{\hat{V}} \\ Z_a \cdot \hat{\hat{I}} \end{pmatrix} = \sqrt{2\hbar\omega Z_a} \begin{pmatrix} -\dfrac{1}{2} & \dfrac{1 + s_{11}}{2s_{21}} \\ \dfrac{s_{22} - 1}{D} & -\dfrac{s_{12}}{D} \end{pmatrix} \begin{pmatrix} \hat{\mathcal{F}}_a \\ \hat{\mathcal{F}}_b \end{pmatrix}. \tag{2.140}$$

Again, all quantities are evaluated at frequency ω.

Equation (2.140) immediately leads to an important conclusion and caveat: *what one calls the "backaction" and "added noise" in the scattering representation (i.e., $\mathcal{F}_a$ and $\mathcal{F}_b$) are not the same as the "backaction" and "added noise" defined in the usual op-amp representation.* For example, the op-amp backaction $\hat{\tilde{I}}$ does not in general coincide with the $\hat{\mathcal{F}}_a$, the backaction in the scattering picture. If we are indeed interested in using our amplifier as a voltage amplifier, we are interested in the total added noise of our amplifier *as defined in the op-amp representation.* As we saw in Section 2.4.7 (see Eq. (2.112)), this quantity involves both the noises $\hat{\tilde{I}}$ and $\hat{\tilde{V}}$.

2.5.2.2 A seemingly ideal two-port amplifier

Scattering versus op-amp quantum limit In this subsection, we demonstrate that an amplifier that is "ideal" and minimally complex when used in the scattering operation mode fails, when used as a voltage op-amp, to have a quantum-limited noise temperature. The system we look at is very similar to the amplifier considered by Grassia [33], though our conclusions are somewhat different than those found there.

In the scattering representation, one might guess that an "ideal" amplifier would be one where there are no reflections of signals at the input and output, and no way for incident signals at the output port to reach the input. In this case, Eq. (2.134) takes the form

$$\begin{pmatrix} \hat{a}_{\text{out}} \\ \hat{b}_{\text{out}} \end{pmatrix} = \begin{pmatrix} 0 & 0 \\ \sqrt{G} & 0 \end{pmatrix} \begin{pmatrix} \hat{a}_{\text{in}} \\ \hat{b}_{\text{in}} \end{pmatrix} + \begin{pmatrix} \hat{\mathcal{F}}_a \\ \hat{\mathcal{F}}_b \end{pmatrix}, \tag{2.141}$$

where we have defined $\sqrt{G} \equiv s_{21}$. All these quantities should be evaluated at the same frequency ω; for clarity, we will omit writing the explicit ω dependence of quantities throughout this subsection.

Turning to the op-amp representation, Eq. (2.141) implies that our amplifier has no reverse gain, and that the input and output impedances are simply given by the impedances of the input and output transmission lines. From Eqs. (2.138), we have

$$\lambda_V = 2\sqrt{\frac{Z_b}{Z_a}G}, \tag{2.142a}$$

$$\lambda'_I = 0, \tag{2.142b}$$

$$Z_{\text{out}} = Z_b, \tag{2.142c}$$

$$\frac{1}{Z_{\text{in}}} = \frac{1}{Z_a}. \tag{2.142d}$$

We immediately see that our amplifier looks less ideal as an op-amp. The input and output impedances are the same as those of the input and output transmission lines. However, for an ideal op-amp, we would have liked $Z_{\text{in}} \to \infty$ and $Z_{\text{out}} \to 0$.

Also of interest are the expressions for the amplifier noises in the op-amp representation:

$$\begin{pmatrix} \hat{\tilde{V}} \\ Z_a \cdot \hat{\tilde{I}} \end{pmatrix} = -\sqrt{2\hbar\omega Z_a} \begin{pmatrix} \frac{1}{2} & -\frac{1}{2\sqrt{G}} \\ 1 & 0 \end{pmatrix} \begin{pmatrix} \hat{\mathcal{F}}_a \\ \hat{\mathcal{F}}_b \end{pmatrix}. \tag{2.143}$$

As $s_{12} = 0$, the backaction noise is the same in both the op-amp and scattering representations: it is determined completely by the noise operator $\hat{\mathcal{F}}_a$. However, the voltage noise (i.e., the intrinsic output noise) involves *both* $\hat{\mathcal{F}}_a$ and $\hat{\mathcal{F}}_b$. We thus have the unavoidable consequence that there will be correlations in $\hat{\tilde{I}}$ and $\hat{\tilde{V}}$.

To make further progress, we note again that commutators of the noise operators $\hat{\mathcal{F}}_a$ and $\hat{\mathcal{F}}_b$ are completely determined by Eq. (2.141) and the requirement that the output operators obey canonical commutation relations. We thus have

$$\left[\hat{\mathcal{F}}_a, \hat{\mathcal{F}}_a^\dagger\right] = 1, \tag{2.144a}$$

$$\left[\hat{\mathcal{F}}_b, \hat{\mathcal{F}}_b^\dagger\right] = 1 - |G|, \tag{2.144b}$$

$$\left[\hat{\mathcal{F}}_a, \hat{\mathcal{F}}_b\right] = \left[\hat{\mathcal{F}}_a, \hat{\mathcal{F}}_b^\dagger\right] = 0. \tag{2.144c}$$

We will be interested in the limit of a large power gain, which requires $|G| \gg 1$. A minimal solution to Eqs. (2.144) would be to have the noise operators determined by two independent (i.e., mutually commuting) auxiliary input mode operators u_{in} and $v_{\text{in}}^\dagger$:

$$\hat{\mathcal{F}}_a = \hat{u}_{\text{in}}, \tag{2.145}$$

$$\hat{\mathcal{F}}_b = \sqrt{|G| - 1}\hat{v}_{\text{in}}^\dagger. \tag{2.146}$$

Further, to minimize the noise of the amplifier, we take the operating state of the amplifier to be the vacuum for both these modes. With these choices, our amplifier is in exactly the minimal form described by Grassia [33]: an input and output line coupled to a negative resistance box and an auxiliary "cold load" via a four-port circulator (see Fig. 2.5). The negative-resistance box is nothing but the single-mode bosonic amplifier discussed in Section 2.5.1; an explicit realization of this element would be a nondegenerate parametric amplifier (see, e.g., Section V.C of [2]). The "cold load" is a semi-infinite transmission line that models dissipation due to a resistor at zero temperature.

Note that, within the scattering picture, one would conclude that our amplifier is ideal: in the large-gain limit, the noise added by the amplifier to $\hat{b}_{\text{out}}$ corresponds to a single quantum at the input:

$$\frac{\langle\{\hat{\mathcal{F}}_b, \hat{\mathcal{F}}_b^\dagger\}\rangle}{|G|} = \frac{|G| - 1}{|G|}\langle\{\hat{v}_{\text{in}}^\dagger, \hat{v}_{\text{in}}\}\rangle \to 1 \tag{2.147}$$

This, however, is *not* the quantity that interests us: as we want to use this system as a voltage op-amp, we would like to know if the noise temperature *defined in the op-amp picture* is as small as possible. We are also usually interested in the case of a signal

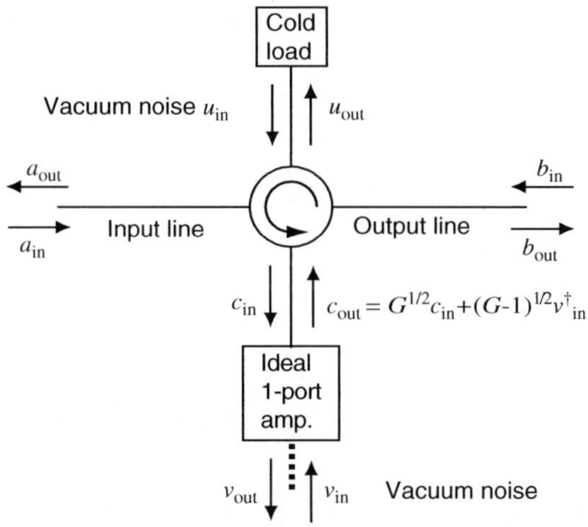

Fig. 2.5 Schematic of a "minimal" two-port amplifier that reaches the quantum limit in the scattering mode of operation, but misses the quantum limit when used as a weakly coupled op-amp. See the text for further description

that is weakly coupled to our amplifier; here, weak coupling means that the input impedance of the amplifier is much larger than the impedance of the signal source (i.e., $Z_{\text{in}} \gg Z_{\text{s}}$). In this limit, the amplifier only slightly increases the total damping of the signal source.

To address whether we can reach the op-amp quantum limit in the weak-coupling regime, we can make use of the results of the general theory presented in Section 2.4.7. In particular, we need to check whether the quantum noise constraint of Eq. (2.117) is optimized, as this is a prerequisite for reaching the (weak-coupling) quantum limit. Thus, we need to calculate the *symmetrized* spectral densities of the current and voltage noises, and their cross-correlation. It is easy to confirm from the definitions of Eqs. (2.132a, b) that these quantities take the form

$$\bar{S}_{VV}[\omega] = \frac{\langle \{\hat{\tilde{V}}[\omega], \hat{\tilde{V}}^{\dagger}[\omega']\} \rangle}{4\pi\delta(\omega - \omega')}, \tag{2.148a}$$

$$\bar{S}_{II}[\omega] = \frac{\langle \{\hat{\tilde{I}}[\omega], \hat{\tilde{I}}^{\dagger}[\omega']\} \rangle}{4\pi\delta(\omega - \omega')}, \tag{2.148b}$$

$$\bar{S}_{VI}[\omega] = \frac{\langle \{\hat{\tilde{V}}[\omega], \hat{\tilde{I}}^{\dagger}[\omega']\} \rangle}{4\pi\delta(\omega - \omega')}. \tag{2.148c}$$

The expectation values here are over the operating state of the amplifier; we have chosen this state to be the vacuum for the auxiliary-mode operators $\hat{u}_{\text{in}}$ and $\hat{v}_{\text{in}}$ to minimize the noise.

Taking $|s_{21}| \gg 1$, and using Eqs. (2.145) and (2.146), we have

$$\bar{S}_{VV}[\omega] = \frac{\hbar\omega Z_a}{4}(\sigma_{uu} + \sigma_{vv}) = \frac{\hbar\omega Z_a}{2}, \tag{2.149a}$$

$$\bar{S}_{II}[\omega] = \frac{\hbar\omega}{Z_a}\sigma_{uu} = \frac{\hbar\omega}{Z_a}, \tag{2.149b}$$

$$\bar{S}_{VI}[\omega] = \frac{\hbar\omega}{2}\sigma_{uu} = \frac{\hbar\omega}{2}, \tag{2.149c}$$

where we have defined

$$\sigma_{ab} \equiv \langle \hat{a}\hat{b}^\dagger + \hat{b}^\dagger\hat{a}\rangle, \tag{2.150}$$

and have used the fact that there cannot be any correlations between the operators u and v in the vacuum state (i.e., $\langle \hat{u}\hat{v}^\dagger\rangle = 0$).

It follows immediately from Eqs. (2.149) that our minimal amplifier does not optimize the quantum noise constraint of Eq. (2.117):

$$\bar{S}_{VV}[\omega]\bar{S}_{II}[\omega] - \left(\text{Im } \bar{S}_{VI}\right)^2 = 2 \times \left(\frac{\hbar\omega}{2}\right)^2. \tag{2.151}$$

The noise product $\bar{S}_{VV}\bar{S}_{II}$ is precisely twice the quantum-limited value. As a result, the general theory of Section 2.4.7 tells us *if one couples an input signal weakly to this amplifier (i.e., $Z_s \ll Z_{in}$), it is impossible to reach the quantum limit on the added noise.* Thus, while our amplifier is ideal in the scattering mode of operation (see Eq. (2.147)), it fails to reach the quantum limit when used in the weak-coupling, op-amp mode of operation. Our amplifier's failure to have "ideal" quantum noise also means that if we tried to use it to do QND qubit detection, the resulting backaction dephasing would be twice as large as the minimum required by quantum mechanics.

2.5.2.3 Why is the op-amp quantum limit is not achieved?

Note that if one substitutes the expressions in Eqs. (2.149) (including the real cross-correlation noise) in the classical lower bound for $T_{\rm N}$ given in Eq. (2.118), one arrives at the disturbing conclusion that $T_N[\omega]$ not only reaches the quantum limit, but surpasses it: one finds $T_N[\omega] = (\sqrt{2}-1)(\hbar\omega/2)$. Before getting too excited (or alarmed) by this, note that this expression was derived for a noise-matched source impedance ($Z_{\rm s} = Z_{\rm N}$), as well as assuming $Z_{\rm s} \ll Z_{\rm in}$. The noise impedance here, however, is given by

$$Z_{\rm N} = \sqrt{\frac{\bar{S}_{VV}}{\bar{S}_{II}}} = \frac{Z_{\rm in}}{\sqrt{2}}. \tag{2.152}$$

Thus, the weak-coupling condition $Z_{\rm s} \ll Z_{\rm in}$ is not fulfilled, and Eq. (2.118) for $T_{\rm N}$ is not valid.

One can instead directly calculate the total added noise of the amplifier (as always, referred back to the input) for a finite ratio of Z_s/Z_{in}. This added noise takes the form

$$\hat{\tilde{V}}_{tot} = -\left(\frac{Z_s Z_a}{Z_s + Z_a}\right)\hat{\tilde{I}} + \hat{\tilde{V}}. \tag{2.153}$$

Note that this classical-looking equation can be rigorously justified within the full quantum theory if one starts with a full description of the amplifier and the signal source (e.g., a parallel LC oscillator attached in parallel to the amplifier input). Plugging in the expressions for $\hat{\tilde{I}}$ and $\hat{\tilde{V}}$, we find

$$\begin{aligned}\hat{\tilde{V}}_{tot} &= \sqrt{\frac{\hbar\omega}{2}}\left[\left(\frac{Z_s Z_a}{Z_s + Z_a}\right)\left(\frac{2}{\sqrt{Z_a}}\hat{u}_{in}\right) - \sqrt{Z_a}(\hat{u}_{in} - \hat{v}^\dagger_{in})\right] \\ &= \sqrt{\frac{\hbar\omega Z_a}{2}}\left[\left(\frac{Z_s - Z_a}{Z_s + Z_a}\right)\hat{u}_{in} - \hat{v}^\dagger_{in}\right].\end{aligned} \tag{2.154}$$

Thus, if one impedance matches the source (i.e., tunes Z_s to $Z_a = Z_{in}$), the mode $\hat{u}_{in}$ does not contribute to the total added noise, and one reaches the appropriately defined quantum limit on the added noise.[5] One is of course very far from the weak-coupling condition needed for the op-amp mode, and is thus effectively operating in the scattering mode.

Returning to the more interesting case of a weak amplifier-signal coupling, for $Z_s \ll Z_a \sim Z_{in}$, one finds that the noise temperature is a large factor Z_{in}/Z_s bigger than the quantum-limited value. It is possible to understand the failure to reach the quantum limit in this weak-coupling limit heuristically. To that end, note again that the amplifier noise cross-correlation $\bar{S}_{IV}$ does not vanish in the large-gain limit (see Eq. (2.149c)). Correlations between the two amplifier noises represent a kind of information, as by making use of them, we can improve the performance of the amplifier. It is easy to take advantage of out-of-phase correlations between $\tilde{I}$ and $\tilde{V}$ (i.e., Im $\bar{S}_{VI}$) by simply tuning the phase of the source impedance (see Eq. (2.116)). However, one cannot take advantage of in-phase noise correlations (i.e., Re $\bar{S}_{VI}$) as easily. To take advantage of the information here, one needs to modify the amplifier itself. By feeding back some of the output voltage to the input, one could effectively cancel out some of the backaction current noise $\tilde{I}$ and thus reduce the overall magnitude of $\bar{S}_{II}$. Hence, the unused information in the cross-correlator Re $\bar{S}_{VI}$ represents a kind of wasted information: had we made use of these correlations via a feedback loop, we could have reduced the noise temperature and increased the information provided by our amplifier. In [2], we discuss this point further, and show how feedback may be implemented in the amplifier considered here by introducing reflections in the transmission lines leading to the circulator.

[5] One must be careful here in defining the added noise, as when Z_s is not much smaller than Z_{in}, the amplifier will appreciably increase the dissipation of the signal source and reduce its intrinsic thermal fluctuations. This reduction should *not* be included in the definition of the added noise. This is analogous to linear position detection in the case where the backaction damping is large; the cooling effect of the backaction damping is not included in the definition of the quantum limit; see. Eq. (2.90).

References

[1] Heisenberg, W. (1930). *The Physical Principles of the Quantum Theory.* University of Chicago Press, Chicago (reprinted by Dover Publications New York 1949).

[2] Clerk, A. A., Devoret, M. H., Girvin, S. M., Marquardt, F., and Schoelkopf, R. J. (2010). Introduction to quantum noise, measurement and amplification. *Rev. Mod. Phys.*, **82**, 1155.

[3] Bruus, H. and Flensberg, K. (2004). *Many-Body Quantum Theory in Condensed Matter Physics: An Introduction.* Oxford University Press, Oxford.

[4] Clerk, A. and Bennett, S. (2005). Quantum nanoelectromechanics with electrons, quasi-particles and Cooper pairs: effective bath descriptions and strong feedback effects. *New J. Phys.*, **7**, 238.

[5] Blencowe, M. P, Imbers, J., and Armour, A. D. (2005). Dynamics of a nanomechanical resonator coupled to a superconducting single-electron transistor. *New J. Phys.*, **7**, 236.

[6] Marquardt, F., Chen, J. P., Clerk, A. A., and Girvin, S. M. (2007). Quantum theory of cavity-assisted sideband cooling of mechanical motion. *Phys. Rev. Lett.*, **99**, 093902.

[7] Wilson-Rae, I., Nooshi, N., Zwerger, W., and Kippenberg, T. J. (2007). Theory of ground state cooling of a mechanical oscillator using dynamical back-action. *Phys. Rev. Lett.*, **99**, 093901.

[8] Blanter, Ya. M. and Büttiker, M. (2000). Shot noise in mesoscopic conductors. *Phys. Rep.*, **336**, 1.

[9] Field, M., Smith, C. G., Pepper, M., Ritchie, D. A., Frost, J. E. F., Jones, G. A. C., and Hasko, D. G. (1993). Measurements of Coulomb blockade with a noninvasive voltage probe. *Phys. Rev. Lett.*, **70**, 1311.

[10] Elzerman, J. M., Hanson, R., van Beveren, L. H. W., Witkamp, B., Vandersypen, L. M. K, and Kouwenhoven, L. P. (2004). Single-shot read-out of an individual electron spin in a quantum dot. *Nature*, **430**, 431.

[11] Petta, J. R., Johnson, A. C., Marcus, C., Hanson, M. P., and Gossard, A. C. (2004). Manipulation of a single charge in a double quantum dot. *Phys. Rev. Lett.*, **93**, 186802.

[12] Poggio, M., Jura, M. P., Degen, C. L., Topinka, M. A., Mamin, H. J., Goldhaber-Gordon, D., and Rugar, D. (2008). An off-board quantum point contact as a sensitive detector of cantilever motion. *Nature Phys.*, **4**, 635.

[13] Gottfried, K. (1966). *Quantum Mechanics*, Volume I: *Fundamentals.* Benjamin, New York.

[14] Braginsky, V. B. and Khalili, F. Y. (1996). Quantum nondemolition measurements: the route from toys to tools. *Rev. Mod. Phys.*, **68**, 1.

[15] Averin, D. (2003). Linear quantum measurements. In *Quantum Noise in Mesoscopic Systems* (ed. Y. Nazarov), p. 229. Kluwer, Amsterdam.

[16] Pathria, R. K. and Beale, P. D. (2011). *Statistical Mechanics*, 3rd edn. Butterworth Heinamann, Oxford.

[17] Clerk, A. A., Girvin, S. M., and Stone, A. D. (2003). Quantum-limited measurement and information in mesoscopic detectors. *Phys. Rev. B*, **67**, 165324.

[18] Devoret, M. H. and Schoelkopf, R. J. (2000). Amplifying quantum signals with the single-electron transistor. *Nature*, **406**, 1039.

[19] Averin, D. V. (2000). Quantum computing and quantum measurement with mesoscopic Josephson junctions. *Fortschr. Phys.*, **48**, 1055.

[20] Korotkov, A. N. and Averin, D. V. (2001). Continuous weak measurement of quantum coherent oscillations. *Phys. Rev. B*, **64**, 165310.

[21] Makhlin, Y., Schön, G., and Shnirman, A. (2001). Quantum-state engineering with Josephson-junction devices. *Rev. Mod. Phys.*, **73**, 357.

[22] Pilgram, S. and Büttiker, M. (2002). Efficiency of mesoscopic detectors. *Phys. Rev. Lett.*, **89**, 200401.

[23] Young, C. E. and Clerk, A. A. (2010). Inelastic backaction due to quantum point contact charge fluctuations. *Phys. Rev. Lett.*, **104**, 186803.

[24] Sprinzak, D., Buks, E., Heiblum, M., and Shtrikman, H. (2000). Controlled dephasing of electrons via a phase sensitive detector. *Phys. Rev. Lett.*, **84**, 5820.

[25] Korotkov, A. N. (1999). Continuous quantum measurement of a double dot. *Phys. Rev. B*, **60**, 5737.

[26] Clerk, A. A. (2004). Quantum-limited position detection and amplification: a linear response perspective. *Phys. Rev. B*, **70**, 245306.

[27] Laflamme, C. and Clerk, A. (2011). Quantum-limited amplification with a nonlinear cavity detector. *Phys. Rev. A*, **83**, 033803.

[28] Caves, C. M. (1982). Quantum limits on noise in linear amplifiers. *Phys. Rev. D*, **26**, 1817.

[29] Haus, H. A. and Mullen, J. A. (1962). Quantum noise in linear amplifiers. *Phys. Rev.*, **128**, 2407.

[30] Heffner, H. (1962). The fundamental noise limit of linear amplifiers. *Proc. IRE*, **50**, 1604.

[31] Haus, H. A. (2000). *Electromagnetic Noise and Quantum Optical Measurements.* Springer-Verlag, New York.

[32] Boylestad, R. and Nashelsky, L. (2006). *Electronic Devices and Circuit Theory*, 9th edn. Prentice Hall, Upper Saddle River, NT.

[33] Grassia, F. (1998). *Fluctuations quantiques et thermiques dans les transducteurs electromecaniques.* PhD. thesis, Université Pierre et Marie Curie.

3
Circuit QED: superconducting qubits coupled to microwave photons

S. M. Girvin

Department of Physics, Yale University,
New Haven, CT 06520, USA

Quantum Machines: Measurement and Control of Engineered Quantum Systems.
Edited by M. Devoret, B. Huard, R. Schoelkopf, and L. F. Cugliandolo.

Chapter Contents

These lecture notes discuss the quantum electrodynamics of superconducting circuits. Some useful papers that students may wish to consult in learning this field are:

- Y. Makhlin, G. Schön, and A. Schnirman. Quantum-state engineering with Josephson-junction devices. *Rev. Mod. Phys.*, **73**, 357 (2001).
- A. Blais, R.-S. Huang, A. Wallraff, S.M. Girvin, and R.J. Schoelkopf. Cavity quantum electrodynamics for superconducting electrical circuits: an architecture for quantum computation. *Phys. Rev. A*, **69**, 062320 (2004).
- A.A. Clerk, M.H. Devoret, S.M. Girvin, F. Marquardt, and R.J. Schoelkopf. Introduction to quantum noise, measurement and amplification. *Rev. Mod. Phys.* (2008). (The full version with pedagogical appendices is most conveniently available at <http://arxiv.org/abs/arXiv:0810.4729>.)

3.1 Introduction to quantum machines

A quantum machine is a device whose degrees of freedom are intrinsically quantum mechanical. Of course, every machine is made of atoms and its microscopic degrees of freedom (the electrons and nuclei) are intrinsically quantum. Atomic physicists have developed remarkable optical techniques to control these microscopic degrees of freedom. In these notes, we will take the condensed matter physics perspective: by degrees of freedom we mean collective variables that are sufficiently "macroscopic" that we can couple to them and control them with externally applied electrical signals or forces. We will extensively study the currents and voltages in superconducting circuits as "macroscopic" quantum degrees of freedom. Another modern example of growing importance would be the center-of-mass position of a movable mirror acting as a mechanical oscillator coupled to the radiation pressure of light in an optomechanical system.

While we do not yet fully understand the properties and capabilities of quantum machines, there is great hope (and some mathematical proof) that such devices will have novel capabilities that are impossible to realize on classical hardware. You might think that quantum machines have already been built. For example, the laser and the transistor would seem to rely on quantum physics for their operation. It is clear that the frequency of a laser cannot be computed without the quantum theory that predicts the excitation energies of the atoms in the laser. Similarly, the optimal bias voltage of a bipolar transistor depends on the electronic band-gap energy of the material from which it is made. Nevertheless, it is only the particular values of the operating *parameters* of these machines that are determined by quantum physics. Once we know the values of these parameters, we see that these are classical machines because their degrees of freedom are purely classical. Indeed, the light output from a laser is special because it is exactly like the radiofrequency output of the classical oscillator that powers a radio station's antenna. Similarly, the currents and voltages in an ordinary transistor circuit need not be treated as noncommuting quantum operators in order to understand the operation of the circuit. Its degrees of freedom are, for all intents and purposes, classical.

These lectures are devoted to understanding the basic components of quantum machines that can be constructed from superconducting electrical circuits. These circuits can be used to create resonators that store individual microwave photons and to create superconducting quantum bits. Both of these circuit elements are intrinsically quantum mechanical. They have quantized energy levels with spacing much greater than temperature (for low enough temperatures!) and they can be placed into quantum superpositions of different energy states. To properly predict their properties, the currents and voltages in such circuits must be represented by noncommuting quantum operators.

The beauty of electrical circuits is that they can be constructed in a modular manner, combining together a few different building blocks in simple ways. The wires connecting these building blocks have to be capable of carrying quantum signals [1, 2], but are still relatively simple, and the problem of spatial mode matching that occurs in optical circuits is largely eliminated (becoming simply a question of impedance matching). Furthermore, with modern lithographic techniques, parallel fabrication of complex structures is relatively straightforward, lending hope that (someday) it will be possible to scale up to processors with large numbers of qubits.

We will study different qubit designs and their relative merits. We will also learn how to control and read out the quantum states of qubits and cavities, and how to entangle their different quantum degrees of freedom. In recent decades, we have come to understand that superposition and entanglement are features of quantum physics that can be used as resources to make powerful quantum computers to process information in ways that are impossible classically. What else we can do with quantum machines is not yet fully understood. The people who invented the laser had no idea that it would be used to play music, communicate over optical fibers, and perform eye surgery. We expect that similar surprises and unexpected applications will be developed once quantum hardware becomes routine enough to play with.

3.2 Quantum electrical circuits

3.2.1 Introduction

Quantum electrodynamics is the theory of interaction between electrons (and atoms) with electromagnetic fields. This chapter discusses the closely related problem of quantization of electrical circuits [1, 3]. Experimental progress over the last decade in creating and controlling quantum coherence in superconducting electrical circuits has been truly remarkable. The quantum electrodynamics of superconducting microwave circuits has been dubbed "circuit QED" by analogy to cavity QED in quantum optics. This chapter will describe the quantum optics approach to microwave circuits, with superconducting qubits playing the role of artificial atoms whose properties can be engineered. Despite being large enough to be visible to the naked eye, these artificial atoms have a very simple discrete set of quantized energy levels that are nearly as well understood [4] as those of the prototypical single-electron atom, hydrogen. Furthermore, it has proven possible to put these atoms into coherent superpositions of different quantum states so that they can act as quantum bits. Through clever

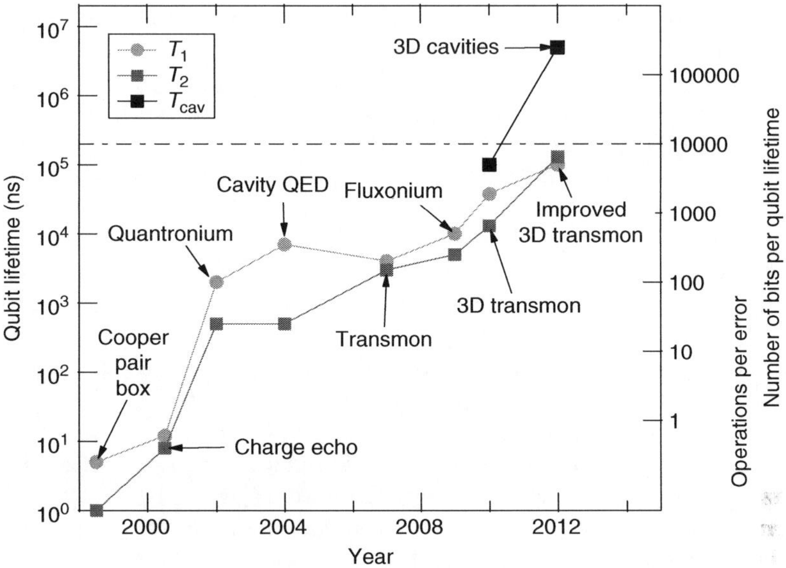

Fig. 3.1 "Schoelkopf's law" plot illustrating the exponential growth for superconducting (charge-) qubit coherence times. Recent experiments [18] with the "fluxonium" qubit design have achieved T_1 times exceeding 1 ms.

engineering, the coherence times of such superposition states has risen more than four orders of magnitude from nanoseconds for the first superconducting qubit created in 1999 [5] up to about 30–150 μs today [6–8]. Recent experiments with the fluxonium qubit design [9] have achieved T_1 times exceeding 1 ms [10]. "Schoelkopf's law" for the exponential growth of coherence time is illustrated in Fig. 3.1.

Simple quantum machines have already been built using superconducting circuits that can manipulate and measure the states of individual qubits [5, 11, 12], as well as individual microwave quanta [13–18], entangle two [19, 20] and three [21, 22] qubits, run simple quantum algorithms [23, 24], and perform rudimentary quantum error correction [25]. Future improved qubit designs, microwave circuit designs, and materials should allow this trend to continue unabated. In addition to being a potentially powerful engineering architecture for building a quantum computer, circuit QED opens up for us a novel new regime to study ultrastrong coupling between "atoms" and individual microwave photons [26]. The concept of the photon is a subtle one, but hopefully this chapter will convince the reader that microwaves, despite their name, really are particles. We will accordingly begin our study with a review of the quantization of electromagnetic fields in circuits and cavities.

The quantization of electrical circuits has been thoroughly addressed in the Les Houches lecture notes of my colleague, Michel Devoret [3], to which I direct the interested reader. The circuit elements that are available to the quantum engineer include those familiar to classical engineers: resistors, capacitors, and inductors. Resistors cause unwanted dissipation and we will attempt to avoid them. See, however, further discussion in the appendix (Section 3.9.2) of spontaneous emission into transmission lines, which act effectively as fixed impedances. Dissipation into a cold resistor can in fact be useful for qubit reset [27] to the ground state, since reset requires removal of entropy to a cold bath.

In addition to these standard circuit elements, there is one special element in superconducting circuits, namely the Josephson tunnel junction. We will be learning more about superconductivity and Josephson junctions later, but for now we simply note the following. With capacitors and inductors, we can build simple LC harmonic oscillators. If we can eliminate all resistors, then the harmonic oscillations will be undamped. The use of superconducting circuits takes us a long way toward this goal of zero dissipation, about which more later. The essential feature of (ordinary) superconductivity is that electrons of opposite spin pair up and condense into a special ground state with a substantial excitation gap 2Δ needed to break one of the pairs and create an excited state. This pair excitation gap is essential to the ability of current to flow in a superconductor without dissipation.[1] A closely related advantage of the excitation gap is that it dramatically reduces the number of effective degrees of freedom in the circuit, allowing us to construct artificial "atoms" that behave like simple single-electron atoms even though they are made up of 10^9–10^{12} aluminum atoms. The extremely powerful force of the Coulomb interactions also plays an essential role in limiting the low-energy degrees of freedom in circuits. When the Coulomb interaction is unscreened, the gapless collective motion of currents is lifted up to the plasma frequency, which is orders of magnitude higher than any relevant frequency scale for the circuits we will consider. (This effect of the long-range Coulomb force occurs in both normal metals and superconductors.) In the presence of screening due to ground planes or shields, the plasma oscillations are "acoustic modes" with a linear dispersion and velocity close to the speed of light in vacuum.[2] When quantized, these will be our propagating photons.

3.2.2 Plasma oscillations

Because the powerful effect of long-range Coulomb interactions plays a crucial role in simplifying the spectrum of quantum electrical circuits, let us begin our analysis by reviewing the plasma oscillations in a bulk metal. Throughout this work, we will use SI units. We will consider infinitesimal density fluctuations δn around the mean electron number density n. In the "jellium" model, the mean charge density is canceled by the ionic background, so the net charge density is

$$\rho(\vec{r}) = -e\,\delta n. \tag{3.1}$$

[1] There do exist gapless superconductors (e.g., d-wave materials such as YBCO) that can carry a dc current without dissipation, but at the microwave frequencies of interest for qubits, the lack of a gap implies significant dissipation.

[2] Flat metallic surfaces and long wires exhibit so-called surface plasmons, which are gapless and have approximately linear dispersion relations due to electrodynamic retardation effects. The purpose of the ground shield surrounding the central wire in a coaxial cable is to prevent radiation losses when the cable is bent into a curve.

The current flowing (to zeroth order in δn) is

$$\vec{J}(\vec{r},t) = -en\vec{v}(\vec{r},t), \tag{3.2}$$

where the local electron mean velocity field obeys Newton's law

$$\frac{\partial}{\partial t}\vec{v} = \frac{-e}{m}\vec{E}, \tag{3.3}$$

where m is the electron (effective) mass. This in turn yields

$$\frac{\partial}{\partial t}\vec{J} = \frac{ne^2}{m}\vec{E}. \tag{3.4}$$

Taking the divergence of both sides of this equation and applying Gauss's law

$$\vec{\nabla}\cdot\vec{E} = \frac{\rho}{\epsilon_0} \tag{3.5}$$

and the continuity equation

$$\vec{\nabla}\cdot\vec{J} + \frac{\partial}{\partial t}\rho = 0 \tag{3.6}$$

yields

$$\frac{\partial^2}{\partial t^2}\rho = -\omega_p^2\rho, \tag{3.7}$$

where the so-called "plasma frequency" is given by[3]

$$\omega_p^2 \equiv \frac{ne^2}{m\epsilon_0}. \tag{3.8}$$

Electromagnetic waves cannot propagate in a plasma at frequencies below the plasma frequency [28]. In the Earth's ionosphere, the typical plasma frequency is in the range of tens of megahertz and varies between night and day, thereby affecting short-wave radio reception. In the typical metals we will be concerned with (e.g., aluminum), the valence electron density is sufficiently high that the plasma frequency is in the ultraviolet region of the optical spectrum. Hence, aluminum (whose plasma frequency $\omega_\mathrm{p}/2\pi \approx 3.6 \times 10^{15}\,\mathrm{Hz}$ corresponds to a photon energy of about 15 eV) is highly reflective in the visible. Essentially, the electrons are so dense and so agile that they screen out any electric fields almost perfectly over a very short screening distance. For frequencies far below the plasma frequency, Maxwell's equations yield

$$\vec{\nabla}\times\vec{\nabla}\times\vec{E} \approx -\lambda_p^{-2}\vec{E}, \tag{3.9}$$

[3] We neglect here the various details of the band structure of Al as well as the possibility that the core electrons in the atoms of the metal contribute a dielectric constant $\epsilon \neq 1$ seen by the valence electrons whose dynamics create the plasma oscillations of the metal.

where the London penetration depth λ_L of the electromagnetic fields is

$$\lambda_{\rm L} = \frac{c}{\omega_p} = \frac{1}{\sqrt{4\pi n r_e}}, \tag{3.10}$$

with the classical radius of the electron given by

$$r_e = \frac{e^2}{4\pi\epsilon_0}\frac{1}{mc^2} \approx 2.818 \times 10^{-15}\,\text{m}. \tag{3.11}$$

For Al, Eq. (3.10) yields[4] $\lambda_L \approx 14\,\text{nm}$ We will be dealing with gigahertz frequency scales many orders of magnitude below the plasma frequency and centimeter wavelength scales relative to which the penetration depth is effectively zero.

Exercise 3.1 Derive Eq. (3.9) in the limit of low frequencies and show that it leads to exponential decay of transverse electromagnetic waves with decay length λ_p.

The above simplified[5] jellium model yields a plasma mode that is completely dispersionless—the mode frequency is independent of wavevector q. The frequency of the bulk collective plasma mode is vastly higher than any microwave frequency that we will be dealing with. From the point of view of quantum mechanics, the amount of energy required to create a bulk plasmon is so large that we can consider these degrees of freedom to be frozen into their quantum mechanical ground state. Hence they can be ignored. The approximations leading to Eq. (3.2) break down at short distances owing to the granularity of the electron charge. At very large wavevectors approaching the Fermi wavevector, the jellium continuous-charge picture breaks down and the plasma oscillation frequency rises and the mode becomes "Landau-damped" owing to the collective charge oscillation mode decaying into single-particle excitations [29]. Conversely for extremely small wavevectors, there is a cutoff associated with the finite size of any sample. This we can take into account by considering the capacitance matrix between different lumps of metal in the circuit we are trying to quantize. In certain circumstances, the capacitance matrix is such that there do exist collective charge-oscillation modes that are down in the microwave range. These will be the important modes that we will quantize. Here, the superconductivity is vital for gapping the single-particle excitations so that the collective charge modes are both simple and extremely weakly damped.

[4] The measured value of the London penetration depth in Al (at zero frequency) is somewhat larger, $\lambda_L \approx 51.5\,\text{nm}$. The difference is presumably due to variation in the core electron dielectric constant with frequency, which has been neglected in our model. It should also be noted that in dirty superconductors, the reduction in the superfluid stiffness causes the penetration depth to increase.

[5] A more careful treatment would have included the change in the Fermi energy as the density oscillates. The resulting Fermi pressure gradients produce a positive quadratic dispersion of the plasma mode with increasing wavevector.

3.2.3 Quantum LC Oscillator

The circuit element with the simplest dynamics is the LC oscillator illustrated schematically in Fig. 3.2(a). Now that we understand that supercurrents can flow essentially without dissipation and that the great strength of the Coulomb interaction lifts density fluctuations up to optical frequencies, we can understand that the LC oscillator has, to a very good approximation, only a single low-energy degree of freedom, namely, uniform divergenceless current flow in the wire of the inductor, which does not build up charge anywhere except on the plates of the capacitor. This is a very good approximation in the "lumped element" limit where the physical size of the LC oscillator is much smaller than the wavelength of electromagnetic waves at the frequency of the oscillator, $\lambda = 2\pi c/\Omega$. (This caveat is associated with the unstated assumption in our discussion of plasma oscillations that we neglected electrodynamic retardation effects. That is, we effectively assumed $c = \infty$.) In terms of the capacitor charge q and the inductor current I, the Lagrangian is readily written

$$\mathcal{L} = \frac{1}{2}LI^2 - \frac{1}{2}\frac{q^2}{C}. \tag{3.12}$$

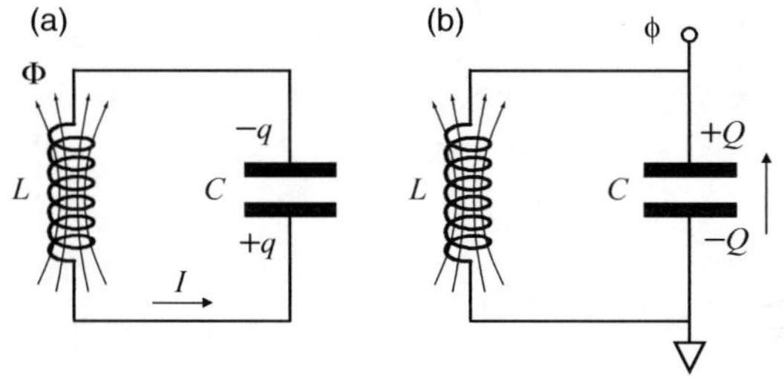

Fig. 3.2 Simple LC electrical oscillator analogous to a mass-and-spring mechanical oscillator. In (a), the position coordinate of the mass is taken to be q, the charge accumulated on the capacitor by the current I flowing through the inductor, and the flux Φ through the inductor is the momentum. The sign convention for the charge is such that $\dot{q} = I$, and therefore the inductance L is analogous to the mass. The role of the spring constant is played by $1/C$ and the potential energy of the capacitor is $(q - q_q)^2/2C$, where q_0 is the offset charge of the capacitor (the equivalent of the equilibrium length of the spring). Hamilton's equation for the time rate of change of the momentum is $\dot{\Phi} = -(q - q_0)/C$. In (b), the position coordinate is now taken to be ϕ, the time integral of the voltage V across the capacitor (i.e., the node flux), and the conjugate momentum is Q, the charge on the capacitor resulting from the electrochemical potential difference between the two plates. The role of the mass is played by C and the spring constant is now $1/L$, with the energy of the inductor given by $(\phi - \phi_0)^2/2L$, where ϕ_0 is the external flux in the loop of the circuit (including the coil of the inductor). Hamilton's equation for the time rate of change of position is $\dot{\phi} = Q/C$. Note the important sign change in the definition of charge, $Q = q_0 - q$, needed to make Hamilton's equations of motion correct in each case. The classical Poisson brackets and the quantum canonical commutation relations between position and momentum are maintained between the two cases: $[\hat{q}, \hat{\Phi}] = [\hat{\phi}, Q] = +i\hbar$.

Using charge conservation, $I = +\dot{q}$, this can be cast into the more familiar form

$$\mathcal{L} = \frac{L}{2}\dot{q}^2 - \frac{1}{2C}q^2. \tag{3.13}$$

Remarkably, we have reduced a complex circuit containing an enormous number of electrons to a system with a *single* degree of freedom q with "mass" L and "spring constant" $1/C$). This is possible only because all but this one degree of freedom are effectively gapped out by a combination of superconductivity (which gaps out the single-particle excitations) and the long-range Coulomb force (which gaps out the collective plasmon (density-fluctuation) degrees of freedom). All that is left is the rigid collective motion of the incompressible electron fluid sloshing back and forth, charging and discharging the capacitor.

Equation (3.13) yields the Euler-Lagrange equation of motion

$$\ddot{q} = -\Omega^2 q, \tag{3.14}$$

where the natural oscillation frequency is

$$\Omega = \frac{1}{\sqrt{LC}}. \tag{3.15}$$

The momentum conjugate to the charge is the flux through the inductor,

$$\Phi = \frac{\delta\mathcal{L}}{\delta\dot{q}} = L\dot{q} = LI. \tag{3.16}$$

Thus, the Hamiltonian can be written

$$H = \Phi\dot{q} - \mathcal{L} = \frac{\Phi^2}{2L} + \frac{1}{2C}q^2. \tag{3.17}$$

Hamilton's equations of motion then give the current through the inductor and the voltage at the node connecting the inductor and the capacitor:

$$\dot{q} = \frac{\partial H}{\partial\Phi} = \frac{\Phi}{L} = I \tag{3.18}$$

$$\dot{\Phi} = -\frac{\partial H}{\partial q} = -\frac{q}{C} = V. \tag{3.19}$$

In the usual way, the coordinate and its conjugate momentum can be promoted to quantum operators obeying the canonical commutation relation

$$[\hat{\Phi}, \hat{q}] = -i\hbar, \tag{3.20}$$

and we can write the Hamiltonian

$$H = \frac{\hbar\Omega}{2}\left(\hat{a}^\dagger\hat{a} + \hat{a}\hat{a}^\dagger\right) = \hbar\Omega\left(\hat{a}^\dagger\hat{a} + \frac{1}{2}\right) \tag{3.21}$$

in terms of raising and lowering operators

$$\hat{a} = +i\frac{1}{\sqrt{2L\hbar\Omega}}\hat{\Phi} + \frac{1}{\sqrt{2C\hbar\Omega}}\hat{q}, \tag{3.22}$$

$$\hat{a}^{\dagger} = -i\frac{1}{\sqrt{2L\hbar\Omega}}\hat{\Phi} + \frac{1}{\sqrt{2C\hbar\Omega}}\hat{q}, \tag{3.23}$$

which obey the usual relation

$$[\hat{a}, \hat{a}^{\dagger}] = 1. \tag{3.24}$$

In the above discussion, we chose the charge q on the capacitor as the natural coordinate of the harmonic oscillator and found that the inductor flux Φ was the momentum conjugate to this flux. In the picture, we interpret the capacitance C as the inverse of the "spring constant," and the inductance L as the "mass." This seems natural from our intuitive view of the capacitance as storing the potential energy and the inductor as storing the kinetic energy. (Actually, the kinetic energy of the electrons makes only a small contribution (called the "kinetic inductance") to the total inductance. It is primarily the energy stored in the magnetic field created by the current that dominates the inductance in most situations.)

When dealing with Josephson junctions, we will start with this same representation but then find that they act as nonlinear inductors, and so it will be more convenient to take the node flux (defined below) to be the coordinate rather than the momentum. In order to get used to this alternative representation, we will practice here on the LC oscillator. Following Devoret [3], let us define the node flux at the point shown in Fig. 3.2(b) by

$$\phi(t) = \int^{t} d\tau\, V(\tau), \tag{3.25}$$

so that $V(t) = \dot{\phi}$. Then, the potential energy stored on the capacitor is

$$U = \frac{1}{2}C\dot{\phi}^2 \tag{3.26}$$

and now looks like the kinetic energy with this choice of coordinate. Similarly, using Faraday's law and the sign convention for the direction of the current defined in Fig. 3.2(b), we have

$$V = L\dot{I} = \dot{\phi}, \tag{3.27}$$

and thus see that the node flux variable ϕ really is the physical magnetic flux Φ winding through the inductor (ignoring any possible external flux applied through the loop of the circuit or the inductor). Hence, the kinetic energy stored in the inductor is

$$T = \frac{1}{2L}\phi^2, \tag{3.28}$$

which now looks like the potential energy. With this choice of coordinate, the Lagrangian becomes

$$\mathcal{L} = \frac{1}{2}C\dot{\phi}^2 - \frac{1}{2L}\phi^2, \tag{3.29}$$

and the momentum conjugate to the flux

$$Q = \frac{\delta\mathcal{L}}{\delta\dot{\phi}} = C\dot{\phi} \tag{3.30}$$

is now the charge as defined with the sign convention in Fig. 3.2(b). Notice the crucial minus sign relative to the previous result. This is necessary to maintain the sign of the commutation relation when we interchange the momentum and coordinate. To reiterate: when the charge is the coordinate and the flux is the conjugate momentum, the commutation relation is

$$[\hat{q}, \hat{\Phi}] = +i\hbar, \tag{3.31}$$

whereas when the flux is the coordinate and the charge is the conjugate momentum, the commutation relation is

$$[\hat{\phi}, \hat{Q}] = +i\hbar. \tag{3.32}$$

Since we have chosen a convention in which $\hat{\Phi} = \hat{\phi}$, we require $\hat{Q} = -\hat{q}$.

Just to be completely explicit, we now repeat the derivation of the Hamiltonian and its quantization for this new choice, which we will be using throughout the remainder of this chapter. Thus, the Hamiltonian can be written

$$H = Q\dot{\phi} - \mathcal{L} = \frac{1}{2C}Q^2 + \frac{\phi^2}{2L}. \tag{3.33}$$

Hamilton's equations of motion are then

$$\dot{\phi} = +\frac{\partial H}{\partial Q} = +\frac{Q}{C}, \tag{3.34}$$

$$\dot{Q} = -\frac{\partial H}{\partial \phi} = -\frac{\phi}{L}. \tag{3.35}$$

Again, in the usual way, the coordinate and its conjugate momentum can be promoted to quantum operators obeying the canonical commutation relation (but note the important position reversal from Eq. (3.20))

$$[\hat{Q}, \hat{\phi}] = -i\hbar, \tag{3.36}$$

and we can write the Hamiltonian

$$H = \frac{\hbar\Omega}{2}\left(\hat{a}^\dagger\hat{a} + \hat{a}\hat{a}^\dagger\right) = \hbar\Omega\left(\hat{a}^\dagger\hat{a} + \frac{1}{2}\right) \tag{3.37}$$

in terms of raising and lowering operators

$$\hat{a} = +i\frac{1}{\sqrt{2C\hbar\Omega}}\hat{Q} + \frac{1}{\sqrt{2L\hbar\Omega}}\hat{\phi}, \tag{3.38}$$

$$\hat{a}^\dagger = -i\frac{1}{\sqrt{2C\hbar\Omega}}\hat{Q} + \frac{1}{\sqrt{2L\hbar\Omega}}\hat{\phi}, \tag{3.39}$$

which obey the usual relation

$$[\hat{a}, \hat{a}^\dagger] = 1. \tag{3.40}$$

The charge and flux operators can be expressed in terms of the raising and lowering operators as

$$\hat{Q} = -iQ_{\text{ZPF}}\,(\hat{a} - \hat{a}^\dagger), \tag{3.41}$$

$$\hat{\phi} = \Phi_{\text{ZPF}}\,(\hat{a} + \hat{a}^\dagger), \tag{3.42}$$

where

$$Q_{\text{ZPF}} = \sqrt{\frac{C\hbar\Omega}{2}} = \sqrt{\frac{\hbar}{2Z}}, \tag{3.43}$$

$$\Phi_{\text{ZPF}} = \sqrt{\frac{L\hbar\Omega}{2}} = \sqrt{\frac{\hbar Z}{2}}, \tag{3.44}$$

where Z is the characteristic impedance of the oscillator

$$Z = \sqrt{\frac{L}{C}}. \tag{3.45}$$

Notice that the notation has been chosen such that the quantum ground-state uncertainties in charge and flux are given by

$$\langle 0|\hat{Q}^2|0\rangle = Q_{\text{ZPF}}^2 \tag{3.46}$$

$$\langle 0|\hat{\phi}^2|0\rangle = \Phi_{\text{ZPF}}^2. \tag{3.47}$$

Exercise 3.2 There is a certain arbitrariness in the choice of phase factors that enter in definition of the raising and lowering operators in Eq. (3.42). We have chosen a convention in which the flux is related to the real part of $\hat{a}$ and the charge is related to the imaginary part of $\hat{a}$. Consider the unitary transformation $U = e^{i\theta\hat{n}}$, where $\hat{n} = a^\dagger a$ is the photon number operator. What does this transformation do to the Fock state $|n\rangle$? How do the raising and lowering operators transform under the action of U? What happens to the expressions for charge and flux under the transformation of U when $\theta = \pi/2$?

Using the superconducting resistance quantum

$$R_{\text{Q}} \equiv \frac{h}{(2e)^2} \approx 6,453.20\,\text{ohm}, \tag{3.48}$$

we can define a dimensionless characteristic impedance

$$z \equiv Z/R_{\rm Q}, \tag{3.49}$$

to obtain

$$Q_{\rm ZPF} = (2e)\sqrt{\frac{1}{4\pi z}}, \tag{3.50a}$$

$$\Phi_{\rm ZPF} = \Phi_0\sqrt{\frac{z}{4\pi}}, \tag{3.50b}$$

where

$$\Phi_0 \equiv \frac{h}{2e} \tag{3.51}$$

is the superconducting flux quantum. Notice that the usual uncertainty product is obeyed:

$$Q_{\rm ZPF}\Phi_{\rm ZPF} = \frac{\hbar}{2}. \tag{3.52}$$

The voltage is an important physical variable, and the voltage operator is given by

$$\begin{aligned} \hat{V} &= \frac{d\hat{\phi}}{dt} = \frac{i}{\hbar}[H, \hat{\phi}] \\ &= \frac{1}{C}\hat{Q} = -i\sqrt{\frac{\hbar\Omega}{2C}}\left(\hat{a} - \hat{a}^\dagger\right) = -iV_{\rm ZPF}\left(\hat{a} - \hat{a}^\dagger\right), \end{aligned} \tag{3.53}$$

where

$$V_{\rm ZPF} = \Omega\Phi_{\rm ZPF} = \Omega\Phi_0\sqrt{\frac{z}{4\pi}}. \tag{3.54}$$

The superconducting flux quantum in convenient units is given by

$$\Phi_0 \approx 2.06783367\,\mu\text{V/GHz} \tag{3.55}$$

which tells us that the vacuum fluctuations of the voltage across the capacitor in a typical 10 GHz, $Z = 100$ ohm impedance resonator circuit will be on the scale of $(1/3)\,\mu$V. Correspondingly the vacuum fluctuations of the current are on the scale of 3 nA. It is remarkable that the quantum fluctuations of currents and voltages in these microwave circuits have the same scales as are routinely measured in the audio range with standard laboratory instruments.

How do we interpret the excitation quanta of this harmonic oscillator? We can think of these as excitations of the collective motion of the electrons in the wire, or we can think of them as photons of the electromagnetic field. Because this is a lumped-element resonator (as opposed to a cavity or other distributed resonator), the electric field appears between the capacitor plates and the magnetic field appears in a separate place, namely, within the coil of the inductor. Nevertheless, it is perfectly acceptable to think of these excitations as photons. The coordinate of the oscillator is the flux in the coil (or, in the first choice we made, the charge on the capacitor plates, which is equivalent to the electric field in the gap between the plates).

One does not normally think about photons in the context of first quantization, but this is also useful for building up intuition and for thinking about things such as the full probability distribution of electric field measurement results. The wavefunction of the vacuum state is a Gaussian in the coordinate ϕ, as shown in Fig. 3.3:

$$\Psi_0(\phi) = \frac{1}{(2\pi\Phi_{\mathrm{ZPF}}^2)^{1/4}} \exp\left(-\frac{1}{4}\frac{\phi^2}{\Phi_{\mathrm{ZPF}}^2}\right). \tag{3.56}$$

If, in the vacuum state, we make a precise measurement of the flux, the resulting value will be random and have a Gaussian probability distribution given by

$$P(\phi) = |\Psi_0(\phi)|^2. \tag{3.57}$$

Hence, the most probable value of the flux is zero. On the other hand, in the one-photon state,

$$\Psi_1(\phi) = \frac{\phi}{\Phi_{\mathrm{ZPF}}} \frac{1}{(2\pi\Phi_{\mathrm{ZPF}}^2)^{1/4}} \exp\left(-\frac{1}{4}\frac{\phi^2}{\Phi_{\mathrm{ZPF}}^2}\right), \tag{3.58}$$

zero flux would never be measured because the wavefunction vanishes at $\phi = 0$. The measured flux is still zero on average. This is true for any (odd) photon Fock state (number eigenstate) from simple parity considerations. On the other hand, if the photon number is uncertain, for example in the coherent superposition state

$$\Psi_+ = \frac{1}{\sqrt{2}}(\Psi_0 + \Psi_1), \tag{3.59}$$

then the centroid of the probability distribution is displaced away from zero as shown in Fig. 3.3 and the average value of the flux will be nonzero. A similar conclusion is readily reached within the second-quantized formulation of Eq. (3.42) by noticing that the flux and charge operators are purely off-diagonal in the photon number basis.

Such superpositions of zero- and one-photon states cannot be achieved by simply weakly driving the oscillator, as this produces a coherent superposition of all photon

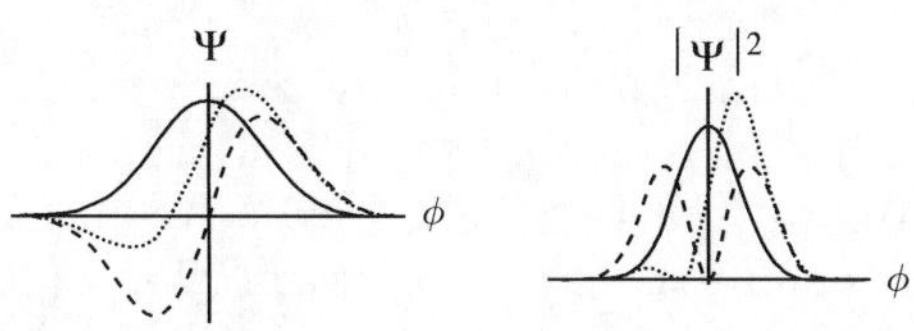

Fig. 3.3 *LC* oscillator wavefunction amplitude (a) and probability density (b) plotted versus the coordinate ϕ. Solid lines, ground state, Ψ_0; long-dashed lines, first excited state, Ψ_1; short-dashed lines, linear combination of the ground and first excited states, $\frac{1}{\sqrt{2}}(\Psi_0 + \psi_1)$.

number states (to be described further below). However they have been achieved experimentally [13–15] by applying control pulses to a qubit to put it into a superposition of the ground state $|g\rangle$ and the excited state $|e\rangle$:

$$|\psi_{\text{initial}}\rangle = \alpha|g\rangle + \beta|e\rangle. \tag{3.60}$$

Allowing the qubit to spontaneously decay (if it is excited) leaves the qubit in the ground state and the electromagnetic field in a superposition of zero and one photon with coefficients α and β inherited from the qubit

$$|\psi_{\text{final}}\rangle = |g\rangle\,[\alpha|0\rangle + \beta|1\rangle]. \tag{3.61}$$

This operation maps a stationary qubit onto a "flying qubit" (the photon) and is an essential step toward communicating quantum information via photons. In the experiment of Houck et al. [13], the photons could be sent into a square-law detector to measure the photon number or into a homodyne detector to measure either quadrature of the electric field (equivalent to measuring $\hat{Q}$ or $\hat{\phi}$ in Eq. (3.42)). The experiment directly showed that the one-photon Fock state had zero electric field on average and that the phase of the electric field for superposition states was determined by the phase imposed initially upon the qubit superposition state. We tend to think of spontaneous emission as an incoherent process, but the above results show that this is not entirely correct. What we really mean by "incoherent" is that the decay of an atom that starts purely in the excited state yields a photon state that varies randomly from shot to shot and which vanishes only on average.

In the UCSB experiments [14, 15], complex superpositions of resonator Fock states were engineered and then measured via the their effect on the state of the qubit, rather than by homodyne measurement of the photon state.

Driven LC oscillators Before continuing, it is useful to return to the classical circuit analysis and think about how we should include a driving force on the oscillator. Returning to Fig. 3.2, let us consider adding a signal source to the circuit at the node labelled ϕ as shown in Fig. 3.4(a). The first question we have to answer is whether we should use a voltage source or a current source. Ideally, the former has zero impedance and the latter has infinite impedance. A voltage source set to zero drive amplitude would short the ϕ node to ground and ruin the oscillator. Conversely, a current source set to zero drive amplitude would have no effect on the oscillator at all since the voltage oscillations would not be damped by the infinite impedance of the current source. Thus, we should use a current source that will minimize the damping. (Generically, resonators will be driven through a coupling capacitor or antenna structure connected to an approximately 50 ohm transmission line, which will introduce some damping.) For the moment, we will assume the drive is classical. (We will say more on the meaning of "classical" later.)

Consider the following modification of the Lagrangian in Eq. (3.29):

$$\mathcal{L} = \frac{1}{2}C\dot{\phi}^2 - \frac{1}{2L}\phi^2 + I_b\phi, \tag{3.62}$$

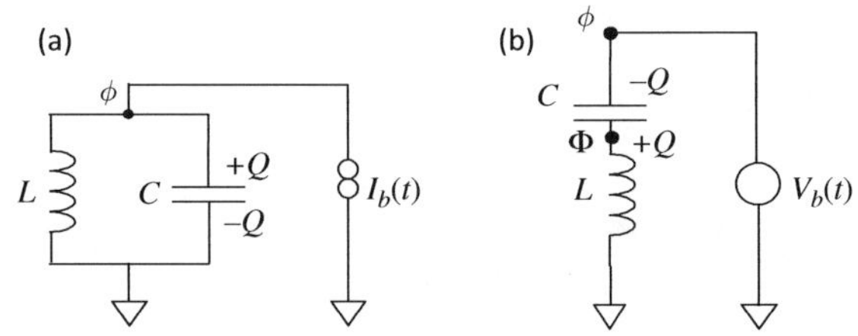

Fig. 3.4 (a) Parallel LC oscillator driven at the node ϕ by a classical external current source with infinite impedance. (b) Series LC oscillator driven at the node ϕ by a classical external voltage source with zero impedance.

where $I_b(t)$ is the (classical) time-dependent bias current delivered by the source. We can think of the third term as a Lagrange multiplier that enforces current conservation. From the Euler-Lagrange equation of motion

$$-\frac{d}{dt}\frac{\delta \mathcal{L}}{\delta \dot{\phi}} + \frac{\delta \mathcal{L}}{\delta \phi} = 0, \tag{3.63}$$

we obtain

$$\dot{Q} + \frac{\phi}{L} = I_b(t), \tag{3.64}$$

which is simply the equation for current conservation at the ϕ node. Converting the Lagrangian to the classical Hamiltonian yields

$$H = \frac{Q^2}{2C} + \frac{\phi^2}{2L} - I_b(t)\phi, \tag{3.65}$$

we see that the bias current acts as a force conjugate to the coordinate ϕ. We can view the current conservation equation (3.64) as Hamilton's equation of motion giving the time rate of change of the momentum in terms of the sum of the oscillator spring force plus the external force:

$$\dot{Q} = -\frac{\phi}{L} + I_b(t). \tag{3.66}$$

So far, we have only considered the parallel LC resonator. We turn now to the series resonator illustrated in Fig. 3.4(b). Clearly, there can be no oscillations unless the node ϕ is connected to ground so that current can flow. This means that the series resonator should be driven by a zero-impedance voltage source instead of a current source. The Lagrangian for this system is

$$\mathcal{L} = \frac{1}{2}C(\dot{\Phi} - \dot{\phi})^2 - \frac{\Phi^2}{2L}, \tag{3.67}$$

from which it follows that the Hamiltonian is

$$H = \frac{Q^2}{2C} + \frac{\Phi^2}{2L} + V_b(t)Q, \tag{3.68}$$

where $\dot{\phi} = V_b(t)$ is fixed by the bias voltage. In this case, the external control parameter is the voltage rather than the drive current, and the internal variable being controlled is the charge rather than the flux.

Exercise 3.3 Rederive the Lagrangian and the Hamiltonian for the series resonator shown in Fig. 3.4(b) except with the capacitor and inductor interchanged so that the external voltage source is attached to the inductor rather than the capacitor. The physics should be identical to the previous case, but the mathematical expressions will look rather different. Can you find a change of coordinates that maps the problem back to the previous form?

Coherent states Now that we understand the classically driven quantum harmonic oscillator, we are in a position to study coherent states of oscillation. A simple way to achieve a superposition of different number states in a quantum oscillator is to drive it with a classical external driving force so that the ground state is displaced and mapped to a so-called "coherent state"

$$\Psi_0(\phi) \longrightarrow \Psi_\Delta(\phi) = \Psi_0(\phi - \Delta). \tag{3.69}$$

Coherent states are discussed below and in further detail in Appendix 3.9.5. In addition to having coherent states displaced in position, one can also have them displaced in momentum. These simply correspond to being in different parts of the classical oscillation cycle. We also discuss below what it means to have a "classical" drive.

Using the Taylor series expansion to all orders, we can write the unitary transformation that displaces the state as

$$\Psi_\Delta(\phi) = e^{-\Delta\partial/\partial\phi}\Psi_0(\phi) \tag{3.70}$$

$$= e^{-(i/\hbar)\Delta\hat{Q}}\Psi_0(\phi), \tag{3.71}$$

which illustrates the fact that the momentum $\hat{Q}$ is the generator of displacements of its conjugate coordinate ϕ. The unitary displacement operator may be written as

$$U_\alpha = e^{-(i/\hbar)\Delta\hat{Q}} = e^{-\alpha(\hat{a}-\hat{a}^\dagger)}, \tag{3.72}$$

where the dimensionless displacement parameter is

$$\alpha \equiv \frac{\Delta Q_{\mathrm{ZPF}}}{\hbar} = \frac{\Delta}{2\Phi_{\mathrm{ZPF}}}. \tag{3.73}$$

Now using the Feynman disentangling theorem [30], derived in the appendix (Section 3.9.4), this can be normal-ordered:

$$U_\alpha = e^{+\alpha\hat{a}^\dagger}e^{-\alpha\hat{a}}e^{-\frac{1}{2}|\alpha|^2}. \tag{3.74}$$

Taking advantage of the fact that $\hat{a}|0\rangle = 0$, we see that in second-quantized notation the coherent state becomes

$$|\alpha\rangle = e^{-\frac{1}{2}|\alpha|^2} e^{\alpha\hat{a}^\dagger}|0\rangle. \tag{3.75}$$

Exercise 3.4 Since U_α is unitary, it must be that $\langle\alpha|\alpha\rangle = 1$. Verify this by direct calculation from Eq. (3.75).

Coherent states have some very nice properties. For example, because they are special coherent superpositions of all possible photon numbers, they are eigenstates of the photon destruction operator:

$$\hat{a}|\alpha\rangle = \alpha|\alpha\rangle. \tag{3.76}$$

You can destroy a photon and still be in the same state! Curiously, coherent states are *not* eigenstates of $\hat{a}^\dagger$. It is clear that $\hat{a}^\dagger|\alpha\rangle$ has no amplitude for zero photons and hence is linearly independent of $|\alpha\rangle$ (and therefore not an eigenstate). One can reach the same conclusion by noting that $\hat{a}$ and $\hat{a}^\dagger$ do not commute:

$$[\hat{a}, \hat{a}^\dagger]|\alpha\rangle = |\alpha\rangle \neq 0. \tag{3.77}$$

On the other hand, it is true that the mean phonon number is given by

$$\bar{N} = \langle\alpha|\hat{a}^\dagger\hat{a}|\alpha\rangle = |\alpha|^2. \tag{3.78}$$

The phonon number distribution in a coherent state is given by the standard Poisson distribution

$$P_n = |\langle n|\alpha\rangle|^2 = \frac{\bar{N}^n}{n!}e^{-\bar{N}}. \tag{3.79}$$

Exercise 3.5 Derive Eqs. (3.76) and (3.79).

Because $\hat{a}^\dagger$ is a raising operator for the energy, the coherent state has a very simple time evolution even though it is itself not an energy eigenstate. The displacement parameter α becomes complex and its phase increases linearly in time. That is, the real and imaginary parts of α simply vary sinusoidally in time, indicating that the displacement alternates between position and momentum:

$$|\alpha(t)\rangle = |e^{-i\Omega t}\alpha(0)\rangle = e^{-\frac{1}{2}|\alpha|^2} e^{\alpha e^{-i\Omega t}\hat{a}^\dagger}|0\rangle. \tag{3.80}$$

This corresponds in the classical limit to the circular motion in phase space of the simple harmonic oscillator.

Rather than working with $\hat{\phi}$ and $\hat{Q}$, we will find it convenient to work with the dimensionless quadrature amplitudes

$$\hat{X} \equiv \frac{1}{2}\left(\hat{a}+\hat{a}^\dagger\right) \tag{3.81}$$

$$\hat{Y} \equiv -i\frac{1}{2}\left(\hat{a}-\hat{a}^\dagger\right). \tag{3.82}$$

These Hermitian operators are effectively the real and imaginary parts of the $\hat{a}$. Like $\hat{\phi}$ and $\hat{Q}$, they are canonically conjugate, with commutator

$$[\hat{X},\hat{Y}] = +\frac{i}{2}, \tag{3.83}$$

and for coherent states obey

$$\langle\alpha|\hat{X}|\alpha\rangle = \mathrm{Re}\,\alpha(t), \tag{3.84}$$

$$\langle\alpha|\hat{Y}|\alpha\rangle = \mathrm{Im}\,\alpha(t), \tag{3.85}$$

$$\langle\alpha|(\hat{X}-\langle\hat{X}\rangle)^2|\alpha\rangle = \langle 0|(\Delta\hat{X})^2|0\rangle = \frac{1}{4}, \tag{3.86}$$

$$\langle\alpha|(\hat{Y}-\langle\hat{Y}\rangle)^2|\alpha\rangle = \langle 0|(\Delta\hat{Y})^2|0\rangle = \frac{1}{4}. \tag{3.87}$$

Equations (3.86) and (3.87) show that there are quantum fluctuations in $\hat{X}$ and $\hat{Y}$ (as there must be, since they do not commute with each other). The resulting uncertainties in the measured values of these quantities play a central role in understanding quantum noise [31]. The energy of the oscillator (in units of $\hbar\Omega$) is

$$\hat{\epsilon} = \hat{X}^2+\hat{Y}^2 = \hat{N}+\frac{1}{2}, \tag{3.88}$$

so the number operator is simply

$$\hat{N} = \hat{X}^2+\hat{Y}^2-\frac{1}{2}. \tag{3.89}$$

To understand the fluctuations in photon number, let us consider a coherent state with amplitude $\alpha = \sqrt{N}$, which is real. As illustrated in Fig. 3.5, fluctuations in $\hat{X}$ lead to photon number fluctuations (fluctuations in the length of the phasor), while fluctuations in $\hat{Y}$ lead to fluctuations in the phase of the coherent state as measured in homodyne detection [31]. As we have seen, the coherent state is nothing more than a displaced vacuum state

$$|\alpha\rangle = U_\alpha|0\rangle. \tag{3.90}$$

Instead of actively displacing the physical system, we can equivalently leave the system alone and displace the coordinate system, transforming all operators according to the usual rule

$$\tilde{a} = U_\alpha^\dagger \hat{a} U_\alpha = \hat{a}+\alpha, \tag{3.91}$$

$$\tilde{a}^\dagger = U_\alpha^\dagger \hat{a}^\dagger U_\alpha = \hat{a}^\dagger+\alpha^*. \tag{3.92}$$

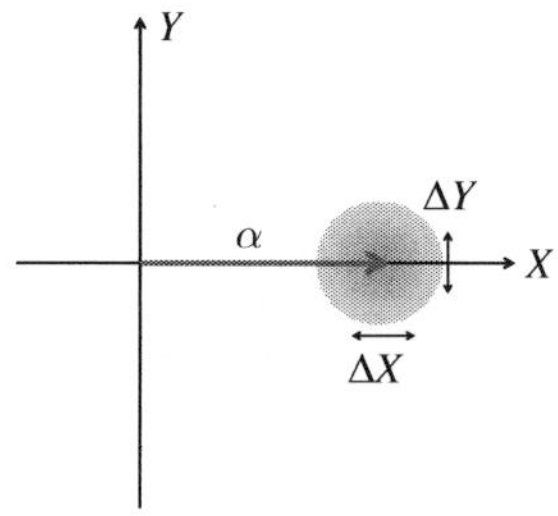

Fig. 3.5 Quantum fluctuations of amplitude and phase quadratures in a coherent state $|\alpha\rangle$.

Note that the analog of Eq. (3.76) is

$$\tilde{a}|0\rangle = \alpha|0\rangle. \tag{3.93}$$

We commonly refer to α as the classical amplitude of the motion, and if $|\alpha| \gg 1$, it dominates over the quantum fluctuations around the classical value of the amplitude. As mentioned earlier, weakly coupling a system to an oscillator mode in a large amplitude coherent state produces what is effectively a classical drive with negligible quantum fluctuations. For example, we might apply a force $\hat{F}$ to an oscillator whose coordinate is $\hat{y} = y_{\mathrm{ZPF}}(\hat{b} + \hat{b}^\dagger)$ via the coupling

$$\hat{V} = -\hat{F}\hat{y}. \tag{3.94}$$

For the case in which the force is supplied by linear coupling to a second "drive" oscillator whose position operator is $\hat{x} = x_{\mathrm{ZPF}}(\hat{a} + \hat{a}^\dagger)$, the Hamiltonian would have the generic form

$$H = \omega_{\mathrm{R}}\hat{b}^\dagger\hat{b} + \omega_{\mathrm{d}}\hat{a}^\dagger\hat{a} + g(\hat{a} + \hat{a}^\dagger)(\hat{b} + \hat{b}^\dagger). \tag{3.95}$$

Changing to a frame rotating with the drive oscillator via the unitary transformation

$$\hat{U} = e^{+i\omega_{\mathrm{d}}t\hat{a}^\dagger\hat{a}}, \tag{3.96}$$

the Hamiltonian becomes

$$H_1 = UHU^\dagger + U\left[-i\frac{d}{dt}, U^\dagger\right] = \omega_{\mathrm{R}}\hat{b}^\dagger\hat{b} + g\left(e^{-i\omega_{\mathrm{d}}t}\hat{a} + e^{-i\omega_{\mathrm{d}}t}\hat{a}^\dagger\right)(\hat{b} + \hat{b}^\dagger). \tag{3.97}$$

If the drive oscillator is initially placed in a high-amplitude coherent state, it is convenient to make the displacement transformation in Eq. (3.92) to obtain the transformed coupling Hamiltonian

$$H_2 = \omega_{\mathrm{R}}\hat{b}^\dagger\hat{b} + g\left(e^{-i\omega_{\mathrm{d}}t}\alpha + e^{-i\omega_{\mathrm{d}}t}\alpha^*\right)(\hat{b} + \hat{b}^\dagger) + H_{\mathrm{Q}}. \tag{3.98}$$

We see in the first two terms that the system oscillator is quantum and subject to a classical drive. The last term describes the quantum fluctuations associated with the drive:

$$H_{\mathrm{Q}} = g\left(e^{-i\omega_{\mathrm{d}}t}\hat{a} + e^{-i\omega_{\mathrm{d}}t}\hat{a}^\dagger\right)(\hat{b} + \hat{b}^\dagger). \tag{3.99}$$

Because (initially at least) the drive oscillator is now in the ground state (in the new frame), the quantum fluctuations of the drive are small compared with the classical part, if (in the original frame) the drive amplitude corresponds to a state with many quanta: $\bar{n} = |\alpha|^2 \gg 1$. This will continue to remain true over time provided that that the drive strength $g|\alpha|$ and the detuning $\omega_{\mathrm{d}} - \omega_{\mathrm{R}}$ are such that the number of quanta transferred from the drive to the system via the action of H_{Q} remains much smaller than $\bar{n}$.

A good example of this physics is provided by a two-port resonator with one weakly coupled port and one strongly coupled port. The damping of the resonator will be controlled by the port strongly coupled to the environment, since most photons will escape through that port. If the system is continuously driven at the weakly coupled port, most photons from the drive line will be reflected, so a relatively large coherent drive from a microwave signal generator is required to excite the resonator cavity. This corresponds to the limit described above of small g and large α, for which the classical approximation is valid. All we require is that the power in the incoming drive wave be mostly reflected so that it greatly exceeds the power emitted by the driven resonator from its strongly coupled port. In the theory of parametric amplifiers, this is known as the "stiff pump" limit. No matter what the driven system does, the pump amplitude stays fixed and essentially classical.

Exercise 3.6 Derive Eqs. (3.91) and (3.92) by differentiating with respect to α and solving the resulting differential equation.

Exercise 3.7 Solve the Heisenberg equation of motion for $\hat{b}$ using the Hamiltonian in Eq. (3.98) but neglecting the quantum fluctuation term H_{Q}. Show that this classical drive applied to an oscillator initially in a coherent state (including possibly the vacuum state) always leaves the system in a coherent state.

Exercise 3.8 Show by direct computation that for the Bose-Einstein number distribution for a thermal photon state,

$$\langle\langle(\hat{N} - \bar{N})^2\rangle\rangle = \bar{N}(\bar{N} + 1). \tag{3.100}$$

If you are familiar with Wick's theorem, use that to achieve the same result.

Exercise 3.9 $\hat{X}^2$ and $\hat{Y}^2$ are clearly Hermitian operators with non-negative eigenvalues. How then can you explain the fact that

$$\langle 0|\hat{X}^2\hat{Y}^2|0\rangle = -\frac{1}{16} \tag{3.101}$$

is negative? Similarly, how can

$$\langle 0|\hat{X}\hat{Y}|0\rangle = \frac{i}{4} \tag{3.102}$$

be complex?

Writing the X quadrature amplitude as

$$\hat{X} = \alpha + \Delta\hat{X}, \tag{3.103}$$

we see that $\Delta\hat{X}$ has the same statistical properties in the coherent state $|\alpha\rangle$ as $\hat{X}$ does in the vacuum state. The number fluctuations are therefore given by the usual Poisson distribution result derived above:

$$\langle\alpha|(\hat{N} - \bar{N})^2|\alpha\rangle = \langle\alpha|\left(2\alpha\Delta\hat{X} + \Delta\hat{X}^2 + \Delta\hat{Y}^2 - \frac{1}{2}\right)^2|\alpha\rangle = \bar{N}. \tag{3.104}$$

Essentially, the above results mean that a coherent laser or microwave beam is as classical as possible. The fluctuations come only from the fact that the photon detection events are discrete and the photons are sprinkled randomly throughout the beam in an uncorrelated manner. A thermal beam has larger fluctuations because the photons tend to bunch together [31].

Fluctuations in the quadrature orthogonal to α cause uncertainty in a measurement of the phase of the coherent state. For the case of α real and in the limit $|\alpha| \gg 1$, we have[6]

$$\Delta\hat{\theta} \approx \frac{\Delta\hat{Y}}{\alpha} \tag{3.105}$$

and

$$\langle\alpha|(\Delta\hat{\theta})^2|\alpha\rangle = \frac{1}{4\bar{N}}. \tag{3.106}$$

Thus, we arrive at the fundamental number-phase uncertainty relation

$$\langle\alpha|(\Delta\hat{\theta})^2|\alpha\rangle^{1/2}\langle\alpha|(\Delta\hat{N})^2|\alpha\rangle^{1/2} \geq \frac{1}{2}. \tag{3.107}$$

Coherent states are minimum-uncertainty Gaussian states that satisfy this relation as an equality. Other non-Gaussian states satisfy this relation only as an inequality.

From the equation of motion of the free oscillator, we see that the quadrature amplitudes obey

$$\hat{X}(t) = \cos(\Omega t)\hat{X}(0) + \sin(\Omega t)\hat{Y}(0), \tag{3.108}$$
$$\hat{Y}(t) = \cos(\Omega t)\hat{Y}(0) - \sin(\Omega t)\hat{X}(0). \tag{3.109}$$

[6] The "phase" operator defined here does not have the angular periodicity of a phase and is only valid for small angles.

In the appendix (Section 3.9.2), we study photons traveling in transmission lines and we again find that the traveling modes are also harmonic oscillators. The above results provide the first hint that the $\sin\Omega t$ and $\cos\Omega t$ quadratures of a quantum electrical signal are canonically conjugate and hence cannot be simultaneously measured with perfect accuracy. Equivalently, even the vacuum contains noise that will appear in any measurement in which one attempts to measure both quadratures of the signal. Equation (3.87) tells us that this uncertainty gives a vacuum "noise energy" (noise power per unit measurement bandwidth) of half a photon [31].

Exercise 3.10 Think through the above statement about noise energy at the classical level. Consider a noise source that is white (i.e., with constant spectral density S) over some large interval. Passing this noise through a filter that transmits a small bandwidth B centered on frequency ω will yield a power $P = SB$. The wider the bandpass, the more power. Thus, we see that the spectral density is power per unit bandwidth, which has units of energy. For a quantum thermal source feeding a photomultiplier (which measures $\hat{a}^\dagger\hat{a}$), this is $S = \hbar\omega\bar{N}$, and we say that "the noise energy is $\bar{N}$ photons." A photomultiplier fed by vacuum noise has zero output. However, listening to the vacuum noise power through a phase-preserving amplifier or (equivalently) using a heterodyne detector that measures the power in the quadrature amplitudes $\langle\hat{X}^2 + \hat{Y}^2\rangle = \frac{1}{2}$ yields a noise energy of half a photon [31].

3.2.4 Coupled LC resonators

Having thoroughly analyzed the simple LC oscillator, it is a useful exercise to consider how to quantize a pair of LC oscillators connected by a coupling capacitor as shown in Fig. 3.6. This will teach us how to handle slightly more complex circuits and will set the stage for understanding the coupling of a qubit to a microwave resonator.

Choosing the fluxes Φ_1 and Φ_2 as the coordinates of the two oscillators, the Lagrangian can be written

$$\mathcal{L} = \frac{1}{2}C_1\dot{\Phi}_1^2 + \frac{1}{2}C_2\dot{\Phi}_2^2 + \frac{1}{2}C_0\left(\dot{\Phi}_1 - \dot{\Phi}_2\right)^2 - \frac{1}{2L_1}\Phi_1^2 - \frac{1}{2L_2}\Phi_2^2. \tag{3.110}$$

It is convenient to use a matrix notation

$$\mathcal{L}\frac{1}{2}\dot{\Phi}C\dot{\Phi} - \frac{1}{2}\Phi L^{-1}\Phi, \tag{3.111}$$

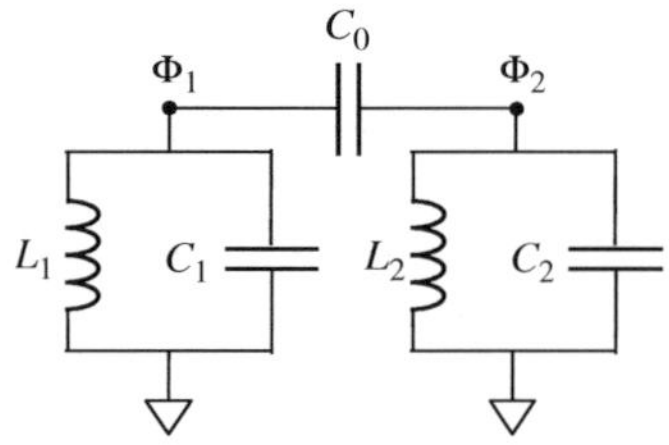

Fig. 3.6 A pair of LC oscillators connected by a coupling capacitor C_0.

where the capacitance matrix is

$$C \equiv \begin{pmatrix} C_1 + C_0 & -C_0 \\ -C_0 & C_2 + C_0 \end{pmatrix} \tag{3.112}$$

and the inverse inductance matrix is

$$L^{-1} \equiv \begin{pmatrix} \frac{1}{L_1} & 0 \\ 0 & \frac{1}{L_2} \end{pmatrix}. \tag{3.113}$$

At this point, there are two ways to proceed, which are described in the following.

3.2.4.1 Method I: find the Hamiltonian, then diagonalize

In the first method, we will use the given coordinates to find the canonical momenta and from there construct the Hamiltonian, which will contain a coupling between the two oscillators.

The canonical momenta are given by

$$Q_i \equiv \frac{\delta \mathcal{L}}{\delta \dot{\Phi}_i} = C_{ij} \dot{\Phi}_j, \tag{3.114}$$

where we employ the Einstein summation convention for repeated indices. In terms of the inverse of the capacitance matrix, we have

$$\dot{\Phi} = C^{-1} Q. \tag{3.115}$$

The Hamiltonian $H = Q_i \dot{\Phi}_i - \mathcal{L}$ now takes the canonical form

$$H = \frac{1}{2} Q C^{-1} Q + \frac{1}{2} \Phi L^{-1} \Phi. \tag{3.116}$$

The inverse of the capacitance matrix is

$$C^{-1} = \frac{1}{C_1 C_2 + C_0 C_1 + C_0 C_2} \begin{pmatrix} C_2 + C_0 & +C_0 \\ +C_0 & C_1 + C_0 \end{pmatrix}, \tag{3.117}$$

It is useful to define two frequencies and a coupling constant:

$$\omega_j^2 == \frac{1}{L_j} \left(C^{-1}\right)_{jj} \tag{3.118}$$

and

$$\beta \equiv \frac{C_0}{\sqrt{(C_1 + c_0)(C_2 + C_0)}}, \tag{3.119}$$

which yields

$$C^{-1} = \begin{pmatrix} L_1\omega_1^2 & +\beta\sqrt{L_1L_2}\omega_1\omega_2 \\ \beta\sqrt{L_1L_2}\omega_1\omega_2 & L_2\omega_2^2 \end{pmatrix}. \tag{3.120}$$

We can now write the Hamiltonian $H = H_0 + V$ in terms of two oscillators with masses L_j and coupled through their momenta:

$$H_0 = \frac{1}{2}L_1\omega_1^2Q_1^2 + \frac{1}{2L_1}\Phi_1^2 + \frac{1}{2}L_2\omega_1^2Q_2^2 + \frac{1}{2L_2}\Phi_2^2 \tag{3.121}$$

$$V = \beta\sqrt{L_1L_2}\omega_1\omega_2Q_1Q_2. \tag{3.122}$$

We quantize as usual by converting to operators with the canonical commutation relation

$$[\hat{Q}_i, \hat{\Phi}_j] = -i\hbar\delta_{ij}. \tag{3.123}$$

Defining creation and annihilation operators in the usual way, we have

$$H_0 = \sum_{j=1}^{2}\hbar\omega_j\left(\hat{a}_j^\dagger\hat{a}_j + \frac{1}{2}\right), \tag{3.124}$$

$$V = -\beta\hbar\sqrt{\omega_1\omega_2}(\hat{a}_1 - \hat{a}_1^\dagger)(\hat{a}_2 - \hat{a}_2^\dagger), \tag{3.125}$$

which can be diagonalized via a Bogoljubov transformation.

Exercise 3.11 Find the Bogoljubov transformation that diagonalizes $H_0 + V$ defined in Eqs. (3.124) and (3.125).

3.2.4.2 Method II: diagonalize the Lagrangian, then the Hamiltonian

The first method used the original coordinates and found their canonical momenta and from there constructed the (nondiagonal) Hamiltonian. In the second method, we will find the normal-mode coordinates that diagonalize the Lagrangian. In terms of these, the Hamiltonian will be automatically diagonal.

When we try to diagonalize the Lagrangian in Eq. (3.111), we are faced with the problem that the capacitance and inductance matrices do not commute and hence cannot be simultaneously diagonalized by a unitary transformation. We can cure this problem by making a *similarity transformation* that maps L^{-1} to the identity matrix. We simply choose scaled coordinates

$$\psi_j = \frac{1}{\sqrt{L_j}}\Phi_j. \tag{3.126}$$

In terms of these, the Lagrangian becomes

$$\mathcal{L} = \frac{1}{2}\dot{\psi}_iA_{ij}\dot{\psi}_j - \frac{1}{2}\psi_i\delta_{ij}\psi_j, \tag{3.127}$$

where

$$A \equiv \begin{pmatrix} \frac{1}{\Omega_1^2} & -\frac{\beta}{\Omega_1\Omega_2} \\ -\frac{\beta}{\Omega_1\Omega_2} & \frac{1}{\Omega_2^2} \end{pmatrix}, \tag{3.128}$$

where we define frequencies (different from those in the previous method)

$$\frac{1}{\Omega_1^2} \equiv L_1(C_1 + C_0), \tag{3.129}$$

$$\frac{1}{\Omega_2^2} \equiv L_2(C_2 + C_0). \tag{3.130}$$

Since A commutes with the identity matrix, we can now proceed as usual to perform a rotation among the coordinates to diagonalize the Lagrangian. Let S be the orthogonal transformation that diagonalizes A. The normal modes and eigenvalues are then given by

$$\tilde{\psi} = S\psi, \tag{3.131}$$

$$\tilde{A} = \begin{pmatrix} \frac{1}{\tilde{\Omega}_1^2} & 0 \\ 0 & \frac{1}{\tilde{\Omega}_2^2} \end{pmatrix} = SAS^T. \tag{3.132}$$

Exercise 3.12 Find the normal modes and eigenfrequencies for Method II. *Hint:* Write $A = \bar{A} + Z\sigma^z + X\sigma^x$ and think of it as a spin problem that has eigenvalues

$$\epsilon_\pm = \bar{A} \pm \sqrt{X^2 + Z^2} \tag{3.133}$$

and eigenfunctions that follow from

$$S = \begin{pmatrix} +\cos(\theta/2) & +\sin(\theta/2) \\ -\sin(\theta/2) & +\cos(\theta/2) \end{pmatrix}, \tag{3.134}$$

where $\tan\theta = X/Z$.

3.2.5 Modes of transmission line resonators

The preceding lengthy discussion of the simple harmonic oscillator has laid the very important groundwork for our next topic which is the quantum modes of transmission lines. We will start with finite-length transmission lines, which have discrete electromagnetic resonances, each of which will turn out to be an independent simple harmonic oscillator. Then, we will move on to the semi-infinite transmission line and discover that it can act like a dissipative bath, even though every one of its electrical elements is nondissipative.

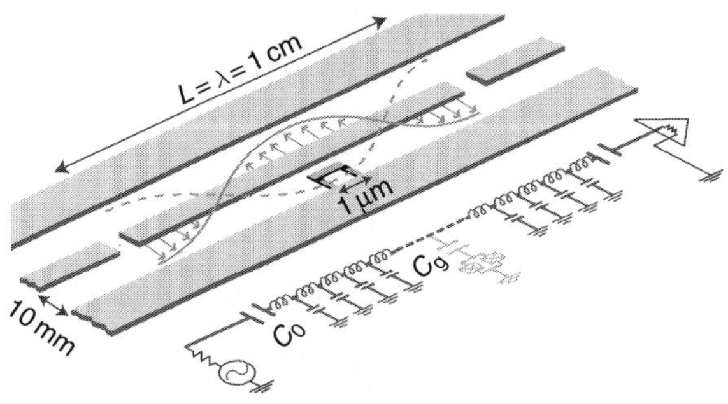

Fig. 3.7 Schematic illustration of a typical coplanar waveguide (CPW) resonator used in circuit QED together with its discretized lumped-element equivalent circuit. The qubit lies between the center pin and the adjacent ground plane and is located at an antinode of the electric field, shown in this case for the full-wave resonance of the CPW. From Blais et al. [32].

Our finite-length transmission line could be a length of ordinary coaxial cable or its two-dimensional equivalent, the coplanar waveguide (CPW), which consists of a superconducting wire evaporated onto an insulating substrate and having superconducting ground planes adjacent to it on the same surface, as shown in Fig. 3.7. Such a system exhibits many standing-wave resonances, and we will soon see that each resonance is an independent harmonic oscillator equivalent to the simple LC oscillator just discussed. The discretized equivalent circuit for the CPW resonator is also shown in Fig. 3.7. In our initial analysis, we will neglect the presence of the qubit and neglect the capacitors C_0 at each end that couple the resonator to the external transmission lines. We can thus assume in this first example open-circuit boundary conditions, for which the current (but not the voltage) vanishes at the ends of the resonator.

It is convenient to define a flux variable analogous to that used above, but now dependent on position [3]:

$$\Phi(x,t) \equiv \int_{-\infty}^{t} d\tau\, V(x,\tau), \tag{3.135}$$

where $V(x,t) = \partial_t \Phi(x,t)$ is the local voltage on the transmission line at position x and time t. The inductance and capacitance per unit length are ℓ and c, respectively. Each segment of the line of length dx has inductance $\ell\, dx$ and the voltage drop along it is $-dx\, \partial_x \partial_t \Phi(x,t)$. The flux through this inductance is thus $-dx\, \partial_x \Phi(x,t)$ and the local value of the current is given by the constitutive equation

$$I(x,t) = -\frac{1}{\ell}\, \partial_x \Phi(x,t). \tag{3.136}$$

The Lagrangian for a system of length L (L is not to be confused with some discrete inductance)

$$\mathcal{L}_g \equiv \int_0^L dx\, \mathcal{L}(x,t) = \int_0^L dx \left[\frac{c}{2} (\partial_t \Phi)^2 - \frac{1}{2\ell} (\partial_x \Phi)^2 \right], \tag{3.137}$$

The Euler-Lagrange equation for this Lagrangian is simply the wave equation

$$v_{\mathrm{p}}^2 \partial_x^2 \Phi - \partial_t^2 \Phi = 0. \tag{3.138}$$

The momentum conjugate to $\Phi(x)$ is simply the charge density

$$q(x,t) \equiv \frac{\delta \mathcal{L}_g}{\delta \partial_t \Phi} = c\partial_t \Phi = cV(x,t), \tag{3.139}$$

and so the Hamiltonian is given by

$$H = \int_0^L dx \left[\frac{1}{2c} q^2 + \frac{1}{2\ell} (\partial_x \Phi)^2 \right]. \tag{3.140}$$

Let us next proceed to consider the classical normal-mode solutions of Eq. (3.138). If we assume a sinusoidal time dependence with angular frequency ω,

$$\Phi(x,t) = e^{-i\omega t} \phi(x), \tag{3.141}$$

we arrive at the Schrödinger-like eigenvalue problem

$$-\partial_x^2 \phi(x) = k^2 \phi(x), \tag{3.142}$$

where $k = \omega / v_{\mathrm{p}}$ and the mode wave velocity is $v_{\mathrm{p}} = 1/\sqrt{\ell c}$. The open-circuit (zero-current) boundary conditions tell us that the eigenfunctions have vanishing derivative at the boundaries. We choose a particular normalization for eigenfunctions which will keep the equations looking as close to those of the single harmonic oscillator as possible

$$\phi_n(x) = \sqrt{2} \cos(k_n x), \tag{3.143}$$

where $n \in \{0, 1, 2, 3, \ldots\}$, $k_n = n\pi/L$. Because for these boundary conditions the operator ∂_x^2 is self-adjoint, and because the eigenvalues are nondegenerate, the eigenfunctions have two helpful properties

$$\int_0^L dx\, \phi_n(x) \phi_m(x) = L\delta_{nm}, \tag{3.144}$$

$$\int_0^L dx\, [\partial_x \phi_n(x)][\partial_x \phi_m(x)] = L k_n^2 \delta_{nm}. \tag{3.145}$$

It follows from this that the Lagrangian can be diagonalized using these (spatial) normal modes as a basis. Let us parameterize the field $\Phi(x,t)$ by

$$\Phi(x,t) = \sum_{n=0}^{\infty} \xi_n(t) \phi_n(x), \tag{3.146}$$

where the ξ_n are *arbitrary* (i.e., not necessarily sinusoidal) functions of time. Substituting into Eq. (3.137) and using Eqs. (3.144) and (3.145),

$$\mathcal{L}_g = \frac{1}{2}Lc\sum_{n=0}^{\infty}[\partial_t\xi_n]^2 - \omega_n^2\xi_n^2, \tag{3.147}$$

we see that each normal mode becomes an independent simple harmonic oscillator. The momentum conjugate to the normal-mode amplitude ξ_n is

$$q_n = \frac{\delta\mathcal{L}_g}{\delta\partial_t\xi_n} = Lc\partial_t\xi_n, \tag{3.148}$$

so the Hamiltonian is

$$H = \frac{1}{2}\sum_{n=0}^{\infty}\left(\frac{1}{Lc}q_n^2 + Lc\omega_n^2\xi_n^2\right), \tag{3.149}$$

which we can quantize as before. Before doing so, let us note that the $n = 0$ mode is a "free particle" rather than a harmonic oscillator because its spring constant vanishes. This mode simply corresponds to a uniform net charge distributed evenly along the transmission line. For a free particle, the momentum (in this case charge) is a constant and the coordinate (flux) increases linearly with time. In most situations, the total charge is indeed simply a constant of the motion (and typically vanishes), and we can ignore the zero mode altogether. We will assume this is the case henceforth.

We end up with a set of independent normal modes with coordinates ξ_n and conjugate momenta q_n that when quantized can be expressed in terms of mode raising and lowering operators in a manner analogous to Eq. (3.42):

$$\hat{\xi}_n = \sqrt{\frac{\hbar}{2\omega_n Lc}}(\hat{a}_n + \hat{a}_n^\dagger), \tag{3.150}$$

$$\hat{q}_n = -i\sqrt{\frac{\hbar\omega_n Lc}{2}}(\hat{a}_n - \hat{a}_n^\dagger), \tag{3.151}$$

where the ladder operators of the different modes obey

$$[\hat{a}_n, \hat{a}_m^\dagger] = \delta_{nm}. \tag{3.152}$$

Note that, just as in the single-mode case in Eq. (3.42), there is a certain arbitrariness in the choice of the phase of the destruction operators (which can be independently varied for each separate mode).

If we are coupling a qubit to a resonator at some particular position x, we need to be able to express the flux and charge density operators at that point in terms of the normal-mode operators. Equation (3.146) is readily extended to the quantum operators,

$$\hat{\Phi}(x) = \sum_n^{\infty}\phi_n(x)\hat{\xi}_n, \tag{3.153}$$

as is Eq. (3.139),

$$\hat{q}(x) = \frac{1}{L}\sum_{n}^{\infty} \phi_n(x)\hat{q}_n. \tag{3.154}$$

Similarly, the analog of Eq. (3.53) for the voltage operator at point x is given by

$$\hat{V}(x) = \frac{1}{c}\hat{q}(x) = \frac{1}{L}\sum_{n=0}^{\infty} \phi_n(x)\hat{q}_n = -i\sum_{n=0}^{\infty}\sqrt{\frac{\hbar\omega_n}{2Lc}}(\hat{a}_n - \hat{a}_n^\dagger)\phi_n(x). \tag{3.155}$$

The total capacitance to ground of the resonator, Lc, enters this expression in a way that is similar to lumped-element oscillator expression in Eq. (3.53). (Recall that L is the length of the resonator, not the inductance.)

Notice that the flux and charge density operators obey the following commutation relation:

$$[\hat{q}(x), \hat{\Phi}(x')] = -i\hbar\frac{1}{L}\sum_{n}^{\infty} \phi_n(x)\phi_n(x'). \tag{3.156}$$

Using the completeness relation (and recalling that the factor of L appears because we did not normalize the eigenfunctions to unity), we end up with the standard field-theoretic relation

$$[\hat{q}(x'), \hat{\Phi}(x)] = -i\hbar\delta(x - x'). \tag{3.157}$$

Expressing the quantum Hamiltonian in Eq. (3.140) in terms of these operators, we have simply

$$\hat{H} = \int_0^L dx \left[\frac{1}{2c}\hat{q}^2 + \frac{1}{2\ell}(\partial_x\hat{\Phi})^2\right]. \tag{3.158}$$

As a “sanity check,” let us look at the Hamilton equations of motion. Using the commutation relation (Eq. (3.157)) and its extension to

$$[\hat{q}(x'), \partial_x\hat{\Phi}(x)] = -i\hbar\partial_x\delta(x - x'), \tag{3.159}$$

we arrive at

$$\partial_t\hat{\Phi}(y) = \frac{i}{\hbar}[\hat{H}, \hat{\Phi}(y)] = \frac{1}{c}\hat{q}(y), \tag{3.160}$$

$$\partial_t\hat{q}(y) = \frac{i}{\hbar}[\hat{H}, \hat{q}(y)] = \frac{1}{\ell}\partial_y^2\hat{\Phi}(y), \tag{3.161}$$

and hence the quantum version of the wave equation (Eq. (3.138)),

$$v_{\mathrm{p}}^2\partial_x^2\hat{\Phi}(x) - \partial_t^2\hat{\Phi}(x) = 0. \tag{3.162}$$

When we studied coherent states of a single oscillator, we found that they were simply the vacuum state displaced in either position (flux) and/or momentum (charge). For a multimode resonator one can coherently displace a linear combination of the of the normal modes. The familiar problem of solving the time evolution of a plucked string is a good classical analog. Suppose that we wish to displace the resonator degrees of freedom so that the local displacement obeys

$$\langle \hat{\Phi}(x) \rangle = \Delta(x), \tag{3.163}$$

where Δ is some specified function. The analog of Eq. (3.72) is simply

$$U_\Delta = e^{-\frac{i}{\hbar}\int_0^L dx\, \Delta(x)\hat{q}(x)}. \tag{3.164}$$

a form that is familiar from the theory of the Luttinger liquid [33, 34]. Using Eq. (3.154), this can be understood in terms of coherent displacement of each of the normal modes:

$$U_\Delta = e^{-\frac{i}{\hbar}\sum_n \Delta_n \hat{q}_n} = \prod_n e^{-\frac{i}{\hbar}\Delta_n \hat{q}_n}. \tag{3.165}$$

Exercise 3.13 In analogy with Eqs. (3.91) and (3.92), show that

$$U_\Delta^\dagger \hat{\Phi}(y) U_\Delta = \hat{\Phi}(y) + \Delta(y). \tag{3.166}$$

Hint: It may be useful to scale $\Delta(x)$ by an overall factor θ and differentiate with respect to θ.

3.2.6 "Black Box" quantization of linear circuits

We have so far studied a single LC oscillator and found that its quantum excitation energy $\hbar\Omega$ is given directly by its classical frequency Ω. We also found in Eq. (3.50) that the characteristic impedance $Z = \sqrt{L/C}$ determines the size of the zero-point fluctuations in flux and charge. The typical circuit that we will study is more complex than a single LC oscillator, and might even be a "black box" whose properties we need to determine. Suppose that we have such a black box and we have access to one port of this structure as shown in Fig. 3.8(a). The only thing we know (or assume) is that all the elements inside the black box are linear and purely reactive; that is, the black box is a network of inductors and capacitors. It might, for example, be a transmission-line resonator such as we studied above. We may ultimately want to connect a qubit or some measurement apparatus to the port of the black box. In order to predict the quantum properties, we need to know each of the normal modes of the box and the size of their zero-point fluctuations as seen at the port. Some modes may be localized inside the box and have very little amplitude at the port. Others may be more strongly coupled to the port.

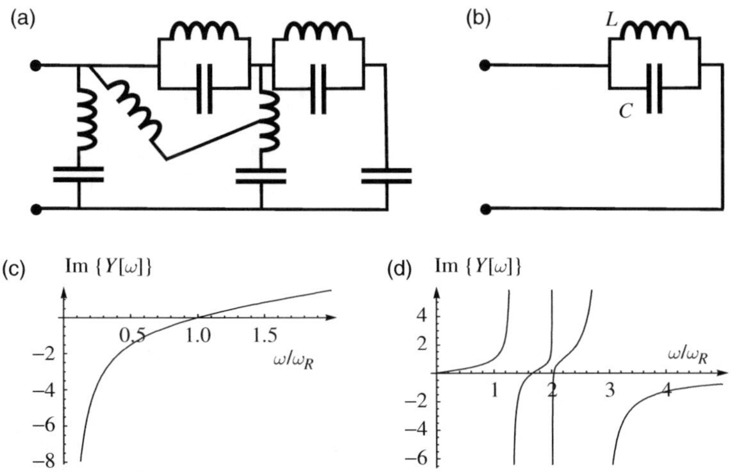

Fig. 3.8 (a) One-port black box containing an arbitrary reactive network. (b) Lumped-element LC resonator. (c) Imaginary part of the admittance of the LC resonator in (b) versus dimensionless frequency, showing that the admittance passes through zero with positive slope at the resonance frequency. (d) Imaginary part of the admittance of a multiresonance circuit with a capacitor in the input line similar to (a). Notice that the slope of the admittance at each of the zeros is different, corresponding to different characteristic impedances of the resonances.

Since the black box is linear, we can probe it by applying a sinusoidal drive and measuring the response. There are two ways to do this. First, one can hook up a current source that forces a current

$$I(t) = i[\omega]e^{j\omega t} + i^*[\omega]e^{-j\omega t} \tag{3.167}$$

through the circuit.[7] The linear response of the circuit is determined by measuring the resulting voltage at the input port,

$$V(t) = v[\omega]e^{j\omega t} + v^*[\omega]e^{-j\omega t}. \tag{3.168}$$

The linear-response coefficient that relates the voltage response to the drive current is known as the impedance:

$$v[\omega] = Z[\omega]i[\omega]. \tag{3.169}$$

Because the box contains only reactive elements (assumed finite in number), the impedance is purely imaginary. The poles of $Z[\omega]$ determine the eigenfrequencies of the

[7] To avoid confusion with the current i we follow the electrical engineering convention of using $j = -\sqrt{-1}$. In addition, to avoid confusion between some function of time and its Fourier transform, we will use the convention that Fourier-transformed quantities have the frequency argument in square brackets.

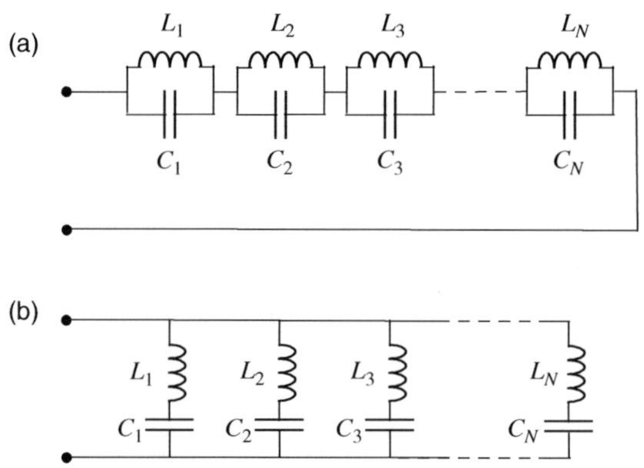

Fig. 3.9 (a) Natural representation of an arbitrary impedance (assuming for simplicity that the impedance vanishes at zero frequency). The jth pole of the impedance occurs at the frequency of the jth collective mode $\omega_j = 1/\sqrt{L_j C_j}$ and can be detected by using an infinite-impedance current source to inject radiofrequency (RF) current into the input port and measuring the resulting RF voltage across the port. (b) Natural representation of an arbitrary admittance (assuming for simplicity that the admittance vanishes at zero frequency). The poles of the admittance determine the natural oscillation frequencies of the circuit when its input is shorted. These can be detected by using a zero-impedance RF voltage source to put a drive voltage across the input port and measuring the resulting RF current that flows into the port.

circuit for which natural oscillations can occur without external input (when the input port is open-circuited). Note that this is consistent with the fact that an ideal current source has infinite internal impedance and hence drives the circuit while effectively keeping the input port open-circuited. The circuit presented in Fig. 3.9(a) is a natural representation of an arbitrary frequency-dependent impedance.[8] It is important to understand that, in general, the circuit elements used in this mathematical representation have no direct correspondence with any of the physical elements in the actual circuit. Note that if there is a pole in the impedance at zero frequency, it corresponds to the "free-particle" Hamiltonian of a capacitor, $H = \hat{Q}^2/2C$, in series with the input (not shown in Fig. 3.9(a)).

The second way to measure the linear response is to attach a zero-impedance voltage source to the input and measure the resulting current response. The linear-response coefficient that relates the current response to the voltage drive is known as the admittance

$$i[\omega] = Y[\omega]v[\omega], \tag{3.170}$$

[8] Note that this particular representation has the property that there is a dc connection through all the inductors to ground. Hence, the impedance vanishes at zero frequency. If this is not the case for the physical circuit, then we must include a series capacitor in the input line. This would be necessary, for example, to represent the impedance of the circuit shown in Fig. 3.8(a).

which is simply the inverse of the impedance:

$$Y[\omega] = Z^{-1}[\omega]. \tag{3.171}$$

The circuit presented in Fig. 3.9(b) is a natural representation of an arbitrary frequency-dependent admittance. The poles of the admittance determine the natural oscillation frequencies of the circuit when its input port is short-circuited. Again, this is consistent with the excitation of these modes, this time using a zero-impedance voltage source. To reiterate, the poles of the admittance (zeros of the impedance) correspond to effective *series* LC resonances that would occur if the input port were short-circuited. These can be important, but for the particular case where nothing is hooked up to the external port, these poles do not correspond to active degrees of freedom. An inductor and capacitor in series cannot oscillate on their own at nonzero frequencies unless the circuit is closed at the input port. Finally, we note that according to Foster's theorem [35], the (imaginary) admittance of a reactive circuit always passes through zero with positive slope, so therefore each zero must be separated from the next by a pole, as shown in Fig. 3.8(d).

Physically, poles of response functions are the most natural thing to consider. However in numerical simulations, zeros are sometimes mathematically easier for a computer to handle than poles. Hence it can be convenient to work with the impedance representation in Fig. 3.9(a) but numerically ascertain the zero-crossings of the admittance rather than the poles of the impedance.

As an example, suppose that the black box contains a single parallel LC oscillator as shown in Fig. 3.8(b). Then the admittance is simply

$$Y[\omega] = j\omega C + \frac{1}{j\omega L} = \frac{+j}{Z_0}\left(\frac{\omega}{\omega_{\mathrm{R}}} - \frac{\omega_{\mathrm{R}}}{\omega}\right), \tag{3.172}$$

where $Z_0 \equiv \sqrt{L/C}$ is the characteristic impedance of the resonance. Note that this is indeed purely imaginary and further that it passes through zero at the resonance frequency $\Omega = 1/\sqrt{LC}$ as shown in Fig. 3.8(c). The admittance is zero because the inductor and capacitor have opposite admittances at the resonance frequency. But this is precisely the condition for self-sustaining oscillation where the currents in the inductor and capacitor are opposite to each other and no external input is needed.

It turns out that knowing the admittance (or impedance) of the box port as a function of frequency completely characterizes the classical and the quantum properties of the black box, as long as it contains only linear elements [36]. We have already seen a hint of this in Eqs. (3.43) and (3.44), where we learned that the characteristic impedance of a resonance determines the zero-point fluctuations of the charge and flux degrees of freedom. Of course, knowing the frequency of an oscillator, we can immediately write down the quantum Hamiltonian (neglecting the zero-point energy)

$$H_0 = \hbar\Omega\hat{a}^\dagger\hat{a}. \tag{3.173}$$

This is not enough, however. If we couple an external circuit to our black box, we need to know the matrix elements of the coupling Hamiltonian. For this, we need to know how to express the charge and flux in terms of a and $a^\dagger$ and hence must know the characteristic impedance of the resonance. Happily, the slope with which the admittance passes through zero determines the characteristic impedance of the resonance,

$$\Omega \left(\frac{\partial Y}{\partial \omega} \right)_\Omega = \frac{2j}{Z_0}, \tag{3.174}$$

so that

$$Z_0 = \frac{2j}{\Omega \left(\partial Y / \partial \omega \right)_\Omega}. \tag{3.175}$$

Using Eqs. (3.43) and (3.44), we can then find any physical quantity we desire.

To see the generality of this result, consider the example of the lumped-element circuit in Fig. 3.10. If $L_1 + L_2 = L$, then this has the same bare resonance frequency Ω, but clearly will have a different coupling to the port. Use of Eq. (3.44) yields

$$\Phi_{\mathrm{ZPF}}^2 = \frac{\hbar \Omega L}{2} \left(\frac{L_1}{L_1 + L_2} \right)^2, \tag{3.176}$$

which is just what we expect from the transformer turns ratio.

Let us suppose for example that we couple to our black box through an inductor L_c as shown in Fig. 3.11. The coupling Hamiltonian is

$$H_1 = \frac{1}{2L_c} (\hat{\Phi} - \hat{\Phi}_{\mathrm{in}})^2. \tag{3.177}$$

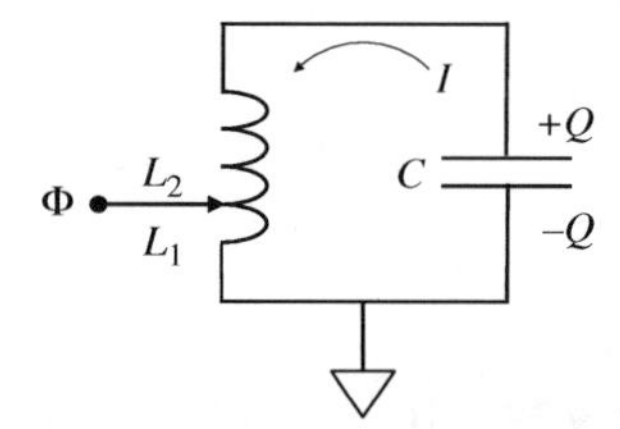

Fig. 3.10 Single-port black box containing a simple LC oscillator with the port connected to an inductive divider with $L_1 + L_2 = L$.

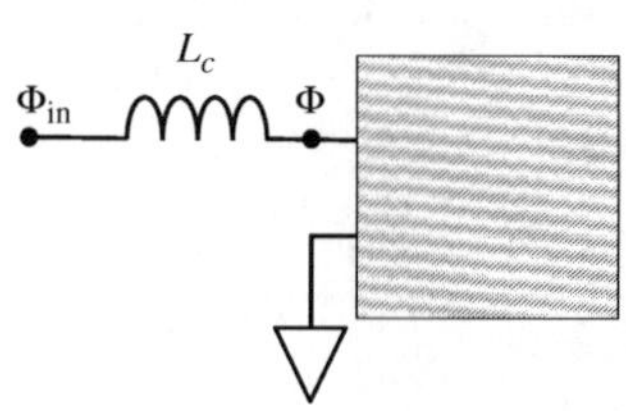

Fig. 3.11 Coupling to a black box via an inductor.

The operator $\hat{\Phi}_{\text{in}}$ is either a classical control field or a quantum operator for whatever system we hook up to our black box. Now that we know the impedance of the resonance, we know how to express $\hat{\Phi}$ using Eq. (3.44), so that we have

$$H_1 = \frac{1}{2L_c}\left[\Phi_{\text{ZPF}}(\hat{a}+\hat{a}^\dagger) - \hat{\Phi}_{\text{in}}\right]^2. \tag{3.178}$$

The case of capacitive rather than inductive coupling is more complex, as can be seen from the example of two capacitively coupled oscillators shown in Fig. 3.6, which we discussed earlier. We found that it was easy to write down the Lagrangian, but finding the Hamiltonian required inverting the capacitance matrix for the entire system. Hence, if we are going to use the flux variable at the input port as the coordinate, it is usually easiest to proceed by treating the coupling capacitor as being inside the black box.

The extension of these results to the case of a multimode black box Hamiltonian is simply

$$H_0 = \sum_m \Omega_m \hat{a}_m^\dagger \hat{a}_m, \tag{3.179}$$

where the summation is over the different modes and the flux operator at the port of the black box is simply

$$\hat{\Phi} = \sum_m \Phi_{\text{ZPF}}^{(m)}\left(\hat{a}_m + \hat{a}_m^\dagger\right). \tag{3.180}$$

This is simply a statement that the voltage across the input port is the sum of the voltages across each of the resonator elements in series as shown in Fig. 3.9(a).

This "black box" formalism will prove useful if it is possible to either measure or use finite-element simulations to compute the admittance as a function of frequency. So far, we have only discussed quantization of linear circuits, which are equivalent to coupled simple harmonic oscillators. Qubits are, of course, not linear circuit elements, but the formalism developed here is especially useful for the study of transmon qubits coupled to resonators, since as we will see in Section 3.4, the transmon qubit is essentially a weakly anharmonic oscillator. The generalization of the discussion above to the coupling of a weakly anharmonic oscillator to a linear black box [36] is discussed in detail in the appendix (Section 3.9.3). The reader should familiarize herself with the discussion of the transmon qubit in Section 3.4 before studying Section 3.9.3.

3.3 Superconductivity

In this chapter, I will not touch upon the microscopic theory of superconductivity (for an introduction, see the classic text by Tinkham [37]). Rather, I will present only the minimal phenomenology needed to understand the Josephson effect in small circuits containing tunnel junctions. A useful overview of Josephson tunnel junctions and various qubit circuits may be found in the papers by Devoret and Martinis [38] and Clarke and Wilhelm [39].

An ordinary (normal-state) tunnel junction consists of two metallic electrodes separated by a thin oxide barrier that allows electrons to tunnel quantum mechanically from one electrode to the other. Because even in rather small (mesoscopic but not nanoscopic) grains, the size of the grain is much larger than the angstrom scale of a typical Fermi wavelength, the "particle-in-a-box" quantum level spacing is extremely tiny and the density of states is essentially a continuum as shown in Fig. 3.12(a). As a result, the tunneling of electrons is an incoherent process that is described well by the irreversible dynamics of Fermi's Golden Rule. Under a voltage bias V, the chemical potential is higher in one grain than the other by an amount eV. Electrons in this energy interval are able to tunnel from "cathode" to "anode" without blocking due to the Pauli exclusion principle. As a result, the tunnel current is linear in the applied voltage (on voltage scales low compared with the tunnel barrier height) and the junction can be characterized as a simple resistor.[9] Because the two electrodes are separated by such a thin barrier, they also form a capacitor; so the equivalent circuit is that shown in Fig. 3.12(b) and the incoherent relaxation of the charge through the junction has the familiar characteristic timescale $\tau = RC$. Obviously, this incoherent behavior is not what we seek in a qubit, and so we must turn to superconductivity to rescue the situation.

Let us begin our discussion of superconductivity by considering a small isolated metallic electrode of some metallic superconductor. Because the electrode is isolated, the

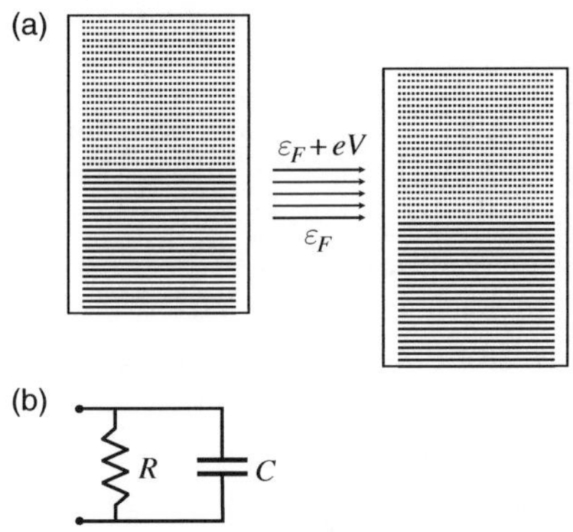

Fig. 3.12 Normal-metal tunnel junction with equilibrium Fermi energy ϵ_F and applied voltage bias V. (a) The single-particle spectrum in each electrode is dense and the occupied (empty) states are indicated by solid (dotted) lines. (b) Lumped-element equivalent circuit. Note that this is an approximate equivalent circuit, only valid when the junction is voltage-biased by a zero-impedance source, so that certain many-body Coulomb interaction effects are negligible.

[9] Strictly speaking, this is not correct. There are certain novel many-body effects associated with the quantum fluctuations of the environment to which the tunnel junction is coupled that can produce singularities in the conductance at low temperatures and low bias voltages [40]. These environmental fluctuation effects become pronounced as the impedance of the source driving the junction increases and begins to approach the quantum of resistance. See [41, 42].

number of electrons in it is fixed and well defined. For the moment, let us assume that the electron number has even parity. The essential physics of an ordinary superconductor is that effective attractive interaction resulting from virtual phonon exchange leads to pairing of electrons of opposite spin into so-called Cooper pairs. If the number of electrons in the electrode is even, then the quantum ground state has all of the electrons paired up into a special nondegenerate low-energy ground state. The system has a finite excitation gap 2Δ, defined by the energy needed to break a Cooper pair into two separate quasiparticles. The scale of this gap is typically several kelvin. As illustrated in Fig. 3.13(a), the quasiparticle states form a near continuum above the gap. (Only if the electrode is extremely tiny (on the scale of a few nanometers) will the single-particle level spacing be appreciable. We will not consider this limit further.)

Recalling that a temperature of 1.0 K corresponds to a frequency of approximately 21 GHz, we will be considering the limit of low temperatures and low frequencies relative to the gap: $k_BT, \hbar\omega \ll 2\Delta$. Hence, to a good approximation, we can say that the primary effect of superconductivity is the reduce the vast Hilbert space of the electrons in the electrode to a single quantum state $|N\rangle$, labeled by the number of pairs, which is a constant of the motion. This simplification is very important and will explain how we can produce macroscopic circuit elements whose quantum energy-level spectrum is as simple as that of a single hydrogen atom. Obviously, however, we have overshot the mark because a quantum system with only one energy level cannot be used to make a two-level qubit. To repair this error, consider a system with *two* metallic electrodes connected by a tunnel junction as shown in Fig. 3.13(b). We will again limit our attention to the quantum ground state of each electrode, assuming

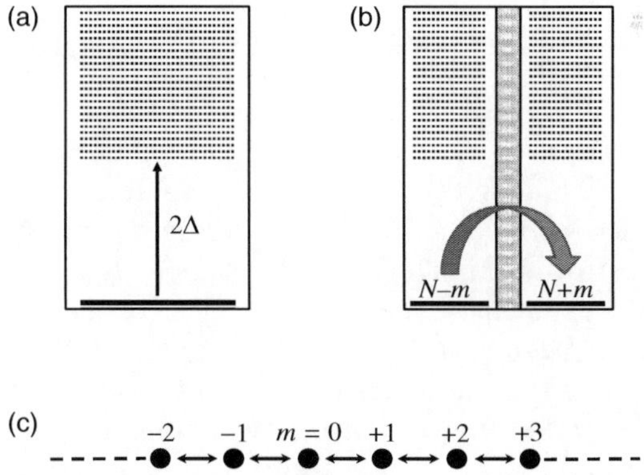

Fig. 3.13 (a) Spectrum of a superconducting Cooper pair box (CPB). For the case of an even number of electrons, there is a unique nondegenerate state separated from the excited states by a gap 2Δ. (b) A pair of CPBs connected by a tunnel barrier to form a Josephson junction. Ignoring the Coulomb energy, there is a large family of degenerate ground states labeled by an integer m representing the number of Cooper pairs transferred from one condensate to the other. (c) "Tight-binding" lattice along which the junction "moves" via Josephson tunneling between "sites" labeled by adjacent values of m.

the electrons in each are fully paired up. Once again, the total number of electron pairs in the system is fixed at some value $N = N_L + N_R$. Unlike the case of the single electrode, the total number of pairs no longer uniquely specifies the quantum state. We must specify the number on each electrode. A useful notation for the set of low-energy states is

$$|m\rangle = |N_L - m, N_R + m\rangle, \tag{3.181}$$

where m defines the number of pairs that have transferred through the tunnel junction from the left electrode to the right electrode, starting from some reference state with pair number $N_L(R)$ on the left (right) electrode.

The two electrodes form a small capacitor, but for the moment we will ignore the Coulomb energy that builds up as Cooper pairs are transferred from one electrode to the other. In this case, we then have a one-dimensional family of states that are labeled by the integer m and that are degenerate in energy.[10] Remarkably, it turns out that the tunnel junction permits pairs of electrons to coherently tunnel together from one side to the other. We will discuss the microscopic details further below, but for now we consider the phenomenological Hamiltonian

$$H_T = -\frac{1}{2}E_J \sum_m \left(|m\rangle\langle m+1| + |m+1\rangle\langle m|\right). \tag{3.182}$$

The parameter E_J is called the Josephson coupling energy and is a measure of the ability of Cooper pairs to tunnel through the junction. To rather good accuracy, E_J is given by the Ambegaokar-Baratoff relation [43]

$$E_J = \frac{1}{2}\frac{h}{(2e)^2}G_N\Delta, \tag{3.183}$$

where $G_N = 1/R_N$ is the normal-state conductance. One can understand this expression by noting that the microscopic tunneling Hamiltonian describes single-electron (not pair) tunneling across the junction. Thus, coherent tunneling of a pair of electrons must be second order in the tunneling matrix element t_{LR}. In the intermediate state after the first tunneling, we have an unpaired electron in each island and so there is an energy of order 2Δ appearing in the denominator. Naively, this suggests that the Josephson coupling will scale inversely with the size of the superconducting gap. However, we note that after the second electron tunnels, we are left with a pair of holes in one island and an extra pair of electrons in the other. Taking into account the quantum amplitudes for both of these pairs to disappear into the condensate brings in two more factors of Δ (and two factors of the density of states to keep the dimensions

[10] Here we need to point out the crucial difference between the superconducting gap and the gap in a band insulator like silicon. The latter gap is tied to a particular density at which the electrons fill up all the valence band (bonding orbitals) and none of the conduction band. In a superconductor, the particular density is not important—the gap follows the Fermi surface as it expands and contracts with density.

correct), which explains why E_J is in fact linearly increasing in the superconducting gap. The normal-state conductance is computed using Fermi's Golden Rule for the tunneling rate, which is of course proportional to the square of the matrix element t_LR. Hence, we see (qualitatively at least) the origin of Eq. (3.183).

You might wonder why E_J is not quartic in the tunneling matrix element since the effective matrix element for the pair tunneling is quadratic and perhaps we should square this as in Fermi's Golden Rule. The answer goes to the heart of the Josephson effect. Our effective Hamiltonian H_T connects discrete states (labeled by m), not a discrete state to a continuum. We will find not the irreversible incoherent dynamics of Fermi's Golden Rule (as we did in the normal-metal case), but rather coherent tunneling that connects a set of degenerate states $\{|m\rangle\}$. We will in effect be doing degenerate perturbation theory (i.e., finding a linear superposition of the degenerate states that diagonalizes H_T) and finding energy eigenvalues (and corresponding dynamics) that is first order in E_J, not second order.

We see that H_T causes m to either increase or decrease by unity, corresponding to the tunneling of a pair to the right or the left. We can gain some intuition by noticing that H_T is identical to that of a one-dimensional tight-binding lattice model with near-neighbor hopping amplitude E_J, as illustrated in Fig. 3.13(c). The (unnormalized) eigenfunctions are plane-wave-like states labeled by a dimensionless "wavevector" $\varphi = ka$, where $a = 1$ is the "lattice constant":

$$|\varphi\rangle = \sum_{m=-\infty}^{+\infty} e^{+im\varphi}|m\rangle. \tag{3.184}$$

Recalling the cosine dispersion of the one-dimensional tight-binding band with near-neighbor hopping, we see that

$$H_T|\varphi\rangle = -E_J \cos\varphi\,|\varphi\rangle. \tag{3.185}$$

Imagine a wavepacket moving to the right on our tight-binding lattice. This corresponds to a net current of Cooper pairs coherently tunneling through the junction, causing the "position" $\langle m\rangle$ to increase linearly with time. The group velocity of the packet is given by the derivative of the energy with respect to wavevector,

$$v_g(\varphi) = \frac{1}{\hbar}\frac{\partial}{\partial\varphi}\left(-E_J\cos\varphi\right), \tag{3.186}$$

so the net current flowing is given by the celebrated (first) Josephson relation

$$I(\varphi) = 2e\,v_g(\varphi) = \frac{2e}{\hbar}E_J\sin\varphi. \tag{3.187}$$

The maximum possible coherent (dissipationless) current occurs at $\varphi = \pi/2$ and is called the critical current

$$I_c = \frac{2e}{\hbar}E_J. \tag{3.188}$$

If more current than this is forced through the junction, the voltage rises from zero to a high value above the excitation gap and our low-energy effective model is no longer applicable.

As an alternative approach to the derivation of the Josephson relation for the current, let us define the operator $\hat{n}$ to be the number operator for the Cooper pairs transferred across the junction:

$$\hat{n} \equiv \sum_m |m\rangle m\langle m|. \tag{3.189}$$

Hamilton's equation of motion gives for the current operator

$$\hat{I} \equiv 2e\frac{d\hat{n}}{dt} = 2e\frac{i}{\hbar}[H_T, \hat{n}] = -i\frac{e}{\hbar}E_J\sum_m \left(|m\rangle\langle m+1| - |m+1\rangle\langle m|\right). \tag{3.190}$$

Next, we simply note that the plane-wave energy eigenfunctions are also eigenfunctions of the current operator obeying

$$\hat{I}|\varphi\rangle = I_c \sin\varphi\,|\varphi\rangle \tag{3.191}$$

which is of course equivalent to Eq. (3.187).

Exercise 3.14 Derive Eq. (3.191).

Equivalently, we note that in the φ representation, the wavefunction is given by

$$\psi(\varphi) = \langle\varphi|\psi\rangle. \tag{3.192}$$

By using Eqs. (3.189) and (3.184), we see that

$$\hat{n}|\varphi\rangle = -i\frac{d}{d\varphi}|\varphi\rangle. \tag{3.193}$$

Thus, it follows that in the φ representation, the number operator is

$$\hat{n} = +i\frac{d}{d\varphi}. \tag{3.194}$$

The confusing sign change comes from considering

$$\langle\varphi|\hat{n}|\psi\rangle = +i\frac{d}{d\varphi}\psi(\varphi). \tag{3.195}$$

It follows from this that

$$\hat{I} = 2e\frac{i}{\hbar}[H_T, \hat{n}] = 2e\frac{i}{\hbar}\left[-E_J\cos\varphi, +i\frac{d}{d\varphi}\right] = I_c\sin\varphi. \tag{3.196}$$

We continue for the moment to ignore the Coulomb interaction, but as a first step toward including it, let us think about the situation where an external electric field is applied and maintained in such a way that there is a fixed voltage drop V across the tunnel junction. This adds to the Hamiltonian a term

$$U = -(2e)V\hat{n}. \tag{3.197}$$

Hamilton's equation of motion yields the equally celebrated (second) Josephson relation[11]

$$\hbar\partial_t\varphi = -\frac{\partial H}{\partial \hat{n}} = 2eV, \tag{3.198}$$

relating the time rate of change of the "momentum" $\hbar\varphi$ to the "force" $2eV$. Equivalently, the solution of the Schrödinger equation is

$$|\Psi(t)\rangle = e^{+\frac{i}{\hbar}E_J\int_0^t d\tau\, \cos\varphi(\tau)}|\varphi(t)\rangle, \tag{3.199}$$

where

$$\varphi(t) = \varphi(0) + \frac{2e}{\hbar}Vt. \tag{3.200}$$

Exercise 3.15 Verify for yourself that Eq. (3.199) does indeed solve the Schrödinger equation

$$i\hbar\partial_t|\Psi(t)\rangle = (H_{\mathrm{T}} + U)|\Psi(t)\rangle. \tag{3.201}$$

The overall phase factor in front of the wavefunction is not particularly important, but the time variation of the "wavevector" $\varphi(t)$ is extremely important because it leads to the ac Josephson effect, namely that dc voltage bias leads to an ac current

$$\langle \hat{I}(t)\rangle = I_{\mathrm{c}}\sin[\varphi(0) + \omega t], \tag{3.202}$$

where the ac Josephson frequency is given by

$$\omega = 2\pi\frac{2e}{h}V. \tag{3.203}$$

The inverse flux quantum in frequency units is

$$\frac{2e}{h} \approx 483.597891(12)\,\mathrm{MHz}/\mu\mathrm{V}. \tag{3.204}$$

Since frequency is the easiest physical quantity to measure with high accuracy, the ac Josephson effect finds great practical use in metrology to maintain (but not define) the SI volt.

[11] As we will discuss in Section 3.4, the fact that the time derivative of the phase variable is proportional to the voltage means that φ is directly proportional to the flux variable introduced in Section 3.2.

To summarize our results, we are using here a representation in which the phase $\varphi(t)$ across the Josephson junction is viewed as a (dimensionless) wavevector. The Hamiltonian of the junction with voltage bias V is

$$H = -E_J \cos\varphi - 2eV\hat{n}, \tag{3.205}$$

where, in the φ representation, the number operator has the form

$$\hat{n} = +i\frac{d}{d\varphi}. \tag{3.206}$$

The pair of Hamilton equations

$$\frac{\partial \hat{n}}{\partial t} = \frac{\partial H}{\partial \hbar\varphi}, \tag{3.207}$$

$$\hbar\partial_t\varphi = -\frac{\partial H}{\partial \hat{n}} \tag{3.208}$$

correspond to the pair of Josephson relations found in Eqs. (3.187) and (3.198).

Stepping away from the particular expression of the Hamiltonian in the φ representation, we can write the general abstract representation of the Hamiltonian as

$$H = -E_J\widehat{\cos\varphi} - 2eV\hat{n}. \tag{3.209}$$

As noted earlier, for the case of an isolated Josephson junction, the number of Cooper pairs transferred through the junction is a well-defined integer and hence the state $|\varphi\rangle$ and the state $|\varphi + 2\pi\rangle$ are identical. Hence, the wavefunction $\psi(\varphi)$ obeys periodic boundary conditions. Because of this, there is no operator $\hat{\varphi}$. Only operators that preserve the periodic boundary conditions (such as $\widehat{\cos\varphi}$ and $\hat{n} = +id/d\varphi$) exist.

3.4 Superconducting qubits

So far, we have studied a single isolated Josephson junction that is able to coherently transfer Cooper pairs from one metallic island to another. Our discussion of this simple structure (known as the Cooper pair box, CPB) has been incomplete because it has neglected the fact that as current flows through the junction, charge builds up on the islands and the Coulomb energy becomes important. Once we include the Coulomb interaction, we will see that this structure makes an excellent artificial atom that can be used as a superconducting qubit. The first evidence that Josephson tunneling causes the CPB to exhibit coherent superpositions of different charge states was obtained in 1998 by Bouchiat *et al.* [44]. This was followed in 1999 by the pioneering experiment of the NEC group [5] demonstrating time-domain control of the quantum state of the CPB using very rapid control pulses to modulate the offset charge.

The remarkable recent progress in creating superconducting quantum bits and manipulating their states has been summarized in several reviews [1, 38, 39, 45–50]. Over 30 years ago Leggett [51] discussed the fundamental issues concerning the collective degrees of freedom in superconducting electrical circuits and the fact that they themselves can behave quantum mechanically. As noted earlier, the essential collective variable in a Josephson junction [38] is the phase difference of the superconducting order parameter across the junction. The first experimental observation of the quantization of the energy levels of the phase "particle" was made by Martinis, Devoret, and Clarke in 1985 [52, 53].

A number of different qubit designs, illustrated in Figs. 3.14 and 3.15, have been developed around the Josephson junction including the CPB [5, 12, 44, 54–58] based on charge, the flux qubit [11, 59, 60], and the phase qubit [61, 62]. Devoret and co-workers have recently introduced the fluxonium qubit [9, 63], in which the small Josephson junction is shunted by a very high inductance created from a string of larger Josephson junctions. Figure 3.16 shows an "evolutionary phylogeny" for these different types of qubits. We will turn now to a discussion of the Hamiltonians of these different types of qubits and subsequently to an analysis of the relative merits in terms of their sensitivity to noise perturbations. Once we understand the Hamiltonians, we will be in a position to classify the qubits according to their location in the "Mendeleev Table" shown in Fig. 3.17. The lectures at this school by Daniel Esteve (see Chapter 6) discuss the different methods by which the state of different qubits can be read out.

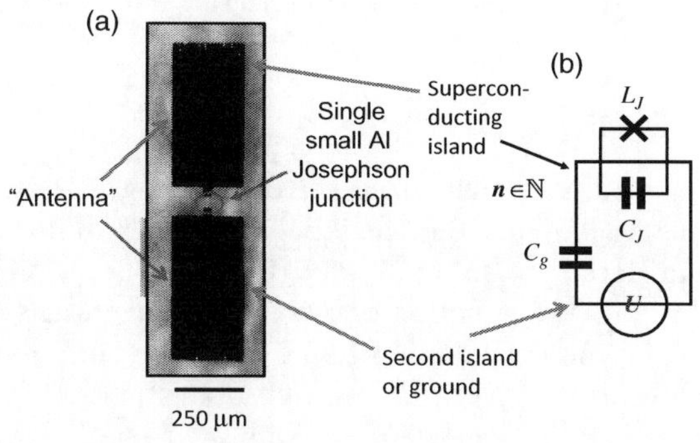

Fig. 3.14 (a) Cooper pair box qubit (from R. Schoelkopf's laboratory) and (b) its equivalent circuit showing a voltage source biasing the box through a coupling ("gate") capacitor c_g. The cross denotes the Josephson junction. The voltage source may represent an intentionally applied bias or be the result of random charges in the insulating substrate supporting the device. The particular device illustrated in (a) is a transmon qubit in a three-dimensional cavity for which there is no dc bias applied (although there may be a random offset voltage due to charges trapped in the substrate).

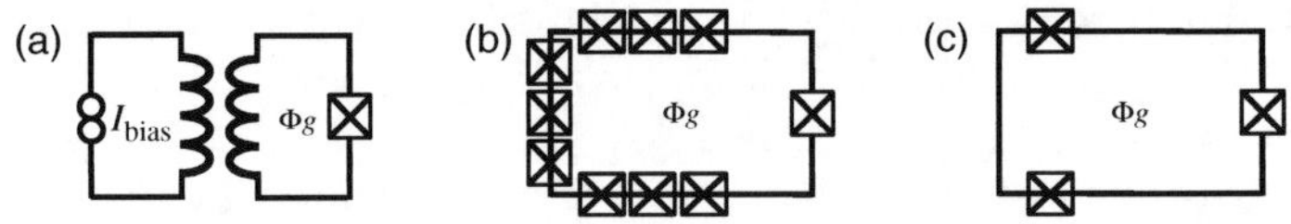

Fig. 3.15 Inductively shunted qubits. (a) Phase qubit with a transformer flux bias circuit driven by current I_{bias}. The Josephson junction is indicated by a box with a cross. (b) Fluxonium qubit. The shunt inductor has been replaced by an array of a large number of Josephson junctions. The array junctions are chosen to have a sufficiently large ratio of Josephson energy E_J to charging energy E_C that phase slips can be neglected and the array is a good approximation to a very large inductor. The flux bias circuit is not shown. (c) Flux qubit consisting of a superconducting loop with three Josephson junctions. The flux bias circuit is not shown.

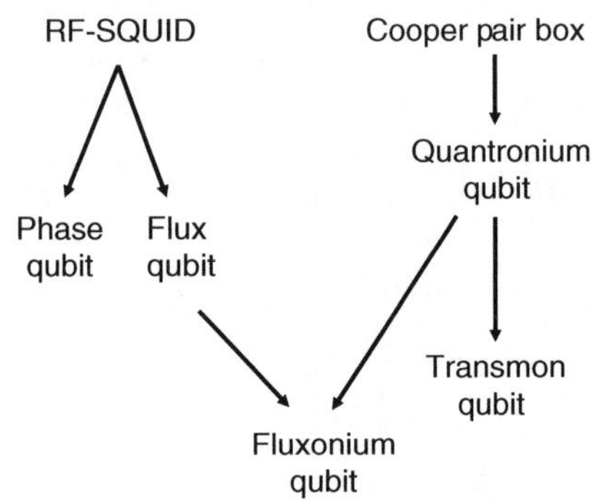

Fig. 3.16 Evolutionary phylogeny of superconducting qubits. (Courtesy M. Devoret.)

3.4.1 The Cooper pair box

The CPB [3] is topologically distinct from the other designs shown in Fig. 3.15 in that it has no wire closing the loop around the junction. The Hamiltonian will be described below and derived in detail in the appendix (Section 3.9.1). The number of Cooper pairs transferred through the junction is a well-defined integer, as we have already discussed in detail. The integer charge implies that the conjugate phase is compact; that is, in the phase representation, the system obeys periodic boundary conditions. As we will see, this implies that charge-based qubits are sensitive to stray electric field noise, but that this can be overcome by putting the CPB in the "transmon" regime where the Josephson tunneling energy dominates over the Coulomb charging energy [57, 64].

Our previous discussion of the Josephson effect neglected the Coulomb interaction. Let us define the charging energy associated with the transfer of a single electron to be

$$E_C = \frac{e^2}{2C_\Sigma}, \tag{3.210}$$

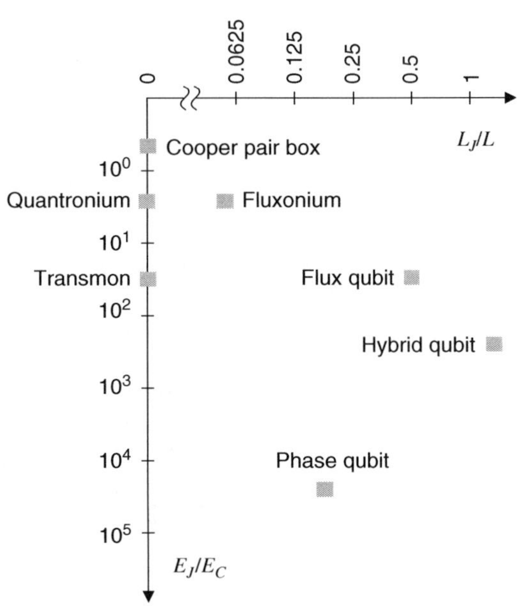

Fig. 3.17 "Periodic Table" of superconducting qubits. E_J is the tunneling Josephson energy, $4E_C$ is the energy cost to charge the junction with one Cooper pair, and $E_L/2$ is the energy cost to "charge" the shunt inductor with one flux quantum. (Courtesy M. Devoret.)

where $C_\Sigma = C_J + C_g$ is the total capacitance between the island electrodes. For the equivalent circuit shown in Fig. 3.14, where the CPB is biased by a low-impedance voltage source [3], the total capacitance connecting the island to ground is the sum of the capacitance across the junction plus the gate capacitance: $C_\Sigma = C_J + C_g$. The Coulomb energy to transfer a Cooper pair is four times larger than for a single electron, and so the Coulomb energy operator is given by

$$\hat{U} = 4E_C(\hat{n} - n_g)^2, \tag{3.211}$$

where

$$n_g \equiv -\frac{C_g V}{2e} \tag{3.212}$$

is called the dimensionless "gate charge" or "offset charge" and represents either the effect of an externally applied electric field or some microscopic junction asymmetry that breaks the degeneracy between positive and negative charge transfers [3]. The number operator $\hat{n}$ has integer eigenvalues, while n_g is a continuous variable, which may either intentionally or randomly and uncontrollably fluctuate. The effects of such fluctuations on the coherence of the qubit will be discussed in Section 3.5.

As we noted earlier, we see from Eq. (3.184) that φ is a compact variable; that is, the state $|\varphi + 2\pi\rangle$ is identified with the state $|\varphi\rangle$. Hence, without loss of generality, we may take the wavevector φ to live in the first Brillouin zone $\varphi \in [-\pi, +\pi]$

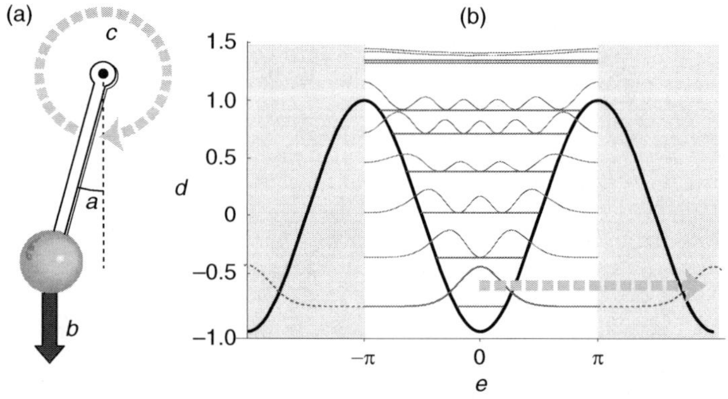

Fig. 3.18 The Cooper pair box Hamiltonian in the phase representation is equivalent to that of a quantum rotor. The offset charge n_g is equivalent to an Aharonov-Bohm flux that produces a Berry phase proportional to the winding number of the rotor trajectory. Unlike other qubit circuit topologies, the rotor wavefunction obeys periodic boundary conditions. From Koch *et al.* [57].

or, equivalently, on a circle. In fact, because of this periodic boundary condition, it is convenient to view φ as an angular coordinate living on a circle, rather than as a quasi-momentum as we have been doing until now. In this picture, we reinterpret the expression in Eq. (3.194) as telling us that the number operator $\hat{n}$ is nothing more than (minus) the angular momentum conjugate to the angle φ. Because φ is compact, the conjugate (dimensionless) angular momentum $\hat{n}$ naturally has integer eigenvalues, which is just the property that we require for the operator representing the number of Cooper pairs that have been transferred through the Josephson junction. Except for the n_g term (to which we will return shortly), the Hamiltonian becomes that of a quantum rotor in a gravitational field (see Fig. 3.18):

$$H = 4E_C\left(\hat{n} - n_g\right)^2 - E_J\widehat{\cos\varphi}, \tag{3.213}$$

where the charging energy determines the (inverse) moment of inertia, and the Josephson energy is a measure of the torque produced by gravity.[12]

We know that for small-amplitude oscillations, the classical quantum rotor is very nearly a simple harmonic oscillator. This can be seen by expanding the cosine to second

[12] This jumping back and forth between representations can easily give beginners a headache. Notice that previously when φ was a momentum and we were studying the pair tunnel Hamiltonian H_T, we were perfectly happy to have a kinetic energy but no potential energy. This is much like the situation of a particle moving freely in space, except here it is hopping on a lattice. Now if we think of φ as a coordinate, we have to think of the Josephson energy as potential energy. Without the Coulomb energy E_C, our rotor has infinite moment of inertia (and hence no kinetic energy and no dynamics). Only when we add the charging energy do we get quantum fluctuations in the phase and dynamical oscillations of our qubit. Confusingly, this situation is sometimes described by saying that without the charging energy, the phase φ is classical.

order (an approximation that is valid in the quantum case in the limit $E_J \gg E_C$ where the zero-point fluctuations in the phase are small). Up to an irrelevant constant in the energy (and ignoring the offset charge for the moment), we have for the classical Hamiltonian

$$H \approx 4E_C n^2 + \frac{1}{2}E_J\varphi^2. \tag{3.214}$$

The replacement of the cosine term, which is correctly periodic in φ, by the quadratic approximation is quite subtle. This is a simple harmonic oscillator only if we extend φ from a compact variable living on a circle to a noncompact variable living on the real line. For the (quantum) harmonic oscillator, we do not have the requirement that the wavefunction be periodic in φ. If, however, E_J/E_C is large enough, the wavefunction will have significant support only near $\varphi = 0$, and hence, we are not making a significant mistake by ignoring the requirement of periodic boundary conditions. The boundary condition of vanishing at infinity is close enough, because the wavefunction is extremely small at $\varphi = \pm\pi$ and so very nearly obeys the required periodic boundary conditions. Further, the small excursions in φ mean that the Taylor series expansion of the cosine is justified. Under these conditions, we obtain a quantum harmonic oscillator with Hamiltonian

$$H \approx 4E_C\hat{n}^2 + \frac{1}{2}E_J\hat{\varphi}^2, \tag{3.215}$$

with $\hat{n}$ now being the ordinary linear (not discrete angular) momentum with continuous spectrum.

It is interesting to try to connect the classical Hamiltonian of this CPB to the classical LC oscillator we studied previously. From the second Josephson relation in Eq. (3.198), we find that the phase angle φ is directly proportional to the flux variable Φ defined in Eq. (3.25) and used in the quantization of the LC oscillator:

$$\varphi = \frac{2e}{\hbar}\Phi = 2\pi\frac{\Phi}{\Phi_0}. \tag{3.216}$$

Thus, each time the flux variable changes by one flux quantum, the superconducting phase variable winds by 2π. The classical Lagrangian for the CPB can be written

$$\mathcal{L} = \frac{1}{2}C\dot{\Phi}^2 + E_J\cos\left(2\pi\frac{\Phi}{\Phi_0}\right), \tag{3.217}$$

and the Hamiltonian becomes

$$H = \frac{1}{2C}Q^2 - E_J\cos\left(2\pi\frac{\Phi}{\Phi_0}\right). \tag{3.218}$$

Expanding the cosine term about $\Phi = 0$ to lowest order (and dropping the zeroth-order term), we have

$$H \approx \frac{1}{2C}Q^2 + \frac{1}{2L_J}\Phi^2, \tag{3.219}$$

where the (small-signal) effective inductance of the Josephson junction is given by

$$L_J = \left(\frac{\hbar}{2e}\right)^2 \frac{1}{E_J}. \tag{3.220}$$

In this approximation, the CPB becomes a simple harmonic oscillator with resonant frequency (known as the Josephson plasma frequency) given by

$$\Omega_J \equiv \frac{1}{\sqrt{L_J C}} = \frac{1}{\hbar}\sqrt{8E_J E_C}. \tag{3.221}$$

The Taylor series expansion of the cosine is justified only if we are discussing small-amplitude motions. Classically, we can always choose to study this limit. Quantum mechanically, we are forced to deal even in the ground state with zero-point fluctuations in Φ. From Eq. (3.44), we see that (in the harmonic approximation used above)

$$2\pi\frac{\Phi_{\mathrm{ZPF}}}{\Phi_0} = \left(2\frac{E_C}{E_J}\right)^{1/4}, \tag{3.222}$$

so the harmonic approximation is, as suggested above, self-consistent in the limit $E_{\mathrm{J}} \gg E_{\mathrm{C}}$. In the quantum rotor picture, this corresponds to strong gravity and large mass (moment of inertia).

For general flux Φ, not necessarily small, we can define the differential (inverse) inductance as

$$L^{-1}(\Phi) \equiv \frac{d^2H}{d\Phi^2} = E_J\left(\frac{2\pi}{\Phi_0}\right)^2 \cos\left(2\pi\frac{\Phi}{\Phi_0}\right), \tag{3.223}$$

and we see that the Josephson junction acts as a nonlinear inductor. It is this key feature that will make the energy levels of the CPB anharmonic.

In the quantum case, if the quantum zero-point fluctuations in Φ are small, then this nonlinear inductor picture can be useful, but in general we need to resort to numerical diagonalization of the CPB Hamiltonian. Let us therefore now return to the full Hamiltonian in Eq. (3.213). In the phase basis, the Schrödinger eigenvalue equation is the Matthieu equation, whose solutions are formally known in terms of Matthieu functions. Numerical diagonalization is more conveniently performed in the charge (number) basis where the Coulomb term is diagonal and the Josephson term is tri-diagonal: $\langle m \pm 1|\cos\varphi|m\rangle = \frac{1}{2}$. The basis states are labeled by the eigenvalue m of the number operator $\hat{n}$ and the Hilbert space must be truncated at some largest $|m| = m_{\mathrm{max}}$. If we are interested only in the first N low-lying states, the size of the Hilbert space needed can be estimated from the zero-point fluctuations of the charge in the harmonic limit given in Eq. (3.43):

$$m_{\mathrm{max}} \gg \sqrt{N}\frac{Q_{\mathrm{ZPF}}}{2e} \sim \sqrt{N}\left(\sqrt{\frac{E_J}{32E_C}}\right)^{1/4}, \tag{3.224}$$

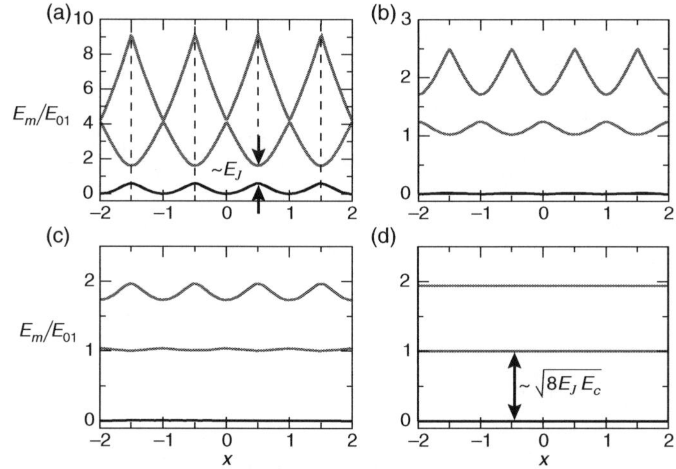

Fig. 3.19 Energy spectrum of the Cooper pair box as a function of offset charge for different values of the dimensionless ratio of Josephson energy to charging energy. The exponential decrease in the charge dispersion is clearly seen. From Koch *et al.* [57].

which, conveniently, is usually not very large.[13]

The qubit spectrum is periodic in the offset charge n_g with unit period, as can be seen in Fig. (3.19). Physically, this simply means that the integer part of the offset charge can always be canceled out by transferring an integer number of Cooper pairs from one island to the other. To understand this mathematically, recall that the wavefunctions obey periodic boundary conditions in the angle φ. Thus, in the φ representation,

$$U_\pm = e^{\pm i\varphi} \tag{3.225}$$

is a "legal" unitary transformation that preserves the boundary conditions (and physically transfers one pair through the junction). Recognizing that such a unitary transformation preserves the spectrum of the Hamiltonian and that it shifts the angular momentum (transferred charge) by one unit,

$$U_\pm \hat{n} U_\pm^\dagger = \hat{n} \mp 1, \tag{3.226}$$

we see that the spectrum must indeed be invariant under unit translations of n_g.

Since the offset term n_g does not appear in the ordinary quantum rotor problem, we need to extend our analogy a bit. It turns out that this term can be viewed as

[13] It is useful at this point to note that for a given tunnel barrier thickness, E_J scales with junction area, but the capacitance C_J (which usually dominates over C_g) and hence the inverse charging energy also scale with junction area. Thus, the Josephson plasma frequency is nearly independent of junction area while the impedance of the Josephson plasma oscillator $Z_J = (L_J/C_\Sigma)^{1/2} = \pi^{-1}h(2e)^{-2}(E_C/E_J)^{1/2}$ scales inversely with the area. Hence, $m_{\max}$ grows with the square root of the area while $\varphi_{\rm ZPF}$ scales inversely with the square root of the junction area.

resulting from the rotor carrying (fake) charge and undergoing an Aharanov-Bohm phase shift as it circles a line of (fake) magnetic flux that is passing through the axis. To see this, let us recall that for a particle with charge q moving in the presence of a vector potential $\vec{A}$, the canonical momentum is replaced by the mechanical momentum

$$\vec{p} \longrightarrow \vec{p} - q\vec{A}(\vec{r}). \tag{3.227}$$

For our quantum rotor turning on the z axis, we are interested in the angular momentum

$$L_z = (\vec{r} \times \vec{p})_z \longrightarrow (\vec{r} \times \vec{p})_z - q(\vec{r} \times \vec{A})_z. \tag{3.228}$$

If the magnetic field is zero everywhere except for an Aharonov-Bohm tube of flux on the z axis, we can choose the following gauge for the vector potential:

$$\vec{A}(\vec{r}) = \Phi_{\mathrm{AB}} \frac{1}{2\pi r} \hat{z} \times \hat{r}, \tag{3.229}$$

which has the correct total flux

$$\oint \vec{A} \cdot d\vec{r} = \Phi_{\mathrm{AB}} \tag{3.230}$$

for any loop with winding number $+1$ around the z axis. The mechanical angular momentum operator thus becomes

$$(\vec{r} \times \vec{p})_z - \frac{q}{2\pi} \Phi_{\mathrm{AB}} = \hbar \left(-i \frac{\partial}{\partial \varphi} - \frac{\Phi_{\mathrm{AB}}}{\Phi_0} \right), \tag{3.231}$$

where Φ_0 is the flux quantum corresponding to charge q (which, since it is arbitrary, we can choose to be $2e$). By comparison with Eq. (3.213), we see that the real offset charge is equivalent to a fake Aharonov-Bohm flux

$$n_g = \frac{\Phi_{\mathrm{AB}}}{\Phi_0}. \tag{3.232}$$

The fact that the spectrum is periodic in Φ_{AB} is simply a reflection of the fact that a flux tube containing an integer number of flux quanta produces an Aharonov-Bohm phase shift of zero (modulo 2π).

In the "charge" limit, $E_C \gg E_J$, the states of our rotator and nearly angular momentum (i.e., charge) eigenstates weakly perturbed by the gravitational torque (Josephson coupling). As can be seen in Fig. 3.19(a), when the offset charge is a half-integer, there are two adjacent charge states that have the same Coulomb energy. This degeneracy is lifted by by the Josephson term. The disadvantage of working in this regime is that E_C is large and the excitation energies are extremely strong functions of the offset charge n_g. As we will see in Section 3.5, environmental noise in n_g will lead to severe qubit decoherence.

In the "transmon" limit, $E_J \gg E_C$, the states of the quantum rotor are nearly harmonic small-amplitude oscillations near the minimum in gravitational potential energy (i.e., Josephson energy). As can be seen in Fig. 3.19(d), the qubit levels are nearly harmonic and very nearly completely insensitive to the value of the offset charge. The reason for this is readily understood within the Aharonov-Bohm analogy of the offset charge. Let us start our discussion in path integral language. We know that the only way that the system can be aware of the value of the Aharonov-Bohm flux is for the rotor to circle completely around the flux tube in order to acquire the Aharonov-Bohm phase shift. The interference between this path and the path where the rotor does not circle the flux modifies the quantum energy. However, in the limit of large E_J (strong gravity), the rotor must tunnel through a very high energy barrier in order to wind the phase from $\varphi = 0$ to $\varphi = 2\pi$. The barrier height is proportional to E_J and the particle "mass" is inversely proportional to E_C. As a result, the contribution of these processes to the energy is exponentially small.

Let us return now to the language of wavefunctions for a more quantitative discussion. Mathematically, we begin by performing the unitary gauge transformation

$$U = e^{-in_g\varphi}, \tag{3.233}$$

which completely removes the offset charge term from the Hamiltonian:

$$U(\hat{n} - n_g)U^\dagger = \hat{n}. \tag{3.234}$$

Notice that the transformed wavefunction $U\Psi$ no longer obeys periodic boundary conditions:

$$\{U\Psi(\varphi + 2\pi)\} = e^{-i2\pi n_g}\{U\Psi(\varphi)\}. \tag{3.235}$$

Thus, while the Hamiltonian becomes independent of n_g,

$$UHU^\dagger = H = 4E_C(\hat{n})^2 - E_J\cos\varphi, \tag{3.236}$$

the change in boundary condition with n_g nevertheless changes the energy eigenvalue spectrum. On the other hand, for large E_J/E_C, the wavefunction is exponentially small at the boundary $\varphi = \pm\pi$, so we do not expect a large change in the spectrum due to this change in boundary condition.

To develop a better intuition for how the change in boundary condition affects the energy eigenvalues, consider the following. Rather than viewing the Hamiltonian as that of a quantum rotor (with compact φ living on the interval $-\pi \le \varphi \le +\pi$), let us view this as the Hamiltonian of a phase "particle" with coordinate (not momentum!) φ moving in the extended cosine potential as shown in Fig. 3.20. Bloch's theorem tells us that the discrete translation symmetry of the potential implies that the wavefunctions must be of the form

$$\Psi_{mk}(\varphi) = e^{ik\varphi}\psi_m(\varphi), \tag{3.237}$$

where ψ_n obeys periodic boundary conditions $\varphi \longrightarrow \varphi + 2\pi$, m is the band index and (because the "lattice constant" of the potential is 2π) the wavevector k lives in the first Brillouin zone $-\frac{1}{2} \le k \le +\frac{1}{2}$. We now simply invoke the boundary condition in

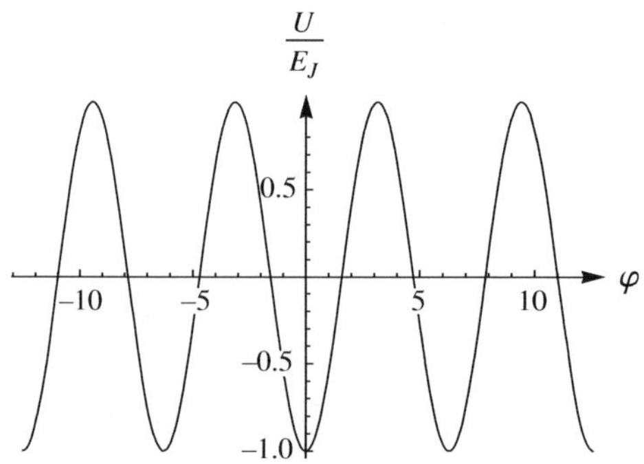

Fig. 3.20 Extended cosine potential $U = -E_J \cos\varphi$.

Eq. (3.235) by choosing the Bloch wavevector to be $k = n_g$. The use of the extended cosine potential is just a crutch to get us to Eq. (3.237). The Bloch-wave solution in the interval $-\pi \leq \varphi \leq +\pi$ solves the quantum rotor problem with compact phase and boundary condition given in Eq. (3.235). Thus, we arrive at the important conclusion that the spectrum of the quantum rotor plotted against offset charge in Fig. 3.19 *is nothing more than the band structure of a one-dimensional solid with a cosine potential in which the offset charge defines the Bloch wavevector.* The excited band energies correspond to the excited states of the rotor. The constraint (or "superselection rule") that the Bloch wavevector is not arbitrary but rather fixed by the offset charge removes the infinite number of extra states that we built into the Hilbert space when we let φ be a noncompact variable.

The band-structure picture is useful because in the limit of large E_J we can invoke (at least for the low-lying bands) the tight-binding approximation to the band structure in which we envision the phase particle in a bound state within a well being able to hop to the corresponding state in an adjacent well by tunneling through the barrier. We know that for near-neighbor hopping the tight-binding model has a simple cosine dispersion as a function of wavevector (which in this case is offset charge). Hence, within this approximation, the energy of the mth level has the form [57]

$$E_m(n_g) \approx \mathcal{E}_m + \epsilon_m \cos(2\pi n_g), \tag{3.238}$$

where the tight-binding bandwidth $2\epsilon_m$ defines the so-called "charge dispersion." Within the WKB approximation for the tight-binding hopping amplitude, the charge dispersion for the mth level is given by [57]

$$\epsilon_m \sim (-1)^m E_C \frac{2^{4m+5}}{m!} \sqrt{\frac{2}{\pi}} \left(\frac{E_J}{2E_C}\right)^{m/2+3/4} e^{-\sqrt{8E_J/E_C}}. \tag{3.239}$$

The charge dispersion is a measure of the sensitivity of the energy to the offset charge. We see that going to large E_J/E_C makes the transmon qubit quite insensitive to (low-frequency) charge noise, which dramatically improves the dephasing time. The fact that the "charge dispersion" is greater for the higher excited states is simply the naturally larger bandwidth of the higher energy states of the band structure. (More energetic particles tunnel through the barrier more readily.)

In the limit of large E_J/E_C, the quantum rotor begins to approach a harmonic oscillator. Fortunately, the anharmonicity defined by

$$A \equiv \omega_{12} - \omega_{01} \sim -E_C \tag{3.240}$$

goes to zero very slowly as the charging energy is reduced and can be easily kept above 100–200 MHz [57], which is adequate to prevent smooth nano-second control pulses from taking the qubit out of the logical subspace (the two lowest levels) [64].

We may perturbatively estimate the anharmonicity (in the limit of negligible charge dispersion) from Eq. (3.236) by assuming that the zero-point fluctuations of the phase are small, allowing us to expand the cosine potential beyond the second-order considered previously to write

$$H \approx H_0 + V, \tag{3.241}$$

where

$$H_0 = 4E_C\hat{n}^2 + \frac{1}{2}E_J\hat{\varphi}^2, \tag{3.242}$$

and

$$V = -\frac{1}{24}E_J\hat{\varphi}^4. \tag{3.243}$$

Using Eq. (3.222), we find

$$\varphi_{\mathrm{ZPF}}^2 = \sqrt{\frac{2E_C}{E_J}}, \tag{3.244}$$

which, as noted earlier in connection with Eq. (3.222), is indeed small for large E_J/E_C, so our assumption is self-consistent in this limit.

Using

$$\hat{\varphi} = \varphi_{\mathrm{ZPF}}(\hat{a} + \hat{a}^\dagger), \tag{3.245}$$

and neglecting off-diagonal terms, we can write the perturbation in a simple form useful for first-order perturbation theory:

$$V = -\frac{1}{12}E_C(\hat{a} + \hat{a}^\dagger)^4 \approx -\frac{E_C}{2}(\hat{a}^\dagger\hat{a}^\dagger\hat{a}\hat{a} + 2\hat{a}^\dagger\hat{a}). \tag{3.246}$$

The second term renormalizes the harmonic oscillator frequency downward slightly,

$$\hbar\tilde{\Omega}_J = \sqrt{8E_JE_C} - E_C, \tag{3.247}$$

and the first introduces an anharmonicity that yields Eq. (3.240).

3.4.2 Inductively shunted qubits

We turn now to the class of superconducting qubits illustrated in Fig. 3.15 that are inductively shunted. In the phase qubit [61, 62], the Josephson junction is shunted by a lumped element inductor. In the fluxonium qubit [9, 63], an extremely large value of inductance is required—so large that it is impossible to achieve with a coiled wire. This is due to parasitic capacitance in any coil and is essentially the result of the small value of the fine structure constant.[14] Thus, instead of a coil, a long chain of Josephson junctions provides the inductance. The flux qubit [11, 59, 60] uses only two junctions in this chain, and hence it turns out that the shunting inductor is not fully linear.

As stated several times previously, the CPB is unique in its topology because there is no connection between the two superconducting islands. Thus, the only way to change the charge is to tunnel an integer number of Cooper pairs through the Josephson junction. The number of Cooper pairs transferred is represented in the Hamiltonian as an angular momentum operator $\hat{n}$ that is conjugate to a compact phase angle φ. We saw from Eq. (3.184) that the state of the CPB $|\varphi + 2\pi\rangle$ is not distinct from (and indeed is identical to) $|\varphi\rangle$. This is what we mean when we say that φ is a compact angular variable living on a circle, or equivalently the wavefunction $\Psi(\varphi)$ obeys periodic boundary conditions. One obvious consequence is that the current flowing through the junction, $I = I_c \sin(\varphi)$, is (from the first Josephson relation) a periodic function of the phase variable.

On the other hand, we know from the second Josephson relation that the phase variable φ is equivalent to the flux variable Φ that we introduced in quantizing the LC oscillator, where we found that the energy stored in the inductor is

$$U = \frac{1}{2L}\Phi^2 = \frac{1}{2}E_L\varphi^2 \tag{3.248}$$

where

$$E_L \equiv \left(\frac{\hbar}{2e}\right)^2 \frac{1}{L} = \left(\frac{\Phi_0}{2\pi}\right)^2 \frac{1}{L}. \tag{3.249}$$

Clearly, the energy U stored in the inductor and the current flowing through it,

$$I = \frac{dU}{d\Phi} = \frac{\Phi_0}{2\pi L}\varphi, \tag{3.250}$$

are *not* periodic in φ. If we shunt our Josephson junction with an external inductor, the change in topology of the circuit has profound consequences on the mathematical description. The states described by φ and $\varphi + 2\pi$ are physically distinct because they differ by how much current is flowing in the inductor. Because charge can move onto the junction capacitor plates continuously through the inductor, the charge variable

[14] The parasitic capacitance in any coil leads to self-resonances. The frequency of the lowest of these self-resonances can be very crudely estimated from matching the corresponding free-space wavelength to the total length of wire in the inductor.

is no longer integer-valued but rather continuous, as expected from the fact that φ is no longer a compact variable and the system no longer obeys periodic boundary conditions. Rather, we expect vanishing boundary conditions because the energy stored in the inductor diverges for large φ: $\Psi(\varphi \longrightarrow \pm\infty) \longrightarrow 0$. Because φ is now an ordinary noncompact coordinate and we will denote this by placing a hat over it when we quantize the system. The Hamiltonian becomes

$$H = 4E_C(\hat{n} - n_g)^2 - E_J \cos\hat{\varphi} + \frac{1}{2}E_L\hat{\varphi}^2. \tag{3.251}$$

Because the charge is now a continuous variable, we expect on physical grounds that a *static* offset charge n_g can be completely screened and should not affect the energy. Mathematically, this can be seen by performing the unitary gauge transformation discussed previously in Eq. (3.233). Notice, however, the important difference that the transformed wavefunction $U\Psi$ still obeys the same vanishing boundary conditions. Hence, unlike the previous case, the spectrum does *not* depend on the static offset charge in any way.

Let us consider the case of a shunt inductor but take the limit $L \longrightarrow \infty$, which means $E_L \longrightarrow 0$. Since the inductor is present, the phase variable is presumably no longer compact and yet it would seem that the inductance term does not change the Hamiltonian. Physically, it seems reasonable to assume that the high-frequency oscillations of the qubit would be unaffected by the enormous reactance of the inductor. This is indeed the case, and the spectrum is the same as before except that now *all* values of the wavevector k are allowed since the phase is noncompact.[15] Thus, there is a continuum of states instead of a single state within each band. The only effect that offset charge has is to shift the k states $k \longrightarrow k + n_g$, but this has no effect on the spectrum, because all values of k are allowed and (in the extended zone scheme) the spectrum is periodic under $k \longrightarrow k + 1$. The continuum of states corresponds to the fact that the inductor can have a fixed dc current flowing in it, which can take on any value (below the the critical current of the junction).

Exercise 3.16 In the limit of $L \longrightarrow \infty$ described above, the eigenstates are plane-wave-like Bloch waves. Using the general periodicity properties of Bloch waves, compute the mean charge (which can be nonzero only because of the infinite inductance)

$$\bar{n} = \langle\Psi_{nk}|\hat{n}|\Psi_{nk}\rangle. \tag{3.252}$$

Strictly speaking, the wavevector k is not gauge-invariant. What is the correct gauge-invariant operator we should have used in this equation?

If the inductive energy E_L is nonzero, then Bloch's theorem no longer applies. The quadratic potential in Eq. (3.251) resulting from the inductive energy leads to the curvature illustrated in Fig. 3.21. The interplay between the quadratic term and

[15] That is, the "superselection rule" discussed previously no longer applies.

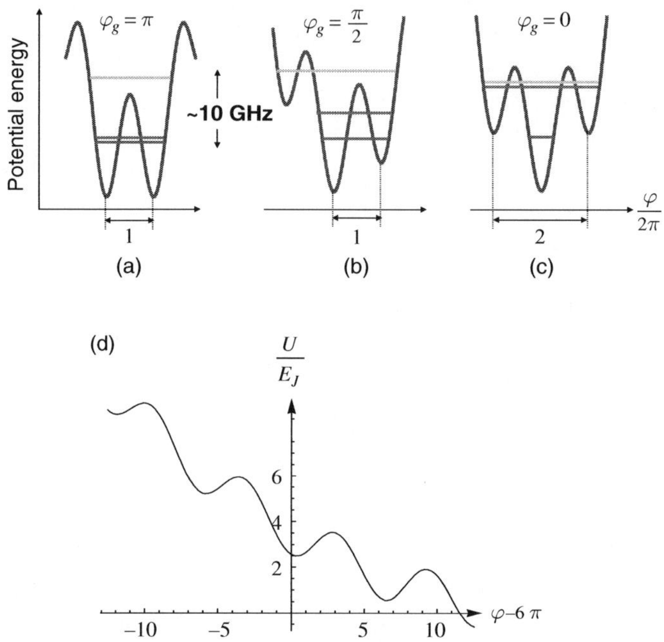

Fig. 3.21 (a) Extended cosine potential $U = -E_J \cos(\varphi + \varphi_g) + E_L\varphi^2$, with dimensionless offset flux $\varphi_g = \pi$. This is used in the fluxonium qubit to produce a "Λ" level configuration; (b) Same as (a) but with $\varphi_g = \pi/2$. (c) Same as (a) but with $\varphi_g = 0$ to produce a "V" level configuration. (d) Same as (a) but displaced a distance $\varphi_g = 6\pi$ to illustrate the current (flux) biased phase qubit. $E_L = 0.01E_J$.

the Josephson cosine term allows us to create a number of different potential energy well shapes and thus generate different interesting qubit spectra. Before delving into this, we need to recognize that there is one more "control knob" at our disposal, namely externally applied flux, which we can view as the inductive analog of the offset charge studied previously. If our inductor is part of a transformer with dc current applied in the other winding, then there is a flux offset and the Hamiltonian becomes

$$H = 4E_C(\hat{n} - n_g)^2 - E_J \cos\hat{\varphi} + \frac{1}{2}E_L(\hat{\varphi} - \varphi_g)^2. \tag{3.253}$$

It is convenient to translate $\hat{\varphi}$ and $\hat{n}$ using the unitary transformation

$$U = e^{i\varphi_g\hat{n}}e^{-in_g\hat{\varphi}}, \tag{3.254}$$

which yields

$$U\hat{\varphi}U^\dagger = \hat{\varphi} + \varphi_g, \tag{3.255}$$

$$U\hat{n}U^\dagger = \hat{n} + n_g. \tag{3.256}$$

The Hamiltonian then becomes

$$H = 4E_C(\hat{n})^2 - E_J \cos(\hat{\varphi} + \varphi_g) + \frac{1}{2}E_L(\hat{\varphi})^2. \tag{3.257}$$

As with the case of offset charge, we see that the resulting spectrum must be periodic in the offset flux.

The ability to control the three energy scales E_J, E_C, and E_L, as well as the offset flux φ_g, gives the experimentalist the opportunity to create a rich variety of level structures in the inductance-shunted family of qubits. The phase qubit [61, 62] is shunted by a lumped-element inductor and typically operates with $E_J/E_C \sim 10^4$, which makes it very nearly a harmonic oscillator. To increase the anharmonicity, it is operated with a large flux bias on the inductor, which drives a steady state current close to the critical current through the junction. In addition to greatly increasing the anharmonicity, this flux bias also lowers the energy-level spacing, so the device is typically constructed with large area junctions having a Josephson plasma frequency $\Omega_J/2\pi = \sqrt{8E_J E_C}/h$ that is very high,[16] about 50 GHz.

In the fluxonium qubit [9, 63], the large inductance (and correspondingly small E_L) is supplied by the kinetic inductance of a Josephson junction array. It also has the advantage that the anharmonicity can be large. Interestingly, the state-dependent polarizability and the excitation spectrum of this design are such that the qubit state can be measured even when the qubit transition frequency is driven down to very low values of order 0.5 GHz. When the external flux φ_g through the closed loop is zero, the potential energy landscape is that shown in Fig. (3.21c). The ground-state wavefunction has most of its support in the deep well near $\varphi = 0$ and there are two relatively closely spaced excited states with support in the two higher wells. This arrangement of three energy levels is known in atomic physics as a "V" configuration. When the external flux is $\varphi_g = \pi$, the situation reverses as shown in Fig. 3.21(a). There are two low-lying states with a large gap to the next excited state. This arrangement is known in atomic physics as a "Λ" configuration. As illustrated in the level scheme of Fig. 3.21(b), the fluxonium qubit has the very nice property that its energy spectrum varies smoothly between these two limits and, unlike the flux qubit, does not have exponential sensitivity to the external flux [9, 63]. A further interesting feature is that, unlike the transmon qubit, the anharmonicity of the level structure can be quite large and it is possible in principle in the "Λ" configuration to have the low energy level splitting much smaller than the frequency of the readout cavity and yet still achieve strong dispersive coupling to the cavity via virtual transitions to the third energy level. This allows efficient readout without suffering from the Purcell effect enhancement of the qubit decay rate via spontaneous fluorescence into the readout cavity [9, 63].

[16] As noted earlier, if the capacitance is dominated by the junction itself, then $E_J E_C$ is independent of junction area. When external and parasitic capacitance is included, the plasma frequency rises with junction area.

3.5 Noise-induced decoherence in qubit circuits

The ideal qubit would have a completely stable transition frequency unperturbed by the external environment and yet would have a port open to the outside world through which its state could be controlled and measured. Unfortunately, things are never ideal and it is our job as quantum engineers to design qubits for the real world. Unlike the Cs and Rb atoms used in atomic clocks, which are (literally) indistinguishable, each qubit we build is a unique individual. The good news is that we can engineer different classes of Hamiltonians and qubit Hamiltonians with different parameters within the same class. The bad news is that, in reality, if we can vary the Hamiltonian parameters, they will tend to vary on their own owing to various random sources of noise. This variation of parameters has two effects. First, it can modulate transition frequencies, which leads to dephasing of superpositions. Second, if the noise (including, as we will see, vacuum noise) is of high frequency, it can cause transitions to occur that change the state of the qubit. A related engineering quandary, which we will explore in more detail in the following, is that a qubit that is completely unperturbed by its environment is also decoupled from our measurement apparatus and hence cannot be read out!

The phenomenological Bloch equation from NMR is a useful starting point to understand the standard parametrization of qubit relaxation rates. In NMR, one typically studies the dynamics of a large collection of spin-1/2 (say) nuclei in a sample by coupling a drive/readout coil to the total magnetic moment. In the absence of any perturbations and assuming there are N spins, all having the same Zeeman energy (no “inhomogeneous broadening”), the Hamiltonian is simply

$$H = \frac{\Omega}{2} \sum_{j=1}^{N} \sigma_j^z. \tag{3.258}$$

The component of the magnetization parallel to the Zeeman field,

$$\hat{m}^z \equiv \frac{1}{N} \sum_{j=1}^{N} \sigma_j^z, \tag{3.259}$$

commutes with the Hamiltonian and so is a constant of the motion. The transverse components of the magnetization precess around the magnetic field direction. If we go to a frame rotating at frequency Ω, these transverse components of the magnetization are also constants of the motion (because the Hamiltonian then vanishes). Inhomogeneous broadening and various fluctuating perturbations will complicate this simple picture and cause the net magnetization to relax to its equilibrium value. This relaxation is characterized in the phenomenological Bloch equations that describe the dynamics of the components of the mean magnetization vector, $m^\mu \equiv \langle \hat{m}^\mu \rangle$, in terms of a longitudinal relaxation time T_1 and a transverse relaxation time T_2:

$$\dot{m}^z = -\frac{1}{T_1}(m^z - \bar{m}^z), \tag{3.260}$$

$$\dot{m}^x = -\frac{1}{T_2} m^x, \tag{3.261}$$

$$\dot{m}^y = -\frac{1}{T_2} m^y, \tag{3.262}$$

The longitudinal relaxation rate can be shown to be given by the sum of the transition rates from down to up and from up to down:

$$\frac{1}{T_1} = \Gamma_\uparrow + \Gamma_\downarrow. \tag{3.263}$$

The equilibrium magnetization is given by

$$\bar{m}^z = \frac{\Gamma_\uparrow - \Gamma_\downarrow}{\Gamma_\uparrow + \Gamma_\downarrow}. \tag{3.264}$$

The (ensemble) transverse relaxation rate can be shown to be given by[17]

$$\frac{1}{T_2^*} = \frac{1}{2T_1} + \frac{1}{T_\varphi}. \tag{3.265}$$

The first term represents the fact that if the spin makes a transition from up to down (or vice versa), the coherent superposition of up and down is destroyed. The second term containing $1/T_\varphi$ represents the rate at which the transverse magnetization decays owing to random fluctuations in the spin precession rate that cause the different spins to get out of phase with each other and thus destroy the mean transverse polarization.

We will be dealing not with a large ensemble of spins that are simultaneously measured (as in NMR), but rather with an individual spin subject to temporal (rather than spatial) fluctuations that cause the precession rate to vary from measurement to measurement. Since it takes many measurements to determine the average magnetization, we must average the effect of the noise across the ensemble of measurements on the single spin.[18] The mathematics is thus similar to the NMR case, but the physical difference in the meaning of the average being taken should be kept in mind.

[17] The notation in the NMR literature is that T_2^* refers to the ensemble transverse relaxation rate, which includes dephasing due to spin-lattice interactions, spatial inhomogeneities of the Zeeman field, and local chemical shifts of the Zeeman splitting. The relaxation time T_2 is that associated only with the homogeneous broadening of the NMR resonance due to spin-lattice interactions. T_2 can be measured using spin-echo methods to cancel out the inhomogeneous broadening. Strictly speaking, there is no unique phase coherence time. One should specify precisely the experiment used to measure the coherence. Thus, it is common to refer to T_2^{Ramsey} and T_2^{echo} in order to be precise. More complex echo sequences (e.g., CPMG) can yield still further measures of coherence.

[18] With a high-fidelity QND measurement of σ^z, one can determine that component of the spin in a single measurement. However, thermal fluctuations (e.g., the initial state of the spin is fluctuating in thermal equilibrium) and quantum fluctuations (e.g., the spin is actually pointing in the x direction, so the measurement result for σ^z is random) imply that one must repeat the experiment many times to determine the quantum/thermal ensemble average magnetization.

Exercise 3.17 Derive Eqs. (3.263) and (3.264).

We turn now to the microscopic noise that leads to the above phenomenology. Suppose that the qubit Hamiltonian contains a set of n parameters $\lambda_j;\, j \in \{1, \ldots, n\}$. Let us take the value of these parameters at the working point to be $\bar{\lambda}_j;\, j \in \{1, \ldots, n\}$, and let the deviation from the nominal working point due to noise in the parameter values be $\eta_j;\, j \in \{1, \ldots, n\}$. The qubit Hamiltonian can then be expressed in a Taylor series expansion

$$H[\vec{\lambda}] = H_0 + \sum_{j=1}^{n} \eta_j V_j + \frac{1}{2} \sum_{j,k=1,n} \eta_j \eta_k V_{jk} + \ldots, \tag{3.266}$$

where $H_0 \equiv H[\vec{\bar{\lambda}}]$ and

$$V_j \equiv \left. \frac{\partial H}{\partial \lambda_j} \right|_{[\vec{\bar{\lambda}}]}, \tag{3.267}$$

$$V_{jk} \equiv \left. \frac{\partial^2 H}{\partial \lambda_j \partial \lambda_k} \right|_{[\vec{\bar{\lambda}}]}. \tag{3.268}$$

The noise amplitudes η_j could represent classical random variables or quantum operators associated with the bath to which the qubit is coupled.

For simplicity, we will initially work only to first order in the expansion. If V_j commutes with H_0, then it is diagonal in the energy eigenbasis and cannot affect the eigenfunctions—it can only modulate the eigenvalues. Such perturbations can dephase superpositions, but cannot cause transitions between levels and hence cannot relax the population. On the other hand, if $[V_j, H_0] \neq 0$, then V_j is not diagonal in the energy basis and can cause both dephasing and relaxation.

As a simple example, consider the Hamiltonian of a two-level system:

$$H = \frac{\Omega}{2} \sigma^z + \vec{\eta}(t) \cdot \vec{\sigma}. \tag{3.269}$$

Let us first ignore the transverse field fluctuations and focus on the η_z longitudinal term, which commutes with H_0. For simplicity, we will take η_z to be a classical random variable. The exact time evolution operator in the interaction representation is

$$U(t_1, t_2) = e^{-\frac{i}{2}\theta(t_1,t_2)\sigma^z}, \tag{3.270}$$

where the random phase accumulated due to the longitudinal noise is given by

$$\theta(t_1, t_2) = 2 \int_{t_1}^{t_2} d\tau\, \eta_z(\tau). \tag{3.271}$$

Assuming η_z has zero mean, then θ also has zero mean and variance given by

$$\langle \theta^2 \rangle = 4 \int_{t_1}^{t_2} d\tau_1 \, d\tau_2 \, \langle \eta_z(\tau_1) \eta_z(\tau_2) \rangle. \tag{3.272}$$

If we assume that η_z itself is Gaussian-distributed, then θ will be Gaussian. Or if η_z is not Gaussian but its autocorrelation function appearing in Eq. (3.272) is short-ranged in time compared with $t_2 - t_1$, then (under rather mild assumptions) the central limit theorem applies because θ is the sum of a large number of independent random variables. If θ is Gaussian-distributed with mean zero, then the cumulant expansion converges at first order and we have the exact result

$$\langle\langle e^{i\theta} \rangle\rangle = e^{-\frac{1}{2}\langle\langle \theta^2 \rangle\rangle} \tag{3.273}$$

where $\langle\langle\rangle\rangle$ refers to the statistical ensemble average over the noise. If the autocorrelation time of the noise is short, we can write Eq. (3.272) as

$$\langle\langle \theta^2 \rangle\rangle \approx 4t S_{\eta_z \eta_z}(0), \tag{3.274}$$

where $S_{\eta_z \eta_z}[\omega]$ is the noise spectral density at frequency ω. The full expressions for the case where the autocorrelation time of the noise is not negligible are given in [65]. The important timescale $t_2 - t_1$ in Eq. (3.272) is given self-consistently by T_2^* itself. Noise that is slow on the timescale of T_2^* produces (the analog of) inhomogeneous broadening. Noise that is of high frequency on the scale of T_2^* makes little contribution to the integral in Eq. (3.272). For sufficiently large T_2^*, it is only low frequencies that count here, and it is generally safe to treat η_z classically and ignore that fact that it typically represents a quantum operator describing a bath variable. (This will not be true for the transverse noise fluctuations, where the high-frequency spectral density is important.)

Let us define the (pure) dephasing rate $1/T_\varphi$ via

$$\langle\langle e^{i\theta} \rangle\rangle = e^{-t/T_\varphi}, \tag{3.275}$$

from which we see that

$$\frac{1}{T_\varphi} = 2S_{\eta_z \eta_z}(0). \tag{3.276}$$

3.5.1 Density matrix description of decoherence

It turns out that the important fluctuations of the transverse fields are the ones resonant with the transition frequency of the qubit, because this is necessary for energy conservation in the transitions between qubit states. For such high-frequency fluctuations, we should more properly use a full quantum theory involving the quantum density matrix of the qubit. In the presence of quantum or classical noise, we desire the ensemble average over the noise of any given physical observable. Let p_j be the probability that an element of the ensemble is in state $|\psi_j\rangle$ (or equivalently that the random noise has driven our system to state $|\psi_j\rangle$). The expectation value of observable $\mathcal{O}$ is

$$\langle\langle \mathcal{O} \rangle\rangle = \sum_j p_j \langle \psi_j | \mathcal{O} | \psi_j \rangle = \mathrm{Tr}\, \rho \mathcal{O}, \tag{3.277}$$

where the density matrix is defined by

$$\rho = \sum_j p_j |\psi_j\rangle\langle\psi_j|. \tag{3.278}$$

Clearly, the density matrix must satisfy

$$\mathrm{Tr}\,\rho = 1, \tag{3.279}$$

$$\mathrm{Tr}\,\rho^2 \leq 1, \tag{3.280}$$

the latter being satisfied as an equality only for a pure state (a state where one of the p_j's is unity and the rest vanish). In fact, for a pure state, the density matrix is idempotent:

$$\rho^2 = \rho. \tag{3.281}$$

The von Neumann entropy is given by an expression closely analogous to the classical one:

$$S = -\mathrm{Tr}\,\{\rho \log \rho\} \tag{3.282}$$

Exercise 3.18 Prove Eqs. (3.280) and (3.281).

The density matrix contains all the information we need to compute the expectation value of any observable, or any moment of the distribution of some observable, for example

$$\langle\langle \mathcal{O}^m \rangle\rangle = \mathrm{Tr}\,\rho\mathcal{O}^m, \tag{3.283}$$

or even the full probability distribution for the measurement results for the observable,

$$P(\lambda) = \mathrm{Tr}\,\{\rho\delta(\lambda - \mathcal{O})\}\,. \tag{3.284}$$

Since a two-level qubit or a spin-1/2 particle has only two independent quantum states, the density matrix is 2×2. The most general such matrix that is Hermitian and has unit trace can be written

$$\rho = \frac{1}{2}\left(1 + \vec{m}\cdot\vec{\sigma}\right), \tag{3.285}$$

where the qubit polarization is given by $\vec{m} = \mathrm{Tr}\,\rho\vec{\sigma}$; that is,

$$m_j = \mathrm{Tr}\,\rho\sigma_j. \tag{3.286}$$

Exercise 3.19 Use Eq. (3.285) to prove Eq. (3.286).

Having now established these statistical results, we can proceed to ensemble-average the density matrix over the noise to obtain the reduced density matrix where the noise-bath degrees of freedom have been integrated out (traced over). In the interaction representation,

$$\langle\langle \rho(t_2)\rangle\rangle = \langle\langle U(t_1,t_2)\rho(t_1)U^\dagger(t_1,t_2)\rangle\rangle, \tag{3.287}$$

where we now take ρ to be the full density matrix for system plus bath and the interaction representation of the time evolution operator is given by the time-ordered exponential of the perturbation $\hat{V}(t)$ that couples the system to the bath:

$$U(t_1,t_2) = T_\tau \exp\left[-\frac{i}{\hbar}\int_{t_1}^{t_2} d\tau\, \hat{V}(\tau)\right] \tag{3.288}$$

Typically, $\hat{V}$ can be expressed as a product (or sum of products) of system operators and bath operators. Notice that we cannot separately average (i.e., trace over the bath) the two time evolution operators appearing in Eq. (3.287), because they are correlated—they depend on the same bath noise. It is convenient to do a perturbative calculation that will be valid for short times $t_2 = t_1 + \Delta t$. We assume the perturbation is weak enough that our perturbative expansion is in fact valid for times longer than the autocorrelation time of the noise, τ_c. In that case, we can safely choose a time step $\Delta t \gg \tau_c$, and the random noise variables in $U(t_1,t_2)$ will be essentially uncorrelated with the past noise variables that led to $\rho(t_1)$. This statistical independence will allow us to make the "no-memory" or Markov approximation [66], which will considerably simplify matters. Again, we can only justify this for noise that is sufficiently weak that low-order perturbation theory is still valid for the chosen value of Δt.

By definition, we choose $\langle\langle \hat{V}(\tau)\rangle\rangle = 0$, so we must (at least in the Markov approximation) take our expansion to second order (just as we do in Fermi's Golden Rule for transition rates):

$$U(t_1,t_2) \approx 1 - \frac{i}{\hbar}\int_{t_1}^{t_2} d\tau\, \hat{V}(\tau) - \frac{1}{2\hbar^2}\int_{t_1}^{t_2} d\tau\, d\tau'\, T_\tau \hat{V}(\tau)\hat{V}(\tau'). \tag{3.289}$$

The time-ordering operator will prove inconvenient, but we can write

$$T_\tau \hat{V}(\tau)\hat{V}(\tau') = \hat{V}(\tau)\hat{V}(\tau') + \theta(\tau'-\tau)[\hat{V}(\tau'),\hat{V}(\tau)]. \tag{3.290}$$

The commutator term does not necessarily vanish, but it turns out that it is possible to lump its effect into the system Hamiltonian, where it leads to small corrections to the system energy levels (e.g., the Lamb shift of atomic levels). We presume that this has already been done and so this commutator term drops out. The reader is directed to [67, 68] for further discussion of these points.

Anticipating being able to drop the terms linear in $\hat{V}$ when we trace out the bath (because they have zero mean and, within the Markov approximation, no correlations with the prior noise that led to $\rho(t_1)$), we are thus led to

$$\begin{aligned}\rho(t_2) = \rho(t_1) + \frac{1}{\hbar^2}\int_{t_1}^{t_2} d\tau\, d\tau' \Big[&\hat{V}(\tau)\rho(t_1)\hat{V}(\tau') \\ &- \frac{1}{2}\hat{V}(\tau)\hat{V}(\tau')\rho(t_1) - \frac{1}{2}\rho(t_1)\hat{V}(\tau)\hat{V}(\tau')\Big].\end{aligned} \tag{3.291}$$

In order to make further progress, we now need to express the perturbation in terms of system and bath variables and trace out the bath variables to obtain an expression for the time evolution of the reduced density matrix describing the system alone. Let us suppose that, in the interaction picture, the bath coupling has the form

$$\frac{1}{\hbar}\hat{V}(\tau) = \hat{\eta}_z(\tau)\hat{\sigma}^z(\tau) + \hat{\eta}_x(\tau)\hat{\sigma}^x(\tau) \tag{3.292}$$

$$= \hat{\eta}_z(\tau)\sigma^z + \hat{\eta}_x(\tau)(e^{i\Omega\tau}\sigma^+ + e^{-i\Omega\tau}\sigma^-). \tag{3.293}$$

To keep the discussion simple, we will assume that the longitudinal and transverse noises are uncorrelated and can therefore be treated separately in our perturbative treatment. We start with the longitudinal noise and will attempt to reproduce our previous semiclassical result in Eq. (3.276). Introducing the notation

$$\mathrm{Tr}_{\,\mathrm{bath}}\rho(t) \equiv \langle\langle \rho(t)\rangle\rangle, \tag{3.294}$$

we can write

$$\langle\langle \rho(t_1+\Delta t)\rangle\rangle = \langle\langle \rho(t_1)\rangle\rangle + \int_{t_1}^{t_2} d\tau\, d\tau'\, \mathrm{Tr}_{\,\mathrm{bath}}\Big\{\hat{\eta}_z(\tau)\sigma^z\rho(t_1)\sigma^z\hat{\eta}_z(\tau')$$
$$-\frac{1}{2}\hat{\eta}_z(\tau)\hat{\eta}_z(\tau')\sigma^z\sigma^z\rho(t_1) - \frac{1}{2}\rho(t_1)\hat{\eta}_z(\tau)\hat{\eta}_z(\tau')\sigma^z\sigma^z\Big\}. \tag{3.295}$$

Using the cyclic property of the trace in the first term, interchanging the dummy labels τ and τ' in the remaining terms, and invoking the Markov approximation, we obtain

$$\langle\langle \rho(t_1+\Delta t)\rangle\rangle = \langle\langle \rho(t_1)\rangle\rangle + \int_{t_1}^{t_2} d\tau\, d\tau'\, \langle\langle \hat{\eta}_z(\tau')\hat{\eta}_z(\tau)\rangle\rangle$$
$$\times\left[\sigma^z\langle\langle \rho(t_1)\rangle\rangle\sigma^z - \frac{1}{2}\sigma^z\sigma^z\langle\langle \rho(t_1)\rangle\rangle - \frac{1}{2}\langle\langle \rho(t_1)\rangle\rangle\sigma^z\sigma^z\right]. \tag{3.296}$$

Of course, $\sigma^z\sigma^z = 1$, but, for notational symmetry, we will not use this. Invoking the Markov approximation as in Eq. (3.274) and defining

$$\gamma_\varphi = 2S_{\eta_z\eta_z}(0), \tag{3.297}$$

we have

$$\langle\langle \rho(t_1+\Delta t)\rangle\rangle = \langle\langle \rho(t_1)\rangle\rangle + \Delta t\frac{\gamma_\varphi}{2}D[\sigma^z]\langle\langle \rho(t_1)\rangle\rangle, \tag{3.298}$$

where the "dissipator" D is called the Lindblad superoperator:[19]

$$D[\mathcal{O}]\rho \equiv \mathcal{O}\rho\mathcal{O}^\dagger - \frac{1}{2}\mathcal{O}^\dagger\mathcal{O}\rho - \frac{1}{2}\rho\mathcal{O}^\dagger\mathcal{O}. \tag{3.299}$$

[19] A "superoperator" is an operator on the N^2-dimensional space of ρ if ρ is written as a column vector of length N^2 instead of an $N \times N$ matrix.

Even though Δt is not supposed to be too small, if we are appropriately cautious about the meaning, we can interpret this difference equation as a differential equation for the time evolution of the reduced density matrix:

$$\frac{d}{dt}\langle\langle\rho(t)\rangle\rangle = \frac{1}{2}\gamma_\varphi D[\sigma^z]\langle\langle\rho(t)\rangle\rangle. \tag{3.300}$$

Let us now try to go from this result to the time evolution of the qubit polarization. Invoking Eq. (3.285), we have

$$\frac{d}{dt}\langle\langle\rho(t)\rangle\rangle = \frac{1}{2}\frac{d\vec{m}}{dt}\cdot\vec{\sigma} = \frac{1}{2}\gamma_\varphi D[\sigma^z]\frac{1}{2}(1+\vec{m}\cdot\sigma). \tag{3.301}$$

Using the fact that the different Pauli matrices anticommute, this can be rewritten

$$\frac{d\vec{m}}{dt}\cdot\vec{\sigma} = -\gamma_\varphi(m_x\sigma^x + m_y\sigma^y). \tag{3.302}$$

From this, we immediately obtain

$$\dot{m}^z = 0, \tag{3.303}$$

$$\dot{m}^x = -\frac{1}{T_\varphi}m^x, \tag{3.304}$$

$$\dot{m}^y = -\frac{1}{T_\varphi}m^y, \tag{3.305}$$

in agreement with the phenomenological Bloch equations (3.260)–(3.262) with $T_1 = \infty$.

In order to obtain a finite energy relaxation rate, we must allow for noise in the transverse field components. Repeating the preceding derivation and defining the up and down transition rates (which agree with the derivation of Fermi's Golden Rule in [31]) in terms of the noise spectral densities

$$\Gamma_\downarrow = S_{\eta_x\eta_x}(+\Omega), \tag{3.306}$$

$$\Gamma_\uparrow = S_{\eta_x\eta_x}(-\Omega), \tag{3.307}$$

Eq. (3.300) becomes the standard master equation [69] in Lindblad form:

$$\frac{d}{dt}\langle\langle\rho(t)\rangle\rangle = \frac{\gamma_\varphi}{2}D[\sigma^z]\langle\langle\rho(t)\rangle\rangle + \Gamma_\uparrow D[\sigma^+]\langle\langle\rho(t)\rangle\rangle + \Gamma_\downarrow D[\sigma^-]\langle\langle\rho(t)\rangle\rangle. \tag{3.308}$$

As discussed extensively in [31], the bath noise spectral density at frequency $+\Omega$ is a measure of the ability of the bath to absorb energy and thereby de-excite the qubit, while the spectral density at $-\Omega$ gives the ability of the bath to emit energy at that frequency, thereby exciting the qubit. Classically, there is no distinction, since the noise correlators are real-valued, but quantum mechanically we must distinguish positive and negative frequency because the noise correlators are complex-valued (which is possible since the noise operators are Hermitian, but their product is not: they do not commute with each other at different times [31]). Evaluating the above, we arrive at the full Bloch equations (3.260)–(3.262).

The Bloch equations in NMR refer to the density matrix for a large ensemble of spins. Here, we are dealing with the ensemble of many repeated experiments on a single "spin." In NMR, inhomogeneous line broadening usually comes from different parts of the sample having slightly different spin precession frequencies. Here, there is only one spin, and inhomogeneous broadening arises from slow temporal (rather than spatial) fluctuations of the qubit splitting over the course of many repeated measurements. Another difference with NMR is that typically in NMR we are in the regime $k_BT \gg \hbar\Omega$ and so $\Gamma_\uparrow \approx \Gamma_\downarrow$. With qubits, we are (typically) in the opposite regime, with high frequency and low temperature where $\Gamma_\downarrow \gg \Gamma_\uparrow$.

Exercise 3.20 Derive Eq. (3.262) in detail from Eq. (3.308).

We emphasize that the Markov approximation is not necessarily valid. For instance, the qubit could be looking at the environmental noise filtered by a resonator or cavity, or, in the case of dephasing noise at low frequencies, the spectrum of the noise may be $1/f$ rather than white. The former can be handled by adding the filter (as a harmonic oscillator) to the Hamiltonian. The latter complicates the analysis and leads to non-exponential decay [65].

Dephasing and read out of charge qubits For generic values of the gate charge, the ground state Ψ_0 and excited state Ψ_1 of a charge qubit differ in their respective static electric "dipole moments." Using Eq. (3.466), we see that the state energy depends on bias voltage V_B fluctuations via the "dipole moment"

$$\tilde{p}_j = \langle\Psi_j|\frac{\partial H}{\partial V_\mathrm{B}}|\Psi_j\rangle \tag{3.309}$$

$$= \beta\langle\Psi_j|\hat{Q}_1|\Psi_j\rangle, \tag{3.310}$$

where we have ignored the last term in Eq. (3.466) since it is not state-dependent. Nakamura *et al.* [5] used the dependence of a certain quasiparticle tunneling rate on p_j to differentiate between the ground state ($j = 0$) and the excited state ($j = 1$) and hence read out the state of the qubit. Aassime *et al.* [70] and Lehnert *et al.* [71] developed an RF single-electron transistor readout scheme for charge-based qubits.

Unfortunately, in the regime where charge-based readout works, a stray electric field $\mathcal{E}$ (fluctuation in bias voltage V_B) causes a first-order perturbation theory shift of the qubit excitation frequency by an amount

$$\delta\omega_{01} = \frac{1}{\hbar}\delta V_B(\tilde{p}_1 - \tilde{p}_0). \tag{3.311}$$

This leads to very rapid dephasing of quantum superpositions at a rate [65]

$$\frac{1}{T_\varphi} = \frac{1}{2}\left(\frac{\tilde{p}_1 - \tilde{p}_0}{\hbar}\right)^2 S_{V_BV_B}, \tag{3.312}$$

where $S_{V_BV_B}$ is the bias voltage spectral density at low frequencies. The total decoherence rate is then given by

$$\frac{1}{T_2^*} = \frac{1}{2T_1} + \frac{1}{T_\varphi}. \tag{3.313}$$

Recognizing this difficulty, Vion *et al.* [12] introduced the notion of operating charge qubits at a "sweet spot" where the transition frequency is an extremum with respect to the voltage and magnetic field control parameters. Several such extrema (known in atomic physics as "clock points") are visible (at half-integer gate-charge points) in the CPB spectrum shown in Fig. 3.19. At these special points, the leading term in the Taylor series expansion presented in Eq. (3.266) vanishes and only the second-order effects of the noise contribute to the dephasing. The qubit coherence time is therefore dramatically increased [12]. Measurements by Devoret's group have demonstrated that these coherence lifetimes are indeed consistent with the expected second-order effects of the noise determined by the curvature of the spectrum at the extremum [72].

We now arrive at an interesting quandary. By tuning the qubit to the sweet spot, the environment is no longer able to detect which state the qubit is in, based on coupling to its electric dipole moment. This is why the coherence time is so dramatically enhanced. But, if the environment cannot measure the state of the qubit by looking at the dipole moment, neither can we! The Saclay group recognized this and developed the concept of reading out the qubit by measuring the state-dependent susceptibility (inductance of the quantronium qubit). Rather than going into the details of this, it is easier in the context of the present discussion to instead think about the closely related state-dependent susceptibility method based on capacitance developed by the Yale group [32, 73], which will be explained in detail in Section 3.6. Because the offset charge is essentially equivalent to an applied voltage, and the potential energy of a capacitor is $\frac{1}{2}CV^2$, the second derivative of the transition energy with respect to n_g, which can be derived from Eq. (3.213) (and is illustrated in Fig. 3.19), is essentially the difference in quantum capacitance [54, 74–78] presented to an external probing field when the qubit is in the ground and excited states. Essentially, this effect was used by the Yale group in developing the dispersive readout [32, 73] based on Eq. (3.326). Working with a low-frequency probe, precisely this effect was measured in [77, 78]. The difference is that the high-frequency dispersive probe depends on the matrix elements related to the quantum capacitance, but, as is clear from (3.326), it also depends on the detuning of the qubit and resonator frequencies. The importance of this difference will become clear in the following.

The invention of the transmon qubit brings us to our next major quandary. As noted in Fig. 3.19, by going to large E_J/E_C, one can dramatically reduce the curvature of the spectrum at the extrema and eventually make the transition frequency exponentially close to constant independent of the gate charge [57, 64]. Operation in this transmon regime leads to even longer lifetimes and long-term qubit frequency stability [6, 64]. If the energy eigenvalues are essentially independent of the offset charge, then neither we nor the environment can read the state of the qubit using either the dipole moment or the susceptibility (quantum capacitance), since neither is dependent on the quantum state. While this explains the even longer coherence times of the transmon, we are left to wonder how it is that the dispersive readout still works even though quantum capacitance is zero in both states. Recall that if the qubit were actually a perfect harmonic oscillator, the transition frequencies would not respond at all to changes in offset charge (displacement of the origin of the oscillator). It is obvious from classical considerations that the susceptibility would be a constant (given by the inverse of the spring constant) independent of the state. The oscillator is highly polarizable and responds strongly to slow variations in offset charge by being displaced, but this displacement to a new equilibrium position has no effect on the excitation spectrum. As noted already, the transmon comes exponentially close to this ideal behavior, and yet the dispersive readout still works. This is because, while the charge dispersion falls off exponentially, the transmon retains its anharmonicity. As can be seen from Eq. (3.326), the cavity pull due to the virtual polarization of the qubit is strongly dependent on the detuning between the qubit and cavity. For the case of the multilevel transmon, the expression for the cavity pull has to be re-derived, but the essential point is that the detuning for the $0 \to 1$ transition is not the same as that for the $1 \to 2$ transition, and so the cavity pull is state-dependent, provided that the cavity frequency is reasonably close to the qubit. For a very low-frequency cavity, we are back in the regime measuring the quantum capacitance where the effect on the cavity is small.

The fluxonium qubit [9, 63] can be operated in a regime that maintains very large anharmonicity even while suppressing the effects of offset charge noise. The fluxonium energy level structure therefore gives the advantage over the transmon that the qubit transition frequency can be far removed from the cavity readout frequency, which prevents Purcell enhancement of the excited-state decay rate.

3.6 Introduction to cavity and circuit QED

Quantum electrodynamics (QED) studies atoms and electrons coupled to the quantum fluctuations of the electromagnetic field. Cavity QED engineers those quantum fluctuations of the vacuum by surrounding the atoms with a resonant cavity that supports only discrete modes of the electromagnetic field. By adjusting the frequencies (and damping) of those resonant modes with respect to the transition frequency of the atoms, one can dramatically alter the coupling of the atoms to their quantum environment (and conversely alter the photon modes via their coupling to the atoms). Because the cavity traps photons, they interact with the atoms repeatedly and one can enter a regime of strong coupling in which the natural excitations of the system are coherent superpositions of atom and photon excitations known as polaritons.

There is a long history of cavity QED studies in the atomic, molecular, and optical physics (AMO) community for both alkali atoms in optical cavities [66, 79–82] and Rydberg atoms in microwave cavities [67, 83–87]. In the optical case, one typically monitors the effect of the atoms on the photons transmitted through (or reflected from) the cavity. It is not possible to measure the state of the atoms after they have fallen through the cavity, because the spontaneous emission lifetime is of the order of nanoseconds at optical frequencies. In the microwave experiments pioneered by the Paris group, it is difficult to directly measure the microwave photons but relatively easy to measure the state of the Rydberg atoms with very high fidelity after they exit the cavity, since they have a lifetime of approximately 30 ms and can be probed with state-selective ionization.

"Circuit QED" uses superconducting qubits as artificial atoms coupled to microwave resonators [1, 26, 32, 73, 88], as illustrated schematically[20] in Fig. 3.22. Measuring the amplitude and phase of microwaves transmitted through (or reflected from) the resonator realizes the equivalent of optical cavity QED at microwave frequencies. It is interesting to note that (typically) there is just a single microwave signal channel running to and from the cavity. Application of microwaves near the cavity frequency excites the cavity and performs a measurement of the state of the qubit. On the other hand, application of microwaves at the qubit transition frequency (assumed to be strongly detuned from the cavity frequency) does not strongly excite the cavity and, most importantly, does *not* perform a measurement. Rather, such microwaves resonant with the qubit can be used to perform single-qubit rotations. Thus, we have a very convenient "frequency multiplexing" in which a single wire carries out different operations depending on the frequency of the microwaves applied to it.

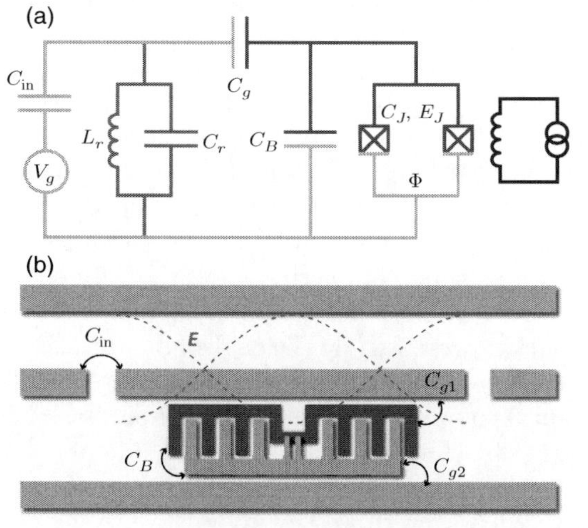

Fig. 3.22 Circuit QED: schematic illustration (not to scale) of a transmon qubit embedded in a coplanar waveguide resonator. (a) shows the lumped-element circuit equivalent to the distributed circuit shown in (b). From Koch *et al.* [57].

[20] There exists a dual geometry in which the Josephson junction qubit is placed in line with the center pin of the resonator and couples directly to the microwave currents flowing in the resonator [26, 89]. In this dual geometry, the fine structure constant is replaced by its *inverse* and the problem is engineering the circuit to *reduce* the coupling to manageable levels.

In recent years, there have been many theoretical proposals for coupling qubits to either three-dimensional cavities or lumped-element resonators and for protocols for analyzing microwave photon states (e.g., [90–103]), and there has been a flurry of experiments on circuit QED [6, 13–17, 21–24, 32, 58, 73, 88, 104–165].

In optical cavity QED, one uses a Fabry-Pérot cavity that (because optical wavelengths are so small) is typically very large compared with a wavelength of light at the frequencies associated with atomic spectra. The frequency of one of the longitudinal modes trapped by the cavity can be adjusted to be close to the transition frequency of the atoms that are dropped into (or, nowadays, trapped within) the cavity. While the coupling to this one special mode can be significantly modified by the cavity, the spontaneous emission rate of the atom into modes not trapped by the cavity is scarcely affected. The mirrors are relatively far apart and the atom sees a lot of "free space." Hence, its spontaneous emission time $\gamma^{-1} \sim 1\,\mathrm{ns}$ remains largely unaffected. The strong-coupling limit where the atom coherently exchanges energy with the one special longitudinal mode more rapidly than it decays into the continuum of other modes requires considerable effort to achieve [80–82].

In circuit QED, extremely strong coupling is trivial to obtain (the atoms can always be made with antennae as large as needed to achieve strong coupling). In fact, it requires care to obtain weak coupling, and the main reason that the cavity is useful is to protect the qubit from the environment by filtering out the vacuum noise which causes spontaneous emission. This physics of using a cavity to modify the coupling to the environment is that of the Purcell effect [166], first observed[21] for a superconducting qubit by [119]. Here, a qubit placed inside a cavity can have its decay rate suppressed if it is far detuned from the cavity resonance or enhanced if the qubit transition frequency is close to a (broad) cavity resonance. The former is useful for protecting quantum superpositions. The latter is useful for providing rapid qubit reset to the ground state. It has also been successfully used to generate single microwave photons on demand and enhance the fidelity of coherent quantum information transfer from a superconducting qubit to a "flying" photon qubit [13]. One can view the Purcell effect as the resonator performing an impedance transformation on the external dissipation presented by the environment to the qubit [119]. The UCSB group [155, 156, 168] has used a tunable transformer coupling to quantitatively explore the role of environmental coupling in a phase qubit circuit over a wide range of coupling strengths and to sculpt the shape of the temporal mode of the photon spontaneously emitted by an excited qubit. Houck's group has utilized a different approach to control the spontaneous emission waveform. They rapidly vary the effective dipole moment of the qubit on timescales short compared with the spontaneous emission time [169]. The ETH group has taken still another approach [170].

[21] In an early pioneering paper, the Saclay group [167] studied a related dissipation time-delay effect through the variation of the macroscopic quantum tunneling rate of a Josephson junction coupled to a transmission line feeding a microwave absorber whose position along the transmission line could be varied *in situ*.

The coupling between the electric field in the cavity and the dipole moment of a (single) atom at position $\vec{R}$ is given by[22]

$$U = -\vec{p} \cdot \vec{E}(\vec{R}). \tag{3.314}$$

As we have seen, each discrete mode of a resonator is an independent harmonic oscillator. We will assume for now that only a single mode (the one closest in frequency to the atom transition) is important. (Of course, it is important to remember that this is not always a good approximation.) The electric field in the cavity (or voltage in a circuit) can be written in terms of the mode polarization direction $\hat{\epsilon}$ and the mode zero-point fluctuation amplitude E_{ZPF} (computed at position $\vec{R}$) and the photon operators as[23]

$$\vec{E} = \hat{\epsilon} E_{\mathrm{ZPF}}(\hat{a} + \hat{a}^\dagger). \tag{3.315}$$

We assume that the atom can be approximated as a two-level system. The dipole moment operator connects the ground and excited states of the atom, and so the interaction Hamiltonian in Eq. (3.314) becomes

$$H_1 = \hbar g(\hat{a} + \hat{a}^\dagger)\sigma^x, \tag{3.316}$$

where σ^x flips the two-level system between states and the "vacuum Rabi coupling" g is given by the dipole matrix element connecting the ground and excited states of the atom:

$$\hbar g = -\langle \psi_1 | \vec{p} \cdot \hat{\epsilon} | \psi_0 \rangle E_{\mathrm{ZPF}}, \tag{3.317}$$

where we again emphasize that E_{ZPF} is the vacuum fluctuation amplitude of the cavity mode at $\vec{R}$, the position of the atom. In atomic cavity QED, the atoms move around and the coupling is strongest at the antinodes of the mode and vanishes at the nodes. In circuit QED, we have the advantage of being able to place the artificial atoms at any desired fixed location within the cavity.

Eq. (3.316) can be rewritten as

$$H_1 = \hbar g(\hat{a}\sigma^+ + \hat{a}^\dagger\sigma^-) + \hbar g(\hat{a}\sigma^- + \hat{a}^\dagger\sigma^+), \tag{3.318}$$

where

$$\sigma^\pm = \frac{1}{2}(\sigma^x \pm i\sigma^y) \tag{3.319}$$

[22] This can be equivalently written in terms of the electromagnetic vector potential and the electron momentum $\sim \vec{P} \cdot \vec{A}$. We neglect the $\vec{A} \cdot \vec{A}$ term, which leads to transitions in which the photon is not absorbed or emitted but rather scattered to a new state (Raman scattering) by the electron. For additional discussion of this topic, see [107, 171–173].

[23] We make a certain gauge choice here and could have used $\vec{E} = -i\hat{\epsilon}E_{\mathrm{ZPF}}(\hat{a} - \hat{a}^\dagger)$ to be closer to the choice made in discussing the voltage of the LC oscillator in Eq. (3.53). Note that one-dimensional coplanar waveguide resonators for microwaves do not have a polarization degree of freedom. Three-dimensional microwave cavities and optical cavities of course do.

are the raising and lowering operators for the qubit energy. If the cavity mode is close in frequency to the atom transition frequency, the first term is important because it nearly conserves the total energy. The second term only mixes states that are far away from each other, or, equivalently, it is rapidly rotating in the interaction picture. Dropping this term is called the rotating-wave approximation (RWA). This is often an excellent approximation, but is not necessarily valid if the coupling is very strong or if the detuning of the atom from the cavity is large so that no one cavity mode is singled out.

In this simplest approximation of a two-level atom coupled to a single cavity mode (within the RWA), the system is described by the Jaynes-Cummings Hamiltonian

$$H = H_0 + V, \tag{3.320}$$

with

$$H_0 = \hbar\omega_{\rm c}\hat{a}^\dagger\hat{a} + \frac{\hbar\omega_{01}}{2}\sigma^z, \tag{3.321}$$

where the single cavity mode is described as a simple harmonic oscillator of angular frequency $\omega_{\rm c}$, the two-level atom is represented as a simple spin-1/2 with excitation frequency ω_{01},

$$V = \hbar g(\hat{a}\sigma^+ + \hat{a}^\dagger\sigma_-), \tag{3.322}$$

and the vacuum Rabi coupling, g represents the dipole matrix element for the process in which the atom absorbs or emits a photon. For the case of a superconducting qubit coupled to an LC resonator, this Hamiltonian is derived in detail in the appendix (Section 3.9.1).

The external driving and damping terms, not written explicitly here, which help control the electromagnetic state of the cavity, are treated using the input-output formalism of quantum optics [31] described in the appendix (Section 3.9.2). The extension of this Hamiltonian to the case of multiple qubits is known as the Tavis-Cummings model [174]. If we had not made the RWA, we would have the so-called Rabi Hamiltonian, which in a certain sense is also integrable [175].

The full Hamiltonian with which we must deal,

$$H = \hbar\omega_{\rm c}\hat{a}^\dagger\hat{a} + \frac{\hbar\omega_{01}}{2}\sigma^z + \hbar g(\hat{a}\sigma^+ + \hat{a}^\dagger\sigma_-) + H_{\rm drive} + H_{\rm damping}, \tag{3.323}$$

includes environmental terms that produce the spontaneous emission of energy from the atom at some rate γ. In addition, in order to excite the atoms with an external drive, one must open a port into the cavity, which then allows spontaneous emission of energy from the cavity at some rate κ (or for a two-sided cavity at rates κ_L and κ_R).

Let us begin our analysis by considering the spectrum of H_0 where the atom and photon are not yet coupled. In Fig. 3.23, we show two ladders of photon states, one for the qubit being in the ground state ($|g\rangle = |\uparrow\rangle$) and the other for the qubit being in the excited state ($|e\rangle = |\downarrow\rangle$). (There is no uniform convention on whether the ground

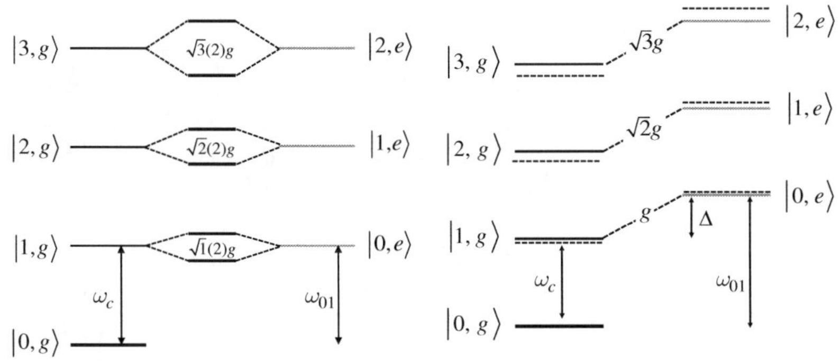

Fig. 3.23 Jaynes-Cummings ladder or "dressed-atom" level structure. (a) Degenerate case $\omega_{01} = \omega_c$. The degenerate levels mix and split by an amount proportional to the vacuum Rabi splitting g. (b) Dispersive case $\omega_{01} = \omega_c + \Delta$. For $\Delta > 0$, the level repulsion causes the cavity frequency to decrease when the qubit is in the ground state and to increase when the qubit is in the excited state.

state should be labeled by spin up or down.) In Fig. 3.23(a), we see the case where the detuning between the qubit frequency and the cavity frequency,

$$\Delta \equiv \omega_{01} - \omega_c, \tag{3.324}$$

vanishes, and in Fig. 3.23(b), we see the case where Δ is positive. The detuning is considered strong if $\Delta \gg g$.

First we consider the case $\Delta = 0$ where the states $|n+1, g\rangle$ and $|n, e\rangle$ are degenerate. As shown in Fig. 3.23(a), this degeneracy is lifted by the dipole coupling matrix element, resulting in an energy splitting $2g$ for the lowest pair and $2g\sqrt{n+1}$ for the higher levels. The splitting $2g$ of the lowest pair of excitations is called the vacuum Rabi splitting [73, 80–82, 87, 107]. The energy eigenstates of $H_0 + V$ are coherent superpositions

$$|\Psi_\pm\rangle = \frac{1}{\sqrt{2}} \left(|n+1, g\rangle \pm |n, e\rangle\right) \tag{3.325}$$

("bonding-antibonding" combinations) of photon excitation and qubit excitation known in the condensed matter and AMO literature as polaritons. The coupling available in circuit QED is now so strong that splittings of about 150 MHz corresponding to about 300 qubit linewidths are easily achieved [117, 124]. The higher-lying excited states form a strongly anharmonic ladder, which can be explored by either strong driving or use of two excitation tones [116, 117, 124].

In the so-called dispersive regime where the qubit is far detuned from the cavity ($|\omega_{01} - \omega_c| \gg g$), we will see below that diagonalization of the Hamiltonian to lowest order in g leads to a second-order dispersive coupling

$$V_{\text{dispersive}} = \hbar \frac{g^2}{\Delta} \left(\hat{a}^\dagger \hat{a} + \frac{1}{2} \right) \sigma^z. \tag{3.326}$$

This coupling is quantum nondemolition (QND) with respect to both photon number and qubit polarization, since it commutes with both. The dispersive coupling can be interpreted either as a shift in the cavity frequency that depends on the state of the qubit, or as the "ac-Stark" or "light" shift (plus the Lamb shift [32, 104, 120]) of the qubit frequency proportional to the number of photons in the cavity. The qubit-state-dependent shift of the cavity frequency leads to changes in the amplitude and phase of photons reflected from or transmitted through the cavity and is the basis of the QND readout of the qubit state in circuit QED [32, 73]. The mean value of the light shift can be used to rapidly tune qubit transition frequencies [104, 106, 114, 176]. The fluctuating part of the light shift can be viewed as the quantum back-action [31] of the qubit measurement. As required by the principles of quantum measurement [31], the photon shot noise [177] in the cavity gradually dephases the qubit superposition as information is gained about σ^z. This backaction effect leads to a broadening of the spectroscopic linewidth of the qubit [104, 106, 178–181]. In the so-called 'strong-dispersive' regime [110], the coupling is so large that the light shift per photon exceeds both the cavity linewidth κ and the atom linewidth γ: $g^2/\Delta > \kappa, \gamma$. In this regime, the qubit spectrum breaks up into a series of separately resolved peaks representing the distribution of photon numbers within the driven cavity [110]. This "photon number" detector was used to distinguish thermal and coherent states in the cavity and could be used to measure number-squeezed states and other non-classical states [110]. Most recently, Paik *et al.* [6] have observed light shifts several orders of magnitude larger than the qubit linewidth. This strong-coupling physics was first beautifully observed in the time domain by the Paris group [83–86].

To derive Eq. (3.326), we need to find a unitary transformation

$$U = e^{\hat{\eta}}, \; U^\dagger = e^{\hat{\eta}^\dagger} = e^{-\hat{\eta}}, \tag{3.327}$$

that removes the off-diagonal term in Eq. (3.322) that is first-order in g. Using the Baker-Campbell-Hausdorff expansion, we have

$$\tilde{H} = UHU^\dagger = H + [\hat{\eta}, H] + \frac{1}{2}[\hat{\eta}, [\hat{\eta}, H]] + \ldots. \tag{3.328}$$

In perturbation theory, we write $H = H_0 + V$, where V is the off-diagonal term linear in g and we expect $\hat{\eta}$ to therefore be of order g. Expanding to second order in g, we have

$$\tilde{H} \approx H_0 + V + [\hat{\eta}, H_0] + [\hat{\eta}, V] + \frac{1}{2}[\hat{\eta}, [\hat{\eta}, H_0]]. \tag{3.329}$$

We need to choose $\hat{\eta}$ to satisfy

$$[\hat{\eta}, H_0] = -V \tag{3.330}$$

in order to remove the lowest-order off-diagonal term. We are then left with only the second-order terms, which conveniently combine to

$$\tilde{H} = H_0 + \frac{1}{2}[\hat{\eta}, V]. \tag{3.331}$$

It is straightforward to verify that the solution to Eq. (3.330) is

$$\hat{\eta} = \frac{g}{\Delta}(\hat{a}\sigma^+ - \hat{a}^\dagger\sigma^-). \tag{3.332}$$

Clearly, the expansion in g is valid only for large enough detuning that $g/\Delta \ll 1$. (On the other hand, if Δ becomes too large, the RWA fails, so caution is required.) Computing the second-order term (and dropping an irrelevant constant), we arrive at the dispersive Hamiltonian

$$\tilde{H} = H_0 + \chi\left(\hat{a}^\dagger\hat{a} + \frac{1}{2}\right)\sigma^z, \tag{3.333}$$

where

$$\chi \equiv \hbar\frac{g^2}{\Delta}. \tag{3.334}$$

For the case of a closed cavity (no damping and no drive), we can do better than this dispersive approximation. Without drive or damping, the Hamiltonian in Eq. (3.320) has the special property (due to the RWA) that it commutes with the excitation number:

$$[H_0 + V, \hat{N}_{\text{ex}}] = 0, \tag{3.335}$$

with

$$\hat{N}_{\text{ex}} \equiv \hat{a}^\dagger\hat{a} + \frac{1+\sigma^z}{2} \tag{3.336}$$

being the sum of the cavity excitation number and the qubit excitation number. This extra symmetry means that the Hamiltonian is block-diagonal. The only state to which $|n,\uparrow\rangle$ is coupled by V is $|n+1,\downarrow\rangle$. Hence, the largest block we have to diagonalize is 2×2. The ground state $|0,\downarrow\rangle$ is (in the RWA) not connected to any other state by V, and hence is an eigenstate on its own. For the higher pairs of states containing a total of $n+1$ excitations, we can write

$$\Psi^{(n+1)} = \alpha|n,\uparrow\rangle + \beta|n+1,\downarrow\rangle, \tag{3.337}$$

for which the 2×2 eigenvalue problem becomes

$$H^{(n+1)}_{2\times 2}\begin{pmatrix}\alpha\\ \beta\end{pmatrix} = E^{(n)}_{\pm}\begin{pmatrix}\alpha\\ \beta\end{pmatrix}, \tag{3.338}$$

where

$$H_{2\times 2}^{(n+1)} = \hbar \begin{pmatrix} n\omega_c + \frac{1}{2}\omega_{01} & g\sqrt{n+1} \\ g\sqrt{n+1} & (n+1)\omega_c - \frac{1}{2}\omega_{01} \end{pmatrix} \tag{3.339}$$

$$= \left(n + \frac{1}{2}\right)\hbar\omega_c + \hbar \begin{pmatrix} \frac{+\Delta}{2} & g\sqrt{n+1} \\ g\sqrt{n+1} & -\frac{\Delta}{2} \end{pmatrix}. \tag{3.340}$$

This has eigenvalues

$$E_{\pm}^{(n+1)} = \left(n + \frac{1}{2}\right)\hbar\omega_c \pm \frac{\hbar}{2}\sqrt{\Delta^2 + 4g^2(n+1)} \tag{3.341}$$

and the eigenfunctions in Eq. (3.337) are given by

$$\Psi_+ = \begin{pmatrix} \alpha_+ \\ \beta_+ \end{pmatrix} = \begin{pmatrix} \cos(\theta/2), \\ \sin(\theta/2) \end{pmatrix}, \tag{3.342}$$

$$\Psi_- = \begin{pmatrix} \alpha_- \\ \beta_- \end{pmatrix} = \begin{pmatrix} -\sin(\theta/2) \\ \cos(\theta/2) \end{pmatrix}, \tag{3.343}$$

where

$$\theta = \tan^{-1}\left(\frac{2g\sqrt{n+1}}{\Delta}\right). \tag{3.344}$$

In the dispersive limit ($g \ll \Delta$), we can perform a Taylor series expansion of the energy eigenvalues in Eq. (3.341) to obtain

$$E_{\pm}^{(n+1)} \approx \left(n + \frac{1}{2}\right)\hbar\omega_c \pm \hbar\Delta\left[\frac{1}{2} + \left(\frac{g}{\Delta}\right)^2 (n+1) - \left(\frac{g}{\Delta}\right)^4 (n+1)^2\right]. \tag{3.345}$$

This can be re-expressed as an effective Hamiltonian

$$\tilde{H} \approx \left(\hat{N}_{\text{ex}} - \frac{1}{2}\right)\hbar\omega_c + \Sigma^z\hbar\Delta\left[\frac{1}{2} + \left(\frac{g}{\Delta}\right)^2 \hat{N}_{\text{ex}} - \left(\frac{g}{\Delta}\right)^4 \hat{N}_{\text{ex}}^2\right], \tag{3.346}$$

where $\Sigma^z = \pm 1$ is a spin label that smoothly connects to σ^z in the limit $g \longrightarrow 0$.

The lowest-order term in the expansion reproduces the result in Eq. (3.333) and the next-order term shows that the cavity inherits some nonlinearity (anharmonicity or self-Kerr effect) from the qubit. The advantage of the exact (within the RWA) solution is that it applies even for small Δ, where the dispersive approximation fails. For $\Delta = 0$, the uncoupled levels occur in degenerate pairs and the degeneracy is lifted by an amount linear in g, giving the simple result that is plotted in Fig. 3.23(a):

$$E_{\pm}^{(n+1)} = \left(n + \frac{1}{2}\right)\hbar\omega_c \pm g\hbar\sqrt{n+1}. \tag{3.347}$$

The single-excitation eigenstates are "polaritons"—coherent superpositions of cavity and qubit excitations. The "vacuum Rabi splitting" [32] of the upper and lower polariton states can be observed spectroscopically [73, 120, 124], provided the system is in the strong-coupling limit ($g \gg \gamma, \kappa$) where the splitting $2g$ exceeds the atom and cavity linewidths γ and κ, respectively. In the strong-coupling limit, the linewidth of each single-polariton excitation is in fact [32] (for $\Delta = 0$)

$$\tilde{\gamma} = \frac{\gamma + \kappa}{2}, \tag{3.348}$$

because the excitation is half qubit and half photon. Because of the peculiar $\sqrt{n+1}$ splitting of the higher excited states, the polariton spectrum is quite anharmonic, so the combination of the ground state and one of the single-polariton excitations can be viewed as a two-level system that (in the strong-coupling limit) can be driven relatively strongly without going up the excitation ladder and therefore can be coherently Rabi-flopped just like a qubit [124], as shown in Fig. 3.24.

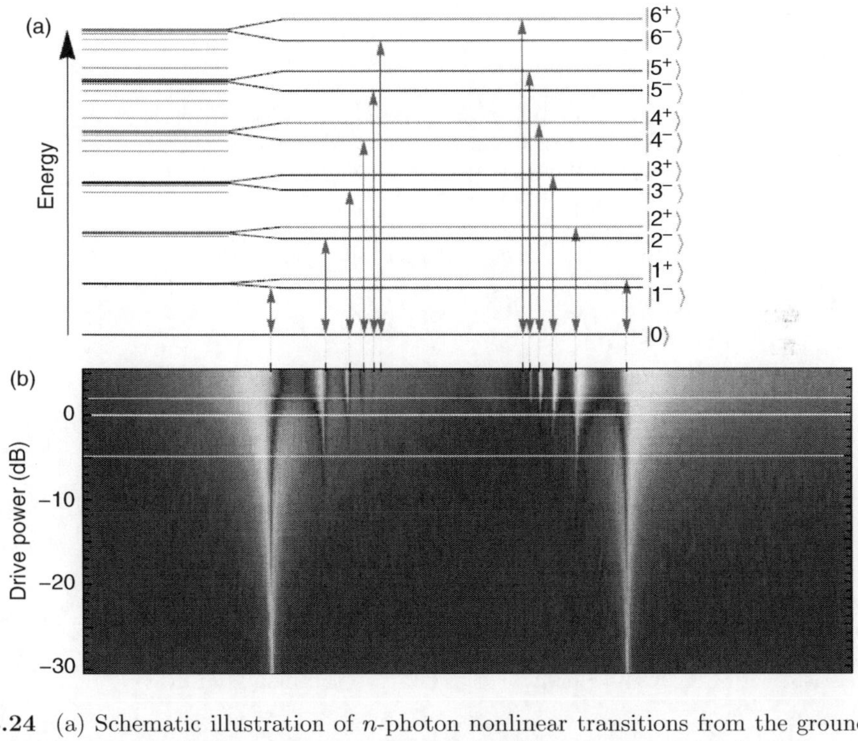

Fig. 3.24 (a) Schematic illustration of n-photon nonlinear transitions from the ground state of qubit and cavity $|0, g\rangle$ to the n excitation manifold $\frac{1}{\sqrt{2}}[|n, g\rangle \pm |n-1, e\rangle$. (b) Driven vacuum Rabi spectrum. The horizontal axis is the drive frequency and the vertical axis is the drive intensity. At low power, one sees the two widely separated vacuum Rabi peaks corresponding to the transition from the ground state to the upper and lower polariton states. At higher powers, one sees multiphoton absorption peaks into the higher excitation manifolds. After [182].

Returning now to the dispersive limit where the qubit is strongly detuned from the cavity, one polariton has primarily qubit character and the other primarily cavity character. For the case of positive detuning, it is $\Psi_+^{(n=0)}$ that is primarily qubit. The rate of energy relaxation of this state is given by the weighted average of the bare qubit and cavity decay rates:

$$\gamma_{\text{tot}} = \cos^2(\theta/2)\,\gamma + \sin^2(\theta/2)\,\kappa. \tag{3.349}$$

For large detuning, where $\sin^2(\theta/2)$ is small, we see that spontaneous emission of a photon via the cavity is very weak. We can say that the qubit must emit its fluorescence photon into the cavity and pay the energy-denominator price of the large detuning before the photon decays out into the continuum. Equivalently, the cavity filters out the vacuum noise at the qubit frequency, which otherwise would have caused fairly rapid spontaneous emission.

Exercise 3.21 The Purcell effect [166] is the modification of the spontaneous emission rate of an atom or qubit due to the presence of a resonator. Depending on circumstances, it can cause the rate of spontaneous fluorescence of the qubit to either be enhanced or reduced. We see from Eq. (3.349) that the fluorescence rate can be reduced by using a long-lived cavity (small κ) and detuning the qubit far from the cavity resonance ($\sin^2(\theta/2) \ll 1$). This equation was derived in the strong-coupling limit $g \gg \kappa$, where it makes sense to first treat the hybridization of the qubit and the cavity and only then add the intrinsic damping of each. In the weak-coupling limit $g \ll \kappa$, it makes sense to treat the density of photon states inside the cavity as a broad continuum (of width κ) and use Fermi's Golden Rule to obtain the fluorescence rate

$$\gamma_\kappa = 2\pi|g|^2\rho(\omega_{01}). \tag{3.350}$$

Show that

$$\rho(\omega_{01}) = -\frac{1}{\pi}\,\text{Im}\,\frac{1}{\omega_{01} - \omega_{\text{c}} + i\kappa/2}, \tag{3.351}$$

and hence that

$$\gamma_\kappa = |g|^2\frac{\kappa}{\Delta^2 + (\kappa/2)^2}. \tag{3.352}$$

Notice that for large detuning $\Delta \gg \kappa$,

$$\gamma_\kappa \approx \frac{g^2}{\Delta^2}\kappa, \tag{3.353}$$

in agreement with Eq. (3.349) (for $N_{\text{ex}} = 1$). On the other hand, for $\Delta = 0$, the rate

$$\gamma_\kappa = \frac{4|g|^2}{\kappa} \tag{3.354}$$

is *inversely* proportional to κ rather than linearly proportional to κ as in the strong-coupling case. This can be understood from the density of states in the resonator being spread out over a greater width and hence having a lower value for large κ as shown in Fig. 3.25. One can view the Purcell effect as the resonator performing an impedance transformation on the external

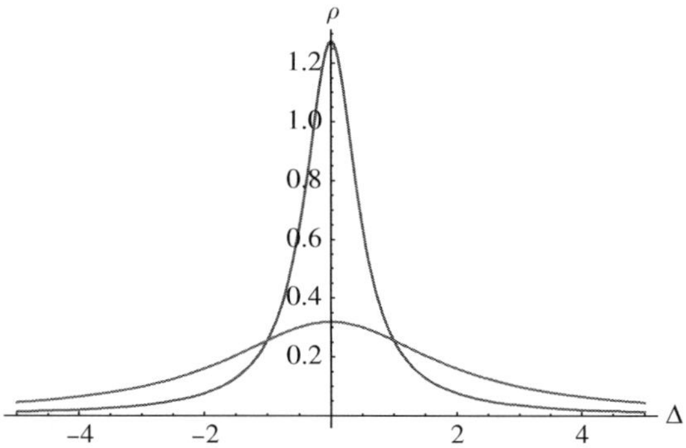

Fig. 3.25 Cavity density of states for two different cavity widths κ. The larger value of κ yields a broader line, but the height is lower at the peak, corresponding to a lower density of states.

dissipation presented by the environment to the qubit [119]. Neeley *et al.* [168] have used a tunable transformer coupling to quantitatively explore the role of environmental coupling in a phase qubit circuit over a wide range of coupling strengths.

The rate of progress in observing novel strong-coupling nonlinear quantum optics effects in superconducting electrical circuits is quite remarkable. As already noted, Houck *et al.* used the Purcell effect [119] to generate nonclassical photon states in a cavity [13]. The states were a superposition of $n = 0$ and $n = 1$ Fock states with controlled amplitude and phase. "Fluorescence tomography" was performed on these states using square-law detection to determine the probability of having a photon. In addition, homodyne measurements were performed to determine the two quadratures of the electric field, which are controlled by the off-diagonal coherence between the $n = 0$ and $n = 1$ Fock states. In particular, Houch *et al.* showed that the mean electric field of the one-photon Fock state was zero.

Higher Fock states up to $n = 6$ were synthesized by the UCSB group [14], who also observed that the decay rate scaled linearly with n as expected [118]. This same effect was seen qualitatively in the frequency domain in the experiment by Schuster *et al.* [110]. The qubit spectrum showed up to six resolved peaks displaying the distribution of photon numbers within the driven cavity and the line width of the peaks increased with n. In a 2009 tour-de-force, Hofheinz *et al.* [15] demonstrated a remarkable method for synthesizing arbitrary photon states (including Fock and various cat states) in a cavity and measuring their Wigner distributions. This level of control now exceeds what has been possible to date with atomic physics methods.

Because microwave photons have 10^4–10^5 times less energy than visible photons, they are much more difficult to detect. The work of Houck *et al.* [13] and Schuster *et al.* [110] showed that individual photons could be detected with low efficiency, and recent work [15, 16] has demonstrated very high-efficiency detection of individual photons in a cavity. However, a general-purpose high-bandwidth "photomultiplier" does not yet exist in the microwave regime. There have been some theoretical proposals for single-photon detection [183, 184], but this remains an important open experimental problem.

Another novel new direction is construction of single artificial atom "lasers" [115, 185, 186], as well as Sisyphus cooling and amplification [121] of an oscillator. The extreme strong coupling available should permit observation of "photon blockade" effects [187] and parametric down-conversion by three-wave mixing [188, 189]. The advances in our understanding and fabrication of Josephson junction circuits motivated by the quest for a quantum computer have led to dramatic advances in the ability to do four-wave mixing, parametric amplification near the quantum limit, and strong squeezing of the vacuum [190–192]. These advances will not only permit much better dispersive readout of qubits, they also open up the possibility of continuous-variable quantum information processing [193, 194], since two-mode squeezed states are an entanglement resource.

3.6.1 Quantum control of qubits in cavities

Suppose that we apply a classical drive with a smooth envelope centered on the qubit transition frequency ω_{01} to the cavity

$$V_d = \{v_R(t)\cos\omega_{01}t + v_I(t)\sin\omega_{01}t\}\,(\hat{a}^\dagger + \hat{a}). \tag{3.355}$$

In the dispersive regime, this drive is far removed from the cavity resonance and only weakly populates the cavity with virtual photons. The vacuum Rabi coupling term of the Jaynes-Cummings model in Eq. (3.322) can then cause coherent rotations of the qubit. This is most easily analyzed by applying the dispersive unitary transformation of Eq. (3.327). To lowest order in g/Δ, we have the original drive on the cavity plus an effective drive directly on the qubit:

$$\tilde{V}_d \approx V_d + V_{dq}, \tag{3.356}$$

where

$$V_{dq} \equiv [\eta, V_d] = \{\lambda_R(t)\cos\omega_{01}t + \lambda_I(t)\sin\omega_{01}t\}\,\sigma^x. \tag{3.357}$$

For large detuning ($\Delta \gg \kappa, \chi$), the complex qubit drive amplitude

$$\lambda_{R(I)} \equiv v_{R(I)}(t)\frac{g}{\Delta} \tag{3.358}$$

can be interpreted as the external drive filtered by the response function of the cavity. (It is important to note that we assume here that since $\Delta \gg \chi$, the filter factor is the same independent of the state of the qubit.) It is convenient to apply a unitary transformation to take us into a frame rotating at the qubit transition frequency,

$$U_{\text{rot}} = e^{\frac{1}{2}i\omega_{01}t\sigma^z}, \tag{3.359}$$

to remove the rapid precession of the "spin." Within the RWA for the drive, we are left with

$$H_{\text{rot}} = U_{\text{rot}} H U_{\text{rot}}^{\dagger} \tag{3.360}$$

$$= \frac{\lambda_R(t)}{2}\sigma^x - \frac{\lambda_I(t)}{2}\sigma^y. \tag{3.361}$$

Thus, we see that the cosine and sine drives produce rotations of the qubit around the x and y axes (in the rotating frame). Rotations about the z axis can be achieved "in hardware" by manipulating the qubit transition frequency to speed up or slow down the precession, or "in software" by mathematically rotating between the cosine and sine components of the drive. This gives us complete quantum control of the qubit state.

We have thus demonstrated that a single input wire to the cavity can be frequency-multiplexed. A drive near the cavity frequency produces a dispersive measurement of the qubit state because the resonance frequency of the cavity and hence the reflection coefficient depend on the state of the qubit. On the other hand, a drive at the qubit frequency is so far detuned from the cavity frequency that the reflection coefficient is independent of the state of the qubit, and so almost no measurement (and almost no measurement-induced dephasing) occurs when applying coherent control pulses to rotate the qubit.

Exercise 3.22 In our derivation, we have assumed a classical drive and implicitly neglected the light shift of the qubit transition frequency that occurs when the cavity contains photons. Consider the derivation more carefully and what constraints the light shift (and the quantum fluctuations in the drive) puts on the fidelity and/or speed of the single-qubit gates.

3.7 Quantum measurements in circuit QED

The theory of quantum measurement has a long history, starting with the founders of quantum mechanics torturing each other with the implications of various gedanken experiments and their interpretation. Today, measurements that are essentially quantum in nature are routinely achieved experimentally. Many of the relevant ideas are better known in the quantum optics community than in the condensed matter community, because the idealized measurement limits were first achieved using quantum optics methods. However, the development of superconducting qubits has also spurred great experimental advances in achieving quantum-limited measurements in condensed matter systems. The author's introduction to the basic facts of quantum measurements came from the review article by Makhlin, Schön, and Shnirman [91]. A recent discussion that bridges the quantum optics/condensed matter domain is given in [31]. The reader is also directed to the discussion by Korotkov of conditional evolution in Chapter 17 of this volume.

Let us start by reviewing the basics. Consider a particle that lives in one dimension and is in quantum state $|\Psi\rangle$. The wavefunction of the particle in the position basis is

$$\psi(x) = \langle x|\Psi\rangle, \tag{3.362}$$

where $|x\rangle$ is the position eigenstate with position eigenvalue x. According to the "Copenhagen interpretation" of quantum mechanics, the probability of finding the particle at position x when the position is measured is given by the Born rule

$$P(x) = \langle\Psi|x\rangle\langle x|\Psi\rangle = |\psi(x)|^2. \tag{3.363}$$

After such a sudden and strong measurement of position, the particle state is said to "collapse" to $|x\rangle$.

Similarly, if we choose instead to measure the particle's momentum (or wavevector k), the probability of distribution of the results is given by

$$\tilde{P}(k) = \langle\Psi|k\rangle\langle k|\Psi\rangle = |\psi[k]|^2, \tag{3.364}$$

where $|k\rangle$ is a momentum eigenstate with momentum eigenvalue $\hbar k$ and the wavefunction in the momentum basis is given by the Fourier transform of the position-basis wavefunction:

$$\psi[k] = \frac{1}{\sqrt{2\pi}} \int_{-\infty}^{+\infty} dx\, e^{-ikx}\psi(x). \tag{3.365}$$

Note that because we have included a factor of $1/\sqrt{2\pi}$ in the Fourier transform, we have

$$\int_{-\infty}^{+\infty} dk\, \tilde{P}(k) = 1. \tag{3.366}$$

One important difference (for a free particle) between measuring position and momentum is that the momentum measurement leaves the system in an eigenstate of the Hamiltonian. Hence, a subsequent measurement of the momentum will yield the same answer. The first result will be a random number drawn from the probability distribution $\tilde{P}(k)$, but all subsequent repeated measurements will yield the same result. This repeatability of the results is the defining characteristic of so-called quantum nondemolition (QND) measurements.

This simple description of a "strong projective" measurement can be physically realized in certain experimental situations. Frequently, however, we need a more sophisticated description of how measurements are done in the laboratory. In practice, we often make experimental measurements on quantum systems by coupling the system to a physical measurement apparatus that itself has a quantum Hamiltonian. We then subsequently make what is effectively a strong projective measurement of some "pointer variable" that is a degree of freedom of the measurement apparatus. Suppose that the measurement pointer variable is described by a position coordinate y. Before the measurement, the wavefunction of the system plus measurement apparatus might be a product state

$$\Psi_0(x, y) = \psi(x)\phi(y), \tag{3.367}$$

where ψ describes the initial state of the system of interest and ϕ describes the initial state of the measurement apparatus. As a result of the coupling between the system and the measurement apparatus, they become correlated (entangled), and the wavefunction is no longer separable. The joint probability distribution for the two variables in terms of the subsequent state Ψ of the combined system is

$$P(x, y) = |\Psi(x, y)|^2, \tag{3.368}$$

while the probability for each separately is

$$P_X(x) = \int dy\, P(x, y), \tag{3.369}$$

$$P_Y(y) = \int dx\, P(x, y). \tag{3.370}$$

Note that these are correctly normalized to unity.

Suppose now that we measure the pointer variable and the result (drawn from the distribution P_Y) is some value y. What can we say from this about the value of x, the quantity we are really interested in? We can say that x has a random value drawn from the *conditional* probability distribution $P(x|y)$. The vertical bar indicates that this is the probability distribution for x given y. Prior to the measurement, x has probability distribution $P_X(x)$. After the measurement of a value y for the pointer variable, we learn that the variable x is drawn from a new probability distribution $P(x|y)$. If this distribution has very narrow support in the vicinity of some value $x_0(y)$, then, by measuring the pointer variable y, we have effectively measured the value of x to be $x_0(y)$. This would be a "strong measurement." In a weak measurement, $P(x|y)$ is broadly distributed, but still slightly narrower than the prior distribution $P_X(x)$. We don't know the precise value of x, but we have learned something about it because the width of its probability distribution has been reduced by conditioning upon the value of y.

From basic probability theory, we know that

$$P(x, y) = P(x|y)P_Y(y). \tag{3.371}$$

That is, the joint probability for x and y to occur is the probability that x occurs given y, times the probability that y occurs. From this, we readily derive

$$P(x|y) = \frac{P(x, y)}{P_Y(y)}, \tag{3.372}$$

which tells us how to update our knowledge of the distribution of x given the new information we have acquired on the value of y. $P_X(x)$ is called the "prior distribution" and $P(x|y)$ is the posterior or updated distribution.

Using Eqs. (3.368) and (3.370), we can express this in terms of the wavefunction:

$$P(x|y) = \frac{|\Psi(x, y)|^2}{\int dx\, |\Psi(x, y)|^2}. \tag{3.373}$$

This simply tells us that the conditional probability of x given y is nothing more than the joint probability evaluated at y and then multiplied by a constant to get the normalization right:

$$\int dx\, P(x|y) = 1. \tag{3.374}$$

Exercise 3.23 Bayes' theorem
We can also write the inverse relation

$$P(y|x) = \frac{P(x,y)}{P_X(x)}, \tag{3.375}$$

from which we can immediately derive Bayes' famous rule relating the two conditional probabilities:

$$P(x|y) = P(y|x)\frac{P_X(x)}{P_Y(y)}. \tag{3.376}$$

Consider one of the examples that Reverend Bayes presented in the eighteenth century. Your friend drops a marble at random somewhere in the interval $[0, L]$. The probability distribution of its position is therefore $P_X(x) = 1/L$. Now your friend drops a second marble at random in the same interval. Suppose your friend updates your knowledge by telling you that the second marble landed to the left of the first marble. How should you update your estimate of the probability distribution of the position of the first marble? Clearly, the new information biases the distribution to the right, but how? Bayes' rule gives us the answer. Let y be a discrete variable that takes values $y = \pm 1$ representing the result that the second marble is left/right of the first. Clearly,

$$P(y = +1|x) = \frac{x}{L}, \tag{3.377}$$

because the further the first marble is to the right, the more likely the second will be to the left. Furthermore,

$$P_Y(y = \pm 1) = \frac{1}{2}, \tag{3.378}$$

because *a priori* (i.e., without prior information about the position of the first marble), the second marble is equally likely to be on the left or right of the first. Thus, from Eq. (3.376),

$$P(x|y = +1) = \frac{2x}{L^2}. \tag{3.379}$$

Given that the second marble landed to the left, there is zero probability that the first marble is at the far left, and the distribution rises linearly from there.

What happens to this conditional probability if we iterate the procedure and the next marble also happens to land to the left of the first one? To the right? Clearly, in the limit of a large number N of such "measurements," the probability distribution for x becomes very narrow and is centered on $x = fL$, where f is the fraction of the marbles that landed to the left of the first. Can you derive the asymptotic form of the probability distribution for large N?

Historical note: Bayes never actually wrote down his rule as an equation. Equation (3.376) was probably first derived later by Pierre-Simon Laplace [195].

Exercise 3.24 Apply Bayes' rule to answer the following questions:

- A certain disease has an incidence rate of 10^{-4} in the general population. The diagnostic test for this disease is highly accurate: the probability of a false negative is only 2%, and the probability of a false positive is only 1%. You are given the test and the result is positive. What is the probability that you have the disease?
- A magician has two unfair coins. Coin A comes up heads with probability $p > \frac{1}{2}$. Coin B comes up heads with probability $1-p$. The magician chooses one of the coins at random and flips it randomly in N trials. It comes up heads every time. What is the probability this is coin A? What is the probability that the next flip will come up heads?
- A resonant cavity coupled to a cold bath has the property that its mean energy decays exponentially at rate κ. You have available a photomultiplier that can detect any photons emitted from the cavity. The cavity is initially prepared in a state that contains one photon with probability p and zero photons with probability $1-p$. After time t, the photomultiplier has not detected a photon. Based on this information, what is the posterior probability that the cavity contains a photon?

3.7.1 Stern-Gerlach measurement of a spin

Ultimately, we will be interested in interrogating the state of a qubit in a resonant cavity using a microwave drive. Before analyzing this somewhat complicated situation, it will be very useful to gain some intuition by reviewing the familiar case of measurement of a spin by using a Stern-Gerlach magnet. The key idea is that we do not directly measure the spin. Rather, we use the magnet to *entangle* the spin with the position of of the particle and then use a detector that measures the position as illustrated in Fig. 3.26.

To keep matters simple, let us stick to a one-dimensional description with a free-particle kinetic energy Hamiltonian H_0 and imagine that the effect of passing through the magnetic field gradient of the Stern-Gerlach magnet is to produce a spin-dependent impulsive force on the particle (assumed to be electrically neutral so that there is no Lorentz force)

$$F(x,t) = +\hbar k_0\, \delta(t)\, \sigma^z, \tag{3.380}$$

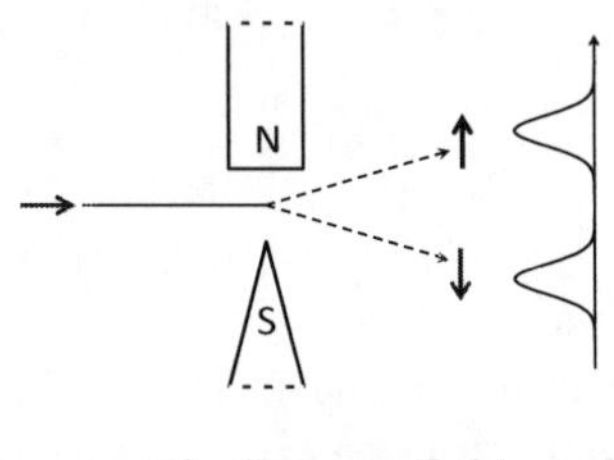

Fig. 3.26 Measurement of a spin using the Zeeman field gradient of a Stern-Gerlach magnet to entangle spin with momentum. The curves on the right show the momentum distribution of each spin component. One can either measure the momentum directly or wait for the momentum difference to evolve into position difference. In the example depicted, the measurement is strong because the momentum distributions of the two components are essentially non-overlapping.

via a potential of the form

$$V(x,t) = -\hbar k_0\,\delta(t)\,\sigma^z x, \tag{3.381}$$

where $\hbar k_0$ is the impulse delivered in some brief interval of time (approximated here by the delta function). Just such spin-dependent impulsive forces are routinely implied in ion-trap quantum computers via application of laser beams with appropriate polarizations.

This coupling term commutes with σ^z, the quantity we want to measure, and hence in principle we can make a QND measurement of the spin. The necessary repeatability is only possible if the quantity in question is a constant of the motion,[24] both respect to the unperturbed Hamiltonian $[H_0, \sigma^z] = 0$ and the perturbation coupling the system to the measurement apparatus, $[V, \sigma^z] = 0$.

Suppose that in the absence of the Stern-Gerlach coupling term, the wavefunction is given by a product of spin and space functions:

$$\psi_0(x) = [a|\uparrow\rangle + b|\downarrow\rangle]\Phi(x), \tag{3.382}$$

where $|a|^2 + |b|^2 = 1$ and $\langle\Phi|\Phi\rangle = 1$. Let us further suppose that the spatial wavepacket Φ is made up of plane waves all of low enough energy so that, for a short time interval around $t = 0$, the packet may be treated as approximately stationary (i.e., $H_0\Phi(x,t) \approx 0$). Then, in the vicinity of time $t = 0$, the solution of the Schrödinger equation will be discontinuous and given approximately by

$$\psi(x,t) \approx e^{ik_0 x\sigma^z\theta(t)}\psi_0(x), \tag{3.383}$$

where $\theta(t)$ is the Heaviside step function. This can be written for $t > 0$ as

$$|\psi\rangle = a|\uparrow\rangle|\Phi_\uparrow\rangle + b|\downarrow\rangle|\Phi_\downarrow\rangle, \tag{3.384}$$

or

$$\psi(x) \approx a|\uparrow\rangle e^{ik_0 x}\Phi(x) + b|\downarrow\rangle e^{-ik_0 x}\Phi(x), \tag{3.385}$$

or equivalently the momentum-basis wavefunction will be

$$\psi[k] = a|\uparrow\rangle\Phi[k - k_0] + b|\downarrow\rangle\Phi[k + k_0]. \tag{3.386}$$

We no longer have a separable wavefunction, because the measuring apparatus has entangled the spin with the momentum. The full time-dependent solution is such that this displacement in momentum space will eventually result in displacement in real space, but, for the purposes of the present discussion, let us imagine we have a mechanism for measuring the momentum directly and immediately. This measurement of the momentum will, via the entanglement, tell us something about the spin.

[24] It is a confusing but common practice to refer to a measurement as being QND even if the quantity being measured is not quite a constant of the motion owing to weak coupling of the system to a dissipative bath. This nonideality is ignored as long as the damping is weak and the quantity would otherwise truly be constant even in the presence of the measurement apparatus.

For simplicity, let us assume that the spatial wavepacket is Gaussian,

$$\Phi(x) = (2\pi\sigma_0^2)^{-1/4} e^{-x^2/4\sigma_0^2}, \tag{3.387}$$

with Fourier transform

$$\Phi[k] = \left(\frac{2\sigma_0^2}{\pi}\right)^{+1/4} e^{-k^2\sigma_0^2}. \tag{3.388}$$

The probability distribution at time $t = 0^+$ for the measured momentum conditioned on the spin being up or down is given by a displaced Gaussian,

$$P(k|\uparrow) = |\Phi_0[k - k_0]|^2, \tag{3.389}$$

$$P(k|\downarrow) = |\Phi_0[k + k_0]|^2, \tag{3.390}$$

and is shown in Fig. 3.27 for different values of the dimensionless impulse $k_0\sigma_0$. Clearly, if $k_0\sigma_0 \gg 1$, there is no ambiguity in the result. If the measured value of k is positive (negative), then the spin is definitely up (down). This corresponds to a strong measurement. However, if $k_0\sigma_0 \ll 1$, we have a weak measurement in which we have acquired only a small amount of information about the spin.

To quantify how much information we have gained, we can make use of Bayes' rule. Given the state of the spin, we can predict the distribution of measured values of k using Eqs. (3.389) and (3.390). In particular, suppose we are given the prior probabilities for the spin to be up, $p(\uparrow) = |a|^2$, and to be down, $p(\downarrow) = |b|^2$. The prior probability distribution for k is thus

$$P(k) = P(k|\uparrow)p(\uparrow) + P(k|\downarrow)p(\downarrow) = |a|^2|\Phi_0[k - k_0]|^2 + |b|^2|\Phi_0[k + k_0]|^2. \tag{3.391}$$

What we want, however, is the inverse conditional probability. That is, given the measured value of k, what can we say about the probability that the spin is up or down? This is given by Bayes' rule:

$$P(\uparrow|k) = P(k|\uparrow)\frac{p(\uparrow)}{P(k)} = \frac{1}{P(k)}|a|^2|\Phi_0[k - k_0]|^2, \tag{3.392}$$

$$P(\downarrow|k) = P(k|\downarrow)\frac{p(\downarrow)}{P(k)} = \frac{1}{P(k)}|b|^2|\Phi_0[k + k_0]|^2. \tag{3.393}$$

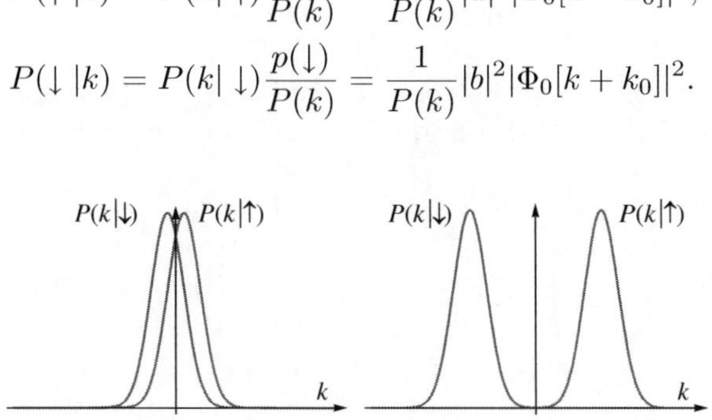

Fig. 3.27 Momentum distribution of the particle after passage through the Stern-Gerlach magnet, conditioned on the spin being up or down. Left panel: weak measurement in which the momentum difference is small relative to the momentum uncertainty. Right panel: strong measurement in which the momentum distributions do not overlap.

For the case of Gaussian wavepackets considered here, this simplifies to

$$P(\uparrow|k) = \frac{|a|^2}{Z}e^{+\lambda k} \tag{3.394}$$

$$P(\downarrow|k) = \frac{|b|^2}{Z}|e^{-\lambda k}, \tag{3.395}$$

where $\lambda \equiv 4k_0\sigma_0^2$ and the normalization simplifies to

$$Z = |a|^2 e^{\lambda k} + |b|^2 e^{-\lambda k}. \tag{3.396}$$

We now specialize to the particular case $|a|^2 = |b|^2 = \frac{1}{2}$, where the prior polarization of the qubit is $m = |a|^2 - |b|^2 = 0$ and the estimate of the polarization following the measurement is

$$m(k) = P(\uparrow|k) - P(\downarrow|k) = \tanh(\lambda k), \tag{3.397}$$

which shows explicitly how the new information from the measurement revises our estimate of the state of the spin. The Shannon entropy for the prior distribution of the value of the spin is

$$S = -\sum_\sigma p_\sigma \ln p_\sigma = -(p_\uparrow \ln p_\uparrow + p_\downarrow \ln p_\downarrow) = \ln 2. \tag{3.398}$$

(This entropy in base e, familiar from statistical mechanics, corresponds to one classical bit of information, which is calculated in base 2.) After the measurement yields a value of k, the Shannon entropy for the spin distribution is

$$S(k) = -\sum_\sigma P(\sigma|k)\ln P(\sigma|k) \tag{3.399}$$

$$= \ln 2 + \ln(\cosh\lambda k) - \lambda k \tanh \lambda k. \tag{3.400}$$

This is reduced from the prior entropy by the amount of information gained by the measurement about the state of the qubit. The information gained is

$$\mathcal{I}(k) = -\ln(\cosh\lambda k) + \lambda k \tanh\lambda k, \tag{3.401}$$

and the average information gained is

$$\bar{\mathcal{I}} = \int dk\, P(k)\mathcal{I}(k). \tag{3.402}$$

For a very weak measurement, we expect λk to be small, in which case, to leading order in the expansion, we have

$$\mathcal{I}(k) \approx \frac{\lambda^2 k^2}{2}, \tag{3.403}$$

and, using Eq. (3.391), the average information gain for weak measurements is

$$\bar{\mathcal{I}} \approx \frac{\lambda^2}{2}\left(\frac{1}{4\sigma_0^2}+k_0^2\right) \approx 2k_0^2\sigma_0^2 = \frac{1}{2}\frac{k_0^2}{(\Delta k)^2}, \tag{3.404}$$

where (for $k_0\sigma_0 \ll 1$) the momentum uncertainty is $(\Delta k)^2 = 1/4\sigma_0^2$. It makes sense that the narrower the momentum uncertainty of the wavepacket, the more information we will gain from the measurement of the spin-dependent impulse imparted to the particle.

The above discussion was couched in the language of classical probability distributions. Let us now repeat it in the language of quantum density matrices.[25] Using the entangled state of the spin and position in Eq. (3.384), the full density matrix is

$$\rho = \begin{pmatrix} |\Phi_\uparrow\rangle aa^*\langle\Phi_\uparrow| & |\Phi_\uparrow\rangle ab^*\langle\Phi_\downarrow| \\ |\Phi_\downarrow\rangle ba^*\langle\Phi_\uparrow| & |\Phi_\downarrow\rangle bb^*\langle\Phi_\downarrow| \end{pmatrix}. \tag{3.405}$$

If the measurement of the momentum yields the result k, then the reduced density matrix for the spin is given by

$$\rho_k = \frac{1}{P(k)}\langle k|\rho|k\rangle = \frac{1}{P(k)}\begin{pmatrix} \langle k|\Phi_\uparrow\rangle aa^*\langle\Phi_\uparrow|k\rangle & \langle k|\Phi_\uparrow\rangle ab^*\langle\Phi_\downarrow|k\rangle \\ \langle k|\Phi_\downarrow\rangle ba^*\langle\Phi_\uparrow|k\rangle & \langle k|\Phi_\downarrow\rangle bb^*\langle\Phi_\downarrow|k\rangle \end{pmatrix} \tag{3.406}$$

$$= \frac{1}{Z}\begin{pmatrix} |a|^2e^{+\lambda k} & ab^* \\ a^*b & |b|^2e^{-\lambda k} \end{pmatrix}, \tag{3.407}$$

where Z is given in Eq. (3.396) and the normalization factor $P(k)$ is

$$P(k) \equiv \mathrm{Tr}\,\langle k|\rho|k\rangle = |a\Phi_0[k-k_0]|^2 + |b\Phi_0[k+k_0]|^2, \tag{3.408}$$

which is nothing but the probability that measurement of the momentum will yield k.

Notice that the diagonal elements of Eq. (3.407) match the classical probabilities in Eqs. (3.394) and (3.395). Eq. (3.407) has a simple interpretation. Suppose for example that the measured value of the momentum turns out to be $k = 0$. Then nothing happens to the spin as a result of the measurement. For $k > 0$, the weight of the up-spin amplitude increases and the down-spin weight decreases by the corresponding factor. The reverse occurs for $k < 0$. Notice that there is no explicit reweighting of the off-diagonal elements of the density matrix because the competing exponential factors cancel. However, the normalization factor indirectly changes the value of the off-diagonal term. The net effect is simply a rotation of the qubit around an axis passing through the equator of the Bloch sphere and lying perpendicular to the qubit polarization. This rotation moves the qubit along a line of constant "longitude" on the Bloch sphere.

[25] Of course, quantum probability amplitudes are not the same thing as classical probabilities, which is why we have to be a bit careful and recall that density matrices have off-diagonal elements that represent superposition effects with no classical analog. Even the interpretation of the density matrix itself can be tricky. It might represent an ensemble of quantum systems or it might represent the observer's incomplete state of knowledge about the state of a quantum system.

If the spin is up, the probability of observing a positive value of k is increased. Conversely, observation of a positive value of k means that the spin is more likely to be up. As already mentioned and as Korotkov discusses in Chapter 17 of this volume, this application of Born's rule from quantum mechanics is consistent with Bayes' rule from classical probability theory. Gaining information from the measured value of k allows us to update our estimate of the spin polarization. Despite the fact that σ^z is a constant of the motion, the polarization of the spin has changed—not owing to any unitary evolution under the action of the Hamiltonian, but rather to the measurement-induced collapse of the momentum state that is entangled with the spin. It is readily verified that the spin is still in a pure state because the eigenvalues of ρ are zero and one. (*Hint*: $\mathrm{Tr}\,\rho = 1$ and $\det \rho = 0$.) Equivalently, the von Neumann entropy (as opposed to the Shannon entropy used above and calculated only from the diagonal elements of ρ) vanishes since the state remains pure:

$$\tilde{S} = -\mathrm{Tr}\,\rho \ln \rho = 0. \tag{3.409}$$

If we ignore the result of the measurement (or equivalently ensemble average over all measurement results) by tracing out the orbital degree of freedom in Eq. (3.405), we obtain the following reduced density matrix for the spin as a result of carrying out (but ignoring) the measurement:[26]

$$\bar{\rho} = \mathrm{Tr}_{\,\mathrm{orbital}}\,\rho = \int_{-\infty}^{+\infty} dk\, P(k)\rho_k = \begin{pmatrix} aa^* & ab^*\langle\Phi_\downarrow|\Phi_\uparrow\rangle \\ ba^*\langle\Phi_\uparrow|\Phi_\downarrow\rangle & bb^* \end{pmatrix}. \tag{3.410}$$

Because we have ignored the measurement results, we see that in this case there is no change in the diagonal elements—we receive no information about the orientation of the spin. Note, however, that the act of measurement has *dephased* the spin because the off-diagonal elements are reduced by a factor given by the overlap of the two pointer states. The stronger the measurement, the greater this reduction will be as the two pointer states become orthogonal and therefore fully distinguishable. This dephasing reduces the expectation value of the transverse components of the spin without affecting the z component. The off-diagonal elements decay as

$$\langle\Phi_\downarrow|\Phi_\uparrow\rangle = \int dx\, e^{2ik_0x}|\Phi(x)|^2 = e^{-2k_0^2(\Delta x)^2}, \tag{3.411}$$

where $(\Delta x)^2 = \sigma_0^2$. As we will see, it makes sense that the dephasing should be proportional to the position uncertainty, because the magnetic field acting on the spin varies linearly with position. The greater the position uncertainty, the greater the magnetic field uncertainty. As we will see below, if we measure the position x, then we know the magnetic field that acted on the spin and we know that this rotated the spin through a definite angle $2k_0x$. The off-diagonal element of the density matrix is thus

[26] Notice the very important fact that the probability measure for k exactly cancels the normalization factor in the denominator of Eq. (3.406), which allows the integral to be trivially carried out.

multiplied by e^{2ik_0x} but the qubit remains in a pure state. Now, however, we have a simple interpretation of Eq. (3.411) telling us that we are ensemble averaging over all the different rotation angles. The average of a bunch of density matrices each of which corresponds to a pure state leads to an impure state. More on this below.

If we assume that the measurement takes time Δt, we can define a dephasing rate Γ_φ via

$$\langle \Phi_\downarrow | \Phi_\uparrow \rangle = e^{-\Gamma_\varphi \Delta t} \tag{3.412}$$

so that

$$\Gamma_\varphi = \frac{2k_0^2(\Delta x)^2}{\Delta t}. \tag{3.413}$$

Similarly, we can define a measurement rate Γ_{meas} from the rate of acquisition of information when we monitor the pointer variable:

$$\Gamma_{\text{meas}} = \frac{\bar{\mathcal{I}}}{\Delta t} = \frac{1}{2}\frac{k_0^2}{(\Delta k)^2}\frac{1}{\Delta t}. \tag{3.414}$$

As expected, the measurement rate is inversely proportional to the momentum uncertainty.

The ratio of these two rates is given by the simple Heisenberg uncertainty expression

$$\frac{\Gamma_\varphi}{\Gamma_{\text{meas}}} = 4(\Delta k)^2(\Delta x)^2 \geq 1. \tag{3.415}$$

The best we can do is have the ensemble-averaged dephasing rate equal the ensemble-averaged measurement rate. In the present example, this equality is actually achieved, meaning that the measurement is quantum-limited. The equality of the measurement rate and the dephasing rate in our quantum-limited detector follows from the uncertainty relation between momentum and position for a minimum-uncertainty Gaussian wavepacket. Quantum-limited measurements are discussed in great detail in [31] and in Chapter 2 in this volume.

Let us now consider the situation where we decide that rather than measure the particle *momentum* (which contains information about the spin through its entanglement), we decide instead to measure the particle *position*, which initially is not affected by the momentum boost and hence is *not* entangled with the spin.[27] We see from Eq. (3.385) that the probability distribution for position is $|\Phi(x)|^2$, independent of the position. Nevertheless, something interesting happens when we make the measurement. Suppose that the measured value of position turns out to be x. Then the analog of Eq. (3.406) for the reduced density matrix for the spin conditioned on having measured x is

[27] After some time, the momentum boost will affect the position of each spin component and position then *is* entangled with the spin.

$$\rho_x = \frac{1}{P_X(x)}\langle x|\rho|x\rangle = \frac{1}{P_X(x)}\begin{pmatrix} \langle x|\Phi_\uparrow\rangle aa^*\langle\Phi_\uparrow|x\rangle & \langle x|\Phi_\uparrow\rangle ab^*\langle\Phi_\downarrow|x\rangle \\ \langle x|\Phi_\downarrow\rangle ba^*\langle\Phi_\uparrow|x\rangle & \langle x|\Phi_\downarrow\rangle bb^*\langle\Phi_\downarrow|x\rangle \end{pmatrix} \quad (3.416)$$

$$= \begin{pmatrix} |a|^2 & e^{-i\varphi}ab^* \\ e^{+i\varphi}a^*b & |b|^2 \end{pmatrix}, \quad (3.417)$$

where $\varphi \equiv 2k_0x$ and

$$P_X(x) = \mathrm{Tr}\,\langle x|\rho|x\rangle = |\Phi_0(x)|^2 \quad (3.418)$$

is the probability distribution for position. Clearly the probabilities to find the spin up and down remain $|a|^2$ and $|b|^2$ and have not been changed by the measurement. Note, however, that the spin has been rotated by an angle φ around the z axis so that it moves along a line of constant latitude. Unlike the previous example of momentum measurement, there is no state collapse that is affecting the spin. Rather, this change in the spin state is the result of the measurement backaction through the unitary time evolution generated by the coupling V. As we can see from Eq. (3.381), the size of the Zeeman field that briefly acts on the spin varies linearly with the position x. Hence, the measured value X determines the rotation angle caused by the coupling V. This is not surprising, because while V commutes with σ^z, it does not commute with the transverse components of the spin and we expect them to be changed by the measurement. Because the measurement of position yields no information about σ^z, the collapse of the position has no effect on σ^z.

The message here is that if we measure the state of the environment presented to the qubit by the measurement apparatus (in this case the pointer variable, which is the particle position or momentum), then we have complete information about what happened to the qubit. What happened may be random, but we know precisely what it is. If, however, we ignore the results of the measurement by ensemble averaging over the measurement results, the random results will lead to *measurement-induced dephasing* of the qubit. In this example, we can ensemble average over all possible results of the momentum measurement or the position measurement. Either way, we should get the same answer because this corresponds to tracing out the environmental degrees of freedom by summing over a complete set of states (in some particular basis).

3.7.2 Dispersive readout of a Qubit in a cavity

Now that we understand Stern-Gerlach measurements, we are ready to study dispersive readout of a qubit in a cavity. This readout works because the cavity frequency, and hence the phase of the reflection coefficient for microwaves, depends on the state of the qubit. In the Stern-Gerlach example, the state of the qubit is entangled with the momentum of the particle wave and we considered the different backaction effects of measuring the particle momentum or the conjugate variable, the position. In this example, the phase of the microwaves reflected from the cavity is entangled with the state of the qubit and will consider the different backaction effects of measuring the phase of the microwaves or its conjugate variable, the photon number.

Suppose that we have the effective Hamiltonian in the dispersive limit given by Eq. (3.333). This is the analog of the Stern-Gerlach Hamiltonian, Eq. (3.381). Instead of the Zeeman splitting varying linearly with position, it now varies linearly with photon number $\hat{N} = a^\dagger a$. The analogy is not perfect because the photon number has discrete integer eigenvalues and for a coherent state obeys a Poisson distribution, while the position is continuous and (in our example) Gaussian-distributed. However, as we shall see, in the limit of weak measurements involving a large number of photons, the Poisson distribution effectively becomes a Gaussian.

If we send a microwave pulse at the bare cavity frequency $\omega_{\rm c}$ in coherent state $|\alpha\rangle$ toward the (one-sided) cavity, then the initial state of the system (before the wavepacket hits the cavity) is a product state

$$|\psi_{\rm in}\rangle = [a|\uparrow\rangle + b|\downarrow\rangle]|\alpha\rangle, \tag{3.419}$$

while the final state has the qubit and cavity entangled,

$$|\psi_{\rm out}\rangle = [a|e^{-i\theta_0}\alpha\rangle|\uparrow\rangle + b|e^{+i\theta_0}\alpha\rangle|\downarrow\rangle], \tag{3.420}$$

where θ_0 is the phase of the complex reflection coefficient given in Eq. (3.517) in the appendix (Section 3.9.2). Using the expression for the dispersive cavity pull in Eq. (3.333) and noting that the drive at the bare cavity frequency is detuned from the actual cavity frequency by an amount $\chi\sigma^z$, the reflection coefficient becomes (with the electrical engineering notation $j = -\sqrt{-1} = -i$)

$$r = \frac{\chi\sigma^z - i\kappa/2}{\chi\sigma^z + i\kappa/2} = e^{-i\theta_0\sigma^z} \tag{3.421}$$

with $\tan(\theta_0/2) = \kappa/2\chi$. As can be seen in Fig. 3.28, for a fixed phase angle difference, the two coherent states become more and more distinguishable as the amplitude increases. Thus, the more photons we use, the stronger the measurement we make.

If, as illustrated in Fig. 3.29, we make a homodyne measurement of the reflected wave, we can choose to measure either of the two quadrature amplitudes defined in Eqs. (3.81) and (3.82). The two quadratures $\hat{X}$ and $\hat{Y}$ are precisely analogous to the

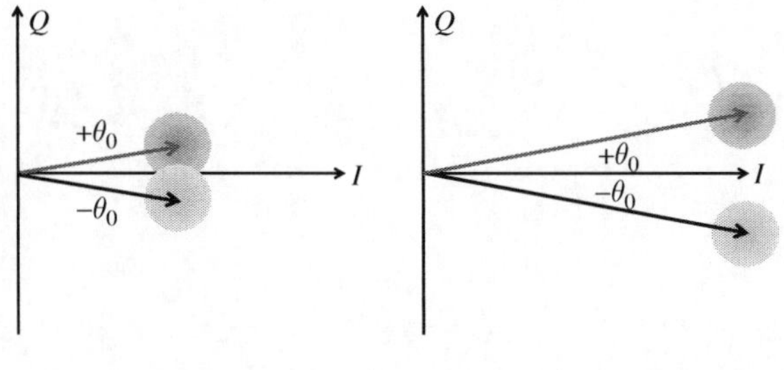

Fig. 3.28 Fresnel "lollipops" showing that two coherent states differing by a fixed phase angle become more and more distinguishable as the amplitude increases. For sufficiently large amplitude, the difference in the Q quadratures exceeds the uncertainty due to the vacuum noise. When they are fully distinguishable, we have a strong quantum measurement of the spin that is entangled with the photon field.

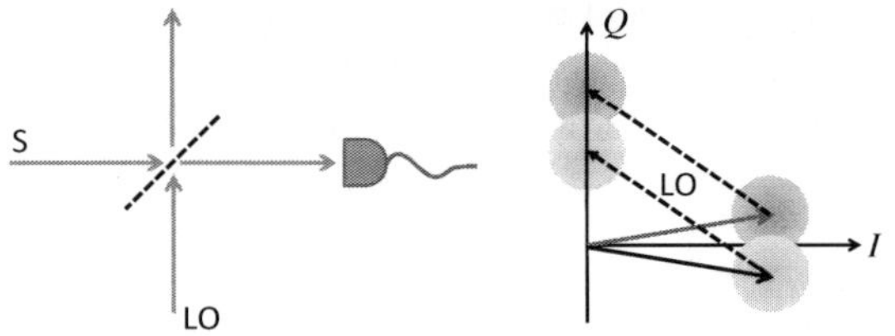

Fig. 3.29 (a) Homodyne measurement in which the signal is mixed with a local oscillator (LO) of the same frequency. The resulting interference allows one to determine the amplitude (rather than simply the power) of the signal. The phase of the LO determines the signal quadrature that is measured by the square-law detector. The beam splitter can be made very strongly transmitting and only weakly reflecting (at the expense of needing more LO power), so that the vacuum noise added by the LO port is negligible. (b) Vector addition of the signal and LO phasors, which converts a phase change of the signal into an amplitude (and hence intensity) change that can be detected by the square-law detector.

position and momentum in the Stern-Gerlach example introduced above.[28] For the coherent state $|e^{-i\theta_0\sigma^z}\alpha\rangle$ with $\alpha = \sqrt{\bar{N}}$ real, we have (see the appendix, Section 3.9.5)

$$\langle \hat{Y} \rangle = -\sigma^z \alpha \sin \theta_0, \tag{3.422}$$

so that, at least for $\theta_0 \ll 1$, measuring $\hat{Y}$ is equivalent to measuring the phase of the wave,[29] and

$$\langle \{\hat{Y} - \langle \hat{Y} \rangle\}^2 \rangle = \frac{1}{4}. \tag{3.423}$$

By comparison with our previous example, we see that the "momentum kick" is

$$Y_0 = -\sqrt{\bar{N}} \sin \theta_0 \tag{3.424}$$

and (as shown in the appendix, Section 3.9.5) the momentum is Gaussian-distributed with variance

$$(\Delta Y)^2 = \frac{1}{4}. \tag{3.425}$$

Hence, by analogy to Eq. (3.414), the measurement rate is

$$\Gamma_{\text{meas}} = \frac{1}{\Delta t} \frac{Y_0^2}{2(\Delta Y)^2} = 2\dot{\bar{N}} \sin^2 \theta_0, \tag{3.426}$$

[28] Note, however, the important fact that, with the normalization we have chosen, their commutator is smaller by a factor of two: $[\hat{X}, \hat{Y}] = \frac{i}{2}$. Hence, the Heisenberg uncertainty relation for the two quadratures is $(\Delta X)^2(\Delta Y)^2 \geq \frac{1}{16}$.

[29] Strictly speaking, there is no "phase operator." Only operators such as $\sin\theta$ that are periodic in θ are well defined.

where $\dot{\bar{N}} = \bar{N}/\Delta t$ is the incident (and outgoing) photon flux. Thus, we see that the measurement rate is maximized when $\theta_0 = \pi/2$, which is achieved when the dispersive shift is half a linewidth: $\chi = \kappa/2$. However, the measurement is weak and continuous only for $\theta_0 \ll 1$, and in this limit we can make the approximation

$$\Gamma_{\text{meas}} \approx 2\dot{\bar{N}}\theta_0^2. \tag{3.427}$$

The corresponding dephasing is most easily determined using the overlap of the pointer coherent states (see the appendix, Section 3.9.5):

$$\langle e^{i\theta_0}\alpha|e^{-i\theta_0}\alpha\rangle = e^{|\alpha|^2(e^{-2i\theta_0}-1)} \approx e^{-i2\theta_0\bar{N}}e^{-2\bar{N}\theta_0^2}. \tag{3.428}$$

The complex exponential simply represents the average phase rotation of the qubit by an angle $2\theta_0$ for each photon that passes through the cavity. The dephasing term represents the ensemble averaging over the fluctuations in the photon number in the coherent state. From this, we deduce

$$\Gamma_\varphi \Delta t = 2\bar{N}\theta_0^2, \tag{3.429}$$

so that again we have a quantum-limited measurement with $\Gamma_\varphi = \Gamma_{\text{meas}}$.

Continuing the Stern-Gerlach analogy, the analog of Eq. (3.407) for the reduced density matrix conditioned on measuring a value y for the $\hat{Y}$ signal quadrature can be found using the results in the appendix (Section 3.9.5):

$$\rho_y = \frac{1}{P(y)}\langle y|\rho|y\rangle = \frac{1}{Z}\begin{pmatrix} |a|^2 e^{+\lambda y} & e^{i\varphi}ab^* \\ e^{-i\varphi}a^*b & |b|^2 e^{-\lambda y}\end{pmatrix}, \tag{3.430}$$

where

$$\lambda \equiv -4\sqrt{\bar{N}}\sin\theta_0 \tag{3.431}$$

and

$$P(y) = |a|^2 e^{-\lambda y} + |b|^2 e^{+\lambda y}, \tag{3.432}$$

and (unlike the Stern-Gerlach case) there is an additional deterministic phase rotation given by

$$\varphi = -\bar{N}\sin 2\theta_0. \tag{3.433}$$

Let us now consider what happens when, instead of measuring the "momentum" (the $\hat{Y}$ signal quadrature), we instead measure the "position" (the $\hat{X}$ signal quadrature). Using the results from the appendix (Section 3.9.5) to project onto a given value x of $\hat{X}$, the analog of Eq. (3.417) is

$$\rho_x = \begin{pmatrix} |a|^2 & e^{i\lambda(x-\bar{x})}e^{i\varphi}ab^* \\ e^{-i\lambda(x-\bar{x})}e^{-i\varphi}a^*b & |b|^2\end{pmatrix}, \tag{3.434}$$

where $\bar{x} = \sqrt{\bar{N}}\cos\theta_0$ is the mean value of x.

Suppose that, instead of measuring the $\hat{X}$ quadrature amplitude via a homodyne measurement, we choose instead to use a photomultiplier to measure $\hat{N}$. Suppose that the result of this measurement is an integer n. Then the reduced density matrix of the spin is readily determined using the following property of coherent states (see the appendix, Section 3.9.5):

$$\langle n|e^{+i\theta_0}\alpha\rangle = \frac{\left(e^{i\theta_0}\alpha\right)^n}{\sqrt{n!}}. \tag{3.435}$$

From this, we have

$$\rho_n = \begin{pmatrix} |a|^2 & ab^* e^{-i2\theta_0 n} \\ ba^* e^{+i2\theta_0 n} & |b|^2 \end{pmatrix}. \tag{3.436}$$

Thus, the qubit is left in a pure state but rotated around the z axis by an angle $\theta_n = -2\theta_0 n$. This simply means that each photon that reflects from the cavity deterministically shifts the phase of the superposition of the two qubit states by a fixed angle, $-2\theta_0$. This is the direct analog of Eq. (3.417) in the Stern-Gerlach example (except that here n has a nonzero mean value rather than fluctuating around zero). For sufficiently small θ_0, we can carry out the ensemble average over the photon number fluctuations using the following approximation:

$$\langle\langle e^{-i2\theta_0 n}\rangle\rangle \approx e^{-i2\theta_0\langle\langle n\rangle\rangle} e^{-\frac{1}{2}\langle\langle[2\theta_0(n-\bar{N})]^2\rangle\rangle} \approx e^{-i2\theta_0\bar{N}} e^{-2\bar{N}\theta_0^2}, \tag{3.437}$$

which correctly reproduces Eq. (3.428).

The connection between measuring the photon number and the $\hat{X}$ signal quadrature can be seen by noting that (for real α) the number of photons is related to the value of $\hat{X}$ whose fluctuations modulate the length of the phasor:

$$\hat{N} = (\alpha + \Delta\hat{X})^2 + (\Delta\hat{Y})^2 - \frac{1}{2} \tag{3.438}$$

$$= \bar{N} + 2\sqrt{\bar{N}}\Delta\hat{X} + [(\Delta\hat{X})^2 + (\Delta\hat{Y})^2 - \frac{1}{2}]. \tag{3.439}$$

For sufficiently small θ_0, the last term in square brackets can be neglected (it vanishes exactly in the state $|\alpha\rangle$). Thus, the off-diagonal phase factor in Eq. (3.436), which contains the photon number measurement result, can be written in terms of the measurement result for the $\hat{X}$ quadrature signal amplitude measurement result:

$$e^{-i2\theta_0 n} \approx e^{-i2\bar{N}\theta_0} e^{-i4\theta_0\sqrt{\bar{N}}(x-\bar{x})} \tag{3.440}$$

which precisely matches the phase factor of the off-diagonal element from Eq. (3.434) in the limit of small θ_0. In this limit, it takes many photons to make the measurement and n becomes Gaussian-distributed and (its discreteness is hard to see) just like the continuous measurement result for x. It is important to emphasize that Eqs. (3.434) and (3.436) are exact for any value of θ_0. We have a simple physical equivalence between them only for small θ_0, however.

Heterodyne detection The above results tell us that the backaction of the homodyne measurement on the qubit depends on which quadrature of the homodyne signal is being measured. In a heterodyne measurement, the frequency of the local oscillator is detuned from that of the signal. One can view this as rotating the measurement between the two quadrature amplitudes at a rate given by the detuning. Thus, we obtain information about both quadratures. This is a bit strange, because the two quadratures are canonically conjugate and hence are incompatible observables—we cannot know them both simultaneously with perfect precision. Thus, just as a phase-preserving amplifier must add some noise [31], so a heterodyne measurement must be subject to some inefficiency. From the quantum noise point of view, this comes about because of additional vacuum noise mixed into the output. If the signal is at frequency f and the local oscillator is at frequency $f-\Delta$, the output will be at frequency Δ. However, vacuum noise at the image frequency $f+\Delta$ will also be mixed down to the same output frequency, increasing the uncertainty in the measurement of the quadrature amplitudes by just the amount required by the Heisenberg uncertainty relation [31].

Alternatively, one can make the equivalent of a heterodyne measurement using two simultaneous homodyne measurements as shown in Fig. 3.30. Here, the additional

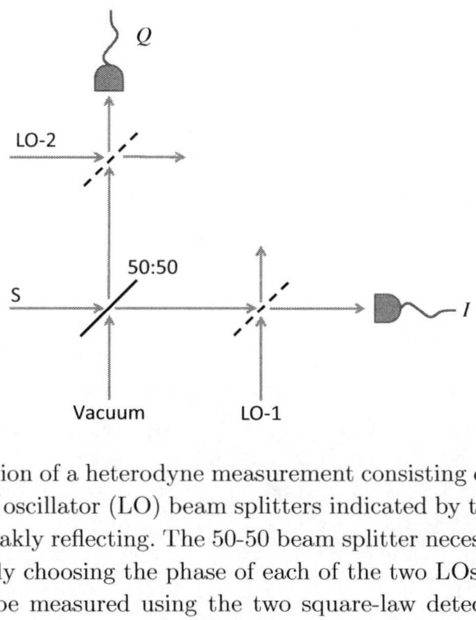

Fig. 3.30 Idealized version of a heterodyne measurement consisting of two separate homodyne measurements. The local oscillator (LO) beam splitters indicated by the dashed lines are highly transmitting and only weakly reflecting. The 50-50 beam splitter necessarily introduces an extra vacuum port. By correctly choosing the phase of each of the two LOs, both quadratures (I,Q) of the input signal can be measured using the two square-law detectors. The added noise of the vacuum port reduces the measurement efficiency by 50% for a given quadrature relative to homodyne measurement of that quadrature. Nevertheless, the state of the qubit remains pure if the two square-law detectors are ideal. In practice, heterodyne measurement is done with a single beam splitter and LO detuned from the signal. The resulting beat note at the intermediate frequency is then amplified. The analog of the added noise from the 50:50 beam splitter is the extra vacuum noise from the "image frequency," which is also mixed down by the LO to the intermediate frequency.

vacuum noise is introduced by the 50-50 beam splitter needed to allow the separate homodyne measurements. Despite the additional noise introduced by the vacuum port of this beam splitter, the qubit remains in a pure state. Measurement of $\hat{X}$ quadrature yields a rotation around the z axis and measurement of the $\hat{Y}$ quadrature pushes the polarization towards $\sigma^z = \pm 1$ as before. However, because only half the signal is used for the $\hat{Y}$ measurement, the measurement rate is only 50% of the previous rate.

Exercise 3.25 Show that if the phases of the two local oscillators in the pseudo-heterodyne setup in Fig. 3.30 are set to measure the same quadrature, then it is possible to recover the homodyne measurement result with no additional noise. *Hint*: First write down the S matrix for the 50-50 beam splitter, making sure that it is unitary. Use this to determine whether the shot noise in the two homodyne signals is correlated. Recall that the shot noise can be viewed as resulting from the interference between the classical input and the vacuum noise that comes with it.

Strong-dispersive limit We have seen that when the qubit transition frequency ω_Q is strongly detuned from the cavity resonance frequency ω_R, the resulting "dispersive" Hamiltonian takes the form in Eq. (3.333):

$$H = \omega_R a^\dagger a + \frac{\omega_Q}{2}\sigma^z + \frac{\chi}{2}a^\dagger a \sigma^z. \tag{3.441}$$

This Hamiltonian is diagonal and deceptively simple-looking, yet we have come to realize that it provides a powerful toolbox for quantum control of the joint states of the cavity and the qubit. The coefficient χ tells us the that the frequency of the cavity shifts by χ when the qubit changes state. The same dispersive coupling term also tells us that the qubit transition frequency changes by χ each time a photon is added to the cavity. Because of the very large transition dipole of superconducting artificial atoms (several electron charges moving about 1 mm), the dispersive shift χ can be some 10^3 times larger than the linewidths of both the cavity and the atom, placing the system deep into what is known as the "strong-dispersive" regime. In this regime, one can easily resolve the different spectral peaks of the qubit corresponding to the different possible integer number of photons in the cavity [6, 16, 110, 196]. That is, the light shift of the qubit is strongly quantized. This opens up new avenues for quantum control by allowing one to apply a Rabi rotation to the qubit that is effective if and only if there is some particular chosen photon number in the cavity.

Conversely, the two possibly cavity frequencies (depending on the state of the qubit) are well resolved. Hence, a drive tone at frequency $\omega_R \pm \chi/2$ will excite (coherently displace) the cavity only if the qubit is in the particular selected state. The combination of a photon-number-dependent qubit rotation and a qubit-state-dependent cavity displacement is an extremely powerful tool set [197–199]. For example, it becomes a relatively simple matter to create entangled states of the qubit and cavity:

$$|\psi\rangle = \frac{1}{\sqrt{2}}\big(|g\rangle|\alpha\rangle \pm |e\rangle|-\alpha\rangle\big) \tag{3.442}$$

where $|e\rangle$ and $|g\rangle$ are the states of the qubit and $|\pm\alpha\rangle$ are coherent states of amplitude $\pm\alpha$ of the cavity.

With a few additional steps, one can disentangle the qubit from the cavity to yield a photon cat state,

$$|\psi\rangle = \frac{1}{\sqrt{2}}|g\rangle\big(|\alpha\rangle \pm |-\alpha\rangle\big), \tag{3.443}$$

or even entangle the qubit with two different cats (one with even photon number parity and one with odd),

$$|\psi\rangle = \frac{1}{2}\left[|g\rangle\left(|\alpha\rangle + |-\alpha\rangle\right) + |e\rangle\left[\left(|\alpha\rangle - |-\alpha\rangle\right)\right]\right]. \tag{3.444}$$

Furthermore, the same tool set allows easy state tomography and direct measurement of the cavity Wigner function as well as the photon number parity [157]. Schrödinger cats of size up to the order of 100 photons can now be produced [158]. The ability to measure the photon number parity opens up the possibility for using Schrödinger cat states as code words in a quantum memory protocol that remarkably is robust against photon loss errors [198].

Experimental progress in circuit QED has now brought us into a new ultrastrong-coupling regime that is well beyond what can be achieved in conventional nonlinear quantum optics. This new regime is opening up new vistas of fundamental quantum physics and affords radically new possibilities for quantum information processing.

Multiport resonators All of the results derived above assumed a single-sided cavity in which there is only one port, so the reflected energy is equal to the incident energy. In this case, the reflection coefficient has unit magnitude, and information about the state of the qubit is encoded only in the phase of the reflection coefficient. For a two-port cavity, there will be a signal reflected from the input port and another signal transmitted through the output port. Depending on the relative coupling strengths of these two ports, information on the qubit state can be contained in one or both signals. This is discussed in more detail in [31].

One interesting case is where the input port is very weakly coupled and the damping of the cavity is dominated by the output port. In this case, there is no information in the signal reflected from the input port. In general, the information in the output port resides in both the amplitude and phase of the signal. For example, if the cavity is irradiated not at the bare cavity frequency ω_c (as in the example discussed above) but rather at $\omega_c - \chi$, then the amplitude will be large for one qubit state and small for the other. This setup is very useful in the strong-dispersive limit $\chi \gg \kappa$. In this limit, the power transmitted through the cavity at frequency ω_c is very small and so not useful for measurement (even though the phase difference between the two qubit states is π; that is, $\theta_0 = \pi/2$).

3.7.3 Multi-qubit dispersive readout

An important idea in circuit QED is the understanding that dispersive coupling to the cavity can be used to perform a simultaneous joint readout of multiple qubits [32, 182].

This joint readout was used for two-qubit state tomography by Majer *et al.* [114]. In its most basic form, the idea is simply that, with two qubits, there are four possible quantum states and four different dispersive frequency pulls of the cavity. If one is in the strong-dispersive coupling regime and all four frequency pulls can be reliably distinguished in a single shot, then one has two bits of classical information and a complete projective measurement of both qubits. Of course, in the presence of qubit decay and amplifier noise, the detector tomography can be complex [182]. A more sophisticated understanding of the situation of imperfect resolution of the four peaks has been developed recently, and Filipp *et al.* [200] have demonstrated that it is possible to reliably measure two-qubit correlations even in the presence of readout noise.

A simplified version of the theory for joint readout of two qubits is the following. While the cavity pull is linear in the qubit polarizations,

$$\delta\omega_c = \chi_1\sigma_1^z + \chi_2\sigma_2^z, \tag{3.445}$$

the corresponding homodyne (transmission) amplitude is *not*,

$$A(\sigma_1^z, \sigma_2^z) = \mathrm{Re}\left\{e^{i\varphi}\frac{\kappa/2}{\Delta - \delta\omega_c + i\kappa/2}\right\}. \tag{3.446}$$

Here, Δ is the detuning of the readout tone from the bare cavity resonance, κ is the cavity linewidth, and φ is the local oscillator phase. Because this can take on only four distinct values (corresponding to two classical bits of information), this expression can always be recast in the form

$$A(\sigma_1^z, \sigma_2^z) = \beta_0 + \beta_1\sigma_1^z + \beta_2\sigma_2^z + \beta_{12}\sigma_1^z\sigma_2^z. \tag{3.447}$$

The joint coefficient β_{12} is in general nonzero (as long as $\Delta \neq 0$) and typically on the same scale as the other coefficients. By using pre-rotations (by angle zero or π) of each of the two qubits prior to making the measurement, it is straightforward to obtain any one or two-qubit correlator in the z basis. Ensemble averaging many such measurements will reduce the statistical uncertainty to arbitrarily low values. For example,

$$\langle\sigma_1^z\sigma_2^z\rangle = \frac{1}{4\beta_{12}}\left\langle A(\sigma_1^z, \sigma_2^z) - A(-\sigma_1^z, \sigma_2^z) - A(\sigma_1^z, -\sigma_2^z) + A(\sigma_1^z, \sigma_2^z)\right\rangle. \tag{3.448}$$

Any other arbitrary correlators (e.g., $\langle\sigma_1^x\sigma_2^y\rangle$) can be achieved by pre-pending rotations through appropriate arbitrary angles. The Yale group has used this multi-qubit readout method to measure values of the CHSH entanglement witness well above the classical bound [20] and to prove three-qubit entanglement in a GHZ state [22].

3.7.4 Nonlinear and latching readouts

We have focused here on dispersive readout using a simple harmonic oscillator resonator parametrically coupled to the qubit. For binary readout of the state of a two-level qubit, other readout methods based on coupling to a nonlinear oscillator have been developed. The Josephson bifurcation amplifier and its cousin the cavity bifurcation amplifier [72, 200, 201] use a large Josephson junction as the nonlinear element. The advantage of a nonlinear cavity is that it can "latch" into one of two stable oscillation modes depending on the state of the qubit. Because of hysteresis, the readout can remain "latched" in its stable state even if the qubit decays, thus providing a high signal-to-noise ratio. A very simple readout scheme has been developed by the Yale group [202] in which the bare cavity is linear but one takes advantage of the anharmonicity inherited from the qubit itself so that a separate nonlinear element is not needed. It is also very robust and easy to implement under a large range of circuit parameter values.

For phase qubits, the UCSB group has developed a high-fidelity readout based on the qubit modulating the macroscopic quantum tunneling rate of a large junction placed in a metastable current-carrying state [61].

3.8 Summary and future directions

The rate of progress in the realization of quantum microwave circuits over the past decade has been truly remarkable and represents both progress towards building novel quantum machines and realizing non-linear quantum optics in a novel strong-coupling regime. Circuit QED is much more than atomic physics with wires. We have a set of modular elements that are readily connected together. Hence, we have the opportunity to assemble large-scale structures from these quantum building blocks and do some real quantum engineering. Further progress will require scaling up the numbers of both qubits and resonators and continuing to advance coherence times and gate fidelities. As the number of qubits grows, it will be important to increase the on-off ratio of the couplings among them. Simply detuning them from each other will probably not be sufficient, and interference between two coupling channels that can null out the net coupling will likely be needed. Houck and co-workers have developed a novel transmon structure in which the vacuum Rabi coupling can be tuned over a wide range using magnetic flux to control the interference between two internal modes of the qubit [203–205].

Another exciting direction involves using multiple physical qubits to realize individual logical qubits to overcome the difficulties of maintaining stable transition frequencies. In particular, the possibility of topological protection [206–210] is beginning to be explored in superconducting qubits [211]. The central idea is that qubits are constructed in which the ground and excited states are degenerate and this degeneracy is robust against local variations in Hamiltonian parameters. Even if the energy levels are not exactly degenerate, it would be very useful to have a qubit with a "Λ" energy-level scheme, that is, two nearly degenerate levels that can be coupled via stimulated Raman pulses through a third level. This would be advantageous both as a robust qubit and for purposes of fundamental quantum optics studies. It seems reasonably certain that this cannot be achieved without applied magnetic flux to frustrate the Josephson couplings (as in a flux qubit or in the fluxonium qubit). Indeed, the fluxonium qubit may turn out to be quite useful as a "Λ" system.

The development of large resonator arrays will be interesting not only as a quantum computation architecture but also for fundamental quantum optics purposes. An array of resonators each containing a qubit that induces a Kerr nonlinearity will be a realization of the boson Hubbard model [212], which exhibits both superfluid and Mott insulator phases. There is now a burgeoning interest in seeing "quantum phase transitions of light" [100, 153, 213–241]. Since the transmon qubit is itself an anharmonic oscillator, one might imagine it would be easier to simply use a lattice of coupled transmons to realize the boson Hubbard model (with negative Kerr coefficient). The advantage of using a lattice of resonators is that their resonance frequencies can be closely matched to a single fixed value. The Kerr coefficient induced by coupling each resonator to an off-resonant qubit will have some variation due to variations in qubit transition frequencies, but this disorder in the Hubbard U will be more tolerable than disorder in the photon "site energies." Just as cold atom systems are now used to simulate condensed matter models, so we may be able to use photons as interacting strongly correlated bosons, which can be probed, measured, and controlled in ways that are impossible in ordinary condensed matter.

The demands placed on classical computers to design and simulate even small numbers of qubits and resonators is already enormous. As the size of our quantum machines grows, the modeling complexity will grow exponentially. Similarly, the experimental measurements and process tomography needed to verify the operation of larger quantum machines will become extremely challenging. We will have to develop calibration and verification protocols that can reliably vet each segment of a quantum processor without the luxury of complete end-to-end process tomography. Today, classical computers are sufficiently complex that we cannot design the next generation by hand. We must use the current generation of computers to design the next. Eventually, this might be true of quantum machines. However, one wonders if we might soon find ourselves in a situation where both existing quantum machines and classical computers are not powerful enough to be used to design and model the next generation of quantum machines. We are rapidly gathering supplies to cross that novel and exciting desert!

3.9 Appendix

3.9.1 Cooper pair box Hamiltonian

The Cooper pair box (CPB) consists of a superconducting island coupled to a superconducting ground via a Josephson junction, or two islands connected by a Josephson junction. The Hamiltonian has the same form in both cases. The equivalent circuit is shown in Fig. 3.14. The bias voltage V could be an intentionally applied dc or ac voltage, or a quantum fluctuating voltage associated with a microwave photon field, or it could be a random voltage representing some local charge asymmetry in the vicinity of the island. For simplicity, we will take the voltage to be supplied by an ideal zero-impedance source. Because figuring out the Hamiltonian of a system connected to a power supply can be confusing, we will consider a physical realization of the voltage source as a very large "buffer" capacitor C_B as shown in Fig. 3.31. In this

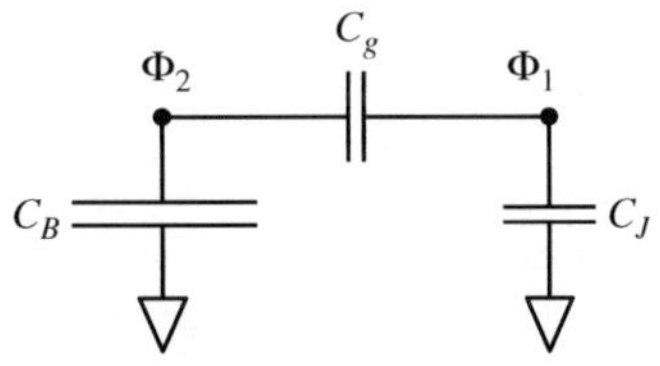

Fig. 3.31 Equivalent circuit for determining the electrostatic energy of a Cooper pair box biased by a voltage source represented by a large capacitor C_B. The Josephson junction capacitance C_J is coupled to the voltage source via the capacitor C_g. There are two node fluxes, Φ_1 and Φ_2.

configuration, the circuit has two islands with corresponding node fluxes Φ_1 and Φ_2. The charging energy can be determined by writing the Lagrangian for the two flux variables,

$$\mathcal{L} = \frac{1}{2}C_B\dot{\Phi}_2^2 + \frac{1}{2}C_g(\dot{\Phi}_2 - \dot{\Phi}_1)^2 + \frac{1}{2}C_J\dot{\Phi}_1^2. \tag{3.449}$$

Defining

$$\Phi \equiv \begin{pmatrix} \Phi_1 \\ \Phi_2 \end{pmatrix}, \tag{3.450}$$

the Lagrangian can be written

$$\mathcal{L} = \frac{1}{2}\dot{\Phi}^T C \dot{\Phi}, \tag{3.451}$$

where the capacitance matrix is given by

$$C = \begin{pmatrix} C_g + C_J & -C_g \\ -C_g & C_g + C_B \end{pmatrix}. \tag{3.452}$$

The electrostatic Hamiltonian is now readily expressed in terms of the charges canonically conjugate to Φ:

$$H = \frac{1}{2}Q^T C^{-1} Q, \tag{3.453}$$

where

$$Q \equiv \begin{pmatrix} Q_1 \\ Q_2 \end{pmatrix}. \tag{3.454}$$

The inverse capacitance matrix is given by

$$C^{-1} = \frac{1}{C_g C_B + C_g C_J + C_B C_J} \begin{pmatrix} C_g + C_B & C_g \\ C_g & C_g + C_J \end{pmatrix}. \tag{3.455}$$

This can be simplified by defining the total capacitance to ground for each of the two islands,

$$C_{1\Sigma} \equiv C_J + C_{2s}, \tag{3.456}$$

$$C_{2\Sigma} \equiv C_B + C_{1s}, \tag{3.457}$$

and the two series capacitances,

$$\frac{1}{C_{1s}} \equiv \frac{1}{C_g} + \frac{1}{C_J}, \tag{3.458}$$

$$\frac{1}{C_{2s}} \equiv \frac{1}{C_g} + \frac{1}{C_B}. \tag{3.459}$$

Using these definitions, the inverse capacitance matrix can be written

$$C^{-1} = \begin{pmatrix} \frac{1}{C_{1\Sigma}} & \frac{\beta}{C_{2\Sigma}} \\ \frac{\beta}{C_{2\Sigma}} & \frac{1}{C_{2\Sigma}} \end{pmatrix}, \tag{3.460}$$

where the dimensionless coupling constant is given by

$$\beta \equiv \frac{C_g}{C_g + C_J}. \tag{3.461}$$

The electrostatic Hamiltonian can then be written

$$H = \frac{Q_1^2}{2C_{1\Sigma}} + \beta \frac{Q_2}{C_{2\Sigma}} Q_1 + \frac{Q_2^2}{2C_{2\Sigma}}. \tag{3.462}$$

We define the "nominal" bias voltage as[30]

$$V_{\mathrm{B}} = \frac{Q_2}{C_{2\Sigma}}. \tag{3.464}$$

In terms of this, we can write the electrostatic Hamiltonian as

$$H = \frac{Q_1^2}{2C_{1\Sigma}} + \beta V_B Q_1 + \frac{1}{2} C_{2\Sigma} V_B^2. \tag{3.465}$$

Including now the Josephson junction energy and quantizing, we arrive at the full CPB Hamiltonian

$$H = \frac{\hat{Q}_1^2}{2C_{1\Sigma}} + \beta V_B \hat{Q}_1 - E_J \cos \frac{2e}{\hbar} \Phi_1 + \frac{1}{2} C_{2\Sigma} V_B^2. \tag{3.466}$$

Note that the last term is a constant of the motion and can be ignored.

Defining the dimensionless offset (or "gate") charge

$$n_g \equiv -\beta \frac{C_{1\Sigma} V_B}{2e} \approx -\frac{C_g V_B}{2e}, \tag{3.467}$$

[30] Note that *actual* voltage on island two is

$$V = \frac{\partial H}{\partial Q_2} = V_{\mathrm{B}} + \beta \frac{Q_1}{C_{2\Sigma}}. \tag{3.463}$$

In the limit of of large C_B, the actual voltage is fully buffered (i.e., becomes independent of Q_1) and is given by $V \approx V(Q_1 = 0) = V_B$.

(with the latter equality only in the limit of large C_B) and defining the charging energy

$$E_C \equiv \frac{e^2}{2C_{1\Sigma}}, \tag{3.468}$$

we can also write the Hamiltonian in terms of the integer-valued number operator

$$\hat{n} \equiv \frac{\hat{Q}_1}{2e} \tag{3.469}$$

representing the excess number of Cooper pairs on island 1. Its conjugate variable is the relative phase angle for the superconducting order parameter across the junction,

$$\varphi = \frac{2e}{\hbar}\Phi_1 = 2\pi\frac{\Phi_1}{\Phi_0}, \tag{3.470}$$

where Φ_0 is the superconducting flux quantum. In terms of this pair of dimensionless charge and phase variables, the Hamiltonian becomes

$$H = 4E_C\left(\hat{n}^2 - 2n_g\hat{n}\right) - E_J\cos\varphi + \frac{1}{2}C_{2\Sigma}V_B^2 \tag{3.471}$$

$$= 4E_C\left(\hat{n} - n_g\right)^2 - E_J\cos\varphi + \frac{1}{2}\frac{C_{1\Sigma}C_{2\Sigma}}{C_g + C_J}V_B^2, \tag{3.472}$$

It is sometimes convenient to work in the (angular) position basis with wavefunction $\Psi(\varphi)$ and $\hat{n} = -i\partial/\partial\varphi$ being represented by the angular momentum conjugate to the angle φ. In other cases, it is more convenient to work in the angular momentum (charge) eigenbasis and recognize that the operator $\cos\varphi$ term is a "torque" that changes the angular momentum by ± 1 unit.

Dropping the last constant term in Eq. (3.472) and assuming the voltage bias is fully buffered (large C_B), we arrive at Eq. (3.211). We again emphasize that n_g is a continuous variable (and often subject to $1/f$ noise), while $\hat{n}$ is integer-valued (i.e., is the angular momentum conjugate to the angular variable φ) and changes by ± 1 each time a Cooper pair tunnels through the Josephson junction connected across C_J.

Cooper pair box coupled to an LC resonator Consider the circuit in Fig. 3.32, which shows a CPB coupled to an LC resonator. For simplicity, we will ignore the

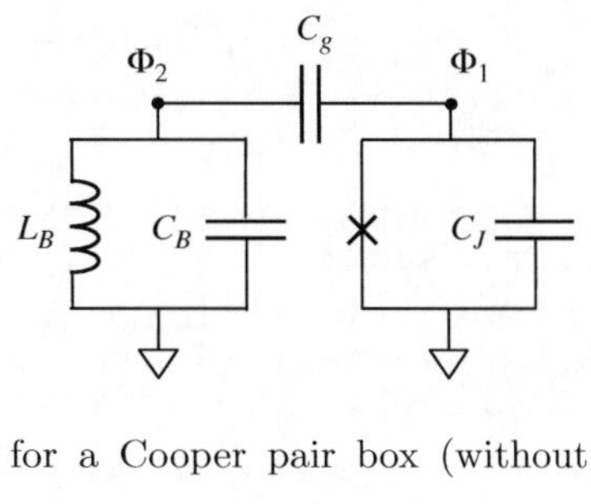

Fig. 3.32 Equivalent circuit for a Cooper pair box (without dc voltage bias) capacitively coupled to an LC resonator.

possibility of a dc bias voltage on the qubit. The Hamiltonian can be immediately written down by analogy with Eq. (3.462):

$$H = H_1 + H_2 + H_{12}, \tag{3.473}$$

$$H_1 = \frac{\hat{Q}_1^2}{2C_{1\Sigma}} - E_J \cos\frac{2e}{\hbar}\hat{\Phi}_1, \tag{3.474}$$

$$H_2 = \frac{\hat{Q}_2^2}{2C_{2\Sigma}} + \frac{1}{2L_B}\hat{\Phi}_2^2, \tag{3.475}$$

$$H_{12} = \frac{\beta}{C_{2\Sigma}}\hat{Q}_1\hat{Q}_2. \tag{3.476}$$

Let the eigenfunctions of H_1 obey

$$H_1|j\rangle = \epsilon_j|j\rangle, \tag{3.477}$$

and let us denote the matrix elements of the charge operator in this basis by

$$Q_{jk} = \langle j|\hat{Q}_1|k\rangle. \tag{3.478}$$

This is the analog of the dipole matrix elements of an atom. The LC resonator Hamiltonian can be written

$$H_2 = \hbar\omega_c\hat{a}^\dagger\hat{a}, \tag{3.479}$$

where $\omega_c = 1/\sqrt{L_B C_{2\Sigma}}$, and the second charge operator can be written following Eq. (3.42),

$$\hat{Q}_2 = -iQ_{2\text{ZPF}}\left(\hat{a} - \hat{a}^\dagger\right), \tag{3.480}$$

where, following Eq. (3.43), we have

$$Q_{2\text{ZPF}} = \sqrt{\frac{C_B\hbar\omega_c}{2}}. \tag{3.481}$$

We now have the full Hamiltonian

$$H = \hbar\omega_c\hat{a}^\dagger\hat{a} + \sum_{k=0}^{\infty}\epsilon_k|k\rangle\langle k| - i\frac{\beta Q_{2\text{ZPF}}Q_{jk}}{2C_{2\Sigma}}|j\rangle(\hat{a} - \hat{a}^\dagger)\langle k|, \tag{3.482}$$

in a form that is convenient for numerical diagonalization.

If the spectrum of the qubit is sufficiently anharmonic, we may be able to restrict our attention to its two lowest states. Projecting our Hamiltonian onto these two states allows us to represent the qubit operators in terms of Pauli spin matrices. Taking the ground state $|0\rangle$ to be represented by spin down $|\downarrow\rangle$ and the excited state $|1\rangle$ to be represented by spin up $|\uparrow\rangle$ we can represent all possible qubit operators within the 2×2 Hilbert space:

$$|0\rangle\langle 0| = \frac{1-\sigma^z}{2}, \tag{3.483}$$

$$|1\rangle\langle 1| = \frac{1+\sigma^z}{2}, \tag{3.484}$$

$$|1\rangle\langle 0| = \sigma^+, \tag{3.485}$$

$$|0\rangle\langle 1| = \sigma^-. \tag{3.486}$$

The Hamiltonian then becomes (dropping an irrelevant constant)

$$H = \hbar\omega_c \hat{a}^\dagger \hat{a} + \frac{\hbar\omega_{01}}{2}\sigma^z - i(\hat{a} - \hat{a}^\dagger)\hbar\left(g_{11}\frac{1+\sigma^z}{2} + g_{00}\frac{1-\sigma^z}{2} + g_{10}\sigma^+ + g_{01}\sigma^-\right), \tag{3.487}$$

where

$$\hbar\omega_{01} \equiv \epsilon_1 - \epsilon_0 \tag{3.488}$$

and

$$\hbar g_{jk} \equiv \frac{\beta Q_{2\mathrm{ZPF}} Q_{jk}}{2C_{2\Sigma}}. \tag{3.489}$$

If the eigenstates of H_1 have a static dipole moment, then the diagonal matrix elements of the charge operator Q_1 will be nonzero. Here we are, for simplicity, ignoring the possibility of a dc bias, which produces an offset charge. In this case, H_1 has a charge-parity symmetry that guarantees that the diagonal matrix elements of the charge operator Q_1 vanish. We are free to choose a gauge (i.e., choose the arbitrary phases of the eigenstates of H_1) so that $g_{01} = g_{10} = g$ is real. We then arrive at the celebrated Jaynes-Cummings Hamiltonian

$$H = \hbar\omega_c \hat{a}^\dagger \hat{a} + \frac{\hbar\omega_{01}}{2}\sigma^z - i\hbar g(\hat{a} - \hat{a}^\dagger)(\sigma^+ + \sigma^-), \tag{3.490}$$

which, when the rotating-wave approximation (RWA) is justified, further simplifies to

$$H = \hbar\omega_c \hat{a}^\dagger \hat{a} + \frac{\hbar\omega_{01}}{2}\sigma^z - i\hbar g(\hat{a}\sigma^+ - \hat{a}^\dagger\sigma^-). \tag{3.491}$$

We can reduce this to the more familiar expression given in Eq. (3.320) by making a rotation of the spin axes via the unitary transformation

$$\hat{U} = e^{i\frac{1}{4}\pi\sigma^z}, \tag{3.492}$$

which yields

$$U\sigma^+ U^\dagger = +i\sigma^+, \tag{3.493}$$

$$U\sigma^- U^\dagger = -i\sigma^+. \tag{3.494}$$

and thus finally

$$H = \hbar\omega_c \hat{a}^\dagger \hat{a} + \frac{\hbar\omega_{01}}{2}\sigma^z + \hbar g(\hat{a}\sigma^+ + \hat{a}^\dagger\sigma^-). \tag{3.495}$$

3.9.2 Semi-infinite transmission lines, dissipation, and input-output theory

This discussion is based on material presented in [31].

So far, we have studied transmission-line resonators, which contain discrete standing-wave resonances. We turn now to the case of infinite and semi-infinite transmission lines, where we will deal with traveling waves rather than standing waves. In order to control the state of quantum bits, we will want to send control pulses down a transmission line to the qubit. However, opening up a port into the qubit to allow control signals in can also allow quantum and thermal noise to enter. Such noises can dephase the coherent superpositions of qubits and also permit the qubits to decay by spontaneously emitting energy out the port and into the transmission line.

Another reason for studying the case of traveling waves in semi-infinite transmission lines is the following. In addition to controlling qubits by sending pulses down transmission lines, we may also wish to *measure* the state of a qubit by sending microwave pulses down a transmission line and seeing how the reflected pulse is modified according to the state of the qubit. Thus, we also need to understand continuum scattering theory for microwaves and how to relate the properties of the reflected wave to those of the incident wave and of the system from which they are reflecting.

You might think that since a transmission line is made of purely reactive elements, it cannot dissipate energy. If, however, it is semi-infinite in length, then a wave launched at one end carries away energy and never returns. As usual in studying irreversible behavior, there is a subtle order of limits here. If time goes to infinity before the transmission-line length goes to infinity, then there is no dissipation. In the opposite order of limits, there is dissipation because the waves travel away and never have time to reach the end of the transmission line and reflect back. We will in fact see that a semi-infinite transmission line acts just like a resistor with resistance equal to its characteristic impedance $R = Z_c = \sqrt{\ell/c}$. Remarkably, however, because the transmission line is made of purely reactive elements, we know, as we have already seen, how to treat it quantum mechanically. Hence, we will be able to develop a fully quantum theory of a dissipative resistor [242].

It is useful to start with a classical model, considering as before a coaxial transmission line modeled as a perfectly conducting wire with inductance per unit length of ℓ and capacitance to ground per unit length c as shown in Fig. 3.33. If the voltage at position x at time t is $V(x,t)$, then the charge density is $q(x,t) = cV(x,t)$. By charge conservation, the current I and the charge density are related by the continuity equation

$$\partial_t q + \partial_x I = 0. \tag{3.496}$$

The constitutive relation (essentially Newton's law) gives the acceleration of the charges:

$$\ell \partial_t I = -\partial_x V. \tag{3.497}$$

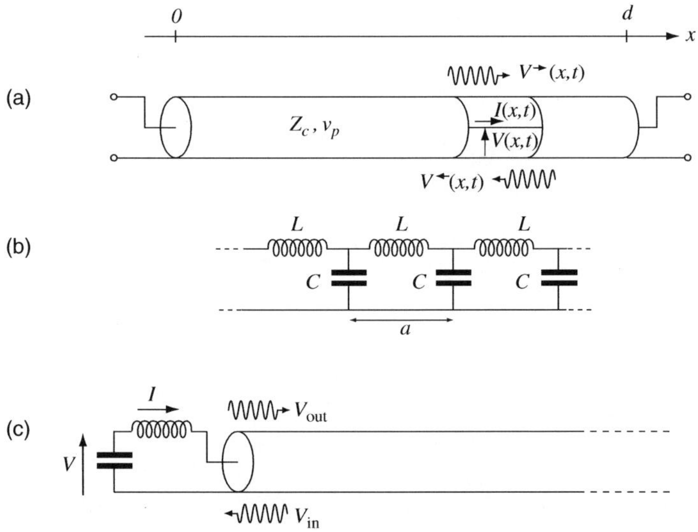

Fig. 3.33 (a) Coaxial transmission line, indicating voltages and currents as defined in the text. (b) Lumped-element representation of a transmission line with capacitance per unit length $c = C/a$ and inductance per unit length $\ell = L/a$. (c) Discrete LC resonator terminating a transmission line.

We can decouple Eqs. (3.496) and (3.497) by introducing left- and right-propagating modes

$$V(x,t) = V^{\rightarrow} + V^{\leftarrow}, \tag{3.498}$$

$$I(x,t) = \frac{1}{Z_c}(V^{\rightarrow} - V^{\leftarrow}). \tag{3.499}$$

In terms of the left- and right-propagating modes, Eqs. (3.496) and (3.497) become

$$v_p \partial_x V^{\rightarrow} + \partial_t V^{\rightarrow} = 0, \tag{3.500}$$

$$v_p \partial_x V^{\leftarrow} - \partial_t V^{\leftarrow} = 0, \tag{3.501}$$

where $v_p \equiv 1/\sqrt{\ell c}$ is the wave phase velocity. These equations have solutions that propagate by uniform translation without changing shape, since the line is (in the present model) dispersionless:

$$V^{\rightarrow}(x,t) = V_{\text{out}}\left(t - \frac{x}{v_p}\right) \tag{3.502}$$

$$V^{\leftarrow}(x,t) = V_{\text{in}}\left(t + \frac{x}{v_p}\right), \tag{3.503}$$

where V_{in} and V_{out} are *arbitrary* functions of their arguments. For an infinite transmission line, V_{out} and V_{in} are completely independent. However, for the case of a semi-infinite line terminated at $x = 0$ (say) by some system S, these two solutions are not independent, but rather related by the boundary condition imposed by the system. We have

$$V(x = 0, t) = V_{\text{out}}(t) + V_{\text{in}}(t), \tag{3.504}$$

$$I(x = 0, t) = \frac{1}{Z_c}[V_{\text{out}}(t) - V_{\text{in}}(t)], \tag{3.505}$$

from which we may eliminate $V(x = 0, t)$ to derive the crucially important "input-output" relation

$$V_{\text{out}}(t) = V_{\text{in}}(t) + Z_c I(x = 0, t). \tag{3.506}$$

The first term on the right-hand side is simply the direct reflection of the input wave, while the second term represents waves radiated into the transmission line by current injected by the system S.

If the system under study is just an open circuit, so that $I(x = 0, t) = 0$, then $V_{\text{out}} = V_{\text{in}}$, meaning that the outgoing wave is simply the result of the incoming wave reflecting from the open-circuit termination. In general, however, there is an additional outgoing wave radiated by the current I that is injected by the system dynamics into the line. In the absence of an incoming wave, we have

$$V(x = 0, t) = Z_c I(x = 0, t), \tag{3.507}$$

indicating that the transmission line acts as a simple resistor that, instead of dissipating energy by Joule heating, carries the energy away from the system as propagating waves. As noted above, the fact that the line can dissipate energy despite containing only purely reactive elements is a consequence of its infinite extent. One must be careful with the order of limits, taking the length to infinity *before* allowing time to go to infinity. In this way, the outgoing waves never reach the far end of the transmission line and reflect back. Since this is a conservative Hamiltonian system, we will be able to quantize these waves and make a quantum theory of resistors [242]. The net power flow carried to the right by the line is

$$P = \frac{1}{Z_c}[V_{\text{out}}^2(t) - V_{\text{in}}^2(t)]. \tag{3.508}$$

The fact that the transmission line presents a dissipative impedance to the system means that it causes damping of the system. It also, however, opens up the possibility of controlling the system via the input field, which partially determines the voltage driving the system. From this point of view, it is convenient to eliminate the output field by writing the voltage as

$$V(x = 0, t) = 2V_{\text{in}}(t) + Z_c I(x = 0, t). \tag{3.509}$$

As we will discuss in more detail, the first term drives the system and the second damps it. From Eq. (3.506), we see that measurement of the outgoing field can be used to determine the current $I(x = 0, t)$ injected by the system into the line and hence to infer the system dynamics that results from the input drive field.

The great benefit of Eq. (3.509) is the following. If the system S is coupled to the transmission line through a capacitor, then the coupling Hamiltonian can be expressed in terms of the voltage at the end, $V(x = 0, t)$. By eliminating V_{out}, we can see how the system is driven by V_{in} and damped by $Z_c I(x = 0, t)$. This classical result will be helpful in understanding the closely analogous quantum expressions, which will be derived further in a rather different manner.

As a simple classical example, consider the system consisting of an LC resonator shown in Fig. 3.33(c). This can be viewed as a simple harmonic oscillator whose coordinate Q is the charge on the capacitor plate (on the side connected to L_0). The current $I(x = 0, t) = \dot{Q}$ plays the role of the velocity of the oscillator. The equation of motion for the oscillator is readily obtained from

$$Q = C_0[-V(x = 0^+, t) - L_0 \dot{I}(x = 0^+, t)]. \tag{3.510}$$

Using Eq. (3.509), we obtain a harmonic oscillator damped by the transmission line and driven by the incoming waves:

$$\ddot{Q} = -\Omega_0^2 Q - \kappa \dot{Q} - \frac{2}{L_0} V_{\text{in}}(t), \tag{3.511}$$

where the resonant frequency is $\Omega_0^2 \equiv 1/\sqrt{L_0 C_0}$. Note that the term $Z_c I(x = 0, t)$ in Eq. (3.509) results in the linear viscous damping rate $\kappa \equiv Z_c/L_0$.

If we solve the equation of motion of the oscillator, we can predict the outgoing field. In the present instance of a simple oscillator, we have a particular example of the general case where the system responds linearly to the input field. We can characterize any such system by a complex, frequency-dependent impedance $Z[\omega]$ defined by

$$Z[\omega] = -\frac{V(x = 0, \omega)}{I(x = 0, \omega)}. \tag{3.512}$$

Note the peculiar minus sign, which results from our definition of positive current flowing to the right (out of the system and into the transmission line). Using Eqs. (3.504), (3.505), and (3.512), we have

$$V_{\text{out}}[\omega] = r[\omega] V_{\text{in}}[\omega], \tag{3.513}$$

where the reflection coefficient r is determined by the impedance mismatch between the system and the line and is given by the well-known result

$$r[\omega] = \frac{Z[\omega] - Z_c}{Z[\omega] + Z_c}. \tag{3.514}$$

If the system is constructed from purely reactive (i.e., lossless) components, then $Z[\omega]$ is purely imaginary and the reflection coefficient obeys $|r| = 1$, which is consistent with Eq. (3.508) and the energy conservation requirement of no net power flow into the lossless system. For example, for the series LC oscillator we have been considering, we have

$$Z[\omega] = \frac{1}{j\omega C_0} + j\omega L_0, \tag{3.515}$$

where, to make contact with the usual electrical engineering sign conventions, we have used $j = -\sqrt{-1} = -i$. If the damping κ of the oscillator induced by coupling it to the transmission line is small, the quality factor of the resonance will be high and we need only consider frequencies near the resonance frequency $\Omega_0 \equiv 1/\sqrt{L_0 C_0}$, where the impedance has a zero. In this case, we may approximate

$$Z[\omega] \approx \frac{2}{jC_0\Omega_0^2}(\Omega_0 - \omega) = 2jL_0(\omega - \Omega_0) \tag{3.516}$$

which yields for the reflection coefficient

$$r[\omega] = \frac{\omega - \Omega_0 + j\kappa/2}{\omega - \Omega_0 - j\kappa/2} \tag{3.517}$$

showing that indeed $|r| = 1$ and that the phase of the reflected signal winds by 2π upon passing through the resonance.[31]

Turning to the more general case where the system also contains lossy elements, one finds that $Z[\omega]$ is no longer purely imaginary, but has a real part satisfying $\operatorname{Re} Z[\omega] > 0$. This in turn implies via Eq. (3.514) that $|r| < 1$. In the special case of impedance matching, $Z[\omega] = Z_c$, all the incident power is dissipated in the system and none is reflected. The other two limits of interest are open-circuit termination with $Z = \infty$ for which $r = +1$ and short-circuit termination $Z = 0$ for which $r = -1$.

Finally, if the system also contains an active device that has energy being pumped into it from a separate external source, it may under the right conditions be described by an effective *negative* resistance $\operatorname{Re} Z[\omega] < 0$ over a certain frequency range. Eq. (3.514) then gives $|r| \geq 1$, implying $|V_{\text{out}}| > |V_{\text{in}}|$. Our system will thus act like the one-port reflection amplifier discussed in great detail in [31].

Quantum input-output theory for a driven cavity The preceding results have been derived on the basis of the classical input-output theory relating the drive, the cavity, and the outgoing waves radiated by the cavity. They can be more formally derived in a full quantum theory of a cavity driven by an external coherent source. The present quantum discussion closely follows standard references on the subject [66, 243, 244]. The crucial feature that distinguishes such an approach from many other treatments of quantum dissipative systems is the goal of keeping the bath modes instead of tracing them out. This is obviously necessary for the situations we have in mind, where the output field emanating from the cavity contains the information acquired during a measurement of the system coupled to the cavity. As we learned from

[31] For the case of resonant *transmission* through a symmetric cavity, the phase shift only winds by π.

the classical treatment, we can eliminate the outgoing waves in favor of a damping term for the system. However, we can recover the solution for the outgoing modes completely from the solution of the equation of motion of the damped system being driven by the incoming waves.

In order to drive the cavity, we must partially open one of its ports, which exposes the cavity both to the external drive and to the vacuum noise outside, permitting energy in the cavity to leak out into the surrounding bath. We will formally separate the degrees of freedom into internal cavity modes and external bath modes. Strictly speaking, once the port is open, these modes are not distinct and we only have "the modes of the universe" [245–247]. However, for high-Q cavities, the distinction is well defined and we can model the decay of the cavity in terms of a spontaneous emission process in which an internal boson is destroyed and an external bath boson is created. We assume a single-sided cavity. For a high-Q cavity, this physics is accurately captured in the Hamiltonian

$$\hat{H} = \hat{H}_{\text{sys}} + \hat{H}_{\text{bath}} + \hat{H}_{\text{int}}. \tag{3.518}$$

The bath Hamiltonian is

$$\hat{H}_{\text{bath}} = \sum_q \hbar\omega_q \hat{b}_q^\dagger \hat{b}_q \tag{3.519}$$

where q labels the quantum numbers of the independent harmonic oscillator bath modes obeying

$$[\hat{b}_q, \hat{b}_{q'}^\dagger] = \delta_{q,q'}. \tag{3.520}$$

Note that since the bath terminates at the system, there is no translational invariance, the normal modes are standing, not running, waves, and the quantum numbers q are not necessarily wavevectors.

The coupling Hamiltonian is (within the rotating-wave approximation)

$$\hat{H}_{\text{int}} = -i\hbar \sum_q \left(f_q \hat{a}^\dagger \hat{b}_q - f_q^* \hat{b}_q^\dagger \hat{a} \right). \tag{3.521}$$

For the moment, we will leave the system (cavity) Hamiltonian to be completely general, specifying only that it consists of a single degree of freedom (i.e., we concentrate on only a single resonance of the cavity with frequency ω_c) obeying the usual bosonic commutation relation

$$[\hat{a}, \hat{a}^\dagger] = 1. \tag{3.522}$$

(Note that this does not imply that it is a harmonic oscillator. We will consider both linear and nonlinear cavities.) The most general linear coupling to the bath modes would include terms of the form $\hat{b}_q^\dagger \hat{a}^\dagger$ and $\hat{b}_q a$, but these are neglected within the rotating-wave approximation because in the interaction representation they oscillate at high frequencies and have little effect on the dynamics.

The Heisenberg equation of motion (EOM) for the bath variables is

$$\dot{\hat{b}}_q = \frac{i}{\hbar}[\hat{H}, \hat{b}_q] = -i\omega_q \hat{b}_q + f_q^* \hat{a}. \tag{3.523}$$

We see that this is simply the EOM of a harmonic oscillator driven by a forcing term due to the motion of the cavity degree of freedom. Since this is a linear system, the EOM can be solved exactly. Let $t_0 < t$ be a time in the distant past before any wavepacket launched at the cavity has reached it. The solution of Eq. (3.523) is

$$\hat{b}_q(t) = e^{-i\omega_q(t-t_0)}\hat{b}_q(t_0) + \int_{t_0}^{t} d\tau\, e^{-i\omega_q(t-\tau)} f_q^* \,\hat{a}(\tau). \tag{3.524}$$

The first term is simply the free evolution of the bath, while the second represents the waves radiated by the cavity into the bath.

The EOM for the cavity mode is

$$\dot{\hat{a}} = \frac{i}{\hbar}[\hat{H}_{\text{sys}}, \hat{a}] - \sum_q f_q \hat{b}_q. \tag{3.525}$$

Substituting Eq. (3.524) into the last term yields

$$\sum_q f_q \hat{b}_q = \sum_q f_q e^{-i\omega_q(t-t_0)}\hat{b}_q(t_0) + \sum_q |f_q|^2 \int_{t_0}^{t} d\tau\, e^{-i(\omega_q-\omega_c)(t-\tau)}[e^{+i\omega_c(\tau-t)}\hat{a}(\tau)], \tag{3.526}$$

where the last term in square brackets is a slowly varying function of τ. To simplify our result, we note that if the cavity system were a simple harmonic oscillator of frequency ω_c, then the decay rate from the $n = 1$ single-photon excited state to the $n = 0$ ground state would be given by the following Fermi Golden Rule expression:

$$\kappa(\omega_c) = 2\pi \sum_q |f_q|^2 \delta(\omega_c - \omega_q). \tag{3.527}$$

It follows from this that

$$\int_{-\infty}^{+\infty} \frac{d\nu}{2\pi}\kappa(\omega_c + \nu) e^{-i\nu(t-\tau)} = \sum_q |f_q|^2 e^{-i(\omega_q-\omega_c)(t-\tau)}. \tag{3.528}$$

We now make the Markov approximation, which assumes that $\kappa(\nu) = \kappa$ is a constant over the range of frequencies relevant to the cavity, so that Eq. (3.528) may be represented as

$$\sum_q |f_q|^2 e^{-i(\omega_q-\omega_c)(t-\tau)} = \kappa\delta(t-\tau). \tag{3.529}$$

Using

$$\int_{-\infty}^{x_0} dx\, \delta(x - x_0) = \frac{1}{2} \tag{3.530}$$

we obtain for the cavity EOM

$$\dot{\hat{a}} = \frac{i}{\hbar}[\hat{H}_{\text{sys}}, \hat{a}] - \frac{\kappa}{2}\hat{a} - \sum_q f_q e^{-i\omega_q(t-t_0)}\hat{b}_q(t_0). \tag{3.531}$$

The second term came from the part of the bath motion representing the wave radiated by the cavity and, within the Markov approximation, has become a simple linear damping term for the cavity mode. Note the important factor of 2. The amplitude decays at half the rate of the intensity (the energy decay rate κ).

Within the spirit of the Markov approximation, it is further convenient to treat $f \equiv \sqrt{|f_q|^2}$ as a constant and define the density of states (also taken to be a constant) by

$$\rho = \sum_q \delta(\omega_c - \omega_q) \tag{3.532}$$

so that the Golden Rule rate becomes

$$\kappa = 2\pi f^2 \rho. \tag{3.533}$$

We can now define the so-called "input mode"

$$\hat{b}_{\rm in}(t) \equiv \frac{1}{\sqrt{2\pi\rho}} \sum_q e^{-i\omega_q(t-t_0)} \hat{b}_q(t_0) \tag{3.534}$$

corresponding to the initial pulse propagating towards the cavity. We finally have for the cavity EOM

$$\dot{\hat{a}} = \frac{i}{\hbar}[\hat{H}_{\rm sys}, \hat{a}] - \frac{\kappa}{2}\hat{a} - \sqrt{\kappa}\, \hat{b}_{\rm in}(t). \tag{3.535}$$

Note that when a wavepacket is launched from the bath toward the cavity, causality prevents it from knowing about the cavity's presence until it reaches the cavity. Hence, the input mode evolves freely as if the cavity were not present, until the time of the collision, at which point it begins to drive the cavity. Since $\hat{b}_{\rm in}(\mathrm{t})$ evolves under the free bath Hamiltonian and acts as the driving term in the cavity EOM, we interpret it physically as the input mode. Eq. (3.535) is the quantum analog of the classical equation (3.511) for our previous example of an LC oscillator driven by a transmission line. The latter would also have been of first order in time if we had worked with the complex amplitude A instead of the coordinate Q.

Eq. (3.534) for the input mode contains a time label just as in the interaction representation. However, it is best interpreted as simply labeling the particular linear combination of the bath modes that is coupled to the system at time t. Some authors even like to think of the bath modes as nonpropagating while the cavity flies along the bath (taken to be one-dimensional) at a velocity v. The system then only interacts briefly with the local mode positioned at $x = vt$ before moving on and interacting with the next local bath mode. We will elaborate on this view further at the end of this subsection.

The expression for the power $P_{\rm in}$ (energy per time) impinging on the cavity depends on the normalization chosen in our definition of $\hat{b}_{\rm in}$. It can be obtained, for example, by imagining the bath modes $\hat{b}_q$ to live on a one-dimensional waveguide with propagation velocity v and length L (using periodic boundary conditions). In that case, we have to sum over all photons to get the average power flowing through a cross-section of the waveguide, $P_{\text{in}} = \sum_q \hbar\omega_q (v_p/L)\langle \hat{b}_q^\dagger \hat{b}_q \rangle$. Inserting the definition of $\hat{b}_{\text{in}}$, Eq. (3.534), the expression for the input power carried by a monochromatic beam at frequency ω is

$$P_{\text{in}}(t) = \hbar\omega \langle \hat{b}_{\text{in}}^\dagger(t) \hat{b}_{\text{in}}(t) \rangle. \tag{3.536}$$

Note that this has the correct dimensions owing to our choice of normalization for $\hat{b}_{\text{in}}$ (with dimensions $\sqrt{\omega}$). In the general case, an integration over frequencies is needed (as will be discussed further below). An analogous formula holds for the power radiated by the cavity, to be discussed now.

The output mode $\hat{b}_{\text{out}}(t)$ is radiated into the bath and evolves freely after the system interacts with $\hat{b}_{\text{in}}(t)$. If the cavity did not respond at all, then the output mode would simply be the input mode reflected off the cavity mirror. If the mirror is partially transparent, then the output mode will also contain waves radiated by the cavity (which is itself being driven by the input mode partially transmitted into the cavity through the mirror) and hence will contain information about the internal dynamics of the cavity. To analyze this output field, let $t_1 > t$ be a time in the distant future after the input field has interacted with the cavity. Then we can write an alternative solution to Eq. (3.523) in terms of the final rather than the initial condition of the bath:

$$\hat{b}_q(t) = e^{-i\omega_q(t-t_1)} \hat{b}_q(t_1) - \int_t^{t_1} d\tau \, e^{-i\omega_q(t-\tau)} f_q^* \hat{a}(\tau). \tag{3.537}$$

Note the important minus sign in the second term associated with the fact that the time t is now the lower limit of integration rather than the upper as it was in Eq. (3.524).

Defining

$$\hat{b}_{\text{out}}(t) \equiv \frac{1}{\sqrt{2\pi\rho}} \sum_q e^{-i\omega_q(t-t_1)} \hat{b}_q(t_1), \tag{3.538}$$

we see that this is simply the free evolution of the bath modes from the distant future (after they have interacted with the cavity) back to the present, indicating that it is indeed appropriate to interpret this as the outgoing field. Proceeding as before, we obtain

$$\dot{\hat{a}} = \frac{i}{\hbar}[\hat{H}_{\text{sys}}, \hat{a}] + \frac{\kappa}{2}\hat{a} - \sqrt{\kappa}\, \hat{b}_{\text{out}}(t). \tag{3.539}$$

Subtracting Eq. (3.539) from Eq. (3.535) yields

$$\hat{b}_{\text{out}}(t) = \hat{b}_{\text{in}}(t) + \sqrt{\kappa}\, \hat{a}(t), \tag{3.540}$$

which is consistent with our interpretation of the outgoing field as the reflected incoming field plus the field radiated by the cavity out through the partially reflecting mirror.

These results are valid for any general cavity Hamiltonian. The general procedure is to solve Eq. (3.535) for $\hat{a}(t)$ for a given input field, and then solve Eq. (3.540) to obtain the output field. For the case of an empty cavity, we can make further progress because the cavity mode is a harmonic oscillator:

$$\hat{H}_{\text{sys}} = \hbar\omega_c \hat{a}^\dagger \hat{a}. \tag{3.541}$$

In this simple case, the cavity EOM becomes

$$\dot{\hat{a}} = -i\omega_c \hat{a} - \frac{\kappa}{2}\hat{a} - \sqrt{\kappa}\,\hat{b}_{\text{in}}(t). \tag{3.542}$$

Eq. (3.542) can be solved by Fourier transformation, yielding

$$\hat{a}[\omega] = -\frac{\sqrt{\kappa}}{i(\omega_c - \omega) + \kappa/2}\,\hat{b}_{\text{in}}\,[\omega] \tag{3.543}$$

$$= -\sqrt{\kappa}\chi_c[\omega - \omega_c]\hat{b}_{\text{in}}[\omega] \tag{3.544}$$

and

$$\hat{b}_{\text{out}}[\omega] = \frac{\omega - \omega_c - i\kappa/2}{\omega - \omega_c + i\kappa/2}\,\hat{b}_{\text{in}}\,[\omega] \tag{3.545}$$

which is the result for the reflection coefficient quoted in Eq. (3.517). For brevity, here and in the following, we will sometimes use the susceptibility of the cavity, defined as

$$\chi_c[\omega - \omega_c] \equiv \frac{1}{-i(\omega - \omega_c) + \kappa/2}. \tag{3.546}$$

For the case of steady driving on resonance where $\omega = \omega_c$, these equations yield

$$\hat{b}_{\text{out}}[\omega] = \frac{\sqrt{\kappa}}{2}\hat{a}[\omega]. \tag{3.547}$$

In steady state, the incoming power equals the outgoing power, and both are related to the photon number inside the single-sided cavity by

$$P = \hbar\omega\langle \hat{b}^\dagger_{\text{out}}(t)\hat{b}_{\text{out}}(t)\rangle = \hbar\omega\frac{\kappa}{4}\langle \hat{a}^\dagger(t)\hat{a}(t)\rangle. \tag{3.548}$$

Note that this does not coincide with the naive expectation, which would be $P = \hbar\omega\kappa\left\langle \hat{a}^\dagger\hat{a}\right\rangle$. The reason for this discrepancy is the interference between the part of the incoming wave that is promptly reflected from the cavity and the field radiated by the cavity. The naive expression becomes correct after the drive has been switched off (where, ignoring the effect of the incoming vacuum noise, we would have $\hat{b}_{\text{out}} = \sqrt{\kappa}\hat{a}$). We note in passing that for a driven two-sided cavity with coupling constants κ_L and κ_R (where $\kappa = \kappa_L + \kappa_R$), the incoming power sent into the left port is related to the photon number by

$$P = \frac{\hbar\omega\kappa^2}{4\kappa_L}\left\langle \hat{a}^\dagger\hat{a}\right\rangle. \tag{3.549}$$

Here, for $\kappa_L = \kappa_R$, the interference effect completely eliminates the reflected beam and we have, in contrast to Eq. (3.548),

$$P = \hbar\omega\frac{\kappa}{2}\left\langle \hat{a}^\dagger \hat{a} \right\rangle . \tag{3.550}$$

Eq. (3.542) can also be solved in the time domain to obtain

$$\begin{aligned} \hat{a}(t) &= e^{-(i\omega_c+\kappa/2)(t-t_0)}\hat{a}(t_0) \\ &\quad -\sqrt{\kappa}\int_{t_0}^{t} d\tau\, e^{-(i\omega_c+\kappa/2)(t-\tau)}\hat{b}_{\text{in}}(\tau). \end{aligned} \tag{3.551}$$

If we take the input field to be a coherent drive at frequency $\omega_L = \omega_c + \Delta$, so that its amplitude has a classical and a quantum part,

$$\hat{b}_{\text{in}}(t) = e^{-i\omega_L t}[\bar{b}_{\text{in}} + \hat{\xi}(t)], \tag{3.552}$$

and if we take the limit $t_0 \to \infty$ so that the initial transient in the cavity amplitude has damped out, then the solution of Eq. (3.551) has the form with

$$\bar{a} = -\frac{\sqrt{\kappa}}{-i\Delta + \kappa/2}\bar{b}_{\text{in}} \tag{3.553}$$

and (in the frame rotating at the drive frequency)

$$\hat{d}(t) = -\sqrt{\kappa}\int_{-\infty}^{t} d\tau\, e^{+(i\Delta-\kappa/2)(t-\tau)}\hat{\xi}(\tau). \tag{3.554}$$

Even in the absence of any classical drive, the input field delivers vacuum fluctuation noise to the cavity. Notice that, from Eqs. (3.534) and (3.552)

$$\begin{aligned} [\hat{b}_{\text{in}}(t), \hat{b}^\dagger_{\text{in}}(t')] &= [\hat{\xi}(t), \hat{\xi}^\dagger(t')] \\ &= \frac{1}{2\pi\rho}\sum_q e^{-i(\omega_q-\omega_L)(t-t')} \\ &= \delta(t-t'). \end{aligned} \tag{3.555}$$

This is the operator equivalent of white noise. Using Eq. (3.551) in the limit $t_0 \to -\infty$ in Eq. (3.554) yields

$$\begin{aligned} [\hat{a}(t), \hat{a}^\dagger(t)] &= [\hat{d}(t), \hat{d}^\dagger(t)] \\ &= \kappa\int_{-\infty}^{t} d\tau \int_{-\infty}^{t} d\tau'\, e^{-(-i\Delta+\kappa/2)(t-\tau)}e^{-(+i\Delta+\kappa/2)(t-\tau')}\delta(\tau-\tau') \\ &= 1, \end{aligned} \tag{3.556}$$

as is required for the cavity bosonic quantum degree of freedom. We can interpret this as saying that the cavity zero-point fluctuations arise from the vacuum noise that enters through the open port. We also now have a simple physical interpretation of the quantum noise in the number of photons in a driven cavity. It is due to the vacuum noise that enters the cavity through the same ports that bring in the classical drive. The interference between the vacuum noise and the classical drive leads to the photon number fluctuations in the cavity.

In thermal equilibrium, $\hat{\xi}$ also contains thermal radiation. If the bath is being probed only over a narrow range of frequencies centered on ω_c (which we have assumed in making the Markov approximation), then we have to a good approximation (consistent with the commutation relation in Eq. (3.556))

$$\langle \hat{\xi}^\dagger(t)\hat{\xi}(t')\rangle = N\delta(t-t'), \tag{3.557}$$

$$\langle \hat{\xi}(t)\hat{\xi}^\dagger(t')\rangle = (N+1)\delta(t-t') \tag{3.558}$$

where $N = n_B(\hbar\omega_c)$ is the thermal equilibrium occupation number of the mode at the frequency of interest. We can gain a better understanding of Eq. (3.557) by Fourier transforming it to obtain the spectral density

$$S[\omega] = \int_{-\infty}^{+\infty} dt\, \langle \hat{\xi}^\dagger(t)\hat{\xi}(t')\rangle e^{i\omega(t-t')} = N. \tag{3.559}$$

As mentioned previously, this dimensionless quantity is the spectral density that would be measured by a photomultiplier: it represents the number of thermal photons passing a given point per unit time per unit bandwidth. Equivalently, the thermally radiated power in a narrow bandwidth B is

$$P = \hbar\omega NB. \tag{3.560}$$

One often hears the confusing statement that the noise added by an amplifier is a certain number N of photons ($N = 20$, say for a good cryogenic high-electron-mobility transistor (HEMT) amplifier operating at 5 GHz). This means that the excess output noise (referred back to the input by dividing by the power gain) produces a flux of N photons per second in a 1 Hz bandwidth, or $10^6 N$ photons per second in 1 MHz of bandwidth.

We can gain further insight into input-output theory by using the following picture. The operator $\hat{b}_{\text{in}}(t)$ represents the classical drive plus vacuum fluctuations that are just about to arrive at the cavity. We will be able to show that the output field is simply the input field a short while later after it has interacted with the cavity. Let us consider the time evolution over a short time period Δt that is very long compared with the inverse bandwidth of the vacuum noise (i.e., the frequency scale beyond which the vacuum noise cannot be treated as constant owing to some property of the environment) but very short compared with the cavity system's slow dynamics. In this circumstance, it is useful to introduce the quantum Wiener increment

$$\widehat{dW} \equiv \int_t^{t+\Delta t} d\tau\, \hat{\xi}(\tau), \tag{3.561}$$

which obeys

$$[\widehat{dW}, \widehat{dW}^\dagger] = \Delta t. \tag{3.562}$$

In the interaction picture (in a displaced frame in which the classical drive has been removed), the Hamiltonian term that couples the cavity to the quantum noise of the environment is, from Eq. (3.521),

$$\hat{V} = -i\hbar\sqrt{\kappa}(\hat{a}^\dagger\hat{\xi} - \hat{a}\hat{\xi}^\dagger). \tag{3.563}$$

Thus, the time evolution operator (in the interaction picture) on the jth short time interval $[t_j, t_j + \Delta t]$ is

$$\hat{U}_j = e^{\sqrt{\kappa}(\hat{a}\, d\widehat{W}^\dagger - \hat{a}^\dagger\, d\widehat{W})}. \tag{3.564}$$

Using this, we can readily evolve the incoming temporal mode forward in time by a small step Δt:

$$d\widehat{W}' = \hat{U}^\dagger d\widehat{W}\hat{U} \approx d\widehat{W} + \sqrt{\kappa}\Delta t\, \hat{a}. \tag{3.565}$$

Recall that in input-output theory we formally defined the outgoing field as the bath field far in the future propagated back (using the free-field time evolution) to the present, which yielded

$$\hat{b}_{\text{out}} = \hat{b}_{\text{in}} + \sqrt{\kappa}\hat{a}. \tag{3.566}$$

Eq. (3.565) is completely equivalent to this. Thus, we confirm our understanding that the incoming field is the bath temporal mode just before it interacts with the cavity and the outgoing field is the bath temporal mode just after it interacts with the cavity.

This leads to the following picture, which is especially useful in the quantum trajectory approach to conditional quantum evolution of a system subject to weak continuous measurement [66, 248]. On top of the classical drive $\bar{b}_{\text{in}}(t)$, the bath supplies to the system a continuous stream of "fresh" harmonic oscillators, each in their ground state (if $T = 0$). Each oscillator with its quantum fluctuation $d\widehat{W}$ interacts briefly for a period Δt with the system and then is disconnected to propagate freely thereafter, never interacting with the system again. Within this picture, it is useful to think of the oscillators arrayed in an infinite stationary line and the cavity flying over them at speed v_p and touching each one for a time Δt.

3.9.3 Coupling a qubit to a linear black box

As we learned in Section 3.4, the transmon qubit is essentially a weakly anharmonic oscillator. In the limit of large Josephson energy and small charging energy ($E_J/E_C \gg 1$), the quantum fluctuations in the phase φ across the Josephson junction are small and we can ignore the fact that φ is a periodic variable. It is important to understand that this is a good approximation only in the limit that the charge dispersion enforced by the periodic boundary conditions on φ can be neglected. One way to make the charging energy small is to make the transmon islands large, which makes the coupling of the transmon to the electromagnetic modes of the cavity large. We therefore do not want to treat this coupling as a perturbation, since the presence of the transmon can significantly alter the cavity modes owing to this strong coupling. It is therefore very useful to treat the harmonic part of the cavity plus qubit Hamiltonian exactly [4, 36].

We then simply have coupled harmonic oscillators whose spectrum is straightforward to understand. In the strong-coupling case, this harmonic oscillator basis can be very efficient for expressing the full Hamiltonian including the anharmonic part of the Hamiltonian. In certain situations, if the anharmonicity is weak, we may be able to treat it analytically in low-order perturbation theory. Even if this is not the case, we have an efficient basis in which to numerically diagonalize the full Hamiltonian. This is particularly true when the cavity has many resonant modes. The mixing and renormalization of these modes by the qubit is largely taken care of by the exact treatment of the harmonic part of H, and so we obtain faster convergence of the numerics in terms of the number of modes we need to keep and the number of quanta in each mode.

The simple harmonic oscillator approximation to the transmon Hamiltonian is given in Eq. (3.219). Let us imagine that we have included this effective LC oscillator in our "black box" whose equivalent circuit is shown in Fig. 3.9(a). The normal modes of the box include a qubit-like mode and many cavity-like modes coupled capacitively via the transmon "antenna." Note that the equivalent circuit describing the input admittance across the terminals of the Josephson junction is still precisely of the form shown in Fig. 3.9(a). The coupling capacitor C_g and the junction capacitance C_J and inductance L_J do *not* appear explicitly in the circuit. Their values help control the resonance frequencies and characteristic impedances of the resonances (i.e., the values of the equivalent circuit elements), but these individual elements from the real circuit do not themselves explicitly appear in the equivalent circuit.

We now turn to the treatment of the anharmonicity. Using Eq. (3.180), the phase across the Josephson junction can be expressed in terms of the normal modes:

$$\varphi = \frac{2e}{\hbar}\Phi = \sum_m \varphi_{\mathrm{ZPF}}^{(m)}(\hat{a}_m + \hat{a}_m^\dagger). \tag{3.567}$$

The Josephson energy is

$$U = -E_J \cos\varphi = -E_J\left(1 - \frac{1}{2}\varphi^2 + \frac{1}{24}\varphi^4 + \dots\right). \tag{3.568}$$

The leading constant is irrelevant and we have already included the quadratic term in the harmonic part of the Hamiltonian. Hence, the anharmonic part of the Hamiltonian analogous to Eq. (3.243) that we need to keep is

$$V = -E_{\mathrm{J}} \cos\varphi + E_{\mathrm{J}}\left(1 - \frac{1}{2}\varphi^2\right) \approx -E_{\mathrm{J}}\frac{1}{24}\varphi^4 + \dots. \tag{3.569}$$

We will make the approximation (valid for large E_J/E_C, i.e., small φ_{ZPF}) of keeping only the quartic term:[32]

$$V \approx -E_J\frac{1}{24}\left[\sum_m \varphi_{\mathrm{ZPF}}^{(m)}(\hat{a}_m + \hat{a}_m^\dagger)\right]^4. \tag{3.570}$$

[32] It is important to note, however, that even if we kept all the terms in the expansion of V, we would still not have the exact result because we are still neglecting the charge dispersion associated with the periodic boundary conditions on φ.

Recall from examination of the transmon anharmonicity that the small parameter in this expansion is E_C, not E_J, because the quantum fluctuations $\varphi_{\rm ZPF}$ are inversely proportional to E_J.

In general, we will have to treat this expression numerically, but we can immediately make some general remarks about the effect of V by normal-ordering the terms and making the rotating-wave approximation. We see that there will be quadratic terms of the form $\hat{b}_m^\dagger \hat{b}_m$ generated that shift the frequencies of the "bare" modes and other quadratic terms of the form $\hat{b}_j^\dagger \hat{b}_k$, which will mix modes that are close in frequency relative to E_C. More significantly, there will be "self-Kerr" terms of the form $\hat{b}_j^\dagger \hat{b}_j^\dagger \hat{b}_j \hat{b}_j$ that make the modes anharmonic and "cross-Kerr" terms of the form $\hat{b}_j^\dagger \hat{b}_j \hat{b}_k^\dagger \hat{b}_k$. The qubit state-dependent shift of the cavity frequency appears here as a cross-Kerr term between the qubit and the cavity mode(s). This gives us a nice physical picture of how to think about the modes, even if we need to turn to numerics to obtain quantitative accuracy.

3.9.4 Feynman's disentangling theorem

Feynman's "disentangling theorem" states that the exponential of the sum of two operators A and B can be simplified in the following manner:

$$e^{(A+B)} = e^A e^B e^{\frac{1}{2}[B,A]}, \tag{3.571}$$

provided that the commutator $[B, A]$ itself commutes with both A and B.

Here is a simple proof that begins with the following lemma. Let us define

$$\tilde{B}(\lambda) = e^{\lambda A} B e^{-\lambda A}, \tag{3.572}$$

where λ is a variable parameter. This object obeys the differential equation

$$\frac{d\tilde{B}}{d\lambda} = e^{\lambda A}[A, B]e^{-\lambda A} = [A, B]. \tag{3.573}$$

This has solution

$$\tilde{B}(\lambda) = B + \lambda[A, B]. \tag{3.574}$$

With this lemma in hand, consider now

$$V(\lambda) \equiv e^{\lambda(A+B)} - e^{\lambda A} e^{\lambda B} e^{\frac{1}{2}\lambda^2[B,A]}. \tag{3.575}$$

Clearly, $V(0) = 0$. If we can prove $V(1)$ also vanishes, then the theorem is proved. Consider the differential equation

$$\frac{dV}{d\lambda} = (A + B)V - AV - e^{\lambda A} B e^{\lambda B} e^{\frac{1}{2}\lambda^2[B,A]} + \lambda[B, A]V \tag{3.576}$$

$$= BV - \tilde{B}(\lambda)V + \lambda[B, A]V = 0, \tag{3.577}$$

where the last equality follows from our first lemma. The solution of this equation is of course

$$V(\lambda) = V(0) = 0, \tag{3.578}$$

which proves our theorem.

3.9.5 Coherent states

A coherent state with complex amplitude α has the defining property of being an eigenstate of the lowering operator,

$$\hat{a}|\alpha\rangle = \alpha|\alpha\rangle, \tag{3.579}$$

from which it follows that

$$\langle\alpha|\hat{a}^\dagger\hat{a}|\alpha\rangle = |\alpha|^2 = \bar{N}, \tag{3.580}$$

where $\bar{N}$ is the mean photon number in the coherent state. It is important to note that a coherent state is *not* an eigenstate of the raising operator. We have

$$\langle\alpha|\hat{a}^\dagger = \langle\alpha|\alpha^*, \tag{3.581}$$

but

$$\hat{a}^\dagger|\alpha\rangle \neq \alpha^*|\alpha\rangle. \tag{3.582}$$

Exercise 3.26 A common source of confusion is to fail to understand the distinction between $e^{i\varphi}|\alpha\rangle$ and $|e^{i\varphi}\alpha\rangle$. If you compute $\langle\hat{a}\rangle$ for the two cases, you will see the difference.

It is readily verified that Eq. (3.579) is solved by the following unitary displacement of the vacuum state:

$$|\alpha\rangle = e^{\alpha\hat{a}^\dagger - \alpha^*\hat{a}}|0\rangle. \tag{3.583}$$

To see this, consider

$$\hat{a}|\alpha\rangle = e^{\alpha\hat{a}^\dagger - \alpha^*\hat{a}}\left\{e^{-(\alpha\hat{a}^\dagger - \alpha^*\hat{a})}\hat{a}e^{\alpha\hat{a}^\dagger - \alpha^*\hat{a}}\right\}|0\rangle. \tag{3.584}$$

Using Eq. (3.574), the term in curly brackets becomes

$$\tilde{a} \equiv \hat{a} + \alpha, \tag{3.585}$$

from which Eq. (3.579) readily follows.

Using Feynman's disentangling theorem from Section 3.9.4, Eq. (3.583) can be written in normal-ordered form with the annihilation operators to the right of the creation operators:

$$|\alpha\rangle = e^{+\alpha\hat{a}^\dagger}e^{-\alpha^*\hat{a}}e^{\frac{1}{2}[-\alpha^*\hat{a}, +\alpha\hat{a}^\dagger]}|0\rangle \tag{3.586}$$

$$= e^{-\frac{1}{2}|\alpha|^2}e^{+\alpha\hat{a}^\dagger}|0\rangle \tag{3.587}$$

$$= e^{-\frac{1}{2}|\alpha|^2}\sum_{n=0}\frac{\alpha^n}{\sqrt{n!}}|n\rangle. \tag{3.588}$$

From this result, it is straightforward to derive the overlap of two coherent states:

$$\langle\beta|\alpha\rangle = e^{-\frac{1}{2}\left(|\alpha|^2 + |\beta|^2 - 2\alpha\beta^*\right)}. \tag{3.589}$$

The following special case appears in the theory of decoherence in the dispersive readout of a qubit:

$$\langle e^{+i\theta_0}\alpha|e^{-i\theta_0}\alpha\rangle = \exp\left[-\bar{N}(1-e^{-2i\theta_0})\right] = e^{-i\bar{N}\sin 2\theta_0}e^{-\bar{N}(1-\cos 2\theta_0)}. \tag{3.590}$$

Let us now examine the properties of the quadrature amplitudes

$$\hat{X} \equiv \frac{\hat{a}+\hat{a}^\dagger}{2}, \tag{3.591}$$

$$\hat{Y} \equiv -i\frac{\hat{a}-\hat{a}^\dagger}{2}, \tag{3.592}$$

which obey the commutation relation

$$[\hat{X},\hat{Y}] = \frac{i}{2}. \tag{3.593}$$

Let us define the projector onto the state with definite quadrature amplitude

$$P_X(x) = \delta(x-\hat{X}), \tag{3.594}$$

$$P_Y(y) = \delta(y-\hat{Y}). \tag{3.595}$$

Using the integral representation of the Dirac delta function, we have

$$P_X(x) = \int_{-\infty}^{\infty}\frac{d\xi}{2\pi}\,e^{i\xi x}e^{-i\xi\hat{X}}, \tag{3.596}$$

$$P_Y(y) = \int_{-\infty}^{\infty}\frac{d\xi}{2\pi}\,e^{i\xi y}e^{-i\xi\hat{Y}}. \tag{3.597}$$

Normal-ordering the operators using the disentangling theorem yields

$$P_X(x) = \int_{-\infty}^{\infty}\frac{d\xi}{2\pi}\,e^{-\frac{1}{8}\xi^2}e^{i\xi x}e^{-\frac{1}{2}i\xi\hat{a}^\dagger}e^{-\frac{1}{2}i\xi\hat{a}}, \tag{3.598}$$

$$P_Y(y) = \int_{-\infty}^{\infty}\frac{d\xi}{2\pi}\,e^{-\frac{1}{8}\xi^2}e^{i\xi y}e^{+\frac{1}{2}\xi\hat{a}^\dagger}e^{-\frac{1}{2}\xi\hat{a}}, \tag{3.599}$$

which allows us to readily compute the matrix elements of these projectors between two coherent states:

$$\langle\beta|P_X(x)|\alpha\rangle = \int_{-\infty}^{\infty}\frac{d\xi}{2\pi}\,e^{-\frac{1}{8}\xi^2}e^{i\xi x}e^{-\frac{1}{2}i\xi\beta^*}e^{-\frac{1}{2}i\xi\alpha}\langle\beta|\alpha\rangle \tag{3.600}$$

$$= \sqrt{\frac{2}{\pi}}e^{-2\left[x-\frac{1}{2}(\beta^*+\alpha)\right]^2}e^{-\frac{1}{2}\left(|\alpha|^2+|\beta|^2-2\alpha\beta^*\right)}, \tag{3.601}$$

$$\langle\beta|P_Y(y)|\alpha\rangle = \int_{-\infty}^{\infty}\frac{d\xi}{2\pi}\,e^{-\frac{1}{8}\xi^2}e^{i\xi y}e^{+\frac{1}{2}\xi\beta^*}e^{-\frac{1}{2}\xi\alpha}\langle\beta|\alpha\rangle \tag{3.602}$$

$$= \sqrt{\frac{2}{\pi}}e^{-2\left[y-\frac{1}{2}i(\beta^*-\alpha)\right]^2}e^{-\frac{1}{2}(|\alpha|^2+|\beta|^2-2\alpha\beta^*)}. \tag{3.603}$$

For the case $\beta = \alpha$, we recognize here the (displaced version of the) Gaussian distribution of position and momentum in the ground state of the harmonic oscillator. For the particular case associated with the dispersive readout of a qubit, we have (for α real)

$$\langle e^{-i\theta_0\sigma^z}\alpha|P_X(x)|e^{-i\theta_0\sigma^z}\alpha\rangle = \sqrt{\frac{2}{\pi}}e^{-2\left(x-\sqrt{\bar{N}}\cos\theta_0\right)^2}, \tag{3.604}$$

$$\langle e^{-i\theta_0\sigma^z}\alpha|P_Y(y)|e^{-i\theta_0\sigma^z}\alpha\rangle = \sqrt{\frac{2}{\pi}}e^{-2\left(y+\sqrt{\bar{N}}\sin\theta_0\sigma^z\right)^2}, \tag{3.605}$$

$$\langle e^{i\theta_0}\alpha|P_X(x)|e^{-i\theta_0}\alpha\rangle = \sqrt{\frac{2}{\pi}}e^{-2\left(x-\sqrt{\bar{N}}\cos\theta_0+i\sqrt{\bar{N}}\sin\theta_0\right)^2}\exp\left[\bar{N}\left(e^{-2i\theta_0}-1\right)\right], \tag{3.606}$$

$$\langle e^{i\theta_0}\alpha|P_Y(y)|e^{-i\theta_0}\alpha\rangle = \sqrt{\frac{2}{\pi}}e^{-2y^2}\exp\left[\bar{N}\left(e^{-2i\theta_0}-1\right)\right]. \tag{3.607}$$

Acknowledgments

This work was supported by Yale University and grants from the National Science Foundation (NSF DMR-1004406) and Army Research Office (ARO W911NF-05-1-0365). I am grateful to my many colleagues at Yale and around the world with whom have collaborated in learning to apply the ideas of quantum optics to electrical circuits. In particular, these lectures are based on work done in collaboration with many students, postdocs, and colleagues, including especially my friends and colleagues Michel H. Devoret and Robert J. Schoelkopf.

References

[1] Schoelkopf, R. J. and Girvin, S. M. (2008). Wiring up quantum systems. *Nature*, **451**, 664.

[2] Devoret, M. H. and Schoelkopf, R. J. (2013). Superconducting circuits for quantum information: an outlook. *Science*, **339**, 1169–1174.

[3] Devoret, M. (1997). Quantum fluctuations in electrical circuits. In *Quantum Fluctuations. Les Houches Session LXIII, June 27–July 28, 1995* (ed. S. Reynaud, E. Giacohiao, and J. Zian-Justin), pp. 351–385. Elsevier, Amsterdam.

[4] Nigg, S. E., Paik, H., Vlastakis, B., Kirchmair, G., Shankar, S., Frunzio, L., Devoret, M. H., Schoelkopf, R. J., and Girvin, S. M. (2012). Black-box superconducting circuit quantization. *Phys. Rev. Lett.*, **108**, 240502.

[5] Nakamura, Y., Pashkin, Yu. A., and Tsai, J. S. (1999). Coherent control of macroscopic quantum states in a single-Cooper-pair box. *Nature*, **398**, 786–788.

[6] Paik, H., Schuster, D. I., Bishop, Lev S., Kirchmair, G., Catelani, G., Sears, A. P., Johnson, B. R., Reagor, M. J., Frunzio, L., Glazman, L. I., Girvin, S. M., Devoret, M. H., and Schoelkopf, R. J. (2011). Observation of high coherence in Josephson junction qubits measured in a three-dimensional circuit QED architecture. *Phys. Rev. Lett.*, **107**, 240501.

[7] Chang, J. B., Vissers, M. R., Corcoles, A. D., Sandberg, M., Gao, J., Abraham, D. W., Chow, J. M., Gambetta, J. M., Rothwell, M. B., Keefe, G. A., Steffen, M., and Pappas, D. P. (2013). Improved superconducting qubit coherence using titanium nitride. *Appl. Phys. Lett.*, **103**, 012602.

[8] Barends, R., Kelly, J., Megrant, A., Sank, D., Jeffrey, E., Chen, Y., Yin, Y., B. Chiaro, J. Mutus, Neill, C., O'Malley, P., Roushan, P., Wenner, J., White, T. C., Cleland, A. N., and Martinis, J. M. (2013). Coherent Josephson qubit suitable for scalable quantum integrated circuits. *Phys. Rev. Lett.*, **111**, 080502.

[9] Manucharyan, V. E., Koch, J., Glazman, L., and Devoret, M. (2009). Fluxonium: single Cooper-pair circuit free of charge offsets. *Science*, **326**, 113–116.

[10] Geerlings, K., Pop, I. M., Masluk, N., Kamal, A., Catelani, G., Glazman, L., and Devoret, M. H. (2013). Fluxonium qubit in a 3D cavity: measurement and analysis. Presented at APS March Meeting 2013. Abstract available at <http://meetings.aps.org/list/BAPS.2013.MAR.c25.9>.

[11] Mooij, J. E., Orlando, T. P., Levitov, L., Tian, L., van der Wal, C. H., and Lloyd, S. (1999). Josephson persistent-current qubit. *Science*, **285**, 1036–1039.

[12] Vion, D., Aassime, A., Cottet, A., Joyez, P., Pothier, H., Urbina, C., Esteve, D., and Devoret, M. H. (2002). Manipulating the quantum state of an electrical circuit. *Science*, **296**, 886–889.

[13] Houck, A. A., Schuster, D. I., Gambetta, J. M., Schreier, J. A., Johnson, B. R., Chow, J. M., Frunzio, L., Majer, J., Devoret, M. H., Girvin, S. M., and Schoelkopf, R. J. (2007). Generating single microwave photons in a circuit. *Nature*, **449**, 328–331.

[14] Hofheinz, M., Weig, E. M., Ansmann, M., Bialczak, R. C., Lucero, E., Neeley, M., O'Connell, A. D., Wang, H., Martinis, J. M., and Cleland, A. N. (2008). Generation of Fock states in a superconducting quantum circuit. *Nature*, **454**, 310–314.

[15] Hofheinz, M., Wang, H., Ansmann, M., Bialczak, R. C., Lucero, E., Neeley, M., O'Connell, A. D., Sank, D., Wenner, J., Martinis, J. M., and Cleland, A. N. (2009). Synthesizing arbitrary quantum states in a superconducting resonator. *Nature*, **459**, 546–549.

[16] Johnson, B. R., Reed, M. D., Houck, A. A., Schuster, D. I., Bishop, Lev S., Ginossar, E., Gambetta, J. M., DiCarlo, L., Frunzio, L., Girvin, S. M., and Schoelkopf, R. J. (2010). Quantum non-demolition detection of single microwave photons in a circuit. *Nature Phys.*, **6**, 663–667.

[17] Mariantoni, M., Wang, H., Bialczak, R. C., Lucero, E., Neeley, M., O'Connell, A. D., Sank, D., Weides, M., Wenner, J., Yamamoto, T., Yin, Y., Zhao, J., Martinis, J. M., and Cleland, A. N. (2011). Photon shell game in three-resonator circuit quantum electrodynamics. *Nature Phys.*, **7**, 287–293.

[18] Wang, H., Mariantoni, M., Bialczak, R. C., Lenander, M., Lucero, E., Neeley, M., O'Connell, A., Sank, D., Weides, M., Wenner, J., Yamamoto, T., Yin, Y., Zhao, J., Martinis, J. M., and Cleland, A. N. (2011). Deterministic entanglement of photons in two superconducting microwave resonators. *Phys. Rev. Lett.*, **106**, 060401.

[19] Ansmann, M., Wang, H., Bialczak, R. C., Hofheinz, M., Lucero, E., Neeley, M., O'Connell, A. D., Sank, D., Weides, M., Wenner, J., Cleland, A. N., and

Martinis, J. M. (2009). Violation of Bell's inequality in Josephson phase qubits. *Nature*, **461**, 504–506.

[20] Chow, J. M., DiCarlo, L., Gambetta, J. M., Nunnenkamp, A., Bishop, L. S., Frunzio, L., Devoret, M. H., Girvin, S. M., and Schoelkopf, R. J. (2010). Detecting highly entangled states with a joint qubit readout. *Phys. Rev. A*, **81**, 062325.

[21] Neeley, M., Bialczak, R. C., Lenander, M., Lucero, E., Mariantoni, M., O'Connell, A. D., Sank, D., Wang, H., Weides, M., Wenner, J., Yin, Y., Yamamoto, T., Cleland, A. N., and Martinis, J. M. (2010). Generation of three-qubit entangled states using superconducting phase qubits. *Nature*, **467**, 570.

[22] DiCarlo, L., Reed, M. D., Sun, L., Johnson, B. R., Chow, J. M., Gambetta, J. M., Frunzio, L., Girvin, S. M., Devoret, M. H., and Schoelkopf, R. J. (2010). Preparation and measurement of three-qubit entanglement in a superconducting circuit. *Nature*, **467**, 574–578.

[23] DiCarlo, L., Chow, J. M., Gambetta, J. M., Bishop, Lev S., Johnson, B. R., Schuster, D. I., Majer, J., Blais, A., Frunzio, L., Girvin, S. M., and Schoelkopf, R. J. (2009). Demonstration of two-qubit algorithms with a superconducting quantum processor. *Nature*, **460**, 240–244.

[24] Mariantoni, M., Wang, H., Yamamoto, T., Neeley, M., Bialczak, R. C., Chen, Y., Lenander, M., Lucero, E., O'Connell, A. D., Sank, D., Weides, M., Wenner, J., Yin, Y., Zhao, J., Korotkov, A. N., Cleland, A. N., and Martinis, J. M. (2011). Implementing the quantum von Neumann architecture with superconducting circuits. *Science*, **334**, 61.

[25] Reed, M. D., DiCarlo, L., Nigg, S. E., Sun, L., Frunzio, L., Girvin, S. M., and Schoelkopf, R. J. (2012). Realization of three-qubit quantum error correction with superconducting circuits. *Nature*, **482**, 382–385.

[26] Devoret, M., Girvin, S., and Schoelkopf, R. (2007). Circuit-QED: How strong can the coupling between a Josephson junction atom and a transmission line resonator be? *Anna. Phys. (Leifzig)*, **16**, 767–779.

[27] Reed, M. D., Johnson, B. R., Houck, A. A., DiCarlo, L., Chow, J. M., Schuster, D. I., Frunzio, L., and Schoelkopf, R. J. (2010). Fast reset and suppressing spontaneous emission of a superconducting qubit. *Appl. Phys. Lett.*, **96**(20), 203110.

[28] Jackson, J. D. (1999). *Electrodynamics*, 3rd edn. Wiley, New York.

[29] Pines, D. (1963). *Elementary Excitations in Solids.* W. A. Benjamin, New York.

[30] Mahan, G. D. (2000). *Many-Particle Physics*, 3rd edn. Springer-Verlag, Berlin.

[31] Clerk, A. A., Devoret, M. H., Girvin, S. M., Marquardt, F., and Schoelkopf, R. J. (2010). Introduction to quantum noise, measurement and amplification. *Rev. Mod. Phys.*, **82**, 1155–1208. A longer version with pedagogical appendices is available at: arXiv:0810.4729.

[32] Blais, A., Huang, R.-S., Wallraff, A., Girvin, S. M., and Schoelkopf, R. J. (2004). Cavity quantum electrodynamics for superconducting electrical circuits: an architecture for quantum computation. *Phys. Rev. A*, **69**, 062320.

[33] Kane, C. L. and Fisher, M. P. A. (1992*b*). Transport in a one-channel Luttinger liquid. *Phys. Rev. Lett.*, **68**, 1220–1223.

[34] Kane, C. L. and Fisher, M. P. A. (1992). Transmission through barriers and resonant tunneling in an interacting one-dimensional electron gas. *Phys. Rev. B*, **46**, 15233–15262.
[35] Foster, R. M. (1924). A reactance theorem. *Bell Syst. Tech. J.*, **3**, 259–267.
[36] Manucharyan, V. E., Boaknin, E., Metcalfe, M., Vijay, R., Siddiqi, I., and Devoret, M. (2007). Microwave bifurcation of a Josephson junction: embedding-circuit requirements. *Phys. Rev. B*, **76**, 014524.
[37] Tinkham, M. (1996). *Introduction to Superconductivity*, 2nd edn. McGraw-Hill, New York (reprinted 2004, Dover Publications, New York).
[38] Devoret, M. H. and Martinis, J. M. (2004). Superconducting qubits. In *Quantum Entanglement and Information Processing. École d'été de Physique des Houches Session LXXIX, 30 July–25 August 2003* (ed. D. Estève, J. M. Raimond, and J. Dalibard), pp. 443–485. Elsevier, Amsterdam.
[39] Clarke, J. and Wilhelm, F. K. (2008). Superconducting quantum bits. *Nature*, **453**, 1031–1042.
[40] Pierre, F., Pothier, H., Joyez, P., Birge, N. O., Esteve, D., and Devoret, M. H. (2001). Electrodynamic dip in the local density of states of a metallic wire. *Phys. Rev. Lett.*, **86**, 1590–1593.
[41] Devoret, M. H., Esteve, D., Grabert, H., Ingold, G.-L., Pothier, H., and Urbina, C. (1990). Effect of the electromagnetic environment on the coulomb blockade in ultrasmall tunnel junctions. *Phys. Rev. Lett.*, **64**, 1824–1827.
[42] Girvin, S. M., Glazman, L. I., Jonson, M., Penn, D. R., and Stiles, M. D. (1990). Quantum fluctuations and the single-junction Coulomb blockade. *Phys. Rev. Lett.*, **64**, 3183–3186.
[43] Ambegaokar, V. and Baratoff, A. (1963). Tunneling between superconductors. *Phys. Rev. Lett.*, **10**, 486. Erratum: *Phys. Rev. Lett.* **11**, 104 (1963).
[44] Bouchiat, V., Vion, D., Joyez, P., Esteve, D., and Devoret, M. H. (1998). Quantum coherence with a single cooper pair. *Physics Scripta*, **T76**, 165.
[45] Esteve, D and Vion, D. (2005). Solid state quantum bit circuits. In *Nanophysics: Coherence and Transport. École d'été de Physique Les Houches Session LXXXI, 28 June–30 July 2004* (ed. H. Bouchiat, Y. Gefen, and S. Guéron, G. Montambaux, and J. Dalibard), pp. 537–575. Elsevier, Amsterdam.
[46] Wendin, G. and Shumeiko, V. S. (2006). In Superconducting circuits, qubits and computing. *Handbook of Theoretical and Computational Nanotechnology*, Volume 3: *Quantum and Molecular Computing, and Quantum Simulations* (ed. M. Rieth and W. Schommers), pp. 223–309. American Scientific Publishers, Los Angeles.
[47] Wendin, G. and Shumeiko, V. S. (2007). Quantum bits with Josephson junctions. *J. Low Temp. Phys.*, **33**, 724.
[48] You, J. Q. and Nori, F. (2005). Superconducting circuits and quantum information. *Phys. Today*, **58**, 42–47.
[49] Nori, F. (2008). Superconducting qubits: atomic physics with a circuit. *Nature Phys.*, **4**, 589–590.
[50] Korotkov, A. N. (2009). Editorial Notes: Special issue on quantum computing with superconducting qubits. *Quantum Info. Process.*, **8**, 51–54.

[51] Leggett, A. J. (1980). Macroscopic quantum systems and the quantum theory of measurement. *Prog. Theor. Phys. Suppl.*, **69**, 80–100.

[52] Martinis, J. M., Devoret, M. H., and Clarke, J. (1985). Energy level quantization in the zero-voltage state of a current-biased Josephson junction. *Phys. Rev. Lett.*, **55**, 1543–1546.

[53] Clarke, J., Cleland, A. N., Devoret, M. H., Esteve, D., and Martinis, J. M. (1988). Quantum mechanics of a macroscopic variable: the phase difference of a Josephson junction. *Science*, **239**, 992–997.

[54] Averin, D. V., Zorin, A. B., and Likharev, K. K. (1985). Bloch oscillations in small Josephson junctions. *Sov. Phys. JETP*, **61**, 407.

[55] Büttiker, M. (1987). Zero-current persistent potential drop across small-capacitance Josephson junctions. *Phys. Rev. B*, **36**, 3548–3555.

[56] Lafarge, P., Joyez, P., Esteve, D., Urbina, C., and Devoret, M. H. (1993). Two-electron quantization of the charge on a superconductor. *Nature*, **365**, 422–424.

[57] Koch, J., Yu, T. M., Gambetta, J., Houck, A. A., Schuster, D. I., Majer, J., Blais, A., Devoret, M. H., Girvin, S. M., and Schoelkopf, R. J. (2007). Charge-insensitive qubit design derived from the Cooper pair box. *Phys. Rev. A*, **76**, 042319.

[58] Schreier, J. A., Houck, A. A., Koch, J., Schuster, D. I., Johnson, B. R., Chow, J. M., Gambetta, J. M., Majer, J., Frunzio, L., Devoret, M. H., Girvin, S. M., and Schoelkopf, R. J. (2008). Suppressing charge noise decoherence in superconducting charge qubits. *Phys. Rev. B*, **77**, 180502.

[59] van der Wal, C. H., ter Haar, A. C. J., Wilhelm, F. K., Schouten, R. N., Harmans, C. J. P. M., Orlando, T. P., Lloyd, S., and Mooij, J. E. (2000). Quantum superposition of macroscopic persistent-current states. *Science*, **290**, 773–777.

[60] Chiorescu, I., Nakamura, Y., Harmans, C. J. P. M., and Mooij, J. E. (2003). Coherent quantum dynamics of a superconducting flux qubit. *Science*, **299**, 1869–1871.

[61] Martinis, J. M., Nam, S., Aumentado, J., and Urbina, C. (2002). Rabi oscillations in a large Josephson-junction qubit. *Phys. Rev. Lett.*, **89**, 117901.

[62] Berkley, A. J., Xu, H., Ramos, R. C., Gubrud, M. A., Strauch, F. W., Johnson, P. R., Anderson, J. R., Dragt, A. J., Lobb, C. J., and Wellstood, F. C. (2003). Entangled macroscopic quantum states in two superconducting qubits. *Science*, **300**, 1548–1550.

[63] Manucharyan, V. E., Koch, J., Brink, M., Glazman, L. I., and Devoret, M. H. (2009). Coherent oscillations between classically separable quantum states of a superconducting loop. arXiv:0910.3039.

[64] Houck, A. A., Koch, J., Devoret, M. H., Girvin, S. M., and Schoelkopf, R. J. (2009). Life after charge noise: recent results with transmon qubits. *Quantum Inf. Process.*, **8**, 105.

[65] Martinis, J. M., Nam, S., Aumentado, J., Lang, K. M., and Urbina, C. (2003). Decoherence of a superconducting qubit due to bias noise. *Phys. Rev. B*, **67**, 094510.

[66] Walls, D. and Milburn, G. (1994). *Quantum Optics*. Spinger-Verlag, Berlin.

[67] Haroche, S. and Raimond, J.-M. (2006). *Exploring the Quantum: Atoms, Cavities and Photons.* Oxford University Press, Oxford.

[68] Meystre, P. and Sargent M. III (1998). *Elements of Quantum Optics.* Springer-Verlag, Berlin.

[69] Carmichael, H. J. (1993). *An Open Systems Approach to Quantum Optics.* Springer-Verlag, Berlin.

[70] Aassime, A., Johansson, G., Wendin, G., Schoelkopf, R. J., and Delsing, P. (2001). Radio-frequency single-electron transistor as readout device for qubits: charge sensitivity and backaction. *Phys. Rev. Lett.*, **86**, 3376–3379.

[71] Lehnert, K. W., Bladh, K., Spietz, L. F., Gunnarsson, D., Schuster, D. I., Delsing, P., and Schoelkopf, R. J. (2003). Measurement of the excited-state lifetime of a microelectronic circuit. *Phys. Rev. Lett.*, **90**, 027002.

[72] Metcalfe, M., Boaknin, E., Manucharyan, V., Vijay, R., Siddiqi, I., Rigetti, C., Frunzio, L., Schoelkopf, R. J., and Devoret, M. H. (2007). Measuring the decoherence of a quantronium qubit with the cavity bifurcation amplifier. *Phys. Rev. B*, **76**, 174516.

[73] Wallraff, A., Schuster, D. I., Blais, A., Frunzio, L., Huang, R.-S., Majer, J., Kumar, S., Girvin, S. M., and Schoelkopf, R. J. (2004). Circuit quantum electrodynamics: coherent coupling of a single photon to a Cooper pair box. *Nature*, **431**, 162–167.

[74] Widom, A., Megaloudis, G., Clark, T. D., Mutton, J. E., Prance, R. J., and Prance, H. (1984). The Josephson pendulum as a nonlinear capacitor. *J. Low Temp. Phys.*, **57**, 651.

[75] Likharev, K. K. and Zorin, A. B. (1985). Theory of the Bloch-wave oscillations in small Josephson junctions. *J. Low Temp. Phys.*, **59**, 347.

[76] Averin, D. V. and Bruder, C. (2003). Variable electrostatic transformer: Controllable coupling of two charge qubits. *Phys. Rev. Lett.*, **91**, 057003.

[77] Duty, T., Johansson, G., Bladh, K., Gunnarsson, D., Wilson, C., and Delsing, P. (2005). Observation of quantum capacitance in the Cooper-pair transistor. *Phys. Rev. Lett.*, **95**, 206807.

[78] Sillanpää, M. A., Lehtinen, T., Paila, A., Makhlin, Y., Roschier, L., and Hakonen, P. J. (2005). Direct observation of Josephson capacitance. *Phys. Rev. Lett.*, **95**, 206806.

[79] Mabuchi, H. and Doherty, A. (2002). Cavity quantum electrodynamics: coherence in context. *Science*, **298**, 1372.

[80] Thompson, R. J., Rempe, G., and Kimble, H. J. (1992). Observation of normal-mode splitting for an atom in an optical cavity. *Phys. Rev. Lett.*, **68**, 1132–1135.

[81] Boca, A., Miller, R., Birnbaum, K. M., Boozer, A. D., McKeever, J., and Kimble, H. J. (2004). Observation of the vacuum Rabi spectrum for one trapped atom. *Phys. Rev. Lett.*, **93**, 233603.

[82] Schuster, I., Kubanek, A., Fuhrmanek, A., Puppe, T., Pinkse, P. W. H., Murr, K., and Rempe, G. (2008). Nonlinear spectroscopy of photons bound to one atom. *Nature Phys.*, **4**, 382–385.

[83] Nogues, G., Rauschenbeutel, A., Osnaghi, S., Brune, M., Raimond, J. M., and Haroche, S. (1999). Seeing a single photon without destroying it. *Nature*, **400**, 239–242.

[84] Guerlin, C., Bernu, J., Deleglise, S., Sayrin, C., Gleyzes, S., Kuhr, S., Brune, M., Raimond, J.-M., and Haroche, S. (2007). Progressive field-state collapse and quantum non-demolition photon counting. *Nature*, **448**, 889–893.

[85] Gleyzes, S., Kuhr, S., Guerlin, C., Bernu, J., Deleglise, S., Busk Hoff, U., Brune, M., Raimond, J.-M., and Haroche, S. (2007). Quantum jumps of light recording the birth and death of a photon in a cavity. *Nature*, **446**, 297–300.

[86] Deleglise, S., Dotsenko, I., Sayrin, C., Bernu, J., Brune, M., Raimond, J.-M., and Haroche, S. (2008). Reconstruction of non-classical cavity field states with snapshots of their decoherence. *Nature*, **455**, 510–514.

[87] Raimond, J., Brune, M., and Haroche, S. (2001). Manipulating quantum entanglement with atoms and photons in a cavity. *Rev. Mod. Phys.*, **73**, 565.

[88] Chiorescu, I., Bertet, P., Semba, K., Nakamura, Y., Harmans, C. J. P. M., and Mooij, J. E. (2004). Coherent dynamics of a flux qubit coupled to a harmonic oscillator. *Nature*, **431**, 159–162.

[89] Bourassa, J., Gambetta, J. M., Jr., Abdumalikov, A. A., Astafiev, O., Nakamura, Y., and Blais, B. (2009). Ultra-strong coupling regime of cavity QED with phase-biased flux qubits. *Phys. Rev. A*, **80**, 032109.

[90] Shnirman, A., Schön, G., and Hermon, Z. (1997). Quantum manipulations of small Josephson junctions. *Phys. Rev. Lett.*, **79**, 2371.

[91] Makhlin, Y., Schön, G., and Shnirman, A. (2001). Quantum-state engineering with Josephson-junction devices. *Rev. Mod. Phys.*, **73**, 357.

[92] Buisson, O. and Hekking, F. (2001). Entangled states in a Josephson charge qubit coupled to a superconducting resonator. In *Macroscopic Quantum Coherence and Quantum Computing* (ed. D. V. Averin, B. Ruggiero, and P. Silvestrini), pp. 137–145. Kluwer/Plenum, New York.

[93] Marquardt, F. and Bruder, C. (2001). Superposition of two mesoscopically distinct quantum states: coupling a Cooper pair box to a large superconducting island. *Phys. Rev. B*, **63**, 054514.

[94] Al-Saidi, W. A. and Stroud, D. (2001). Eigenstates of a small Josephson junction coupled to a resonant cavity. *Phys. Rev. B*, **65**, 014512.

[95] Plastina, F. and Falci, G. (2003). Communicating Josephson qubits. *Phys. Rev. B*, **67**, 224514.

[96] Blais, A., Maassen van den Brink, A., and Zagoskin, A. (2003). Tunable coupling of superconducting qubits. *Phys. Rev. Lett.*, **90**, 127901.

[97] Yang, C.-P., Chu, S.-I., and Han, S. (2003). Possible realization of entanglement, logical gates, and quantum-information transfer with superconducting-quantum-interference device qubits in cavity QED. *Phys. Rev. A*, **67**, 042311.

[98] You, J. Q. and Nori, F. (2003). Quantum information processing with superconducting qubits in a microwave field. *Phys. Rev. B*, **68**, 064509.

[99] Menzel, E. P., Deppe, F., Mariantoni, M., Caballero, M. A. Araque, Baust, A., Niemczyk, T., Hoffmann, E., Marx, A., Solano, E., and Gross, R. (2010).

Dual-path state reconstruction scheme for propagating quantum microwaves and detector noise tomography. *Phys. Rev. Lett.*, **105**, 100401.

[100] Koch, J. and Le Hur, K. (2009). Superfluid–Mott insulator transition of light in the Jaynes-Cummings lattice. *Phys. Rev. A*, **80**, 023811.

[101] Koch, J., Houck, A. A., Hur, Le K., and Girvin, S. M. (2010). Time-reversal symmetry breaking in circuit-QED based photon lattices. *Phys. Rev. A*, **82**, 043811.

[102] Nunnenkamp, A., Koch, J., and Girvin, S. M. (2011). Synthetic gauge fields and homodyne transmission in Jaynes-Cummings lattices. *New J. Phys.*, **13**, 095008.

[103] Viehmann, O., von Delft, J., and Marquardt, F. (2011). Superradiant phase transitions and the standard description of circuit QED. *Phys. Rev. Lett.*, **107**, 113602.

[104] Schuster, D. I., Wallraff, A., Blais, A., Frunzio, L., Huang, R. S., Majer, J., Girvin, S. M., and Schoelkopf, R. J. (2005). ac Stark shift and dephasing of a superconducting qubit strongly coupled to a cavity field. *Phys. Rev. Lett.*, **94**, 123602.

[105] Wallraff, A., Schuster, D. I., Blais, A., Frunzio, L., Majer, J., Devoret, M. H., Girvin, S. M., and Schoelkopf, R. J. (2005). Approaching unit visibility for control of a superconducting qubit with dispersive readout. *Phys. Rev. Lett.*, **95**, 060501.

[106] Schuster, D. I., Wallraff, A., Blais, A., Frunzio, L., Huang, R.-S., Majer, J., Girvin, S. M., and Schoelkopf, R. J. (2007). Erratum: ac Stark shift and dephasing of a superconducting qubit strongly coupled to a cavity field [*Phys. Rev. Lett.* **94**, 123602 (2005)]. *Phys. Rev. Lett.*, **98**, 049902.

[107] Johansson, J., Saito, S., Meno, T., Nakano, H., Ueda, M., Semba, K., and Takayanagi, H. (2006). Vacuum Rabi oscillations in a macroscopic superconducting qubit LC oscillator system. *Phys. Rev. Lett.*, **96**, 127006.

[108] Siddiqi, I., Vijay, R., Metcalfe, M., Boaknin, E., Frunzio, L., Schoelkopf, R. J., and Devoret, M. H. (2006). Dispersive measurements of superconducting qubit coherence with a fast latching readout. *Phys. Rev. B*, **73**, 054510.

[109] Boulant, N., Ithier, G., Meeson, P., Nguyen, F., Vion, D., Esteve, D., Siddiqi, I., Vijay, R., Rigetti, C., Pierre, F., and Devoret, M. (2007). Quantum nondemolition readout using a Josephson bifurcation amplifier. *Phys. Rev. B*, **76**, 014525.

[110] Schuster, D. I., Houck, A. A., Schreier, J. A., Wallraff, A., Gambetta, J., Blais, A., Frunzio, L., Johnson, B., Devoret, M. H., Girvin, S. M., and Schoelkopf, R. J. (2007). Resolving photon number states in a superconducting circuit. *Nature*, **445**, 515–518.

[111] Wallraff, A., Schuster, D. I., Blais, A., Gambetta, J. M., Schreier, J., Frunzio, L., Devoret, M. H., Girvin, S. M., and Schoelkopf, R. J. (2007). Sideband transitions and two-tone spectroscopy of a superconducting qubit strongly coupled to an on-chip cavity. *Phys. Rev. Lett.*, **99**, 050501.

[112] Sillanpää, M. A., Park, J. I., and Simmonds, R. W. (2007). Coherent quantum state storage and transfer between two phase qubits via a resonant cavity. *Nature*, **449**, 438–442.

[113] Leek, P. J., Fink, J. M., Blais, A., Bianchetti, R., Goeppl, M., Gambetta, J. M., Schuster, D. I., Frunzio, L., Schoelkopf, R. J., and Wallraff, A. (2007). Observation of Berry's phase in a solid-state qubit. *Science*, **318**, 1889–1892.

[114] Majer, J., Chow, J. M., Gambetta, J. M., Koch, J., Johnson, B. R., Schreier, J. A., Frunzio, L., Schuster, D. I., Houck, A. A., Wallraff, A., Blais, A., Devoret, M. H., Girvin, S. M., and Schoelkopf, R. J. (2007). Coupling superconducting qubits via a cavity bus. *Nature*, **449**, 443–447.

[115] Astafiev, O., Inomata, K., Niskanen, A. O., Yamamoto, T., Pashkin, Yu. A., Nakamura, Y., and Tsai, J. S. (2007). Single artificial-atom lasing. *Nature*, **449**, 588–590.

[116] Deppe, F., Mariantoni, M., Menzel, E. P., Marx, A., Saito, S., Kakuyanagi, K., Tanaka, H., Meno, T., Semba, K., Takayanagi, H., Solano, E., and Gross, R. (2008). Two-photon probe of the Jaynes-Cummings model and controlled symmetry breaking in circuit QED. *Nature Phys.*, **4**, 686–691.

[117] Fink, J. M., Goeppl, M., Baur, M., Bianchetti, R., Leek, P. J., Blais, A., and Wallraff, A. (2008). Climbing the Jaynes-Cummings ladder and observing its root n nonlinearity in a cavity QED system. *Nature*, **454**, 315–318.

[118] Wang, H., Hofheinz, M., Ansmann, M., Bialczak, R. C., Lucero, E., Neeley, M., O'Connell, A. D., Sank, D., Wenner, J., Cleland, A. N., and Martinis, J. M. (2008). Measurement of the decay of Fock states in a superconducting quantum circuit. *Phys. Rev. Lett.*, **101**, 240401.

[119] Houck, A. A., Schreier, J. A., Johnson, B. R., Chow, J. M., Koch, Jens, Gambetta, J. M., Schuster, D. I., Frunzio, L., Devoret, M. H., Girvin, S. M., and Schoelkopf, R. J. (2008). Controlling the spontaneous emission of a superconducting transmon qubit. *Phys. Rev. Lett.*, **101**, 080502.

[120] Fragner, A., Goeppl, M., Fink, J. M., Baur, M., Bianchetti, R., Leek, P. J., Blais, A., and Wallraff, A. (2008). Resolving vacuum fluctuations in an electrical circuit by measuring the Lamb shift. *Science*, **322**, 1357–1360.

[121] Grajcar, M., van der Ploeg, S. H. W., Izmalkov, A., Il'ichev, E., Meyer, H.-G., Fedorov, A., Shnirman, A., and Schon, G. (2008). Sisyphus cooling and amplification by a superconducting qubit. *Nature Phys.*, **4**, 612–616.

[122] Sandberg, M., Wilson, C. M., Persson, F., Johansson, G., Shumeiko, V., Bauch, T., Duty, T., and Delsing, P. (2008). Fast tuning of superconducting microwave cavities. In *Solid-State Quantum Computing. Proceedings 2nd International Workshop on Solid-State Quantum Computing/Mini-School on Quantum Information Science, Taipei, Taiwan, June 23–27, 2008.* (ed. H. S. Goan, and Y. N. Chen) *AIP Conf. Proc.*, **1074**, 12–21.

[123] Il'ichev, E., van der Ploeg, S. H. W., Grajcar, M., and Meyer, H. G. (2009). Weak continuous measurements of multiqubits systems. *Quantum Inf. Process.*, **8**, 133–153.

[124] Bishop, Lev S., Chow, J. M., Koch, J., Houck, A. A., Devoret, M. H., Thuneberg, E., Girvin, S. M., and Schoelkopf, R. J. (2009). Nonlinear response of the vacuum Rabi resonance. *Nature Phys.*, **5**, 105–109.

[125] Chow, J. M., Gambetta, J. M., Tornberg, L., Koch, J., Bishop, L. S., Houck, A. A., Johnson, B. R., Frunzio, L., Girvin, S. M., and Schoelkopf, R. J. (2009).

Randomized benchmarking and process tomography for gate errors in a solid-state qubit. *Phys. Rev. Lett.*, **102**, 090502.

[126] Fink, J. M., Bianchetti, R., Baur, M., Göppl, M., Steffen, L., Filipp, S., Leek, P. J., Blais, A., and Wallraff, A. (2009). Dressed collective qubit states and the Tavis-Cummings model in circuit QED. *Phys. Rev. Lett.*, **103**, 083601.

[127] Baur, M., Filipp, S., Bianchetti, R., Fink, J. M., Göppl, M., Steffen, L., Leek, P. J., Blais, A., and Wallraff, A. (2009). Measurement of Autler-Townes and Mollow transitions in a strongly driven superconducting qubit. *Phys. Rev. Lett.*, **102**, 243602.

[128] Leek, P. J., Filipp, S., Maurer, P., Baur, M., Bianchetti, R., Fink, J. M., Göppl, M., Steffen, L., and Wallraff, A. (2009). Using sideband transitions for two-qubit operations in superconducting circuits. *Phys. Rev. B*, **79**, 180511(R).

[129] Leek, P. J., Baur, M., Fink, J. M., Bianchetti, R., Steffen, L., Filipp, S., and Wallraff, A. (2010). Cavity quantum electrodynamics with separate photon storage and qubit readout modes. *Phys. Rev. Lett.*, **104**, 100504.

[130] Bianchetti, R., Filipp, S., Baur, M., Fink, J. M., Lang, C., Steffen, L., Boissonneault, M., Blais, A., and Wallraff, A. (2010). Control and tomography of a three level superconducting artificial atom. *Phys. Rev. Lett.*, **105**, 223601.

[131] Niemczyk, T., Deppe, F., Huebl, H., Menzel, E. P., Hocke, F., Schwarz, M. J., Garcia-Ripoll, J. J., Zueco, D., Hümmer, T., Solano, E., Marx, A., and Gross, R. (2010). Circuit quantum electrodynamics in the ultrastrong-coupling regime. *Nature Phys.*, **6**, 772–776.

[132] da Silva, M. P., Bozyigit, D., Wallraff, A., and Blais, A. (2010). Schemes for the observation of photon correlation functions in circuit QED with linear detectors. *Phys. Rev. A*, **82**, 043804.

[133] Fink, J. M., Steffen, L., Studer, P., Bishop, Lev S., Baur, M., Bianchetti, R., Bozyigit, D., Lang, C., Filipp, S., Leek, P. J., and Wallraff, A. (2010). Quantum-to-classical transition in cavity quantum electrodynamics. *Phys. Rev. Lett.*, **105**, 163601.

[134] Astafiev, O., Zagoskin, A. M., Abdumalikov, A. A. Jr., Pashkin, Yu. A., Yamamoto, T., Inomata, K., Nakamura, Y., and Tsai, J. S. (2010). Resonance fluorescence of a single artificial atom. *Science*, **327**, 840–843.

[135] Hoffman, A. J., Srinivasan, S. J., Schmidt, S., Spietz, L., Aumentado, J., Türeci, H. E., and Houck, A. A. (2011). Dispersive photon blockade in a superconducting circuit. *Phys. Rev. Lett.*, **107**, 053602.

[136] Gambetta, J. M., Houck, A. A., and Blais, A. (2011). Superconducting qubit with Purcell protection and tunable coupling. *Phys. Rev. Lett.*, **106**, 030502.

[137] Srinivasan, S. J., Hoffman, A. J., Gambetta, J. M., and Houck, A. A. (2011). Tunable coupling in circuit quantum electrodynamics using a superconducting charge qubit with a V-shaped energy level diagram. *Phys. Rev. Lett.*, **106**, 083601.

[138] Wilson, C. M., Johansson, G., Pourkabirian, A., Simoen, M., Johansson, J. R., Duty, T., Nori, F., and Delsing, P. (2011). Observation of the dynamical Casimir effect in a superconducting circuit. *Nature*, **479**, 376–379.

[139] Bylander, J., Gustavsson, S., Yan, F., Yoshihara, F., Harrabi, K., Fitch, G., Cory, D. G., Nakamura, Y., Tsai, J.-S., and Oliver, W. D. (2011). Noise

spectroscopy through dynamical decoupling with a superconducting flux qubit. *Nature Phys.*, **7**, 565–570.

[140] Ong, F. R., Boissonneault, M., Mallet, F., Palacios-Laloy, A., Dewes, A., Doherty, A. C., Blais, A., Bertet, P., Vion, D., and Esteve, D. (2011). Circuit QED with a nonlinear resonator: ac-Stark shift and dephasing. *Phys. Rev. Lett.*, **106**, 167002.

[141] Vijay, R., Slichter, D. H., and Siddiqi, I. (2011). Observation of quantum jumps in a superconducting artificial atom. *Phys. Rev. Lett.*, **106**, 110502.

[142] Bozyigit, D., Lang, C., Steffen, L., Fink, J. M., Eichler, C., Baur, M., Bianchetti, R., Leek, P. J., Filipp, S., da Silva, M. P., Blais, A., and Wallraff, A. (2011). Antibunching of microwave-frequency photons observed in correlation measurements using linear detectors. *Nature Phys.*, **7**, 154–158.

[143] Eichler, C., Bozyigit, D., Lang, C., Steffen, L., Fink, J., and Wallraff, A. (2011). Experimental state tomography of itinerant single microwave photons. *Phys. Rev. Lett.*, **106**, 220503.

[144] Eichler, C., Bozyigit, D., Lang, C., Baur, M., Steffen, L., Fink, J. M., Filipp, S., and Wallraff, A. (2011). Observation of two-mode squeezing in the microwave frequency domain. *Phys. Rev. Lett.*, **107**, 113601.

[145] Lang, C., Bozyigit, D., Eichler, C., Steffen, L., Fink, J. M., A. A. Abdumalikov, Jr., Baur, M., Filipp, S., da Silva, M. P., Blais, A., and Wallraff, A. (2011). Observation of resonant photon blockade at microwave frequencies using correlation function measurements. *Phys. Rev. Lett.*, **106**, 243601.

[146] Hoi, I.-C., Wilson, C. M., Johansson, G., Palomaki, T., Peropadre, B., and Delsing, P. (2011). Demonstration of a single-photon router in the microwave regime. *Phys. Rev. Lett.*, **107**, 073601.

[147] Kim, Z., Suri, B., Zaretskey, V., Novikov, S., Osborn, K. D., Mizel, A., Wellstood, F. C., and Palmer, B. S. (2011). Decoupling a Cooper-pair box to enhance the lifetime to 0.2 ms. *Phys. Rev. Lett.*, **106**, 120501.

[148] Dewes, A., Lauro, R., Ong, Florian R., Schmitt, V., Milman, P., Bertet, P., Vion, D., and Esteve, D. (2012). Quantum speeding-up of computation demonstrated in a superconducting two-qubit processor. *Phys. Rev. B*, **85**, 140503.

[149] Baur, M., Fedorov, A., Steffen, L., Filipp, S., da Silva, M. P., and Wallraff, A. (2012). Benchmarking a quantum teleportation protocol in superconducting circuits using tomography and an entanglement witness. *Phys. Rev. Lett.*, **108**, 040502.

[150] Eichler, C., Lang, C., Fink, J. M., Govenius, J., Filipp, S., and Wallraff, A. (2012). Observation of entanglement between itinerant microwave photons and a superconducting qubit. *Phys. Rev. Lett.*, **109**, 240501.

[151] Murch, K. W., Vool, U., Zhou, D., Weber, S. J., Girvin, S. M., and Siddiqi, I. (2012). Cavity-assisted quantum bath engineering. *Phys. Rev. Lett.*, **109**, 183602.

[152] Johnson, J. E., Macklin, C., Slichter, D. H., Vijay, R., Weingarten, E. B., Clarke, J., and Siddiqi, I. (2012). Heralded state preparation in a superconducting qubit. *Phys. Rev. Lett.*, **109**, 050506.

[153] Houck, A. A., Tureci, H., and Koch, J. (2012). On-chip quantum simulation with superconducting circuits. *Nature Phys.*, **8**, 292–299.

[154] Underwood, D. L., Shanks, W. E., Koch, J., and Houck, A. A. (2012). Low-disorder microwave cavity lattices for quantum simulation with photons. *Phys. Rev. A*, **86**, 023837.

[155] Yin, Y., Chen, Y., Sank, D., O'Malley, P. J. J., White, T. C., Barends, R., Kelly, J., Lucero, E., Mariantoni, M., Megrant, A., Neill, C., Vainsencher, A., Wenner, J., Korotkov, A. N., Cleland, A. N., and Martinis, J. M. (2013). Catch and release of microwave photon states. *Phys. Rev. Lett.*, **110**, 107001.

[156] Sete, E. A., Galiautdinov, A., Mlinar, E., Martinis, J. M., and Korotkov, A. N. (2013). Catch-disperse-release readout for superconducting qubits. *Phys. Rev. Lett.*, **110**, 210501.

[157] Kirchmair, G., Vlastakis, B., Leghtas, Z., Nigg, S. E, Paik, H., Ginossar, E., Mirrahimi, M., Frunzio, L., Girvin, S. M., and Schoelkopf, R. J. (2013). Observation of quantum state collapse and revival due to the single-photon Kerr effect. *Nature*, **495**, 205–209.

[158] Vlastakis, B., Kirchmair, G., Leghtas, Z., Nigg, S. E., Frunzio, L., Girvin, S. M., Mirrahimi, M., Devoret, M. H., and Schoelkopf, R. J. (2013). Deterministically encoding quantum information in 100-photon Schrödinger cat states. *Science*, **342**, 607–610. (in press).

[159] Hoi, I.-C., Palomaki, T., Lindkvist, J., Johansson, G., Delsing, P., and Wilson, C. M. (2012). Generation of nonclassical microwave states using an artificial atom in 1D open space. *Phys. Rev. Lett.*, **108**, 263601.

[160] Hoi, I.-C., Kockum, A. F., Palomaki, T., Stace, T. M., Fan, B., Tornberg, L., Sathyamoorthy, S. R., Johansson, G., Delsing, P., and Wilson, C. M. (2013). Giant cross-kerr effect for propagating microwaves induced by an artificial atom. *Phys. Rev. Lett.*, **111**, 053601.

[161] Hatridge, M., Shankar, S., Mirrahimi, M., Schackert, F., Geerlings, K., Brecht, T., Sliwa, K. M., Abdo, B., Frunzio, L., Girvin, S. M., Schoelkopf, R. J., and Devoret, M. H. (2013). Quantum back-action of an individual variable-strength measurement. *Science*, **339**, 178–181.

[162] Woolley, M. J., Lang, C., Eichler, C., Wallraff, A., and Blais, A. (2013). Signatures of Hong-Ou-Mandel interference at microwave frequencies. *New J. Phys.*, **15**, 105025.

[163] Córcoles, A. D., Gambetta, J. M., Chow, J. M., Smolin, J. A., Ware, M., Strand, J., Plourde, B. L. T., and Steffen, M. (2013, Mar). Process verification of two-qubit quantum gates by randomized benchmarking. *Phys. Rev. A*, **87**, 030301.

[164] Murch, K. W., Weber, S. J., Beck, K. M., Ginossar, E., and Siddiqi, I. (2013). Reduction of the radiative decay of atomic coherence in squeezed vacuum. *Nature*, **499**, 62–65.

[165] Shanks, W. E., Underwood, D. L., and Houck, A. A. (2013). A scanning transmon qubit for strong coupling circuit quantum electrodynamics. *Nature Commun.*, **4**, 1991.

[166] Purcell, E. M. (1946). Spontaneous emission probabilities at radio frequencies. *Phys. Rev.*, **69**, 681.

[167] Turlot, E., Esteve, D., Urbina, C., Martinis, J. M., Devoret, M. H., Linkwitz, S., and Grabert, H. (1989). Escape oscillations of a Josephson junction switching out of the zero-voltage state. *Phys. Rev. Lett.*, **62**, 1788–1791.

[168] Neeley, M., Ansmann, M., Bialczak, R. C., Hofheinz, M., Katz, N., Lucero, E., O'Connell, A., Wang, H., Cleland, A. N., and Martinis, J. M. (2008). Transformed dissipation in superconducting quantum circuits. *Phys. Rev. B*, **77**, 180508.

[169] Srinivasan, S. J., Sundaresan, N. M., Sadri, D., Liu, Y., Gambetta, J. M., Yu, T., Girvin, S. M., and Houck, A. A. (2013). Time-reversal symmetrization of spontaneous emission for high fidelity quantum state transfer. arXiv:1308.3471.

[170] Pechal, M., Eichler, C., Zeytinoglu, S., Berger, S., Wallraff, A., and Filipp, S. (2013). Microwave-controlled generation of shaped single photons in circuit quantum electrodynamics. arXiv:1308.4094.

[171] Nataf, P. and Ciuti, C. (2010). No-go theorem for superradiant quantum phase transitions in cavity QED and counter-example in circuit QED. *Nature Commun.*, **1**, 72.

[172] Ciuti, C. and Nataf, P. (2012). Comment on "Superradiant phase transitions and the standard description of circuit QED". *Phys. Rev. Lett.*, **109**, 179301.

[173] Viehmann, O., von Delft, J., and Marquardt, F. (2012). Reply to comment on "Superradiant phase transitions and the standard description of circuit QED." arXiv:1202.2916.

[174] Tavis, M. and Cummings, F. W. (1968). Exact solution for an n-molecule radiation-field Hamiltonian. *Phys. Rev.*, **170**, 379–384.

[175] Braak, D. (2011). Integrability of the rabi model. *Phys. Rev. Lett.*, **107**, 100401.

[176] Gambetta, J., Blais, A., Schuster, D. I., Wallraff, A., Frunzio, L., Majer, J., Devoret, M. H., Girvin, S. M., and Schoelkopf, R. J. (2006). Qubit-photon interactions in a cavity: measurement-induced dephasing and number splitting. *Phys. Rev. A*, **74**, 042318.

[177] Bertet, P., Chiorescu, I., Burkard, G., Semba, K., Harmans, C. J. P. M., DiVincenzo, D. P., and Mooij, J. E. (2005). Dephasing of a superconducting qubit induced by photon noise. *Phys. Rev. Lett.*, **95**, 257002.

[178] Ithier, G, Collin, E., Joyez, P., Meeson, P. J., Vion, D., Esteve, D., Chiarello, F., Shnirman, A., Makhlin, Y., Schriefl, J., and Schon, G. (2005). Decoherence in a superconducting quantum bit circuit. *Phys. Rev. B*, **72**, 134519.

[179] Lupascu, A., Harmans, C. J. P. M., and Mooij, J. E. (2005). Quantum state detection of a superconducting flux qubit using a dc-SQUID in the inductive mode. *Phys. Rev. B*, **71**, 184506.

[180] Boissonneault, M., Gambetta, J. M., and Blais, A. (2008). Nonlinear dispersive regime of cavity QED: the dressed dephasing model. *Phys. Rev. A*, **77**, 060305.

[181] Boissonneault, M., Gambetta, J. M., and Blais, A. (2009). Dispersive regime of circuit QED: photon-dependent qubit dephasing and relaxation rates. *Phys. Rev. A*, **79**, 013819.

[182] Bishop, L. S, Tornberg, L., Price, D., Ginossar, E., Nunnenkamp, A., Houck, A. A., Gambetta, J. M., Koch, J., Johansson, G., Girvin, S. M., and Schoelkopf, R. J. (2009). Proposal for generating and detecting multi-qubit ghz states in circuit qed. *New J. Phys.*, **11**, 073040.

[183] Helmer, F., Mariantoni, M., Solano, E., and Marquardt, F. (2009). Quantum nondemolition photon detection in circuit QED and the quantum Zeno effect. *Phys. Rev. A*, **79**, 052115.

[184] Romero, G., Garcia-Ripoll, J. J., and Solano, E. (2009). Microwave photon detector in circuit QED. *Phys. Rev. Lett.*, **102**, 173602.

[185] Marthaler, M., Schön, G., and Shnirman, A. (2008). Photon-number squeezing in circuit quantum electrodynamics. *Phys. Rev. Lett.*, **101**, 147001.

[186] Ashhab, S., Johansson, J. R., Zagoskin, A. M., and Nori, F. (2009). Single-artificial-atom lasing using a voltage-biased superconducting charge qubit. *New J. Phys.*, **11**, 023030.

[187] Birnbaum, K. M., Boca, A., Miller, R., Boozer, A. D., Northup, T. E., and Kimble, H. J. (2005). Photon blockade in an optical cavity with one trapped atom. *Nature*, **436**, 87–90.

[188] Moon, K. and Girvin, S. M. (2005). Theory of microwave parametric down-conversion and squeezing using circuit QED. *Phys. Rev. Lett.*, **95**, 140504.

[189] Marquardt, F. (2007). Efficient on-chip source of microwave photon pairs in superconducting circuit QED. *Phys. Rev. B*, **76**, 205416.

[190] Castellanos-Beltran, M. A., Irwin, K. D., Hilton, G. C., Vale, L. R., and Lehnert, K. W. (2008). Amplification and squeezing of quantum noise with a tunable Josephson metamaterial. *Nature Phys.*, **4**, 929–931.

[191] Bergeal, N., Schackert, F., Metcalfe, M., Vijay, R., Manucharyan, V. E., Frunzio, L., Prober, D. E., Schoelkopf, R. J., Girvin, S. M., and Devoret, M. H. (2010). Analog information processing at the quantum limit with a Josephson ring modulator. *Nature Phys.*, **6**, 296–302.

[192] Bergeal, N., Vijay, R., Manucharyan, V. E., Siddiqi, I., Schoelkopf, R. J., Girvin, S. M., and Devoret, M. H. (2010). Phase-preserving amplification near the quantum limit with a Josephson ring modulator. *Nature*, **465**, 64–68.

[193] Gottesman, D., Kitaev, A., and Preskill, J. (2001). Encoding a qubit in an oscillator. *Phys. Rev. A*, **64**, 012310.

[194] Braunstein, S. L. and van Loock, P. (2005). Quantum information with continuous variables. *Rev. Mod. Phys.*, **77**, 513–577.

[195] McGrayne, S. B. (2011). *The Theory That Would Not Die: How Bayes' Rule Cracked the Enigma Code, Hunted Down Russian Submarines, and Emerged Triumphant from Two Centuries of Controversy.* Yale University Press, New Haver, CT.

[196] Sears, A. P., Petrenko, A., Catelani, G., Sun, L., Paik, Hanhee, Kirchmair, G., Frunzio, L., Glazman, L. I., Girvin, S. M., and Schoelkopf, R. J. (2012). Photon shot noise dephasing in the strong-dispersive limit of circuit QED. *Phys. Rev. B*, **86**, 180504.

[197] Leghtas, Z., Kirchmair, G., Vlastakis, B., Devoret, M. H., Schoelkopf, R. J., and Mirrahimi, M. (2013). Deterministic protocol for mapping a qubit to coherent state superpositions in a cavity. *Phys. Rev. A*, **87**, 042315.

[198] Leghtas, Z., Kirchmair, G., Vlastakis, B., Schoelkopf, R., Devoret, M., and Mirrahimi, M. (2013). Hardware-efficient autonomous quantum error correction. *Phys. Rev. Lett.*, **111**, 120501.

[199] Nigg, S. E. and Girvin, S. M. (2013). Stabilizer quantum error correction toolbox for superconducting qubits. *Phys. Rev. Lett.*, **110**, 243604.

[200] Siddiqi, I., Vijay, R., Pierre, F., Wilson, C. M., Metcalfe, M., Rigetti, C., Frunzio, L., and Devoret, M. H. (2004). Rf-driven Josephson bifurcation amplifier for quantum measurement. *Phys. Rev. Lett.*, **93**, 207002.

[201] Mallet, F., Ong, F. R., Palacios-Laloy, A., Nguyen, F., Bertet, P., Vion, D., and Esteve, D. (2009). Single-shot qubit readout in circuit quantum electrodynamics. *Nature Phys.*, **5**, 791.

[202] Reed, M. D., DiCarlo, L., Johnson, B. R., Sun, L., Schuster, D. I., Frunzio, L., and Schoelkopf, R. J. (2010). High-fidelity readout in circuit quantum electrodynamics using the Jaynes-Cummings nonlinearity. *Phys. Rev. Lett.*, **105**, 173601.

[203] S. Srinivasan, A. Hoffman, Y. Liu, J. Gambetta, and A. Houck, "Tunable coupling cavity QED with a superconducting artificial atom," in Conference on Lasers and Electro-Optics 2012, OSA Technical Digest (online) (Optical Society of America, 2012), paper JW1I.2.

[204] Gambetta, J. M., Houck, A. A., and Blais, A. (2011). Superconducting qubit with Purcell protection and tunable coupling. *Phys. Rev. Lett.*, **106**, 030502.

[205] Hoffman, A. J., Srinivasan, S. J., Gambetta, J. M., and Houck, A. A. (2011). Coherent control of a superconducting qubit with dynamically tunable qubit-cavity coupling. *Phys. Rev. B*, **84**, 184515.

[206] Kitaev, A. Yu. (2003). Fault-tolerant quantum computation by anyons. *Ann. Phys. (NY)*, **303**, 2–30.

[207] Ioffe, L. B. and Feigelman, M. V. (2002). Possible realization of an ideal quantum computer in Josephson junction array. *Phys. Rev. B*, **66**, 224503.

[208] Ioffe, L. B., Feigel'man, M. V., Ioselevich, A., Ivanov, D., Troyer, M., and Blatter, G. (2002). Topologically protected quantum bits using Josephson junction arrays. *Nature*, **415**, 503–506.

[209] Doucot, B., Feigelman, M. V., and Ioffe, L. B. (2003). Topological order in the insulating Josephson junction arrays. *Phys. Rev. Lett.*, **90**, 107003.

[210] Doucot, B., Feigelman, M. V., Ioffe, L. B., and Ioselevich, A. S. (2005). Protected qubits and Chern-Simons theories in Josephson junction arrays. *Phys. Rev. B*, **71**, 024505.

[211] Gladchenko, S., Olaya, D., Dupont-Ferrier, E., Doucot, B., Ioffe, L. B., and Gershenson, M. E. (2009). Superconducting nanocircuits for topologically protected qubits. *Nature Phys.*, **5**, 48–53.

[212] Fisher, M. P. A., Weichman, P. B., Grinstein, G., and Fisher, D. S. (1989). Boson localization and the superfluid-insulator transition. *Phys. Rev. B*, **40**, 546–570.

[213] Greentree, A. D., Tahan, C., Cole, J. H., and Hollenberg, L. C. L. (2006). Quantum phase transitions of light. *Nature Phys.*, **2**, 856–861.

[214] Illuminati, F. (2006). Quantum optics: light does matter. *Nature Phys.*, **2**, 803–804.

[215] Hartmann, M. J. and Plenio, M. B. (2007). Strong photon nonlinearities and photonic Mott insulators. *Phys. Rev. Lett.*, **99**, 103601.

[216] Jarrett, T. C., Olaya-Castro, A., and Johnson, N. F. (2007). Optical signatures of quantum phase transitions in a light-matter system. *EPL*, **77**, 34001.

[217] Rossini, D. and Fazio, R. (2007). Mott-insulating and glassy phases of polaritons in 1D arrays of coupled cavities. *Phys. Rev. Lett.*, **99**, 186401.

[218] Angelakis, D. G., Santos, M. F., and Bose, S. (2007). Photon-blockade-induced mott transitions and xy spin models in coupled cavity arrays. *Phys. Rev. A*, **76**, 031805.

[219] Hartmann, M. J., Brandao, F. G. S. L., and Plenio, M. B. (2008). A polaritonic two-component Bose-Hubbard model. *New J. Phys.*, **10**, 033011.

[220] Makin, M. I., Cole, J. H., Tahan, C., Hollenberg, L. C. L., and Greentree, A. D. (2008). Quantum phase transitions in photonic cavities with two-level systems. *Phys. Rev. A*, **77**, 053819.

[221] Aichhorn, M., Hohenadler, M., Tahan, C., and Littlewood, P. B. (2008). Quantum fluctuations, temperature, and detuning effects in solid-light systems. *Phys. Rev. Lett.*, **100**, 216401.

[222] Cho, J., Angelakis, D. G., and Bose, S. (2008). Simulation of high-spin Heisenberg models in coupled cavities. *Phys. Rev. A*, **78**, 062338.

[223] Cho, J., Angelakis, D. G., and Bose, S. (2008). Fractional quantum Hall state in coupled cavities. *Phys. Rev. Lett.*, **101**, 246809.

[224] Zhao, J., Sandvik, A. W., and Ueda, K. (2008). Insulator to superfluid transition in coupled photonic cavities in two dimensions. arXiv:0806.3603.

[225] Na, N., Utsunomiya, S., Tian, L., and Yamamoto, Y. (2008). Strongly correlated polaritons in a two-dimensional array of photonic crystal microcavities. *Phys. Rev. A*, **77**, 031803.

[226] Lei, S.-C. and Lee, R.-K. (2008). Quantum phase transitions of light in the Dicke-Bose-Hubbard model. *Phys. Rev. A*, **77**, 033827.

[227] Cho, J., Angelakis, D. G., and Bose, S. (2008). Fractional quantum Hall state in coupled cavities. *Phys. Rev. Lett.*, **101**, 246809.

[228] Ji, A.-C., Sun, Q., Xie, X. C., and Liu, W. M. (2009). Josephson effect for photons in two weakly linked microcavities. *Phys. Rev. Lett.*, **102**, 023602.

[229] Hartmann, J., Brandao, F. G. S. L., and Plenio, M. B. (2009). Quantum many-body phenomena in coupled cavity arrays. *Laser Photon. Rev.*, **2**, 527–556.

[230] Grochol, M. (2009). Quantum phase transitions in an array of coupled nanocavity quantum dots. *Phys. Rev. B*, **79**, 205306.

[231] Dalidovich, D. and Kennett, M. P. (2009). Bose-Hubbard model in the presence of Ohmic dissipation. *Phys. Rev. A*, **79**, 053611.

[232] Carusotto, I., Gerace, D., Tureci, H. E., Liberato, S. De, Ciuti, C., and Imamoglu, A. (2009). Fermionized photons in an array of driven dissipative nonlinear cavities. *Phys. Rev. Lett.*, **103**, 033601.

[233] Schmidt, S. and Blatter, G. (2009). Strong coupling theory for the Jaynes-Cummings-hubbard model. *Phys. Rev. Lett.*, **103**, 086403.

[234] Lieb, M. and Hartmann, M. J. (2010). Bose-Hubbard dynamics of polaritons in a chain of circuit quantum electrodynamics cavities. *New J. Phys.*, **12**, 093031.

[235] Hartmann, M. J. (2010). Polariton crystallization in driven arrays of lossy nonlinear resonators. *Phys. Rev. Lett.*, **104**, 113601.

[236] Schmidt, S., Gerace, D., Houck, A. A., Blatter, G., and Türeci, H. E. (2010). Nonequilibrium delocalization-localization transition of photons in circuit quantum electrodynamics. *Phys. Rev. B*, **82**, 100508.

[237] Hayward, A. L. C., Martin, A. M., and Greentree, A. D. (2012). Fractional quantum Hall physics in Jaynes-Cummings-Hubbard lattices. *Phys. Rev. Lett.*, **108**, 223602.

[238] Nissen, F., Schmidt, S., Biondi, M., Blatter, G., Türeci, H. E., and Keeling, J. (2012). Nonequilibrium dynamics of coupled qubit-cavity arrays. *Phys. Rev. Lett.*, **108**, 233603.

[239] Petrescu, A., Houck, A. A., and Le Hur, K. (2012). Anomalous Hall effects of light and chiral edge modes on the Kagomé lattice. *Phys. Rev. A*, **86**, 053804.

[240] Hwang, M.-J. and Choi, M.-S. (2013). Large-scale maximal entanglement and Majorana bound states in coupled circuit quantum electrodynamic systems. *Phys. Rev. B*, **87**, 125404.

[241] Schmidt, S. and Koch, J. (2013). Circuit QED lattices: towards quantum simulation with superconducting circuits. *Ann. Phys. (Leipzig)*, **525**, 395–412.

[242] Caldeira, A. O. and Leggett, A. J. (1983). Quantum tunneling in a dissipative system. *Ann. Phys. (NY)*, **149**, 374–456.

[243] Yurke, B. (1984). Use of cavities in squeezed state generation. *Phys. Rev. A*, **29**, 408.

[244] Yurke, B. and Denker, J. S. (1984). Quantum network theory. *Phys. Rev. A*, **29**, 1419.

[245] Lang, R., Scully, M. O., and Lamb, W. E. Jr. (1973). Why is the laser line so narrow? A theory of single-quasimode laser operation. *Phys. Rev. A*, **7**, 1788.

[246] Gea-Banacloche, J., Lu, N., Pedrotti, L. M., Prasad, S., Scully, M. O., and Wodkiewicz, K. (1990). Treatment of the spectrum of squeezing based on the modes of the universe. I. Theory and a physical picture. *Phys. Rev. A*, **41**, 369.

[247] Gea-Banacloche, J., Lu, N., Pedrotti, L. M., Prasad, S., Scully, M. O., and Wodkiewicz, K. (1990). Treatment of the spectrum of squeezing based on the modes of the universe. II. Applications. *Phys. Rev. A*, **41**, 381.

[248] Gardiner, C. W., Parkins, A. S., and Zoller, P. (1992). Wave-function quantum stochastic differential equations and quantum simulation methods. *Phys. Rev. A*, **46**, 4363.

4

Quantum logic gates in superconducting qubits

J. M. Martinis

Department of Physics, University of California,
Santa Barbara, CA 93111, USA

Quantum Machines: Measurement and Control of Engineered Quantum Systems.
Edited by M. Devoret, B. Huard, R. Schoelkopf, and L. F. Cugliandolo.

Chapter Contents

4.1 Introduction

Successful operation of a quantum computer will require unprecedented control of quantum systems. The basic qubit operations, quantum logic gates, are described by the linear Schrödinger equation: the "analog" nature of quantum state evolution makes these logic gates fundamentally sensitive to imperfections in control and loss of energy. In contrast, conventional digital logic can correct errors due to built-in gain and nonlinearity. In a quantum computer, these imperfections fortunately can be removed with error-correction protocols, which work as long as the probability for the production of errors is small enough.

The performance specifications for error correction depend on details of the quantum computer architecture. Rough estimates for conventional gate-based architectures give limits below $\sim 10^{-4}$ [1], whereas more recent proposals based on surface codes may allow errors in the 10^{-2} range [2].

Much research in superconducting qubits has been directed toward improving the coherence of qubits and demonstrating quantum logic gates, both for single and coupled qubits. I am optimistic that quantum gates can eventually meet performance requirements needed for error correction.

Here, I focus on several important issues concerning the high-level design of quantum logic gates. In particular, I will review the need to effectively turn on and off coupling interactions between qubits to produce scalable controlled-not (CNOT) gates. This is an important topic for superconducting qubits, since they typically use fixed coupling elements set by fabrication.

4.2 Transition logic gates

To illustrate some design issues, I first discuss a simple example of logic gates defined by inducing transitions between quantum states that are selected by their transition frequency. Figure 4.1(a) shows the energy-level diagram for two uncoupled qubits with frequencies ω_1 and ω_2. A single-qubit gate is generated by applying a pulse of microwaves, at an excitation frequency ω_1 for changing the first qubit state or frequency ω_2 for the second. As the two qubits are not coupled, the energy-level diagram has equal transition frequencies for pairs of states, given by the dashed and dotted arrows.

Although single-qubit gates are simple to generate when they are uncoupled, it is not possible to also have CNOT logic. As illustrated in Fig. 4.1(b), the CNOT gate must swap the state amplitudes between $|10\rangle$ and $|11\rangle$ while other state amplitudes remain unchanged. A simple solution would be to produce a π-pulse transition between these two states, as shown in Fig. 4.1(a). However, this transition frequency is degenerate with that for the states $|00\rangle \leftrightarrow |01\rangle$, and such action would produce a pair of transitions that are, of course, equivalent to a single-qubit gate.

This problem can be overcome by adding coupling between the two qubits, such that the four transition frequencies are all different [3]. Now, an applied π-pulse at the $|10\rangle \leftrightarrow |11\rangle$ frequency gives a CNOT, since all other transitions are off resonant. However, this solution adds significant complexity: now a *pair* of frequencies have to be applied to make the single-qubit transitions, and their pulse strength has to be

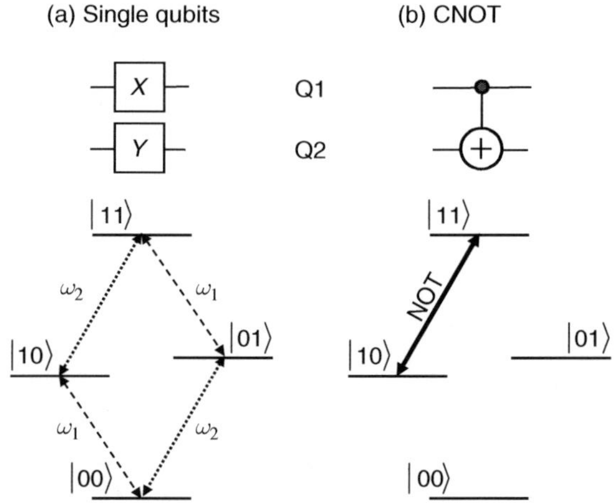

Fig. 4.1 Single and CNOT logic gates, in the transition picture, for two uncoupled qubits. Plotted is the qubit energy (vertical) for the four possible states. (a) Here, the transition frequency ω_1 for the first qubit (dashed lines) is the same for the pair of transitions $|00\rangle \leftrightarrow |10\rangle$ and $|01\rangle \leftrightarrow |11\rangle$. There is a similar pair (dotted lines) at frequency ω_2 for the second qubit. (b) The CNOT logic gate must swap only the states $|10\rangle$ and $|11\rangle$, which cannot be accomplished because of the degeneracy with the transition frequency $|00\rangle \leftrightarrow |01\rangle$.

matched carefully in order to maintain symmetric transition strengths and produce what looks like a single-qubit logic operation.

Although such transition logic is often discussed in the literature as "CNOT gates for quantum computation," in reality these probably cannot be used in a real quantum computer because they are not *scalable* gates. What happens to these logic gates as the number of qubits n increases? First, the number of transition frequencies that one must track grows as 2^n, which implies that simply calibrating the qubit system has exponential overhead. The exponentially large number of transition frequencies implies there can be significant problems with frequency overlap and crowding. The number of applied frequencies for single and CNOT gates grows as 2^{n-1} and 2^{n-2}, respectively, so that such gates become increasing more complex to generate and control accurately. Even at a modest size of $n = 8$, keeping track of 128 transition frequencies for a single-qubit logic gate would certainly be taxing. The fundamental problem is that exponential growth of classical resources is needed to control logic built with transition frequencies, which makes the system unscalable for constructing a real quantum computer.

4.3 Tunable frequency logic

A solution to this scaling problem is to turn the qubit coupling on and off. Although adjustable couplers have been demonstrated in superconducting qubits [4–8], the circuitry is somewhat complex, and the more favored approach is to use fixed capacitive coupling and a tunable qubit frequency. Here, circuit design is simpler, and the interaction strength can be effectively turned up and down by tuning the qubits into and out of resonance. In Fig. 4.1(a), "on coupling" would correspond to adjusting the qubit frequencies $\omega_1 = \omega_2$ so that the $|10\rangle$ and $|01\rangle$ states are resonant, with the resulting swapping [9] between the two gates combined with single-qubit gates [10, 11] producing a CNOT.

The figure of merit for tunable logic is given through the fixed coupling energy g and the off detuning energy Δ. When the qubits have zero detuning, the swapping frequency [9] is given by $2g/\hbar$, whereas, off resonance, the effective interaction energy between the qubits is given by the dispersive interaction [12], which scales as g^2/Δ. In this simple picture, the off-to-on strength is the ratio of these energies: $(g^2/\Delta)/g = g/\Delta$. Note here that the effective coupling is never entirely turned off: with typical values for the phase qubit given by $g/h = 30\,\text{MHz}$ and $\Delta/h = 300\,\text{MHz}$, the off-coupling detuning is not large enough to neglect the coupling.

There are presently two general ways to design such gates. At UCSB, we use direct swapping transitions between the $|10\rangle$ and $|01\rangle$ states, as well as between the $|11\rangle$ and $|20\rangle$ states for controlled-phase and CNOT [13]. The Yale group uses a dispersive interaction between these states [14], which requires about five times larger g than for the coupling interaction, so the gates can remain adiabatic in similar gate times; the off coupling is similar, since they use greater detuning.

Logical errors from finite dispersion g^2/Δ can be understood in a simple model. Defining the transition energy between the ground and excited states of qubit 1 as E_1, this energy depends on whether qubit 2 is in the state $\Psi_2 = |0\rangle$ or $|1\rangle$:

$$E_1 = \begin{cases} E_c & \text{for } \Psi_2 = |0\rangle, \\ E_c + g^2/\Delta & \text{for } \Psi_2 = |1\rangle, \end{cases} \tag{4.1}$$

where E_c is a constant. The dispersion energy of qubit 1 depends on the state of the coupled qubit 2. Defining a rotating frame for qubit 1 at frequency E_1/h, the time dependence of an equal superposition of ground and excited states is

$$\Psi_1(t) = \frac{1}{\sqrt{2}} \times \begin{cases} |0\rangle + |1\rangle & \text{for } \Psi_2 = |0\rangle, \\ |0\rangle + e^{-itg^2/\hbar\Delta}|1\rangle & \text{for } \Psi_2 = |1\rangle, \end{cases} \tag{4.2}$$

which shows a dependence on the state of qubit 2. For uncoupled qubits ($\Delta \to \infty$), the state of qubit 1 does not depend at all on the state of qubit 2: for finite Δ, the phase change in the second equation in Eq. (4.2) can be understood as a phase error from nonzero coupling. Note that this phase error is zero at time $t = 0$, increases quadratically with time, and then has a maximum magnitude of unity at $tg^2/\hbar\Delta = \pi$.

Although small at first, the potentially large magnitude from this phase error implies that small dispersive coupling cannot be ignored. It can be neglected for present experiments because the simple algorithms do not need to store data for long times, but as algorithm complexity grows, this error growing as t^2 will be increasingly important.

4.4 Refocusing

The phase error of Eq. (4.2) may be reduced through refocusing techniques [1], as developed for nuclear magnetic resonance (NMR). The basic idea is to periodically change the state of a qubit with a π-pulse so that the phase accumulation is effectively balanced out between the $|0\rangle$ and $|1\rangle$ states. For best cancelation, the phase reversal should occur halfway during the qubit storage time.

Although simple for two qubits, refocusing becomes increasing complex as the number of qubits increases, since refocusing should optimally be placed between every pair of qubits that are coupled together. As the number of pairs scales as $n!$, significant control overhead is expected for large n. For example, in the experiment to factor 15 in NMR, most of the time of the algorithm was spent turning off the coupling between spins via refocusing [15]. It will be interesting to see if this technique will be a scalable solution for a large number of qubits.

4.5 Quantum von Neumann architecture and RezQu protocol

Another solution to the off-coupling problem uses what we call the quantum von Neumann architecture [16], as illustrated in Fig. 4.2. Here, each qubit is coupled to a memory resonator as well as the resonator bus, the latter providing inter-qubit coupling. The qubit and resonator bus has residual coupling as described previously, so a state in the resonator bus will disperse the state of each qubit, causing phase errors.

This source of error is minimized in the von Neumann architecture by storing the qubit state in the memory resonator. In the RezQu (resonator zero-qubit) protocol, the qubit state is swapped into the memory resonator when it is not being manipulated by single- or coupled-qubit logic. When a quantum state is stored in memory, the qubit is

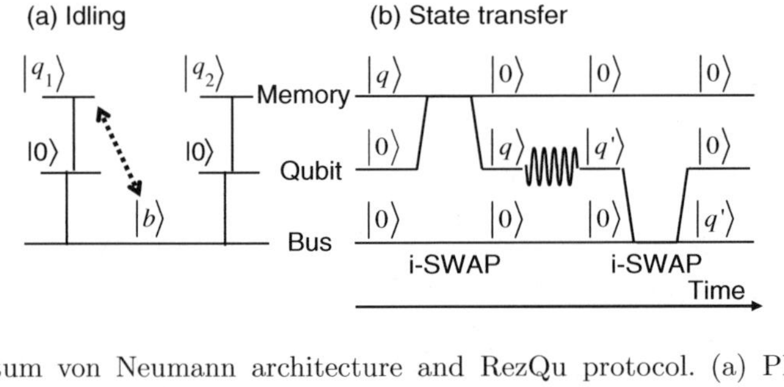

Fig. 4.2 Quantum von Neumann architecture and RezQu protocol. (a) Plot of qubit and resonator frequency (vertical) for two qubits and their memory resonator, along with a bus resonator. Qubits and resonators are coupled (vertical thin lines), and the qubit frequency may move up and down to the bus and memory frequencies. Only small (fourth-order) residual coupling remains between the memory $|q_1\rangle$ and bus $|b\rangle$ states (dashed line). (b) Plot of qubit frequency versus time showing memory-to-qubit transfer, single-qubit logic via a microwave pulse, and qubit-to-bus transfer.

in the ground state $|0\rangle$. Although there is a frequency shift in the memory qubit and the bus from the qubit ground state, no error is generated, since the state is known and always produces a constant shift.

Phase errors remain between the memory and bus resonators, but they are quite small because they proceed through a fourth-order process that produces a virtual state in the qubit by memory-qubit coupling, then a frequency shift in the bus through qubit-bus coupling. It is interesting to note that this coupling from memory to bus goes to zero when the qubit frequency is placed halfway between the memory and bus frequencies (A. Korotkov, personal communication, 2012).

An additional advantage of this architecture is the use of resonators as qubit memories, since these elements presently have the longest coherence time. Additionally, resonators require no control signals, so they are inexpensive in terms of control lines and electronics. Using memory resonators with closely tuned frequencies might also minimize the difficulty of accurately tracking differential phases between the many qubits.

4.6 Conclusions

Now that long coherence times have been demonstrated in superconducting qubits, achieving high-fidelity gates is an important topic for future research. In this chapter, I have discussed two fundamental issues that confront the designer of quantum logic: scalability and qubit errors from residual off-coupling. As solutions are available, either with refocusing or using the RezQu architecture, much progress is expected in the next few years.

References

[1] Nielsen, M. A., and Chuang, I. L. (2000). *Quantum Computation and Quantum Information.* Cambridge University Press, Cambridge.

[2] Fowler, A. G., Stephens, A. M., and Groszkowski, P. (2009). High-threshold universal quantum computation on the surface code. *Phys. Rev. A*, **80**, 052312.

[3] Plantenberg, J. H., de Groot, P. C., Harmans, C. J. P. M., and Mooij, J. E. (2007). Demonstration of controlled-NOT quantum gates on a pair of superconducting quantum bits. *Nature*, **447**, 836.

[4] Hime, T., Reichardt, P. A., Plourde, B. L. T., Robertson, T. L., Wu, C.-E., Ustinor, A. V., and Clorhe, J. (2006). Solid state qubits with current-controlled coupling. *Science*, **314**, 1427.

[5] Niskanen, A. O., Harrabi, K., Yoshihara, F., Nakamwa, Y., Lloyd, S., and Tsai, J. S. (2007). Quantum coherent tunable coupling of superconducting qubits. *Science*, **316**, 723.

[6] Auman, M. S., Attorare, F., Whittaher, J. D., Citrak, K., Li, D., Sirois, A. Strong, J., Teuful, J. D., and Simmonds, R. W. (2010). rf-SQUID-mediated coherent tunable coupling between a superconducting phase qubit and a lamped-element resonator. *Phys. Rev. Lett.*, **104**, 177004.

[7] Harris, R., Berkley, A. J., Johnson, M. W., Bunyk, P., Govorkov, S., Thom, M. C., Uchaikin, S., Wilson, A. B., Chung, J., Holtham, E., Biamonte, J. D., Smirnov, A. Yu., Amin, M. H. S., and Alec Maassen van den Brink (2007). Sign- and magnitude-tunable coupler for superconducting flux qubits. *Phys. Rev. Lett.* **98**, 177001.

[8] Bialczak, R. C., Ansmann, M., Hofheinz, M., Lenander, M., Lucero, E., Neeley, M., O'Connell, A. D., Sank, D., Wang, H., Weides, M., Wenner, J., Yamamoto, T., Cleland, A. N., and Martinis, J. M. (2011). Fast tunable coupler for superconducting qubits. *Phys. Rev. Lett.* **106**, 060501.

[9] McDermott, R., Simmonds, R. W., Steffen, M., Cooper, K. B., Cicak, K., Osborn, K., Oh, S., Pappas, D. P., and Martinis, J. M. (2005). Simultaneous state measurement of coupled Josephson phase qubits. *Science*, **307**, 1299.

[10] Geller, M. R., Pritchett, E. J., Galiautdinov, A., and Martinis, J. M. (2010). Quantum logic with weakly coupled qubits. *Phys. Rev. A*, **81**, 012320.

[11] Dewes, A., Ong, F. R., Schmitt, V., Lauro, R., Boulant, N., Bertet, P., Vion, D., and Esteve, D.(2012). Characterization of a two-transmon processor with individual single-shot qubit readout. *Phys. Rev. Lett.* **108**, 057002.

[12] Wallraff, A., Schuster, D. I., Blais, A., Frunzio, L., Huang, R.-S., Majer, J., Kumar, S., Girvin, S. M., and Schoelkopf, R. J. (2004). Circuit quantum electrodynamics: coherent coupling of a single photon to a Cooper pair box. *Nature*, **431**, 162.

[13] Yamamoto, T., Neeley, M., Lucero, E., Bialczak, R. C., Kelly, J., Lenander, M., Mariantoni, M., O'Connell, A. D., Sank, D., Wang, H., Weides, M., Wenner, J., Yin, Y., Cleland, A. N., and Martinis, J. M. (2010). Quantum process tomography of two-qubit controlled-Z and controlled-NOT gates using superconducting phase qubits. *Phys. Rev. B*, **82**, 184515.

[14] DiCarlo, L., Chow, J. M., Gambetta, J. M., Bishop, L. S., Johnson, B. R., Schuster, D. I., Majer, J., Blais, A., Frunzio, L., Girvin, S. M., and Schoelkopf, R. J. (2009). Demonstration of two-qubit algorithms with a superconducting quantum processor. *Nature*, **460**, 240.

[15] Vandersypen, L. M., Steffen, M., Breyta, G., Yannomi, C. S., Sherwood, M. H., and Chuang, I. L. (2001). Experimental realization of Shor's aquantum factoring algorithm using nuclear magnetic resonance. *Nature*, **414**, 883.

[16] Mariantoni, M., Wang, H., Yamamoto, E., Neeley, M., Bialczak, R. C., Chen, Y., Lenander, M., Lucero, E., O'Connell, A. D., Sank, D., Weides, M., Wenner, J., Yin, Y., Zhao, J., Korotkov, A. N., Cleland, A. N., and Martinis, J. M. (2011). Implementing the quantum von Neumann architecture with superconducting circuits. *Science*, **334**, 61.

5

Exploring quantum matter with ultracold atoms

I. BLOCH

Max-Planck-Institute of Quantum Optics
85748 Garching, Germany
Ludwig-Maximilians-University of Munich
80799 Munich, Germany

Quantum Machines: Measurement and Control of Engineered Quantum Systems.
Edited by M. Devoret, B. Huard, R. Schoelkopf, and L. F. Cugliandolo.

Chapter Contents

The course was aimed at giving a basic introduction to the physics of ultracold quantum gases in optical lattices. It started out with an introduction to the generation of ultracold quantum gases using laser and evaporative cooling techniques. Then, the collisional interactions between atoms were discussed, with a special focus on revisiting scattering theory and an elementary lecture on Feshbach resonances. Through such resonances, the interactions between atoms can be tuned over a wide range using an adjustable magnetic field.

The different detection techniques for probing ultracold quantum matter were then introduced, covering a wide range of topics ranging from standard absorption imaging, over noise-correlation measurements to the more recently achieved high-resolution imaging of ultracold atoms. It was also shown how, with the latter method, nonlocal order parameters and quantum fluctuations of a many-body system can be directly measured.

Next, the course then discussed many-body effects in ultracold atoms, with a special focus on optical-lattice-based experiments. Ultracold quantum gases in optical lattices can be seen as prototypical realizations of Hubbard model systems, and the different quantum phases present in these systems for weak and strong interactions were discussed.

The course then considered the control of few-body effects in optical lattices, and highlighted experiments in which superexchange interactions and resonating valence bond states on single plaquettes in optical lattices were realized. Different topologies of the states were outlined, and it was shown how the dynamical evolution of the quantum states could be revealed in experiments.

The course then turned to the example of where ultracold quantum gases can provide a new test bed for many-body dynamics far from equilibrium. The simple model case of Landau-Zener transitions between coupled one-dimensional quantum gases was discussed, as well as the relaxation of an isolated quantum system towards equilibrium.

Finally, an outlook on the possibility of investigating many-body systems with long-range interactions using polar molecules or Rydberg atoms was given and applications for quantum information and quantum optics were outlined.

The course content largely overlapped with a previous course given by the same lecturer at Les Houches, and detailed lecture notes can be found in the proceedings of that course:

- I. Bloch. Ultracold atoms in optical lattices. In *Many-Body Physics with Ultracold Gases. Lecture Notes of the Les Houches Summer School: Volume 94, July 2010* (ed. C. Salomon, G.V. Shlyapnikov, and L.F. Cugliandolo), pp. 71–108. Oxford University Press, Oxford, 2012.

For the interested reader, the following additional references may be helpful as a starting point for further studies:

- I. Bloch, J. Dalibard, and W. Zwerger. Many-body physics with ultracold gases. *Rev. Mod. Phys.*, **80**, 885–964 (2008).

- I. Bloch, J. Dalibard, and S. Nascimbene. Quantum simulations with ultracold atoms. *Nature Phys.*, **8**, 267–276 (2012).
- D. Jaksch and P. Zoller. The cold atoms Hubbard toolbox. *Ann. Phys. (NY)*, **315**, 52–79 (2005).
- M. Lewenstein, A. Sanpera, V. Ahufinger, B. Damski, A. Sen(De), and U. Sen. Ultracold atomic gases in optical lattices: mimicking condensed matter physics and beyond. *Adv. Phys.*, **56**, 243–379 (2007).
- T. Esslinger. Fermi-Hubbard physics with atoms in an optical lattice. *Annu. Rev. Condens. Matter Phys.*, **1**, 129–152 (2010).

6

Readout of superconducting qubits

D. ESTEVE

Quantronics Group
Service de Physique de l'Etat Condensé/IRAMIS/DSM (CNRS URA 2464)
CEA Saclay, 91191 Gif-sur-Yvette, France

Quantum Machines: Measurement and Control of Engineered Quantum Systems.
Edited by M. Devoret, B. Huard, R. Schoelkopf, and L. F. Cugliandolo.

Chapter Contents

Performing the readout of a quantum register with high fidelity and with no more backaction than requested by a projective quantum measurement is one of the main challenges to the development of a scalable processor architecture based on superconducting qubits.

First, qubit readout consists in entangling it with a pointer with two distinguishable states, the pointer itself often being a mesoscopic system. Readout involves two fidelities: the outcome fidelity, and the projection fidelity associated with the qubit state after the measurement. Ideally, readout is faithful and leaves the qubit in the state corresponding to the measured outcome. In these lectures, qubit readout was discussed within the general framework of quantum measurements as now found in textbooks (see, e.g., [1]). The positive operator-valued measure was introduced in the case of the readout of a phase qubit through its tunneling out of a metastable well.

Quantum mechanics states that qubit readout at least brings in as much decoherence as the amount of information extracted from the qubit. Readout backaction was briefly discussed for a few readout schemes, and in more depth for the linear dispersive readout of a transmon through the qubit-state-dependent frequency pull of the cavity in which it is embedded (see Chapter 3 of this volume). In this scheme, the signal-to-noise ratio is limited by the readout circuitry, which can be alleviated by using microwave amplifiers operated close to the quantum limit [2] (see also Chapter 16). It was also explained how dispersive readout of a continuously driven transmon was used to demonstrate the violation of the Leggett-Garg inequality, proving thus that this "macroscopic" electrical circuit does not fulfill the hypotheses of macrorealism and is thus unquestionably quantum [3].

The readout of multi-qubit circuits was discussed in the context of running quantum algorithms. Here, achieving high-fidelity single-shot readout of the qubit register is mandatory for demonstrating the quantum speed-up expected from quantum algorithms. It was shown how the Josephson bifurcation amplifier [4] can provide a high-fidelity readout of the transmon with induced decoherence comparable to the amount of information taken from the transmon [5, 6]. The two-transmon processor developed in quantronics was discussed, including the process tomography of a universal quantum gate [7]. The demonstration of quantum speed-up for the Grover search algorithm on four objects was finally presented [8].

References

[1] Wiseman, H., and Milburn, G. (2009). *Quantum Measurement and Control.* Cambridge University Press, Cambridge.

[2] Vijay, R., Macklin, C., Slichter, D. H., Weber, S. J., Murch, K. W., Naik, R., Korotkov, A. N., and Siddiqi, I. (2012). Quantum feedback control of a superconducting qubit: persistent Rabi oscillations. *Nature*, **490**, 77.

[3] Palacios-Laloy, A., Mallet, F., Nguyen, F., Bertet, P., Vion, D., Esteve, D., and Korotkov, A. (2010). *Nature Phys.*, **6**, 442.

[4] Vijay, R., Devoret, M. H., and Siddiqi, I. (2009). *Rev. Sci. Instrum.*, **80**, 111101.

[5] Mallet, F., Ong, F. R., Palacios-Laloy, A., Nguyen, F., Bertet, P., Vion, D., and Esteve, D. (2009). *Nature Phys.*, **5**, 791.
[6] Boissonneault, M., Doherty, A. C., Ong, F. R., Bertet, P., Vion, D., Esteve, D., and Blais, A. (2012) *Phys. Rev. A*, **85**, 022305.
[7] Dewes, A., Ong, F. R., Schmitt, V., Lauro, R., Boulant, N., Bertet, P., Vion, D., and Esteve, D. (2012) *Phys. Rev. Lett.*, **108**, 057002.
[8] Dewes, A., Lauro, R., Ong, F. R., Schmitt, V., Milman, P., Bertet, P., Vion, D., and Esteve, D. (2012) *Phys. Rev. B*, **85**, 140503.

7
Quantum error correction

I. L. Chuang

Massachusetts Institute of Technology
Cambridge, Massachusetts 02139, USA

Quantum Machines: Measurement and Control of Engineered Quantum Systems.
Edited by M. Devoret, B. Huard, R. Schoelkopf, and L. F. Cugliandolo.

Chapter Contents

Quantum error correction is perhaps the deepest and most surprising concept to arise from the union of quantum physics, computer science, and information theory. How can the effects of an intrinsically irreversible process, such as spontaneous emission, be reversed, so that a quantum state could in principle stay essentially perfectly intact, forever?

Here, we describe the principles of quantum error corrections, and the applications of quantum error correction in the design and operation of simple quantum machines, over the course of five lectures:

- Lecture 1: Unraveling open quantum system dynamics;
- Lecture 2: Reversing quantum noise;
- Lecture 3: Quantum error correction;
- Lecture 4: Modern quantum codes;
- Lecture 5: Fault-tolerant quantum computation.

One additional lecture on applications of quantum error correction is omitted from this chapter. The reader is assumed to be familiar with (but not necessarily an expert at) ideas from quantum circuits and methods from advanced quantum mechanics, including qubits, basic quantum gates, Pauli matrices, density matrices, tensor products, partial traces, Bloch equations, and master equations (see, e.g., the lectures by Mabuchi and Girvin Chapters 1 and 3 of this volume).

7.1 Unraveling open quantum system dynamics

Quantum error correction is a concept predicated upon a clear understanding of the dynamics of quantum systems interacting with an environment. We begin by describing several different approaches to studying such *open* quantum system dynamics, based on the idea of *unraveling* dynamics into potential single-system views. This then leads to the quantum Monte Carlo wavefunction method, with which we illustrate various equivalent (and some unexpected!) interpretations of the dynamics of a model open quantum system—a two-level atom interacting with the vacuum.

7.1.1 Concepts of open quantum systems

The evolution of a *closed* quantum system is governed by the Schrödinger equation

$$|\psi(t)\rangle = e^{-iHt/\hbar}|\psi(0)\rangle \,, \tag{7.1}$$

where $|\psi(t)\rangle$ is the state of the system at time t and H is a fixed, time-independent Hamiltonian. Note that if we were to want to allow H to be time-varying, this would necessarily imply a mechanism driving the time dependence, meaning that the system would no longer be closed. Also note that because H is a Hermitian operator, $U = e^{-iHt/\hbar}$ is a unitary transformation. Thus, all closed quantum systems evolve unitarily.

How does one describe dynamics such as friction, which entails loss of energy from the system? Classical dynamical open systems are described by modeling a coupling between the system and an *environment* (Fig. 7.1). Often, a variety of assumptions

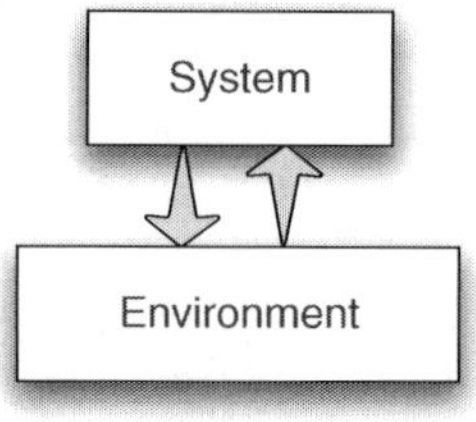

Fig. 7.1 System-environment model for open system dynamics.

are made to make this model tractable—for example, the coupling may be assumed to be weak, or the environment may be approximated as being largely unchanging, and memoryless. But the basic principles are that properties of the system, such as its energy, may change owing to coupling to the environment, and this coupling typically introduces elements of randomness into the behavior of the system.

Open quantum systems can be modeled similarly. One key step is to replace the pure-state description of the system $|\psi\rangle$ with a density matrix description ρ. If the state is $|\psi\rangle$, then $\rho = |\psi\rangle\langle\psi|$, but the state in general need not be pure—statistical mixtures such as $\rho = \sum_k p_k |\psi_k\rangle\langle\psi_k|$ (where p_k are probabilities) may also occur. The density matrix analogue of the Schrödinger equation is the von Neumann equation

$$\dot{\rho} = -i[H, \rho], \tag{7.2}$$

but more generally (i.e., for open quantum systems), the evolution need not be Hamiltonian.

The evolution, however, must respect necessary and sufficient conditions for ρ to remain a legitimate description of a quantum state, and thus the evolution is constrained to what is known as the "Lindblad form":

$$\dot{\rho} = -i[H, \rho] + \sum_k L_k \rho L_k^\dagger - \frac{1}{2} L_k^\dagger L_k \rho - \frac{1}{2} \rho L_k^\dagger L_k \,. \tag{7.3}$$

L_k are known as *jump operators*, and this equation is known as a quantum master equation.

The dynamics predicted of open quantum systems modeled with such master equations are in excellent agreement with experimental observations. It is therefore interesting to closely study what these predictions suggest, if full quantum control of the system were available.

7.1.2 Single quantum systems and unravelings

If you could watch the state of a single quantum system undergoing dynamics described by a quantum master equation, what would you see? The answer to this is very interesting, and is even historically controversial.

Ensembles and Icthysauria Implicit in the density matrix formalism is the notion of an *ensemble* description of quantum dynamics. In fact, the concept of ensembles is vital to the foundations of quantum mechanics; only ensemble averages can be measured, and one might even say that in a sense, only ensemble averages are "real." Schrödinger, in 1952, wrote (and we quote here from [1]):

...we never experiment with just one electron or atom or (small) molecule. In thought experiments, we sometimes assume that we do; this invariably entails ridiculous consequences... In the first place, it is fair to state that we are not experimenting with single particles, any more than we can raise Ichthysauria in the zoo.

Despite this rather extreme opinion, modern physics experiments are now broadly moving in the direction of coherently addressing and manipulating single quantum systems, such as single atoms and single molecules, in the laboratory. Experiments have been successful with both (hot and cold) neutral atoms and ions, and with a variety of single (hot) molecules. These date back to Dehmelt's 1975 proposal to trap single ions to provide an atomic time standard.

Such experiments with single quantum systems motivate a description of open quantum system dynamics in terms of statistical averages over single realizations, as illustrated by the following experiment. Consider a single three-level ion, such as strontium, with an 8 ns lifetime 422 nm S-P transition, and a ~400 ms lifetime 674 nm S-D transition (Fig. 7.2). Experimental observation of the blue fluorescence from this atom, when strongly excited at 422 nm, and weakly excited at 674 nm, shows a very interesting signature: quantum jumps. When the atom is in the metastable D state, the 422 nm fluorescence ceases, because no photons can be scattered from the strong transition until it transitions back to the ground S state. A typical observation record might look as shown in Fig. 7.3.

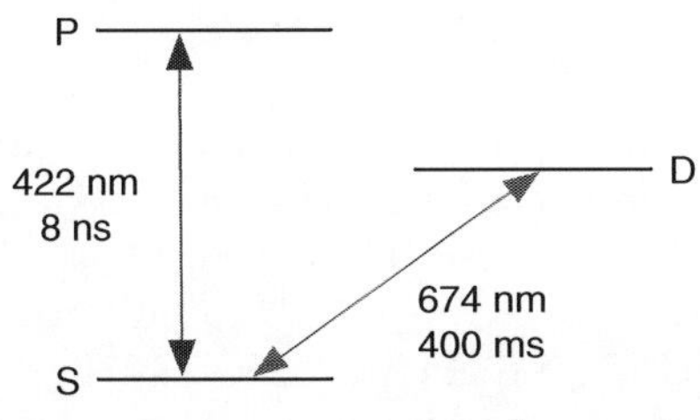

Fig. 7.2 Energy levels of Sr^+, a typical ion used in quantum computation experiments.

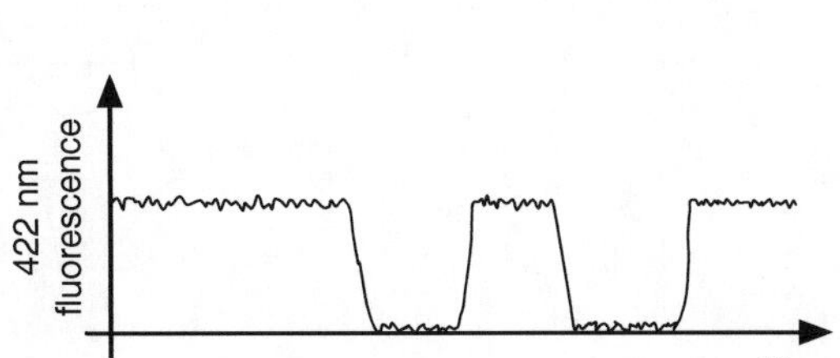

Fig. 7.3 Quantum jumps observed in the fluorescence spectrum of ions such as Sr^+.

Naturally, the *average* fluorescence rate will be consistent with that expected by the relative strengths of the red and blue excitations, but this would be just a constant value. Such quantum jumps, however, are ubiquitous in experiments involving single atoms and other single quanta, and it would be nice to have a theoretical prescription for describing such observations. In addition, it would be nice to have available possible *interpretations* of how the individual systems might be evolving to give some observed ensemble behavior, such as an exponential decay.

Unravelings In general, we may interpret the dynamics of a system described by a density matrix in terms of various *unravelings* into ensembles of different statistical mixtures of pure states. While in principle all of the unravelings are equivalent, and none of them can be distinguished by any experimental measurement, these unravelings nevertheless provide insightful interpretations, particularly in view of understanding when and how quantum error correction works.

Consider the two-qubit state

$$|\psi_{AB}\rangle = \sqrt{\frac{3}{4}}|00\rangle + \sqrt{\frac{1}{4}}|11\rangle\,, \tag{7.4}$$

where the labels denote qubits A and B, respectively, from left to right. If B were measured (in the computational basis, as shown in Fig. 7.4(a), but the outcome left unknown to the party holding qubit A), then A would be found to be in state $|0\rangle$ with probability 3/4, and in the state $|1\rangle$ with probability 1/4, so we would then describe the state of qubit A alone with the density matrix

$$\rho_A = \frac{1}{4}\begin{bmatrix} 3 & 0 \\ 0 & 1 \end{bmatrix}\,. \tag{7.5}$$

What if the measurement of qubit B were done in a different basis, for example, as shown in Fig. 7.4(b)? The Hadamard gate H (not to be confused with a Hamiltonian) maps $|0\rangle \to (|0\rangle + |1\rangle)/\sqrt{2}$, and $|1\rangle \to (|0\rangle - |1\rangle)/\sqrt{2}$, turning $|\psi_{AB}\rangle$ into

$$\sqrt{\frac{3}{4}}|0\rangle\frac{|0\rangle + |1\rangle}{\sqrt{2}} + \sqrt{\frac{1}{4}}|1\rangle\frac{|0\rangle - |1\rangle}{\sqrt{2}}\,. \tag{7.6}$$

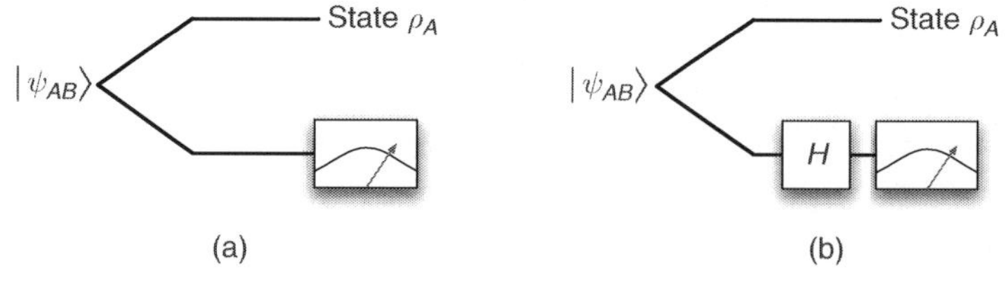

Fig. 7.4 A bipartite quantum state $|\psi_{AB}\rangle$, subject to two different measurements on subsystem B, resulting in subsystem state ρ_A.

Now if the measurement of B gives 0, then A is found to be in the state $\sqrt{\frac{3}{4}}|0\rangle + \sqrt{\frac{1}{4}}|1\rangle$, whereas if B were measured to be 1, then A is found to be in the state $\sqrt{\frac{3}{4}}|0\rangle - \sqrt{\frac{1}{4}}|1\rangle$. This seems very different from the first case, without the Hadamard gate. However, the statistical mixture of these states, weighted by the equal probability for each measurement to occur, gives the density matrix for A

$$\rho_A = \frac{1}{4}\begin{bmatrix} 3 & 0 \\ 0 & 1 \end{bmatrix} . \tag{7.7}$$

This is exactly the same as in the first case.

In fact, there are in general an infinite number of possible ways in which a density matrix such as ρ_A can be expressed as a statistical mixture $\rho_A = \sum_k p_k|\psi_k\rangle\langle\psi_k|$ of pure states. This is a well-known fact, and these different unravelings are all equivalent, in the sense that they cannot be experimentally distinguished from each other by any quantum measurement on A.

What is less well known, and conceptually vital to the idea of quantum error correction, is that *the dynamical evolution of open quantum systems can also be unraveled into an infinite number of possible evolution trajectories*, in general. In the example above, we saw that different unravelings of the density matrix state ρ_A resulted from different measurement bases employed in measuring the B subsystem of the "purified" state $|\psi_{AB}\rangle$. The different possible unravelings of evolution trajectories also arise from a similar degree of freedom, which is in a sense the basis in which the environment sees the system.

7.1.3 Quantum Monte Carlo Wavefunction Method

Consider a single two-level atom evolving under the Hamiltonian H_0, while also undergoing spontaneous emission. This is a scenario often used to formulate the optical Bloch equations (OBEs). The system's dynamics can be modeled using the quantum Monte Carlo wavefunction (QMCWF) technique, using the following steps.

Let us assume the atom begins in some state $|\psi\rangle$, and let the atomic levels be $|e\rangle$ and $|g\rangle$. Let Γ be the decay rate of the excited state, as determined, for example, from Fermi's Golden Rule. The QMCWF technique introduces a new *non-Hermitian* "Hamiltonian"

$$H = H_0 - i\frac{\Gamma}{2}|e\rangle\langle e| , \tag{7.8}$$

which governs evolution of the state, according to the following. For each differential time step dt, the state $|\psi\rangle$ evolves as

1. Compute $dp = \Gamma\, dt\, |\langle e|\psi\rangle|^2$.
2. Let $0 \le \epsilon \le 1$ be a uniformly distributed random number.
3. If $\epsilon < dp$, then $|\psi\rangle \leftarrow |g\rangle$.
4. If $\epsilon \ge dp$, then $|\psi\rangle \leftarrow \left(e^{-iH\,dt}/\sqrt{1-dp}\right)|\psi\rangle$.
5. Go to **1.**

Physically, dp is the probability of the atom jumping from $|e\rangle$ to $|g\rangle$, that is, spontaneously emitting a photon, during the time interval dt. Note that typically $dp \ll 1$, and

$$dp = 1 - \langle\psi|e^{iH^\dagger dt}e^{-iHdt}|\psi\rangle\,, \tag{7.9}$$

so $\sqrt{1-dp}$ is the normalization of the state after dt of evolution under the non-Hermitian Hamiltonian H. Essentially, the QMCWF scheme models an imaginary observer watching for a spontaneously emitted photon coming from the atom. If a photon is emitted, then the atom transitions immediately into $|g\rangle$; this is the rule given in step 3, and such a transition is known as a *quantum jump*.

Otherwise, if no photon is emitted, the state changes nevertheless, albeit by only a small amount—because the observer saw no photon, and thus the probability of the atom being in $|e\rangle$ is slightly diminished. Specifically, if $H_0 = 0$, and $|\psi\rangle = \alpha|g\rangle + \beta|e\rangle$, then, according to the rule in step 4,

$$|\psi\rangle \leftarrow \frac{\alpha|g\rangle + \beta e^{-\Gamma dt/2}|e\rangle}{\sqrt{|\alpha|^2 + |\beta|^2 e^{-\Gamma dt}}}\,. \tag{7.10}$$

This is exactly the same physics as in the OBEs.

Equivalence of QMCWF and the optical Bloch equations The equivalence of QMCWF and the OBEs is shown by demonstrating that the evolution of the density matrix

$$\rho(t) = \overline{|\psi(t)\rangle\langle\psi(t)|} \tag{7.11}$$

satisfies the OBEs, where the average (denoted by the overline) is taken over instances of running the QMCWF procedure. This follows from computing the density matrix for the state after one QMCWF procedure step:

$$\begin{aligned}
\rho(t+dt) &= dp\,|g\rangle\langle g| + (1-dp)\frac{e^{-iH\,dt}|\psi\rangle\langle\psi|e^{+iH^\dagger\,dt}}{1-dp} && (7.12)\\
&\approx dp\,|g\rangle\langle g| + (1-iH\,dt)\rho(t)(1+iH^\dagger\,dt) && (7.13)\\
&\approx dp\,|g\rangle\langle g| + \rho(t) - i\{H\rho(t) - \rho(t)H^\dagger\}\,dt && (7.14)\\
&= \Gamma\,dt\,\langle e|\rho(t)|e\rangle\,|g\rangle\langle g| + \rho(t) - i\{H\rho(t) - \rho(t)H^\dagger\}\,dt && (7.15)\\
&= \Gamma\,dt\,|g\rangle\langle e|\rho(t)|e\rangle\langle g| + \rho(t)\\
&\quad -dt\left\{i\,[H_0,\rho(t)] - \frac{\Gamma}{2}\left(|e\rangle\langle e|\rho(t) + \rho(t)|e\rangle\langle e|\right)\right\}. && (7.16)
\end{aligned}$$

Writing this as a coarse-grained differential equation, taking the limit of small dt, we find

$$\frac{d}{dt}\rho(t) \approx \frac{\rho(t+dt)-\rho(t)}{dt} \tag{7.17}$$

$$= -i\left[H_0,\rho(t)\right] - \frac{\Gamma}{2}\Big(|e\rangle\langle e|\rho(t) + \rho(t)|e\rangle\langle e|\Big) + \Gamma|g\rangle\langle e|\ \rho(t)\ |e\rangle\langle g|\,. \tag{7.18}$$

This is a quantum master equation of the form of Eq. (7.3), with jump operator $L = \sqrt{\Gamma}|g\rangle\langle e|$. It is also the OBE.

7.1.4 Quantum Monte Carlo wavefunction technique: general case

The method of the QMCWF technique can be applied not just to the atom + vacuum scenario, but also to model any open quantum system dynamics. The relaxation part of a master equation may be described by a Lindblad operator

$$\mathcal{L}(\rho) = -\frac{1}{2}\sum_k C_k^\dagger C_k\rho + \rho C_k^\dagger C_k + \sum_k C_k\rho C_k^\dagger\,, \tag{7.19}$$

where C_k are known as "quantum jump" operators. The corresponding QMCWF procedure employs the non-Hermitian Hamiltonian

$$H = H_0 - \frac{i}{2}\sum_k C_k^\dagger C_k \tag{7.20}$$

and the steps

1. Compute $dp = \sum_k dp_k$, and $dp_k = dt\,\langle\psi|C_k^\dagger C_k|\psi\rangle$.
2. Let $0 \le \epsilon \le 1$ be a uniformly distributed random number.
3. If $\epsilon < dp$, then $|\psi\rangle \leftarrow C_k|\psi\rangle/\sqrt{dp_k/dt}$, with k randomly chosen with probability dp_k/dp.
4. If $\epsilon \ge dp$, then $|\psi\rangle \leftarrow \left(e^{-iH\,dt}/\sqrt{1-dp}\right)|\psi\rangle$.
5. Go to **1.**

There are some computational advantages to the QMCWF approach over numerical solution of the differential equations normally obtained with master equations. In particular, an N-dimensional Hilbert space described by a density matrix requires $O(N^2)$ variables, whereas a pure-state wavefunction requires only $O(N)$ variables. Of course, calculating expectation values means that the stochastic evolution of the QMCWF technique must be repeated $\sim N$ times, so the overall effect is a tradeoff between storage space and computational time. However, the QMCWF technique is immediately parallelizable, and often (but not always) desired observables converge quickly.

For more about this, see, for example, the nice article by Molmer, Castin, and Dalibard [2].

7.1.5 Open-system dynamics have infinitely many equivalent unravelings

The various physical origins of phase damping are interesting to consider, because they can teach us an important fact about models of decoherence for single quantum systems: just as ρ can be *unraveled* in infinitely many ways as $\sum_k p_k|\psi_k\rangle\langle\psi_k|$, decoherence *processes* may also be unraveled in an infinite number of ways, each equally equivalent and equally physically meaningful.

We consider three different physical models of phase damping, on a two-level atom.

Random phase noise A two-level atom of frequency ω_0 excited by far off-resonance light, $\omega \gg \omega_0$, experiences an AC Stark shift of amount proportional to the light intensity. If this intensity fluctuates, then the atom's phase is randomly modulated, causing the evolution

$$|e\rangle \rightarrow e^{-i\theta}|e\rangle\,, \tag{7.21}$$

where θ is the phase imparted by the AC Stark shift. Suppose θ is modeled as a Gaussian-distributed random variable, with mean zero and variance $2\lambda t$, such that

$$\text{prob}(\theta) = \frac{1}{\sqrt{4\pi\lambda t}} e^{-\theta^2/4\lambda t}\,. \tag{7.22}$$

If the initial state of the atom is $\psi\rangle = a|g\rangle + b|e\rangle$, then, after time t, it evolves into the average state described by the density matrix

$$\rho(t) = \int_{-\infty}^{+\infty} \begin{bmatrix} |a|^2 & ab^* e^{i\theta} \\ a^* b e^{-i\theta} & |b|^2 \end{bmatrix} \text{prob}(\theta)\, d\theta\,, \tag{7.23}$$

which is found to be

$$\rho(t) = \begin{bmatrix} |a|^2 & ab^* e^{-\lambda t} \\ a^* b e^{-\lambda t} & |b|^2 \end{bmatrix}. \tag{7.24}$$

If the atomic Hamiltonian is $H_0 = \omega_0|e\rangle\langle e|$, and the atom's initial state is $|\psi\rangle = (|g\rangle + |e\rangle)/\sqrt{2}$, then the dipole moment of the atom shows a simple Rabi oscillation (Fig. 7.5). When random phase noise is imposed on the atom, then its dipole moment decays with time. The evolution of a *single* atom, according to the "trajectory" described by the random walk of Eq. (7.21), is a noisy Rabi oscillation, as shown in Fig. 7.6 (plotted for an illustrative random walk timestep).

Elastic collisions Another physical origin for phase damping is elastic collisions. Assume the two-level atom bounces along a waveguide, interacting with the walls without losing kinetic energy, but changing its trajectory slightly at each bounce, in a manner depending on the state of the atom. This can be modeled by a Hamiltonian interacting the atom with a single mode environment,

$$H_{SE} = |e\rangle\langle e| \otimes (\gamma|0\rangle\langle 1| + \gamma^*|1\rangle\langle 0|)\,, \tag{7.25}$$

with coupling constant γ (this parameterizes the strength of the noise, much like λ in the model above). During a small differential time step dt, an initial atomic state $a|g\rangle + b|e\rangle$ coupled to an environment $|0\rangle$ evolves to become

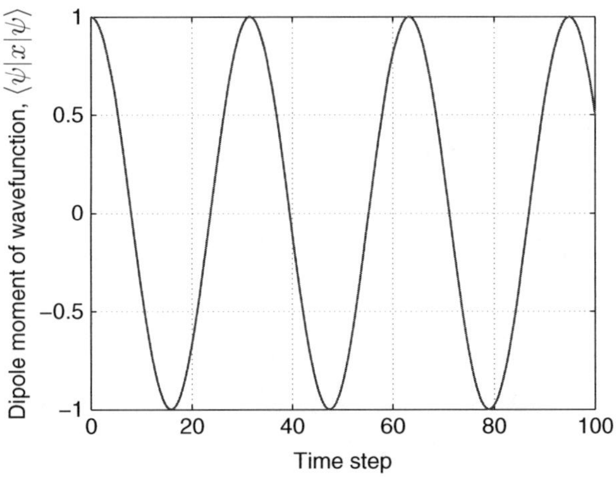

Fig. 7.5 Quantum Monte Carlo simulation of the Rabi oscillation of a two-level atom, without noise.

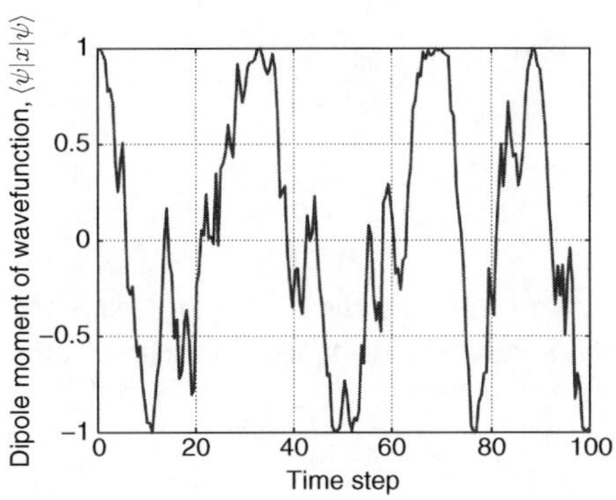

Fig. 7.6 Quantum Monte Carlo simulation of the evolution of two-level atom, with random phase noise.

$$(a|g\rangle + b|e\rangle) \otimes |0\rangle \to a|g\rangle|0\rangle + b|e\rangle(\cos\theta|0\rangle + \sin\theta|1\rangle) \tag{7.26}$$

$$= (a|g\rangle + b\cos\theta|e\rangle)\,|0\rangle + (b\sin\theta|e\rangle)\,|1\rangle\,, \tag{7.27}$$

where $e^{-\gamma dt} = \cos\theta$. This expression is very similar to that obtained for the gedanken experiment used in the QMCWF model of spontaneous emission; the difference is that when a photon is observed in the environment, the atom does not collapse into $|g\rangle$, but rather into $|e\rangle$. In other words, it does not lose energy; it only loses *information* about what state it was in, before the collapse.

Just as in the proof of the equivalence of QMCWF to the OBEs, we can compute the density matrix evolution which this model gives rise to, by writing down an expression for $\rho(t)$, based on Eq. (7.27):

$$\rho(t) = (a|g\rangle + b\cos\theta|e\rangle)\left(\langle g|a^* + \langle e|b^* \cos\theta\right) + |b|^2 \sin^2\theta|e\rangle\langle e| \tag{7.28}$$

$$= \begin{bmatrix} |a|^2 & ab^* e^{-\lambda t} \\ a^* b e^{-\lambda t} & |b|^2 \end{bmatrix} . \tag{7.29}$$

Note that this is exactly the same evolution as we obtained for the random phase noise model, Eq. (7.24).

Despite the density matrix evolution being identical to that of the random phase model, the elastic collision model implies a different single-particle evolution trajectory. In contrast to the noisy Rabi oscillations previously seen, for the elastic collisions, the atomic state initially decays, then *jumps* into the $|e\rangle$ state at some random time; this is illustrated by the sample trajectory shown in Fig. 7.7.

Phase flips A third physical model for phase damping is based on quantum jumps. The Lindblad operator for phase damping is evidently

$$\mathcal{L}(\rho) = -\frac{1}{T_2}\left(|e\rangle\langle e|\rho|g\rangle\langle g| + |g\rangle\langle g|\rho|e\rangle\langle e|\right) . \tag{7.30}$$

Equivalently, it may be rewritten in standard form as

$$\mathcal{L}(\rho) = -\frac{1}{2}\left(C^\dagger C\rho + \rho C^\dagger C\right) + C\rho C^\dagger , \tag{7.31}$$

where

$$C = \frac{1}{\sqrt{2T_2}}\left(|g\rangle\langle g| - |e\rangle\langle e|\right) . \tag{7.32}$$

Note that $C^\dagger C$ is proportional to the identity, and thus $H = H_0$, such that no evolution occurs owing to this relaxation process except when a quantum jump occurs.

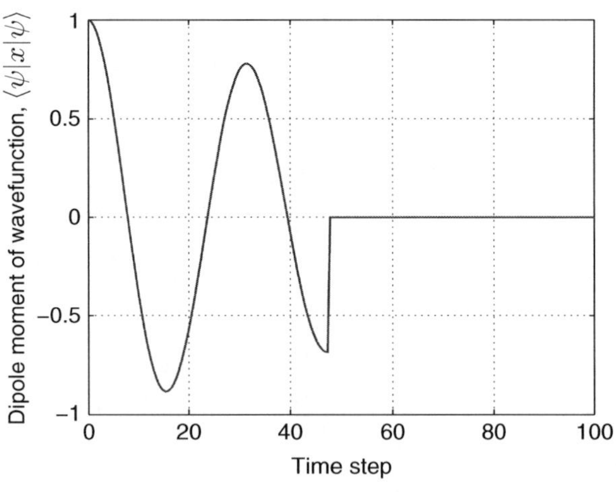

Fig. 7.7 Quantum Monte Carlo simulation of the evolution of two-level atom with noise from elastic collisions.

Moreover, the effect of a quantum jump is to flip the phase of the atom by -1, changing $|e\rangle \to -|e\rangle$.

If such a flip happens with probability $(1 - e^{-\lambda t})/2$ at time t, then the density matrix for this evolution is thus

$$\rho(t) = \frac{1+e^{-\lambda t}}{2}\left(a|g\rangle + b|e\rangle\right)\left(\langle g|a^* + \langle e|b^*\right) + \frac{1-e^{-\lambda t}}{2}\left(a|g\rangle - b|e\rangle\right)\left(\langle g|a^* - \langle e|b^*\right) \tag{7.33}$$

$$= \begin{bmatrix} |a|^2 & ab^*e^{-\lambda t} \\ a^*be^{-\lambda t} & |b|^2 \end{bmatrix}. \tag{7.34}$$

This is again the same density matrix dynamics as previously obtained for the random phase noise and elastic collision models. However, the trajectories of individual evolutions are different; at each moment in time, either the two-level atom remains *completely unchanged* or its excited state flips sign, inverting its dipole moment, as shown in Fig. 7.8

Discussion We have seen three models of phase damping, all of which produce the same density matrix evolution, but each of which has very different microscopic trajectories for individually evolving systems.

Which is correct?

The answer is that all of them are correct, and yet none are. Any of the three can be used for physical intuition and interpretation, but only as long as the only conclusions drawn depend on statistical averages. In fact, in the absence of control over the environment, *no experiment can distinguish between phase damping processes described by these three models, even in principle.* This strong statement arises from

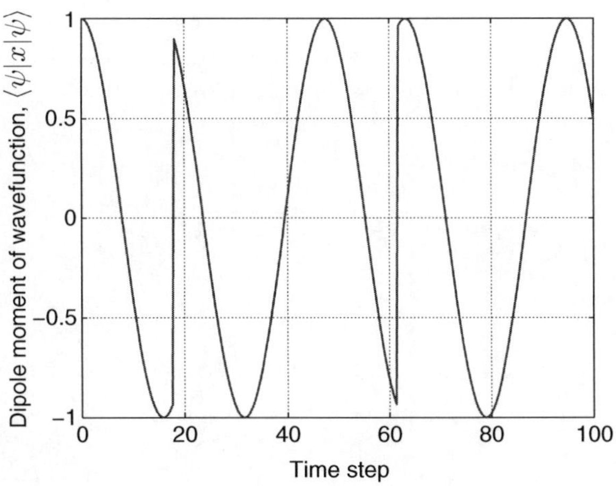

Fig. 7.8 Quantum Monte Carlo simulation of the evolution of excited two-level atom under phase flip noise model.

the fact that there are an infinite number of ways a (mixed) density matrix can be written as a statistical mixture of pure states; correspondingly, there are an infinite number of "unravelings" of density matrix time evolutions into statistical evolutions of pure-state wavefunctions.

The freedom of *interpretation* that arises in studying decoherence processes originates from a unitary degree of freedom. For example, in the QMCWF model, a gedanken experiment is performed in which the atom is allowed to decay and the emitted photon is captured. The evolution of the atom cannot depend on what measurement basis is used for the photodetection. This basis choice is a unitary transform, which can be chosen arbitrarily in the gedanken experiment, and different choices lead to the different unravelings of the master equation into trajectories.

7.2 Reversing quantum noise

Error correction builds on the idea of errors as being discrete events in time. In contrast, the traditional master-equation model of open quantum system dynamics is based on continuous-time evolution. We now build bridges between these two viewpoints, by introducing the *operator-sum representation* (OSR), a discrete-time model for open quantum system evolution. We examine how different unravelings of open quantum system dynamics arise due to the unitary degree of freedom in the operator-sum representation. We then show how to map back and forth between master equations and the OSR, then conclude with an example of how quantum error detection can be described with the OSR formalism.

7.2.1 The operator-sum representation

Here is an operational definition of the OSR that is simple and useful in analysis of experiments. Let the system, initially in state $|\psi\rangle$ (or, more generally, ρ), interact with an environment, initially in state $|e\rangle$, as shown in Fig. 7.9. The system and environment are initially separable. The interaction is modeled as a unitary operator U acting on the joint Hilbert space of system and environment, producing $U|e\rangle|\psi\rangle$ as the output.

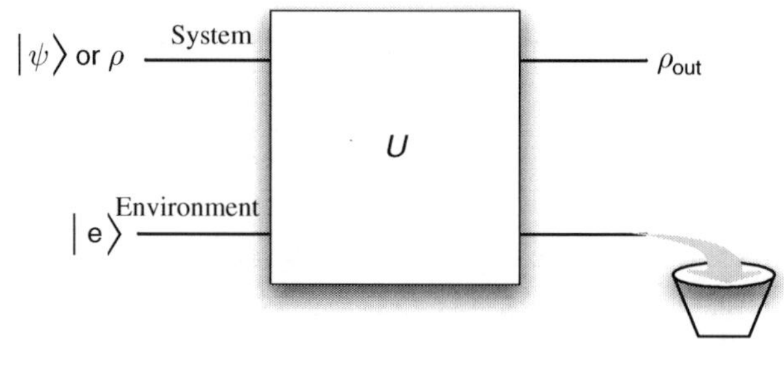

Fig. 7.9 System-environment interaction model used to define the operator-sum representation.

The resulting environment state is disregarded; this can be modeled by tracing over the environment, with an orthonormal basis of states $|e_k\rangle$.

The resulting system state ρ_{out} is thus a statistical mixture of pure states. We may denote this (using a somewhat nonstandard but pedagogically clear and useful notation) as

$$\rho_{\text{out}} = \langle e_0|U|e\rangle|\psi\rangle \oplus \langle e_1|U|e\rangle|\psi\rangle \oplus \cdots \tag{7.35}$$

$$= \bigoplus_k \langle e_k|U|e\rangle|\psi\rangle\,, \tag{7.36}$$

where "$\oplus$" means "or" and the density matrix is expressed as an unraveling into a statistical mixture of probabilistically weighted pure states.

The expression $\langle e_k|U|e\rangle$ may be identified as operators E_k, known as "operation elements," or more colloquially as "Kraus operators," which act on the Hilbert space of the system. In terms of the E_k, the output state is given as

$$\rho_{\text{out}} = \mathcal{E}(\rho) = \sum_k E_k \rho E_k^\dagger\,, \tag{7.37}$$

where we have substituted $\rho = |\psi\rangle\langle\psi|$ (it may be a mixed state, in general), and where it follows from the definition of E_k that

$$\sum_k E_k^\dagger E_k = \sum_k \langle e|U^\dagger|e_k\rangle\langle e_k|U|e\rangle \tag{7.38}$$

$$= I\,, \tag{7.39}$$

owing to the completeness of the basis $|e_k\rangle$.

Equation (7.37) is known as the operator-sum representation for $\mathcal{E}$. The OSR is defined by its operation elements. It has three defining properties:

(1) $\text{Tr}[\mathcal{E}(\rho)] = 1$ (probability-preserving).
(2) $\mathcal{E}(\sum_k p_k\rho_k) = \sum_k p_k\mathcal{E}(\rho_k)$ (convex and linear).
(3) if $\rho > 0$, then $\mathcal{E}(\rho) > 0$. Also, if $\rho = \text{Tr}_B(|\psi_{AB}\rangle)$ (i.e., is part of a composite system), then $(I \otimes \mathcal{E})(|\psi_{AB}\rangle) > 0$, where I acts on B and $\mathcal{E}$ acts on A (positive and completely positive).

More about these conditions, and their physical origins and interpretations, can be found in [3]. More important to us here are the facts that (A) any quantum operation $\mathcal{E}$ satisfying these three properties always has an operator sum representation, and (B) any set of operation elements E_k such that $\sum_k E_k^\dagger E_k = I$ is a valid OSR.

7.2.1.1 OSR examples

Since it is straightforward to check that $\sum_k E_k^\dagger E_k = I$, it is easy to construct sets of operation elements giving valid OSR models. Thus, we may quickly explore a wide range of examples of open quantum system dynamics, described in terms of the OSR.

Example 1

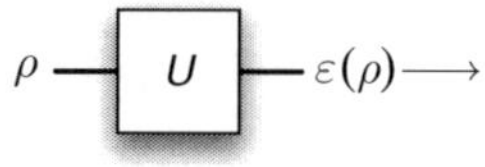

$\longrightarrow E(\rho) = U\rho U^\dagger.$

Example 2

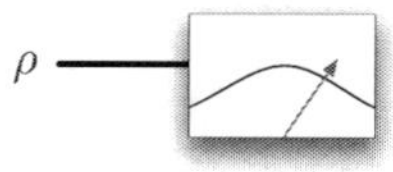

$\longrightarrow E(\rho) = \sum_k P_k \rho P_k,$

where P_k are the projectors onto the measurement basis. For example, the standard computational basis measurement has $P_0 = |0\rangle\langle 0|$ and $P_1 = |1\rangle\langle 1|$.

Example 3

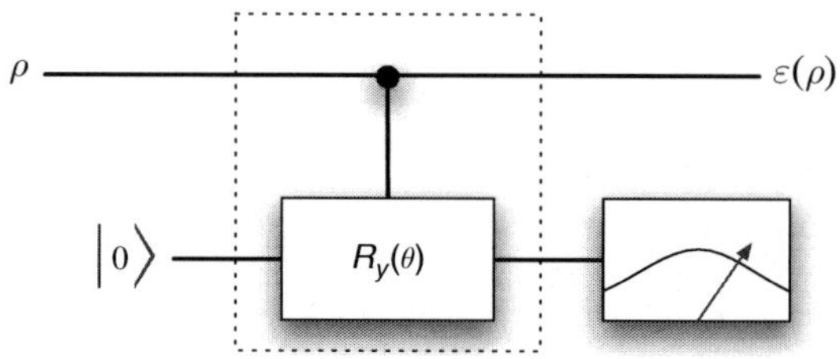

Here the single-qubit rotation is

$$R_y(\theta) = \begin{bmatrix} \cos(\theta/2) & -\sin(\theta/2) \\ \sin(\theta/2) & \cos(\theta/2) \end{bmatrix}, \tag{7.40}$$

so the *controlled* rotation is

$$U = \begin{bmatrix} 1 & 0 & . & . \\ 0 & \cos(\theta/2) & . & . \\ 0 & 0 & . & . \\ 0 & \sin(\theta/2) & . & . \end{bmatrix}, \tag{7.41}$$

where the first two rows and columns are labeled by $|00\rangle$, $|01\rangle$ with the assignment $|\text{env}, \text{sys}\rangle$; these are the only elements of the matrix that we are concerned with, because the environment starts in the $|0\rangle$ state. We can thus read off the operation elements for an operator-sum description of this quantum process:

$$E_0 = \begin{bmatrix} 1 & 0 \\ 0 & \sqrt{1-\lambda} \end{bmatrix}, \tag{7.42}$$

$$E_1 = \begin{bmatrix} 0 & 0 \\ 0 & \sqrt{\lambda} \end{bmatrix}, \tag{7.43}$$

where we have defined $\sqrt{1-\lambda} = \cos(\theta/2)$ just for convenience. The quantum operation is

$$\mathcal{E}(\rho) = E_0 \rho E_0^\dagger + E_1 \rho E_1^\dagger, \tag{7.44}$$

and to understand better what it does, let us apply it to a generic input density matrix $\rho = \begin{bmatrix} a & b \\ c & d \end{bmatrix}$. We find

$$\mathcal{E}(\rho) = \begin{bmatrix} a & \sqrt{1-\lambda}b \\ \sqrt{1-\lambda}c & d \end{bmatrix}. \tag{7.45}$$

This is very interesting, because if we identify $\sqrt{1-\lambda}$ as being e^{-t/T_2} (e.g., evolution for some fixed time interval t), then this quantum operation is immediately identifiable as being phase damping, with time constant T_2. In NMR, this process is known as transverse relaxation.

Physically, we may interpret E_0 as being the Hamiltonian evolution that happens in the absence of a quantum jump; since λ is typically small, this means that the evolution is largely unitary. Occasionally, however, with small probability of order λ, a quantum jump happens, which leaves the system in the state $|1\rangle$, annihilating $|0\rangle$, and in the process discarding any relative phase between $|0\rangle$ and $|1\rangle$. It is as if the qubit were being slightly measured, continuously in time. Eventually, the system collapses into a fixed point, which is a statistical mixture of $|0\rangle$ and $|1\rangle$, with probabilities that reflect the original state amplitudes. Specifically, for example, $a|0\rangle + b|1\rangle$ evolves under repeated phase damping to become $\rho = |a|^2|0\rangle\langle 0| + |b|^2|1\rangle\langle 1|$. This is exactly the process that we described as being "elastic collisions" in Section 7.1.5.

Example 4

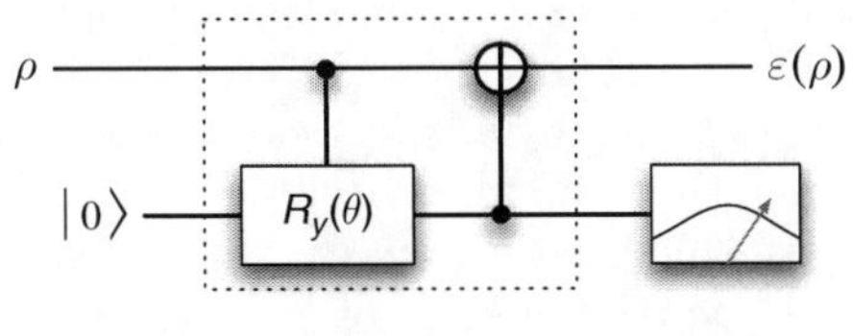

The operation elements for this system-environment model can be derived from those for the phase damping model, by observing that they are the same except when the environment is measured to be 1, when an extra X flip is applied to the system. Thus, the E_1 of phase damping (Eq. (7.43)) is replaced by

$$E_1 = X \begin{bmatrix} 0 & 0 \\ 0 & \sqrt{\lambda} \end{bmatrix} \tag{7.46}$$

$$= \begin{bmatrix} 0 & 1 \\ 1 & 0 \end{bmatrix} \begin{bmatrix} 0 & 0 \\ 0 & \sqrt{\lambda} \end{bmatrix} \tag{7.47}$$

$$= \begin{bmatrix} 0 & \sqrt{\lambda} \\ 0 & 0 \end{bmatrix}. \tag{7.48}$$

The operation elements are thus

$$E_0 = \begin{bmatrix} 1 & 0 \\ 0 & \sqrt{1-\lambda} \end{bmatrix}, \tag{7.49}$$

$$E_1 = \begin{bmatrix} 0 & \sqrt{\lambda} \\ 0 & 0 \end{bmatrix}, \tag{7.50}$$

Applying this to a generic input density matrix $\rho = \begin{bmatrix} a & b \\ c & d \end{bmatrix}$ (where $a = 1 - d$), we find

$$\mathcal{E}(\rho) = \begin{bmatrix} 1 - d(1-\lambda) & b\sqrt{1-\lambda} \\ c\sqrt{1-\lambda} & d(1-\lambda) \end{bmatrix}, \tag{7.51}$$

and, identifying $1 - \lambda = e^{-t/T_1}$, we obtain

$$\mathcal{E}(\rho) = \begin{bmatrix} 1 - de^{-t/T_1} & be^{-t/2T_1} \\ ce^{-t/2T_1} & de^{-t/T_1} \end{bmatrix}. \tag{7.52}$$

This is a process known as amplitude damping, or, in atomic physics, spontaneous emission, and, in NMR, longitudinal relaxation. Note how this derivation neatly shows that T_1 processes have an effective $T_2 = 2T_1$, in agreement with standard derivations from the OBEs.

7.2.2 The unitary degree of freedom in the OSR

We have seen how a set of operators $\{E_k\}$ satisfying $\sum_k E_k^\dagger E_k = I$ defines an OSR. However, they do not define it *uniquely*. In fact, in general, an infinite set of operation elements can define the same OSR. Why?

The reason is because the state of the system must be independent of the basis in which the environment is traced over. Consider the system-environment model, but now allow a new unitary transform U' to be performed on the state of the environment just before it is discarded (Fig. 7.10). $\mathcal{E}(\rho)$ must be independent of U', and thus there is a unitary degree of freedom in the OSR.

This unitary degree of freedom is explicitly captured in the following theorem.

Theorem *Let $\mathcal{E}(\rho) = \sum_k E_k \rho E_k^\dagger$ and $\mathcal{F}(\rho) = \sum_k F_k \rho F_k^\dagger$. Then $\mathcal{E} = \mathcal{F}$ iff $\exists$ a unitary U with elements u_{jk} such that $E_j = \sum_k u_{jk} F_k$.*

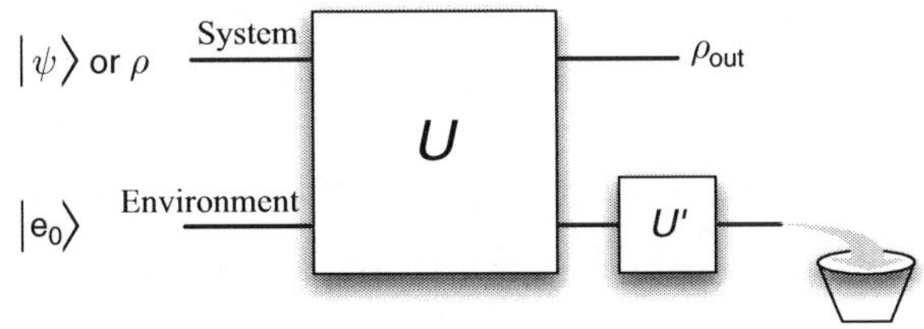

Fig. 7.10 Unitary degree of freedom in the definition of the operator-sum representation, due to the possibility in change of basis of the environment enacted by U'.

Phase damping unraveled in another basis A specific example of how this unitary degree of freedom is relevant to quantum error correction comes from studying phase damping. Recall that phase damping of a qubit is defined by the operation elements

$$E_0 = \begin{bmatrix} 1 & 0 \\ 0 & \sqrt{1-\lambda} \end{bmatrix}, \tag{7.53}$$

$$E_1 = \begin{bmatrix} 0 & 0 \\ 0 & \sqrt{\lambda} \end{bmatrix}, \tag{7.54}$$

Let

$$U = \begin{bmatrix} \sqrt{\alpha} & -\sqrt{1-\alpha} \\ \sqrt{1-\alpha} & \sqrt{\alpha} \end{bmatrix}. \tag{7.55}$$

This gives the new operation elements

$$F_0 = \sqrt{\alpha}E_0 + \sqrt{1-\alpha}E_1, \tag{7.56}$$

$$F_1 = -\sqrt{1-\alpha}E_0 + \sqrt{\alpha}E_1. \tag{7.57}$$

Now, if we choose $\alpha = (1+\sqrt{1-\lambda})/2$, so that $1-\alpha = (1-\sqrt{1-\lambda})/2$, then it turns out that

$$F_0 = \sqrt{\alpha} \begin{bmatrix} 1 & 0 \\ 0 & 1 \end{bmatrix} = \sqrt{\alpha} I\,, \tag{7.58}$$

$$F_1 = \sqrt{1-\alpha} \begin{bmatrix} 1 & 0 \\ 0 & -1 \end{bmatrix} = \sqrt{1-\alpha} Z\,. \tag{7.59}$$

This is a very different physical interpretation of phase damping!

Physically, it may be interpreted as a process in which, at every point in time, the system either stays exactly the same (F_0) or is modified by $|1\rangle \to -|1\rangle$ (i.e., F_1). This is the process we described in Section 7.1.5 as being random phase flips. The point is that this is also phase damping, and in fact is *experimentally indistinguishable from phase damping.* However, it is a very *useful* interpretation of phase damping. While the traditional model of phase damping as a continuous incremental process does not seem to admit a simple procedure for reversing errors, the phase flip model of phase damping does, because it happens to coincide with the binary idea of "there is an error" or "there is no error," used in the traditional conceptualization of error correction.

7.2.3 Equivalence between OSR and master equations

The dynamics of an open quantum system can be described equivalently by an OSR map or by a master equation of the Lindblad form. Here, we present an explicit conversion between the two forms, following a beautiful treatment of this given in the marvelous text *Exploring the Quantum*, by Haroche and Raimond [4].

Recall that a master equation for ρ in Lindblad form is the differential equation

$$\dot{\rho} = -i[H, \rho] + \sum_k L_k \rho L_k^\dagger - \frac{1}{2} L_k^\dagger L_k \rho - \frac{1}{2} \rho L_k^\dagger L_k \,. \tag{7.60}$$

This originates from a coarse-graining process, in which the approximation

$$\frac{d\rho}{dt} \approx \frac{\mathcal{E}(\rho) - \rho}{\tau} \tag{7.61}$$

is made, where τ parameterizes the strength of the interaction and $\mathcal{E}$ may be taken to be the map

$$\mathcal{E}(\rho) = \sum_k E_k \rho E_k^\dagger \tag{7.62}$$

given by the OSR of the same process.

Without loss of generality, we may say that

$$E_0 = I - iK\tau + O(\tau^2) \,, \tag{7.63}$$

where K is independent of τ, so that E_0 is the single dominant "order of unity" operation element[1] in $\mathcal{E}$. We want to identify E_0 with Hamiltonian evolution, but recognize that it also includes some anti-Hermitian evolution arising from the damping, as observed in Eq. (7.20), of the quantum Monte Carlo wavefunction technique. We thus extract those two pieces of evolution from K by defining

$$H = \frac{K + K^\dagger}{2} \,, \tag{7.64}$$

$$J = i\frac{K - K^\dagger}{2} \,, \tag{7.65}$$

where H is now the Hermitian Hamiltonian, J is the anti-Hermitian damping term, and $K = H - iJ$.

[1] If any other operation element is also of order unity (e.g., $E_1 = \alpha_1 I - iK_1\tau + O(\tau^2)$), then there exists a unitary transform by which E_1 can be combined with E_0 to produce a new E_0' of the form of Eq. (7.63), and a new $E_1' = O(\tau)$.

We now wish to express J in terms of just operation elements $\{E_k\}$; this may be done by studying the contribution of E_0 to the OSR. The contribution from the E_0 "order-of-unity" term to the OSR is, to leading order in τ,

$$E_0\rho E_0^\dagger \approx (I - iK\tau)\rho(I + iK^\dagger\tau) \tag{7.66}$$

$$\approx \rho - i(H\rho - \rho H)\tau - (J\rho + \rho J)\tau \tag{7.67}$$

$$\approx \rho - i\tau[H,\rho] - \tau(J\rho + \rho J)\,. \tag{7.68}$$

Recall that $\sum_k E_k^\dagger E_k = I$. Thus,

$$\sum_{k\neq 0} E_k^\dagger E_k = I - E_0^\dagger E_0 \tag{7.69}$$

$$\approx I - (I + iK^\dagger\tau)(I - iK\tau) \tag{7.70}$$

$$\approx -i(K^\dagger - K)\tau = 2J\tau\,. \tag{7.71}$$

It follows from this that

$$J = \frac{1}{2\tau}\sum_{k\neq 0} E_k^\dagger E_k = \frac{1}{2}\sum_{k\neq 0} L_k^\dagger L_k\,, \tag{7.72}$$

where we now make the identification $L_k = E_k/\sqrt{\tau}$ as being the jump operators for evolution during the discrete time interval determined by τ. This is justified by noting that $E_k\rho E_k^\dagger = O(\tau)$ for $k \neq 0$.

Finally, putting together all the pieces, we have

$$\mathcal{E}(\rho) = \sum_k E_k\rho E_k^\dagger \tag{7.73}$$

$$\approx \rho - i\tau[H,\rho] - \tau\sum_{k\neq 0}\left(\frac{1}{2}L_k^\dagger L_k\rho + \frac{1}{2}\rho L_k^\dagger L_k\right) + \tau\sum_{k\neq 0} L_k\rho L_k^\dagger\,. \tag{7.74}$$

Plugging this into the coarse-grained differential equation (7.61), we obtain

$$\dot{\rho} \approx -i[H,\rho] - \sum_{k\neq 0}\frac{1}{2}L_k^\dagger L_k\rho + \frac{1}{2}\rho L_k^\dagger L_k - L_k\rho L_k^\dagger\,, \tag{7.75}$$

where, to summarize, H is the Hermitian part of E_0/τ, $L_k\rho L_k^\dagger$ describes quantum jumps, and $L_k^\dagger L_k\rho + \rho L_k^\dagger L_k$ describes the anti-Hermitian part of E_0/τ.

7.2.4 Example: quantum error detection

The utility of the OSR for quantum error correction can be made apparent from a simple example, illustrating the process of quantum error detection.

Consider the action of a quantum operation $\mathcal{E}$ defined by operation elements $\{E_k\}$ on a Hilbert subspace C. It is helpful to distinguish four distinct behaviors:

- $E_k = U$: unitary rotation.
- $E_k = P$: projection.
- $E_k = \alpha P$: uniform shrinking (projection into subspace with scaling).
- $E_k \sim \begin{bmatrix} 1 & 0 \\ 0 & \sqrt{1-\lambda} \end{bmatrix}$: deformation.

The images $E_k \rho E_k^\dagger$ of each ρ may also collide with each other, meaning they are not in orthogonal subspaces for different k, or they may not.

Let us call the operation elements quantum *errors*. We say that a set of errors $\{E_k\}$ on a subspace C is *distinguishable* if $E_k C E_k^\dagger$ is disjoint for all k. In the case that we have a disjoint set of errors, the errors can be detected using an error detection code.

Example: Detection code for phase damping on two qubits Consider a two-qubit system, and let the Hilbert (sub)space C be defined by

$$C = \text{span}\,\{a(|00\rangle + |11\rangle) + b(|01\rangle + |10\rangle)\} \tag{7.76}$$

over the range of legal complex values of a and b. We shall call C the *code space.* Let E_{jk} be the operation elements for phase damping on two qubits. Recall for one qubit that phase damping may be described by the operation elements

$$E_0 = \sqrt{\alpha} I\,, \tag{7.77}$$
$$E_1 = \sqrt{1-\alpha} Z\,. \tag{7.78}$$

In terms of these, we may define the two-qubit operation elements as

$$E_{00} = E_0 \otimes E_0 = \alpha II\,, \tag{7.79}$$
$$E_{01} = E_0 \otimes E_1 = \sqrt{\alpha(1-\alpha)} IZ\,, \tag{7.80}$$
$$E_{10} = E_1 \otimes E_0 = \sqrt{\alpha(1-\alpha)} ZI\,, \tag{7.81}$$
$$E_{11} = E_1 \otimes E_1 = (1-\alpha) ZZ\,. \tag{7.82}$$

Note, however, that we shall leave out E_{11} in the set of errors that we wish to be able to detect; this is because it is a higher-order error, $O(\alpha)$ rather than $O(\sqrt{\alpha})$.

Now we work out what the effect of the error is on the code space C:

$$E_{00} \circ C = E_{00} C E_{00}^\dagger = \alpha^2 C\,, \tag{7.83}$$
$$E_{01} \circ C = E_{10} C E_{10}^\dagger = \alpha(1-\alpha)\left[a(|00\rangle - |11\rangle) - b(|01\rangle - |10\rangle)\right]\,, \tag{7.84}$$
$$E_{10} \circ C = E_{01} C E_{01}^\dagger = \alpha(1-\alpha)\left[a(|00\rangle - |11\rangle) + b(|01\rangle - |10\rangle)\right]\,. \tag{7.85}$$

Note that the case of no error, E_{00}, results in an image subspace that is orthogonal to both the subspaces from error E_{01} and from error E_{10}. Thus, there exists a projection operator that can distinguish these two cases from each other—that is, when an error has occurred, versus when it has not.

Example: Detection code for amplitude damping on two qubits Consider a two-qubit system, and let the Hilbert (sub)space C be defined by

$$C = \text{span}\{a|01\rangle + b|10\rangle\} \tag{7.86}$$

over the range of legal values of a and b. Let E_{jk} be the operation elements for amplitude damping on two qubits. Recall for one qubit that amplitude damping may be described by the operation elements

$$E_0 = \begin{bmatrix} 1 & 0 \\ 0 & \sqrt{1-\lambda} \end{bmatrix}, \tag{7.87}$$

$$E_1 = \begin{bmatrix} 0 & \sqrt{\lambda} \\ 0 & 0 \end{bmatrix}. \tag{7.88}$$

Just as for the phase damping example, we may define the two-qubit operation elements for amplitude damping as $E_{jk} = E_j \otimes E_k$, from which it follows that their action on C is

$$E_{00} \circ C = E_{00} C E_{00}^\dagger = \sqrt{1-\lambda}\,(a|01\rangle + b|10\rangle)\ , \tag{7.89}$$

$$E_{01} \circ C = E_{01} C E_{01}^\dagger = a\sqrt{\lambda}|00\rangle\,, \tag{7.90}$$

$$E_{10} \circ C = E_{10} C E_{10}^\dagger = b\sqrt{\lambda}|00\rangle\,. \tag{7.91}$$

Error detection is thus realized by checking to see if the state is $|00\rangle$. If so, then an error has occurred. Otherwise, it is likely that an error has not occurred, since the probability of E_{11} occurring is small compared with the probability of E_{01} or E_{10} occurring.

7.3 Quantum error correction

Given the discrete error model for open quantum systems provided by the OSR, we are now well positioned to describe in detail how quantum error correction works. We begin by reviewing concepts of classical coding, then extend this into a simple initial example of a rudimentary three-qubit quantum code. We show how operator measurement and error syndromes play an integral role in the correction procedure, then generalize to the nine-qubit code, which can correct for any single-qubit error. We end this section by presenting the quantum error correction criteria.

Portions of this section are excerpted from [5].

7.3.1 Classical coding

A classical code is defined as being a set of bit strings with certain properties. Specifically, we say that an $[n, k, d]$ code is a set of 2^k n-bit strings with minimum Hamming distance d. The Hamming distance $d(x, y)$ between bit strings x and y is the weight of (number of 1's in) $x \oplus y$, where $\oplus$ denotes bitwise addition modulo two.

Example The set of bit strings $\{000, 111\}$ is a $[3, 1, 3]$ code. If each of the three bits is sent through a communication channel, in which, with probability p, a bit is flipped, then on transmitting 000, the following bit strings are received:

RECEIVE	PROBABILITY	DECODE
000	$(1-p)^3$	0
001 010 100	$p(1-p)^2$	0
011 101 110	$p^2(1-p)$	1
111	p^3	1

Decoding is done by computing the majority vote of the bits received. With no coding (sending a single-bit message), the probability of error is p. Using the triple redundancy code, the probability of the decoded bit being wrong is $3p^2(1-p)+p^3 = 3p^2+2p^3 = O(p^2)$. To leading order, the error probability now depends quadratically instead of linearly on p; this is the key signature of error correction in action.

7.3.2 Quantum coding

The existence of a quantum analogue to classical error correction is perhaps one of the most surprising and fundamentally interesting results of quantum information. Before 1995, it was believed that the concept of error correction could not apply to quantum systems, for the following reasons:

- Quantum states collapse when measured.
- Errors are continuous.
- Quantum states cannot be cloned

However, it turns out there are ways around all of these objections. The part of the Hilbert space containing the quantum information to be preserved need not be measured; only the effect of the environment need be determined by a measurement. And using entangled states allow errors to be made orthogonal and distinguishable. Entanglement also replaces the role played by redundant copies in classical error correction.

These ideas can be illustrated by the three qubit bit-flip code.

3-qubit bit-flip code Suppose errors occur to a qubit as described by the quantum operation $\mathcal{E}(\rho) = (1-p)\rho + pX\rho X$, where p is a probability. This process is known as a "bit-flip" channel, because the qubit is left alone with probability $1-p$, and flipped by X with probability p. We will provide an encoding for a quantum state to allow bit-flip errors to be corrected.

Definition An $[[n,k]]$ quantum code C is a k-qubit subspace of an n-qubit Hilbert space.

Example For $k=1$, $n=3$, we can define logical zero as $|0_L\rangle = |000\rangle$ and $|1_L\rangle = |111\rangle$. Much like the three-bit code example in Section 7.3.1, we find that the effect of the quantum noise process (acting independently and identically on the three qubits in the code) performs the transformation

$$a|0_L\rangle + b|1_L\rangle \xrightarrow{\mathcal{E}} \begin{cases} a|000\rangle + b|111\rangle & (1-p)^3 \\ a|001\rangle + b|110\rangle & p(1-p)^2 \\ a|010\rangle + b|101\rangle & p(1-p)^2 \\ a|100\rangle + b|011\rangle & p(1-p)^2 \\ a|011\rangle + b|100\rangle & p^2(1-p) \\ a|101\rangle + b|010\rangle & p^2(1-p) \\ a|110\rangle + b|001\rangle & p^2(1-p) \\ a|111\rangle + b|000\rangle & p^3 \end{cases} \tag{7.92}$$

where the probabilities of each pathway are given in the column on the right. How can we determine what error occurred?

7.3.3 Operator measurement and error syndromes

Operator measurement Consider the quantum circuit shown in Fig. 7.11. Say that U has eigenvalues ± 1 and corresponding eigenvectors $|u_\pm\rangle$. Then, a single-qubit state $|\psi\rangle$ can be expressed as $c_0|u_+\rangle + c_1|u_-\rangle$. Plugging this state into the circuit gives the following sequence of transformations, following circuit elements from left to right:

$$(|0\rangle)(c_0|u_+\rangle + c_1|u_-\rangle) \tag{7.93}$$

$$\to \left(\frac{|0\rangle + |1\rangle}{\sqrt{2}}\right)(c_0|u_+\rangle + c_1|u_-\rangle) \tag{7.94}$$

$$\to \frac{|0\rangle(c_0|u_+\rangle + c_1|u_-\rangle)}{\sqrt{2}} + \frac{|1\rangle(c_0|u_+\rangle - c_1|u_-\rangle)}{\sqrt{2}} \tag{7.95}$$

$$\to \frac{|0\rangle + |1\rangle}{2}(c_0|u_+\rangle + c_1|u_-\rangle) + \frac{|0\rangle - |1\rangle}{2}(c_0|u_+\rangle - c_1|u_-\rangle) \tag{7.96}$$

$$= c_0|0\rangle|u_+\rangle + c_1|1\rangle|u_-\rangle\,. \tag{7.97}$$

Thus, when the measurement result is 0, the circuit outputs $|u_+\rangle$ and when 1, it outputs $|u_-\rangle$.

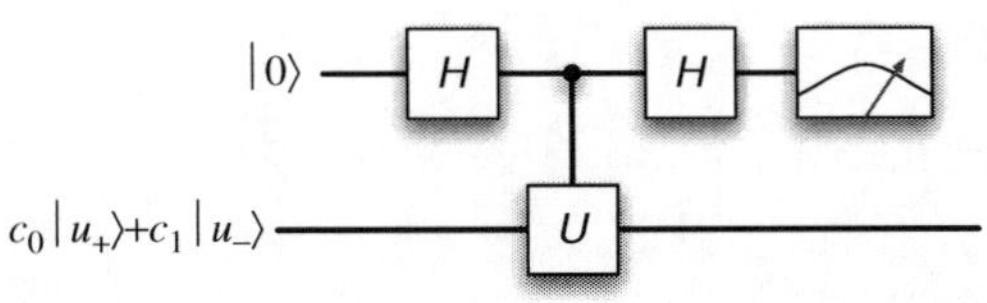

Fig. 7.11 Quantum circuit for measurement of an operator acting on a single qubit.

We can use this operator measurement procedure to determine the error caused by the environment, without measuring the encoded qubit. Returning to the state in Eq. (7.92), let us measure the two unitary operators

$$U_1 = Z_1 Z_2 = ZZI \qquad \text{and} \qquad U_2 = Z_2 Z_3 = IZZ\,. \tag{7.98}$$

These are operators on more than one qubit, but because they both have order 2, they have only two distinct eigenvalues, and can be measured by exactly the same circuit as in Fig. 7.11, albeit with more qubits entering and leaving the controlled-U operation. Considering just the four high-probability output states, we find

STATE	U_1 RESULT	U_2 RESULT	RECOVERY OPERATION
$a\|000\rangle + b\|111\rangle$	0	0	I
$a\|001\rangle + b\|110\rangle$	0	1	X_3
$a\|010\rangle + b\|101\rangle$	1	1	X_2
$a\|100\rangle + b\|011\rangle$	1	0	X_1

(7.99)

where the second and third columns give the measurement results (either 0 or 1) for U_1 and U_2, known as the *syndrome.* Note that none of the measurement results distinguish between a and b, and thus the encoded qubit is left intact. After doing this measurement, we can recover the initial state $a|0_L\rangle + b|1_L\rangle$ by applying the unitary transformation given in the fourth column.

The final output state ρ after performing this correction procedure is not perfect, because of the probability of two or more errors occurring. Quantitatively, the fidelity of the reconstructed state is

$$F(\rho, |\psi\rangle) = \sqrt{\langle\psi|\rho|\psi\rangle} = \sqrt{1 - 3p^2 + 2p^3}\,. \tag{7.100}$$

Interestingly, this scheme also corrects a small rotational error, even though the error is continuous! Let the error be described by the quantum operation

$$\mathcal{E}(\rho) = e^{-i\sqrt{\epsilon}X_1}\rho e^{i\sqrt{\epsilon}X_1}\,. \tag{7.101}$$

The rotation gate produces the output $|\psi'\rangle = R_{x_1}(2\sqrt{\epsilon})|\psi\rangle = |\psi\rangle - i\sqrt{\epsilon}X_1|\psi\rangle$, so the syndrome measurement result collapses this state into either $|\psi\rangle$ (with probability $\approx 1-\epsilon$) or $X_1|\psi\rangle$ (with probability $\approx \epsilon$). The fidelity of the resulting output, after recovery $\mathcal{R}$, is

$$F(\mathcal{R}(\mathcal{E}(|\psi\rangle)), |\psi\rangle) \approx \sqrt{1-\epsilon^2} \approx 1 - \frac{\epsilon^2}{2}\,. \tag{7.102}$$

In comparison, if no encoding had been done, then the output fidelity would have been $\approx 1 - \epsilon$. This reduction of error from $O(\epsilon)$ to $O(\epsilon^2)$ is a signature of the success of the error correction, even for continuous errors.

3-qubit phase-flip code A similar procedure can be used to correct phase-flip errors. Let $\mathcal{E}(\rho) = (1-p)\rho + pZ\rho Z$. Recall that $HZH = X$ and $HXH = Z$. Thus, $H\mathcal{E}(H\rho H)H = \mathcal{E}_{\text{bit flip}}$. Thus, we can correct phase-flip errors by the encoding $|0_L\rangle = |{+}{+}{+}\rangle$, $|1_L\rangle = |{-}{-}{-}\rangle$, where $|\pm\rangle \equiv (|0\rangle \pm |1\rangle)/\sqrt{2}$, and we perform the syndrome measurements XXI and IXX.

The reason phase-flip errors are significant and interesting is because of the following.

Claim Arbitrary qubit errors are a combination of bit-flip (X), phase-flip (Z), and bit-phase flip (XZ) errors.

Proof. sketch Recall that a general quantum operation can be written as $\mathcal{E}(\rho) = \sum_k E_k \rho E_k^\dagger$. Let ρ be a single qubit, and define $E_k = \sum_j c_{jk}\sigma_j$, where $\sigma_j \in \{I, X, Y, Z\}$. Then

$$\mathcal{E}(\rho) = \sum_{kjj'} c_{kj}{c^*}_{kj'}\sigma_j \rho \sigma_{j'} \tag{7.103}$$

$$= \sum_{kl} \chi_{jl}\sigma_j \rho \sigma_l\,, \tag{7.104}$$

where $\chi_{jl} = \sum_k c_{kj}c^*_{kl}$ is a matrix of complex numbers. □

For example, for $R_x(2\sqrt{\epsilon})|\psi\rangle \approx |\psi\rangle - i\sqrt{\epsilon}X|\psi\rangle$, we have $\mathcal{E}(\rho) = \rho - i\sqrt{\epsilon}(X\rho - \rho X) + \epsilon X\rho X$. The syndrome measurement then projects the environment into definite error states, removing the off-diagonal terms in this expression.

7.3.4 Nine-qubit code

A combination of the bit-flip and phase-flip codes allows correction of *any* single-qubit error. This code encodes a single qubit as

$$|0_L\rangle = \frac{(|000\rangle + |111\rangle)^{\otimes 3}}{\sqrt{8}}\,, \tag{7.105}$$

$$|1_L\rangle = \frac{(|000\rangle - |111\rangle)^{\otimes 3}}{\sqrt{8}}\,. \tag{7.106}$$

The syndrome measurement operators are

$$Z_1Z_2 \quad Z_2Z_3 \quad Z_4Z_5 \quad Z_5Z_6 \quad Z_7Z_8 \quad Z_8Z_9 \tag{7.107}$$

for bit-flip errors and

$$X_1X_2X_3X_4X_5X_6 \qquad X_4X_5X_6X_7X_8X_9 \tag{7.108}$$

for phase-flip errors. It is simple to see that any single-qubit phase-flip or bit-flip or bit–phase-flip error gives a unique syndrome measurement result, by direct examination.

7.3.5 QEC criteria

So far, we have given examples of specific quantum codes constructed by analogy from classical codes. More generally, there is a set of criteria that all quantum codes must satisfy to be able to correct for errors caused by a given quantum operation. Let the error (e.g., in a quantum channel) be described by $\mathcal{E}(\rho) = \sum_k E_k \rho E_k^\dagger$.

Theorem *Let C be a quantum code defined by the orthonormal states $\{|\psi_\ell\rangle\}$. There exists a quantum operation $\mathcal{R}$ correcting $\mathcal{E}$ on C iff the following two conditions are satisfied:*

- *Orthogonality: $\langle\psi_\ell|E_j^\dagger E_k|\psi_\ell\rangle = 0$, $\forall j \neq k$, $\forall \ell$.*
- *Nondeformation: $\langle\psi_\ell|E_k^\dagger E_k|\psi_\ell\rangle = d_k$, $\forall \ell$.*

The first condition stipulates that orthogonal basis states remain orthogonal, and the second that code space basis vectors stay of equal length. The important observation is that these are a set of algebraic conditions that allow a quantum code to be found, in principle, for any error process. Of course, the size of the code space required may be nontrivial, so it is not efficient to search randomly for solutions.

Proof. Let $P = \sum_\ell |\psi_\ell\rangle\langle\psi_\ell|$ be the projector onto the code space C. We shall assume the condition $\langle\psi_m|E_j^\dagger E_k|\psi_\ell\rangle = 0$; this is stronger than the orthogonality condition, but captures the physically reasonable expectation that errors are not so severe that they change one codeword to another. Note that

$$PE_k^\dagger E_j P = d_k \delta_{jk} P\,. \tag{7.109}$$

The polar decomposition gives $E_k P = U_k \sqrt{PE_k^\dagger E_k P} = \sqrt{d_k} U_k P$, where U_k is unitary.

First, construct the syndrome measurement. Let

$$P_k = U_k P U_k^\dagger = \frac{E_k P U_k^\dagger}{\sqrt{d_k}} = \frac{U_k P E_k^\dagger}{\sqrt{d_k}}\,. \tag{7.110}$$

By Eq. (7.109), the P_k are orthogonal, for $k \neq j$:

$$P_k P_j \propto U_k P E_k^\dagger E_j P U_j^\dagger = 0\,. \tag{7.111}$$

So let us measure P_k and obtain k as the syndrome measurement result.

Second, perform a recovery operation to undo the error, using $U_k^\dagger$. Let

$$\mathcal{R}(\rho) = \sum_k U_k^\dagger P_k \rho P_k U_k \tag{7.112}$$

be the recovery operation. Note that for $|\psi\rangle \in C$,

$$U_k^\dagger P_k E_j|\psi\rangle = \frac{1}{\sqrt{d_k}} U_k^\dagger U_k P E_k^\dagger E_j P|\psi\rangle \tag{7.113}$$

$$= \sqrt{d_k}\,\delta_{jk}|\psi\rangle\,. \tag{7.114}$$

Thus,

$$\mathcal{R}(\mathcal{E}(|\psi\rangle)) = \mathcal{R}\left(\sum_j E_j|\psi\rangle\langle\psi|E_j^\dagger\right) \tag{7.115}$$

$$= \sum_{kj} U_k^\dagger P_k k E_j|\psi\rangle\langle\psi|E_j^\dagger P_k^\dagger U_k \tag{7.116}$$

$$= \sum_{kj} d_k \delta_{jk}|\psi\rangle\langle\psi| \tag{7.117}$$

$$= |\psi\rangle\langle\psi|\sum_k d_k\,, \tag{7.118}$$

which is proportional to the original state, as desired. □

7.4 Modern quantum codes

The central importance of quantum error correction lies not merely in its possibility, but, more crucially, in the existence of efficiently describable families of *good* quantum codes. These are good because they encode a number k of logical qubits that is a nonvanishing proportion of the number n of physical qubits employed, as $n \to \infty$. The most important family of such quantum codes are known as *stabilizer* codes, discovered and defined by Gottesman [6]. We begin our study of these codes by defining stabilizer states, then exploring stabilizer codes and stabilizer code criteria. We then define the CSS family of codes, which are stabilizer codes derived from classical codes, before ending by touching on the family of truly quantum code-word stabilized quantum codes.

7.4.1 Stabilizer states

Let us define the Pauli group G_n on n qubits as being all the n-fold tensor products of $\{X, Y, Z\}$ (recall that these are the Pauli matrices) and $\{\pm 1, \pm i\}$. For example, for $n = 1$, $G_1 = \{X, Y, Z, I, -X, \ldots\}$. Note that for $g, h \in G_n$, either $gh = hg$ (they commute) or $gh = -hg$ (they anticommute). When the meaning of n is implied, we may write G instead of G_n.

A *stabilizer* for the Hilbert space spanned by basis states $V_s \doteq \{|\psi_\ell\rangle\}$ (where $\doteq$ means "spanned by") is defined as the set

$$S = \{g \in G \mid g|\psi\rangle = |\psi\rangle,\ \forall\ |\psi\rangle \in V_s\}\,. \tag{7.119}$$

In order for S to be nontrivial, we stipulate that $-I \notin S$. Note that S is Abelian. More properties of and theorems applying to stabilizers and stabilizer states are presented in [3]. Here, we summarize some essential ideas, and go through some helpful examples.

Example 1 $V_s \doteq \{|00\rangle\} \longrightarrow S = \{ZZ, IZ, ZI, II\}$. Note that $ZZ = Z \otimes Z$; we often suppress the tensor product in this notation. S may also be written in terms of its generators, as $S = \langle ZI, IZ\rangle$.

Example 2 $V_s \doteq \{(|00\rangle + |11\rangle)/\sqrt{2}\} \longrightarrow S = \{XX, ZZ, II, -YY\} = \langle XX, ZZ\rangle$. Note that the dimension of V_s is 2^k, where $k = n -$ (min #generators of S).

Example 3 $S = \{X, Y\} \longrightarrow$ What is V_s?

Example 4 $V_s = \{|000\rangle, |111\rangle\} \longrightarrow S = \langle IZZ, ZZI\rangle$.

Example 5 $V_s = \{(|000\rangle + |111\rangle)^{\otimes 3}, (|000\rangle - |111\rangle)^{\otimes 3}\} \longrightarrow n = 9$, $k = 1$, minimum # generators $= 8$.

Example 6 $S = \langle XX\rangle \longrightarrow V_s = \{|00\rangle + |11\rangle, |01\rangle + |10\rangle\}$.

Example 7 $V_s = \{|1\rangle\} \longrightarrow S = \langle -Z\rangle$.

Example 8 $V_s = \{|0\rangle\} \longrightarrow S = \langle Z\rangle$.

Example 9 $V_s = \{(|01\rangle + |10\rangle)/\sqrt{2}, |11\rangle\} \longrightarrow S = ?$

7.4.2 Stabilizer codes and criteria

An $[[n, k]]$ stabilizer code $C(S)$ may be defined as the vector space stabilized by $S \subset G_n$, where $S = \langle g_1, g_2, \cdots, g_{n-k}\rangle$.

Theorem *Given errors $\{E_a\} \in G_n$, if $\exists g \in S$ such that $\forall E = E_a^\dagger E_b$ $(a \neq b)$, $Eg = -gE$, then $\{E_a\}$ can be corrected by the code $C(S)$.*

Proof. The theorem may be proved by showing that the criterion that errors anticommute with at least one stabilizer generator implies the orthogonality and nondeformation conditions of the QEC criteria discussed in Section 7.3.5.

For orthogonality, we desire that $\forall|\psi\rangle \in C(S)$, $\langle\psi|E_a^\dagger E_b|\psi\rangle = 0$ when $a \neq b$. Let g be the generator that anticommutes with $E = E_a^\dagger E_b$. Then

$$\langle\psi|E|\psi\rangle = \langle\psi|Eg|\psi\rangle \tag{7.120}$$
$$= -\langle\psi|gE|\psi\rangle \tag{7.121}$$
$$= -\langle\psi|E|\psi\rangle \tag{7.122}$$
$$= 0\,, \tag{7.123}$$

so the orthogonality requirement is satisfied.

For nondeformation, we desire that $\forall|\psi\rangle \in C(S)$, $\langle\psi|E_a^\dagger E_a|\psi\rangle$ is constant and independent of $|\psi\rangle$. For stabilizer codes, $E_a^\dagger E_a = I$, so this is manifestly satisfied. □

Stabilizer codes also admit a nice means to obtain the error syndrome. For $S = \langle g_1, g_2, \cdots, g_{n-k}\rangle$ and error E, the error syndrome is a k-bit string s where $s_j = 0$ if $Eg_j = g_jE$, and $s_j = 1$ otherwise. Each error has a unique syndrome for a "nondegenerate" code.

7.4.2.1 Stabilizer code examples

Example 1 $S = \langle IZZ, ZZI\rangle \longrightarrow E = IXI$ has syndrome 11, but $E = XXX$ is not an error, because it does not anticommute with any stabilizer generator.

Example 2 $S = \langle XX \rangle \longrightarrow C(S) = \{|00\rangle + |11\rangle, |01\rangle + |10\rangle\}$. This code may seem like it could correct errors ZI and IZ, but, in fact, the product of those two errors, ZZ, commutes with S, and thus it cannot. It can *detect* phase-flip errors, but not correct them. This is the code that we studied in Section 7.2.4.

Example 3

$$S = \left\langle \begin{matrix} \mathtt{XZIZ} \\ \mathtt{ZXZI} \\ \mathtt{IZXZ} \\ \mathtt{ZIZX} \end{matrix} \right\rangle \longrightarrow n = 4,\ k = 0.$$

This code can "correct" any single qubit error, but in fact it encodes no qubits.

Example 4

$$S = \left\langle \begin{matrix} \mathtt{IIIXXXX} \\ \mathtt{IXXIIXX} \\ \mathtt{XIXIXIX} \\ \mathtt{IIIZZZZ} \\ \mathtt{IZZIIZZ} \\ \mathtt{ZIZIZIZ} \end{matrix} \right\rangle \longrightarrow n = 7 \text{ and } k = 1.$$

This code can correct any single qubit error (it is easy to check that any single-qubit Pauli operator anticommutes with at least one of these generators). There are $2^6 = 64$ possible syndrome values, but only $22 = 7 \cdot 3 + 1$ possible errors, so some other multi-qubit errors are also correctable. This code is known as the 7-qubit Steane code.

Example 5

$$S = \left\langle \begin{matrix} \mathtt{XZZXI} \\ \mathtt{IXZZX} \\ \mathtt{XIXZZ} \\ \mathtt{ZXIXZ} \end{matrix} \right\rangle \longrightarrow n = 5 \text{ and } k = 1.$$

This code can correct any single-qubit error (it is easy to check that any single-qubit Pauli operator anticommutes with at least one of these generators). There are $2^4 = 16$ possible syndrome values, and $16 = 5 \cdot 3 + 1$ possible errors, and since these two match exactly, the code is known as being a *perfect* code. It is the smallest possible perfect quantum code.

7.4.2.2 Stabilizer code distance

Let $\mathrm{wt}(E)$ be the number of nonidentity Pauli operators in the n-qubit operator $E \in G_n$. A stabilizer code of distance d has the property that for all $E \in G_n$ of weight $\mathrm{wt}(E) < d$, either $E \in S$, or $\exists g \in S$ such that $gE = -gE$.

A distance $d = 2t + 1$ code can correct t errors.

7.4.3 Calderbank-Shor-Steane codes

The stabilizer codes formalism is very neat and simple to employ. However, what we wish is to efficiently and systematically describe an entire family of quantum codes, with known $[[n,k]]$ (and, better yet, $[[n,k,d]]$, for a good measure of distance d) parameters. One effective approach to this goal is to build on the vast knowledge of good classical codes. Given families of $[n,k,d]$ classical codes, how can one construct families of $[[n,k,d]]$ quantum codes?

The CSS codes, discovered separately by Robert Calderbank and Peter Shor [7] and by Andrew Steane [8], accomplish this by combining two classical codes to create one quantum code. One classical code, C_1, corrects for bit-flip errors and the other, C_2, is used to correct for phase-flip errors. The combination thus allows any single-qubit error to be corrected. The key is that C_1 and C_2 must have a specific relationship to each other, in order that they do not conflict when combined to produce the single quantum code. CSS codes are stabilizer codes, and the essence of this restriction is that the stabilizer generators produced from C_1 and from C_2 must all commute with each other, while keeping the code space size the same. We describe how this works, beginning with a very brief review of classical codes.

Classical linear codes Recall that a $[n,k,d]$ code C encodes k bits in n bits with minimum Hamming distance d. The $n \times k$ generator matrix G for C is defined such that it encodes an input bitstring $\vec{x}$ into a codeword $\vec{v} = G\vec{x}$. Note that we use vector notation to denote classical bitstrings in this section; by convention, these will be column vectors. Columns of G are basis codewords. For example, if

$$G = \begin{bmatrix} 1 \\ 1 \\ 1 \end{bmatrix}, \tag{7.124}$$

then

$$G[0] = \begin{bmatrix} 0 \\ 0 \\ 0 \end{bmatrix} \tag{7.125}$$

is an encoded 0 and

$$G[1] = \begin{bmatrix} 1 \\ 1 \\ 1 \end{bmatrix} \tag{7.126}$$

is an encoded 1.

The parity check matrix H for C is defined to be an $(n-k) \times n$ matrix such that $H\vec{v} = 0$ for all $\vec{v} \in C$. Equivalently, $HG = 0$. For example, for the triple-redundancy code,

$$H = \begin{bmatrix} 1 & 1 & 0 \\ 0 & 1 & 1 \end{bmatrix}. \tag{7.127}$$

Rows of H are orthonormal to columns of G. H is particularly useful in decoding data and determining the error syndrome. If an error $\vec{e}$ occurs to $\vec{v}$, producing $\vec{v}+\vec{e}$, then $H(\vec{v}+\vec{e}) = H\vec{e}$ is the syndrome for error $\vec{e}$.

Note that we may think of codewords as being constraints on the columns of H. For example,

$$\begin{bmatrix} 1 & 1 & 0 \\ 0 & 1 & 1 \end{bmatrix} \begin{bmatrix} 1 \\ 1 \\ 1 \end{bmatrix} = 0\,, \tag{7.128}$$

showing that the logical 1 of the triple-redundancy code enforces a constraint that the sum of each of the columns of H is zero. It thus follows that if C has distance d, then H has at least d linearly independent columns.

The Hamming code is a well-known family of classical linear codes, parameterized by an integer r. For $r \geq 2$, the code is determined by its parity check matrix H, which has $2^r - 1$ columns. The columns are the bitstring representations of the integers 1 to $2^r - 1$, from left to right. For example, when $r = 2$, the triple-redundancy code is obtained. When $r = 3$, the parity check matrix is

$$H = \begin{bmatrix} 0 & 0 & 0 & 1 & 1 & 1 & 1 \\ 0 & 1 & 1 & 0 & 0 & 1 & 1 \\ 1 & 0 & 1 & 0 & 1 & 0 & 1 \end{bmatrix}. \tag{7.129}$$

The generator matrix is

$$G = \begin{bmatrix} 1\,1\,1\cdots \\ H \end{bmatrix}^T = \begin{bmatrix} 1 & \\ 1 & \\ 1 & H^T \\ 1 & \\ \vdots & \end{bmatrix}. \tag{7.130}$$

For $r = 3$, this is

$$G = \begin{bmatrix} 1 & 0 & 0 & 1 \\ 1 & 0 & 1 & 0 \\ 1 & 0 & 1 & 1 \\ 1 & 1 & 0 & 0 \\ & & \vdots & \end{bmatrix}. \tag{7.131}$$

In the case $r = 3$, the code parameters may be recognized to be $[7, 4, 3]$. For general r, the code parameters are $[2^r - 1, 2^r - 1 - r, 3]$. What is nice about this family of codes is that the *rate* $R = k/n$ is $1 - r/(2^r - 1)$, which is the highest possible for distance-3 codes of $n = 2^r - 1$ bits.

For a code C, the generator G and parity check matrix H are kinds of mirror images of each other, in the sense that G creates codewords and H annihilates them. If their roles are flipped, then a new "dual" code $C^\perp$ can be constructed, whose codewords are orthogonal to those of the original code C. For $C^\perp$, the generator is $G^\perp = H^T$, and the parity check matrix is $H^\perp = G^T$.

The CSS construction The CSS code is a quantum code constructed from two classical linear codes, C_1 and C_2. Let C_1 be an $[n, k_1]$ code, and let C_2 be an $[n, k_2]$ code, such that $C_2 \subset C_1$, and C_1 and $C_2^\perp$ both correct t errors (a code of distance d corrects $(d-1)/2$ errors, for d odd). The code CSS(C_1, C_2) is defined by the stabilizer

$$S = \left\langle X[G_2^T], Z[H_1] \right\rangle , \tag{7.132}$$

where $X[\cdot]$ means the set of Pauli group elements given by taking the rows of the argument and replacing 0 with I and 1 with X, and similarly $Z[\cdot]$ maps 0 to I and 1 to Z. And G_2 is the generator of C_2, and H_1 is the parity check matrix of C_1.

We claim that CSS(C_1, C_2) corrects t single-qubit errors.

First, let us prove that S is a valid stabilizer. For this, we need all the elements of $\{X[G_2^T], Z[H_1]\}$ to commute with each other. It is clear by construction that the elements of $X[G_2^T]$ commute with each other, since the rows of G_2^T are basis codewords, and we are working with linear codes. The rows of $Z[H_1]$ similarly commute with each other. Moreover, the rows of $X[G_2^T]$ and $Z[H_1]$ commute with each other, since $G_2^T H_1^T = (H_1 G_2)^T = 0$.

Next, recall that a stabilizer code can correct an error $E \in G_n$ if E anticommutes with at least one stabilizer generator element. Suppose E has weight t, meaning that it has t nonidentity elements. Then CSS(C_1, C_2) corrects E if and only if one of the elements of $\{X[G_2^T], Z[H_1]\}$ anticommutes with it.

Suppose the errors in E are all X errors (i.e., bit flips), at positions given by bitstring $\vec{e}_1$. E will commute with all $\{X[G_2^T]\}$, but will anticommute with at least one row of $\{Z[H_1]\}$ if $H_1\vec{e}_1 \neq 0$. Since C_1 corrects t errors, then $H_1\vec{e}_1 \neq 0$ if $\vec{e}_1$ has weight no more than t. Thus, CSS(C_1, C_2) corrects up to t bit-flip qubit errors.

Similarly, we may suppose the errors in E are all Z errors (i.e., phase flips), at positions given by bitstring $\vec{e}_2$. E will commute with all $\{Z[H_1]\}$, but will anticommute with at least one row of $\{X[G_2^T]\}$ if $G_2^T\vec{e}_2 \neq 0$. Note that G_2^T is the parity check matrix of $C_2^\perp$, and in the CSS construction it is stipulated that $C_2^\perp$ correct at least t errors. Thus, it follows that CSS(C_1, C_2) corrects up to t phase-flip qubit errors.

7.4.4 Graph states and codeword-stabilized quantum codes

Stabilizer codes, and especially CSS codes, are probably the most important family of codes to understand, because of their use in fault-tolerant quantum computation constructions. However, there are limitations to stabilizer codes. For example, it is known that they are not the most efficient codes, in terms of the ratio of the size of the encoded space to the size of the Hilbert space used by the code. It is also known that stabilizer codes admit no universal set of naturally fault-tolerant "transversal" gates, without departing the code space; this therefore substantially complicates fault-tolerant quantum computation constructions. Finally, stabilizer codes are inherently built on an Abelian group structure (in coding theory terminology, they are known as *additive* codes). There is no fundamental reason why quantum codes need be restricted to such Abelian constructions.

A very interesting and promising new approach to quantum codes, known as *codeword-stabilized* quantum codes (CSSZ09), is based on combining quantum states and classical codes to produce quantum codes. Here is a very brief tour through the basic ideas behind these codes, beginning with graph states and graph codes.

Graph states The graph states are an interesting and important subset of stabilizer states, defined by having a stabilizer generator set $\{g_k\}$ with three properties:

- First, the number of generators n equals the number of qubits acted upon. Thus, the stabilized vector space has $k = 0$ (it is a single state).
- Second, $(g_k)_k = X$. In other words, if the generators are presented as an $n \times n$ matrix of Pauli operators, then the diagonal elements are X operators.
- Third, $(g_k)_j \in \{Z, I\}$ and $(g_k)_j = (g_j)_k$ for all $j \neq k$. In other words, if the generators are presented as an $n \times n$ matrix of Pauli operators, then all the off-diagonal elements are either I or Z, and they are symmetric about the diagonal.

The *adjacency matrix* A for a graph state is given by the $n \times n$ binary-valued matrix with elements

$$A_{kj} = \begin{cases} 1 & \text{if } (g_k)_j = Z\,, \\ 0 & \text{otherwise}\,. \end{cases} \tag{7.133}$$

Note that A is symmetric, and has zeros along its diagonal.

Despite graph states being a subset of stabilizer states, it turns out that all stabilizer states are equivalent to graph states under conjugation by local Clifford operators. The local Clifford group is generated by the Hadamard operation H, and the phase gate S, and it is helpful to recall how these act by conjugation on Pauli operators:

	H	S
X	Z	Y
Y	-Y	-X
Z	X	Z

Any stabilizer generator set $\mathcal{S}$ (of full rank) can thus be transformed into a graph state generator set $G(\mathcal{S})$ by suitably relabeling qubits, recombining generators, and conjugating operators acting on some of the qubits by H and S.

Graph codes A *graph code* $C(G, \{s^1, s^2, \cdots\})$ is defined by an n-qubit graph state $|G\rangle$ and a set of n-bit strings $\{s^1, s^2, \cdots\}$. The code is the vector space spanned by $\{|G_{s^1}\rangle, |G_{s^2}\rangle, \cdots\}$, where

$$|G_s\rangle = \prod_{k=0}^{n-1} (Z_k)^{s_k} |G\rangle\,, \tag{7.134}$$

s_k is the kth bit of s, and Z_k is the Pauli Z operator acting on the kth qubit.

For example, we may construct a graph code defined by $C(G, \{0000, 0011, 1100, 1111\})$ with G being a square (Fig. 7.12). The four codewords for this code are $|G\rangle$, $IIZZ|G\rangle$, $\texttt{ZZII}|G\rangle$, and $\texttt{ZZZZ}|G\rangle$.

Here is another interesting example, known as the kite graph code. Let the 6-qubit graph code $C(G, \{000000, 011100\})$ have G as this "kite" graph (Fig. 7.13). Note that

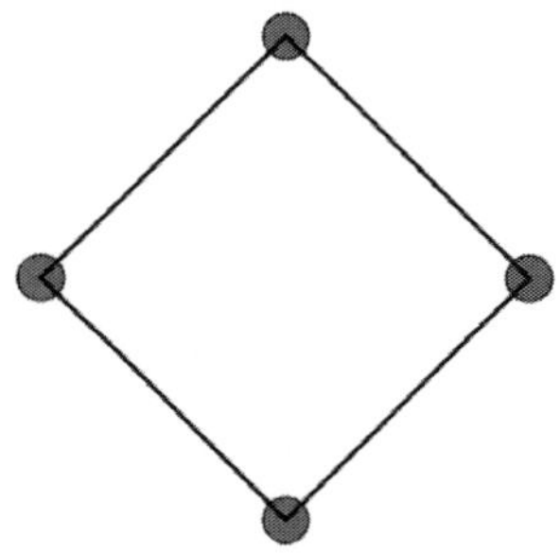

Fig. 7.12 Square graph.

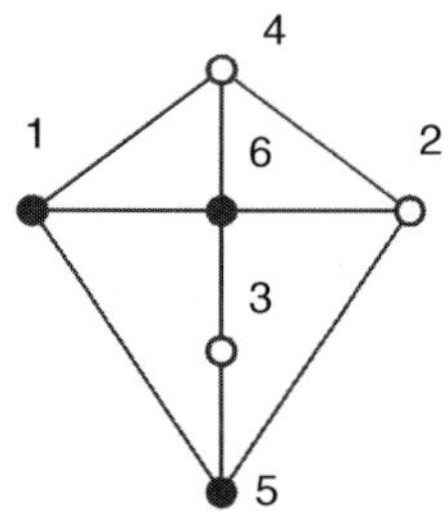

Fig. 7.13 Kite graph.

the nodes with the Z operations applied, corresponding to the 1's in 011100, are the nodes labeled 2, 3, 4 in the diagram above. Thus, if we number the qubits starting from 1 on the left of each Pauli operator string, the codewords are $|G\rangle$ and $\texttt{IZZZII}|G\rangle$.

Graph state construction circuits Graph states can be constructed using simple quantum circuits U_{graph} that act upon a set of qubits $|0\rangle^n$ using Hadamard and controlled-Z gates. First $|+\rangle$ states are created for each qubit (node), then controlled-Z gates are applied to every pair of qubits linked by an edge in the graph. This gives

$$U_{\text{graph}} = U_{\text{edges}} H^{\otimes n}. \tag{7.135}$$

Graph code construction circuits For quantum error correction, however, it is valuable to go further, and have a systematic method for constructing quantum circuits U_{enc} that act on an arbitrary input qubit, plus ancilla qubits, and produces an encoded graph code state, that is,

$$U_{\text{enc}} \left[(a|0\rangle + b|1\rangle) \otimes |0\rangle^{n-1} \right] = a|0_L\rangle + b|1_L\rangle \,, \tag{7.136}$$

where $|0_L\rangle$ and $|1_L\rangle$ are the graph code codeword states.

Consider the 5-qubit graph code $C(G, \{00000, 11111\})$, where G is a pentagon, as shown in Fig. 7.14. The two codewords are $|0_L\rangle = |G\rangle$ and $|1_L\rangle = \texttt{ZZZZZ}|G\rangle$.

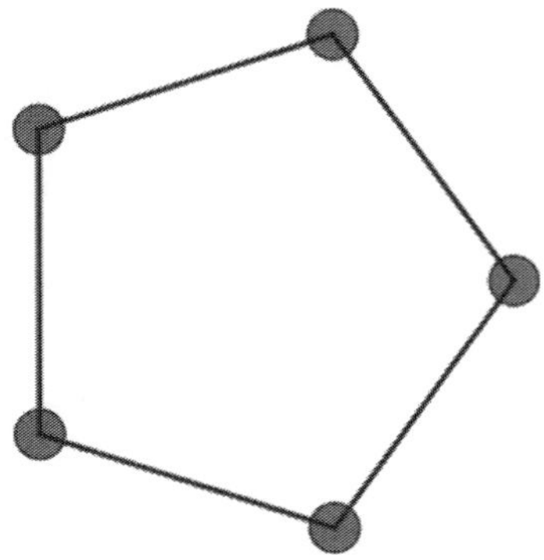

Fig. 7.14 Pentagon graph.

An encoding circuit for this code may be constructed in two parts. Let $U_{\text{enc}} = U_{\text{graph}} U_{\text{prep}}$, where $U_{\text{graph}} = U_{\text{edges}} H^{\otimes n}$ is the standard circuit for creating $|G\rangle$. Since single-qubit Z gates commute with controlled-Z gates, and $ZH = HX$, it follows that

$$
\begin{aligned}
a|0_L\rangle + b|1_L\rangle &= a|G\rangle + b\mathtt{ZZZZZ}|G\rangle & (7.137)\\
&= (a\mathtt{IIIII} + b\mathtt{ZZZZZ})|G\rangle & (7.138)\\
&= (a\mathtt{IIIII} + b\mathtt{ZZZZZ})U_{\text{edges}} H^{\otimes n}|0\rangle^{\otimes n} & (7.139)\\
&= U_{\text{edges}}(a\mathtt{IIIII} + b\mathtt{ZZZZZ})H^{\otimes n}|0\rangle^{\otimes n} & (7.140)\\
&= U_{\text{edges}} H^{\otimes n}(a\mathtt{IIIII} + b\mathtt{XXXXX})|0\rangle^{\otimes n} & (7.141)\\
&= U_{\text{edges}} H^{\otimes n} U_{\text{prep}} \left[|0\rangle^{\otimes n-1} \otimes ((a\mathtt{I} + b\mathtt{X})|0\rangle)\right] . & (7.142)
\end{aligned}
$$

where U_{prep} is a simple circuit that transforms IIIIX to XXXXX.

Codeword stabilized quantum codes A codeword-stabilized (CWS) quantum code $Q(G, C)$ is built from two ingredients:

- a graph state $|G\rangle$;
- a classical n-bit code C.

The codewords of $Q(G, C)$ are the states

$$|w_i\rangle = w_i|G\rangle = Z^{C_i}|G\rangle \tag{7.143}$$

for all $C_i \in C$. The conceptual idea behind this construction is that, for a graph state, an X error happening to qubit k is equivalent to two or more Z errors. Specifically, the Z errors occur to the qubits adjacent in the graph to k. Thus, quantum X and Z errors can be corrected by harnessing a classical code to correct all the equivalent *patterns* of Z errors induced on the graph state: only one kind of error need be corrected.

The $[[n, k, d]]$ notation for distance-d quantum codes encoding k qubits with n qubits can be generalized to $((n, K, d))$, where K now denotes the *dimension* of the encoded Hilbert space (so, for k qubits, $K = 2^k$). In terms of this more general notation, one central achievement of CWS codes is a set of systematic recipes for their construction [9, 10], which have resulted in the discovery of new families of quantum codes that outperform any possible stabilizer code. This includes one-off codes with parameters $((5,6,2))$, $((9,12,3))$, $((10,24,3))$, and families such as the quantum Goethals-Preparata codes [11], with parameters $((2^m, 2^{2^m-5m+1}, 8))$.

7.5 Fault-tolerant quantum computation

Error correction provides the means by which essentially noiseless communication is made possible today. The foundation for this possibility is Shannon's noisy coding theorem, which holds that the probability of error in transmitting a message through a noisy channel can be *infinitesimally close to zero*, as long as the transmission rate is below a certain *threshold*, known as the *capacity* of the channel.

Remarkably, it turns out that a similar theorem holds for computation, as was first investigated by von Neumann in the 1950s [12]. The basic idea is that a collection of noisy gates can be assembled to perform essentially arbitrarily noiseless computation, as long as the error probability of each gate is below a certain threshold. This defines a *computational capacity* of a set of noisy gates, analogous to the communication capacity of a noisy channel.

Historically, this *threshold theorem* for fault-tolerant computation was important in the early days, of unreliable vacuum-tube-based computers [13], but modern electronic computers eschew this principle in favor of simply realizing nearly perfect integrated circuits. Quantum computation, however, does not have such luxury, because of the inherent universality of quantum noise. Thus, constructions for fault-tolerant quantum computation have an essential role in the existence of realistic scenarios for large-scale quantum information processing systems.

Here, we describe the ideas of fault-tolerant quantum computation, beginning with the threshold theorem for classical computation, moving from there to the basic principles for fault tolerance, and ending with a threshold estimate for reliable quantum computation.

7.5.1 The threshold theorem

The *threshold theorem* for classical computation may be stated as follows:

Theorem 7.1 *A (uniform) circuit comprising N error-free gates can be simulated with probability of error less than ϵ, using*

$$O\left(poly\left(N\log\frac{N}{\epsilon}\right)\right) \tag{7.144}$$

gates, which fail with probability p, provided that $p < p_{th}$, where p_{th} is independent of N and ϵ.

The proof of this theorem follows from constructing a circuit that computes on encoded data, without ever decoding. Let us illustrate the ideas with an example. Suppose we wish to construct a fault-tolerant NAND gate, using unreliable two-input NAND gates and three-input MAJ (majority) gates. Let each unreliable gate fail with

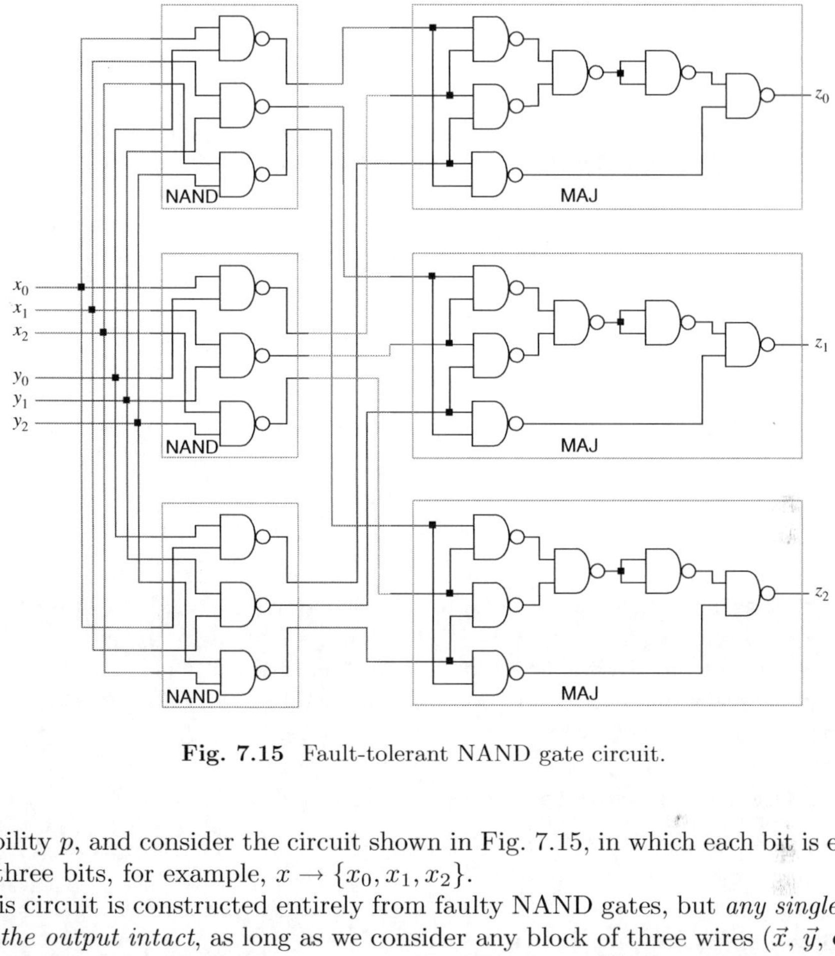

Fig. 7.15 Fault-tolerant NAND gate circuit.

probability p, and consider the circuit shown in Fig. 7.15, in which each bit is encoded using three bits, for example, $x \to \{x_0, x_1, x_2\}$.

This circuit is constructed entirely from faulty NAND gates, but *any single failure leaves the output intact,* as long as we consider any block of three wires ($\vec{x}$, $\vec{y}$, or $\vec{z}$) to be correct if it has no more than one error. In fact, the probability of the output block $\vec{z}$ having two or more errors is approximately $167p^2$, because there are 167 possible pairs of single errors that lead to the output being incorrect. Each of these errors may be a single NAND gate failure or an error in an input block. The circuit thus has lower failure probability than a single gate if $167p^2 < p$, meaning that $p < 1/167 \approx 0.006$.

The construction of this fault-tolerant NAND gate may be appreciated as generally being a recursive structure of NAND and MAJ gates. If MAJ gates were available as primitives, then the circuit would comprise three NAND gates and three MAJ gates. And if we ignored errors on the input blocks for simplicity, then the output would be incorrect only if two or more gates failed. There are (6 choose 2) = 15 possible choices of two gates, and thus $p_{\text{fail}} \sim 15p^2$. Now recursively construct a circuit using 9 bits to encode one bit, using the three-bit fault-tolerant NAND gate as the building block. Assume the MAJ gate is also recursively constructed. The resulting circuit fails if two of its blocks fail, and each of its blocks fails with probability $15p^2$. There are again six blocks, so

$$p_{\text{fail}} \approx 15\left(15p^2\right)^2 = \frac{(15p)^4}{15}\,. \tag{7.145}$$

In general, we may define each configuration of errors that causes an output error as being a *fault path*. Let c be the number of fault paths, and denote the initial 3-bit encoded circuit as a "level 1" encoding, the 9-bit encoded circuit as a "level 2" encoding, and so forth. We find that

$$\text{Level 1}: \; c\,p_{\text{fail}} = (cp)^2, \tag{7.146}$$

$$\text{Level 2}: \; c\,p_{\text{fail}} = (cp)^4, \tag{7.147}$$

$$\text{Level k}: \; c\,p_{\text{fail}} = (cp)^{2^k}. \tag{7.148}$$

Thus, if $cp < 1$, then as $k \to \infty$, we have $p_{\text{fail}} \to 0$. Therefore, we may identify the *threshold for fault tolerance* (often also referred to as the *accuracy threshold*) as

$$p_{\text{th}} = \frac{1}{c} = \frac{1}{\text{number of fault paths in a universal FT gate}}\,. \tag{7.149}$$

Note that this number is independent of N and ϵ. It is important to appreciate that p_{th} is actually just a property of the error correction code and the family of gates available as primitives. It is the inverse of the number of fault paths in the circuit used to construct a universal gate acting on the data encoded in the code. The challenge is to minimize this number, through clever circuit design, intelligent choice of codes, and optimal selection of physical gates employed.

7.5.2 Principles of fault tolerance

How can the number of fault paths in a circuit be minimized? We define a procedure (i.e., a quantum circuit, with gates, measurement, and state preparation) as being *fault-tolerant* if a single component failure causes at most one error in each encoded block of bits in the output.

Figure 7.16 shows a set of examples illustrating good (i.e., fault-tolerant) and bad (i.e., non-fault-tolerant) classical and quantum circuits. The first key idea behind realizing a fault-tolerant procedure is to control the extent to which one error can propagate to become more than one error. One approach to ensuring control over such error propagation is to employ only *transversal* gates. Such gates act (qu)bitwise on (qu)bits in a circuit. Specifically, for a quantum circuit, a gate U is transversal if $U = U_1 \otimes U_2 \otimes \cdots \otimes U_n$, for an n-qubit encoding.

The second key idea is to systematically employ error correction. Error correction terminates the propagation paths of possible errors. But one must be careful, since error correction circuits themselves must be built from faulty parts, and thus the error correction circuit is usually constructed using transversal gates (though that is not strictly necessary).

What is the cost of making a procedure fault-tolerant? Let d be the number of elementary gates in a given fault-tolerant procedure for some logical fault-tolerant gate ("logical" means that it acts on encoded (qu)bits). For k levels of recursion, the

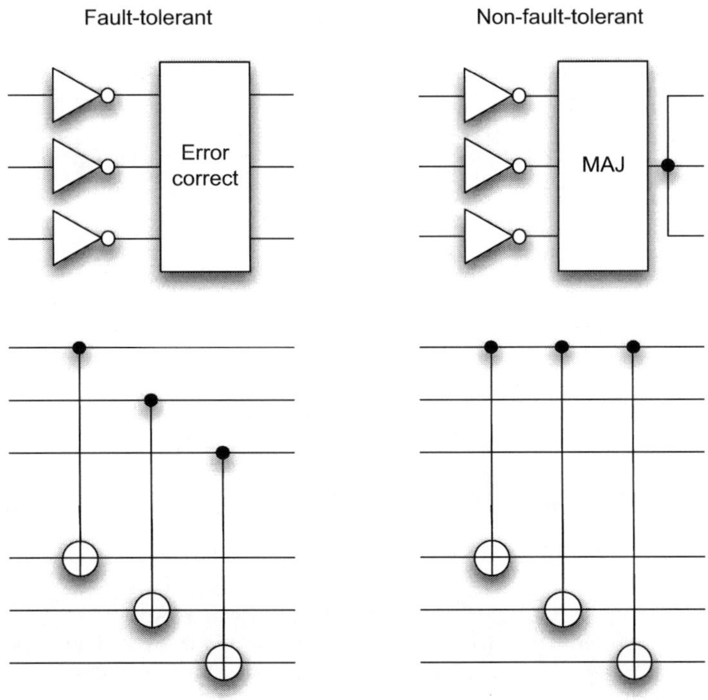

Fig. 7.16 Fault-tolerant and non-fault-tolerant procedures.

circuit size required for one gate is d^k. This exponential growth may give rise to some concern, but the rate at which errors decrease acts to alleviate this.

Specifically, suppose the goal is to simulate an N-gate circuit, with probability of error less than ϵ. We thus require each gate in the fault-tolerant procedure to have an error less than ϵ/N. It is convenient to combine Eq. (7.148) with Eq. (7.149), giving the failure probability of the k-level procedure as

$$\frac{p_{\text{fail}}}{p_{\text{th}}} = \left(\frac{p}{p_{\text{th}}}\right)^{2^k}. \tag{7.150}$$

Note that this is doubly exponential in k.

It follows that the size of circuit required to obtain a failure probability ϵ will only be logarithmic in $1/\epsilon$. We may see this explicitly by first computing the recursion level k necessary to satisfy $p_{\text{fail}} < \epsilon/N$. This gives

$$2^k \approx \frac{\log \dfrac{\epsilon}{N p_{\text{th}}}}{\log \dfrac{p}{p_{\text{th}}}}. \tag{7.151}$$

Since the total number of gates in the fault-tolerant procedure is Nd^k, we thus find that

$$Nd^k = N \left(\frac{\log \frac{N}{\epsilon} p_{\text{th}}}{\log \frac{p_{\text{th}}}{p}} \right)^{\log_2 d} \tag{7.152}$$

$$= \text{poly}\left(\log \frac{N}{\epsilon}\right) N \,. \tag{7.153}$$

This completes the proof of Theorem 7.1. □

The two principles for fault tolerance to keep in mind are

- compute on encoded data;
- control error propagation using fault-tolerant procedures.

7.5.3 Threshold for quantum computation

What is the fault-tolerance threshold for quantum computation? Since this is determined by the number of fault paths in the optimal fault-tolerant procedure for a universal gate, we need to begin by understanding how to compute on encoded quantum data. We study how this is done for stabilizer codes, and give an estimate for p_{th} based on the fault-tolerant procedure for a controlled- NOT gate acting on logical qubits encoded in the 7-qubit Steane code.

Normalizers Let S be a stabilizer. Define the *normalizer* of S as being

$$N(S) = \{g \in G \mid ghg^\dagger \in S,\ \forall\, h \in S\} \,. \tag{7.154}$$

Another way to express the normalizer is as the set of all Pauli operators that commute with all elements of S:

$$N(S) = \{g \in G \mid [g,h] = 0,\ \forall\, h \in S\} \,. \tag{7.155}$$

Example 1 $S = \langle IZ, ZI \rangle$. $N(S)$ includes II and ZZ, as well as IZ and ZI. So it is clear that $S \subseteq N(S)$.

Example 2 $S = \langle XX \rangle$. $N(S)$ includes XI and IX (and more) in addition to S.

Example 3 $S = \langle IZZ, ZIZ \rangle$. $N(S)$ includes XXX and ZZZ (and more) in addition to S.

Generalized normalizers and the Clifford group The *generalized normalizer* of S is the same idea as the normalizer, but instead of just being restricted to elements of the Pauli group G, the generalized normalizer may contain arbitrary unitary U. The elements of the generalized normalizer must all commute with all the elements of S. Note that sometimes, when the context is clear, the word "generalized" is dropped, so that writing the normalizer of S, or $N(S)$, implicitly refers to the generalized normalizer.

The generalized normalizer of G is defined as being the *Clifford group*. For one qubit, the Clifford group is generated by the Hadamard gate H and the phase gate S (recall that $S = \sqrt{Z}$). For more than one qubit, the Clifford group is generated by H, S, and the controlled-NOT gate. We shall refer to this group as being C_2, for reasons relating to the teleportation hierarchy of fault-tolerant gates [14].

Universal gates on encoded quantum data The conceptual importance of normalizer operations for fault-tolerant quantum computation is the fact that they represent all the gates that can be performed on a stabilizer code state, without leaving the code space. Ideally, therefore, there would exist a stabilizer code such that its normalizer would admit any unitary operation on its logical qubits.

However, two important facts prohibit this ideal. First is the Gottesman-Knill theorem [3], according to which any quantum circuit composed of gates from the Clifford group can be efficiently simulated when acting upon stabilizer states and when employing measurements in the computational basis. One might thus hope that there exists a stabilizer code with transversal non-Clifford gates. Indeed, such codes turn out to exist [15], but unfortunately they only admit a restricted set of gates, which are not universal. And, in fact, it is now known that no stabilizer code can exist that has a transversal set of gates that is universal for quantum computation [16].

Despite this severe constraint, fault-tolerant quantum computation is still possible, but at the added expense of having to exit the code space to perform certain operations. The general technique for accomplishing this employs a process that "teleports" a gate into the circuit, consuming a pre-made ancilla quantum state that encodes the gate. This gate may be, for example, a T gate ($T = \sqrt{S}$, also known as a $\pi/8$ gate), which is universal when combined with H and controlled-NOT. The ancilla state may be prepared by a variety of fault-tolerant procedures, such as *magic state distillation* [17].

The teleportation procedure itself only requires Clifford group gates, and thus can be accomplished in a fault-tolerant manner on an encoded quantum state. The limiting step in the teleportation is the controlled- NOT gate required, owing to the complexity of its two-qubit operation.

Thus, we may obtain a good estimate of the threshold for fault-tolerant quantum computation by considering the threshold for performing a fault-tolerant controlled-NOT gate.

Fault-tolerance threshold for a controlled-not gate We began Section 7.5.1 with an estimate of the threshold for a fault-tolerant NAND gate, based on counting the number of fault paths in a fault-tolerant procedure for a NAND gate circuit acting on bits encoded in a triple-redundancy code. An analogous estimation can be done for a fault-tolerant controlled-NOT gate, acting on qubits encoded in the 7-qubit Steane code.

The fault-tolerant procedure for a controlled-NOT gate is conceptually simple when implemented on qubits encoded in the 7-qubit Steane code; it is seven controlled-NOT gates performed transversally, followed by fault-tolerant quantum error correction, as shown in Fig. 7.17. This is possible because all CSS codes have a transversal controlled-NOT gate (and, in fact, all quantum codes with a transversal controlled-NOT gate are CSS codes).

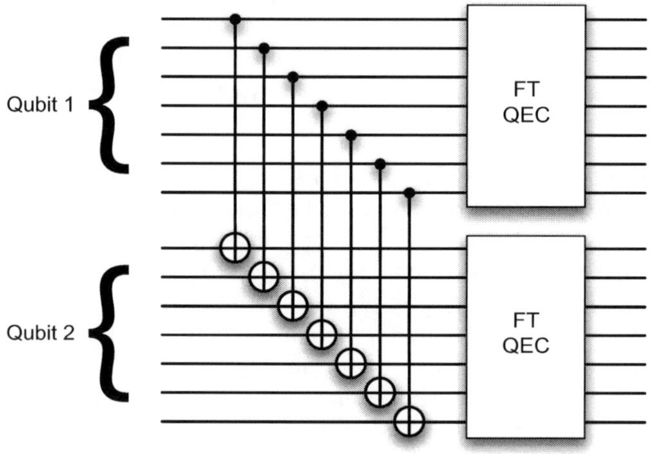

Fig. 7.17 Failt-tolerant controlled-not gate circuit.

The challenge is in the fault-tolerant quantum error correction. Recall from Section 7.3.3 that quantum error correction is realized by performing a series of syndrome measurements, followed by some classical processing, then application of recovery operations. The challenge is the measurements, which are implemented by operator measurement circuits. The Steane code has six syndrome operators to be measured. Suppose one of them is the operator $IZZIIZZ$, for the sake of example. The standard procedure for operator measurement of this syndrome would use the circuit shown in Fig. 7.18.

However, this circuit is manifestly non-fault-tolerant. Any error on the top qubit (the one being measured) would propagate through the controlled-NOT operators into becoming four errors on the encoded qubit block. That is unacceptable.

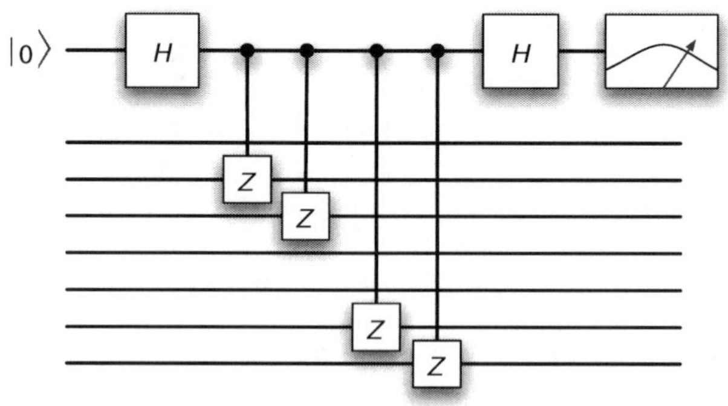

Fig. 7.18 Non-fault-tolerant syndrome operator measurement circuit.

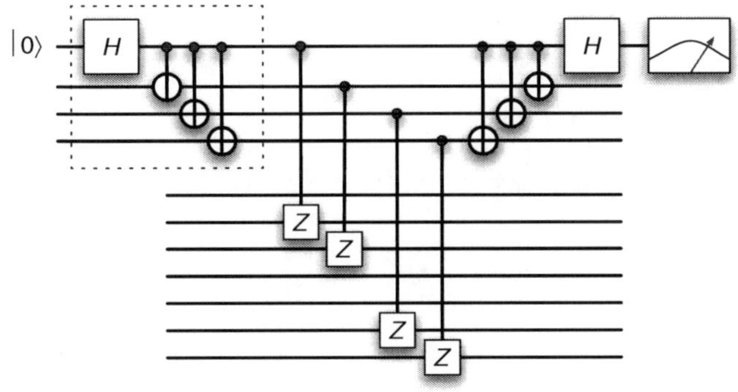

Fig. 7.19 Fault-tolerant syndrome operator measurement circuit.

There are several clever ways to solve this problem and provide means for fault-tolerant syndrome operator measurement. One employs "cat" states. Note that the operator measurement circuit uses a control qubit in the state $(|0\rangle + |1\rangle)/\sqrt{2}$, before the controlled-$Z$ gates. In place of this, one can instead use the four-qubit entangled state $(|0000\rangle + |1111\rangle)/\sqrt{2}$, as shown in Fig. 7.19.

As long as this "cat" state can be prepared (offline) with sufficiently low error probability, then error propagation is limited, since each controlled-Z operation now comes from a separate control qubit. This approach has some additional nice properties, due to the asymmetry of X and Z error propagation across the controlled Pauli gates.

Another approach employs encoded logical $|0_L\rangle$ and $|+_L\rangle = (|0_L\rangle + |1_L\rangle)/\sqrt{2}$ qubit states, and transversal controlled-NOT and controlled-Z gates, instead of physical qubit states, in the operator measurement. And a third approach employs teleportation, which simultaneously performs error correction and movement of a logical qubit value from one block into a "fresh" block of qubits, which can have fewer errors, owing to checks performed during the state preparation.

The essence of each of these approaches is to take advantage of the ability to prepare certain "ancilla" states with low error probability offline (i.e., before they are needed). These states then assist in the measurement process, to lower the probability of measurement error, or to limit the propagation of errors during the measurement. Nevertheless, it is still necessary to repeat each measurement three times, because the final measurement may fail with probability p. Classical majority voting is then done on the measurement results, to obtain the bits of the syndrome.

So how many gates are involved, all together? Table 7.1 gives a breakdown of the gates needed for quantum error correction. Overall, we obtain a subtotal of 72 gates needed for the error correction. Add to this 7 gates needed for the controlled-NOT gate, and up to 2 gates needed for the classically controlled recovery operation, and the total is 81 gates. The number of fault paths is bounded above by (81 choose 2) = 3240, so the threshold is

$$p_{\text{th}} < \frac{1}{3240} \approx 3 \times 10^{-4} \,. \tag{7.156}$$

Table 7.1 Approximate count of gates needed for fault-tolerant quantum error correction on the 7-qubit Steane code

No. of gates	Task
6	Syndrome operators
$\times$ 4	Gates each
$\times$ 3	Repetitions of measurement

By the arguments given above, this is the threshold for fault-tolerant quantum computation using the 7-qubit Steane code, assuming the standard quantum circuit model.

Recent work, using topological quantum codes (all of them stabilizer codes), has reached fault tolerance thresholds that appear to be higher than this—possibly as high as 10^{-1} [18, 19] (though the "code threshold" numbers come from an optimistic error model that assumes no measurement errors). However, one must be careful to make sure such thresholds do not come at the cost of much higher overhead costs. This overhead is the physical (space and time) expense of creating, testing, storing, and moving around the ancilla states necessary for the fault-tolerant procedures employed.

One reasonable measure of overhead is the number of physical controlled-NOT gates needed per unit fault-tolerant procedure for a standard gate (like a logical controlled-NOT). For the 7-qubit Steane code example given here, the overhead cost is of the order of 10^2 gates, as long as the error probability p is sufficiently below p_{th} to allow just one level of recursion. On the other hand, for typical topological code parameters [20], the overhead is between 10^3 and 10^5, to achieve similar overall failure probabilities. Of course, when p is higher than the p_{th} for the Steane code, then fault tolerance is unachievable for any amount of overhead, whereas, in contrast, it is still theoretically achievable using a topological code.

Recent directions in the literature bring together topological codes, subsystem codes [21], and graphs [22], with the potential for substantially lowering overhead requirements into a regime closer to experimental practicality. Many interesting questions also remain open, including the possibility of fault-tolerant quantum computation based on nonstabilizer codes such as the codeword-stabilized quantum codes. Moving forward, there is a great deal of promise for realistic fault-tolerant quantum computation with new approaches to quantum error correction.

References

[1] Gerry, C. C. and Knight, P. L. (2005). *Introductory Quantum Optics*. Cambridge University Press, Cambridge.

[2] Molmer, K., Castin, Y. and Dalibard, J. (1993). Monte Carlo wavefunction method in quantum optics. *J. Opt. Soc. Am. B*, **10**, 524.

[3] Nielsen, M. A. and Chuang, I. L. (2000). *Quantum Computation and Quantum Information*. Cambridge University Press, Cambridge.

[4] Haroche, S. and Raimond, J.-M. (2006). *Exploring the Quantum: Atoms, Cavities, and Photons.* Oxford University Press, Oxford.

[5] Chuang, I. (2004). Principles of quantum computation. In *Quantum Entanglement and Information Processing. Ecole d'été de Physique les Houches Session LXXIX, 30 July – 25 August 2003* (ed. D. Esteve, J.-M. Raimond, and J. Dalibard), P. I. Elsevier, Amsterdam.

[6] Gottesman, D. (1997) *Stabilizer Codes and Quantum Error Correction.* PhD thesis, California Institute of Technology.

[7] Calderbank, A. R. and Shor, P. W. (1996). Good quantum error-correcting codes exist. *Phys. Rev. A*, **54**, 1098.

[8] Steane, A. M. (1996). Simple quantum error correcting codes. *Phys. Rev. A*, **54**, 4741.

[9] Cross, A., Smith, G., Smolin, J. A. and Zeng, B. (2009). Codeword stabilized quantum codes. *IEEE Trans. Inf. Theory*, **55**, 433.

[10] Chuang, I. L., Cross, A. W., Smith, G., Smolin, J. A. and Zeng, B. (2009). Codeword stabilized quantum codes: algorithm and structure. *J. Math. Phys.*, **50**, 042109.

[11] Grassl, M. and Roetteler, M. (2008). Non-additive quantum codes from Goethals and Preparata codes. In *Proceedings of 2008 IEEE Information Theory Workshop*, p. 396. arXiv: 0801.2144.

[12] von Neumann, J. (1956). Probabilistic logics and synthesis of reliable organisms from unreliable components. In *Automata Studies* (eds. C. Shannon and J. McCarthy), p. 43. Princeton University Press, Princeton, NJ.

[13] Winograd, S. and Cowan, J. D. (1963). *Reliable Computation in the Presence of Noise.* MIT Press, Cambridge, MA.

[14] Gottesman, D. and Chuang, I. L. (1999). Quantum teleportation is a universal computational primitive. *Nature*, **402**, 390.

[15] Rains, E. M., Hardin, R. H., Shor, P. W. and Sloane, N. J. A. (1997). A nonadditive quantum code. *Phys. Rev. Lett.*, **79**, 953.

[16] Zeng, B., Cross, A. W. and Chuang, I. L. (2011). Transversality versus universality for additive quantum codes. *IEEE Trans. Inf. Theory*, **57**, 6272.

[17] Bravyi, S. and Kitaev, A. (2005). Universal quantum computation with ideal Clifford gates and noisy ancillas. *Phys. Rev. A*, **71**, 022316.

[18] Ohzeki, M. (2009). Accuracy thresholds of topological color codes on the hexagonal and square-octagonal lattices. *Phys. Rev. E*, **80**, 011141.

[19] Sarvepalli, P. and Raussendorf, R. (2012). Efficient decoding of topological color codes. *Phys. Rev. A*, **85**, 022317.

[20] Cross, A. W., DiVincenzo, D. P. and Terhal, B. M. (2007). A comparative code study for quantum fault-tolerance. *Quantum Inf. Compat.*, **9**, 0541.

[21] Andrist, R. S., Bombin, H., Katzgraber, H. G. and Martin-Delgado, M. A. (2012). Optimal error correction in topological subsystem codes. *Phys. Rev. A*, **85**, 050302.

[22] Sarvepalli, P. and Brown, K. R. (2012) Topological subsystem codes from graphs and hypergraphs. *Phys. Rev. A*, **86**, 042336.

8
Quantum optomechanics

F. MARQUARDT

University of Erlangen-Nuremberg, Institute of Theoretical Physics
Staudtstr. 7, 91058 Erlangen, Germany
Max Planck Institute for the Science of Light
Günther-Scharowsky-Straße 1/Bau 24, 91058 Erlangen, Germany

Quantum Machines: Measurement and Control of Engineered Quantum Systems.
Edited by M. Devoret, B. Huard, R. Schoelkopf, and L. F. Cugliandolo.

Chapter Contents

8.1 Introduction

This chapter is a basic introduction to what is now known as (cavity) optomechanics, a field at the intersection of nanophysics and quantum optics that has developed over the past few years. This field deals with the interaction between light and micro- or nanomechanical motion. A typical setup may involve a laser-driven optical cavity with a vibrating end-mirror, but many different setups exist by now, even in superconducting microwave circuits (see Chapter 9 of this volume) and cold-atom experiments. The field has developed rapidly during the past few years, starting with demonstrations of laser-cooling and sensitive displacement detection. For short reviews with many relevant references, see [1–4]. In the present chapter, I have only picked a few illustrative references, but the discussion is hopefully more didactic and also covers some very recent material not found in those reviews. I will emphasize the quantum aspects of optomechanical systems, which are now becoming important. Related chapters in this volume are those by Konrad Lehnert (Chapter 9: specific implementation in superconducting circuits, direct classical calculation) and Aashish Clerk (Chapter 2: quantum limits to measurement), as well as Jack Harris (Chapter 14).

As of 2011, the first experiments have reported laser cooling down to near the quantum ground state of a nanomechanical resonator. These achievements have unlocked the door toward the quantum regime of optomechanics. The quantum regime will now start to be explored, and the present chapter should provide the basics needed for understanding those developments. In general, the field is driven by a variety of different goals and promising aspects, both fundamental and applied:

- Applications in ultrasensitive measurements (of small displacements, forces, mass changes etc.), with precisions down to the limits set by quantum mechanics.
- Possible fundamental tests of quantum mechanics in a new regime of massive objects ($\sim 10^{14}$ atoms!) being placed into superposition states or getting entangled with other quantum objects. This might be used to test speculations about novel decoherence mechanisms, such as gravitationally induced collapse of the wavefunction.
- Hybrid systems: mechanical motion couples to many different systems (light, atoms, superconducting qubits, microwave resonators, spins, etc.), and can therefore be used as an interface between quantum systems. Optomechanics may provide the most efficient way of converting quantum information at gigahertz frequencies (such as in superconducting qubits) up to optical frequencies, into flying qubits.
- Classical signal processing: optomechanics provides tunable nonlinearities (for the light field and the mechanical motion) that may be exploited for on-chip signal processing, potentially integrated with sensing.
- Quantum information processing: qubits (or continuous variable quantum states) may be stored in relatively long-lived mechanical excitations that can be manipulated and read out by the light field.
- Optomechanical arrays and circuits: a recent experimental development, in which vibrating photonic crystal structures are fabricated on a chip, will allow for more complex setups with many optical and vibrational modes, coupling to optical and phononic waveguides.

8.1.1 Interaction between light and mechanics

Consider the most elementary light-matter interaction, a light beam scattering off any arbitrary object (atom, molecule, small glass sphere, nanobeam, etc.). There is always elastic light scattering, with the outgoing light frequency identical to the incoming frequency: $\omega' = \omega$. Inelastic scattering, in contrast, will be accompanied by excitation or de-excitation of the material object. For example, internal atomic transitions may be excited. However, independent of the internal electronic details of the atoms or molecules, it is always possible to have Raman scattering due to the object's mechanical vibrations: $\omega' = \omega \pm \Omega$, where Ω is the vibrational frequency.

The vibrations gain or lose energy, respectively, for these Stokes/anti-Stokes processes ($\omega' = \omega \mp \Omega$). If both of these processes occur at an equal rate, the vibrations will merely heat up. This is not a very surprising outcome from shining light on an object.

However, one may use an optical cavity to suppress the Stokes process. Then, the photons are preferentially backscattered blue-shifted, $\omega' = \omega + \Omega$, carrying away energy. This is the principle of cavity-assisted laser cooling! Although the same can be (and is) done with internal atomic resonances, using a cavity means independence of the internal structure of the atoms that constitute the object. Thus, we arrive at the basic optomechanical setup: a laser-driven optical cavity coupled to the mechanical vibrations of some object. This is a very generic setting. The purpose of the cavity is to select optical frequencies (e.g., to suppress the Stokes process), to resonantly enhance the light intensity (leading to much stronger radiation forces), and to enhance the sensitivity to the mechanical vibrations.

We will see that this setup displays features of a true two-way interaction between light and mechanics. This is in contrast to optical tweezers, optical lattices, or vibrational spectroscopy, where the light field controls the mechanics (or vice versa) but the loop is not closed.

The general goal in optomechanics, based on these ideas, is to cool, control, and readout (nano)mechanical motion via the light field, and to use the mechanics to produce interesting states of the light as well.

8.1.2 Basic optomechanical Hamiltonian

Consider a laser-driven cavity, whose optical resonance frequency is controlled by the displacement $\hat{x}$ of some object (e.g., a movable end-mirror as in Fig. 8.1):

$$\hat{H} = \hbar\omega(\hat{x})\hat{a}^\dagger\hat{a} + \hbar\Omega\hat{b}^\dagger\hat{b} + \dots, \tag{8.1}$$

where "..." stand for terms such as laser driving, photon, and phonon decay, which will be taken into account later. Here $\hat{x} = x_{\mathrm{ZPF}}(\hat{b} + \hat{b}^\dagger)$ is the displacement of the one vibrational mode we pick for consideration (neglecting all the other normal modes for the moment). $x_{\mathrm{ZPF}} = (\hbar/2m\Omega)^{1/2}$ is the size of the mechanical zero-point fluctuations (the width of the mechanical ground-state wavefunction). For many setups, this is of the order of 10^{-15} m!

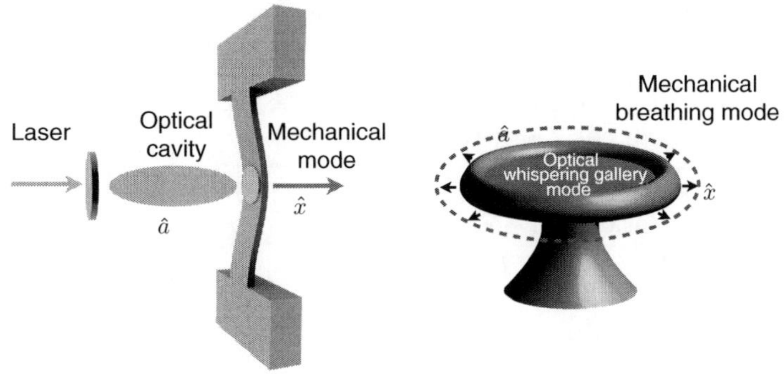

Fig. 8.1 (a) The standard schematic optomechanical setup, with a mirror on a nanobeam. Many other versions exist, for example in the form of a vibrating microtoroidal whispering gallery mode resonator (b).

The optical resonance frequency can be expanded in the displacement, with L the cavity length and ω_{cav} the optical resonance frequency for $x = 0$:

$$\omega(\hat{x}) = \omega_{\text{cav}} \left(1 - \frac{\hat{x}}{L} + \dots\right). \tag{8.2}$$

Thus, $\hat{H}$ contains a term of the type $-\hat{F}\hat{x}$, where we identify $\hat{F}$ as the radiation pressure force:

$$\hat{F} = \frac{\hbar\omega_{\text{cav}}}{L}\hat{a}^\dagger\hat{a}\,. \tag{8.3}$$

In summary, the standard optomechanical Hamiltonian reads

$$\hat{H} = \hbar\omega_{\text{cav}}\hat{a}^\dagger\hat{a} + \hbar\Omega\hat{b}^\dagger\hat{b} - \hbar g_0\hat{a}^\dagger\hat{a}(\hat{b} + \hat{b}^\dagger) + \dots\,. \tag{8.4}$$

Here we have identified

$$g_0 = \omega_{\text{cav}}\frac{x_{\text{ZPF}}}{L} \tag{8.5}$$

as the optomechanical single-photon coupling strength. It represents the optical frequency shift produced by a zero-point displacement. Therefore, the optomechanical coupling term to be used in the following is

$$-\hbar g_0\hat{a}^\dagger\hat{a}(\hat{b} + \hat{b}^\dagger)\,. \tag{8.6}$$

In essence, we are facing a parametric coupling where a mechanical displacement controls the frequency of some (driven) resonance. It is no surprise, then, that this is a rather generic situation that can be obtained in many different setups.

8.1.3 Many setups

A wide variety of optomechanical systems have now been investigated, (almost) all of them described by the same Hamiltonian. In each case, we are able only to cite a few examples, to provide an entry into the literature:

- movable mirrors on a cantilever or nanobeam [5–7];
- microtoroids with optical whispering-gallery modes [8];
- membranes/nanowires inside or next to a cavity [9–11];
- atomic clouds [12, 13];
- nanobeams or vibrating plate capacitors in superconducting microwave resonators [14, 15];
- photonic crystals patterned into free-standing nanobeams [16, 17].

8.1.4 Typical scales

These setups span a large range of scales, of frequencies Ω in the kilohertz to gigahertz range and masses m ranging from 10^{-15} kg to 1 g. The typical values of g_0 for real mechanical setups (as opposed to atomic clouds, whose small masses give a large x_{ZPF}) are rather small; for example, for an $L \sim 1\,\mathrm{cm}$ cavity, we would get $g_0 \sim 100\,\mathrm{Hz}$. Reducing the size of the cavity to a wavelength, $L \sim 1\,\mu\mathrm{m}$, increases g_0 up to the megahertz region. This is achieved in the photonic crystal setups.

For mechanical damping, the typical decay rates Γ are in the hertz to megahertz range, with good setups reaching quality factors $Q = \Omega/\Gamma$ of 10^5 up to 10^7. The optical decay rate κ depends on the cavity size, and typically ranges from megahertz to gigahertz.

8.1.5 Basic optomechanics: classical picture

Plotting the circulating light intensity versus displacement x, we get a series of Fabry-Perot resonances, spaced by $\lambda/2$. The intensity is proportional to the radiation pressure force $F(x)$, and we can convert this force into a potential, $V'_{\mathrm{rad}}(x) = -F(x)$, which displays a series of rounded steps (Fig. 8.2(a)). Adding this to the mechanical harmonic oscillator potential reveals the possibility of having several stable minima ("static multistability"). When sweeping external parameters back and forth, this can lead to hysteresis, as the system follows first one and then another branch (first seen in [18]). For example, this effect is routinely observed in recording the optical transmission while sweeping the laser detuning, provided the laser intensity is strong enough (Fig. 8.2(b)).

In addition, the slope of $F(x)$ defines a modification of the mechanical spring constant,

$$D = D_0 - \frac{dF}{dx}\,, \tag{8.7}$$

a feature known as the "optical spring effect."

However, this picture is incomplete, as it neglects retardation effects introduced by the finite cavity photon decay rate κ. The force follows the motion only with some

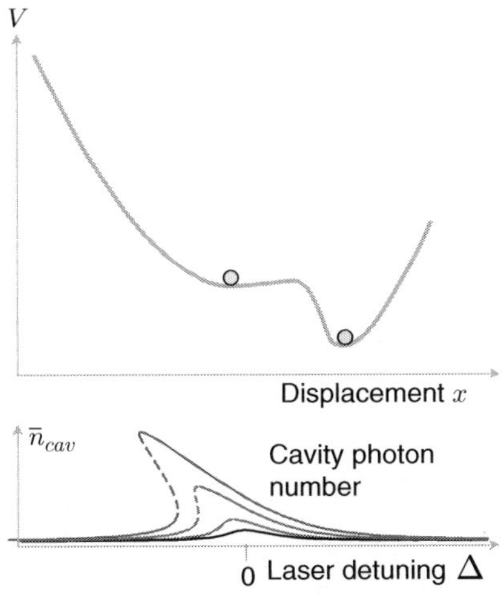

Fig. 8.2 (a) Effective mechanical potential resulting from the radiation force, with static bistability. (b) Corresponding hysteresis observed in sweeping the laser detuning.

time delay. This leads to effects such as friction. For example, assume the equilibrium position to sit somewhere on the rising slope of the resonance. In thermal equilibrium, there will be oscillations around this position. We want to ask what is the effect of the delayed radiation force during one cycle of oscillation. It turns, out that the work done is negative,

$$\oint F dx < 0; \tag{8.8}$$

that is, the radiation force extracts mechanical energy: there is extra, light-induced damping. Radiation-induced damping of this kind was first observed in pioneering experiments with a microwave cavity by Braginsky and co-workers over 40 years ago [19]. This can be used to cool down the mechanical motion, which is important for reaching the quantum regime. More on this later.

8.1.6 Basic optomechanics: quantum picture

Most of the experiments to date can be described using the regime of "linearized optomechanics." Assume a strong laser drive, such that

$$\hat{a} = (\alpha + \delta\hat{a})\, e^{-i\omega_L t}\,, \tag{8.9}$$

where $\alpha = \sqrt{n_{\text{phot}}}$ is the amplitude of the light field inside the cavity (here assumed real-valued without loss of generality). Thus, when expanding the photon number, we get

$$\hat{a}^\dagger \hat{a} = \alpha^2 + \alpha(\delta\hat{a} + \delta\hat{a}^\dagger) + \delta\hat{a}^\dagger \delta\hat{a}\,. \tag{8.10}$$

We neglect the last term (much smaller than the second for large drive), and omit the first (which only gives rise to an average classical force). Thus, we obtain the Hamiltonian of linearized optomechanics:

$$\hat{H} \approx -\hbar g_0 \alpha (\delta\hat{a}^\dagger + \delta\hat{a})(\hat{b} + \hat{b}^\dagger) \\ +\hbar\Omega\hat{b}^\dagger\hat{b} - \hbar\Delta\delta\hat{a}^\dagger\delta\hat{a} + \ldots \, . \tag{8.11}$$

Here we have gone into a frame rotating at the laser frequency (see the definition of $\delta\hat{a}$ in Eq. (8.9)), which introduces the laser detuning

$$\Delta \equiv \omega_L - \omega_{\text{cav}} \, . \tag{8.12}$$

The omitted terms $(\ldots)$ in Eq. (8.11) no longer contain the laser drive, since that has been taken care of by shifting $\hat{a}$. The Hamiltonian (8.11) will be the basis for much of our discussion. We can introduce the (driving-enhanced) optomechanical coupling strength

$$g = g_0 \alpha \, , \tag{8.13}$$

which can be tuned by the amplitude of the incoming laser drive. All experiments to date rely on this driving-enhanced coupling g. Later, we will even see that time-dependent modulation of g can be exploited. In summary, the linearized optomechanical interaction is

$$-\hbar g(\delta\hat{a} + \delta\hat{a}^\dagger)(\hat{b} + \hat{b}^\dagger) \, . \tag{8.14}$$

We thus arrive at the picture of two coupled oscillators: a mechanical oscillator $\hat{b}$ and a laser-driven cavity mode $\delta\hat{a}$, at frequencies Ω and $-\Delta$, respectively (Fig. 8.3). It is important to remember, though, that $\delta\hat{a}$ describes the excitations on top of the strongly driven cavity field. The shifted oscillator $\delta\hat{a}$ itself is in its ground state (since we have subtracted the large coherent-state amplitude α). Putting a single photon

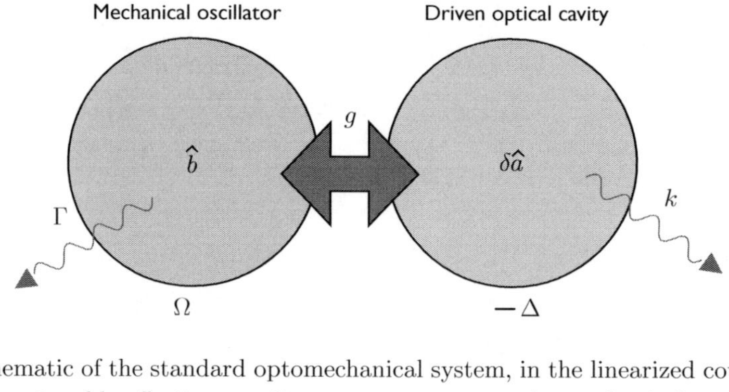

Fig. 8.3 Schematic of the standard optomechanical system, in the linearized coupling approximation. A laser-tunable effective coupling $g = g_0\alpha$ connects the mechanical oscillator and the laser-driven cavity, with the detuning $\Delta = \omega_L - \omega_{\text{cav}}$.

into the oscillator $\delta\hat{a}$ (i.e., $\delta\hat{a}^\dagger\delta\hat{a} = 1$) would not be equivalent to adding a photon to the cavity field.

Starting from this Hamiltonian, we can briefly give an overview of the most important optomechanical effects (only some of which have already been observed experimentally until now). These are selected by picking the right laser detuning Δ. We will later discuss many of these effects in more detail.

Red-detuned regime $\Delta = -\Omega$ Omitting all off-resonant terms ("rotating-wave approximation"), the coupling reduces to

$$\delta\hat{a}^\dagger\hat{b} + \delta\hat{a}\hat{b}^\dagger\,, \tag{8.15}$$

which describes state exchange between two resonant oscillators (or a beamsplitter, in quantum optics language). This can be used for

- state transfer between phonons and photons (this needs $g > \kappa$, in which case one observes hybridization between $\delta\hat{a}$ and $\hat{b}$, i.e., a peak splitting in the mechanical or optical response, the so-called "strong-coupling regime" of optomechanics);
- cooling (note that $\delta\hat{a}$ is an oscillator at zero temperature, which can take up the surplus energy of the mechanical mode; Fermi's Golden Rule yields a cooling rate $\sim g^2/\kappa$).

Blue-detuned regime $\Delta = +\Omega$ The resonant terms are

$$\hat{b}\delta\hat{a} + \hat{b}^\dagger\delta\hat{a}^\dagger\,, \tag{8.16}$$

which is a "two-mode squeezing" Hamiltonian. The following features may be observed:

- two-mode squeezing, entanglement;
- "mechanical lasing" instability (self-sustained optomechanical oscillations), if the growth of the mechanical energy overwhelms the intrinsic losses due to mechanical friction.

On-resonance regime $\Delta = 0$ We get

$$(\delta\hat{a} + \delta\hat{a}^\dagger)\hat{x}\,,$$

which means that the optical state experiences a shift proportional to the displacement. This translates into a phase shift of the reflected/transmitted light beam, which can be used to read out the mechanical motion. In this regime, the cavity is simply operated as an interferometer.

Large detuning $|\Delta| \gg \Omega, \kappa$ We can use second-order perturbation theory to eliminate the driven cavity mode, which leads to

$$\hat{H}_{\text{eff}} = \ldots + \hbar\frac{g^2}{\Delta}(\hat{b} + \hat{b}^\dagger)^2\,. \tag{8.17}$$

This is, once again, the optical spring effect. We realize that the spring constant can be varied by changing the detuning or the laser drive.

8.2 Basic linearized dynamics of optomechanical systems

8.2.1 Light-induced change of mechanical response

Both the optical spring effect and the optomechanical damping rate can be understood as aspects of the light-induced modification of the mechanical response.

Start from the classical equations of motion (e.g., derive the Heisenberg equations from $\dot{\hat{a}} = [\hat{a}, \hat{H}]/i\hbar$, etc. and then consider the classical limit, with $\alpha = \langle\hat{a}\rangle$ and $x = \langle\hat{x}\rangle$):

$$\ddot{x} = -\Omega^2(x - x_0) - \Gamma\dot{x} + \frac{F_{\text{rad}} + F_{\text{ext}}}{m}, \tag{8.18}$$

$$\dot{\alpha} = \left[i(\Delta + Gx) - \frac{\kappa}{2}\right]\alpha + \frac{\kappa}{2}\alpha_{\max} \tag{8.19}$$

Here, we have set $G = g_0/x_{\text{ZPF}}$ and introduced the laser drive amplitude via $\alpha_{\max}$, which is the light amplitude obtained when driving at resonance. $F_{\text{rad}} = (\hbar\omega_{\text{cav}}/L)\,|\alpha|^2$ is the radiation pressure force. These coupled nonlinear equations contain all the dynamics of a single optomechanical system, including the instability and even chaotic motion. They can easily be solved numerically, although a full overview of the intricate dynamics in the nonlinear regime is still lacking even now.

However, here we want to linearize, to figure out the mechanical response to some small external force F_{ext}. We set

$$x(t) = \bar{x} + \delta x(t), \tag{8.20}$$

$$\alpha(t) = \bar{\alpha} + \delta\alpha(t). \tag{8.21}$$

Then we expand the equations to linear order and eliminate $\delta\alpha$, in Fourier space. We obtain the frequency-dependent mechanical response (where $\delta x[\omega] = \int dt\, e^{i\omega t}\delta x(t)$):

$$\delta x[\omega] = \frac{F_{\text{ext}}[\omega]/m}{\Omega^2 - \omega^2 - i\omega\Gamma + \Sigma_{\text{opt}}(\omega)}. \tag{8.22}$$

Here all the light-induced modifications of the mechanical response are summarized in the "self-energy"

$$\Sigma_{\text{opt}}(\omega) \approx -2ig^2\omega[\chi_c(\omega) - \chi_c^*(-\omega)]\,, \tag{8.23}$$

where $\chi_c(\omega) = [-i\omega - i\Delta + \kappa/2]^{-1}$ is the cavity response function.

If the coupling g is weak, there are just two effects, which can be quantitatively obtained by evaluating $\Sigma_{\text{opt}}(\omega)$ at the unperturbed resonance $\omega = \Omega$.

Light-induced frequency shift

$$\delta\Omega \approx \frac{\text{Re}\,\Sigma_{\text{opt}}(\omega = \Omega)}{2\Omega}. \tag{8.24}$$

Again, this is the optical spring effect, but now for arbitrary detuning.

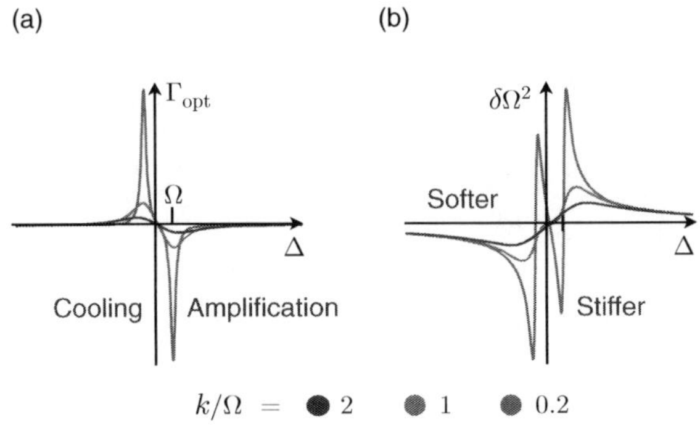

Fig. 8.4 (a) Optomechanical damping rate and (b) optical spring effect (mechanical frequency shift) versus laser detuning, for various values of κ/Ω.

Optomechanical damping rate

$$\Gamma_{opt} \approx -\frac{\text{Im}\,\Sigma_{opt}(\omega = \Omega)}{\Omega}. \tag{8.25}$$

These two quantities are strongly dependent on detuning and on the ratio κ/Ω (Fig. 8.4). For red detuning $\Delta < 0$, one obtains extra damping ($\Gamma_{opt} > 0$), while there is antidamping, leading to heating, for $\Delta > 0$. The frequency shift $\delta\Omega$ is negative for red detuning and positive for blue detuning, in the limit $\Omega \ll \kappa$. It shows more complicated dispersive behaviour for $\Omega \gg \kappa$.

Strong-coupling regime Interestingly, we can still make use of the expression for the mechanical response, Eq. (8.22), even when $g > \kappa$ (strong-coupling regime). However, we then have to keep the full frequency dependence in $\Sigma_{opt}(\omega)$. One observes that in this regime the mechanical response has two distinct peaks (rather than one), and they are separated by $2g$. This is the peak splitting characteristic of hybridization between two oscillator modes (here the mechanical mode and the laser-driven cavity mode). As we mentioned before, this is also the regime in which coherent excitations can be transferred between the laser-driven mode and the mechanical mode.

8.2.2 Quantum limit of optomechanical cooling

The damping/cooling rate Γ_{opt} can be increased, in principle, to arbitrary values by ramping up the laser intensity. In a classical theory, the effective temperature of the mechanical mode would then go to zero, according to $T_{\text{eff}} = T\Gamma/(\Gamma_{opt} + \Gamma)$. However, the shot noise fluctuations of the photon field prevent this simple picture from being true. We will briefly indicate the very generally applicable "quantum noise" approach to this problem (which reiterates our treatment in [20]).

Assume any quantum system coupled weakly to an arbitrary bath, via $\hat{V} = -\hat{F}\hat{x}$, where $\hat{F}$ refers to the bath and $\hat{x}$ to the system. One can show that Fermi's Golden Rule yields a transition rate between system levels i and f given by

$$\Gamma_{f \leftarrow i} = \frac{1}{\hbar^2} S_{FF} \left(\omega = \frac{E_i - E_f}{\hbar} \right) |\langle f | \hat{x} | i \rangle|^2 , \tag{8.26}$$

with the quantum noise spectrum

$$S_{FF}(\omega) \equiv \int dt \, e^{i\omega t} \langle \hat{F}(t) \hat{F}(0) \rangle . \tag{8.27}$$

Here the expectation value and the evolution of $\hat{F}$ are taken with respect to the uncoupled system (i.e., S_{FF} is a property of the bath only). In general, $S_{FF}(\omega)$ is not symmetric in ω, but it is always real-valued and positive.

If the system is a harmonic oscillator, the spectrum need be evaluated only at $\omega = \pm\Omega$. Then we can attach an effective temperature to the bath at this frequency, which is also the temperature that the oscillator settles into if it is only coupled to that bath (and nothing else). Detailed balance (the "Stokes relation") yields

$$e^{\hbar\Omega/k_B T_{\text{eff}}} \equiv \frac{\Gamma_{0 \leftarrow 1}}{\Gamma_{1 \leftarrow 0}} = \frac{S_{FF}(\omega = \Omega)}{S_{FF}(\omega = -\Omega)} . \tag{8.28}$$

Now the Bose-Einstein distribution means that the Stokes factor on the left-hand side is also equal to $\bar{n}/(\bar{n}+1)$, where $\bar{n}$ is the steady-state average phonon number that the oscillator settles into. This means that we get

$$\bar{n} = \left[\frac{S_{FF}(\omega = \Omega)}{S_{FF}(\omega = -\Omega)} - 1 \right]^{-1} . \tag{8.29}$$

This is still completely general (for weak coupling to any bath). Now, for the optomechanical case, $\hat{F} = (\hbar\omega_{\text{cav}}/L)\hat{a}^\dagger\hat{a} = (\hbar g_0/x_{\text{ZPF}})\hat{a}^\dagger\hat{a}$ is the radiation force. We can calculate its correlator and its spectrum by using $\hat{a} = (\alpha + \delta\hat{a})e^{-i\omega_L t}$ and $\langle \delta\hat{a}(t)\delta\hat{a}^\dagger(0) \rangle = \exp[i\Delta t - (\kappa/2)\,|t|]$. We find

$$\frac{x_{\text{ZPF}}^2}{\hbar^2} S_{FF}(\omega) = g_0^2 \frac{\kappa n_{\text{phot}}}{(\omega + \Delta)^2 + (\kappa/2)^2} .$$

Per Eq. (8.29), this leads to a minimum phonon number reachable in optomechanical cooling, which is (in the "resolved sideband regime" $\Omega \gg \kappa$, at $\Delta = -\Omega$)

$$\bar{n}_{\min} = \left(\frac{\kappa}{4\Omega} \right)^2 . \tag{8.30}$$

Apparently, in the resolved sideband regime, the phonon number can become smaller than 1 and the system gets close to the mechanical ground state [20, 21]. The first optomechanical cooling experiments started in 2004 with a photothermal force [22] and then in 2006 with true radiation pressure forces [5, 6, 8]. However, many challenges had to be overcome to cool to the ground state. This was achieved only in 2011 for an optomechanical setup with optical radiation [17] and also for a microwave setup [15]. These experiments have opened the door to the quantum regime of optomechanics.

Incidentally, we note that the optomechanical damping rate can also be expressed via S_{FF}, as

$$\Gamma_{\rm opt} = \frac{1}{\hbar^2} x^2_{\rm ZPF} [S_{FF}(\omega = \Omega) - S_{FF}(\omega = -\Omega)] \,. \tag{8.31}$$

This coincides with the expression from the classical linearized analysis, when taking into account $g^2 = g_0^2 n_{\rm phot}$.

8.2.3 Displacement readout

We briefly discuss the measurement of $x(t)$, whose precision is limited by the so-called "standard quantum limit" (SQL) of displacement detection. A much more extended discussion on the quantum limits of measurement and amplification can be found in Chapter 2 of this volume, as well as in the review [23] and the book [24]. An alternative discussion, geared towards electromechanical circuits, can be found in Chapter 9.

The phase shift of the reflected light is linear in the displacement (for small displacements). It is set by the optical frequency shift multiplied by the photon lifetime: $\theta \sim (\omega_{\rm cav}/\kappa)(x/L)$. A direct observation of $x(t)$ in the time domain, for an oscillator in thermal equilibrium, would exhibit an oscillation at Ω, whose amplitude and phase slowly fluctuate on a timescale $1/\Gamma$ (Fig. 8.5). The total size of the fluctuations is determined by the equipartition theorem, that is, $m\Omega^2 \langle x^2 \rangle /2 = k_B T/2$.

However, since the phase-shift measurement is sensitive to all mechanical normal modes simultaneously, it is better to go into frequency space and look at the noise spectrum, as obtained experimentally via a spectrum analyzer from the phase-shift signal. The symmetrized displacement noise spectrum $\bar{S}_{xx}(\omega) = [S_{xx}(\omega) + S_{xx}(-\omega)]/2$ is related to the susceptibility of the harmonic oscillator via the fluctuation-dissipation theorem:

$$\bar{S}_{xx}(\omega) = 2\hbar \coth\left(\frac{\hbar\omega}{k_B T}\right) \operatorname{Im} \chi_{xx}(\omega) \,, \tag{8.32}$$

where $\delta x[\omega] = \chi_{xx}(\omega) F_{\rm ext}[\omega]$ defines the susceptibility, that is, $\chi_{xx}(\omega) = (\Omega^2 - \omega^2 - i\omega\Gamma)^{-1} m^{-1}$. $\bar{S}_{xx}(\omega)$ consists of two Lorentzians, centered at $\omega = \pm\Omega$, with a width set by Γ and an area proportional to the temperature:

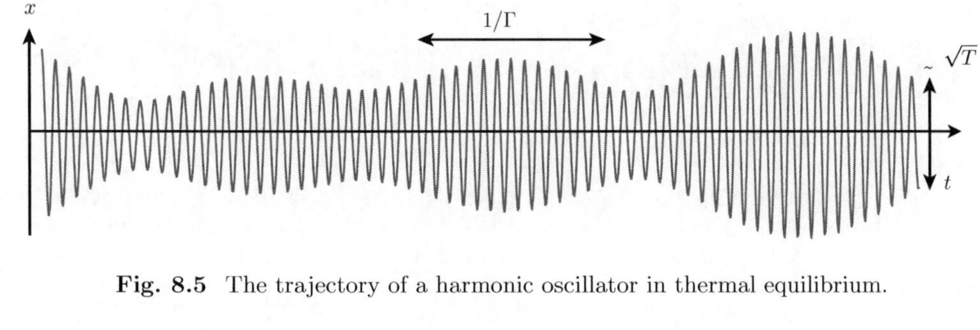

Fig. 8.5 The trajectory of a harmonic oscillator in thermal equilibrium.

$$\int_{-\infty}^{+\infty} d\omega \, \bar{S}_{xx}(\omega) = \langle x(0)^2 \rangle = k_B T / m\Omega^2 \,, \tag{8.33}$$

with the last equality evaluated in the classical limit.

Now the measured signal also contains contributions from the noise in the phase-shift determination ("imprecision noise"), and from some additional noisy component of the motion that is driven by the fluctuating radiation pressure force ("backaction noise"), whose noise spectrum has been calculated in Section 8.2.2:

$$x^{\rm meas}(t) = x(t) + \delta x^{\rm imp}(t) + \delta x^{\rm backaction}(t) \,. \tag{8.34}$$

Thus, the total measured noise spectrum $\bar{S}_{xx}^{\rm meas}(\omega)$ contains both of these extra noise contributions (Fig. 8.6). One may try to reduce the imprecision noise by increasing the laser intensity. However, this will enhance the backaction noise. There is always an optimum at intermediate laser intensity, and it is set by the SQL:

$$\bar{S}_{xx}^{\rm added}(\omega) \geq \bar{S}_{xx}^{T=0}(\omega) \,; \tag{8.35}$$

that is, the minimum added noise is equal to the zero-point fluctuations. When measuring at the mechanical resonance, this means that one can resolve $x(t)$ down to the mechanical zero-point fluctuations during the damping time Γ^{-1} (but no further).

The reason for this limit is ultimately Heisenberg's uncertainty relation: The weak, continuous measurement of the full mechanical trajectory $x(t)$ that is performed in this interferometric setup cannot be infinitely precise, because this would amount to a simultaneous determination of position and momentum at any given time. This can be seen most easily when decomposing the trajectory into "quadratures":

$$\hat{x}(t) = \hat{x}(0)\cos(\Omega t) + \frac{\hat{p}(0)}{m\Omega}\sin(\Omega t) \,. \tag{8.36}$$

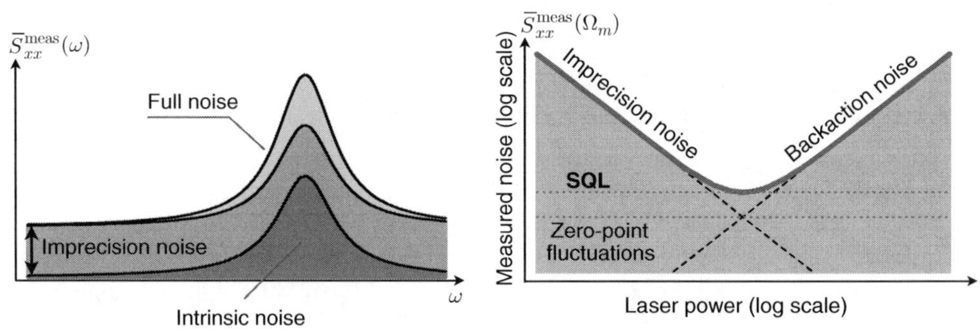

Fig. 8.6 The measured noise spectrum, with extra contributions from the imprecision noise (phase-shift fluctuations) and from the backaction noise (induced by the fluctuating radiation force). There is an optimum at intermediate laser power (i.e., intermediate "measurement rate") and it is set by the SQL.

Optomechanical experiments have already reached the regime where the imprecision noise alone would fall below the SQL (which means that the backaction noise already has to be appreciable). There is more about this in Chapter 9 by Konrad Lehnert, whose group has pioneered an optomechanical setup involving a microwave resonator coupled to nanomechanical motion.

8.3 Nonlinear dynamics

Recall the detuning dependence of the light-induced optomechanical damping rate $\Gamma_{\mathrm{opt}}(\Delta)$. At blue detuning, Γ_{opt} becomes negative, such that the full damping rate of the mechanical system is reduced below its intrinsic value Γ:

$$\Gamma_{\mathrm{full}} = \Gamma + \Gamma_{\mathrm{opt}}(\Delta) . \tag{8.37}$$

What happens if Γ_{full} itself becomes negative, which will eventually happen for sufficiently large laser drive? Then the mechanical system is unstable, and any small initial fluctuation of $x(t)$ will grow exponentially. The energy will increase as $\exp(|\Gamma_{\mathrm{full}}|\,t)$. Finally, this growth saturates owing to nonlinear effects, and the mechanical oscillator settles into a regime of "self-sustained oscillations" (see, e.g., [25–27] for experiments, and [28, 29] for theory). The power input needed to sustain these oscillations (against the intrinsic mechanical friction) is provided by the radiation force. This is completely analogous to the working of a laser, where some pump provides always replenishes the coherent lasing oscillations. For that reason, the phenomenon encountered here is also sometimes called "mechanical lasing" (or, in the quantum regime, "phonon lasing").

The oscillation amplitude A depends on all the microscopic parameters, such as detuning Δ and laser drive power. A plot of A versus Δ (or any other parameter) would reveal that it grows like $\sqrt{\Delta - \Delta_c}$ just above the threshold Δ_c. Mathematically, this behaviour is known as a "Hopf bifurcation." Interestingly, there are also attractors (stable values of A) that are not smoothly connected to the $A = 0$ solution, starting out immediately at a finite amplitude. The presence of multiple stable attractors for a given set of external parameters is known as "dynamical multistability."

The amplitude A can be found from the "power balance" condition

$$\langle F(t)\dot{x}\rangle_t = m\Gamma\langle \dot{x}^2\rangle_t \tag{8.38}$$

(Fig. 8.7), where one makes the ansatz (supported by numerics) of simple sinusoidal oscillations at the original mechanical eigenfrequency:

$$x(t) = x_0 + A\cos(\Omega t) . \tag{8.39}$$

The radiation force $F \propto |\alpha(t)|^2$ is then calculated from the solution for the light amplitude $\alpha(t)$, which can be obtained exactly as a Fourier series, given this particular form of $x(t)$.

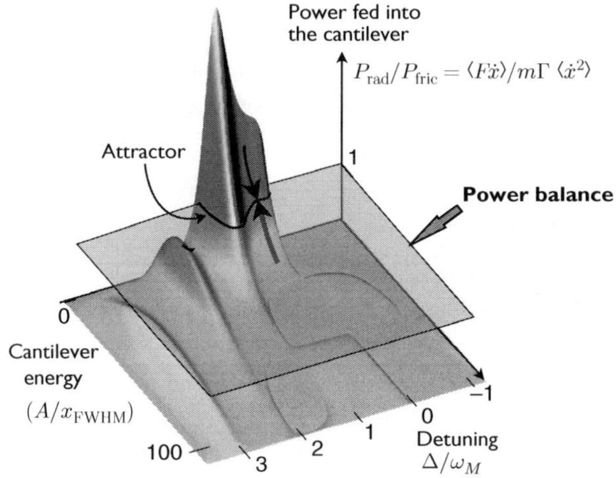

Fig. 8.7 The attractor diagram is obtained by demanding that the power fed into the mechanical oscillation by the radiation force be equal to the power lost owing to mechanical friction.

We note several interesting aspects of the nonlinear dynamical regime:

- At larger laser drive, the system displays chaotic motion, where the simple ansatz for $x(t)$ ceases to be a good approximation [30]. This is still mostly unexplored.
- The phase of the oscillation is picked at random, but it is fixed in time in the ideal case. Any noise (e.g., thermal Langevin forces) will lead to a slow phase diffusion that becomes more pronounced for smaller amplitudes, that is, close to the threshold (as in a laser).
- Noise forces can also lead to stochastic jumps between attractors. One can ascribe an effective potential to the slow motion of the amplitude, and one finds Arrhenius-type expressions for the noise-induced tunneling between minima in this potential. Again, this has yet to be observed.
- Coupling many such optomechanical self-sustained oscillators (little "clocks") can lead to synchronization [31], an interesting and universal collective phenomenon. This is also helpful in reducing the effects of noise, making the clocks more stable.
- In the quantum regime, quantum fluctuations are amplified strongly just below threshold, which is no longer sharp. A phase-space plot of the mechanical Wigner density reveals the co-existence of several attractors. The same is also found in the phonon number distribution, which therefore displays large fluctuations.

8.4 Basic quantum state manipulations

8.4.1 Squeezing the light field

One of the simplest nontrivial quantum states of a harmonic oscillator is a squeezed state, where the position and momentum uncertainty still achieve the Heisenberg bound, but are not equal to the ground-state uncertainties. For example, the position uncertainty may be reduced, at the expense of the momentum uncertainty. For a light beam, it is better to talk about uncertainties of amplitude and phase. Squeezing these fluctuations in a light beam is important for being able to perform interferometric measurements with a precision that goes beyond the standard quantum limit, as was first recognized for the case of laser-interferometric gravitational wave observatories.

An optomechanical system can be viewed as an effective Kerr medium (i.e., providing a nonlinearity for the photon field) with some time-delayed response [32]. In the simplest case, for large $\kappa \gg \Omega$, we can write $\omega_{\rm cav} = \omega_{\rm cav}(x(I))$ for the optical resonance of the cavity depending on intensity I. This is analogous to having a medium inside the cavity with an intensity-dependent refractive index.

Thus, an incoming laser beam with intensity fluctuations may have these fluctuations reduced after backreflection from an optomechanical system ("amplitude squeezing"). Note that this can happen only at finite frequencies of the intensity fluctuations, since at zero frequency the intensity going into a one-sided cavity must equal exactly the intensity going out of the cavity (neglecting losses). When plotting the intensity noise spectrum $S_{II}(\omega)$ against fluctuation frequency ω, one finds at low but finite frequencies a dip below the constant "white" noise spectrum that characterizes the shot noise of a coherent laser beam. The details depend on detuning, laser power, mechanical frequency, cavity decay rate, etc.

The squeezing spectrum can be obtained by solving the linearized Heisenberg equations of motion and applying input-output theory, which relates the backscattered light field to the incoming light field and the light amplitude inside the cavity:

$$\hat{a}_{\rm out}(t) = \hat{a}_{\rm in}(t) + \sqrt{\kappa}\hat{a}(t)\,. \tag{8.40}$$

We will not go through this lengthy but straightforward exercise here. However, for reference, we display the Heisenberg equations of motion (still in their full, nonlinearized, form):

$$\dot{\hat{b}} = (-i\Omega - \Gamma/2)\hat{b} + ig_0\hat{a}^\dagger\hat{a} - \sqrt{\Gamma}\hat{b}_{\rm in}(t), \tag{8.41}$$

$$\dot{\hat{a}} = \{i[\Delta + g_0(\hat{b}+\hat{b}^\dagger)] - \kappa/2\}\hat{a} - \sqrt{\kappa}\hat{a}_{\rm in}(t)\,, \tag{8.42}$$

where $-\sqrt{\kappa}\,\langle\hat{a}_{\rm in}\rangle = (\kappa/2)\alpha_{\rm max}$ is set by the laser drive amplitude. The fields $\hat{b}_{\rm in}$ and $\hat{a}_{\rm in}$ contain the (quantum and thermal) noise acting on the mechanical and on the optical mode. One has $\langle\hat{a}_{\rm in}(t)\hat{a}^\dagger_{\rm in}(0)\rangle = \delta(t)$, $\langle\hat{a}^\dagger_{\rm in}(t)\hat{a}_{\rm in}(0)\rangle = 0$, $\langle\hat{b}_{\rm in}(t)\hat{b}^\dagger_{\rm in}(0)\rangle = (n_{\rm th}+1)\delta(t)$, and $\langle\hat{b}^\dagger_{\rm in}(t)\hat{b}_{\rm in}(0)\rangle = n_{\rm th}\delta(t)$.

8.4.2 Squeezing the mechanical motion

A very simple way to produce squeezing in any oscillator is to suddenly change its frequency. Then, the former ground-state wavefunction will start to expand and contract. This process can be made more efficient by periodically modulating the frequency, $\Omega(t)$. This gives rise to what is known as parametric amplification, and it is the trick employed by a child on a swing to increase the oscillation amplitude.

Let us briefly look into the quantum theory of this process. We start from the Hamiltonian of a parametrically driven harmonic oscillator:

$$\hat{H} = \hbar\Omega\hat{b}^\dagger\hat{b} + \hbar\epsilon\cos(2\Omega t)(\hat{b}+\hat{b}^\dagger)^2\,. \tag{8.43}$$

Here Ω is the initial frequency. The second term stems from the time-dependent modulation of $\Omega(t)$, as in $\Omega(t)^2\hat{x}^2$. It can be expanded into several terms, many Ø of which (like $e^{-i2\Omega t}\hat{b}^\dagger\hat{b}$) are off-resonant, which can be seen by inserting the unperturbed Heisenberg evolution: $\hat{b}(t) = \hat{b}(0)e^{-i\Omega t}$. We keep only the resonant terms ("rotating-wave approximation") and switch into a rotating frame, where $\hat{b}^{\rm old}(t) = \hat{b}^{\rm new}(t)e^{-i\Omega t}$. We will again omit the superscript in the following. Then we get

$$\hat{H} = \frac{\hbar\epsilon}{2}(\hat{b}^2 + (\hat{b}^\dagger)^2)\,, \tag{8.44}$$

which is known as a "squeezing Hamiltonian". The Heisenberg equations of motion, $\dot{\hat{b}} = [\hat{b}, \hat{H}]/i\hbar = -i\epsilon\hat{b}^\dagger$ and $\dot{\hat{b}}^\dagger = i\epsilon\hat{b}$, are solved by

$$\hat{b}(t) = \hat{b}(0)\cosh(\epsilon t) - i\hat{b}^\dagger(0)\sinh(\epsilon t)\,. \tag{8.45}$$

Thus, we find that the particular quadratures

$$(1 \pm i)\hat{b} + (1 \mp i)\hat{b}^\dagger \sim \hat{x} \mp \frac{\hat{p}}{m\Omega} \tag{8.46}$$

grow (shrink) like $e^{\pm\epsilon t}$ at long times. This then produces a state that is squeezed in these directions, as can best be observed in a phase-space plot.

Coming back to the optomechanical system, we can obviously exploit the optical spring effect to obtain a time-dependent effective mechanical frequency, just by modulating the laser intensity. Let us briefly recall how the optical spring effect would be derived quantum mechanically. Starting from the linearized Hamiltonian

$$\hat{H} = -\hbar\Delta\delta\hat{a}^\dagger\delta\hat{a} - \hbar g(\delta\hat{a} + \delta\hat{a}^\dagger)(\hat{b} + \hat{b}^\dagger) + \ldots\,, \tag{8.47}$$

we can apply second-order perturbation theory to eliminate the laser-driven cavity. This yields

$$\hat{H}_{\rm eff} = \hbar\frac{g^2}{\Delta}(\hat{b} + \hat{b}^\dagger)^2 + \ldots\,. \tag{8.48}$$

Now $g = g_0\sqrt{n_{\rm phot}(t)}$ can be modulated via the laser drive (equivalently, the detuning may be modulated) [33].

We will later see that the optical spring effect (and its modulation) are also valuable tools for producing entanglement.

8.4.3 Quantum nondemolition (QND) readout of mechanical quadratures

Suppose we have produced an interesting mechanical state. How would we go about characterizing it? Full knowledge of its density matrix would be the optimal goal, because it allows us to calculate any desired characteristic. Unfortunately, the simple interferometric displacement measurement will not be sufficient, because its precision is limited by the standard quantum limit, thus "blurring" all the finer features of the mechanical quantum state.

As the SQL is enforced due to Heisenberg's uncertainty relation between the quadratures, there is a way out: just measure only one of the quadratures at a time, to infinite precision, for example $\hat{x}_\varphi$ in the decomposition

$$\hat{x}(t) = \hat{x}_\varphi \cos(\Omega t - \varphi) + \hat{y}_\varphi \sin(\Omega t - \varphi)\,. \tag{8.49}$$

This is a quantum nondemolition (QND) measurement, as it can be repeated with the same result; that is, it does not destroy the quantum state. Then repeatedly produce the same state in many runs, eventually obtaining the probability density of that particular quadrature, $p(x_\varphi)$. Finally, repeat the same procedure for all angles φ. The resulting set of probability densities can be used to reconstruct the Wigner phase-space density

$$W(x,p) \equiv \int \frac{dy}{2\pi\hbar}\, e^{-ipy/\hbar} \rho\left(x + \frac{y}{2}, x - \frac{y}{2}\right). \tag{8.50}$$

The $p(x_\varphi)$ are projections of the Wigner density along all possible directions. The inverse Radon transform can be used to reconstruct a two-dimensional density from all its one-dimensional projections, a method known as tomography (Fig. 8.8). "Quantum state tomography" has already been successfully applied to states of the electromagnetic field, such as Fock states of a well-defined photon number, as well as to a nanomechanical resonator coupled to a superconducting qubit [49] using a method different from quadrature measurements.

Again, returning to optomechanics, how would we measure only a single selected mechanical quadrature [34]? First, imagine the effect of a "stroboscopic" measurement of position, at periodic times $t + n2\pi/\Omega$. The x measurement will perturb the momentum, but this will have no effect on the x fluctuations a full period later (a special property of the harmonic oscillator). Now, it turns out that the same goal can be achieved just as well by periodically modulating the amplitude of the incoming measurement laser (like $\cos(\Omega t)$). The phase of these amplitude modulations will then select the phase of the mechanical quadrature to be measured. All the noisy backaction forces only affect the other quadrature, not the measured one.

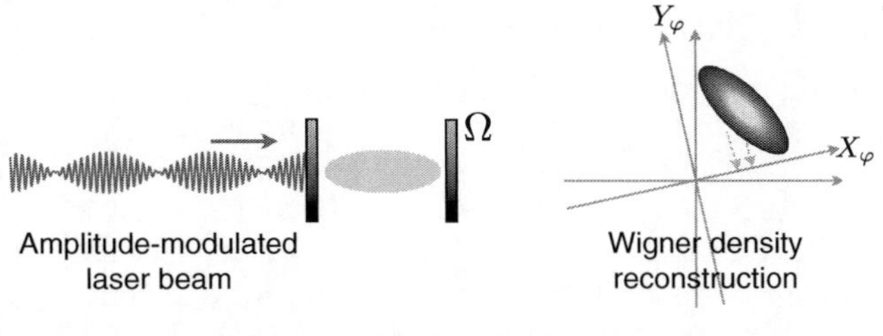

Fig. 8.8 A simple amplitude modulation of the measurement laser beam yields a selected mechanical quadrature. Repeating the measurement many times, for different quadrature phases, will enable full tomographical reconstruction of the mechanical Wigner density.

8.5 Optomechanical entanglement

8.5.1 Mechanical entanglement via light-induced coupling

Suppose the optical mode couples to two mechanical modes (e.g., two membranes or nanowires inside a cavity, or two normal modes of one object). Then the linearized Hamiltonian will be (assuming identical coupling strength)

$$\hat{H} = \ldots - \hbar g(\delta\hat{a} + \delta\hat{a}^\dagger)(\hat{X}_1 + \hat{X}_2)\,, \tag{8.51}$$

where $\hat{X}_j = \hat{b}_j + \hat{b}_j^\dagger$. Again, using second-order perturbation theory, we arrive at the "collective" optical spring effect, which couples the different oscillators:

$$\hat{H}_{\text{eff}} = \ldots + \frac{\hbar g^2}{\Delta}(\hat{X}_1 + \hat{X}_2)^2\,. \tag{8.52}$$

This is a laser-tunable mechanical coupling. Now, if the oscillators are in their ground state, this coupling will automatically lead to entanglement. At finite temperatures, one will have to employ laser cooling. There are two alternatives. One can cool quickly into the ground state, then switch off the cooling beams and see some transient entanglement. Otherwise, one can have the cooling laser beam switched on all the time and obtain steady-state entanglement (e.g. [35]). The cooling has to be strong enough to make the coupled system go near its ground state. However, if it is too strong, the entanglement will be destroyed again [36].

An even more efficient alternative is to periodically modulate $g(t)$ at a frequency $\Omega_1 + \Omega_2$, to engineer a two-mode squeezing Hamiltonian [37]

$$\hat{H}_{\text{eff}} = \ldots + \hbar J(\hat{b}_1\hat{b}_2 + \hat{b}_1^\dagger\hat{b}_2^\dagger)\,, \tag{8.53}$$

written in the rotating-wave approximation, in the rotating frame, where the bare time evolution of $\hat{b}_1$ and $\hat{b}_2$ has been transformed away. If $g(t)$ is modulated at $\Omega_1 - \Omega_2$, we get a beamsplitter Hamiltonian that can be used for state transfer between the mechanical modes:

$$\hat{H}_{\text{eff}} = \ldots + \hbar J(\hat{b}_1^\dagger\hat{b}_2 + \hat{b}_2^\dagger\hat{b}_1)\,. \tag{8.54}$$

8.5.2 Light-mechanics entanglement

The linearized optomechanical Hamiltonian

$$\hat{H} = \hbar\Omega\hat{b}^\dagger\hat{b} - \hbar\Delta\delta\hat{a}^\dagger\delta\hat{a} - \hbar g(\delta\hat{a} + \delta\hat{a}^\dagger)(\hat{b} + \hat{b}^\dagger) + \ldots \tag{8.55}$$

describes two coupled harmonic oscillators, in resonance for $\Delta = -\Omega$. In principle, this is already enough to generate light-mechanics entanglement, and various schemes have been analyzed. Characterizing the entanglement involves doing homodyne measurements on the light field and single-quadrature measurements on the mechanical oscillator (as already discussed), such that correlators between the light and mechanical quadratures can be extracted. Incidentally, we note that this continuous variable entanglement (nowadays routinely generated for intense light beams) is the type originally envisioned by Einstein, Podolsky, and Rosen. They had considered an extreme two-mode squeezed state, $\psi \sim \delta(x_1 - x_2)\delta(p_1 + p_2)$.

Light-mechanics entanglement in an optomechanical setup can also be analyzed using the full optomechanical interaction,

$$-\hbar g_0 \hat{a}^\dagger \hat{a}(\hat{b} + \hat{b}^\dagger)\,. \tag{8.56}$$

This corresponds to a photon-number-dependent force, which shifts the mechanical potential by a discrete amount $\delta\hat{x} = \hbar g_0 \hat{a}^\dagger \hat{a}/(x_{\mathrm{ZPF}} m \Omega^2)$ ("force over spring constant"). Assume for simplicity that the cavity is initially filled with a superposition of different photon numbers, $\sum_n c_n |n\rangle$ (e.g., a coherent state of the light field). Then, for each photon number, the mechanical oscillation will proceed in a harmonic potential that is shifted by a different amount. The oscillation will be described by a coherent state of the mechanical motion, $|\beta_n(t)\rangle$, with $\hat{b}\,|\beta_n(t)\rangle = \beta_n(t)\,|\beta_n(t)\rangle$. Thus, after some time, we have an entangled state [38, 39]

$$|\psi\rangle\,(t) = \sum_{n=0}^{\infty} c_n\,|n\rangle \otimes |\beta_n(t)\rangle\, e^{i\varphi_n(t)}\,. \tag{8.57}$$

It is interesting to study the ratio between the displacement per photon and the mechanical zero-point fluctuations:

$$\frac{\delta\hat{x}/\hat{a}^\dagger\hat{a}}{x_{\mathrm{ZPF}}} = 2\frac{g_0}{\Omega}\,. \tag{8.58}$$

If this is larger than one, then the overlap between the different coherent states almost vanishes after half a cycle: $|\langle\beta_n|\beta_{n'}\rangle| \ll 1$. One may term this state a "Schrödinger cat state" in an optomechanical system.

8.6 Fundamental tests of quantum mechanics

As the mechanical objects we are dealing with are relatively large (e.g., about 10^{14} atoms in a typical cantilever), they might be used to test quantum mechanics in a new regime, namely, that of large masses. Note that the experiments by Arndt *et al.*, [40] sending buckyballs and heavier molecules through interferometers, are dealing only with hundreds of atoms. It should be stated clearly that so far there is no indication whatsoever (in all of physics) that quantum mechanics will break down in any regime. Nevertheless, some researchers worry about the measurement process or about the classical limit emerging on larger length/mass scales, or about the cosmological implications of a wavefunction. There have been proposals that the Schrödinger equation might need to be modified by small extra terms (maybe arising due to as-yet mysterious effects of quantum gravity). These terms would describe an extra decoherence mechanism that would fundamentally prevent superpositions from being established on a macroscopic scale. While all of this is highly speculative, it is true that there have been no tests of quantum mechanics in the regime of large masses, and optomechanical systems may offer a way of testing it.

One early idea [41, 42] is to extract the decoherence of a mechanical object via the light field. A kind of optomechanical "which-path" setup could be realized by having a Michelson interferometer with two cavities, one in each arm, and one of them with a movable end-mirror. As we have seen in Section 8.5, the resulting superposition between a 0-photon and a 1-photon state will have its coherence reduced by the overlap of the corresponding mechanical states: $|\rho_{10}(t)| = |\langle\beta_0(t)|\beta_1(t)\rangle|$. This coherence can be extracted in the photon interference. Neglecting mechanical dissipation and decoherence, the photon coherence displays periodic revivals, at integer multiples of the mechanical period. This is even true at higher temperatures. At these times, the mechanical oscillator comes back to its initial state, thereby erasing the "which-path information" about the photon going through the optomechanical cavity. Only mechanical dissipation (or the extra "collapse of the wavefunction" speculated about in extensions of quantum mechanics) will suppress these revivals. The challenge, however, will be to distinguish the standard sources of decoherence from the hypothetical extra effects.

For this purpose, it helps to have some idea of the parametric dependence of such extra decoherence sources. For example, in the proposal on gravitationally induced decoherence (made independently by Penrose and Diósi [43, 44]), the decoherence rate is given determined by the gravitational self-energy of a configuration where a mass is in a superposition of either here or there. At large distances, $\Gamma_\varphi \sim GM^2/\hbar R$, where R is the size of the object, M its mass, and G Newton's gravitational constant.

Any such experiment will benefit greatly from suppressing the ordinary decoherence, as set by the mechanical damping rate Γ that quantifies the coupling to the mechanical thermal environment. Therefore, recent proposals for employing levitating objects (e.g., dielectric spheres trapped in an optical dipole potential [45–47]) aim to reduce Γ essentially down to zero.

8.7 Hybrid systems

Light couples to atoms or quantum-dot excitons, but it also couples to mechanical motion, which in turn can be made to couple to spins, superconducting qubits, microwave resonators, or quantum dots.

For example, the motion of one of those mechanical resonators can be made to couple to a cloud of cold atoms or even to the motion of a single atom inside an optical lattice potential [48].

Superconducting qubits have been shown to couple strongly to a piezoelectric nanoresonator [49]. This nanoresonator then in turn could couple to the light field, ultimately allowing transfer between gigahertz quantum information and flying qubits. One potential tool to be used in this task is the photon-phonon translator to be discussed in Section 8.9.2.

8.8 Ultrastrong coupling

We have seen the mechanical mode and the laser-driven optical mode hybridize in the "strong-coupling" regime $g = g_0\sqrt{\bar{n}_{\text{cav}}} > \kappa$. However, the physics there is still described by two coupled harmonic oscillators. We now want to ask what happens when even g_0, the single-photon coupling rate, gets large by itself.

Effects of "ultrastrong coupling" g_0 (or "single-photon strong coupling") were investigated in early work that mostly completely neglected decay [38, 39] and more realistically recently for the optomechanical instability [29], for the cavity spectrum [50], and for optomechanical photon blockade [51].

If $g_0 > \kappa$, then the strong-coupling regime exists even for, on average, 1 photon inside the cavity. The optical frequency shift produced by the zero-point displacement becomes large in that case. The linearized Hamiltonian is no longer applicable, and we have to go back to the full Hamiltonian

$$\hat{H} = -\hbar g_0 \hat{a}^\dagger \hat{a}(\hat{b} + \hat{b}^\dagger) + \hbar\Omega\hat{b}^\dagger\hat{b} - \hbar\Delta\hat{a}^\dagger\hat{a} + \dots . \tag{8.59}$$

As explained before, there is a photon-number-dependent shift of the mechanical oscillator equilibrium position. We can incorporate this by shifting the mechanical operators, $\hat{b}^{\text{new}} = \hat{b}^{\text{old}} - (g_0/\Omega)\hat{a}^\dagger\hat{a}$, which is the form suggested by writing the first two terms in the combined form $\hbar\Omega(\hat{b}^\dagger)^{\text{new}}\hat{b}^{\text{new}}$. We will drop the superscript "new" in the following. We obtain

$$\hat{H}' = \hbar\Omega\hat{b}^\dagger\hat{b} - \hbar\Delta\hat{a}^\dagger\hat{a} - \frac{\hbar g_0^2}{\Omega}(\hat{a}^\dagger\hat{a})^2 + \dots . \tag{8.60}$$

Apparently, there is an effective attractive interaction between photons, mediated by the mechanics. Note that the transformation $\hat{H}' = \hat{U}^\dagger\hat{H}\hat{U}$ can be accomplished by what in condensed matter physics would be called the "polaron transformation": $\hat{U} = \exp[(\hat{b}^\dagger - \hat{b})(g_0/\Omega)\hat{a}^\dagger\hat{a}]$. This accomplishes the shift of the oscillator's equilibrium position. Although the terms displayed in $\hat{H}'$ do not have any interaction between $\hat{b}$ and $\hat{a}$ left, there is another term that we have not explicitly mentioned so far. The laser driving $\hbar\epsilon_L(\hat{a}+\hat{a}^\dagger)$ becomes $\hbar\epsilon_L(\hat{a}^\dagger e^{(\hat{b}-\hat{b}^\dagger)g_0/\Omega}+\text{h.c.})$ under the unitary transformation. Physically, this means that, on putting another photon inside the cavity, the oscillator potential is displaced. This creates a more complicated superposition of phonon states out of what may have been a phonon eigenstate in the old potential. The overlap

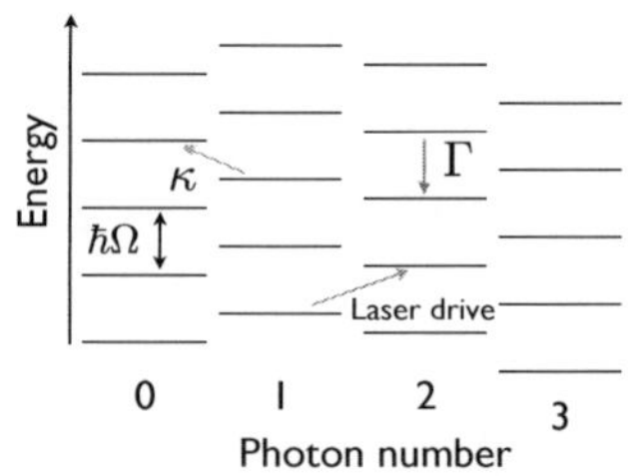

Fig. 8.9 Level scheme (in the rotating frame) for an optomechanical system with a large g_0. In this plot, $\Delta < 0$.

matrix elements between the old and new phonon eigenstates are known as "Franck-Condon factors" in molecular physics (where the role of the photons would be played by electrons undergoing transitions between different orbitals).

The energy-level scheme shows a phonon ladder for each photon number N being displaced by an amount $-\hbar\Delta N - \hbar(g_0^2/\Omega)N^2$. Different levels are connected by the laser drive, and by the decay processes (κ and Γ) (Fig. 8.9).

If

$$\frac{g_0^2}{\Omega} > \kappa\,, \tag{8.61}$$

then the photon anharmonicity induced by the mechanics can be resolved. As a consequence, one may encounter a situation where the transition between 0 and 1 photons is resonant (horizontal in this level scheme), but the transition between 1 and 2 photons is off-resonant (by more than the linewidth). This would be a case of optomechanical "photon blockade" [51], leading to antibunching of transmitted photons.

8.9 Multimode optomechanical systems

Interesting additional functionality becomes available when one can couple multiple optical and mechanical modes in some suitable way.

8.9.1 "Membrane in the middle"

One of the simplest such multimode setups is a membrane sitting inside an optical cavity, as pioneered in Jack Harris's laboratory at Yale [9, 52] (Fig. 8.10). If the membrane is highly reflective, the setup can be viewed as two optical modes. These are tunnel-coupled (photon transmission through the membrane), and their optical frequencies are pulled up/down upon movement of the membrane. The Hamiltonian is

$$\begin{aligned}\hat{H} = &-\hbar\Delta(\hat{a}_L^\dagger\hat{a}_L + \hat{a}_R^\dagger\hat{a}_R) \\ &-\hbar g_0(\hat{a}_L^\dagger\hat{a}_L - \hat{a}_R^\dagger\hat{a}_R)(\hat{b} + \hat{b}^\dagger) \\ &+\hbar J(\hat{a}_L^\dagger\hat{a}_R + \hat{a}_R^\dagger\hat{a}_L)\,.\end{aligned} \tag{8.62}$$

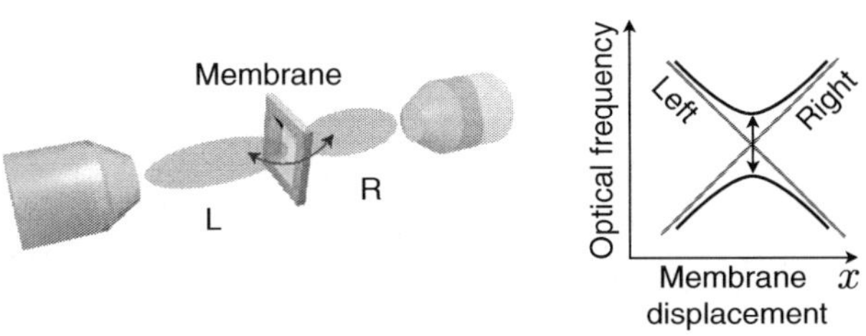

Fig. 8.10 The "membrane-in-the-middle" setup, where photons tunnel between two optical resonances (left and right of the membrane). The optical spectrum displays an avoided crossing, with an x^2 dependence on displacement at the degeneracy point.

If one diagonalizes $\hat{H}$ at fixed $\hat{b}+\hat{b}^\dagger = X$, then one finds the two optical resonances show an avoided crossing at $X = 0$, with a splitting set by $2J$. Crucially, this implies the optical resonance frequencies depend quadratically on displacement at the degeneracy point (compare the "sweet spot" of a Cooper pair box qubit). Focusing on one of the resonances, we encounter a new "quadratic" type of optomechanical coupling:

$$\hbar \frac{d^2\omega}{dx^2} x_{\mathrm{ZPF}}^2 (\hat{b} + \hat{b}^\dagger)^2 \hat{a}^\dagger \hat{a}. \tag{8.63}$$

Now, in the rotating-wave approximation we can keep only the non-oscillating terms in $(\hat{b}+\hat{b}^\dagger)^2$, which amounts to $2\hat{b}^\dagger\hat{b}+1$. This is the phonon number, coupling to the photon number. As a consequence, the optical phase shift will becomes a measure of phonon number, permitting a QND detection of mechanical Fock states. After cooling to the mechanical ground state, one will be able to observe a series of quantum jumps to higher phonon numbers, as thermal phonons leak back into the mechanical oscillator. Note that this would be a very challenging experiment. If the second mirror is leaky or there is absorption inside the setup, then extra noise quickly destroys the ability to perform QND detection. One needs $g_0 > \kappa_{\mathrm{abs}}$, where κ_{abs} represents the unwanted decay channels (not including the coupling through the front mirror, which is there even in the idealized setup) [53].

8.9.2 The phonon-photon translator

Here, we want to explain the basic idea behind a very nice application of multimode setups: the phonon-photon translator (Fig. 8.11), first suggested by Painter's group [54]. The goal is to convert single optical photons into single gigahertz-frequency phonons (and back again). When realized, this could be crucial for quantum information processing.

The first question is how to realize the frequency conversion. Here, two optical modes come into play. Indeed, we can start from the setup already described, although it would be realized inside a photonic crystal to obtain gigahertz mechanical frequency and be able to engineer photonic and phononic waveguides. We set

$$\hat{a}_{1/2} = \frac{\hat{a}_L \pm \hat{a}_R}{\sqrt{2}}, \tag{8.64}$$

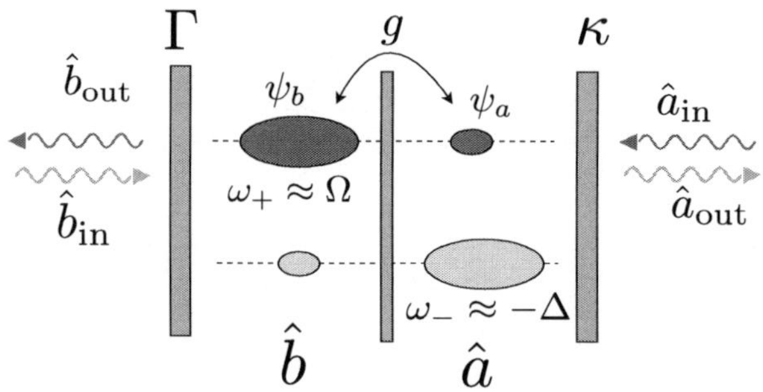

Fig. 8.11 The photon-phonon translator. An essential ingredient consists in matching the effective coupling to the incoming photon field and the outgoing phonon field, by demanding $\Gamma|\psi_b|^2 = \kappa|\psi_a|^2$. Then, incoming photons will be converted 100% into outgoing phonons (traveling along a phonon waveguide).

in which case the Hamiltonian becomes (at the degeneracy point)

$$\hat{H} = \hbar\omega_1 \hat{a}_1^\dagger \hat{a}_1 + \hbar\omega_2 \hat{a}_2^\dagger \hat{a}_2 - \hbar g_0(\hat{a}_1^\dagger \hat{a}_2 + \hat{a}_2^\dagger \hat{a}_1)(\hat{b} + \hat{b}^\dagger) + \dots . \tag{8.65}$$

Now, if we drive the $\hat{a}_2$ mode strongly, we can replace the operator by the mean amplitude, and recover the following Hamiltonian:

$$\hat{H} \approx -\hbar\Delta \hat{a}_1^\dagger \hat{a}_1 - \hbar g(\hat{a}_1^\dagger + \hat{a}_1)(\hat{b} + \hat{b}^\dagger) + \dots , \tag{8.66}$$

where $g = g_0\alpha_2$. The idea is similar to the standard approach of the linearized optomechanical Hamiltonian. However, physically, there is an important difference: here the $\hat{a}_1$ operator still refers to real photons (not a shifted field mode). This is essential for the application in mind. Physically, frequency conversion will be achieved by taking a photon $\hat{a}_1$, dumping most of its energy into the coherent stream of photons passing through $\hat{a}_2$, and placing the rest into the phonon field.

In the following, for brevity, we will set $\hat{a}_1 \mapsto \hat{a}$. Then, we have the following Heisenberg equations of motion, including the damping and the in-coupling fields:

$$\dot{\hat{a}} = (i\Delta - \kappa/2)\hat{a} - \sqrt{\kappa}\hat{a}_{in} + ig(\hat{b} + \hat{b}^\dagger), \tag{8.67}$$

$$\dot{\hat{b}} = (-i\Omega - \Gamma/2)\hat{b} - \sqrt{\Gamma}\hat{b}_{in} + ig(\hat{a} + \hat{a}^\dagger). \tag{8.68}$$

In contrast to the usual case, we will assume that the setup involves a phononic waveguide coupled to $\hat{b}$. This waveguide transports away the phonons created inside the mode, and it provides the dominant decay channel (ideally, all of Γ is due to this coupling). Note that we will drop the off-resonant terms $\hat{b}^\dagger$ and $\hat{a}^\dagger$ on the right-hand-side.

Applying input-output theory, we have

$$\hat{b}_{out} = \hat{b}_{in} + \sqrt{\Gamma}\hat{b} = \dots = S_{bb}\hat{b}_{in} + S_{ba}\hat{a}_{in} . \tag{8.69}$$

Here, the strategy will be to eliminate $\hat{a}$ and $\hat{b}$ and write the result fully in terms of the incoming fields. In other words, we want to obtain the scattering matrix of this setup. Our goal will be to make the prefactor S_{bb} equal to zero. Then, there will be perfect transmission from the photon field into the phonon field (and $|S_{ba}| = 1$). As S_{bb} is complex-valued, these are two equations. However, we can choose two parameters, (e.g., g and Δ) to satisfy both of these equations.

Instead of going through the mathematics and writing down the full expression for S_{bb}, we can also appeal to physical intuition. We are in fact facing an impedance-matching problem: usually, $\Gamma \ll \kappa$, and we have to find a way of still transmitting 100%.

The equations of motion correspond to the case of a cavity with unequal end-mirrors (transmissions characterized by Γ and κ). Usually, one cannot get perfect transmission in that case, except if one applies a trick: inserting another mirror into the middle of the cavity (transmission set by g) gives a hybridization of modes in the left and right half of this fictitious cavity (not to be confused with the original membrane in the middle). If the original frequencies of the left (right) mode are Ω ($-\Delta$), then the hybridized frequencies are

$$\omega_\pm = \frac{\Omega - \Delta}{2} \pm \sqrt{\left(\frac{\Omega + \Delta}{2}\right)^2 + g^2}\,. \tag{8.70}$$

The "molecular orbital" of the upper mode ($\omega_+ \approx \Omega$, by assumption) will have amplitudes ψ_b and ψ_a in the left (right) cavity, with ψ_b dominating. We want to match the decay rates, by demanding

$$\Gamma \left|\psi_b\right|^2 = \kappa \left|\psi_a\right|^2\,. \tag{8.71}$$

On the other hand, the eigenvalue problem tells us that the ratio of probabilities is just given by

$$\frac{\left|\psi_a\right|^2}{\left|\psi_b\right|^2} \approx \frac{g^2}{(\Omega + \Delta)^2} = \frac{\Gamma}{\kappa}\,, \tag{8.72}$$

where we have assumed $\Omega + \Delta \gg g$. This is, finally, the condition we have to fulfill to obtain perfect transmission when irradiating the setup at the upper frequency ω_+. Then, indeed, one obtains the desired photon-phonon translator (which can be run both ways). Having such a scheme allows for several interesting applications [54]:

- phonon/photon conversion;
- single-phonon detection via single-photon detection;
- storing light/delaying light in a long-lived phonon memory;
- frequency-filtering optical signals (the bandwidth is set by Γ).

8.9.3 Collective dynamics of optomechanical arrays

The photonic crystal design, with strongly localized optical and vibrational modes, can be engineered to build arrays of many coupled modes [31, 55].

Even in the classical regime, one may be able to observe very interesting (and useful) collective dynamics, by coupling many optomechanical self-induced oscillations. This gives rise to synchronization [31] between the oscillations, possibly important for metrology and timekeeping.

In the quantum regime, one might use arrays to transfer quantum states between different optical and mechanical modes, or entangle them selectively. If one reaches the ultrastrong-coupling regime (say, $g_0^2/\Omega > \kappa$), then one will be able to observe nonlinear quantum many-body physics in these optomechanical arrays. The investigations into these possibilities are presently only starting to emerge.

References

[1] Kippenberg, T. J. and Vahala, K. J. (2008). *Science*, **321**, 1172–1176.

[2] Marquardt, F. and Girvin, S. M. (2009). *Physics*, **2**, 40.

[3] Favero, I. and Karrai, K. (2009). *Nature Photon.*, **3**, 201–205.

[4] Genes, C., Mari, A., Vitali, D., and Tombesi, P. (2009). *Adv. At. Mol. Opt. Phys.*, **57**, 33–86.

[5] Arcizet, O., Cohadon, P.-F., Briant, T., Pinard, M., and Heidmann, A. (2006). *Nature*, **444**, 71–74.

[6] Gigan, S., Böhm, H. R., Paternostro, M., Blaser, F., Langer, G., Hertzberg, J. B., Schwab, K. C., Bäuerle, D., Aspelmeyer, M., and Zeilinger, A. (2006). *Nature*, **444**, 67–70.

[7] Favero, I., Metzger, C., Camerer, S., König, D., Lorenz, H., Kotthaus, J. P., and Karrai, K. (2007). *Appl. Phys. Lett.*, **90**, 104101.

[8] Schliesser, A., Del'Haye, P., Nooshi, N., Vahala, K. J., and Kippenberg, T. J. (2006). *Phys. Rev. Lett.*, **97**, 243905.

[9] Thompson, J. D., Zwickl, B. M., Jayich, A. M., Marquardt, F., Girvin, S. M., and Harris, J. G. E. (2008). *Nature*, **452**, 72–75.

[10] Favero, I., Stapfner, S., Hunger, D., Paulitschke, P., Reichel, J., Lorenz, H., Weig, E. M., and Karrai, K. (2009). *Opt. Express*, **17**, 12813–12820.

[11] Anetsberger, G., Arcizet, O., Unterreithmeier, Q. P., Rivière, R., Schliesser, A., Weig, E. M., Kotthaus, J. P., and Kippenberg, T. J. (2009). *Nature Phys.*, **5**, 909–914.

[12] Murch, K. W., Moore, K. L., Gupta, S., and Stamper-Kurn, D. M. (2008). *Nature Phys.*, **4**, 561–564.

[13] Brennecke, F., Ritter, S., Donner, T., and Esslinger, T. (2008). *Science*, **322**, 235–238.

[14] Regal, C. A., Teufel, J. D., and Lehnert, K. W. (2008). *Nature Phys.*, **4**, 555–560.

[15] Teufel, J. D., Donner, T., Li, D., Harlaw, J. W., Allman, M. S., Cicak, K., Sirois, A. J., Whittaker, J. D., Lehnert, K. W., and Simmonds, R. W. (2011). *Nature*, **475**, 359–363.

[16] Eichenfield, M., Camacho, R., Chan, J., Vahala, K. J., and Painter, O. (2009). *Nature*, **459**, 550–555.

[17] Chan, J., Alegre, T. P. M., Safavi-Naeini, A. H., Hill, J. T., Krause, A., Gröblacher, S., Aspelmeyer, M., and Painter, O. (2011). *Nature*, **478**, 89–92.

[18] Dorsel, A., McCullen, J. D., Meystre, P., Vignes, E., and Walther, H. (1983). *Phys. Rev. Lett.*, **51**, 1550–1553.

[19] Braginsky, V. B., Manukin, A. B., and Tikhonov, M. Yu. (1970). *Sov. Phys. JETP*, **31**, 829.

[20] Marquardt, F., Chen, J. P., Clerk, A. A., and Girvin, S. M. (2007). *Phys. Rev. Lett.*, **99**, 093902.

[21] Wilson-Rae, I., Nooshi, N., Zwerger, W., and Kippenberg, T. J. (2007). *Phys. Rev. Lett.*, **99**, 093901.

[22] Höhberger Metzger, C. and Karrai, K. (2004). *Nature*, **432**, 1002–1005.

[23] Clerk, A. A., Devoret, M. H., Girvin, S. M., Marquardt, F., and Schoelkopf, R. J. (2010). *Rev. Mod. Phys.*, **82**, 1155–1208. Longer version at arXiv:0810.4729.

[24] Braginsky, V. B. and Khalili, F. Y. (1995). *Quantum Measurements.* Cambridge University Press, Cambridge.

[25] Höhberger, C. and Karrai, K. (2004). In *Proceedings of the 4th IEEE Conference on Nanotechnology.* IEEE, New York.

[26] Kippenberg, T. J., Rokhsari, H., Carmon, T., Scherer, A., and Vahala, K. J. (2005). *Phys. Rev. Lett.*, **95**, 033901.

[27] Metzger, C., Favero, I., Ortlieb, A., and Karrai, K. (2008). *Phys. Rev. B*, **78**, 035309.

[28] Marquardt, F., Harris, J. G. E., and Girvin, S. M. (2006). *Phys. Rev. Lett.*, **96**, 103901.

[29] Ludwig, M., Kubala, B., and Marquardt, F. (2008). *New J. Phys.*, **10**, 095013.

[30] Carmon, T., Cross, M. C., and Vahala, K. J. (2007). *Phys. Rev. Lett.*, **98**, 167203.

[31] Heinrich, G., Ludwig, M., Qian, J., Kubala, B., and Marquardt, F. (2011). *Phys. Rev. Lett.*, **107**, 043603.

[32] Fabre, C., Pinard, M., Bourzeix, S., Heidmann, A., Giacobino, E., and Reynaud, S. (1994). *Phys. Rev. A*, **49**, 1337–1343.

[33] Mari, A. and Eisert, J. (2009). *Phys. Rev. Lett.*, **103**, 213603.

[34] Clerk, A. A., Marquardt, F., and Jacobs, K. (2008). *New J. Phys.*, **10**, 095010.

[35] Hartmann, M. J. and Plenio, M. B. (2008). *Phys. Rev. Lett.*, **101**, 200503.

[36] Ludwig, M., Hammerer, K., and Marquardt, F. (2010). *Phys. Rev. A*, **82**, 012333.

[37] Wallquist, M., Hammerer, K., Zoller, P., Genes, C., Ludwig, M., Marquardt, F., Treutlein, P., Ye, J., and Kimble, H. J. (2010). *Phys. Rev. A*, **81**, 023816.

[38] Mancini, S., Man'ko, V. I., and Tombesi, P. (1997). *Phys. Rev. A*, **55**, 3042–3050.

[39] Bose, S., Jacobs, K., and Knight, P. L. (1997). *Phys. Rev. A*, **56**, 4175–4186.

[40] Arndt, M., Nairz, O., Voss-Andreae, J., Keller, C., van der Zouw., G., and Zeilinger, A. (1999). *Nature*, **401**, 680–682.

[41] Bose, S., Jacobs, K., and Knight, P. L. (1999, 5). *Phys. Rev. A*, **59**, 3204–3210.

[42] Marshall, W., Simon, C., Penrose, R., and Bouwmeester, D. (2003). *Phys. Rev. Lett.*, **91**, 130401.

[43] Penrose, R. (2000). In *Quantum [Un]Speatables: From Bell or Quantum Information* (ed. R. A. Bertlmann and A. Zeilinger), pp. 319–331. Springer-Verlag, Berlin.

[44] Diósi, L. (2000). In *Decoherence: Theoretical, Experimental, and Conceptual Problems* (ed. P. Blanchard, D. Gintini, E. Joos, C. Kiefer, and I.-O. Stamatescu), pp. 243–250. Lecture Notes in Physics, Vol. 538. Springer-Verlag, Berlin.
[45] Romero-Isart, O., Juan, M. L., Quidant, R., and Cirac, J. I. (2010). *New J. Phys.*, **12**, 033015.
[46] Chang, D. E., Regal, C. A., Papp, S. B., Wilson, D. J., Ye, J., Painter, O., Kimble, H. J., and Zoller, P. (2010). *Proc. Natl. Acad. Sci. USA*, **107**, 1005–1010.
[47] Barker, P. F. and Schneider, M. N. (2010). *Phys. Rev. A*, **81**, 023826.
[48] Hammerer, K., Wallquist, M., Genes, C., Ludwig, M., Marquardt, F., Treutlein, P., Zoller, P., Ye, J., and Kimble, H. J. (2009). *Phys. Rev. Lett.*, **103**, 063005.
[49] O'Connell, A. D., Hofheinz, M., Ansmann, M., Bialczak, R. C., Lenander, M., Lucero, E., Neeley, M., Sank, D., Wang, H., Weides, M., Wenner, J., Martinis, J. M., and Cleland, A. N. (2010). *Nature*, **464**, 697–703.
[50] Nunnenkamp, A., Borkje, K., and Girvin, S. M. (2011). *Phys. Rev. Lett.*, **107**, 063602.
[51] Rabl, P. (2011). *Phys. Rev. Lett.*, **107**, 063601.
[52] Jayich, A. M., Sankey, J. C., Zwickl, B. M., Yang, C., Thompson, J. D., Girvin, S. M., Clerk, A. A., Marquardt, F., and Harris, J. G. E. (2008). *New J. Phys.*, **10**, 095008.
[53] Miao, H., Danilishin, S., Corbitt, T., and Chen, Y. (2009). *Phys. Rev. Lett.*, **103**, 100402.
[54] Safavi-Naeini, A. H. and Painter, O. (2011). *New J. Phys.*, **13**, 013017.
[55] Chang, D. E., Safavi-Naeini, A. H., Hafezi, M., and Painter, O. (2011). *New J. Phys.*, **13**, 2011.

9

Micromechanics and superconducting circuits

K. W. Lehnert

JILA, National Institute of Standards and Technology
Physics Department, The University of Colorado
Boulder, CO 80309, USA

Quantum Machines: Measurement and Control of Engineered Quantum Systems.
Edited by M. Devoret, B. Huard, R. Schoelkopf, and L. F. Cugliandolo.

Chapter Contents

9.1 Introduction

The field of nano- and micromechanics investigates the mechanical motion of micron-sized structures. Although this field has already found wide application in commercial technology, it remains an active area of scientific inquiry. In particular, micromechanics provides a method to sense tiny forces in nanoscale structures [1], to search for deviations in standard quantum theory [2], and to coax larger more tangible objects into exhibiting quantum behavior [3].

One of the most promising strategies for observing and exploiting the quantum motion of a macroscopic mechanical oscillator is realized by coupling a high-Q lower-resonance-frequency mechanical oscillator to a superconducting, higher-frequency, electrically resonant circuit [4]. When the circuit has a resonance frequency greater than a few gigahertz (the domain of microwave electricity) and is operated at a temperature well below 1 K, the electrical circuit is naturally prepared in a pure quantum state. In contrast, the much lower-frequency mechanical oscillator will have many thermal excitations even for temperatures much less than 1 K. However, if the parametric interaction between motion and electricity is strong enough, it is possible to prepare the mechanical oscillator in a pure quantum state. The mechanical oscillator essentially inherits its quantum behavior through interaction with the electrical circuit [5, 6].

Although micro-electromechanics (MEMS) provides a route to manipulating the quantum state of a mechanical oscillator, the theoretical work investigating this possibility has been expressed primarily in the language of optics [7, 8] rather than electromechanics. This situation has arisen because the experimental effort in quantum electromechanics has been closely analogous to the topic of cavity optomechanics. Cavity optomechanics also exploits parametric coupling between low-frequency mechanical motion and high-frequency electromagnetic energy to pursue similar goals. In cavity optomechanical structures, the motion of a mechanical oscillator alters the resonance frequency of an optical cavity rather than a microwave resonant circuit. Indeed, to explicitly relate these two efforts [9, 10], recent work with mechanical oscillators coupled to microwave resonant circuits has been called "microwave cavity optomechanics."

The analogy between cavity optomechanics and electromechanics is strong, but there are pitfalls in relying exclusively on this analogy. Because the mathematical tools of optics are naturally quantum, it is not immediately clear if a particular phenomenon is genuinely quantum or is in the correspondence limit. In addition, approximations valid in the optical domain may not be correct in the electrical domain. Furthermore, measurements of electricity and of light rely on very different technologies. For example, one can relate the concept of an optical photon directly to the outcome of an experiment, whereas the tools of electrical measurement detect field amplitude rather than photon number. Finally, the optical and electrical technologies offer their own distinct possibilities for extending research outside of the regime where the analogy between optomechanics and electromechanics is valid (see, e.g., [3]).

This chapter is intended to encourage a quantum theory of electromechanics built from a foundation of classical electromechanics. It is organized as follows. In Section 9.2, I introduce a simple electromechanical model that nevertheless contains interesting electromechanical phenomena. In Section 9.3, I discuss measurement within this model, including semiclassical notions of quantum-limited measurement. Finally, I show in Section 9.4 that the model can be quantized to yield the cavity optomechanical Hamiltonian. To maintain the connection to electricity, I work in conventional electrical variables (such as charge and voltage), not in dimensionless fields. I also use analogies to classical feedback theory to interpret various electromechanical phenomena.

9.2 Electromechanics of superconducting circuits

Electromechanical effects in the structures that comprise the field of nano- and micro-electromechanics (NEMS and MEMS) are generally associated with the deformation of mechanically compliant structures (Fig. 9.1). This type of motion is distinct from the motion in some larger-scale electromechanical devices, such as motors and generators, where shafts spin on bearings. Spinning and sliding structures are less practical at micron scales, where the forces of friction usually overwhelm any inertial forces. The electromechanics of flexural motion can often be modeled as a capacitor $C(x)$ (or inductor $L(x)$) with a capacitance (inductance) that depends on the deflection x of the flexural mode.[1] In principle, the shape of the flexural mode itself depends on electrical forces as well as the elastic forces. As a first approximation, one restricts the electrical energy such that modification to the flexural modes is small. One can then find the flexural modes of a compliant structure in the absence of electricity and use the shapes of those modes to calculate $C(x)$ or $L(x)$.

When considering a system that has interacting electrical and mechanical degrees of freedom, standard linear circuit analysis is poorly adapted to finding the coupled equations of motion. Specifically, the interaction between motion and electricity is usually nonlinear in the electrical variables (e.g., the force between the electrodes of a capacitor depends of the square of the separated charge). Within the steady-state ac theory of currents and voltages, it is difficult to linearize this interaction in a systematic

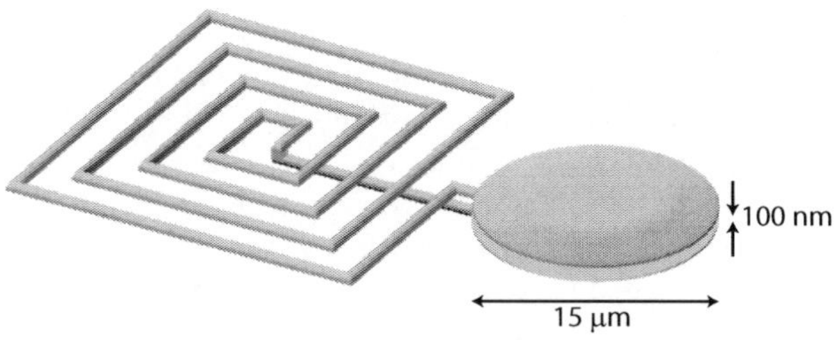

Fig. 9.1 A sketch of an electromechanical device. The structure comprises a spiral inductor and a parallel-plate capacitor where the separation between plates is a dynamical variable. With the indicated scales, the circuit would resonate at a few gigahertz.

[1] In contrast, piezoelectric, magnetostrictive, and piezoresistive phenomena are not well modeled in this way.

way. To solve electromechanical problems systematically, one treats the electrical and mechanical variables in an equivalent way by using the Lagrangian formalism. The coupling between electrical and mechanical degrees of freedom then emerges directly from the function $C(x)$ or $L(x)$. Although it is not widely used in electrical engineering (except in electromechanics), there is a fully developed Lagrangian method for finding the equations of motion for a circuit.

9.2.1 Electromechanical model

In order to illustrate this method, I introduce a simple electromechanical model that nevertheless exhibits the following phenomena: electromechanical spring effect, electromechanical damping and amplification of motion, and electromechanical strong coupling. From a semiclassical analysis of this model, one can trace the origin of the quantum limit on displacement sensitivity and identify a quantum backaction force associated with measurement itself.

The model (Fig. 9.2) is a series RLC circuit driven with a voltage source of source impedance Z_0, where the capacitor is the mechanically compliant element described by a function $C(x)$. This model is both more general and more relevant than it may appear. Near resonance, any electrical resonator can be recast as an effective RLC circuit. Because the circuit includes two resistors, it can separately model dissipation internal to the resonator and dissipation from energy coupled out of the resonator. Drawn in this way, one can find the behavior of the lumped-element circuit and later interpret the results as the behavior of a particular microwave network used in many experiments.

To find the Lagrangian of this model, I use the charge q passing through the circuit elements as one generalized coordinate, and take x as the other. The Lagrangian is then $\mathcal{L} = (1/2)[L\dot{q}^2 + m\dot{x}^2 - q^2/C(x) - k_s x^2] + qV + F_{\text{ext}}x$, where V is a generalized force acting on q and F_{ext} is an external force acting on x. If one knew the function $C(x)$, then the resulting nonlinear Lagrange equations of motion could be solved numerically.

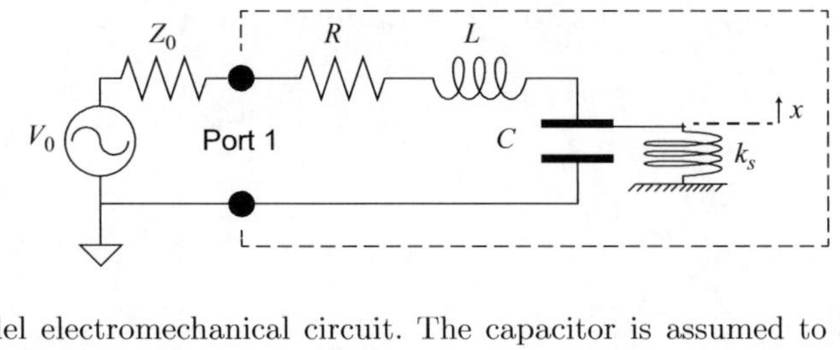

Fig. 9.2 A model electromechanical circuit. The capacitor is assumed to be a mechanically compliant element whose capacitance depends on a coordinate x. That coordinate has a mass m; it experiences both a spring force $-k_s x$ and a dissipative force $-\Gamma_M m\dot{x}$. In addition, it experiences a force F_{ext} exerted by some external agency. The dashed box separates portions of the circuit that will later be considered as part of the electromechanical resonant circuit (inside the box) and part of the external circuitry (outside the box).

9.2.2 Dynamics of an electromechanical circuit

In most NEMS and MEMS experiments, $C(x)$ is not known with much accuracy. Even if it were, a numerical solution to a specific situation wouldn't provide much insight. Instead, one can approximate the *coupling* to linear order, expanding $1/C(x) \approx 1/C(\bar{x}) - [1/C(\bar{x})^2][\partial C(\bar{x})/\partial x](x - \bar{x})$, where $\bar{x}$ is the static value of the coordinate x. The Lagrangian with linearized coupling yields the following equations of motion:

$$L\ddot{q} = V - \frac{q}{C}\left(1 - \frac{1}{C}\frac{\partial C}{\partial x}x\right) - \dot{q}(R + Z_0), \tag{9.1}$$

$$m\ddot{x} = F_{\text{ext}} - k_s(x + \bar{x}) + \frac{q^2}{2C^2}\frac{\partial C}{\partial x} - \Gamma_M m\dot{x}, \tag{9.2}$$

where I have added by hand dissipative terms $\dot{q}(R + Z_0)$ and $\Gamma_M m\dot{x}$, and redefined x as the displacement from the static value $(x - \bar{x} \to x)$. In addition, I have defined $C \equiv C(\bar{x})$. Although these equations of motion are still nonlinear, approximate solutions can be found perturbatively. While one might be able to guess linearized equations of motion for this simple example, for more complex electromechanical circuits the Lagrangian offers a systematic way to analyze the electromechanical interaction.

To solve Eqs. (9.1) and (9.2), I use experiments to guide the approximations. For typical experiments in this field, the mechanical oscillator's resonance frequency $\Omega_M = \sqrt{k_s/m}$ is much lower than the cavity resonance frequency ω_{cir}, both the mechanical oscillator and the electrical circuits have quality factors $Q > 1000$, and coupling between motion and electricity is weak unless the circuit is strongly driven. Consistent with these limits, I linearize the equations of motion around a strong voltage drive near resonance, $V(t) = (1/2)\{[V_0 + V_1(t)]e^{i\omega_D t} + \text{c.c.}\}$ and seek solutions of the form $q(t) = (1/2)\{[q_0 + q_1(t)]e^{i\omega_D t} + \text{c.c.}\}$ and $x = x(t)$. The quantity $V_1(t)$ can be used to model voltages imposed at frequencies other than ω_D, including the Johnson-Nyquist noise of the resistors themselves. I first find static solutions q_0 and $\bar{x}$; I then work to first order in x and the small parameters associated with a strong drive, $|q_1|/|q_0| \ll 1$, and a slowly varying response, $|\dot{q}_1(t)|/|\omega_D q_1(t)| \ll 1$. In addition, I drop terms oscillating at $2\omega_D$, as neither the circuit nor the mechanical oscillator will respond at these frequencies. The linear equations of motion are

$$-G\frac{q_0 q_1^*(t) + q_0^* q_1(t)}{2C\omega_{\text{cir}}} + F_{\text{ext}} = (\ddot{x} + \Gamma_m\dot{x} + \Omega_M^2 x)m, \tag{9.3}$$

$$\frac{V_1(t)}{L} + 2Gq_0\omega_{\text{cir}}x(t) = q_1(t)(\omega_D^2 - \omega_{\text{cir}}^2 - i\omega_D\kappa) + \dot{q}_1(t)(2i\omega_D + \kappa), \tag{9.4}$$

$$\frac{V_1^*(t)}{L} + 2Gq_0^*\omega_{\text{cir}}x(t) = q_1^*(t)(\omega_D^2 - \omega_{\text{cir}}^2 + i\omega_D\kappa) + \dot{q}_1^*(t)(-2i\omega_D + \kappa), \tag{9.5}$$

where $\kappa = \kappa_I + \kappa_E$, $\kappa_I = R/L$, and $\kappa_E = Z_0/L$ are the total, internal, and coupled rates of energy loss from the circuit, and the electromechanical coupling $G = -\omega_{\text{cir}}(1/2C)(\partial C/\partial x)$. I have explicitly written Eq. (9.5) to emphasize that

$q_1(t)$ and $V_1(t)$ are complex quantities. The solutions to these equations are easily found in the frequency domain. To postpone some of the confusion related to the different definitions of the Fourier transform, I solve these equations by ansatz, writing $F_{\rm ext}(t) = (1/2)[F_{\rm ext}(\omega)e^{i\omega t} + \text{c.c.}]$ and $V_1(t) = (1/2)[V_1(\omega)e^{i\omega t} + V_1(-\omega)e^{-i\omega t}]$, and seeking solutions of the form $q_1(t) = (1/2)[q_1(\omega)e^{i\omega t} + q_1(-\omega)e^{-i\omega t}]$ and $x(t) = (1/2)[x(\omega)e^{i\omega t} + \text{c.c.}]$. The resulting frequency-domain equations are

$$x(\omega) = \chi_M(\omega)\left\{F_{\rm ext}(\omega) - \frac{G[q_0 q_1^*(-\omega) + q_0^* q_1(\omega)]}{2C\omega_{\rm cir}}\right\}, \tag{9.6}$$

$$q_1(\omega) = \chi_c(\omega)[CV_1(\omega)\omega_{\rm cir} + q_0 G x(\omega)], \tag{9.7}$$

$$q_1^*(-\omega) = \chi_c^*(-\omega)[CV_1^*(-\omega)\omega_{\rm cir} + q_0^* G x(\omega)], \tag{9.8}$$

where I have defined the bare mechanical susceptibility $\chi_M^{-1} = m(\Omega_M^2 - \omega^2 + i\Gamma_M\omega)$ and the circuit susceptibility $\chi_c(\omega) = (1 - \Delta/2\omega_D)/[\Delta' - \omega + i(\kappa/2)(1 - \omega/\omega_D)]$. I have written the circuit susceptibility in terms of the detuning of the drive from circuit resonance $\Delta = \omega_D - \omega_{\rm cir}$ and defined $\Delta' = \Delta - \Delta^2/2\omega_D$. Note that, in my notation, the coefficients of the conjugate terms rotating at $+\omega$ are $q_1^*(-\omega)$ and $V_1^*(-\omega)$. For high-Q circuits and mechanical oscillators probed near resonance, the susceptibilities can be approximated by Lorentzian functions to find simpler forms $\chi_M^{-1} \approx 2m\Omega_M(\Omega_M - \omega + i\Gamma_M/2)$ and $\chi_c^{-1} \approx (\Delta - \omega + i\kappa/2)$.

These frequency-domain equations determine the susceptibility of the state variables x and q_1 to control fields $F_{\rm ext}$ and V_1. One can define the dressed mechanical and electrical susceptibilities as $\widetilde{\chi_{\rm M}}(\omega) = x(\omega)/F_{\rm ext}(\omega)$ with $V_1 = 0$ and $\widetilde{\chi_c}(\omega) = C\omega_{\rm cir} q_1(\omega)/V_1(\omega)$ with $F_{\rm ext} = 0$. In addition, one can find the mixed susceptibilities $q_1/F_{\rm ext}$ with $V_1 = 0$ and x/V_1 with $F_{\rm ext} = 0$. All four of these quantities play a prominent role in superconducting electromechanics. For example, the notion of cooling mechanical motion by electromechanical interactions can be understood as a consequence of the modification of mechanical susceptibility by the electromechanical interaction [11]. The mechanical susceptibility in the presence of the strong drive is modified as

$$\widetilde{\chi_M} = \frac{\chi_M}{1 + \chi_M A[\chi_c(\omega) + \chi_c^*(-\omega)]}, \tag{9.9}$$

and the mixed electromechanical susceptibility is closely related,

$$\frac{q_1}{F_{\rm ext}} = \widetilde{\chi_M}\chi_c q_0 G. \tag{9.10}$$

where $A = G^2|q_0|^2/(2C\omega_{\rm cir})$ parameterizes the strength of the electromechanical coupling enhanced by the strong drive.

9.2.3 Dynamical backaction in electromechanics

The form of Eq. (9.9) is familiar from feedback control theory. This equation describes an endogenous feedback effect that alters the "open-loop" bare response of the oscillator to give the "closed-loop" or dressed response. In other words, the force on the oscillator's coordinate x depends on x itself through the electromechanical force $F_{\text{rad}}(\omega) = -x(\omega)\eta(\omega)$ (so named because it is the analog of radiation pressure in cavity optomechanics), where $\eta(\omega) = \chi_c(\omega) + \chi_c^*(-\omega)$. This phenomenon is sometimes known as dynamical backaction.

Owing to dynamical backaction, the mechanical oscillator may resonate at a higher or lower frequency or appear to be more or less strongly damped. Dynamical backaction modifies the Lorentzian susceptibility of a mechanical oscillator, but under many experimentally realized conditions, it remains of Lorentzian form. Consider that the electrical circuits typically resonate at frequencies of a few gigahertz while the mechanical structures oscillate below 10 MHz, yet both the electrical resonators and mechanical oscillators have similar quality factors. Consequently, the Lorentzian function $\chi_{\text{M}}(\omega)$ is much more sharply peaked than $\chi_c(\omega)$ and $\eta(\omega)$. If A is not too large $(A/(2\Omega_M m\kappa^2) \ll 1)$, one can approximate Eq. (9.9) by evaluating $\chi_c(\omega)$ at $\omega = \Omega_M$, the peak in $\chi_M(\omega)$. Within this approximation, the dressed response is another Lorentzian with a new width and resonance frequency. To see this, rewrite Eq. (9.9) as $\widetilde{\chi_M}(\omega)^{-1} \approx 2m\Omega_M(\Omega_M - \omega + i\Gamma_M/2) + \eta(\Omega_M)$. The shifts in resonance and damping are found from the real and imaginary parts of $\eta(\Omega_M)$ as

$$\delta\Omega_M = \frac{\text{Re}[\eta(\Omega_M)]}{2\Omega_M m} = \frac{2A}{\Omega_M m}\left[\frac{\Delta + \Omega_M}{4(\Delta + \Omega_M)^2 + \kappa^2} + \frac{\Delta - \Omega_M}{4(\Delta - \Omega_M)^2 + \kappa^2}\right], \quad (9.11)$$

$$\delta\Gamma_M = \frac{\text{Im}[\eta(\Omega_M)]}{\Omega_M m} = \frac{2A}{\Omega_M m}\left[\frac{\kappa}{4(\Delta + \Omega_M)^2 + \kappa^2} - \frac{\kappa}{4(\Delta - \Omega_M)^2 + \kappa^2}\right]. \quad (9.12)$$

These functions have very different characters, depending on the ratio Ω_M/κ. For the case in which this so-called sideband factor is large, $\Omega_M/\kappa \gg 1$ (resolved sideband limit), the substantial changes in response occur at detunings $|\Delta| \approx \Omega_M$. The changes in the real and imaginary parts of the mechanical response just reflect the real and imaginary parts of the circuit response. The largest change in dissipation is four times greater than the maximum resonance frequency shift. In contrast, when the sideband factor is small (unresolved sideband limit) the electrical spring (or anti-spring) and the electrical damping (or amplification) are strongest when $|\Delta| \approx \kappa/2$, and the change in dissipation is much smaller than the shift in resonance frequency (Fig. 9.3).

The approximate evaluation of $\eta(\omega) \approx \eta(\Omega_M)$ in the expression for $\widetilde{\chi_M}(\omega)$ will fail to give accurate results if $\delta\Gamma_M > \kappa$. The full frequency-dependent expression for $\widetilde{\chi_M}(\omega)$ is correct, but it will no longer be a simple Lorentzian. When the strong drive is detuned from cavity resonance for optimum damping, $\Delta = -\Omega_M$, the dressed response has a particularly interesting behavior. As A is increased, the dressed mechanical response first broadens, but as $\delta\Gamma_M$ approaches κ, the response splits into a two-peaked structure. The two peaks can be separated by an amount much larger than κ itself (Fig. 9.4). The transition from a simple Lorentzian to a split response is just the transition from over- to underdamped behavior of the linear equations of motion. The marginal case of critical damping, which occurs when $A/(2\Omega_M m) = \Gamma_M\kappa/4$, corresponds to a complete conversion of mechanical energy into energy dissipated in the electrical loss, and vice versa. For larger values of A, the split spectrum can

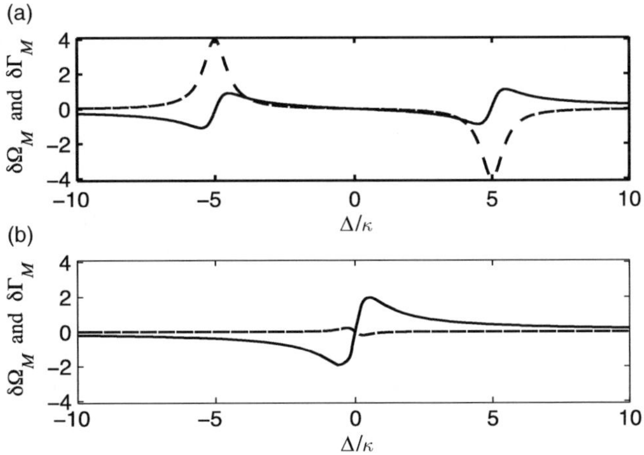

Fig. 9.3 The electromechanical spring and damping effects. (a) $\delta\Omega_M$ (solid line) and $\delta\Gamma_M$ (dashed line) for $\kappa = \Omega_M/5$. (b) $\delta\Omega_M$ (solid line) and $\delta\Gamma_M$ (dashed line) for $\kappa = 5\Omega_M$. The frequency shifts are all plotted in units of $A/(2\Omega_M m\kappa)$. Four dynamical backaction phenomena are apparent. The mechanical oscillator may behave as if bound by a stiffer or softer spring, and it may be more or less damped owing to electromechanical feedback.

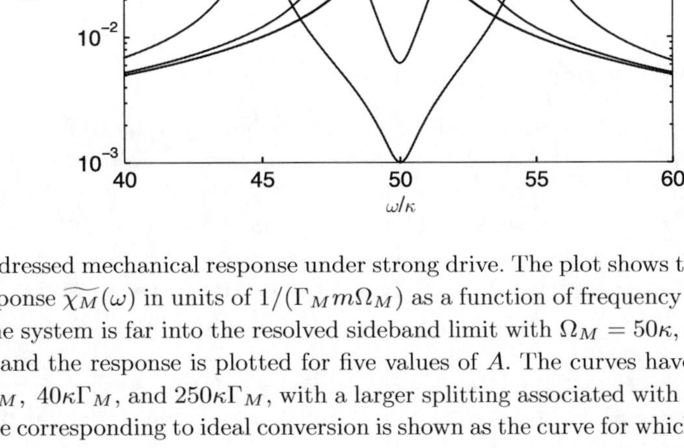

Fig. 9.4 The dressed mechanical response under strong drive. The plot shows the magnitude of the dressed response $\widetilde{\chi_M}(\omega)$ in units of $1/(\Gamma_M m\Omega_M)$ as a function of frequency ω for increasing values of A. The system is far into the resolved sideband limit with $\Omega_M = 50\kappa$, $\Delta = -\Omega_M$, and $\Gamma_M = 0.1\Omega_M$, and the response is plotted for five values of A. The curves have $A/(2m\Omega_M) = 0$, $\kappa\Gamma_M/4$, $\kappa\Gamma_M$, $40\kappa\Gamma_M$, and $250\kappa\Gamma_M$, with a larger splitting associated with a larger value of A. The response corresponding to ideal conversion is shown as the curve for which $A/(2m\Omega_M) = \kappa\Gamma_M/4$.

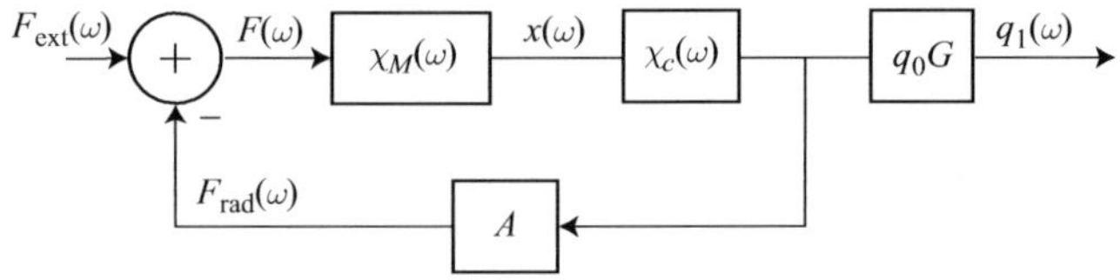

Fig. 9.5 A diagram expressing the electromechanical backaction as a type of built- in feedback. The feedback diagram pictorially represents the linearized equations of motion in the limit $\Delta = -\Omega_M$ and $\kappa \ll \Omega_M$. In this limit, the feedback strength is simply A.

be interpreted as energy sloshing between the mechanical and electrical degrees of freedom. If the energy leaves the mechanical oscillator more rapidly than it is dissipated, it builds up in the electrical resonator and re-enters the mechanical oscillator. Because the rate of energy transfer between the mechanical oscillator and the electrical circuit is $2\sqrt{A/2\Omega_M m}$, the condition for well-separated peaks in the response is $2\sqrt{A/2\Omega_M m} > \max(\kappa, \Gamma_M)$.

The conditions for ideal conversion and for observing a split response evoke the physics of coupled quantum systems. Indeed, if one had control over the electrical and mechanical degrees of freedom at the level of a single quantum, this physics would correspond to exchanging electrical photons for mechanical phonons. In the classical description, there is a more mundane interpretation of the structure in the oscillator's dressed response. If the electromechanical damping is viewed as resulting from a feedback force, the split spectrum appears when the feedback strength has exceeded the servo bandwidth. Near optimal damping in the resolved sideband limit, $\chi_c^*(-\omega) \ll \chi_c(\omega)$. In this limit, one can describe the dressed response as resulting from feedback, where the feedback strength is just A itself (Fig. 9.5). The peaks in the split spectrum are just the "servo bumps" that appear in the response of many feedback control loops.

9.3 Measuring oscillator motion

Electromechanics offers an appealing way to measure the motion of the oscillator itself. When operating at temperatures below 1 K, microwave circuits can be prepared in pure quantum states, and recent advances in amplification allow microwave signals to be measured with sensitivity near the quantum limit [12, 13]. If mechanical motion creates microwave electrical signals and those signals can be measured with quantum-limited sensitivity, is the mechanical motion also measured with quantum-limited sensitivity? To answer this question, one should refine the notion of a quantum limit on measurement.

Quantum limits arise in essence because more information is required to specify a quantum state than can be extracted in a single measurement. If an arbitrary quantum state could be determined in a single measurement, one could prepare many copies of that same state, violating basic theorems of quantum information [14]. When specialized to the case of linear measurements of a harmonic oscillator, the consequence of this fact can be expressed as a minimum amount of noise that must be added when measuring the state of the oscillator. If one attempts to determine the state of the mechanical oscillator by measuring both the oscillator's coordinate x and conjugate momentum p, the measurement must introduce noise at least as large as the ground-state fluctuations of x or p. Expressed as an energy, this minimum added noise has a variance $\hbar\Omega_M/2$ for the mechanical oscillator [15]. As is usually the case with such limits, knowing the limit provides little insight into how to reach it. It is the experimentalists task to design a measurement capable of reaching this limit.

It is evident from Eq. (9.7) that the charge modulation serves as a quantity from which the oscillator's position can be inferred. Rather than measure directly the charge signal $q_1(t)$, one instead usually detects the voltage made by that charge signal on a resistor remote from the circuit itself. Conceptually, one could just monitor the voltage across Z_0 in Fig. 9.2. In experiments, a microwave network is usually connected to port 1 in Fig. 9.2. As such, it is possible to separate the incident and reflected waves at port 1. If the network is arranged appropriately (as shown in Fig. 9.6), the charge dynamics are unaffected. In Fig. 9.6, the signal voltage at the amplifier input is $V_s(t) = V(t)/2 - \dot{q}(t)Z_0$.

For frequencies near the circuit resonance $\omega_D \pm \omega \approx \omega_{\text{cir}}$, the Fourier components of the voltage can be expressed as $V_s(\omega_D \pm \omega) = (1/2)[V_1(\pm\omega) - iq_1(\pm\omega)Z_c\kappa_E]$

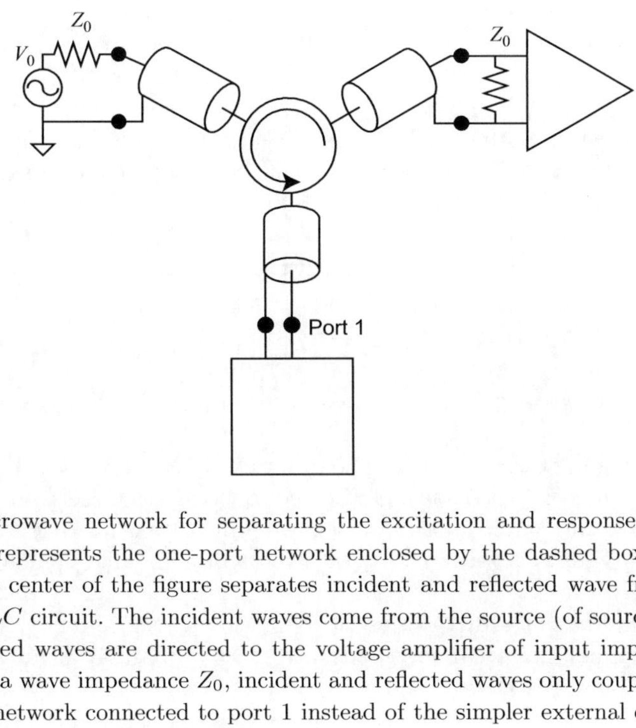

Fig. 9.6 A microwave network for separating the excitation and response of a circuit. The square element represents the one-port network enclosed by the dashed box in Fig. 9.2. The circulator in the center of the figure separates incident and reflected wave from each other at port 1 of the RLC circuit. The incident waves come from the source (of source impedance Z_0), while the reflected waves are directed to the voltage amplifier of input impedance Z_0. If the cables also have a wave impedance Z_0, incident and reflected waves only couple at port 1. With this microwave network connected to port 1 instead of the simpler external circuit, the charge dynamics of the RLC circuit are unchanged.

by using $\omega_{\rm cir} = 1/\sqrt{LC}$ and defining $Z_c = \sqrt{L/C}$. Putting this together with the electromechanical susceptibilities gives the measured voltage for a particular x expressed as the complex amplitude of the measured upper or lower sidebands:

$$V_s(\omega_D + \omega) = \frac{V_1(\omega)}{2} - \frac{i\kappa_E/2}{\Delta - \omega + i(\kappa/2)}[q_0 Z_c G x(\omega) + V_1(\omega)] \tag{9.13}$$

$$V_s(\omega_D - \omega) = \frac{V_1(-\omega)}{2} - \frac{i\kappa_E/2}{\Delta + \omega + i(\kappa/2)}[q_0 Z_c G x^*(\omega) + V_1(-\omega)]. \tag{9.14}$$

From these expressions, one can infer x from V_s if $V_1 = 0$.

9.3.1 Measurement in the unresolved sideband limit

In the unresolved-sideband limit, the optimum strategy for measurement is apparent. For $\omega \ll \kappa$, one can operate at zero detuning ($\Delta = 0$) and the denominators in Eqs. (9.13) and (9.14) are near their minima where the Fourier components of the upper and lower sidebands are $V_s(\omega_D + \omega) = [V_1(\omega)/2](1 - 2\kappa_E/\kappa) - (\kappa_E/\kappa) q_0 Z_c x(\omega)$ and $V_s(\omega_{\rm D} - \omega) = [V_1(-\omega)/2](1 - 2\kappa_E/\kappa) - (\kappa_E/\kappa) q_0 Z_c x^*(\omega)$. Both sidebands contain the same information and have the same amplitude. Which sideband should one measure?

When noise is added to the model, it is clear that the best measurement sensitivity is achieved by measuring a particular linear combination of both sidebands. Introducing a noise process into the amplifier, the measured voltage at the amplifier output—expressed as a voltage at its input—can be written as $V_{\rm meas}(\omega_D + \omega) = V_s(\omega_D + \omega) + V_{n,\rm amp}(\omega_D + \omega)$, where $V_{n,\rm amp}(\omega)$ is a random complex value drawn from a Gaussian distribution with spectral density $S_{V,\rm amp}(\omega)$. The inference of x from $V_{\rm meas}$ is imperfect because of this noise process. The influence of the noise can be reduced if information about $x(\omega)$ is extracted from both sidebands. For example, as shown in Fig. 9.7, the amplifier output may be mixed with a local oscillator tone $\sin(\omega_D t)$, creating a microwave interferometer. The two sidebands appear at the output of the mixer summed as $V_{\rm IF}(\omega) = V_{\rm meas}(\omega_D + \omega) - V^*_{\rm meas}(\omega_D - \omega)$. The mixer output (for $V_1 = 0$) then depends on x as

$$V_{\rm IF}(\omega) = xG\frac{\kappa_E}{\kappa} Z_c(\omega)(q_0 - q_0^*) + V_n(\omega_D + \omega) - V_n^*(\omega_D - \omega). \tag{9.15}$$

If q_0 is purely imaginary (i.e., the charge oscillations at ω_D are $\pi/2$ radians out of phase with the local oscillator), then the signal power quadruples (signal amplitude doubles) while the noise power only doubles as it is a sum of two uncorrelated noise amplitudes. Indeed, if V_0 is real (assumed without loss of generality), then q_0 is imaginary for $\Delta = 0$.

For the measurement strategy described in this section, where $\Delta = 0$, the influence of the amplifier noise can be made arbitrarily small. If q_0 increases, then the signal power increases relative to the noise power. The limit on measurement sensitivity seems to be set by the largest practical value of q_0, the value of $S_{V,\rm amp}$, and the ratio κ_E/κ. Perhaps more surprisingly, $S_{V,\rm amp}$ itself can be made arbitrarily small,

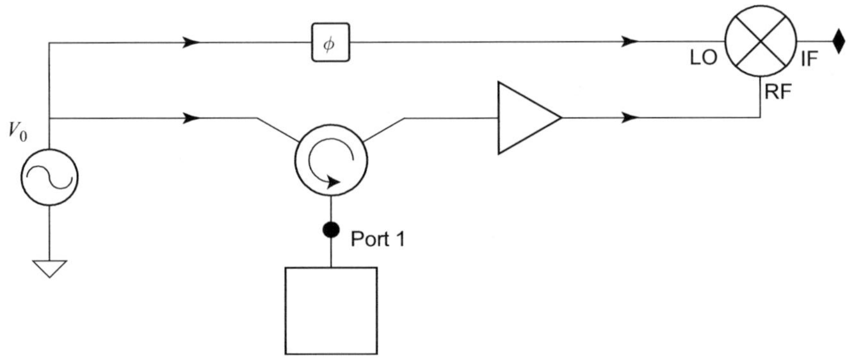

Fig. 9.7 A microwave interferometer. This schematic diagram shows the signal flow in a microwave interferometer designed to recover the phase modulation created by an electromechanical circuit. The tone created by a microwave signal generator is split into two arms. The upper arm provides a phase reference to the local oscillator (LO) port of the mixer; the tone in the lower arm acquires a phase modulation on reflection by the electromechanical circuit. After amplification, the reflected signal $V_s(t)$ is passed to the radiofrequency (RF) port of the mixer and the phase modulation is recovered at the intermediate-frequency (IF) port. The voltage at the IF port is $V_{\rm IF} \propto \langle V_s(t)\cos(\omega_D t + \phi)\rangle$, where the angular brackets indicate a time average over times long compared with $1/\omega_D$ but short compared with $1/\Omega_M$. A phase shift $\phi = -\pi/2$ applied to the LO tone selects the phase modulation to appear at the IF port.

even when considering quantum limits on voltage measurement or amplification. The real and imaginary parts of a voltage tone, which are canonically conjugate quantum observables, cannot both be amplified (and therefore measured) without the addition of a minimum amount of noise. However, there is no quantum limit on amplification of just one of the canonically conjugate observables [15]. Indeed, the linear combination of sibebands expressed by Eq. (9.15) corresponds to measuring the ω spectral component of the phase-quadrature fluctuations of the drive tone. By using a Josephson parametric amplifier as the first stage of amplification, it is indeed possible to measure only one quadrature of an electrical signal [12, 16].

When applying the drive tone on resonance, the quantum limit on measurement of the mechanical oscillator does not arise from a quantum limit on *measurement* of microwave voltages. Rather, the limit arises from fluctuations that are present in the drive voltage $V(t)$. A complete description of the quantum statistical properties of harmonic oscillators (including electrical modes in transmission lines) requires a fully quantum mechanical treatment. Nevertheless, a semiclassical description does provide some intuition. Namely, each dissipative element in the model must introduce a Langevin force as required by the fluctuation-dissipation theorem. The fluctuations associated with the electrical dissipation (Fig. 9.6) can be modeled by letting V_1 be a small random voltage where $V_1(\omega)$ is drawn from a Gaussian distribution with spectral density $S_{V,\rm src} = \hbar\omega Z_0 \coth[\hbar(\omega_D + \omega)/2k_BT] \approx \hbar\omega Z_0 \coth(\hbar\omega_D/2k_BT)$. With this spectral density, the average energy in each transmission-line mode will be the sum of a (one-dimensional) Planck distribution plus a half-quantum of vacuum energy.

For temperatures $T \ll \hbar\omega_D/k_B$, there are still voltage fluctuations. These quantum fluctuations can influence a measurement of x in two ways. First, some component of these fluctuations will appear at the amplifier input and interfere with the inference of x, contributing to an imprecision in the measurement of x similar to added noise by an amplifier. The value of $x_{\text{meas}} = V_{\text{IF}}/2GZ_c|q_0|$ inferred from measurement can be written as

$$x_{\text{meas}}(\omega) = x + \frac{V_1(\omega_{\text{D}} + \omega) - V_1^*(\omega_{\text{D}} - \omega)}{2GZ_c|q_0|}\left(1 - \frac{\kappa}{2\kappa_E}\right), \tag{9.16}$$

a sum of the "true" coordinate and a purely apparent fluctuation, the imprecision. Second, a component of the voltage fluctuations will introduce a fluctuating force F_{BA} through the radiation pressure term on the right-hand side of Eq. (9.6) as

$$F_{\text{BA}}(\omega) = \frac{|q_0|G[V_1(\omega_D + \omega) + V_1^*(\omega_D - \omega)]}{\kappa}. \tag{9.17}$$

This is a kind of backaction force that will create real fluctuations in x. If $\Delta = 0$, the imprecision and backaction are derived from two orthogonal quadratures of the input field. It is the amplitude fluctuations of $V(t)$ that exert a force on the oscillator but the phase fluctuations that contribute to imprecision. Although these fluctuations have the same origin and the same variance, they are independent. For a continuous measurement of x, one measures both the amplitude and phase of the oscillations at frequency Ω_M; consequently, one can determine both the position and momentum of the oscillator. For such a measurement, one expects to encounter a quantum limit [17].

Understanding the origin of the imprecision and backaction provides a semiclassical picture of this quantum limit. The backaction force and imprecision have inverse dependence on q_0: the imprecision shrinks with increasing q_0 while the backaction grows. Because the noise introduced by the measurement of x is the sum of the imprecision and backaction, the added noise will have a minimum value at a particular optimum value of the q_0. The added noise at this optimum will not vanish, owing to the nonzero quantum noise of the source $S_{V,\text{src}}(T = 0) = \hbar\omega Z_0$. Written as a spectral density, the total added noise is

$$\begin{aligned} S_{x,\text{tot}} &= S_{x,\text{Im}} + S_{F,\text{BA}}|\chi_M|^2 \\ &= \hbar\omega_{\text{cir}} Z_0 \left[\frac{1}{2(GZ_c|q_0|)^2}\left(1 - \frac{\kappa}{2\kappa_E}\right)^2 + \frac{2|q_0|^2G^2}{\kappa^2}|\chi_M|^2\right]. \end{aligned} \tag{9.18}$$

Specializing to a measurement at the mechanical oscillator's resonance frequency and assuming that $R = 0 \Rightarrow \kappa = \kappa_E$, one finds that the minimum total noise is $S_{x,\text{tot}} = \hbar|\chi_M(\Omega_M)| = \hbar/(m\Gamma_M\Omega_M)$. This minimum value is reached at the optimum drive amplitude $q_{0,\text{opt}} = \sqrt{\kappa_E/(4G^2Z_c|\chi_M|)}$. The spectral density of the ground-state zero-point motion $S_{x,\text{zpf}} = S_F(T = 0, \omega = \Omega_M)|\chi_M|^2$ is also $S_{x,\text{zpf}} = \hbar/(m\Gamma_M\Omega_M)$, where $S_F(T, \omega) = \hbar\omega m\Gamma_M \coth(\hbar\omega/2k_BT)$ is the fluctuation-dissipation force associated with the mechanical dissipation. If the voltage measurement adds no noise, $S_{V,\text{amp}}(\omega) = 0$, the internal loss of the circuit is negligible in comparison with

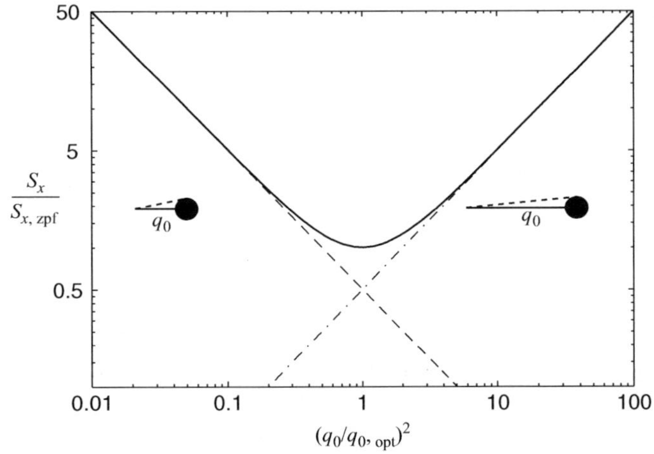

Fig. 9.8 Displacement noise spectral density versus drive power. The total noise added by a measurement as a function of drive power reaches the quantum limit for the ideal case described in the text. The total noise (solid line) is the sum of the imprecision (dashed line) and the backaction motion (dashed dotted line). The displacement noise is plotted in units of $S_{x,\text{zpf}}$. The annotations illustrate schematically a phasor representation of the drive tone composed of a deterministic amplitude (solid lines) and the quantum fluctuations (circles). They show that as the drive amplitude increases, the phase of the drive tone can be measured with improved precision (angle between solid and dashed lines).

the external loss, $\kappa_I \ll \kappa_E$, and the voltage source has no more noise than quantum noise, $S_{V,\text{src}}(\omega) = \hbar\omega Z_0$, then this measurement strategy reaches the quantum limit (Fig. 9.8).

9.3.2 Measurement in the resolved-sideband limit

In the resolved-sideband limit, the measurement strategy changes. In particular, a detuning of $\Delta = 0$ is unfavorable because both sidebands are far from the circuit resonance frequency ω_{cir}. If technical factors limit the size of q_0 instead of V_0, optimum detuning, $\Delta = -\Omega_M$, provides a better measurement of x than $\Delta = 0$. For this detuning, the high-frequency sideband is then resonant with the circuit, $\omega_D + \Omega_M = \omega_{\text{cir}}$.

Measuring the position of the oscillator with a detuning $\Delta = -\Omega_M$ differs from a measurement with $\Delta = 0$ in several important ways. Most significantly, the susceptibility of the oscillator χ_M will be modified by the dynamical backaction described in Section 9.2.3. The oscillator is subject to a dissipative force caused by the measurement itself. The position x can still be inferred from a measurement of $V_s(\omega_D + \omega)$, but the oscillator will not have the same position for a given force.

In addition to the modified susceptibility, the notion of measuring the oscillator's position changes. A measurement of x with temporal resolution τ requires measuring a range of frequencies $\delta\omega \approx \pi/\tau$. In the unresolved-sideband limit, it is possible to resolve the position of the oscillator at times shorter than its period. This resolution is not possible far into the resolved-sideband limit, because $\delta\omega$ would have to be much greater than κ. Rather than measuring the position of the oscillator, in the resolved-sideband limit, it is more accurate to say that one measures the slowly varying quadrature amplitudes X and Y of the oscillator, writing the oscillator's position $x(t) = X(t)\cos(\Omega_M t) + Y(t)\sin(\Omega_M t)$.

Finally, the quantum limit on the measurement of the oscillator's state arises from the electrical measurement in a different manner. In contrast to the $\Delta = 0$ case, only the higher-frequency sideband $V_s(\omega_D + \omega)$ will have appreciable size when $\Delta = -\Omega_M$. The real and imaginary parts of $V_s(\omega_D + \Omega_M)$ encode X and Y. One must measure both quadratures of the signal voltage to infer both canonically conjugate observables X and Y. Consequently, the quantum limit for position measurement does arise partially from a quantum limit on voltage measurement. Backaction also contributes to the quantum limit, but in a different manner than for the $\Delta = 0$ case. The additional damping $\delta\Gamma_M$ associated with the dynamical backaction is not dissipation so much as a transfer of energy from mechanical to electrical degrees of freedom. In a full quantum description of the optimally detuned and resolved-sideband case, the microwave drive creates a coupling that deterministically maps the state of the mechanical oscillator onto the state of the microwave field emitted by port 1 (Fig. 9.6). Simultaneously, the state of the voltage source $V_1(\omega_{\mathrm{cir}})$ incident on port 1 is mapped onto the state of the oscillator. In the time necessary to estimate X and Y with sufficient precision to resolve energy at the single-quantum level, the oscillator's state will be substantially altered by dynamical backaction, and one cannot improve the measurement precision by measuring for a longer period of time.

9.4 Quantum description of an electromechanical circuit

In attempting to describe the quantum limit of measurement of the mechanical oscillator, I have reached the end of the applicability of a classical description of the electromechanical circuit in Section 9.2.1. An accurate description of many quantum effects is possible within a semiclassical model where the circuit and mechanical oscillator are treated classically, but the Langevin forces associated with dissipation are given symmetrized quantum spectral densities as $S_{V,\mathrm{src}}(\omega) = \hbar\omega Z_0 \coth(\hbar\omega/2k_BT)$ and $S_F(\omega) = \hbar\omega m\Gamma_M \coth(\hbar\omega/2k_BT)$. This powerful correspondence is a feature of linear equations of motion and linear measurement of field amplitudes. Nevertheless, to satisfactorily account for all features of electromechanics requires a quantum description.

The starting point for most quantum treatments of electromechanical circuits is the cavity optomechanical Hamiltonian [4, 5, 7, 8]. A straightforward application of canonical quantization shows that the linearized Lagrangian

$$\mathcal{L} = \frac{1}{2}\left[L\dot{q}^2 + m\dot{x}^2 - \frac{q^2}{C}\left(1 - x\frac{1}{C}\frac{\partial C}{\partial x}\right) - k_s x^2\right] + qV + F_{\mathrm{ext}}x \qquad (9.19)$$

almost corresponds to the cavity optomechanical Hamiltonian.

Canonical quantization begins by finding the classical Hamiltonian using the formal definition $\mathcal{H} = \dot{q}\Phi + \dot{x}p - \mathcal{L}$, where the canonical momenta are $p \equiv \partial\mathcal{L}/\partial\dot{x} = m\dot{x}$ and $\Phi \equiv \partial\mathcal{L}/\partial\dot{q} = L\dot{q}$. Writing $\mathcal{H}$ as a function of q, x, p, and Φ gives

$$H = \frac{p^2}{2m} + \frac{k_s x^2}{2} + \frac{\Phi^2}{2L} + \frac{q^2}{2C}\left(1 - \frac{1}{C}\frac{\partial C}{\partial x}x\right) - qV - xF_{\text{ext}}. \tag{9.20}$$

Having written the classical Hamiltonian in terms of system coordinates and their canonical momenta, one simply writes the quantum Hamiltonian by replacing each coordinate or momentum by its operator (e.g., $q \to \hat{q}$) and requiring that they obey the canonical commutations relations $[\hat{x}, \hat{p}] = i\hbar$, and $[\hat{q}, \hat{\Phi}] = i\hbar$. Defining the creation and annihilation operators for the two harmonic oscillators,

$$\hat{A} = \hat{q}\sqrt{Z_c/2\hbar} + i\hat{\Phi}\sqrt{1/2Z_c\hbar},$$

$$\hat{B} = \hat{x}\sqrt{Z_M/2\hbar} + i\hat{p}\sqrt{1/2Z_M\hbar},$$

yields the quantum Hamiltonian

$$\begin{aligned}\hat{H} = {} & \hbar\omega_{\text{cir}}\left(\hat{A}^\dagger\hat{A} + \frac{1}{2}\right) + \hbar\Omega_M\left(\hat{B}^\dagger\hat{B} + \frac{1}{2}\right) + \hbar G x_{\text{ZPF}}(\hat{B} + \hat{B}^\dagger)\hat{A}^\dagger\hat{A} \\ & -(\hat{A} + \hat{A}^\dagger)\sqrt{\frac{\hbar}{2Z_c}}V - (\hat{B} + \hat{B}^\dagger)\sqrt{\frac{\hbar}{2Z_M}}F_{\text{ext}} \\ & + \frac{\hbar G}{2}\left\{\hat{A}\hat{A} + \hat{A}^\dagger\hat{A}^\dagger + 1\right\}x_{\text{ZPF}}(\hat{B} + \hat{B}^\dagger),\end{aligned} \tag{9.21}$$

where $Z_M = \sqrt{k_s m}$ and $x_{\text{ZPF}} = \sqrt{\hbar/2Z_M}$.

The terms in curly brackets in Eq. (9.21) are usually dropped from the cavity optomechanical Hamiltonian. Indeed, there is a good physical argument to ignore them for systems where $\Omega_M \ll \omega_{\text{cir}}$. The force associated with these terms oscillates at about $2\omega_{\text{cir}}$, where the susceptibility of the mechanical oscillator is likely to be vanishingly small. I have emphasized them here because it is possible [3] to build mechanical structures that respond at $2\omega_{\text{cir}}$ when $2\omega_{\text{cir}}$ is a microwave rather than an optical frequency.

Heisenberg equations of motion can readily be found from the Hamiltonian; however, these quantum equation of motion won't correspond to the classical equations of motion in Section 9.2.2. In particular, the dissipative terms added to Eqs. (9.1) and (9.2) are absent. Including dissipation is more complex in a quantum description than in a classical description, as one must add new quantum degrees of freedom to keep track of the modes of the environment. Fortunately, for the high-Q resonant circuits and mechanical oscillators considered here, the so-called formalism of input-output theory is valid [18]. Indeed, the strategy used in the theory of cavity optomechanical effects applies input-output theory to add dissipation to the Hamiltonian in Eq. (9.21). From the linearized Heisenberg-Langevin equations of motion that are derived from this approximate method, one can find exactly the same phenomena described in Section 9.2.3 [7].

Acknowledgments

This chapter was developed through many discussions with Jennifer Harlow, Tauno Palomaki, Cindy Regal, John Teufel, and John Price.

References

[1] Degen, C. L., Poggio, M., Mamin, H. J., Rettner, C. T., and Rugar, D. (2009). *Proc. Nat. Acad. Sci. USA*, **106**, 1313–1317.

[2] Kleckner, D., Pikovski, I., Jeffrey, E., Ament, L., Eliel, E., van den Brink, J., and Bouwmeester, D. (2008). *New J. Phys.*, **10**, 095020.

[3] O'Connell, A. D., Hofheinz, M., Ansmann, M., Bialczak, R. C., Lenander, M., Lucero, E., Neeley, M., Sank, D., Wang, H., Weides, M., Wenner, J., Martinis, J. M., and Cleland, A. N. (2010). *Nature*, **464**, 697–703.

[4] Regal, C. A., Teufel, J. D., and Lehnert, K. W. (2008). *Nature Phys.*, **4**, 555–560.

[5] Teufel, J. D., Regal, C. A., and Lehnert, K. W. (2008). *New J. Phys.*, **10**, 095002.

[6] Teufel, J. D., Donner, T., Li, D., Harlow, J. W., Allman, M. S., Cicak, K., Sirois, A. J., Whittaker, J. D., Lehnert, K. W., and Simmonds, R. W. (2011). *Nature*, **475**, 359–363.

[7] Marquardt, F., Chen, J. P., Clerk, A. A., and Girvin, S. M. (2007). *Phys. Rev. Lett.*, **99**, 093902.

[8] Wilson-Rae, I., Nooshi, N., Zwerger, W., and Kippenberg, T. J. (2007). *Phys. Rev. Lett.*, **99**, 093901.

[9] Marquardt, F. (2008). *Nature Phys.*, **4**, 513–514.

[10] Kippenberg, T. J. and Vahala, K. J. (2008). *Science*, **321**, 1172–1176.

[11] Teufel, J. D., Harlow, J. W., Regal, C. A., and Lehnert, K. W. (2008). *Phys. Rev. Lett.*, **101**, 197203.

[12] Castellanos-Beltran, M. A., Irwin, K. D., Hilton, G. C., Vale, L. R., and Lehnert, K. W. (2008). *Nature Phys.*, **4**, 929.

[13] Bergeal, N., Schackhert, F., Metcalfe, M., Vijay, R., Manucharyan, V. E., Frunzio, L., Prober, D. E., Schoelkopf, R. J., Girvin, S. M., and Devoret, M. H. (2010). *Nature*, **465**, 64–68.

[14] Wootters, W. K. and Zurek, W. H. (1982). *Nature*, **299**, 802–803.

[15] Caves, C. M. (1982). *Phys. Rev. D*, **26**, 1817–1839.

[16] Teufel, J. D., Donner, T., Castellanos-Beltran, M. A., Harlow, J. W., and Lehnert, K. W. (2009). *Nature Nanotechnol.*, **4**, 820–823.

[17] Clerk, A. A. (2004). *Phys. Rev. B*, **70**, 245306.

[18] Clerk, A. A., Devoret, M. H., Girvin, S. M., Marquardt, F., and Schoelkopf, R. J. (2010). *Rev. Mod. Phys.*, **82**, 1155–1208. Longer version at arXiv:0810.4729.

10
Two-electron spin qubits in GaAs: control and dephasing due to nuclear spins

A. YACOBY and H. BLUHM

Department of Physics, Harvard University, Cambridge, MA 02138, USA

Quantum Machines: Measurement and Control of Engineered Quantum Systems.
Edited by M. Devoret, B. Huard, R. Schoelkopf, and L. F. Cugliandolo.

Chapter Contents

10.1 Introduction

The potential realization of quantum computers has attracted much attention because of their promise to perform certain calculations that are practically intractable for classical computers. While a classical bit attains only two values (0 and 1), the phase space of a quantum bit (a two-level system) is in one-to-one correspondence with the points on the surface of a three-dimensional sphere, known as the Bloch sphere [1], where the basis states (corresponding to the classical 0 and 1) are represented at the north and south poles (Fig. 10.1(a)). A generic manipulation of the qubit needed to implement universal gate operations requires the ability to perform rotations around two axes in the Bloch sphere [2–5] (e.g., the z and x axes). In the work described in this chapter, the two-level quantum bit (the smallest logical unit of the quantum computer) is encoded in the spin state of two electrons confined in a double-well potential. This semiconductor-based system has potential for good scalability, and its manipulation is all-electrical and potentially fast enough to enable 10^4 universal gate operations within the coherence time, an essential requirement for conventional quantum error correction schemes and well within the threshold for surface codes.

10.2 Two-electron logical spin qubits

The double-well potential that confines the electrons is formed by applying a negative voltage to metal gates deposited on top of a two-dimensional electron gas (2DEG) embedded in a GaAs/AlGaAs heterostructure. The negative potential depletes the electrons underneath the metal gates, creating two isolated puddles of electrons (a double quantum dot; Fig. 10.1(b)). The number of electrons in the dots can be controlled by tuning the potentials on the gates. We restrict the total occupation of the double quantum dot to two electrons, and describe their spatial separation by the parameter ε: for $\varepsilon \gg t_c$, where t_c is the interdot tunnel coupling, both electrons are in the right quantum dot (the (0,2) configuration) and for $\varepsilon \ll t_c$, one electron occupies each dot (the (1,1) configuration). The parameter ε and hence the dots' charge state can be continuously swept through intermediate configurations by varying the voltages on the metal gates. In the (0,2) charge configuration, the only energetically accessible spin configuration is the singlet state $S(0,2) = (|\uparrow\downarrow\rangle - |\downarrow\uparrow\rangle)/\sqrt{2}$ (the arrows indicate the direction of the electron spins). As we separate the electrons, the wavefunctions' overlap decreases and four spin configurations become energetically degenerate: the singlet $S(1,1)$ and three triplets $T_0 = (|\uparrow\downarrow\rangle + |\downarrow\uparrow\rangle)/\sqrt{2}$, $T_- = |\downarrow\downarrow\rangle$, and $T_+ = |\uparrow\uparrow\rangle$ (Fig. 10.1(d)). We select the states $S(1,1)$ and T_0, both having zero z component of the spin angular momentum, as the basis states of our logical qubit [6, 7] and lift the degeneracy with the states T_- and T_+ by applying an external magnetic field B_{ext}. The Zeeman energy $E_z = g\mu_B B$ (where $g = -0.4$ is the g-factor for GaAs and μ_B the Bohr magneton) shifts the T_+ state to lower energies, creating a crossing point with the singlet (marked by the circle in Fig. 10.1(d)) at a value of ε that depends on $\mathbf{B} = \mathbf{B}_{\text{ext}} + \mathbf{B}_{\text{nuc}}$, where $\mathbf{B}_{\text{nuc}} \equiv (\mathbf{B}_{\text{nuc},L} + \mathbf{B}_{\text{nuc},R})/2$ is the average hyperfine field and $\mathbf{B}_{\text{nuc},L}$ and $\mathbf{B}_{\text{nuc},R}$ are the nuclear fields felt by the electron in the left and right dots, respectively (Fig. 10.1(c)). Within this logical subspace, rotations around the z-axis are controlled by the energy splitting between $S(1,1)$ and T_0, denoted by $J(\varepsilon)$.

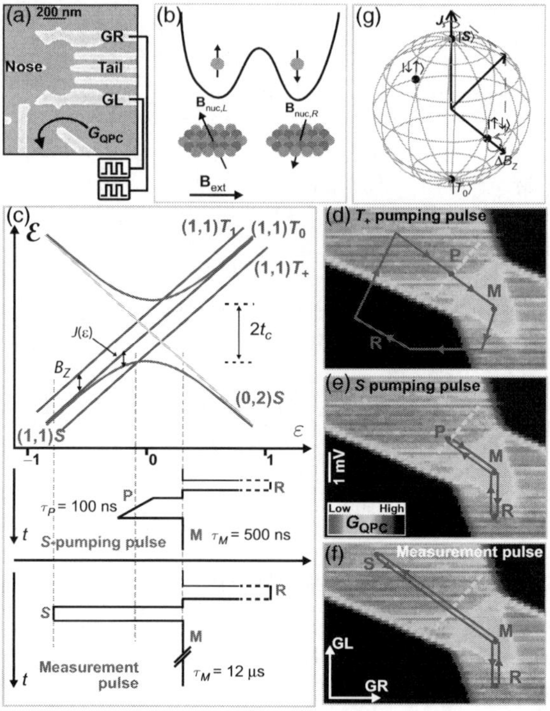

Fig. 10.1 Pump and measurement schemes. (a) SEM micrograph of a typical device. Voltages on gates GL and GR control the charge configuration of the two dots, and the central gates (nose and tail) control the tunnel coupling t_c between the two dots. The average charge configuration is detected by measuring the conductance G_{QPC} through a capacitively coupled quantum point contact. (b) $\mathbf{B}_{\text{nuc},L}$ and $B_{\text{nuc},R}$ are the local magnetic fields experienced by the electrons in the left and right dots through hyperfine coupling with the Ga and As nuclei. (c) Schematic representation of the energy levels at the (0,2)–(1,1) charge transition for finite external magnetic field. The detuning ε from the degeneracy point is controlled by the voltages on GL and GR. Two pulse cycles are presented: (1) *nuclear pumping*: the system is moved to point P, where S and T_+ are degenerate and can mix; (2) *measurement pulse*: the system is moved to large negative detuning where the states S and T_0 can mix. (d) The T_+-pumping pulse scheme, (e) the S-pumping pulse scheme, and (f) the measurement pulse scheme, all shown as functions of GL and GR. (g) Geometrical representation (Bloch sphere) of the two-level system (S and T_0) and the two rotation axes (J and ΔB^z_{nuc}) allowing the implementation of universal single-qubit gates.

This evolution amounts to a coherent exchange of the two electrons. Rotations around the x axis are controlled by the z component of a magnetic field gradient across the two electrons, $\Delta\mathbf{B}_{\text{nuc}} = \mathbf{B}_{\text{nuc},L} - \mathbf{B}_{\text{nuc},R}$. If we thus let a state evolve around a combined axis $J\mathbf{z} + g\mu_B\Delta B^z_{\text{nuc}}\mathbf{x}$, the precession frequency will be given by $f = \sqrt{J^2 + (g\mu_B\Delta B^z_{\text{nuc}})^2}/h$ (h being Planck's constant).

10.3 Single-qubit gates

While controlled z rotations have been previously shown [8], controlled rotations around the x axis of the two-electron logical qubit are much harder to implement. The challenge is to provide a stable magnetic field gradient across the two dots that exceeds the intrinsic nuclear fluctuations due to the hyperfine interaction. While such a gradient have been obtained using permanent magnets [9], here we present two polarization schemes that take advantage of the nuclear polarization in the host lattice by which the gradient can be increased to values significantly exceeding its fluctuations and very efficiently stabilized. Both pumping schemes make use of the degeneracy point between $S(1,1)$ and T_+. Transitions between the two states that are driven by the transverse component of $\Delta\mathbf{B}_{\mathrm{nuc}}$ [10] are accompanied by a spin flip of the nuclei in order to conserve the total angular momentum. Our first pumping scheme follows a standard recipe [11, 12] of initializing the system in the $S(0,2)$ state followed by a 50 or 100 ns long sweep across the S–T_+ degeneracy point. This process ideally transfers one unit of angular momentum into the nuclear subsystem. In addition, we have developed an alternative pumping scheme whereby we initialize the system in a $T_+(1,1)$ state followed by a similar slow passage through the S–T_+ degeneracy point. This new T_+-pumping scheme allows us to polarize the nuclear subsystem in a direction opposite to the S-pumping scheme. The T_+-pumping scheme works only when the Zeeman energy exceeds the electron temperature in the reservoirs: the system is swept slowly into (0,1) and subsequently reloaded into the (1,1) charge state (Fig. 10.1(g)). First the right and then the left electron align with the external field owing to large Zeeman energy ($\approx 12.5\,\mu$eV at 500 mT), which preferentially loads a T_+ state.

While these nuclear pumping schemes should produce nuclear polarization, it is not at all obvious that this nuclear polarization should be different across the two dots [12]. Since the mixing between the $S(1,1)$ and $T_0(1,1)$ is only sensitive to the field gradient, we use a pulse cycle that monitors the coherent evolution around the x axis in order to measure this gradient. The system is first reset into a $S(0,2)$ state. ε is then abruptly set to point S in (1,1) for an evolution time τ_S (see Fig. 10.1(e)). Here $\Delta B^z_{\mathrm{nuc}} \gg J(\varepsilon)/g\mu_B$ drives coherent oscillations between $S(1,1)$ and T_0 and the probability of being in a singlet state oscillates in time as $p(S) = \cos^2(g\mu_B \Delta B^z_{\mathrm{nuc}}\tau_S/2\hbar)$. When the system is brought back to the measurement point M, only transitions from $S(1,1)$ to $S(0,2)$ are allowed, while T_0, remains blocked in the (1,1) charge configuration. This spin-blockade effect allows us to map the spin configuration of the state onto a charge configuration [13], which is measured by a charge sensor [13]. Here we use a quantum point contact (QPC) positioned next to one of the the two quantum dots (Fig. 10.1(b)) in order to detect changes in the double-dot charge configuration. The QPC signal, averaged over many gradient-probing cycles, is proportional to the probability of being in a singlet state.

A steady-state nuclear field can be achieved by continuously alternating between a pump cycle that runs for a time t_{pump}, and a gradient probing cycle that runs for 1 s, as schematically visualized in Fig. 10.2(a). In each measurement stage, a gradient probing pulse with a different separation time τ_S is used ($0 \leq \tau_S \leq 30$ ns for each τ_S sweep). The outcome of a measurement repeating τ_S sweeps 40 times and using $t_{\mathrm{pump}} = 60$ ms

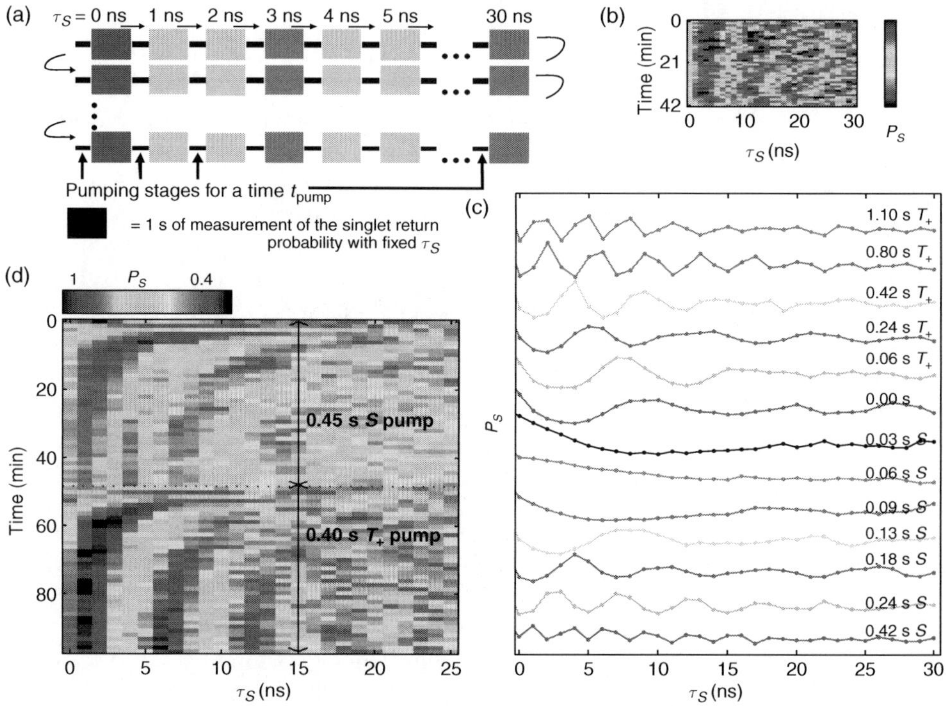

Fig. 10.2 Build-up of a gradient with two different pumping cycles. (a) Schematics explaining the measurement of the singlet return probability as a function of τ_S. Each measurement point (rectangle) is an average over 1 s and preceded by a pump cycle that runs for a time t_{pump}. Each line is a repetition of the same measurement procedure. (b) Measurement of the singlet return probability as a function of τ_S for $t_{\text{pump}} = 60\,\text{ms}$. (c) Singlet return probability as a function of separation time τ_S for different t_{pump}. Each data-set line is an average over 30 τ_S sweeps. P_s is normalized by the size of the dc charge transition Traces are displaced for clarity. (d) Crossover between S and T_+ pumping. The plot shows an average of nine subsequent repetitions of the same measurement. The disappearance of the oscillations on changing the pump pulse followed by a recovery (around 0 and 50 min) suggests that S and T_+ pumping produce ΔB^z_{nuc} of opposite signs. $B_{\text{ext}} = 1.5\,\text{T}$ for these data sets.

is shown in Fig. 10.2(b). An oscillatory signal with a frequency fluctuating around a steady mean is clearly visible. In the present measurement, the gradient is kept in a steady state for 40 minutes, but this time could have been extended indefinitely. Each curve in Fig. 10.2(c) shows an average over 30 τ_S sweeps and the different values of t_{pump} control the steady-state value of the gradient in each data set. We observe that the oscillations vanish (corresponding to a ΔB^z_{nuc} fluctuating around 0) at moderate S pumping rather than $t_{\text{pump}} = 0$. This appears to reflect a small polarization effect from the measurement pulses that can be compensated with S pumping. To compare the S- and T_+-pumping schemes, we have taken a measurement where we have switched

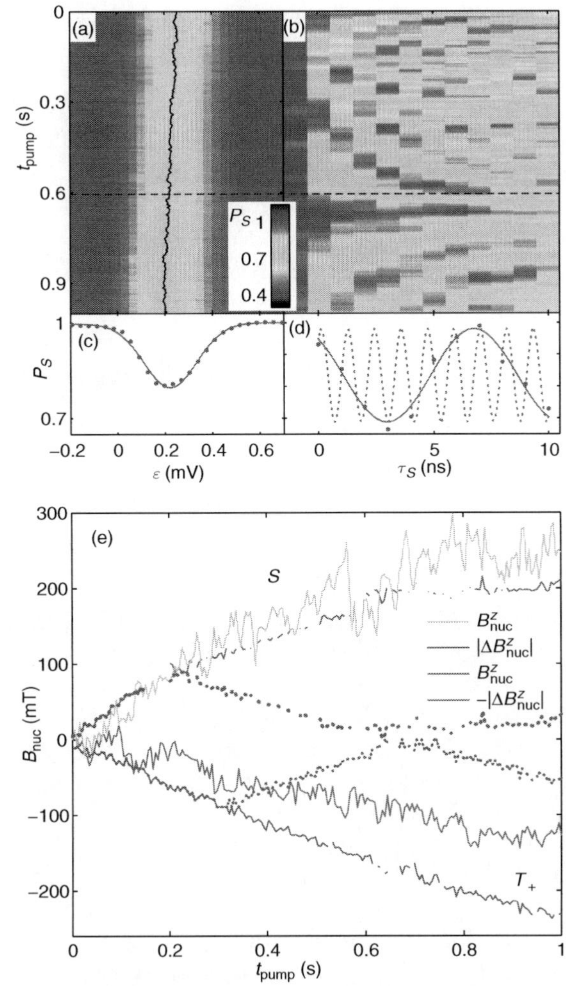

Fig. 10.3 Comparison of the gradient and the average nuclear field. (a) Line scans of the position of the S–T_+ transition along the ε axis as a function of pump time: the T_+-pumping cycle, applied for a duration t_{pump}, precedes each 2 s long measurement interval. (b) Singlet return probability versus τ_S under the same pumping conditions as in (a). The repeated increase and decrease of the apparent oscillation frequency is a consequence of aliasing of a monotonically increasing frequency due to the 1 ns sampling interval. (c) Single line scan from the data in (a) at $t_{\text{pump}} = 0.6$ s. (d) The corresponding τ_S sweep from (b). The continuous line is a sinusoidal fit and the dashed line reconstructs the actual, non-aliased time dependence of the singlet probability. (e) B_{nuc} and ΔB^z_{nuc} extracted from fits as shown in (c) and (d) as functions of t_{pump} for both T_+ and S pumping. The shift of the S–T_+ transition was converted to B^z_{nuc} using its measured dependence on B_{ext}; ΔB^z_{nuc} was obtained from the fitted oscillation frequency corrected for aliasing. The dotted lines shows the same data before this correction. The fluctuations in the B^z_{nuc} curves reflect measurement noise. $B_{\text{ext}} = 500$ mT for these data sets.

between S pumping and T_+ pumping every 40 τ_s sweeps (see Fig. 10.2(c)). The data show that, on changing the pump cycle, the oscillations disappear and then recover after a few minutes. This observation suggests that the two pump cycles pump nuclei in opposite directions and generate gradients of opposite signs.

While the magnitude of the gradient is determined via coherent x rotations, we can also measure the average value of the nuclear field by monitoring the position of the S–T_+ transition [11]. This should clarify whether spins are flipped only in one or in both dots. Figures 10.3(a,b) show interleaved measurements of the position of the S–T_+ transition [12] and the oscillatory S–T_0 mixing as a function of t_{pump} using the T_+-pumping cycle. A shift of the S–T_+ transition to more negative ε corresponds to the buildup of an average field B^z_{nuc} oriented opposite to the external magnetic field, consistent with spin flips from down to up in the nuclear system [14]. Figure 10.3(e) shows that at $B_{\text{ext}} = 500\,\text{mT}$, ΔB^z_{nuc} reaches 230 mT while B^z_{nuc} is about 130 mT. The ratio of nearly 2 indicates that the nuclei are polarized predominantly in one of the two dots. Data obtained using the S-pumping cycle (Fig. 10.3(e)) show a ΔB^z_{nuc} that tends to be slightly smaller than the average field.

10.4 Universal single-qubit rotations

Combining our slowly tunable x-rotation gate with the electrically controllable exchange gate allows single-qubit rotations around an axis that can be rapidly tilted to any desired angle between 0 and nearly $\pi/2$ away from the x axis (angle θ in Fig. 10.4(d)). Concatenating rotations around different axis allows us to implement universal quantum control. We demonstrate and characterize the rotation around an arbitrary axis using state tomography, consisting of three independent measurements of the probability of being in an $|S\rangle \equiv |Z\rangle$, in an $|S\rangle + |T_0\rangle \equiv |\uparrow\downarrow\rangle \equiv |X\rangle$, and in an $|S\rangle + i|T_0\rangle \equiv |Y\rangle$ state [1], with pulses shown in Fig. 10.4(a). This allows us to fully reconstruct the time evolution of the state vector. For each of the measurements, we first prepare an $|\uparrow\downarrow\rangle$ state by loading an $S(0{,}2)$ and adiabatically switching off $J(\varepsilon)$ in (1,1). The desired rotation is performed by quickly setting J to a finite value for a time τ_{rot}. Rapidly returning to $S(0,2)$ allows us to measure $p(|Z\rangle) \equiv |\langle Z|\psi\rangle|^2$, whereas slowly increasing J brings $|\uparrow\downarrow\rangle$ onto $|S\rangle$ and $|\downarrow\uparrow\rangle$ onto $|T_0\rangle$, thus allowing the readout of $p(|X\rangle) \equiv |\langle X|\psi\rangle|^2$. To obtain $p(|Y\rangle) \equiv |\langle Y|\psi\rangle|^2$, J is turned off for a time corresponding to a $\pi/2$ rotation around the x axis before rapidly returning to M. Results of this procedure for a particular choice of J and ΔB^z_{nuc} are shown in Fig. 10.4(b) as functions of τ_{rot}. For ideal pulses, one would expect $p(|X\rangle)$ to oscillate sinusoidally between 1 and $(1 + \cos 2\theta)/2$ and $p(|Z\rangle)$ between 1/2 and $(1 + \sin 2\theta)/2$, whereas $p(|Y\rangle)$ should vary symmetrically around 1/2. Deviations from this behavior can be attributed to a finite pulse rise time and high-pass filtering of the pulses. Fits to a model incorporating these effects and inhomogeneous broadening due to fluctuations in ΔB^z_{nuc} give a good match with the data (Fig. 10.4(b)). In Fig. 10.4(d), the data and fits are displayed in the Bloch sphere representation.

The mechanism responsible for the large gradient due to pumping is currently not clearly established. One possible cause is an asymmetry in the size of the two dots due to local disorder. Both the probability to flip a nuclear spin in one of the dots and

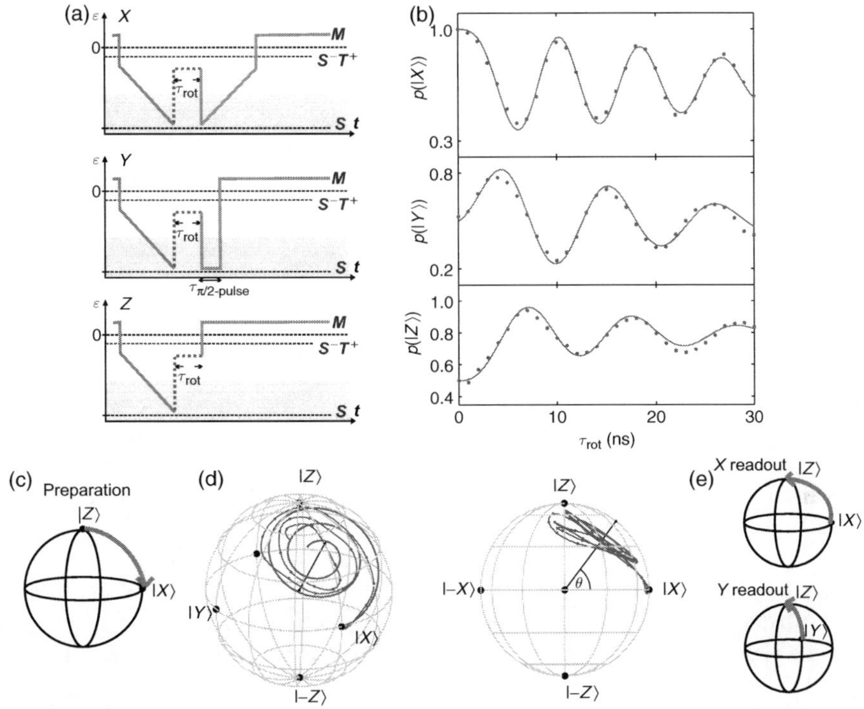

Fig. 10.4 State tomography and universal gate. (a) Pulse schemes to measure the singlet probability $p(|S\rangle) \equiv p(|Z\rangle)$, the $|\uparrow\downarrow\rangle$ probability $p(|\uparrow\downarrow\rangle) \equiv p(|X\rangle)$, and the $|S\rangle + i|T_0\rangle$ probability $p(|S\rangle + i|T_0\rangle) \equiv p(|Y\rangle)$ after rotation around a tilted, tunable axis. (b) Measurements taken with the X, Y, and Z pulses (dots) and fits (line) to a numerical solution of the Schrödinger equation of the S–T_0 Hamiltonian incorporating the finite pulse rise time and inhomogeneous broadening due to fluctuations in ΔB^z_{nuc}. (c) An $|\uparrow\downarrow\rangle = |X\rangle$ state is prepared by adiabatically turning off J. (d) Representation of the measured and fitted trajectories in the Bloch sphere. In order to eliminate phase shifts due to the slightly different frequencies (see text), the timescales for the X data have been rescaled using spline interpolation so that the expected phase relations are maintained. The heavier line is a spline interpolation of the data points in (b). (e) Visualization of the X- and Y-readout schemes on the Bloch sphere.

the change in hyperfine field due to that flip are inversely proportional to the number of nuclei N over which the electron wavefunction extends [15]. The overall $1/N^2$ dependence results in the smaller dot being polarized more rapidly. It was shown in numerical simulations that different dot sizes can indeed lead to an asymmetric dynamics polarization [16].

10.5 Controlling the nuclear environment

Interaction of the qubit spin(s) with nearby nuclear spins is a significant source of decoherence in several systems [8, 17–20]. It is therefore very attractive to prepare the spin environment of the electron in a way that mitigates this decoherence. One approach would be to polarize it [21], but the nearly complete polarization required for improved coherence [22] is difficult to achieve. In this section, we present a method to narrow the distribution of the fluctuating nuclear hyperfine field while maintaining a weak polarization. In addition to the reduced decoherence, such a narrowed state is of interest for studying the long-time quantum dynamics arising from the spin-bath interaction [23]. Our method relies on first letting the qubit evolve under the influence of the bath. Its final state is a sensitive measure of the instantaneous nuclear gradient. Subsequent dynamic nuclear polarization steps depend on this final state of the qubit. Therefore, the qubit itself is an integral part of a feedback loop that probes the gradient and then pumps it to a desired value.

In recent years, several similar qubit-nuclear feedback schemes have been employed, such as resonance locking experiments under microwave [24] and optical [25–27] irradiation. Vink *et al.*, [14] and Latta *et al.*, [26] inferred a narrowing of the hyperfine field from the observed bidirectional polarization, keeping the system on resonance, but this was not experimentally verified. Xu *et al.*, [25] detected narrowing spectroscopically, and Greilich *et al.*, [27] studied an ensemble of optically controlled quantum dots. Here, we directly measure the narrowed distribution of the hyperfine field and the dephasing time T_2^* of a single, electrically controlled qubit. T_2^* is enhanced by nearly an order of magnitude. In contrast to previous experiments, where the feedback mechanism is intrinsic to the polarization dynamics [28, 29], here we implement feedback by manipulating the qubit.

The spin qubit studied in this work employs the $m = 0$ subspace of two electron spins in a double quantum dot. The energy splitting between the two basis states depends on the hyperfine field gradient between the two dots. Reilly *et al.*, [12] reported a complete elimination of this gradient and an associated enhancement of T_2^* in similar devices. However, there is now a more likely interpretation of that experiment in terms of a loss of readout contrast due to a *large* hyperfine field gradient that accelerates inelastic decay [30]. Furthermore, maintaining a nonzero average field gradient as done here is essential for universal fast electrical control of the qubit [31].

Throughout this work, $B_{\text{ext}} = 0.7\,\text{T}$ was applied along the z axis, parallel to the 2DEG. The encoding of the qubit in two spins enables fast electrical control via the energy difference ε between states with both electrons in one dot and one electron in each dot, respectively [2, 8, 31]. The energies of the four spin states depend on ε as shown in Fig. 10.5(a).

We probe the hyperfine field gradient ΔB^z_{nuc}, by measuring this free precession using a standard prepare-evolve-measure cycle [8, 31, 32], relying on spin-to-charge conversion and a QPC for readout. The probability P_S of finding the electron in $|S\rangle$ varies sinusoidally with the evolution time τ_S (inset to Fig. 10.6(a)), with a frequency given by $f = |g\mu_B \Delta B^z_{\text{nuc}}|/h$, where $g \approx -0.4$ is the g factor for a confined electron in GaAs. Fitting a sine curve every time a τ_S sweep is completed yields a time trace of $|\Delta B^z_{\text{nuc}}|$ with a sampling rate of about 1 Hz, which is fast enough to probe slow random variations of the nuclear polarization [33].

This near-real-time measurement of ΔB^z_{nuc} allows us to use pump cycles discussed in detail in [11, 12, 31] to compensate the fluctuations of the spin bath. These pump cycles use the degeneracy point of $|S\rangle$ and $|T_+\rangle$ (point P in Fig. 10.5(b)) to exchange

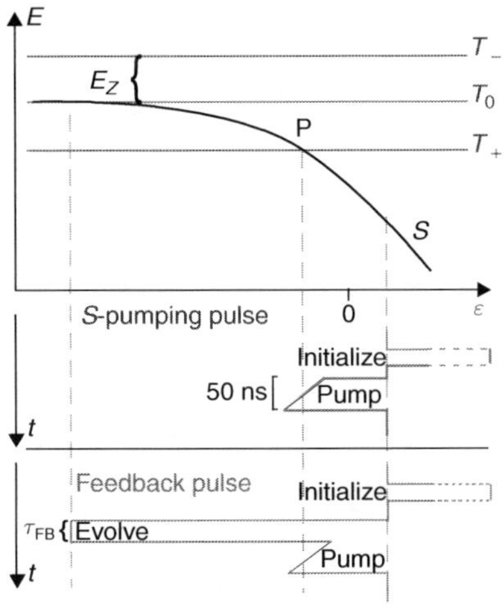

Fig. 10.5 Schematic of $\varepsilon(t)$ for the S and feedback pump pulses (bottom). The top pannel shows the corresponding portion of the energy diagram from Fig. 10.1(c).

spin between the electrons and the nuclei. In the so-called S (T_+) pumping cycle, we prepare an $|S\rangle$ ($|T_+\rangle$) and then sweep through the S–T_+ transition, which builds up a polarization of the same (opposite) sign as the applied field in each of the dots. How these pump cycles affect ΔB^z_{nuc} depends on the imbalance of the polarization rates in the two dots, for example arising from different dot sizes [16] due to disorder. Experimentally, we can probe their effect by running on the order of 10^6 cycles (at a 4 MHz repetition rate) between measurements of ΔB^z_{nuc}. This alternation between measuring and pumping was applied throughout the remainder of our experiment. We find that the two pump cycles always change ΔB^z_{nuc} in opposite directions. If the gradient reaches zero while pumping, it immediately increases again, which suggests a sign change. This behavior is consistent with [31], where the T_+ cycle was first introduced.

To quantify the effect of pumping, we switch between S and T_+ pumping whenever ΔB^z_{nuc} reaches one of two predetermined limits. This leads to a sawtooth-like time dependence of ΔB^z_{nuc}, as shown in Fig. 10.6(b). Averaging over many such cycles yields the mean rate of change of the gradient, $\Delta B^z_{\text{nuc}}/dt$, as a function of its value ΔB^z_{nuc} (Fig. 10.6(d)). The approximately linear relation for S and T_+ pumping reflects the relaxation of the polarization due to spin diffusion. For a fixed pump time and pulse, ΔB^z_{nuc} saturates once pumping and relaxation balance each other, but continues to fluctuate on timescales of up to minutes (Fig. 10.6(a)), with a root-mean-square (rms) amplitude $\delta\Delta B^z_{\text{nuc}}$ of about 3 mT (Fig. 10.7c). These fluctuations lead to a Gaussian decay of coherent S–T_0 oscillations after a time $T_2^* = \hbar\sqrt{2}/(g^*\mu_B\delta\Delta B^z_{\text{nuc}}) = 14$ ns when averaging over many τ_S sweeps with different oscillation frequencies (Fig. 10.7(a)).

Pumping with the S and T_+ pulses for a fixed time generally does not change $\delta\Delta B^z_{\text{nuc}}$ appreciably. However, our ability to rapidly measure and manipulate ΔB^z_{nuc}

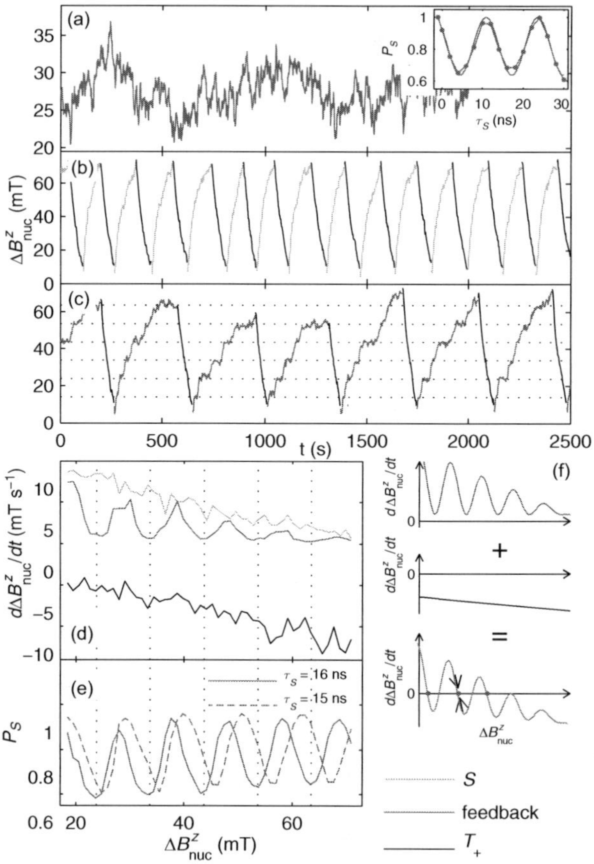

Fig. 10.6 Polarization rates. (a) Time trace of ΔB^z_{nuc} showing the fluctuations of the nuclear hyperfine field. The nonzero mean of ΔB^z_{nuc} is due to pumping (without feedback) between measurements. Each data point reflects a fit to the S–T_0 precession rate as shown in the inset. (b) Same, but alternating between 0.4 s S and T_+ pumping between measurements. The legend at the lower right is valid for (b)–(f). (c) Same for a feedback pulse with $\tau_{FB} = 15$ ns and T_+. (d) Pump rates extracted by averaging over time traces such as those in (b) and (c), but much longer. $d\Delta B^z_{\text{nuc}}/dt$ is normalized by the time spent pumping. (e) Singlet probability P_S after a fixed mixing time τ_S as a function of ΔB^z_{nuc}, extracted from the same data as the 'red and bright portions of the curve in (c) and the middle curve in (d)'. P_S is normalized by the change of the charge signal when sweeping gate voltages across the charge transition and does not decay all the way to zero, because of inelastic decay of the metastable T_0 during the measurement process. The dotted lines in (c)–(e) indicate where the polarization stagnates because the qubit is swept through the S–T_+ transition in $|T_0\rangle$. The best match between P_S and $d\Delta B^z_{\text{nuc}}/dt$ is obtained for $\tau_S = 16$ ns, whereas $\tau_{FB} = 15$ ns. This difference probably reflects finite pulse rise times. (f) Schematic illustration of the summing of the pump rates as shown in (d) from S (top) and T_+ pumping (middle) to obtain stable fixed points (filled circles, bottom). Fluctuations of ΔB^z_{nuc} away from these points are compensated by a restoring pump effect (arrows).

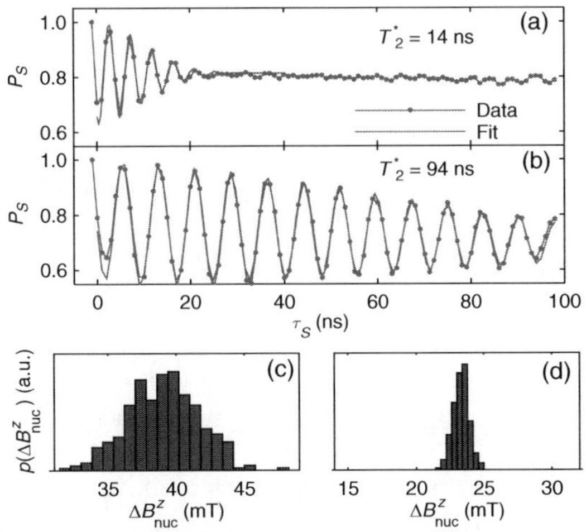

Fig. 10.7 T_2^* enhancement. (a) Ensemble average of S–T_0 oscillations for fixed pumping (0.7 s T^+ pumping and 0.25 s S pumping repeated every 2.2 s). The pumping maintains a nonzero $\Delta B^z_{\text{nuc}} = 39\,\text{mT}$, whose fluctuations lead to a dephasing time of $T_2^* = 14\,\text{ns}$. Other pump times give different oscillation frequencies, but similar dephasing times. P_S is normalized as explained in the caption of Fig. 10.6. (b) Same measurement, but for 30 ms of feedback pumping with $\tau_{FB} = 30\,\text{ns}$ and 31 ms of T_+ pumping after each 39 ms long measurement. The feedback extends T_2^* to 94 ns. The thin black lines show the fits to $1-P_S \propto 1+\sin(g^*\mu_B\Delta B^z_{\text{nuc}}\tau_S/\hbar)e^{-(\tau_S/T_2^*)^*}+\alpha\tau_S$ used to extract T_2^* and $\Delta B^z_{\text{nuc}} = 23\,\text{mT}$. The linear term in the fit model accounts for crosstalk of the gate pulses to the charge sensor. (c, d) Corresponding distributions of ΔB^z_{nuc}, obtained by histogramming instantaneous values of ΔB^z_{nuc} before ensemble averaging without (c) and with (d) pulse feedback.

enables us to narrow its distribution using software-based feedback. A proportional-integral feedback loop determines the type and duration of pumping between measurements from the time trace of ΔB^z_{nuc}. This procedure reduced $\delta\Delta B^z_{\text{nuc}}$ by about a factor of 2, corresponding to $T_2^* \approx 30\,\text{ns}$. It was limited by the ~1 Hz sampling rate of ΔB^z_{nuc} and could thus be improved with a faster readout technique [32, 34].

While this software feedback method already uses the same qubit to measure and polarize the nuclei, these two tasks are linked via a relatively slow readout process and the measurement computer. In order to speed up the feedback response, we have bypassed this connection by combining both operations into a single pulse derived from the S cycle (Fig. 10.5, bottom). In such a pulse feedback cycle, the qubit first probes its nuclear environment and then polarizes it depending on the result: after initialization in $|S\rangle$, the qubit is allowed to evolve at $\varepsilon \ll 0$ for a time τ_{FB}. As in the probe cycle, its state oscillates between $|S\rangle$ and $|T_0\rangle$, and the probability of ending in $|S\rangle$ is given by $[1+\cos(g^*\mu_B\Delta B^z_{\text{nuc}}\tau_{FB}/\hbar)]/2$. On sweeping ε past the S–T_+ transition, a nuclear spin can only be flipped by the $|S\rangle$ component of the qubit's state emerging from the evolution. Thus, the pump rate should be proportional to the ΔB^z_{nuc}-dependent singlet probability when averaged over many cycles. We have verified this behavior by characterizing the feedback pulse in the same way as the S and T_+ pulses. The measured mean pump rate $d\Delta B^z_{\text{nuc}}/dt$ oscillates as a function of ΔB^z_{nuc} between 0 and the value corresponding to S pumping (Fig. 10.6(d)). As expected, this modulation follows the singlet probability P_S extracted from the same data (Fig. 10.6(e)). Its period is given by $h/g^*\mu_B\tau_{FB}$.

In order to obtain a stable fixed point for ΔB^z_{nuc}, the pump rate has to cross zero with a negative slope. Fluctuations of ΔB^z_{nuc} away from the fixed point are then corrected by an opposing pump effect (Fig. 10.6(f), bottom). However, the feedback cycle alone pumps nuclei in one direction only (Fig. 10.6(d) and (f), bottom). At the minima of P_S, $d\Delta B^z_{\text{nuc}}/dt$ approaches zero because the qubit is swept past the S–T_+ transition in a $|T_0\rangle$. This stagnation of the polarization results in step-like structures in the time traces in Fig. 10.6(c). On either side of these points, $d\Delta B^z_{\text{nuc}}/dt$ remains positive. Thus, ΔB^z_{nuc} continues to grow once a fluctuation pushes it past one of these unstable fixed points. The required sign change can be engineered by preceding the feedback pulse with some amount of T_+ pumping. The resulting mean pump rate is the sum of a weakly ΔB^z_{nuc}-dependent negative T_+-pulse component and a positive contribution oscillating with ΔB^z_{nuc} from the feedback pulse (Fig. 10.6(f)).

To test the stabilizing effect of this pulse combination, we applied it for a fixed time between measurements of P_S for different τ_S. Figures 10.7(b) and (d) demonstrate an enhancement of T_2^* from 16 ns to 94 ns, and the corresponding narrowing of the distribution of ΔB^z_{nuc} around a fixed point. Here, the pump pulses were applied for 61 ms per 100 ms interval. The remaining 39 ms were spent measuring P_S for a single τ_S, and the data were averaged over 232 sweeps of τ_S. We estimate that the pump rate achieved here limits the feedback response time to about 1 s.

In a quantum processor, the measurements could be replaced by a sequence of gate operations, whose fidelity would be substantially improved by the reduction of fluctuations. For a 1 ns π rotation generated by a gradient of 100 mT [31] with rms fluctuations of $\delta\Delta B^z_{\text{nuc}} = 0.5\,\text{mT}$ as demonstrated here, the fidelity is $\pi^2(\delta\Delta B^z_{\text{nuc}})^2/4(\Delta B^z_{\text{nuc}})^2 \lesssim 10^{-4}$ [35]. However, owing to the slowness of the nuclear bath, an error of order unity accumulates after only 100 such gates. This limitation could be overcome by making gates insensitive to $\delta\Delta B^z_{\text{nuc}}$ to first order [36, 37]. A pulse angle error of order $(\delta\Delta B^z_{\text{nuc}})^2/(\Delta B^z_{\text{nuc}})^2 \sim 10^{-4}$ would allow 10^4 operations per error. In either case, the improvement in gate fidelity is at least quadratic in the narrowing ratio. Thus, narrowing procedures are very effective at overcoming the limitations imposed by a fluctuating nuclear spin bath. The flexibility of our approach should allow an adaptation to other systems or other protocols on the same system.

10.6 Decoupling from the nuclear bath

The promise of quantum-dot spin qubits as a solid state approach to quantum computing is demonstrated by the successful realization of initialization, control, and single-shot readout of electron spin qubits in GaAs quantum dots using optical [4], magnetic [38], and fully electrical [31, 32, 39] techniques. To further advance spin-based quantum computing, it is vital to mitigate decoherence due to the interaction of the electron spin with the spins of nuclei of the host material. Understanding the dynamics of this system is also of great fundamental interest [23, 40].

Through the hyperfine interaction, an electron spin in a GaAs quantum dot is subjected to an effective magnetic field produced by the nuclear spins. Under typical experimental conditions, this so-called "Overhauser field" has a random magnitude and direction. Typically, measurements of the coherent electron spin precession involve averaging over many experimental runs, and thus over many Overhauser field configurations. As a result, the coherence signal is suppressed for evolution times $\tau \gtrsim T_2^* \approx 10\,\mathrm{ns}$ [11]. However, the nuclear spins evolve much more slowly than the electron spins, so that the Overhauser field is nearly static for a sufficiently short duration of electron spin evolution. Therefore, it is possible to partially eliminate the effect of the random nuclear field by flipping the electron spin halfway through an interval of free precession [41], a procedure known as Hahn echo. The random contributions of the Overhauser field to the electron spin precession before and after the spin reversal then approximately cancel out. For longer evolution times, the effective field acting on the electron spin generally changes between the two halves of the precession interval. This change leads to an eventual loss of coherence on a timescale determined by the details of the nuclear spin dynamics.

Previous Hahn-echo experiments in GaAs quantum-dot spin qubits have demonstrated spin dephasing times of around $1\,\mu\mathrm{s}$ at relatively low magnetic fields up to $100\,\mathrm{mT}$ [8, 42]. Recent theoretical studies of decoherence due to the hyperfine interaction [43–45] are generally consistent with these experimental results, but predict revivals of the echo signal after several microseconds, as also seen in other systems [17]. This prediction already indicates that the initial decay of the echo does not reflect irreversible decoherence, but is a consequence of the coherent Larmor precession of the nuclei. Theoretical work also predicted much longer coherence times at higher external magnetic fields [46] or when using more advanced pulse sequences [47, 48]. The classic example is the Carr-Purcell-Meiboom-Gill (CPMG) sequence [8, 49], but several alternatives have recently been developed [50, 51] and demonstrated [52, 53]. The performance of such schemes is expected to improve as more control pulses are added [48]. Here, we provide direct experimental confirmations for all the above predictions.

We initialize the system at a large detuning, where the ground state is a spin singlet with both electrons residing in a single dot. As ε is swept to negative values, the electrons separate into different dots, thus preparing the singlet state $S \equiv (|\uparrow\downarrow\rangle - |\downarrow\uparrow\rangle)/\sqrt{2}$. For very large negative detunings, the electron spins in the two dots are decoupled, and each individually experiences a Zeeman field composed of the homogeneous external field and a fluctuating local hyperfine field. A difference $\Delta B^z_{\mathrm{nuc}}$ between the z components of the hyperfine fields in the two dots leads to an energy splitting between the basis states $|\uparrow\downarrow\rangle$ and $|\downarrow\uparrow\rangle$. This splitting causes precession between the singlet S and the triplet $T_0 \equiv (|\uparrow\downarrow\rangle + |\downarrow\uparrow\rangle)/\sqrt{2}$, and its fluctuations lead to dephasing of the qubit. We implement the echo π pulses by pulsing to small negative detunings, where inter-dot tunneling leads to an exchange splitting between S and T_0. This splitting

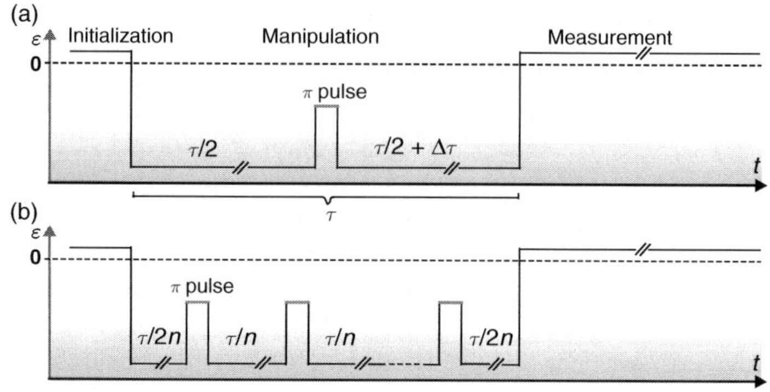

Fig. 10.8 Pulse sequences. (a) Hahn-echo sequence. After evolving for a time $\tau/2$, the two electrons are exchanged with a π pulse. The singlet state is recovered after further evolution for another $\tau/2$, independent of $\Delta B^z_{\mathrm{nuc}}$. (b) Carr-Purcell-Meiboom-Gill sequence: in this higher-order decoupling sequence, n π-pulses at time intervals τ/n are applied.

drives coherent oscillations between the states $|\uparrow\downarrow\rangle$ and $|\downarrow\uparrow\rangle$. The pulse profiles for the Hahn-echo and CPMG sequence are shown in Figs. 10.8(a,b).

Figure 10.9(a) shows the Hahn-echo signals for different magnetic fields B_{ext}. At high fields, we find a monotonic decay of the Hahn-echo signal with the total duration of free precession τ. The echo initially decays very slowly (approximately proportionally to τ^4) and is decays to $1/e$ after 30 μs. As the magnetic field is reduced, the echo signal develops oscillations with a timescale of microseconds. For even lower fields, the oscillations evolve into full collapses of the signal, with revivals at later times on a ten microsecond time scale. These revivals were predicted by Cywinski *et al.*, [44, 45] based on a quantum mechanical treatment of the hyperfine interaction between electron and nuclear spins. In the following, we outline a semiclassical model [54] that can reproduce the lowest-order result of Cywinski *et al.*, [44, 45] and accounts for additional effects that are essential for fitting our data. This model provides the theoretical echo signal, $C(\tau) \equiv 2p(S) - 1 = -\mathrm{Re}\langle\uparrow\downarrow|\rho(\tau)\,|\downarrow\uparrow\rangle$, where $p(S)$ is the probability of finding the electron in a singlet state and $\rho(\tau)$ is the qubit's density matrix at the end of the evolution time. We have used this model to produce the quantitatively accurate fits displayed in Fig. 10.9(a).

For each electron spin, the Zeeman energy splitting is proportional to the total magnetic field (Fig. 10.9(b)). Time dependence of the parallel and transverse nuclear

$$B_{\mathrm{tot}} = \sqrt{(B_{\mathrm{ext}} + B^z_{\mathrm{nuc}})^2 + {B^{\perp}_{\mathrm{nuc}}}^2} \approx B_{\mathrm{ext}} + B^z_{\mathrm{nuc}} + (B^{\perp}_{\mathrm{nuc}})^2/2B_{\mathrm{ext}}$$

components B^z_{nuc} and $(B^{\perp}_{\mathrm{nuc}})^2$ can lead to dephasing of the electron spin. Assuming statistical independence between B^z_{nuc} and $(B^{\perp}_{\mathrm{nuc}})^2$, the theoretical echo signal can be written as a product, $C(\tau) = A_z(\tau)R_{\perp}(\tau)$, where $A_z(\tau)$ and $R_{\perp}(\tau)$ account for the contributions of B^z_{nuc} and $B^{\perp}_{\mathrm{nuc}}$ to the electron spin precession. In the experimental range of the magnetic fields, the time dependence of B^z_{nuc} is mainly caused by spectral

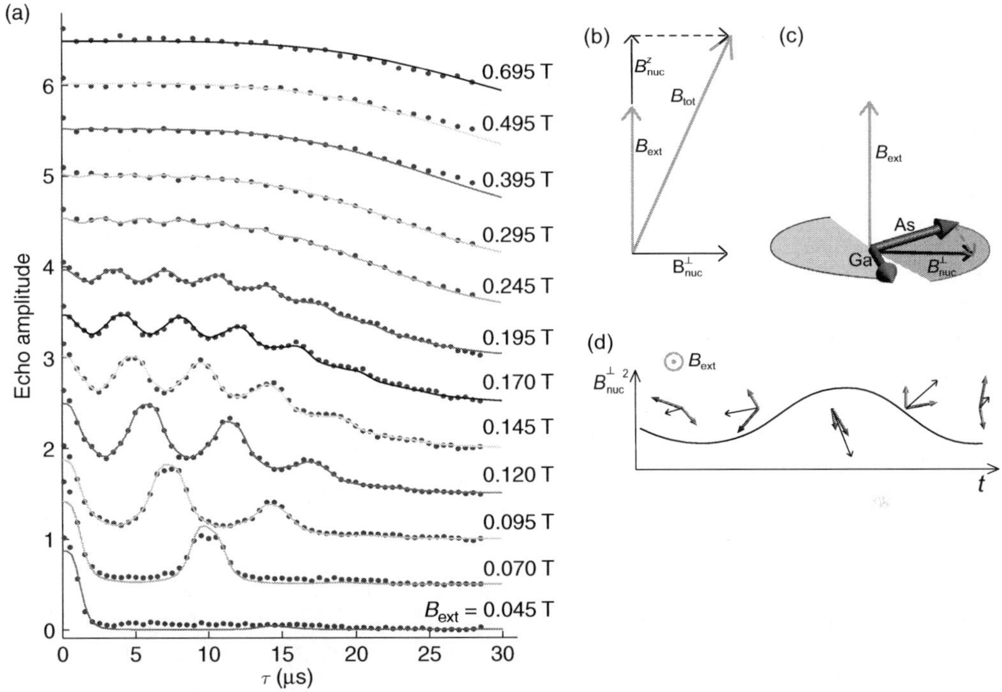

Fig. 10.9 Echo amplitude. (a) Echo signal as a function of the total evolution time τ for different values of the magnetic field. The fits to the data are obtained from the model discussed in the text. Curves are offset for clarity and normalized to unity visibility at short time. (b) The total Zeeman field seen by the electron is the vector sum of the external field and the Overhauser fields parallel and perpendicular to it. (c) The three nuclear species (only two shown for clarity) contributing to the Overhauser field precess at different Larmor frequencies in the external field. (d) As a result of the relative precession of the nuclear species, the total transverse nuclear field oscillates at the Larmor frequency difference(s).

diffusion due to the dipole-dipole interaction between nuclear spins. This process is predicted to lead to a decay of the form $A_z(\tau) = \exp[-(\tau/T_{SD})^4]$ [43, 46]. As we will now discuss, $R_\perp(\tau)$ has a more interesting nonmonotonic structure, which arises from the relative precession of nuclear spins in the external field with different Larmor frequencies.

The transverse hyperfine field $\vec{B}^{\perp}_{\rm nuc}$ is a vector sum of contributions from the three nuclear species ^{69}Ga, ^{71}Ga, and ^{75}As. Due to the different precession rates of these species (Fig. 10.9(c)), $[B^{\perp}_{\rm nuc}(t)]^2$ contains harmonics at the three relative Larmor frequencies (Fig. 10.9(d)), in addition to a constant term. The contribution of the constant term to the singlet return probability is eliminated by the echo pulse. For a general free precession period, the time dependence leads to a suppression of the echo signal. However, if the precession interval $\tau/2$ is a multiple of all three relative Larmor periods, the oscillatory components contribute no net phase to the electron spin evolution. As a result, the echo amplitude revives whenever the commensurability condition is met. Averaging the singlet return probability over initial Overhauser field configurations [54] leads to the collapse-and-revival behavior predicted by Cywinski *et al.*, [44, 45].

At low fields, the echo envelope decays more quickly than at high fields (see Fig. 10.9(a)). This field dependence can be accounted for by including a spread of the Larmor precession frequencies for each nuclear species. Such a variation is also manifest in the width of NMR lines and naturally arises from dipolar and other interactions between nuclei [55]. We model it as a shift of the magnetic field acting on each individual nuclear spin by an amount B_{loc}, where B_{loc} is a Gaussian random variable with standard deviation δB_{loc}. This spread of precession frequencies leads to an aperiodic time-dependence of $(B^{\perp}_{\text{nuc}})^2$, which cannot be removed by the electron spin echo.

Using the model outlined above and described in more detail in Section 10.7, we have fit all the data in Fig. 10.9(a) with a single set of field-independent parameters that were chosen to obtain a good match with all data sets: the number of nuclei in each of the two dots, N, the spectral diffusion time constant T_{SD}, and δB_{loc}. In addition, the scale factor for each data set was allowed to vary to account for the imperfect normalization of the data. The value of N determines the depths of the dips between revivals. The best fit yields $N = 4.4 \times 10^6$, which is in good agreement with an independent determination from a measurement of $T_2^* = \sqrt{N}\hbar/(g^*\mu_B \cdot 4.0\,\text{T})$, giving $N = 4.9 \times 10^6$ [10]. From the fit, we also obtain $T_{SD} \approx 37\,\mu\text{s}$ and $\delta B_{\text{loc}} = 0.3\,\text{mT}$. The measured NMR linewidth in pure GaAs is about 0.1 mT [55]. A possible origin for the larger field inhomogeneity found here is the quadrupole splitting arising from the presence of the two localized electrons [56]. The inhomogeneity of the Knight shift is expected to have a similar but quantitatively negligible effect for our parameters. The value of T_{SD} is consistent with theoretical estimates [46]. Interestingly, the spread of nuclear Larmor frequencies, captured by δB_{loc}, contributes significantly to the echo decay even at the highest fields investigated. We have also verified that the Hahn-echo lifetime is not significantly affected by dynamic nuclear polarization, which can be used to increase T_2^* [57].

In order to measure the long Hahn-echo decay times of up to 30 μs, it was necessary to systematically optimize the pulses. Small differences in the gate voltages before and after the π pulse shift the electronic wavefunction relative to the inhomogeneously polarized nuclei. Such shifts cause the electrons to sample different Overhauser fields at different times, and thus lead to an imperfect echo. We have minimized this effect by compensating for a systematic drift of ε over the course of each pulse sequence.

Substantially longer coherence times are expected for more elaborate decoupling sequences [48]. We implemented the CPMG sequence [49], which consists of an n-fold repetition of the Hahn echo, thus requiring n π pulses, as shown in Fig. 10.8(d). Figure 10.10 shows data for $n = 6$, 10, and 16. For $n = 16$, the echo signal clearly persists for more than 200 μs. To verify the interpretation of the data, we have measured the dependence of the echo on small changes in the final free-precession time and the duration of the exchange pulses for $n = 10$, $\tau = 5$ and $120\,\mu$s. Because of the large number of potential tuning parameters, we have not optimized these CPMG pulses.

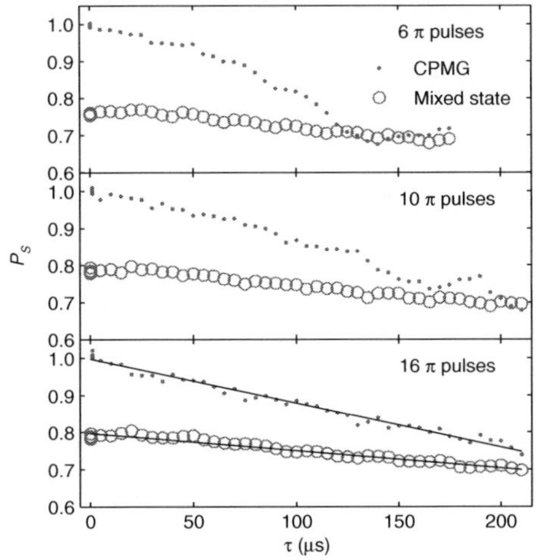

Fig. 10.10 CPMG decoupling experiments with 6, 10 and 16 π pulses at $B_{\text{ext}} = 0.4\,\text{T}$. The dots show the readout signal of the CPMG pulses, the circles represent reference measurements with the same evolution time without any π pulses (equivalent to T_2^* measurements), which produce a completely dephased state. P_S is the sensor signal normalized by the dc contrast associated with the transfer of an electron from one dot to the other, so that a singlet corresponds to $P_S = 1$. Inelastic decay during the readout phase and possibly other visibility loss mechanisms increase P_S for the mixed state above the ideal value of 0.5. The linear trends in the reference measurement and the initial decay of the CPMG signal likely reflect leakage out of the logical subspace. The linear fits to the 16-pulse data (lines) intersect at $\tau = 276\,\mu\text{s}$, which can be taken as a rough estimate of or lower bound on the coherence time.

We expect that, with improved pulses, the same extension of the coherence time could be achieved with fewer pulses. The linear initial decay of the signal in Fig. 10.10 is not well understood. The similar variation of the reference signal corresponding to a completely mixed state is suggestive of a single-electron T_1 process causing leakage into the T^+ and T^- states. The decay time constant sets a lower bound for the largest achievable coherence time.

Our measurements demonstrate coherence times of GaAs spin qubits of at least $200\,\mu\text{s}$, two orders of magnitude larger than previously shown. The duration of each of the π pulses could easily be reduced below the 6 ns used here. Thus, more than 10^5 operations could be carried out within the coherence time, well in excess of the commonly quoted error correction threshold of order 10^4 for conventional quantum error correction schemes and well within the threshold required for surface codes. Furthermore, one may hope to achieve millisecond scale coherence times with improved decoupling sequences [48] without adding complexity. The excellent agreement with the model for the field and time dependence of the Hahn-echo revivals shows that many aspects of the dephasing of electron spins due to the nuclear hyperfine interaction are now well understood. The insight gained may also help pave the way toward probing macroscopic quantum effects in a mesoscopic ensemble of a few million nuclear spins.

10.7 Semiclassical picture of dephasing

A model for decoherence of a single electron spin due to the hyperfine interaction was derived in [44], based on a quantum mechanical treatment of electron-mediated nuclear flip-flops. Here, we derive the same result by treating the Overhauser fields acting on the electron as semiclassical variables, which provides a simple intuition for the collapses and revivals seen in Fig. 10.9. The heteronuclear terms in the expansion $(B_{\text{nuc}}^{\perp})^2 = \sum_{kl} \sum_{i=x,y} A_k A_l I_k^i I_l^i$ (i.e, those where nuclei k and l are of different species) oscillate at the relative Larmor frequencies $\omega_k - \omega_l$. Averaging over all initial nuclear spin states amounts to averaging over the phase and amplitude of these oscillations, which randomizes the electronic phase accumulated during an echo experiment and thus causes the collapse of the echo amplitude. However, if the free-evolution time $\tau/2$ is a multiple of each of the three relative Larmor periods, the contribution of those oscillations to the electron phase vanishes regardless of the nuclear state, which leads to the revival peaks. Note that it is a fortunate coincidence that the three Larmor frequencies are nearly equidistant, so that the commensurability condition is easily fulfilled to a good approximation for all three pairs of species simultaneously.

We extend our model to allow for a spatial variation δB of the local magnetic field determining the nuclear Larmor frequencies, $B_{\text{ext}} + \delta B$. Such a variation phenomenologically accounts for the random dipolar and electron-mediated exchange field of neighboring nuclei and for possible quadrupolar shifts from the electric field associated with the electron. It leads to fluctuations of the transverse field from each nuclear species beyond a rigid rotation, which contribute to the envelope decay of the echo amplitude.

The fact that the two electrons are decoupled and interact with two independent nuclear ensembles during the free evolution allows us to begin by considering dephasing in a single dot, and then to apply the result to the relevant case of a double dot. Our notation largely follows that of [44]. A detailed justification of our semiclassical treatment is given elsewhere [54].

Because of their smaller g factor, the nuclei evolve three orders of magnitude more slowly than the electrons. Thus, the precession of the electron spin adiabatically follows the instantaneous eigenstates aligned with the total field $\mathbf{B}_{\text{tot}} = \mathbf{B}_{\text{ext}} + \mathbf{B}_{\text{nuc}}$, and we only need to consider the level splitting $g^*\mu_B B_{\text{tot}} = g^*\mu_B\sqrt{(B_{\text{ext}} + B_{\text{nuc}}^z)^2 + (B_{\text{nuc}}^{\perp})^2} \approx g^*\mu_B[B_{\text{ext}} + B_{\text{nuc}}^z + (B_{\text{nuc}}^{\perp})^2/2B_{\text{ext}}]$. We will thus work with the electronic Hamiltonian $\hat{H}(t) = g^*\mu_B\{B_{\text{ext}} + B_{\text{nuc}}^z(t) + [B_{\text{nuc}}^{\perp}(t)]^2/2B_{\text{ext}}\}\hat{S}_z$. In [44], this effective Hamiltonian was formally derived from a Schrieffer-Wolff transformation.

We account for fluctuations in B_{nuc}^z by assuming that they are uncorrelated with those of $B_{\text{nuc}}^{\perp}$ and lead to a field-independent echo decay factor $\exp[(-\tau/\tau_{SD})^4]$, as computed in [46]. In this section, we focus on the effect of $[B_{\text{nuc}}^{\perp}(t)]^2$, which we write as a sum of semiclassical fields $[B_{\text{nuc}}^{\perp\sigma}(t)]^2 = \sum_{k,l} B_k^\sigma(t) B_l^\sigma(t)^*$. The summation indices k and l run over groups of (approximately) equivalent nuclei. Each group k includes only nuclei of the same species $\alpha(k)$, experiencing the same local field inhomogeneity δB_k and having the same hyperfine coupling strength A_k. The $B_k^\sigma = B_k^{\sigma x} + iB_k^{\sigma y}$ are complex-valued, with the real and imaginary parts representing the x and y component of the total transverse field from group k. $\sigma = \pm 1$ specifies the initial $\hat{S}^z$ eigenstate $|\sigma\rangle$ of the electron, which is important because it determines the sign of the Knight shift. For the time evolution of the nuclear field, we assume that nuclei in group k precess at a constant Larmor frequency $\omega_k + \delta\omega_k + \sigma c(t) A_k/2$. Here $\omega_k = \gamma_{\alpha(k)} B_{\text{ext}}$ is the species $\alpha(k)$-dependent Larmor frequency, and $\delta\omega_k = \gamma_{\alpha(k)} \delta B_k$ its change due to a local field δB_k. $c(t) = \pm 1$ describes the pulse sequence and changes at every π pulse, thus switching the sign of the Knight-shift term $\sigma c(t) A_k/2$. For the Hahn echo, $c(t) = 1$ (-1) for $t < \tau/2$ $(t > \tau/2)$. The time dependence is thus

$$B_k^\sigma(t) = B_k^\sigma(0) \exp\left\{ i \left[\omega_k t + \delta\omega_k t + \frac{\sigma A_k}{2} \int_0^t c(t)\, dt \right] \right\}.$$

The relative phase Φ picked up between the two electronic states $|\sigma = \pm\rangle$ over a pulse sequence of duration τ is then given by

$$\begin{aligned}
\Phi &= \frac{g^* \mu_B}{4\hbar B_{\text{ext}}} \sum_{\sigma=\pm 1} \int_0^\tau c(t)\, dt \sum_{k,l} B_k^\sigma(t) B_l^{\sigma *}(t) \\
&= \frac{g^* \mu_B}{4\hbar B_{\text{ext}}} \sum_{k,l} \overline{B}_l \overline{B}_k z_k z_l^* \sum_{\sigma=\pm 1} \times \int_0^\tau c(t)\, dt \exp\left[i(\omega_{kl} + \delta\omega_{kl}) t + i\sigma A_{kl} \int_0^t dt'\, c(t') \right] \\
&\equiv \sum_{k,l} T_{kl} \frac{z_k z_l^*}{2}, \qquad (10.1)
\end{aligned}$$

where we have expressed the initial conditions for $B_k^\sigma(0)$ by the dimensionless complex variables z_k and the rms transverse field values, $\overline{B}_k = \sqrt{a_{\alpha(k)} N_k n_{\alpha(k)}/2} A_k$. Furthermore, $\omega_{kl} = \omega_k - \omega_l$, $\delta\omega_{kl} = \delta\omega_k - \delta\omega_l$, $A_{kl} = (A_k - A_l)/2$, $a_{\alpha(k)} = 2/3(I_{\alpha(k)}+1)$, $I_{\alpha(k)} = 5/2$, and the number of nuclei of species α per unit cell, n_α, are defined as in [44]. N_k is the number of unit cells contributing to group k. The $\{T_{kl}\}$ are identical to the coarse-grained $\tilde{T}$ matrix of [44], except for our extension to account for the local field variation δB.

We now carry out an ensemble average over the initial conditions by treating $\{z_k\}$ as uncorrelated Gaussian-distributed random variables (justified by the large number of nuclear spins involved) with zero mean and unit variance, that is, with probability density function

$$p(z, z^*)\, dz\, dz^* = \frac{i}{4\pi} \exp\left(-\frac{z z^*}{2} \right) dz\, dz^*.$$

$$\begin{aligned}
\langle e^{-i\Phi} \rangle &= \int \left(\prod_j dz_j\, dz_j^*\, p(z_j, z_j^*) \right) \exp\left(-i \sum_{k,l} T_{kl} \frac{z_k^* z_l}{2} \right) \\
&= \prod_j \int dz\, dz^*\, p(z, z^*) \exp\left(-i\lambda_j \frac{|z|^2}{2} \right) \\
&= \prod_j \frac{1}{1 + i\lambda_j}, \qquad (10.2)
\end{aligned}$$

where the $\{\lambda_j\}$ are the eigenvalues of the T matrix, and we have used the fact that the distribution of the z_k is invariant under the basis change associated with the diagonalization of the T matrix.

So far, we have only considered the decoherence of a single electron spin. Since the nuclei in the two halves of our double dot can be assumed to be statistically independent and the phase difference between the $|\uparrow\downarrow\rangle$ and $|\downarrow\uparrow\rangle$ states of the double dot is the sum of the single-dot phases, the two-electron decoherence function is a product of two single-dot functions: $\langle e^{i\Phi_{\text{double-dot}}}\rangle = \langle e^{i(\Phi_{\text{left}}-\Phi_{\text{right}})}\rangle = \langle e^{i\Phi_{\text{left}}}\rangle\langle e^{-i\Phi_{\text{right}}}\rangle = |\langle e^{i\Phi}\rangle|^2$, where the subscripts denote the two halves of the double dot and the last equality holds for identical dots.

A quantitative comparison of T_{SD} with theory [46] is difficult because the latter depends sensitively on the shape of the electronic wavefunction, but a rough comparison gives good agreement. From a self-consistent Poisson-Schrödinger model, we estimate a full width at half-maximum of the transverse electronic charge density in our heterostructure of 12 nm. We thus compare with the curve for a quantum well width $z_0 = 20$ nm from Fig. 14(b) of [46]. The fitted $T_{SD} = 37$ μs has to be multiplied by a factor $2^{1/4}$ to account for dephasing occurring for both electrons in the double dot [58]. Under these assumptions, theory and experiment match for a Fock-Darwin radius of $\ell = 25$ nm, which would correspond to $N = (32/3)\pi z_0 \ell^2/a_0^3 = 2.3 \times 10^6$, where $a_0 = 0.565$ nm is the lattice constant of GaAs.

In pure GaAs, the measured NMR linewidth is about 0.1 mT. The broadening arises from a fluctuating effective magnetic field due to dipolar and indirect exchange interactions between nuclei [55, 56]. The correlation time of this field is given by the timescale for dipolar nuclear flip-flops. Assuming a typical dipolar field of at most 0.1 mT, the correlation time is at least 100 μs, substantially longer than the duration of Hahn-echo experiments. Thus, the broadening can be described by the static distribution of δB_k used in our model to a good approximation. Note that the correlation time of B^z_{nuc} is even longer (up to minutes), yet the $\dot{B}^z_{\text{nuc}}$ fluctuations are responsible for part of the electron coherence. Here, a quasistatic approximation would yield no echo decay at all.

In addition to the interactions, the nuclei will experience a static quadrupole splitting due to electric field gradients originating from the electron. We proceed to estimate the resulting second moment of the NMR line, which we identify with δB^2. We crudely model the electron as a charge density of 50 nm lateral and 10 nm vertical extent. Its direct contribution to the electric field gradient, $\partial E_z/\partial z \sim e/[\epsilon_r\epsilon_0(10\,\text{nm}\cdot 50\,\text{nm}^2)] \approx 7\times 10^{13}\,\text{V/m}^2$, is negligible. In addition, the electric field of the electron induces a field gradient at the site of the nuclei by deforming the atomic orbitals. Following Eq. (6) of [56], we obtain $\partial E_z/\partial z = R_{14}e/(\epsilon_r\epsilon_0 \cdot 50\,\text{nm}^2)$, where $R_{14} \approx 1 \times 10^{12}\,\text{m}^{-1}$ is the relevant response tensor element. The resulting rms linewidth broadening for the magnetic field along the [110] direction is $\sqrt{9(4I(I+1)-3)/5}/(4I(2I-1)\gamma_\alpha)Q_\alpha\partial E_z/\partial z$ (see Eq. (1) of [56]), where Q_α is the quadrupole moment of species α. Substituting literature values, we obtain rms linewidths of 0.6–2.5 G for the different species. The quadrupole shift is thus of the correct order of magnitude to explain the larger δB of about 3 G needed to fit our Hahn-echo data. Interestingly, the quadrupolar shift should have a pronounced dependence on the direction of the applied magnetic field.

According to our model, it vanishes for B_{ext} along the [100] direction and is largest along the [111] direction. In our experiments, B_{ext} was applied in the [110] direction. Thus, a strong enhancement of the envelope decay time can be expected for B_{ext} along [100].

References

[1] Nielsen, M. A., and Chuang, I. L. (2000). *Quantum Computation and Quantum Information.* Cambridge University Press, Cambridge.

[2] Levy, J. (2002). Universal quantum computation with spin-1/2 pairs and Heisenberg exchange. *Phys. Rev. Lett.*, **89**, 147902.

[3] Coish, W. A., and Loss, D. (2007). Exchange-controlled single-electron-spin rotations in quantum dots. *Phys. Rev. B*, **75**, 161302.

[4] Press, D., Ladd, T. D., Zhang, B. Y., and Yamamoto, Y. (2008). Complete quantum control of a single quantum dot spin using ultrafast optical pulses. *Nature*, **456**, 218.

[5] Hanson, R. and Burkard, G. (2007). Universal set of quantum gates for double-dot spin qubits with fixed interdot coupling. *Phys. Rev. Lett.*, **98**, 050502.

[6] Lidar, D. A., Chuang, I. L., and Whaley, K. B. (1998). Decoherence-free subspace for quantum computation. *Phys. Rev. Lett.*, **81**, 2594.

[7] Taylor, J. M., Engel, H. A., Dur, W., Yacoby, A., Marcus, C. M., Zoller, P., and Lukin, M. D. (2005). Fault-tollerant architecture for quantum computation using electrically controlled semiconductor spins. *Nature Phys.*, **1**, 177.

[8] Petta, J. R., Johnson, A. C., Taylor, J. M., Laird, E. A., Yacoby, A., Lukin, M. D., Marcus, C. M., Hanson, M. P., and Gossard, A. C. (2005). Coherent manipulation of coupled electron spins in semiconductor quantum dots. *Science*, **309**, 2180.

[9] Obata, T., Pioro-Ladrière, M., Tokura, Y., Shin, Y.-S., Kubo, T., Yoshida, K., Taniyama, T., and Tarucha, S. (2010). Coherent manipulation of individual electron spin in a double quantum dot integrated with a micromagnet. *Phys. Rev. B*, **81**, 085317.

[10] Taylor, J. M., Petta, J. R., Johnson, A. C., Yacoby, A., Marcus, C. M., and Lukin, M. D. (2007). Relaxation, dephasing, and quantum control of electron spins in double quantum dots. *Phys. Rev. B*, **76**, 035315.

[11] Petta, J. R., Taylor, J. M., Johnson, A. C., Yacoby, A., Lukin, M. D., Marcus, C. M., Hanson, M. P., and Gossard, A. C. (2008). Dynamic nuclear polarization with single electron spins. *Phys. Rev. Lett.*, **100**, 067601.

[12] Reilly, D. J., Taylor, J. M., Petta, J. R., Marcus, C. M., Hanson, M. P., and Gossard, A. C. (2008). Suppressing spin qubit dephasing by nuclear state preparation. *Science*, **321**, 817.

[13] Johnson, A. C., Petta, J. R., Taylor, J. M., Yacoby, A., Lukin, M. D., Marcus, C. M., Hanson, M. P., and Gossard, A. C. (2005). Triple-singlet spin relaxation via nuclei in a double quantum dot. *Nature*, **435**, 925.

[14] Paget, D., Lampel, G., and Sapoval, B. (1977). Low field electron-nuclear spin coupling in gallium arsenide under optical pumping conditions. *Phys. Rev. B*, **15**, 5780–5796.

[15] Stopa, M., Krich, J. J., and Yacoby, A. (2010). Inhomogeneous nuclear spin flips: feedback mechanism between electronic states in a double quantum dot and the underlying nuclear spin bath. *Phys. Rev. B*, **81**(4), 041304.
[16] Gullans, M., Krich, J. J., Taylor, J. M., Bluhm, H., Halperin, B. I., Marcus, C. M., Stopa, M., Yacoby, A., and Lukin, M. D. (2010). Dynamic nuclear polarization in double quantum dots. *Phys. Rev. Lett.*, **104**, 226807.
[17] Childress, L., Gurudev Dutt, M. V., Taylor, J. M., Zibrov, A. S., Jelezko, F., Wrachtrup, J., Hemmer, P. R., and Lukin, M. D. (2006). Coherent dynamics of coupled electron and nuclear spin qubits in diamond. *Science*, **314**, 281.
[18] Koppens, F. H. L., Buizert, C., Tielrooij, K. J., Vink, I. T., Nowack, K. C., Meunier, T., Kouwenhoven, L. P., and Vandersypen, L. M. K. (2006). Driven coherent oscillations of a single electron spin in a quantum dot. *Nature*, **442**, 766.
[19] Fischer, J., Coish, W. A., Bulaev, D. V., and Loss, D. (2008). Spin decoherence of a heavy hole coupled to nuclear spins in a quantum dot. *Phys. Rev. B*, **78**, 155329.
[20] Tyryshkin, A. M., Lyon, S. A., Astashkin, A. V., and Raitsimring, A. M. (2003). Electron spin relaxation times of phosphorus donors in silicon. *Phys. Rev. B*, **68**, 193207.
[21] Takahashi, S., Hanson, R., van Tol, J., Sherwin, M. S., and Awschalom, D. D. (2008). Quenching spin decoherence in diamond through spin bath polarization. *Phys. Rev. Lett.*, **101**, 047601.
[22] Coish, W. A. and Loss, D. (2004). Hyperfine interaction in a quantum dot: non-Markovian electron spin dynamics. *Phys. Rev. B*, **70**, 195340.
[23] Coish, W. A., Fischer, J., and Loss, D. (2008). Exponential decay in a spin bath. *Phys. Rev. B*, **77**, 125329.
[24] Vink, I. T., Nowack, K. C., Koppens, F. H. L., Danon, J., Nazarov, Yu. V., and Vandersypen, L. M. K. (2009). Locking electron spins into magnetic resonance by electron-nuclear feedback. *Nature Phys.*, **5**, 764.
[25] Xu, X., Yao, W., Sun, B., Steel, D. G., Bracker, A. S., Gammon, D., and Sham, L. J. (2009). Optically controlled locking of the nuclear field via coherent dark-state spectroscopy. *Nature*, **459**, 1105.
[26] Latta, C., Högele, A., Zhao, Y., Vamivakas, A. N., Maletinsky, P., Kroner, M., Dreiser, J., Carusotto, I., Badolato, A., Schuh, D., Wegscheider, W., Atature, M., and Imamoglu, A. (2009). Confluence of resonant laser excitation and bi-directional quantum dot nuclear spin polarization. *Nature Phys.*, **5**, 758.
[27] Greilich, A., Shabaev, A., Yakovlev, D. R., Efros, Al. L., Yugova, I. A., Reuter, D., Wieck, A. D., and Bayer, M. (2007). Nuclei-induced frequency focusing of electron spin coherence. *Science*, **317**, 1896.
[28] Danon, J., Vink, I. T., Koppens, F. H. L., Nowack, K. C., Vandersypen, L. M. K., and Nazarov, Yu. V. (2009). Multiple nuclear polarization states in a double quantum dot. *Phys. Rev. Lett.*, **103**, 046601.
[29] Rudner, M. S. and Levitov, L. S. (2007). Electrically driven reverse Overhauser pumping of nuclear spins in quantum dots. *Phys. Rev. Lett.*, **99**(24), 246602.

[30] Barthel, C., Medford, J., Bluhm, H., Yacoby, A., Marcus, C. M., Hanson, M. P., and Gossard, A. C. (2012). Relaxation and readout visibility of a singlet-triplet qubit in an overhauser field gradient. *Phys. Rev. B*, **85**, 035306.

[31] Foletti, S., Bluhm, H., Mahalu, D., Umansky, V., and Yacoby, A. (2009). Universal quantum control in two-electron spin quantum bits using dynamic nuclear polarization. *Nature Phys.*, **5**, 903.

[32] Barthel, C., Reilly, D. J., Marcus, C. M., Hanson, M. P., and Gossard, A. C. (2009). Rapid single-shot measurement of a singlet-triplet qubit. *Phys. Rev. Lett.*, **103**, 160503.

[33] Reilly, D. J., Taylor, J. M., Laird, E. A., Petta, J. R., Marcus, C. M., Hanson, M. P., and Gossard, A. C. (2008). Measurement of temporal correlations of the overhauser field in a double quantum dot. *Phys. Rev. Lett.*, **101**, 236803.

[34] Reilly, D. J., Marcus, C. M., Hanson, M. P., and Gossard, A. C. (2007). Fast single-charge sensing with rf quantum point contact. *Appl. Phys. Lett.*, **91**, 162101.

[35] Vandersypen, L. M. K. and Chuang, I. L. (2004). NMR techniques for quantum control and computation. *Rev. Mod. Phys.*, **76**, 1037–1069.

[36] Khodjasteh, K. and Viola, L. (2009). Dynamically error-corrected gates for universal quantum computation. *Phys. Rev. Lett.*, **102**, 080501.

[37] Khodjasteh, K., Bluhm, H., and Viola, L. (2012). Automated synthesis of dynamically corrected quantum gates. *Phys. Rev. A*, **86**, 042329.

[38] Koppens, F. H. L., Folk, J. A., Elzerman, J. M., Hanson, R., Willems van Beveren, L. H., Vink, I. T., Tranitz, H. P., Wegschneider, W., Kouwenhoven, L. P., and Vandersypen, L. M. K. (2005). Control and detection of singlet-triplet mixing in a random nuclear field. *Science*, **309**, 1346.

[39] Nowack, K. C., Koppens, F. H. L., Nazarov, Yu. V., and Vandersypen, L. M. K. (2007). Coherent control of a single electron spin with electric fields. *Science*, **318**, 1430.

[40] Chen, G., Bergman, D. L., and Balents, L. (2007). Semiclassical dynamics and long-time asymptotics of the central-spin problem in a quantum dot. *Phys. Rev. B*, **76**, 045312.

[41] Hahn, E. L. (1950). Spin echoes. *Phys. Rev.*, **80**, 580.

[42] Koppens, F. H. L., Nowack, K. C., and Vandersypen, L. M. K. (2008). Spin echo of a single spin in a quantum dot. *Phys. Rev. Lett.*, **100**, 236802.

[43] Yao, W., Liu, R. B., and Sham, L. J. (2006). Theory of electron spin decoherence by interacting nuclear spins in a quantum dot. *Phys. Rev. B*, **74**, 195301.

[44] Cywinski, L., Witzel, W. M., and Das Sarma, S. (2009). Pure quantum dephasing of a solid-state electron spin qubit in a large nuclear spin bath coupled by long-range hyperfine mediated interactions. *Phys. Rev. B*, **79**, 245314.

[45] Cywinski, L., Witzel, W. M., and Das Sarma, S. (2009). Electron spin dephasing due to hyperfine interactions with a nuclear spin bath. *Phys. Rev. Lett.*, **102**, 057601.

[46] Witzel, W. M. and Das Sarma, S. (2006). Quantum theory for electron spin decoherence induced by nuclear spin dynamics in semiconductor quantum computer architecture. *Phys. Rev. B*, **74**, 035322.

[47] Witzel, W. M. and Das Sarma, S. (2007). Multiple-pulse coherence enhancement of solid state spin qubits. *Phys. Rev. Lett.*, **98**, 077601.

[48] Lee, B., Witzel, W. M., and Das Sarma, S. (2008). Universal pulse sequence to minimize spin dephasing in the central spin decoherence problem. *Phys. Rev. Lett.*, **100**, 160505.

[49] Meiboom, S. and Gill, D. (1958). Modified spin-echo method for measuring nuclear relaxation times. *Rev. Sci. Instrum.*, **29**, 688.

[50] Khodjasteh, K. and Lidar, D. A. (2007). Performance of deterministic dynamical decoupling schemes: Concatenated and periodic pulse sequences. *Phys. Rev. A*, **75**, 062310.

[51] Uhrig, G. S. (2007). Keeping a quantum bit alive by optimizing π-pulse sequences. *Phys. Rev. Lett.*, **98**, 100504.

[52] Biercuk, M. J., Uys, H., VanDevender, A. P., Shiga, N., Itano, W. M., and Bollinger, J. J. (2009). Optimized dynamical decoupling in a model quantum memory. *Nature*, **458**, 996–1000.

[53] Du, J., Rong, X., Zhao, N., Wang, Y., Yang, J., and Liu, R. B. (2009). Preserving electron spin coherence in solids by optimal dynamical decoupling. *Nature*, **461**, 1265–1268.

[54] Neder, I., Rudner, M. S., Bluhm, H., Foletti, S., Halperin, B. I., and Yacoby, A. (2011). Semiclassical model for the dephasing of a two-electron spin qubit coupled to a coherently evolving nuclear spin bath. *Phys. Rev. B*, **84**, 035441.

[55] Sundfors, R. K. (1969). Exchange and quadrupole broadening of nuclear acoustic resonance line shapes in the III-V semiconductors. *Phys. Rev.*, **185**, 458–472.

[56] Hester, R. K., Sher, A., Soest, J. F., and Weisz, G. (1974). Nuclear-magnetic-resonance detection of charge defects in gallium arsenide. *Phys. Rev. B*, **10**, 4262–4273.

[57] Bluhm, H., Foletti, S., Mahalu, D., Umansky, V., and Yacoby, A. (2010). Enhancing the coherence of a spin qubit by operating it as a feedback loop that controls its nuclear spin bath. *Phys. Rev. Lett.*, **105**, 216803.

[58] Witzel, W. M. and Das Sarma, S. (2008). Wavefunction considerations for the central spin decoherence problem in a nuclear spin bath. *Phys. Rev. B*, **77**, 165319.

11

Exploring the quantum world with photons trapped in cavities and Rydberg atoms

J.-M. Raimond

Laboratoire Kastler Brossel, ENS, CNRS, UPMC-Paris 6
24 rue Lhomond, 75231 Paris Cedex 05, France

Quantum Machines: Measurement and Control of Engineered Quantum Systems.
Edited by M. Devoret, B. Huard, R. Schoelkopf, and L. F. Cugliandolo.

Chapter Contents

11.1 Introduction

Experiments on fundamental aspects of quantum physics are the focus of considerable interest. On the one hand, this interest is motivated by the intriguing quantum description of reality. State superpositions, quantum measurements, entanglement, and nonlocality are quite incompatible with our classical intuition and deserve to be observed in a variety of contexts. On the other hand, the interest in quantum physics is triggered by the fascinating perspectives of quantum information. Harnessing the weird quantum logic makes it possible to achieve information treatment functions far beyond the reach of classical systems [1].

Cavity quantum electrodynamics (cQED) is a favourite system for explorations of quantum physics [2]. cQED experiments implement the simplest matter-field system: a single two-level emitter, a "spin-1/2", coupled to a single quantum field mode represented by a one-dimensional harmonic oscillator, a "spring." In the "strong-coupling" regime, the coherent interaction between the spin and the spring overwhelms the dissipative processes responsible for decoherence. The results of these experiments thus illustrate directly the basic postulates of quantum physics.

The history of cQED originates in a seminal remark by Purcell in 1946 [3]. He noted that the spontaneous emission of a spin can be modified by changing the boundary conditions imposed on the vacuum field. The first cQED experiments in the 1980s observed modifications of the emission rate—either an enhancement [4] or an inhibition [5, 6].

Since these early studies, cQED has thrived. It reached strong coupling with the first micromaser experiments, using transitions between Rydberg levels coupled to millimeter-wave superconducting cavities [7]. It has been extended to the optical domain, with atomic transitions coupled to optical Fabry-Perot cavities [8], then with solid state emitters (e.g., quantum dots) coupled to monolithic cavities using Bragg mirrors [9] or photonic bandgap structures [10]. It even has industrial applications. The widespread VCSEL lasers, for instance, rely on cQED effects in the weak-coupling regime. A recent and fascinating avatar of cQED is circuit QED, with superconducting qubit circuits coupled to high-quality stripline cavities. The experimental design is extremely flexible and leads to quite remarkable achievements. This subject is described in detail in other chapters in this volume.

This chapter will focus on the microwave cQED experiments realized in our laboratory, using circular Rydberg atoms coupled to superconducting millimeter-wave Fabry-Perot cavities. Both the spin and the spring systems have in this context an extremely long lifetime, of the order of a second for the spring. The excellent controllability of both systems and the slow pace of the experiments, with a typical time step in the few tens of microseconds range, allow us to realize complex sequences.

We first present (Section 11.2) an overview of the experimental techniques and of the theoretical tools describing the atom-cavity interaction and cavity relaxation. We then show in Section 11.3 that this system is ideal for the illustration of the quantum measurement postulates. It can be used for an ideal, projective, quantum nondemolition (QND) measurement of the number of photons in the cavity. Atoms repeatedly probing the cavity count the photons without absorbing them. By performing repeated measurements on the cavity field, we witness for the first time the quantum jumps of light, as photons are lost one by one owing to cavity damping.

With the addition of controlled displacements of the field, this QND measurement can be turned into a complete determination of the cavity quantum state, providing an unprecedented insight into its quantum nature (Section 11.4). Reconstructing the state of "Schrödinger cats," quantum superpositions of mesoscopic components with different classical properties, we observe their decoherence. They are rapidly transformed into mere statistical mixtures by their coupling to the cavity's environment.

This experiment sheds light on the quantum-to-classical transition, one of the main open questions in contemporary quantum physics. Why is it that we observe a tiny fraction of all possible quantum states at our scale? Why is it that we never see a cat [11] in a superposition of being dead and alive at the same time? Why is it that our macroscopic meters, while measuring quantum systems in state superpositions, never evolve into absurd superpositions of different indications? Why is it that the classical world, ruled by quantum physics, is classical at all? Environment-induced decoherence [12] provides partial answers to these important questions.

In order to harness the quantum for useful purposes, we have to face the decoherence obstacle, which must be combated efficiently. We discuss in Section 11.5 a quantum feedback experiment. Transposing to the quantum realm the concept of feedback, it prepares on demand nonclassical states with a well-defined photon number (Fock states). It also protects them against decoherence, by rapidly reverting the quantum jumps due to relaxation.

Another strategy against decoherence uses reservoir engineering. We propose in Section 11.7 to prepare and protect nonclassical field states, by coupling strongly the cavity to a carefully designed environment. We then conclude (Section 11.7) by discussing a few perspectives for these microwave cQED experiments.

11.2 A spin and a spring

11.2.1 Experimental setup

The scheme of our microwave cQED set-up is presented in Fig. 11.1 (for a detailed description see [2, 13]). Circular Rydberg atom samples are prepared in box B at precisely defined time intervals, with a selected velocity v ($v = 250\,\mathrm{m\,s^{-1}}$ in most experiments described here), out of a thermal rubidium beam by laser and microwave excitation. They propagate through the experiment, cooled down to 0.8 K by an ^{3}He refrigerator. They interact with the superconducting Fabry-Perot cavity C and are finally detected by field ionization in D.

Circular Rydberg states have a large principal quantum number and maximum orbital and magnetic quantum numbers. Their wavefunction is located around a circular Bohr orbit. As a consequence of the correspondence principle, their properties can be computed classically. However, they are extremely well suited for experiments unveiling the quantum.

We mainly use two adjacent states, $|e\rangle$ and $|g\rangle$, connected by a dipole transition at a frequency $\omega_a/2\pi = 51.1\,\mathrm{GHz}$. These states have no residual degeneracy, making

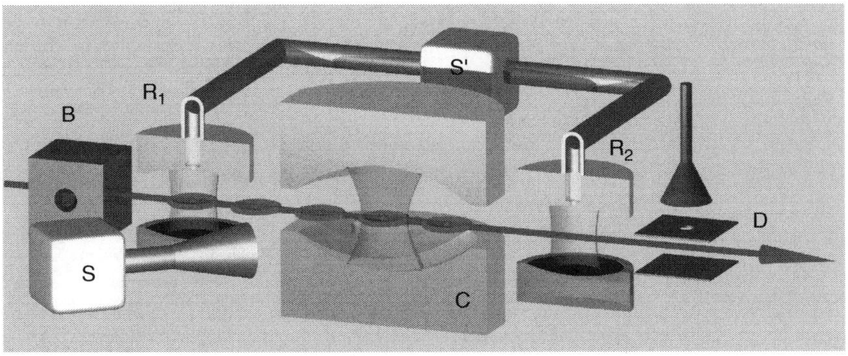

Fig. 11.1 Scheme of a microwave cavity QED set-up.

them a nearly perfect implementation of a two-level system (qubit), whose quantum state is pictorially described as a unit vector evolving on the Bloch sphere (with level $|e\rangle$ and $|g\rangle$ at the north and south poles respectively). The atomic Hamiltonian is simply $H_a = \hbar\omega_a\sigma_z/2$, where σ_z is the standard Pauli operator for the z axis of the Bloch sphere.

Circular states have a long lifetime, about 30 ms, when placed in a small homogeneous static electric field defining a preferred axis for their highly anisotropic wavefunction and lifting their degeneracy with the huge hydrogenic manifold. They are very strongly coupled to resonant microwave radiation, owing to the large size of the electron orbit (0.12 μm radius), resulting in a huge transition dipole matrix element.

Highly efficient field-ionization detection provides a measurement of the atomic state in the $\{|e\rangle, |g\rangle\}$ basis. Finally, the atomic transition frequency can be tuned, with a nanosecond time resolution, by the Stark effect in a static electric field. The selective preparation of these levels is complex, using laser, radiofrequency, and static fields [2]. The number of atoms in each sample is random and obeys Poisson statistics, and its average value is kept low, minimizing the effect of samples containing more than one atom.

In order to get information on the atom-cavity interaction, we use a Ramsey atomic interferometry technique. Before and after their interaction with C, the atoms, initially prepared in $|g\rangle$, undergo two classical $\pi/2$ resonant pulses in the low-Q cavities R_1 and R_2 driven by the source S$'$. Using a proper phase reference, the first pulse realizes the transformation $|g\rangle \to (|g\rangle + |e\rangle)/\sqrt{2}$ and prepares a state represented by a vector along the Ox axis in the equatorial plane of the Bloch sphere. In a proper interaction representation, the second pulse realizes the transformations $|g\rangle \to [|g\rangle + \exp(i\phi_r)\,|e\rangle]/\sqrt{2}$ and $|e\rangle \to [|e\rangle - \exp(-i\phi_r)\,|g\rangle]/\sqrt{2}$, where the phase ϕ_r of the Ramsey interferometer can be tuned by adjusting the relative phase of the pulses in R_1 and R_2 or by applying a transient Stark shift on the atomic transition between R_1 and R_2. The final probability $\pi_e(\phi_r)$ for observing the atom in $|e\rangle$,

$$\pi_e(\phi_r) = (1 + \cos\phi_r)/2\ , \tag{11.1}$$

exhibits "Ramsey fringes," oscillating between 0 and 1 as a function of ϕ_r.

The small static electric field required for the stabilization of the circular states makes them incompatible with closed metallic cavities. We use thus a Fabry-Perot design, with two mirrors facing each other. A field across the mirrors stabilizes the atoms and makes it possible to tune their transition via the Stark effect while they interact with C.

The photon (or energy) storage time T_c is the most critical parameter in these experiments. A long T_c requires a very high mirror conductivity, which can only be provided by superconducting metals at cryogenic temperature. It also requires an excellent surface state to minimize losses induced by diffraction on defects. The quest for high-quality cavities has been a long process, since these two requirements are somewhat incompatible. We finally developed a fabrication technique based on diamond-machined copper substrates, with an extremely smooth surface (10 nm roughness), covered with a thin (12 μm) layer of high-purity niobium deposited by cathode sputtering [14].

The mirrors have a 50 mm radius and are 27 mm apart. They have a toroidal shape in order to lift the polarization degeneracy. They sustain a nondegenerate TEM_{900} Gaussian mode with a linear polarization orthogonal to the cavity axis and a waist $w = 6$ mm. The frequency of the mode is adjusted by piezoelectric elements changing the cavity length. The damping times of the cavities used in recent experiments range from 65 ms to 0.13 s, a macroscopic time interval. The latter corresponds to a quality factor $Q = 4.5 \times 10^{10}$ and to a finesse $\mathcal{F} = 4.9 \times 10^9$, a thousand times better than that of the best optical cavities.

The mode is a quantum harmonic oscillator, with Hamiltonian $H_c = \hbar\omega_c(N+1/2)$, where ω_c is the field's angular frequency and $N = a^\dagger a$ is the photon number operator (a is the photon annihilation operator). The eigenstates of H_c are the nonclassical Fock states $|n\rangle$, with a well-defined number n of photons. The vacuum $|0\rangle$ is the ground state.

A classical source S, weakly coupled to the mode via the diffraction loss channels (Fig. 11.1), can be used to inject in C a coherent semiclassical state $|\alpha\rangle$ defined by the complex amplitude α. This injection is represented by the unitary displacement operator $D(\alpha) = \exp\left(\alpha a^\dagger - \alpha^* a\right)$, with $|\alpha\rangle = D(\alpha)|0\rangle$. The coherent state is an eigenstate of the annihilation operator a (with eigenvalue α). It can be expanded on the Fock states basis as $|\alpha\rangle = \sum_n c_n |n\rangle$ with $c_n = \exp(-|\alpha|^2/2)\alpha^n/\sqrt{n!}$. Its photon number distribution $P(n) = |c_n|^2$ is Poissonian with an average $\overline{n} = |\alpha|^2$. Note that the displacement operator describes a global translation in phase space. In particular, it acts on an initial coherent state $|\beta\rangle$ according to $D(\alpha)|\beta\rangle = \exp(\alpha\beta^* - \beta\alpha^*)|\beta + \alpha\rangle$, a quantum version of the addition of classical fields in the Fresnel (phase) plane.

11.2.2 Atom-field interaction

The atom-field interaction is described by the Jaynes-Cummings model [15]. The complete Hamiltonian $H = H_a + H_c + H_{ac}$. The interaction term H_{ac} is

$$H_{ac} = -i\hbar\frac{\Omega_0}{2} f(vt)(a\sigma_+ - a^\dagger\sigma_-) \; , \tag{11.2}$$

where $\sigma_+ = |e\rangle\langle g|$ and $\sigma_- = |g\rangle\langle e|$ are the atomic raising and lowering operators.

The "vacuum Rabi frequency" Ω_0 measures the strength of the atom-field coupling when the atom is located at the cavity centre, where the electric field amplitude is maximal. It is proportional to the dipole matrix element of the atomic transition and to the amplitude of a single photon field stored in the cavity. In our experiments, $\Omega_0/2\pi = 50\,\text{kHz}$. The function $f(vt)$ reflects the variation of the atom-field coupling with time while the atom crosses the Gaussian mode at right-angle to the cavity axis. Taking the origin of time when the atom reaches the axis, f is simply given by $f(vt) = \exp(-v^2t^2/w^2)$.

The eigenstates of H, the atom-field "dressed states," can be straightforwardly expressed in the basis of the uncoupled atom-cavity states $|e,n\rangle$ and $|g,n\rangle$, eigenstates of $H_a + H_c$ [2]. We will only consider here either exact atom-cavity resonance $\delta = \omega_a - \omega_c = 0$ or the nonresonant dispersive case $|\delta| \gg \Omega_0$.

At resonance ($\delta = 0$), the uncoupled states $\{|e,n\rangle\}$ and $\{|g,n+1\rangle\}$ have the same energy, $(n+3/2)\hbar\omega_c$. Note that $|g,0\rangle$ is separate and is not affected by the atom-field coupling. The eigenstates of $H_a + H_c$ form a ladder of twofold degenerate multiplicities. The coupling H_{ac} lifts this degeneracy. The dressed states with $n+1$ excitations are $|\pm,n\rangle = (|e,n\rangle \pm i\,|g,n+1\rangle)/\sqrt{2}$, with energies separated at the cavity center ($f = 1$) by $\hbar\Omega_n$ where $\Omega_n = \Omega_0\sqrt{n+1}$.

An atom initially prepared in $|e\rangle$ with an n-photon Fock state in the cavity corresponds to an initial quantum superposition of the two nondegenerate dressed states $|\pm,n\rangle$. The later evolution is thus a quantum Rabi oscillation between $|e,n\rangle$ and $|g,n+1\rangle$ at frequency Ω_n, the atom periodically emitting and reabsorbing a photon in C.

Note that, at most times during this evolution, the atom-cavity system is in an entangled state. Quantum Rabi oscillations have been used in a variety of cQED experiments to create and engineer atom-field entanglement, culminating in the generation of a three-particle entangled state of the GHZ type, with three two-qubit quantum gate actions [13, 16].

Note that the atomic motion through the mode is simply taken into account in this regime. A complete crossing of the mode at a constant velocity v is equivalent to a constant coupling at the cavity center during an effective interaction time $t_r = \sqrt{\pi}w/v$.

Far from resonance ($|\delta| \gg \Omega_0$), the dressed states nearly coincide with the uncoupled levels. Energy exchange between the atom and the field is prohibited by energy conservation. The only effect of the mutual interaction is a slight shift of the joint atom-cavity levels.

A simple second-order calculation shows that the atomic frequency is shifted at the cavity center ($f = 1$) in the field of an n-photon state by $\Delta\omega_a = 2(n+1/2)s_0$, where $s_0 = \Omega_0^2/4\delta$. This energy shift includes a constant part s_0, which corresponds to the Lamb shift of $|e\rangle$ induced by the vacuum fluctuations in C. The other part, $2ns_0$, proportional to the field intensity, is the few-photon limit of the light shifts usually observed in strong laser fields.

When the atom crosses the mode containing n photons in a state superposition $(|e\rangle + |g\rangle)/\sqrt{2}$ (produced in the Ramsey zone R_1), the transient shift of the atomic frequency results in a phase shift of this superposition, by an amount $\phi_0(n+1/2)$, where $\phi_0 = 2s_0t_d$. The effective dispersive interaction time is $t_d = \sqrt{\pi/2}(w/v)$. The

Ramsey fringe signal is accordingly shifted. The probabilities $\pi_e(\phi_r|n)$ and $\pi_g(\phi_r|n)$ for detecting the atom in $|e\rangle$ or $|g\rangle$ conditioned to the presence of n photons in the cavity are

$$\pi_e(\phi_r|n) = 1 - \pi_g(\phi_r|n) = \frac{1}{2}\{1 + \cos\left[\phi_r + \phi_0(n + 1/2)\right]\} . \tag{11.3}$$

The detection of an atom thus carries information on the photon number. Since this atom is unable to absorb the cavity field, this information can be used for a QND measurement of n (Section 11.3).

As a reciprocal effect, the cavity frequency is shifted when an atom is located in the mode. An atom in $|e\rangle$ at the cavity center shifts the cavity resonance by $\Delta_e\Omega_c = s_0$. An atom in $|g\rangle$ shifts it by an opposite amount $\Delta_g\Omega_c = -s_0$. This is the single-atom limit of a refractive-index effect. The nonresonant atom, unable to absorb or emit light, behaves as a piece of transparent dielectric that modifies by its index the effective cavity length and thus its resonance frequency. With the strongly coupled circular atoms, this effect is large even though there is only a single atom in the mode.

This index effect results in a global phase shift for a coherent state $|\alpha\rangle$ injected in C before the atom crosses it. The dispersive interaction then realizes the transformations

$$|e,\alpha\rangle \rightarrow \exp(-i\phi_0/2)\,|e, \exp(-i\phi_0/2)\alpha\rangle , \tag{11.4}$$
$$|g,\alpha\rangle \rightarrow |g, \exp(+i\phi_0/2)\alpha\rangle . \tag{11.5}$$

Interaction with an atom in a state superposition results in an entangled atom-cavity state, where the two atomic energy eigenstates are correlated with mesoscopic coherent fields having the same energy but different classical phases. This situation is quite reminiscent of Schrödinger's famous cat [11], cast in a superposition of its "dead" and "alive" states through its interaction with a single atom. In this context, cQED experiments are particularly well suited for studying the decoherence of mesoscopic quantum states.

11.2.3 Field relaxation

The atom-cavity effective interaction time is at most in the 100 μs range, since the atoms are crossing the millimeter-sized cavity mode at thermal velocities. This is much shorter than the lifetime of the atomic levels (30 ms). Atomic relaxation can be safely neglected, and cavity damping is the main source of decoherence in these experiments.

The damping of a cavity mode, of a spring, has been described since the early days of quantum relaxation theory. When C is coupled linearly to a large environment, including many degrees of freedom, whose eigenfrequencies span continuously a large interval around ω_c, the master equation ruling the evolution of the field density operator ρ can be cast in the general Lindblad form [17]

$$\frac{d\rho}{dt} = \frac{1}{i\hbar}[H_c, \rho] + \sum_{i=1}^{2}\left\{L_i\rho L_i^\dagger - \frac{1}{2}\left(L_i^\dagger L_i\rho + \rho L_i^\dagger L_i\right)\right\} . \tag{11.6}$$

The operators L_i are simply $L_1 = \sqrt{\kappa(1+n_{\text{th}})}\,a$ and $L_2 = \sqrt{\kappa n_{\text{th}}}\,a^\dagger$, where $\kappa = 1/T_c = \omega_c/Q$ is the cavity energy damping rate and $n_{\text{th}} = 1/[\exp(\hbar\omega_c/k_BT) - 1]$ the mean number of blackbody photons per mode at the mirror temperature T, as given by Planck's law (k_B is Boltzmann's constant). Note that L_2 vanishes at zero temperature. In our experiment, $T = 0.8\,\text{K}$ and $n_{\text{th}} = 0.05$. The operators L_1 and L_2, proportional to a and $a^\dagger$ respectively, describe the modifications, the "quantum jumps," of the cavity state when a photon leaks out from the cavity into the environment or when a thermal photon is created.

Coherent states, being eigenstates of the annihilation operator, do not change when a photon escapes into the environment. They are thus rather immune to relaxation. A direct solution of Eq. (11.6) at zero temperature shows that an initial coherent state $|\alpha_0\rangle$ remains coherent, its amplitude being exponentially damped with time constant $T_c/2$. Thus, the average photon number decays as the classical field energy, with the time constant T_c. This feature qualifies the coherent states as (approximate) "pointer states" for the cavity-environment coupling, according to the definition by Zurek [12].

For the photon number distribution $P(n) = \langle n|\rho|n\rangle$, the master equation reduces to

$$\frac{dP(n)}{dt} = \kappa(1+n_{\text{th}})(n+1)P(n+1)+\kappa n_{\text{th}}nP(n-1)-[\kappa(1+n_{\text{th}})n+\kappa n_{\text{th}}(n+1)]\,P(n)\,, \tag{11.7}$$

whose steady state can be obtained by a detailed-balance argument. It coincides obviously with the equilibrium blackbody state. If we consider an initial Fock state $|n_0\rangle$ and the $T = 0\,\text{K}$ case, $P(n_0)$ is ruled at short times by

$$\frac{dP(n_0)}{dt} = -\kappa n_0 P(n_0)\,, \tag{11.8}$$

showing that the lifetime of the $|n_0\rangle$ state is of the order of T_c/n_0.

The larger the Fock state, the smaller its lifetime. This is a decoherence effect. Like the Schrödinger's cat, the Fock states are extremely fragile—more and more so when their energy increases. In fact, any Fock state with a large n_0 can be expressed as a superposition of n_0 non-overlapping coherent components with the same average energy. Each coherent component is rather immune to decoherence, but their quantum superposition is reminiscent of Schrödinger's cat. It is fragile and decays rapidly into a statistical mixture, spoiling the Fock state.

11.3 Ideal QND measurement of the photon number: the quantum jumps of light

Most actual measurements are quite far from the ideal, projective measurement described in elementary quantum physics textbooks. This is particularly true when it comes to counting the number of photons in a light field, for instance in a laser pulse. Modern photodetectors can count photons with high reliability. However, instead of projecting the field on the Fock state corresponding to the result obtained, they cast it onto the vacuum, since all detected photons are absorbed, their energy being used to create the measurable electronic signal.

This total demolition of the light quantum state is not a requirement of quantum physics, which allows QND measurement processes [18]. They are simply ideal projective measurements of an observable that is a constant of motion in the evolution ruled by the system's Hamiltonian. Repeated QND measurements thus always give the same result as the first one in the series. A sudden transition between two eigenvalues of the measured observable, a "quantum jump," can be traced to an extra source of evolution for the system—a relaxation process for instance. Projective measurements of the photon number are obviously QND, since the photon number operator N is a constant of motion.

QND measurements of a light beam intensity have been realized with nonlinear optical systems (for a review, see [19]). The intensity of the signal beam changes the index of a nonlinear material (e.g., an atomic vapour). This refractive-index modification is read out by a weak meter beam in an interferometric arrangement. Owing to the weak optical nonlinearities, these measurements only operate with macroscopic signal and meter beams, whose quantum fluctuations become correlated. They are not adapted for QND measurements at the single-photon level, nor of course for the observation of the quantum jumps of the light intensity.

We have realized a QND measurement at the single-photon level with our cQED setup [20, 21]. We probe the cavity field with nonresonant atoms in a state superposition, which, as discussed above, provide information on the photon number n in the cavity without absorbing or emitting light.

Figure 11.2 presents the Bloch vector describing the atomic state at successive stages in this QND process. The atom is originally prepared in $|g\rangle$ (south pole). The first Ramsey pulse in R_1 casts it in a superposition of $|e\rangle$ and $|g\rangle$ represented by a vector along the Ox axis (Fig. 11.2(a)). The atom then interacts dispersively with the cavity mode, resulting in a phase shift of the atomic coherence. We have represented in Fig. 11.2(b) the $\phi_0 = \pi/4$ situation, using an implicit interaction representation such that the atomic state evolution is only due to the interaction with C. After crossing the mode in the Fock state $|n\rangle$, the Bloch vector has rotated by an angle $(n + 1/2)\phi_0$.

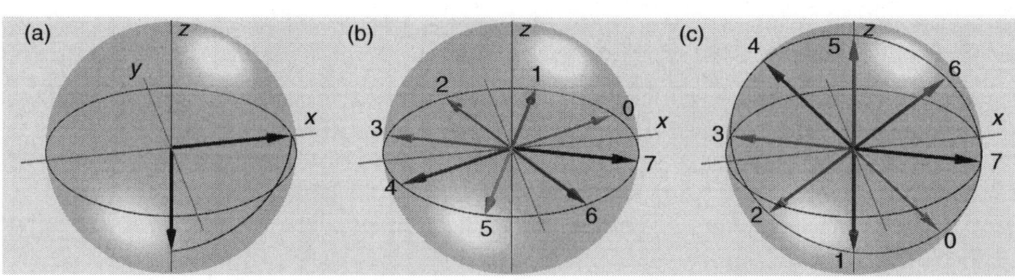

Fig. 11.2 Evolution of the atomic Bloch vector in a QND photon number measurement process. (a) First Ramsey $\pi/2$ pulse in R_1. (b) After the dispersive interaction with C, 8 orientations of the Bloch vector are correlated to the photon numbers from 0 to 7, for $\phi_0 = \pi/4$. (c) After the second Ramsey pulse in R_2.

Eight different orientations of the Bloch vector at the exit of the cavity are entangled with the photon numbers from 0 to 7 (eight photons produce the same rotation as the vacuum, $\phi_0/2$). A single atomic detection is clearly not sufficient to pin down the photon number. After a second $\pi/2$ rotation in R_2 (Fig. 11.2(c)), which maps one state of the equatorial plane onto $|e\rangle$, the atom is detected in the $\{|e\rangle, |g\rangle\}$ basis, providing a single bit of information. This does not allow for a complete discrimination of the eight non-orthogonal atomic states at the cavity exit. In simple terms, a single bit is not enough to count from 0 to 7.

A single-atom detection thus realizes a weak measurement of the cavity field, which nevertheless changes the cavity state. Since the atom cannot emit or absorb photons, the cavity state modification is only due to its entanglement with the atom. The corresponding (completely positive) map for the field density matrix ρ is

$$\rho \longrightarrow \rho_j = \frac{M_j \rho M_j^\dagger}{\pi_j(\phi_r|\rho)} , \tag{11.9}$$

where the index $j = \{e, g\}$ indicates the measured atomic state.

The measurement operators M_e and M_g are

$$M_g = \sin\left[\frac{\phi_r + \phi_0(N + 1/2)}{2}\right] , \tag{11.10}$$

$$M_e = \cos\left[\frac{\phi_r + \phi_0(N + 1/2)}{2}\right] , \tag{11.11}$$

and define a positive operator-valued measurement (POVM) with elements $E_j = M_j^\dagger M_j$, with $E_e + E_g = \mathbb{1}$. The denominator $\pi_j(\phi_r|\rho)$ in Eq. (11.9) is the probability for detecting the atom in state j conditioned by the field state ρ:

$$\pi_j(\phi_r|\rho) = \mathrm{Tr}\left(M_j \rho M_j^\dagger\right) . \tag{11.12}$$

Since the M_j's are diagonal in the Fock-state basis, this probability can also be written

$$\pi_j(\phi_r|\rho) = \sum_n P(n)\pi_j(\phi_r|n) , \tag{11.13}$$

$P(n) = \langle n|\rho|n\rangle$ being the photon number distribution. The conditional probabilities $\pi_j(\phi_r|n)$ are given by Eq. (11.3).

The new photon number distribution after a detection in state j for the Ramsey interferometer setting ϕ_r, $P(n|j, \phi_r)$, is thus [21]

$$P(n|j, \phi_r) = \frac{\pi_j(\phi_r|n)}{\pi_j(\phi_r|\rho)} P(n) . \tag{11.14}$$

We recover here the usual Bayes law of conditional probabilities. The probability for having n photons, conditioned to an atomic detection in j with the Ramsey interferometer phase set at ϕ_r, is, up to a normalization factor (denominator on the right-hand side of Eq. (11.14)), the initial photon number distribution $P(n)$ multiplied by the conditional probability for detecting the atom in j with this Ramsey phase setting when there are n photons in the cavity. Note that if the atom escapes detection, owing to the finite efficiency of D, it does not change the photon distribution, since the two density matrices corresponding to a detection in $|e\rangle$ or $|g\rangle$ should then be summed, weighted by the respective detection probabilities.

After each atomic detection, $P(n)$ is thus multiplied by a sinusoidal function of n, proportional to $\pi_j(\phi_r|n)$. Some photon numbers are less likely to be found after the measurement (those corresponding to the highest probability for detecting the atom in the other state). Sending N_a atoms one by one through C iterates this photon number "decimation" operation. The final $P(n)$ is

$$P_{N_a}(n) = \frac{P_0(n)}{Z} \prod_{i=1}^{N_a} \pi_{j_i}(\phi_{r,i}|n) , \tag{11.15}$$

where j_i is the detected state for the atom labelled i in the series, $\phi_{r,i}$ is the Ramsey interferometer phase used for this atom, and $P_0(n)$ is the initial photon number distribution.

Each factor in the product on the right-hand side of Eq. (11.15) decimates some photon numbers. Numerical simulations and mathematical arguments [22] show that, after many atoms have been detected, the final photon number distribution reduces to a single photon number n_m, $P_{N_a}(n) = \delta_{n,n_m}$, provided that the initial distribution $P_0(n)$ spans a photon number range such that no two photon numbers lead to the same detected atomic state (e.g., the $\{0, 7\}$ range when $\phi_0 = \pi/4$). The accumulation of partial information provided by weak measurements ends up in a complete, projective measurement of the photon number.

The final outcome is independent of the initial distribution $P_0(n)$. We can thus choose it freely, provided it does not cancel in the relevant photon number range. A possible choice is a flat distribution, reflecting our lack of knowledge of the photon number distribution in the interval 0–7.

We realized such a measurement on a coherent field with an average photon number $\overline{n} = 3.82$ (the probability for having 8 photons or more in this field is a few percent only). We record the detection of $N_t = 110$ atoms, in a time of the order of 26 ms, with $\phi_0/\pi = 0.233$ rad. The Ramsey phase ϕ_r is randomly chosen among four values differing by about $\pi/4$ to speed up convergence. We use a slightly modified decimation law to incorporate the finite contrast of the Ramsey interferometer (for details, see [21]).

Figure 11.3 presents the evolution of the photon number distribution after the detection of N_a atoms ($N_a = 0, \ldots, N_t$) for two realizations of the experiment. Let us discuss the one shown in (a). The initial $P_0(n)$ is flat. After one atomic detection, it has a sine shape. After about 20 atomic detections, the decimation process leaves only two possible photon numbers, 5 and 4. For a short time interval, 4 dominates, but, after about 70 atomic detections, only 5 is left, completing the measurement. The other realization (b) exhibits a similar behaviour, the final photon number being then 7. We

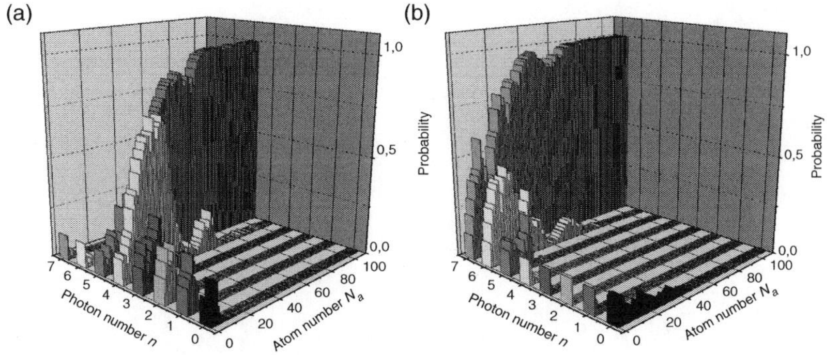

Fig. 11.3 (a, b) Two individual realizations of the QND measurement of a coherent field. The histograms show the evolution of the photon number distribution with the number N_a of detected atoms. From Guerlin *et al.* [21]. Reprinted by permission from Macmillan Publishers Ltd.

observe here the progressive collapse of the field state due to an accumulation of weak measurements.

The final photon number is selected by the random atomic detection outcomes. Each realization of the experiment leads, after recording N_t atoms, to a different, random photon number according to the basic postulates of quantum physics. Once again, "God is playing dice" in a quantum measurement. The probability for obtaining the final value n_m must be ideally given by the initial photon number distribution $P_0(n_m)$.

We have recorded 2000 individual realizations of the detection sequence and computed, for each of them, the average photon number $\langle n\rangle$ for the final distribution $P_{N_t}(n)$. The histogram of the measured $\langle n\rangle$ values is shown in Fig. 11.4. In about 80% of the cases, we do finally obtain an integer photon number, showing that the decimation process has converged. The background in the histogram corresponds to 20% of the experiments for which either N_t atoms have not been sufficient to grant convergence or a quantum jump due to cavity relaxation occurred during the measurement process itself. When convergence is obtained, the observed probabilities fit with a Poisson law (open circles). Its average photon number, 3.46, coincides nicely with an independent calibration of the average photon number in the cavity at the time when the measurement is complete.

Since the probe atoms do not absorb the cavity field, we can follow the photon number evolution over a long time interval. We keep sending dispersive atoms in C. At each time t, we infer a photon number distribution and its average $\langle n\rangle(t)$ from information provided by the last N_t atomic detections. Figure 11.5(a) presents the time evolution of the inferred average photon number over 0.7 s for two realizations of the experiment whose initial phase (first N_t atoms) is shown in Fig. 11.3.

The initial phase corresponds to the approximately 26 ms time interval required to perform the state collapse. Then, we observe a plateau with a constant photon

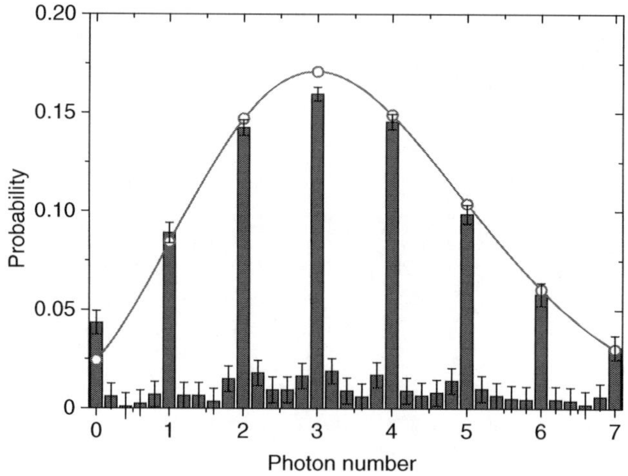

Fig. 11.4 Reconstructed photon number distribution in the initial cavity field: histogram of the average value $\langle n \rangle$ of the final photon distribution $P_{N_t}(n)$. The error bars reflect the statistical uncertainty (2000 realizations of the experiment have been used). The open circles (and the continuous line joining them for visual convenience) represent a Poisson law with 3.46 photons on the average. From Guerlin *et al.* [21]. Reprinted by permission from Macmillan Publishers Ltd.

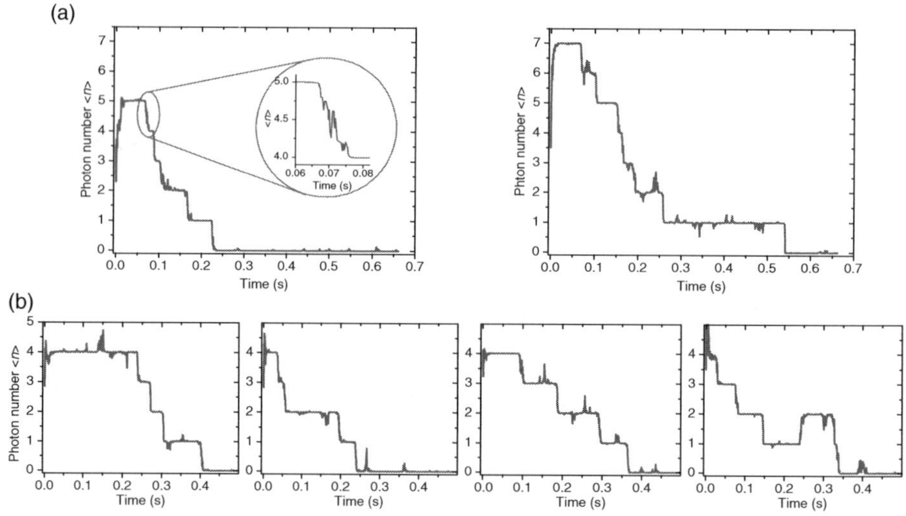

Fig. 11.5 Repeated measurements. (a) Time evolution of the mean photon number $\langle n \rangle(t)$ for the two sequences whose collapse is shown in Fig. 11.3. (b) Four other trajectories with an initial collapse in the $n_m = 4$ Fock state. From Guerlin *et al.* [21]. Reprinted by permission from Macmillan Publishers Ltd.

number (5 or 7). In both cases, the duration of this plateau is long enough to allow for two measurements of the photon number with two independent samples of $N_t = 110$ atoms. That these two measurements lead to the same result is a clear indication of the QND character of this process.

Our photon-counting procedure thus illustrates all the basic postulates for an ideal projective quantum measurement. It provides a quantized result (an integer photon number), whose value is randomly chosen with a statistical distribution given by the initial cavity state. The repeatability of the measurement shows that it does project the field state onto an eigenstate of the measured observable, the $|n_m\rangle$ Fock state, a situation radically different from that of all standard photodetections.

After the initial plateau, we observe a stepwise relaxation of the photon number toward zero. We monitor here the quantum jumps of light, while the photons escape one by one from the cavity into the loss channels. The inset in the first curve shows a zoom on the quantum jump between the 5- and 4-photon states. Recording it requires the accumulation of information from a few tens of atomic detection and is performed in about 10 ms. Note that the 1-photon state in the trajectory on the right has an exceptionally long lifetime (about 300 ms, nearly three average lifetimes T_c).

Figure 11.5(b) presents four other trajectories, with an initial collapse in the 4-photon state, exhibiting the randomness of the quantum jump occurrences. Note on the rightmost trajectory an upwards quantum jump corresponding to the creation of a thermal excitation in C.

The master equation (11.6) predicts a smooth relaxation of the average energy, the photon number decaying exponentially with a time constant T_c. This evolution is quite different from the sudden jumps in Fig. 11.5. This illustrates the difference between a single realization of a quantum experiment and the ensemble average predicted by the density operator. Averaging thousands of quantum jump trajectories, starting from randomly selected photon numbers, all exhibiting quantum jumps at different times, we indeed recover a smooth evolution, in excellent agreement with the predictions of the master equation.

We have performed a detailed study of field relaxation [23]. By a careful analysis of all quantum jump trajectories, we reconstruct the damping-induced evolution of the photon number distribution, starting from all Fock states with 0–7 photons. We get clear evidence of the fast decay of the high-lying Fock states, whose lifetime is T_c/n. By a fit of the complete data, we extract the coefficients of the most general linear master equation governing the evolution of $P(n,t)$. The results of this partial quantum process tomography are in excellent agreement with Eq. (11.7). Similar measurements have been performed simultaneously in the context of circuit QED [24].

Finally, the QND measurement process leads naturally to observation of the quantum Zeno effect [25]. Whereas frequently repeated measurements do not affect the dynamics of incoherent damping, they inhibit the coherent evolution of a quantum system. An observed system does not evolve under the action of its Hamiltonian, in clear correspondence with the second Zeno paradox negating motion. The quantum Zeno effect has been observed in the evolution of a variety of two-level systems, first for a trapped ion [26].

We have observed it for the coherent classical evolution of the cavity field [27]. We realize a long series of identical coherent injections with a very small amplitude in the initially empty cavity. These amplitudes add up, and the field is finally in a mesoscopic coherent state whose average photon number grows quadratically with the number of injections, or with time. When we perform, between injections, QND measurements of n, we project repeatedly, with a high probability, the field back onto $|0\rangle$, since the probability for finding $n = 1$ after a single injection is low. We observe that the field energy effectively remains nearly zero, exhibiting the Zeno effect for the classical runaway process of field injection.

11.4 Monitoring the decoherence of mesoscopic quantum superpositions

The short lifetime of the Fock states is a first indication of the fragility of nonclassical states when they reach the mesoscopic scale. An even more striking example is provided by a Schrödinger cat's decay.

A cat state, a quantum superposition of two mesoscopic coherent fields in C with the same average photon number but different classical phases, can be prepared by a single atom interacting dispersively with a coherent state γ, initially injected into C by the source S (Fig. 11.1). The atom, initially in $|g\rangle$, is prepared in $(|e\rangle + |g\rangle)/\sqrt{2}$ in R_1. It interacts dispersively with the cavity field, resulting in the atom-cavity entangled state $\exp(-i\phi_0/2)\,|e\rangle\,|\gamma\exp(-i\phi_0/2)\rangle + |g\rangle\,|\gamma\exp(+i\phi_0/2)\rangle$. This is a situation quite reminiscent of the Schrödinger cat metaphor, with a two-level quantum system entangled to a "macroscopic" degree of freedom (the classical phase of the field).

Setting $\phi_r = -\phi_0/2$, a final atomic detection after R_2 projects (within an irrelevant global phase) the field onto

$$|\Psi_\pm\rangle = \frac{1}{\mathcal{N}}(|\gamma e^{+i\phi_0/2}\rangle \pm |\gamma e^{-i\phi_0/2}\rangle) , \qquad (11.16)$$

where the $+(-)$ sign applies for a detection in e (g), $\mathcal{N}$ being a normalization factor. When $\phi_0 = \pi$, the cat state with the $+$ $(-)$ sign expands only on even (odd) photon number states. In reference to this simple situation, we call even (odd) cat the state with the $+$ $(-)$ sign for all ϕ_0 values.

The cat state is, when $|\gamma|$ is large, a quantum superposition of two quite distinct classical states. It is expected to decay rapidly owing to its coupling to the environment via the cavity losses. In an early experiment [28], we performed a first investigation of this decoherence process, showing that the decay rate of the coherence between the superposed states increases rapidly with their separation in phase space, in good agreement with theoretical expectations.

A much more detailed insight of the cat coherence can be obtained through a complete reconstruction of the field density operator as a function of time.

As discussed in Section 11.3, the QND measurement of the photon number leads to a partial reconstruction. Many measurements performed on the same quantum state allow us to reconstruct the photon number distribution, that is, the diagonal of the density operator in the Fock-state basis.

We have no access, however, to the nondiagonal elements, which contain information on the field's phase distribution. We are unable, for instance, to distinguish a statistical mixture of Fock states with no phase information from a coherent state, with a well-defined phase. Since we use nonresonant probe atoms, they cannot extract information about the field phase directly.

A simple modification of the QND scheme allows us to reconstruct the full density operator [29]. Before sending the QND probe atoms, we first perform a displacement of the cavity field, by letting the source S inject into C a coherent amplitude β. The source S has of course a well-defined, controlled phase with respect to the initial cavity field amplitude γ. This "homodyning" procedure turns the cavity field density operator into $\rho_\beta = D(\beta)\rho D(-\beta)$. We then send many QND probe atoms through C. By repeating the experiment many times, we measure the probability $\pi_e(\phi_r|\rho_\beta)$, conditioned to the translated cavity state. It simply reads $\pi_e(\phi_r|\rho_\beta) = \mathrm{Tr}\,[G(\beta,\phi_r)\rho]$, where $G(\beta,\phi_r) = D(-\beta)M_e^\dagger M_e D(\beta)$. We thus obtain finally, within statistical noise, the average of the observable $G(\beta,\phi_r)$ in the initial cavity state.

Resuming the experiment for very many (up to 600) different translation amplitudes β carefully chosen in phase space, we obtain the average value of many different observables $G(\beta,\phi_r)$ in the state ρ. We infer from these results all the matrix elements of ρ, using an approach based on the maximum-entropy principle [30]. This procedure determines the density matrix that best fits the data, while having a maximum entropy $-\mathrm{Tr}\rho\ln\rho$. We thus make sure that the reconstructed state does not contain more information than that provided by the experimental data.

We represent the reconstructed state by its Wigner function, which is more pictorial than the density operator. It is a quasi-probability distribution $W(\alpha)$ defined over the phase plane by

$$W(\alpha) = \frac{2}{\pi}\mathrm{Tr}\left[D(-\alpha)\rho D(\alpha)e^{iN\pi}\right] \; . \tag{11.17}$$

In other words, $W(\alpha)$ is the expectation value of the parity operator $\exp(iN\pi)$ (eigenvalue $+1$ for even Fock states, -1 for odd Fock states) in the cavity field translated by $-\alpha$. The Wigner function contains all information on the field state, since ρ can be directly computed from W [2]. For all ordinary states (coherent, thermal, squeezed), W is a positive Gaussian. Negativities in the Wigner function are a clear indication of the nonclassicality of a state. Note that another method for measuring W has been used in trapped ions [31] and circuit QED [32].

Figure 11.6(a) presents the measured Wigner function of the even cat (preparation atom detected in $|e\rangle$), for $|\gamma| = \sqrt{3.5}$ and $\phi_0 = 0.74\,\pi$. The Wigner function exhibits two positive bumps centered on the superposed classical amplitudes, $\gamma\exp(\pm i\phi_0/2)$ (the cat's "ears"). In between these two classical features, we observe a high contrast interference pattern that reveals the quantum nature of the superposition (the cat's "smile"). The observed Wigner function is quite close to the theoretical expectation (shown in the inset).

The "size" of this cat is conveniently measured by the square of the distance $\mathcal{D}$ between the two classical components in phase space: $\mathcal{D}^2 = 11.8$ photons, since the decoherence timescale is expected to be $T_d = 2T_c/\mathcal{D}^2$ at zero temperature [2].

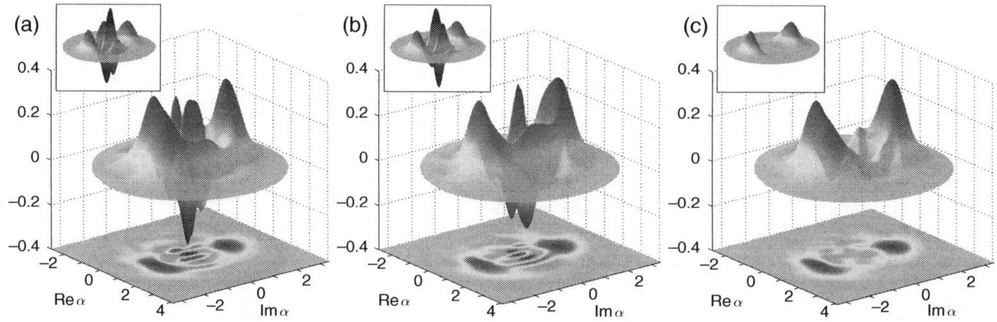

Fig. 11.6 Experimental Wigner functions of an even cat (a) an odd cat (b), and a statistical mixture of two coherent components (c). The insets present the theoretical expectations. The average photon number is 3.5 and the phase shift is $\phi_0 = 0.74\pi$. From Deléglise *et al.* [29]. Reprinted by permission from Macmillan Publishers Ltd.

Note that the ears of our cat are not exactly Gaussian (as they should be for a superposition of coherent states). This is not due to an imperfection in the reconstruction procedure, but to the cat preparation stage. Since we are not deeply into the dispersive regime (the atom-cavity detuning is $\delta/2\pi = 51\,\text{kHz}$, which is not very large compared with $\Omega_0/2\pi$), the phase shift of a Fock state depends in a nonlinear way on the photon number, leading to a slight deformation of the coherent components.

Figure 11.6(b) presents the Wigner function of the odd cat. The classical components are the same, but the interference pattern is phase-shifted by π compared with that of the even cat. Finally, Fig. 11.6(c) presents the Wigner function of a statistical mixture of the two coherent components. It is obtained by mixing data corresponding to different detected states of the preparation atom. The cavity state is then a statistical mixture of the odd and the even cat, or, equivalently, a statistical mixture of the two coherent components. The classical ears are still there, but the smile has gone, as expected.

For the reconstruction of ρ, we can repeat the experiment many times for each β value. We thus use only a few QND atoms in each realization and still achieve good statistics. We chose to detect about 20 atoms in a 4 ms time interval. We are thus able to measure the time evolution of the field state (or of its Wigner function) with a decent resolution.

Figure 11.7 presents four snapshots of the even-cat Wigner function evolution (the conditions are those of Fig. 11.6(a)). The quantum feature, the cat's smile, decays much faster than the energy (the latter decay corresponds to a slow motion of the ears towards the origin, with a time constant $2T_c$). After about 23 ms, the contrast of the interference pattern is considerably reduced and we are left with a mere statistical mixture. From these data, we deduce the decoherence timescale T_d, defined as the damping time of the nondiagonal elements of the density matrix in the coherent-state basis. We get $T_d = 17 \pm 3\,\text{ms}$, in excellent agreement with the theoretical expectation, 19.5 ms, taking into account the finite mode temperature.

Fig. 11.7 Four snapshots of the Wigner function of a decaying Schrödinger cat state. The initial state is the even cat of Fig. 11.6(a). The snapshots are taken 1.3, 4.3, 15.8 and 22.9 ms (from front to back) after cat preparation.

We get in this way a detailed insight into the cat state decoherence. More precise measurements, with considerably larger cats, could allow us to realize a full quantum process tomography of the cavity relaxation. Experiments on "decoherence metrology" are interesting to test alternative interpretations of the conspicuous lack of quantum superpositions at the macroscopic scale [33].

11.5 An experiment on quantum feedback

It is fairly easy in our cQED experiments to prepare nonclassical states: Schrödinger cats conditioned to the detection of the preparation atom or Fock states conditioned to the outcome of the QND measurement process. However, this preparation is nondeterministic (being conditioned to an initial random atomic detection) and the obtained state suffers rapid decoherence, typically $\overline{n}$ times faster than the energy damping, where $\overline{n}$ is the average photon number.

If nonclassical states are to be used as a resource, it would be quite interesting to prepare them on demand and to protect them from relaxation. A possible route toward this goal is quantum feedback [34, 35], which extends to the quantum realm the feedback circuits present in every complex classical system.

As in any feedback operation, a "sensor" gets information on the system, a "controller" estimates a proper distance between the present state of the system and the "target" one, and drives accordingly an "actuator," which steers the system toward the required operating point.

The main difficulty that quantum feedback has to face is that the measurement by the sensor changes the system's state, making quantum feedback loops more subtle than their classical counterparts. A review of the general quantum feedback problem and associated references can be found in Chapter 1 of this volume.

We have realized a quantum feedback experiment preparing on demand Fock states and protecting them against decoherence by inverting rapidly the adverse effects of quantum jumps [36]. The principle of this experiment [37], inspired by [38], is sketched in Fig. 11.8.

We prepare initially in C a coherent state with a real amplitude $\alpha = \sqrt{n_t}$, where $|n_t\rangle$ is the target Fock state. We send a stream of atomic samples, prepared at a $T_a = 82\,\mu\text{s}$ time interval and interacting one after the other with C in the dispersive regime. The detection of the atoms performs a weak measurement of the field intensity, described by the operators M_e and M_g (see Eq. (11.11)). Setting the dispersive phase shift ϕ_0 near $\pi/4$ allows us to distinguish the photon numbers between 0 and 7.

Instead of piling up information provided by many atoms to perform a complete QND measurement of n, which would prepare a randomly selected Fock state, we act on the cavity field after each atomic detection. Information provided by the atom is used by the controller K (a fast real-time classical computer ADwin-ProII by Jäger Messtechnik) to infer the cavity field density operator ρ. The controller then computes

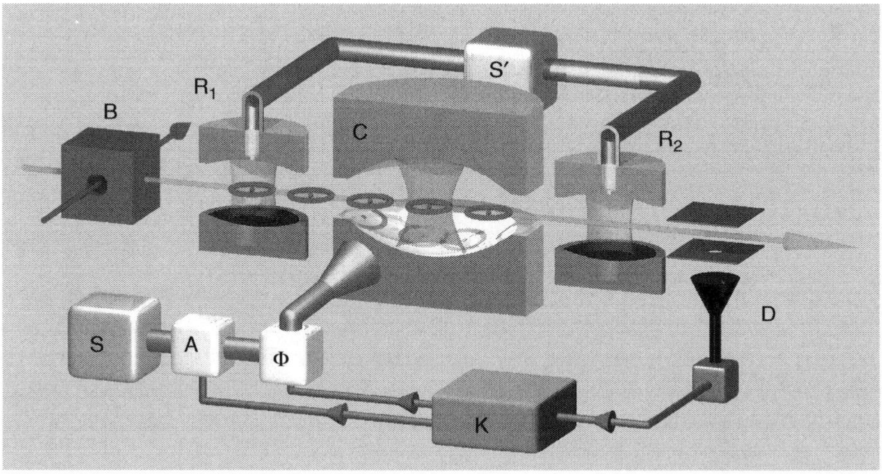

Fig. 11.8 Principle of the quantum feedback experiment for the preparation and protection of Fock states.

the amplitude α of a displacement minimizing a proper distance, $d(n_t, \rho_\alpha)$, between the translated density operator $\rho_\alpha = D(\alpha)\rho D(-\alpha)$ and the target Fock state. It then sets the amplitude and phase controls A and Φ acting on S to apply this displacement. Numerical simulations show that iterations of this elementary feedback loop drive efficiently the cavity state towards the target [37].

The atomic samples should efficiently distinguish the Fock states around n_t. A proper choice for the Ramsey interferometer phase is $\phi_r(n_t) = (\pi/2) - \phi_0(n_t + 1/2)$, corresponding to a mid-fringe setting when C contains n_t photons. We have shown by numerical simulations that this choice is well adapted for $n_t = 1, 2$.

For larger target photon numbers, however, quantum jumps due to dissipation are more frequent. We then alternate between two values of the Ramsey phase: $\phi_r(n_t)$ providing a good sensitivity and $\phi_r(n_t - 1)$ providing a fast detection of the most frequent downward quantum jumps.

The state estimation is a critical part of the feedback loop. It must include all available information on the cavity field. It must take into account our initial knowledge of the field state, all information provided by the atomic detections, and all feedback actions so far. It must also include all the known experimental imperfections, carefully calibrated before each experimental run. They include cavity relaxation in the time interval between two atomic detections (the cavity damping time is $T_c = 65\,\text{ms}$ in this experiment) and the finite contrast of the Ramsey interferometer.

The estimation must also include the imperfections of the atomic detector D, which has a 39% detection efficiency and a finite state attribution error rate. The actual atomic sample composition can thus differ from the detection results. In fact, we attribute the finite (80%) Ramsey fringe contrast to detector errors only. This simplification of the state estimation has little effect on the feedback efficiency, as shown by numerical simulations.

The average number of atoms in each sample is about 0.6, with a negligible probability for having more than 2 atoms. Since we operate far into the dispersive interaction regime ($\delta/2\pi = 245\,\text{kHz} \gg \Omega_0/2\pi$), two atoms crossing C at the same time probe its field independently. The detection of a single atom in $|e\rangle$, for instance, may correspond to five actual compositions of the atomic sample and hence five density operator mappings.

We could have had one atom $|e\rangle$ with no detection error, the density operator mapping being defined by the measurement operator M_e. We could have had one atom in $|g\rangle$ (measurement operator M_g) with an erroneous detection, or two atoms in $|e\rangle$ (M_e^2) with one atom failing to be detected, or one atom in $|e\rangle$ and the other in $|g\rangle$ ($M_e M_g$) with or without error detections and with one atom undetected, or finally two atoms in $|g\rangle$ (M_g^2) with one atom escaping detection and one detection error. The final density matrices corresponding to all these actual sample compositions must be averaged, weighted by the occurrence probabilities of the corresponding events, which are obtained by a straightforward but tedious Bayesian inference analysis [36].

The state estimator must also take care of the 344 μs time interval between the interaction of an atomic sample with C and its detection in D. When a sample is finally detected, four other ones have already interacted with C and are on their way toward D. State estimation takes into account the yet unread measurements performed by these samples [37].

All the state estimation computations should be performed in a time shorter than the time interval T_a between atomic samples. We perform them with a carefully optimized code in a truncated Fock-state basis, with 8 elements for $n_t \leq 2$ ($|0\rangle, \ldots, |7\rangle$) and 9 for $n_t = 3$ and 4. Moreover, since the initial amplitude is real, the initial density operator is real in the Fock-state basis, as well as the target. We may thus restrict the amplitude α to real values at each step, and hence the estimated density operator to real matrix elements.

The controller adjusts the value of the real injected field amplitude α in order to minimize the distance $d(n_t, \rho_\alpha)$ of the estimated density matrix after translation, $\rho_\alpha = D(\alpha)\rho D(-\alpha)$, with the target state $|n_t\rangle\langle n_t|$. A simple distance definition is $d = 1 - F_t(\rho_\alpha)$, where $F_t(\rho_\alpha) = \text{Tr}(\rho_\alpha |n_t\rangle\langle n_t|)$ is the fidelity with respect to the target state. However, this distance reaches its maximum value, 1, for all states orthogonal to $|n_t\rangle$. It does not provide clear information on the actual difference between the estimated state and the target. We use instead [36]

$$d(n_t, \rho_\alpha) = 1 - \text{Tr}(\Lambda^{(n_t)}\rho_\alpha) , \tag{11.18}$$

where $\Lambda^{(n_t)}$ is a diagonal matrix in the Fock state basis, with $\langle n_t|\Lambda^{(n_t)}|n_t\rangle = 1$, so that the distance cancels in the target state. The diagonal elements of this matrix are chosen so that d has a global minimum on the target state and local maxima on all other Fock states. They are further fine-tuned by optimizing the feedback convergence speed in numerical simulations.

An explicit optimization of the amplitude α is far too time-consuming to be performed at each iteration of the loop. In order to perform computations faster, we use a second-order approximation of the displacement operators:

$$\rho_\alpha = D(\alpha)\rho D(-\alpha) \approx \rho - \alpha[\rho, a^\dagger - a] + \frac{\alpha^2}{2}[[\rho, a^\dagger - a], a^\dagger - a] , \tag{11.19}$$

leading to

$$d(n_t, \rho_\alpha) \approx d(n_t, \rho) - a_1\alpha - a_2\frac{\alpha^2}{2} , \tag{11.20}$$

with $a_1 = \text{Tr}[\Lambda^{(n_t)}, a^\dagger - a]\rho$ and $a_2 = \text{Tr}[[\Lambda^{(n_t)}, a^\dagger - a], a^\dagger - a]\rho$.

Each step should reduce d, and hence $(a_1\alpha + a_2\alpha^2/2)$ must be large and positive. If $a_2 < 0$, then $d(n_t, \rho_\alpha)$ is concave and the optimum is achieved for $\alpha = -a_1/a_2$. Since we have used a second-order approximation in α, we moreover limit the amplitude modulus to $|\alpha| < \alpha_{\text{max}} = 0.1$. For mere experimental convenience, we arbitrarily set to zero all amplitudes whose modulus is lower than 10^{-3}. When $a_2 \geq 0$, d is a convex function of α and we can thus expect to be far from its global minimum. We therefore apply the largest allowed displacement with the proper sign: $\alpha = \alpha_{\text{max}}$ for positive a_1 and $\alpha = -\alpha_{\text{max}}$ for nonpositive a_1.

The controller K sets the amplitude α by pulsing the continuous-wave source S with the switch A. The injected amplitude modulus is a precisely calibrated linear function of the injection time t_i. Typically, the injection of the maximum amplitude $\alpha_{\max}$ is performed in 63 μs, a time shorter than the interval T_a between samples (note that the injection pulse overlaps with the independent calculation by K of the amplitude for the next iteration of the feedback loop). The sign of α is controlled by the programmable $\{0, \pi\}$ phase-shifter Φ.

Figure 11.9 presents one feedback run lasting 160 ms, aiming at the preparation of a two-photon state ($n_t = 2$). Initially, the distance to the target is large and the feedback active, with repeated injections. After a while, the cavity state converges toward the target. The probability $P(2)$ for having two photons reaches a high value, in the 80% range and the estimated density matrix is quite close to that of a Fock state. We have thus prepared on demand the prescribed nonclassical state.

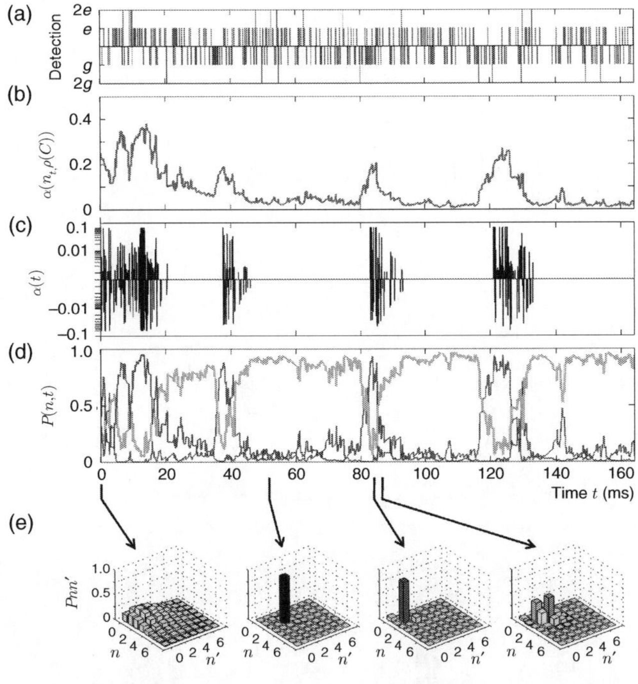

Fig. 11.9 One feedback run lasting 160 ms with $|n_t\rangle = |2\rangle$. (a) Atomic detection events (the two-atom detections appear as double-length bars). (b) Distance $d(n_t, \rho)$ between the estimated density matrix and the target state. (c) Applied injection amplitude (on a logarithmic scale, as $\mathrm{sgn}(\alpha)\log(|\alpha|)$). (d) Photon number distribution with $P(2)$ by the thick light gray line, $P(n < 2)$ by the thin black line, and $P(n > 2)$ by the thin gray line. (e) Four snapshots of the estimated density matrix, at time $t = 0$ (initial coherent state), at a time when convergence is realized, immediately after a quantum jump and finally during quantum jump recovery. The precise times are indicated by arrows from the above timescale. From Sayrin *et al.* [36]. Reprinted by permission from Macmillan Publishers Ltd.

After a while, a quantum jump toward $n = 1$ occurs. The feedback activity resumes, until the target state is restored. We observe, in Fig. 11.9(e), that the field has no defined phase immediately after the jump, since the density matrix is diagonal. During recovery, the injections create transient nondiagonal elements, which are required to pull the state back toward the target and which finally vanish when convergence is achieved.

We have measured the average convergence time toward the target state. The feedback procedure leads to a Fock state preparation noticeably faster than a "fail and resume" approach based on QND measurement (we perform a QND measurement of n, restart the sequence if n is found to be different from n_t, until we finally find $n = n_t$). For $n_t = 3$, 63% of the feedback trajectories have reached $P(3) > 80\%$ after 50 ms only, whereas 250 ms are required for the "fail and resume" brute force approach. We have also studied the dynamics of the recovery after a quantum jump, from $n = n_t = 3$ down to $n = 2$. The estimator needs about 3 ms (7 detected atoms) to "realize" that a jump has occurred and the target-state restoration lasts for an additional 8 ms.

We estimate the fidelity of the feedback procedure by interrupting the sequence and realizing an independent measurement of the photon number distribution with QND atoms, using a fast procedure outlined in [23]. Figure 11.10 presents the results obtained for n_t from 1 to 4 in (a) to (d), respectively. We use for these data two feedback interruption criteria. For the light-gray histograms, we interrupt the feedback

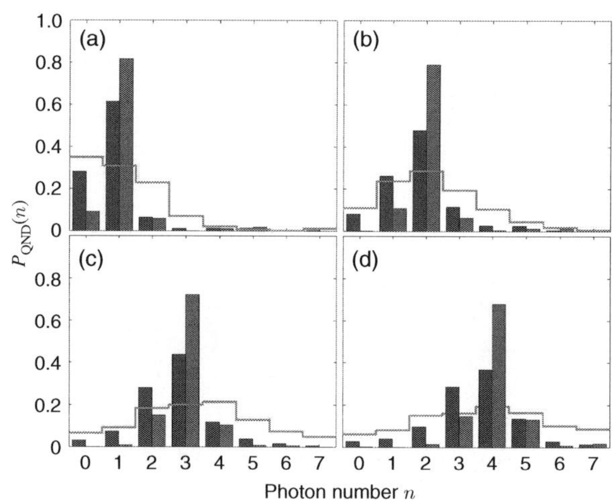

Fig. 11.10 (a)–(d) show photon number distributions after feedback operation for $n_t = 1$ to 4, respectively. The dark-gray histograms are obtained when the feedback operation is interrupted at a fixed 164 ms time. If the feedback is interrupted when the estimated probability for having n_t photons is larger than 80%, we get the light-gray histograms. Finally, the line represents the measured photon number distribution in the initial coherent field. From Sayrin *et al.* [36]. Reprinted by permission from Macmillan Publishers Ltd.

when the probability of having n_t photons in the estimated density matrix, $\langle n_t|\rho|n_t\rangle$, reaches 80% for three consecutive atomic samples. We observe that the state estimation is correct, since the independently measured $P(n_t)$ is then close to 80%.

For the dark-gray histograms, we interrupt the feedback sequence after a fixed amount of time (164 ms), large enough to reaching a steady state on the average. The fidelity with respect to the target state is lower than for the light-gray histograms, because the field resides for a fraction of time in states with $n \neq n_t$ owing to relaxation-induced quantum jumps. The photon number distribution is, however, clearly nonclassical, with $P(1) > 60\%$ for $n_t = 1$ and $P(3) \approx 45\%$ for $n_t = 3$. This shows that our feedback procedure does protect the nonclassical Fock states against decoherence.

We have implemented a real-time quantum feedback operating in the steady state. The slow pace of the Rydberg-atoms cQED experiments is a considerable asset, since it allows K to perform complex state estimation procedures in the time interval between two measurements of the cavity field. This would be much more difficult in optical or circuit QED experiments, in which the timescale is typically 1000 times faster. During the preparation of this chapter, we have realized a new feedback experiment with quantum actuators made up of Rydberg atoms interacting resonantly with C and delivering (or subtracting) single photons [39].

11.6 Reservoir engineering

Quantum feedback protects nonclassical states against decoherence. Reservoir engineering is another route toward decoherence control [40, 41]. The principle is to couple the system to an "engineered" bath whose pointer states [12] are the nonclassical states of interest. When the coupling to the engineered reservoir is much stronger than that to the standard loss channels, the target state is efficiently protected against decoherence. Many proposals for reservoir engineering have been put forth, mostly in the ion trap or cQED context. However, they imply rather complex operations.

We propose a simple engineered reservoir for the cavity field mode made up of atoms crossing it one after the other [42]. It stabilizes interesting nonclassical states, including cat states, squeezed states, and superpositions of multiple coherent components.

Each atom of the reservoir is initially prepared in R_1 in the state superposition $|u_a\rangle = \cos(u/2)\,|g\rangle + \sin(u/2)\,|e\rangle$, represented by a vector in the xOz plane of the Bloch sphere, at an angle u with the north-south vertical axis. The atom undergoes a composite interaction with the cavity mode. It first interacts dispersively with it, with a positive detuning δ (single-photon phase shift ϕ_0). The atom is then set in resonance with the cavity mode for a short time interval t_r when it is close to the cavity axis. It undergoes a resonant Rabi oscillation in the cavity field. The detuning is finally set to $-\delta$ for a second dispersive interaction, corresponding to the phase shift $-\phi_0$, opposite to that of the first dispersive interaction. The atom is finally discarded. The Stark effect produced by an electric field applied across the cavity mirror makes it possible to control easily in real time the atomic transition frequency and to achieve this sequence of operations.

Let us first get an insight into the reservoir's operation by considering only the resonant interaction, described by the evolution operator $U_r(\Theta)$:

$$U_r(\Theta) = |g\rangle\langle g| \cos\frac{\Theta\sqrt{N}}{2} + |e\rangle\langle e| \cos\frac{\Theta\sqrt{N+\mathbb{1}}}{2}$$
$$-|e\rangle\langle g|\, a\frac{\sin\Theta\sqrt{N}/2}{\sqrt{N}} + |g\rangle\langle e| \frac{\sin\Theta\sqrt{N}/2}{\sqrt{N}} a^\dagger \ ,$$

where $\Theta = \Omega_0 t_r$. If $u = 0$ (atoms sent in $|g\rangle$), the resonant reservoir obviously stabilizes the vacuum state. The resonant atoms in their lower state absorb any initial field present in C. Incidentally, we use this operation routinely to empty the cavity rapidly at the end of all experimental sequences.

When the atom is injected with a small u value, $0 < u \ll 1$, we realize in C a micromaser pumped below population inversion [43]. The balance between atomic emission and absorption stabilizes a nearly coherent state $|\alpha\rangle$. The real amplitude α is given by $\alpha = 2u/\Theta$ in the $\Theta \ll 1$ limit.

This steady-state value can be understood in simple terms. When it is reached, the resonant interaction corresponds to a rotation of the Bloch vector by an angle $-\Theta\alpha$ in the xOz plane of the Bloch sphere. The Bloch vector, initially at an angle u with the vertical axis, evolves first toward the south pole, corresponding to an atomic emission, and continues its rotation (atomic absorption) until it finally reaches an angle $u - \alpha\Theta = -u$ with the vertical axis. The atom exits the cavity with the same average energy as it had entering it. The cavity energy is unchanged and the field amplitude is stable.

Numerical simulations of the atom-cavity interaction show that this picture remains valid even when u and Θ are of the order of 1. Of course, stabilizing a coherent state is not utterly interesting (a classical source does it).

Let us now consider the full interaction, described by the evolution operator $U = U_d^\dagger(\phi_0)U_rU_d(\phi_0)$, where

$$U_d(\phi_0) = |g\rangle\,\langle g|\, e^{-i\phi_0 N/2} + |e\rangle\,\langle e|\, e^{i\phi_0(N+\mathbb{1})/2} \tag{11.21}$$

is the dispersive evolution operator corresponding to the phase shift per photon ϕ_0. After some algebra [42], making use of the relation $af(N) = f(N+\mathbb{1})a$, we cast the total evolution operator in the form

$$U = e^{-ih_0(N)}U_r e^{ih_0(N)}, \tag{11.22}$$

with

$$h_0(N) = -\phi_0 N(N+\mathbb{1})/2 \tag{11.23}$$

The resonant interaction of a single atom with the mode leaves the coherent state $|\alpha\rangle$ unchanged. It is thus clear from the above equations that the composite interaction leaves invariant the state $U_0\,|\alpha\rangle$, where $U_0 = \exp[-ih_0(N)]$ is the evolution operator in the fictitious Hamiltonian h_0 for a unit time interval.

The pointer state of the composite interaction thus results from the action on the coherent state $|\alpha\rangle$ of a photon-number-conserving Hamiltonian including a term proportional to N^2. This corresponds to the propagation of a coherent state through a Kerr medium, with a refractive index containing a term proportional to the intensity. The Kerr effect has been widely analyzed in the theoretical literature as a way to produce nonclassical fields.

During its propagation through the Kerr medium, an initial coherent state $|\alpha\rangle$ is first distorted into a squeezed state, with quantum fluctuations on one quadrature reduced below the initial ones. Since the photon number distribution is conserved, the squeezing is limited, and the state gradually evolves into a "banana-shaped" state spreading in the phase plane along a circle with radius $|\alpha|$. When the spreading extends over the whole circle, quantum interference phenomena appear, which are conspicuous on a plot of the Husimi-Q or of the Wigner function of the field [2]. For definite interaction times, the state evolves into a quantum superposition of 6, 5, 4, 3, and finally 2 coherent components with opposite phases—a cat state of the field. By tuning the phase shift ϕ_0, we can thus stabilize with the composite reservoir any of those highly nonclassical states produced by the fictitious Kerr Hamiltonian h_0.

In order to confirm this insight, we have performed numerical simulations, using the parameters of our actual cQED experiment. In particular, we include the competition of the engineered reservoir with the standard cavity losses, with $T_c = 65\,\text{ms}$ and $n_{\text{th}} = 0.05$. The dispersive and resonant interactions are computed exactly, integrating the Jaynes–Cummings Hamiltonian (11.2) with the help of the quantum optics package for Matlab [44].

In these computations, the atomic samples are prepared at regular time intervals T_a with velocity v such that they are separated by $3w$ along the beam. They thus interact separately with C. They contain 0.3 atoms on average and we neglect the effect of two-atom samples. The resonant and dispersive interaction periods are symmetric with respect to the time when the samples cross the cavity axis.

Figure 11.11(a) presents the Wigner function $W(\xi)$ associated with the cavity field state ρ_{200} after its interaction with 200 atomic samples for $v = 70\,\text{m}\,\text{s}^{-1}$,

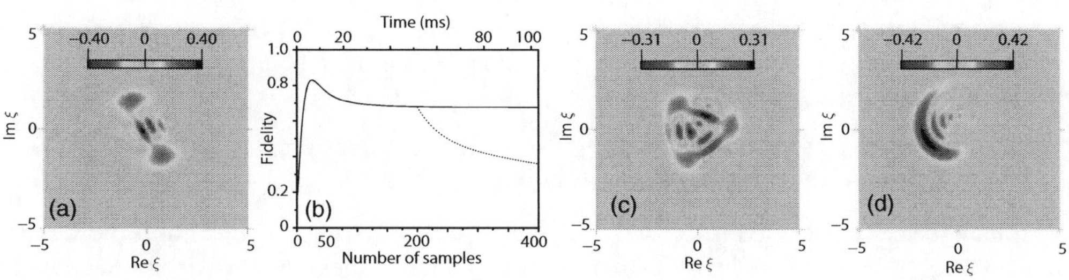

Fig. 11.11 Nonclassical state stabilization. (a) Wigner function of the cavity field after interaction with 200 atoms. Parameters (see text) are optimized to produce a state close to a two-component cat state. (b) Solid line: fidelity of the generated state against the closest cat as a function of number of interactions (bottom axis) or of time (upper axis). Dashed line: reservoir is switched off after 200 interactions. (c) and (d) show Wigner functions of stabilized cavity fields close to a three-component cat and to a "banana state," respectively.

$T_a = 257\,\mu s$, $\Delta = 2.2\,\Omega_0$, $t_r = 5\,\mu s$, $\Theta \approx \pi/2$, and $u = 0.45\pi$. The (irrelevant) initial cavity state is the vacuum. We get a state quite similar to a Schrödinger's cat (see Section 11.4), with an average photon number $\overline{n} = 2.72$ and $\mathcal{D}^2 = 10.9$, where the mesoscopicity parameter $\mathcal{D}$ is the distance in phase space between the classical components.

The purity $P = \mathrm{Tr}(\rho_{200}^2)$ is 51%. We estimate the fidelity $F = \mathrm{Tr}[\rho_{200}\rho_{\mathrm{cat}}]$ of this state with respect to an ideal cat, optimized by adjusting in the reference state ρ_{cat} the phase and amplitude of the coherent components and their relative quantum phase. We get $F = 69\%$. We have checked that cavity relaxation is the main cause of imperfection, F being 98% with an ideal cavity.

Figure 11.11(b) shows as a solid line the fidelity F of the prepared state with respect to the ideal cat state as a function of atomic sample number (i.e., as a function of time). The transient reflects the competition between the fast buildup of the state, the fidelity raising over a few atomic samples only, and the decoherence, whose timescale $T_d = T_c/(2\overline{n})$ becomes relevant once a large average photon number $\overline{n}$ has been produced. The steady-state fidelity is reached after about 100 samples. The dashed line shows F when we switch off the reservoir after 200 interactions. It initially drops rapidly, illustrating the efficient protection offered by the engineered reservoir.

For a slightly larger detuning $\Delta = 3.7\,\Omega_0$ (all other parameters being unchanged), we obtain a superposition of three components (Fig. 11.11(c)), with $\overline{n} = 2.70$ photons, $P = 56\%$, and $F = 73\%$ with respect to the closest ideal three-coherent-component superposition. Figure 11.11(d) shows $W(\xi)$ obtained after 200 samples with $v = 150\,\mathrm{m\,s^{-1}}$, $t_i = 120\,\mu s$, $t_r = 5\,\mu s$, that is, $\Theta \approx \pi/2$, $u = \pi/2$, and $\Delta = 7\,\Omega_0$. These parameters correspond to a weak Kerr interaction, leading to the production of a "banana" state, with interesting negativities in its Wigner function. The field has $\overline{n} = 3.52$ and $P = 91\%$.

This last setting is already within reach of our setup. The other two require a moderate atomic-beam laser cooling to achieve proper velocities. We have checked that the scheme is not sensitive to experimental imperfections (a few percent variation of the interaction parameters does not appreciably modify the steady state), as long as the symmetry between the two dispersive interactions is accurate. We hope to be able to try this reservoir engineering procedure in the near future.

11.7 Conclusions and perspectives

We have briefly described microwave cQED experiments exploring some fundamental features of the quantum world. Atoms interacting dispersively with the cavity extract information about the photon number at a rate that is compared with the field damping. This information can be used for a QND measurement, an ideal projective measurement of the photon number that illustrates directly all the basic postulates of quantum theory. It leads to a detailed study of the relaxation of the photon number states and to the direct observation of the quantum Zeno effect in a new environment.

These experiments are also extremely well suited for exploration of the boundary between the quantum and classical worlds. The interaction of a single atom with a coherent state prepares a mesoscopic equivalent of Schrödinger's famous cat, a quantum superposition of classically distinct states. A full quantum state reconstruction procedure, also based on the QND atomic probes, allows us to get detailed insight into the decoherence of this cat state.

Finally, the slow pace of these experiments and the exquisite controllability of the atomic system make it possible to test strategies to combat decoherence. Quantum feedback uses information extracted by QND probes to deterministically steer the cavity field toward a prescribed Fock state. It also protects these nonclassical resources from decoherence by rapidly reversing the adverse effects of quantum jumps. Quantum reservoir engineering opens the way to the steady-state generation of nonclassical states.

All these achievements open promising perspectives, which are nicely complementary to those of circuit QED. In order to achieve an even higher degree of control, we are now developing our experiment in two directions.

The first aims at the deterministic preparation of circular Rydberg atoms on demand. The random excitation used so far, leading to a Poisson atom number distribution, is difficult for experiments involving a fixed number of atoms, such as those on quantum information [16].

The dipole blockade mechanism [45] leads to a deterministic preparation of individual Rydberg atoms. We plan to use a dense sample of about 100 cold atoms, with a micrometer-sized extension. A resonant laser excites a transition from the ground state to a low-angular-momentum Rydberg level ($60S$ state). Once the first atom has been excited to the Rydberg state, it shifts all the others far out of resonance with the laser, owing to the extremely strong dipole-dipole interaction between the very excited Rydberg levels. Numerical simulations show that the probability for exciting two atoms at the same time is low, in the fraction of a percent range.

The strong confinement required is obtained with an atom-chip device, which should moreover operate at cryogenic temperatures. We have therefore developed a superconducting atom chip experiment [46]. This context is quite appropriate for the preparation of a small Bose-Einstein condensate [47]. In the near future, we plan to investigate the dipole blockade mechanism. The prepared Rydberg state could easily be transferred to the circular levels and used on-chip [48], possibly coupled to a stripline microwave resonator, or launched in high-Q cavities using the Stark effect for the acceleration at a macroscopic velocity.

In the present setup, the atom-cavity interaction time is limited to about 100 μs by the thermal velocity of the atomic beam. This is an asset when we need to extract information rapidly out of the cavity, but it is a severe limitation for other experiments. We are therefore developing a new setup, represented in Fig. 11.12, where the atoms will be prepared inside the cavity itself, out of a slow atomic beam in a fountain arrangement. Atoms excited near their turning point interact with the mode for times in the millisecond range, only limited by their free fall through the mode's waist. The limited level lifetime makes it mandatory to perform the field ionization detection also in the cavity structure.

With these long interaction times, we could generate large cat states, containing up to a few tens of photons with high fidelities, and monitor their decoherence. We

Fig. 11.12 Experimental setup under construction, featuring a slow atomic beam in a fountain arrangement to prepare circular atoms nearly at rest in a high-Q cavity (only one mirror is shown). The electrodes around the cavity are used for the circular state preparation and field-ionization detection. A fast horizontal atomic beam with its Ramsey zones and detectors is used for cavity diagnostics.

could realize quantum random walks for the phase of the cavity field, driven by a single atom [49].

Finally, the atom-cavity interaction time is long enough to resolve the anharmonicity of the dressed-levels ladder and to address selectively a transition corresponding to one precise photon number. We have shown recently that we could realize quantum Zeno dynamics (QZD) in this context [50].

QZD [51] generalizes the quantum Zeno effect to measurements of an observable with degenerate eigenspaces. Under such frequently repeated measurements, the system evolution is confined in one of these subspaces, and proceeds under the restriction of the Hamiltonian in the subspace.

Frequent interrogation of a photon-number-selective transition in the dressed levels implements such a dynamics, restricting the evolution to photon numbers smaller or larger than the addressed one. This leads to nontrivial dynamics and to the efficient generation of nonclassical states.

Combining these interrogations with global displacements, we have proposed "phase space tweezers," able to pick out a single coherent component in a complex cat-like superposition and to move it at will, independently from the others. Moreover, these tweezers can be adapted to prepare such superpositions rapidly from the initial vacuum state—a rather fascinating perspective.

References

[1] Nielsen, M. A. and Chuang, I. L. (2000). *Quantum Computation and Quantum Information*. Cambridge University Press, Cambridge.

[2] Haroche, S. and Raimond, J.-M. (2006). *Exploring the Quantum: Atoms, Cavities, and Photons*. Oxford University Press, Oxford.

[3] Purcell, E. M. (1946). *Phys. Rev.*, **69**, 681.
[4] Goy, P., Raimond, J.-M., Gross, M., and Haroche, S. (1983). *Phys. Rev. Lett.*, **50**, 1903.
[5] Gabrielse, G. and Dehmet, H. (1985). *Phys. Rev. Lett.*, **55**, 67.
[6] Hulet, R. G., Hilfer, E. S., and Kleppner, D. (1985). *Phys. Rev. Lett.*, **55**, 2137.
[7] Meschede, D., Walther, H., and Müller, G. (1985). *Phys. Rev. Lett.*, **54**, 551.
[8] Thompson, R. J., Rempe, G., and Kimble, H. J. (1992). *Phys. Rev. Lett.*, **68**, 1132.
[9] Reithmaier, J. P., Sek, G., Löffler, A., Hofmann, G., Kuhn, S., Reitzenstein, S., Keldysh, L. V., Kulakovskii, V. D., Reinecke, T. L., and Forchel, A. (2004). *Nature*, **432**, 197.
[10] Badolato, A., Hennessy, K., Atatüre, M., Dreiser, J., Hu, E., Petroff, P. M., and Imamoglu, A. (2005). *Science*, **308**, 1158.
[11] Schrödinger, E. (1935). *Naturwissenschaften*, **23**, 807, 823, 844.
[12] Zurek, W. H. (2003). *Rev. Mod. Phys.*, **75**, 715.
[13] Raimond, J.-M., Brune, M., and Haroche, S. (2001). *Rev. Mod. Phys.*, **73**, 565.
[14] Kuhr, S., Gleyzes, S., Guerlin, C., Bernu, J., Hoff, U. B., Deléglise, S., Osnaghi, S., Brune, M., Raimond, J. M., Haroche, S., Jacques, E., Bosland, P., and Visentin, B. (2007). *Appl. Phys. Lett.*, **90**, 164101.
[15] Jaynes, E. T. and Cummings, F. W. (1963). *Proc. IEEE*, **51**, 89.
[16] Rauschenbeutel, A., Nogues, G., Osnaghi, S., Bertet, P., Brune, M., Raimond, J.-M., and Haroche, S. (2000). *Science*, **288**, 2024.
[17] Lindblad, G. (1976). *Commun. Math. Phys.*, **48**, 119.
[18] Braginsky, V. B. and Vorontosov, Y. I. (1974). *Usp. Fiz. Nauk*, **114**, 41. [*Sov. Phys. Usp.*, **17**, 644 (1975)].
[19] Grangier, P., Levenson, J. A., and Poizat, J.-P. (1998). *Nature*, **396**, 537.
[20] Gleyzes, S., Kuhr, S., Guerlin, C., Bernu, J., Deléglise, S., Hoff, U. Busk, Brune, M., Raimond, J. M., and Haroche, S. (2007). *Nature*, **446**, 297.
[21] Guerlin, C., Bernu, J., Deléglise, S., Sayrin, C., Gleyzes, S., Kuhr, S., Brune, M., Raimond, J. M., and Haroche, S. (2007). *Nature*, **448**, 889.
[22] Bauer, M. and Bernard, D. (2011). *Phys. Rev. A*, **84**, 044103.
[23] Brune, M., Bernu, J., Guerlin, C., Deléglise, S., Sayrin, C., Gleyzes, S., Kuhr, S., Dotsenko, I., Raimond, J. M., and Haroche, S. (2008). *Phys. Rev. Lett.*, **101**, 240402.
[24] Wang, H., Hofheinz, M., Ansmann, M., Bialczak, R. C., Lucero, E., Neeley, M., O'Connell, A. D., Sank, D., Wenner, J., Cleland, A. N., and Martinis, J. M. (2008). *Phys. Rev. Lett.*, **101**(24), 240401.
[25] Misra, B. and Sudarshan, E. C. G. (1977). *J. Math. Phys.*, **18**, 756.
[26] Itano, W. M., Heinzen, D. J., Bollinger, J. J., and Wineland, D. J. (1990). *Phys. Rev. A*, **41**, 2295.
[27] Bernu, J., Deléglise, Sayrin, C., Kuhr, S., Dotsenko, I., Brune, M., Raimond, J. M., and Haroche, S. (2008). *Phys. Rev. Lett.*, **101**, 180402.
[28] Brune, M., Hagley, E., Dreyer, J., Maître, X., Maali, A., Wunderlich, C., Raimond, J. M., and Haroche, S. (1996). *Phys. Rev. Lett.*, **77**(24), 4887–4890.

[29] Deléglise, S., Dotsenko, I., Sayrin, C., Bernu, J., Brune, M., Raimond, J. M., and Haroche, S. (2008). *Nature*, **455**, 510.
[30] Buzek, V. and Drobny, G. (2000). *J. Mod. Opt.*, **47**, 2823.
[31] Leibfried, D., Meekhof, D. M., King, B. E., Monroe, C., Itano, W. M., and Wineland, D. J. (1996). *Phys. Rev. Lett.*, **77**, 4281.
[32] Hofheinz, M., Wang, H., Ansmann, M., Bialczak, R. C., Lucero, E., Neely, M., O'Connel, A. D., Sank, D., Wenner, J., Martinis, J. M., and Cleland, A. N. (2009). *Nature*, **459**, 546–549.
[33] Adler, S. L. and Bassi, A. (2009). *Science*, **325**, 275–276.
[34] Wiseman, H. M. (1994). *Phys. Rev. A*, **49**, 2133–2150.
[35] Doherty, A. C., Habib, S., Jacobs, K., Mabuchi, H., and Tan, S. M. (2000). *Phys. Rev. A*, **62**, 012105.
[36] Sayrin, C., Dotsenko, I., Zhou, X., Peaudecerf, B., Rybarczyk, T., Gleyzes, S., Rouchon, P., Mirrahimi, M., Amini, H., Brune, M., Raimond, J.-M., and Haroche, S. (2011). *Nature*, **477**, 73.
[37] Dotsenko, I., Mirrahimi, M., Brune, M., Haroche, S., Raimond, J. M., and Rouchon, P. (2009). *Phys. Rev. A*, **80**, 013805.
[38] Geremia, J. M. (2006). *Phys. Rev. Lett.*, **97**, 073601.
[39] Zhou, X., Dotsenko, I., Peaudecerf, B., Rybarczyk, T., Sayrin, C., Gleyzes, S., Raimond, J. M., Brune, M., and Haroche, S. (2012). *Phys. Rev. Lett.*, **108**, 243602.
[40] Poyatos, J. F., Cirac, J. I., and Zoller, P. (1996). *Phys. Rev. Lett.*, **77**, 4728.
[41] Carvalho, A. R. R., Milman, P., de Matos Filho, R. L., and Davidovich, L. (2001). *Phys. Rev. Lett.*, **86**, 4988–4991.
[42] Sarlette, A., Raimond, J.-M., Brune, M., and Rouchon, P. (2011). *Phys. Rev. Lett.*, **107**, 010402.
[43] Slosser, J. J., Meystre, P., and Braunstein, S. L. (1989). *Phys. Rev. Lett.*, **63**, 934–937.
[44] Tan, S. M. (1999). *J. Opt. B*, **1**, 424.
[45] Lukin, M. D., Fleischhauer, M., Cote, R., Duan, L. M., Jaksch, D., Cirac, J. I., and Zoller, P. (2001). *Phys. Rev. Lett.*, **87**, 037901.
[46] Nirrengarten, T., Qarry, A., Roux, C., Emmert, A., Nogues, G., Brune, M., Raimond, J.-M., and Haroche, S. (2006). *Phys. Rev. Lett.*, **97**, 200405.
[47] Roux, C., Emmert, A., Lupascu, A., Nirrengarten, T., Nogues, G., Brune, M., Raimond, J. M., and Haroche, S. (2008). *EPL*, **81**, 56004.
[48] Hyafil, P., Mozley, J., Perrin, A., Tailleur, J., Nogues, G., Brune, M., Raimond, J.-M., and Haroche, S. (2004). *Phys. Rev. Lett.*, **93**, 103001.
[49] Sanders, B. C., Bartlett, S. D., Tregenna, B., and Knight, P. L. (2003). *Phys. Rev. A*, **67**, 042305.
[50] Raimond, J. M., Sayrin, C., Gleyzes, S., Dotsenko, I., Brune, M., Haroche, S., Facchi, P., and Pascazio, S. (2010). *Phys. Rev. Lett.*, **105**, 213601.
[51] Facchi, P. and Pascazio, S. (2002). *Phys. Rev. Lett.*, **89**(8), 080401.

12
SQUID amplifiers

J. Clarke[1], M. H. Devoret[2, 3],
and A. Kamal[3, 4]

[1]Department of Physics, University of California, Berkeley, California 94720, USA
[2]College de France, II Place Marcelin Berthelot, F-75231 Paris Cedex, France
[3]Department of Applied Physics, Yale University, New Haven, CT 06520, USA
[4]Research Laboratory of Electronics, Massachusetts Institute of Technology, Cambridge, MA 02139, USA

Quantum Machines: Measurement and Control of Engineered Quantum Systems.
Edited by M. Devoret, B. Huard, R. Schoelkopf, and L. F. Cugliandolo.

Chapter Contents

12.1 Introduction and overview

Superconducting quantum interference devices (SQUIDs) are exceedingly sensitive detectors of magnetic flux. They are amazingly versatile, and are able to measure any physical quantity that can be converted to a flux, for example, magnetic field, magnetic field gradient, current, voltage, displacement, magnetic susceptibility, far-infrared radiation, the density of axions (if they exist), and the state of a superconducting qubit. As a result, the applications of SQUIDs are wide ranging, from the detection of tiny magnetic fields produced by the human brain and the measurement of fluctuating geomagnetic fields in remote areas to the detection of gravity waves and the observation of spin noise in an ensemble of magnetic nuclei. Hundreds of thousands of SQUIDs are in operation today.

SQUIDs combine two physical phenomena: flux quantization (i.e., the fact that the flux in a closed superconducting loop is quantized [1] in units of the flux quantum $\Phi_0 \equiv h/2e \approx 2.07 \times 10^{-15}$ Wb) and Josephson tunneling [2, 3]. There are two kinds of SQUIDs. The first, the dc SQUID [4], consists of two Josephson junctions connected in parallel on a superconducting loop, and is so named because it can be operated with a static current bias. The second, the rf SQUID [5–7], involves a single Josephson junction interrupting the current flow around a superconducting loop, and is operated with a radiofrequency flux bias. In both cases, the output from the SQUID is periodic with period Φ_0 in the magnetic flux applied to the loop. Typically, above a few hertz, the flux noise is of the order of $10^{-6}\Phi_0\,\mathrm{Hz}^{-1/2}$, although for some devices the noise may be substantially lower. We confine our attention to dc SQUIDs, fabricated from low-transition-temperature (T_c) superconductors. A detailed description of high-T_c SQUIDs can be found in the article by Koelle *et al.* [8]. The *SQUID Handbooks* [9, 10] give a comprehensive account of SQUIDs and their applications.

In this chapter, we describe the principles and operation of the dc SQUID, with an emphasis on its application to amplifiers in both the classical and quantum regimes. We begin, in Section 12.2, with a brief review of the resistively shunted Josephson junction, with particular emphasis on the effects of noise and the observation of quantum fluctuations. In Section 12.3, we discuss the equations of motion for the dc SQUID, the current-voltage characteristics, and noise in the classical and quantum regimes. Section 12.4 is concerned with SQUID amplifiers in the classical regime, and Sections 12.5 and 12.6 describe SQUID amplifiers in the quantum regime. Section 12.7 contains our conclusions.

12.2 The resistively shunted Josephson junction

12.2.1 Equation of motion: the classical Langevin equation

A Josephson junction consists of two superconductors separated by a thin insulating barrier. Cooper pairs of electrons tunnel through the barrier, maintaining phase coherence in the process. The applied current I controls the difference $\delta = \phi_1 - \phi_2$ between the phases of the two superconductors according to the current-phase relation

$$I = I_0 \sin\delta, \tag{12.1}$$

where I_0 is the critical current, that is, the maximum supercurrent the junction can sustain. When the current is increased from zero, there is initially no voltage across the junction; for $I > I_0$, a voltage V appears, and δ evolves with time according to the voltage-frequency relation

$$\dot{\delta} = \frac{2eV}{\hbar} = \frac{2\pi V}{\Phi_0}. \tag{12.2}$$

Equations (12.1) and (12.2) imply that the Josephson junction is intrinsically a nonlinear inductance. Differentiating Eq. (12.1) with respect to time gives $\dot{I} = I_0(\cos\delta)\dot{\delta} = (2\pi V I_0 \cos\delta)/\Phi_0$. Comparing this result with Faraday's law $L\dot{I} = V$, where L is an inductance, yields the Josephson inductance

$$L_J = \frac{\Phi_0}{2\pi I_0 \cos\delta} = \frac{L_{J0}}{\left(1 - I^2/I_0^2\right)^{1/2}} \qquad (I < I_0), \tag{12.3}$$

where

$$L_{J0} \equiv \frac{\Phi_0}{2\pi I_0} \tag{12.4}$$

is the Josephson inductance for $I = 0$. For $I_0 = 1\,\mu\text{A}$, $L_{J0} = 329\,\text{pH}$. We have followed the usual sign convention for the direction of the current and voltage in a branch of a circuit.

A Josephson tunnel junction has a hysteretic current-voltage ($I-V$) characteristic. As the current is increased from zero, the voltage switches abruptly to a nonzero value when I exceeds I_0, returning to zero only when I is reduced to a value much less than I_0. This hysteresis must be eliminated for dc SQUIDs operated in the conventional manner, and one does so by shunting the junction with an external shunt resistance. The "resistively shunted junction" (RSJ) model [11, 12] is shown in Fig. 12.1(a). The junction has a critical current I_0 and is in parallel with its self-capacitance C and its shunt resistance R, which has a current noise source $I_N(t)$ associated with it. The equation of motion is

$$C\dot{V} + I_0 \sin\delta + \frac{V}{R} = I + I_N(t). \tag{12.5}$$

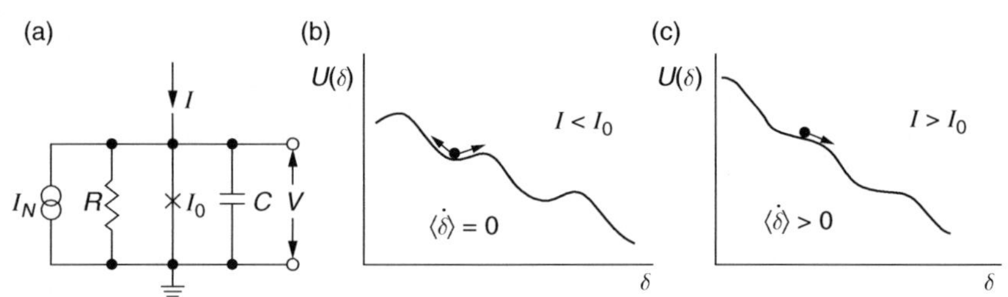

Fig. 12.1 (a) The resistively shunted Josephson junction. (b) and (c) show the tilted washboard model for $I < I_0$ and $I > I_0$.

Neglecting the noise term for now and setting $V = \hbar\dot{\delta}/2e$, we obtain

$$\frac{\hbar C}{2e}\ddot{\delta} + \frac{\hbar}{2eR}\dot{\delta} = I - I_0 \sin\delta = -\frac{2e}{\hbar}\frac{\partial U}{\partial \delta}, \tag{12.6}$$

where

$$U = -\frac{\Phi_0}{2\pi}(I\delta + I_0 \cos\delta). \tag{12.7}$$

For $\delta = 0$,

$$U_0 = -\frac{I_0 \Phi_0}{2\pi}, \tag{12.8}$$

the coupling energy of the junction [2].

One obtains considerable insight into the dynamics of the junction by realizing that Eq. (12.6) also describes the motion of a ball moving on the "tilted washboard" potential U. The term involving C represents the mass of the particle, the $1/R$ term represents the damping of the motion, and the average "tilt" of the washboard is proportional to $-I$. For values of $I < I_0$, the particle is confined to one of the potential wells (Fig. 12.1(b)), where it oscillates back and forth at the plasma frequency [2, 3] $\omega_p = (2\pi I_0/\Phi_0 C)^{1/2}[1-(I/I_0)^2]^{1/4}$. In this state $\langle\dot{\delta}\rangle = 0$ and hence the average voltage across the junction is zero ($\langle\rangle$ represents a time average). As the current is increased to I_0, the tilt increases, and when I exceeds I_0, the particle rolls down the washboard; in this state, $\langle\dot{\delta}\rangle$ is nonzero and a voltage appears across the junction (Fig. 12.1(c)). As the current is increased further, $\langle\dot{\delta}\rangle$ increases, as does V. For the nonhysteretic case, as soon as I is reduced below I_0 the particle becomes trapped in one of the wells, and V returns to zero. In this, the overdamped case, we require [11, 12]

$$\beta_C \equiv \left(\frac{2\pi I_0 R}{\Phi_0}\right) RC = \omega_J RC \leq 1. \tag{12.9}$$

Here, ω_J is the Josephson frequency corresponding to the voltage $I_0 R$.

We introduce the effects of noise by restoring the noise term in Eq. (12.6) to obtain the classical Langevin equation

$$\frac{\hbar C}{2e}\ddot{\delta} + \frac{\hbar}{2eR}\dot{\delta} + I_0 \sin\delta = I + I_N(t). \tag{12.10}$$

In the thermal-noise limit, the spectral density of $I_N(t)$ is given by the Nyquist formula

$$S_I(f) = \frac{4k_B T}{R}, \tag{12.11}$$

where f is the frequency. It is evident that $I_N(t)$ causes the tilt in the washboard to fluctuate with time. This fluctuation has two effects on the junction. First, when $I < I_0$, from time to time fluctuations cause the total current $I + I_N(t)$ to exceed I_0, enabling the particle to roll out of one potential minimum into the next. For

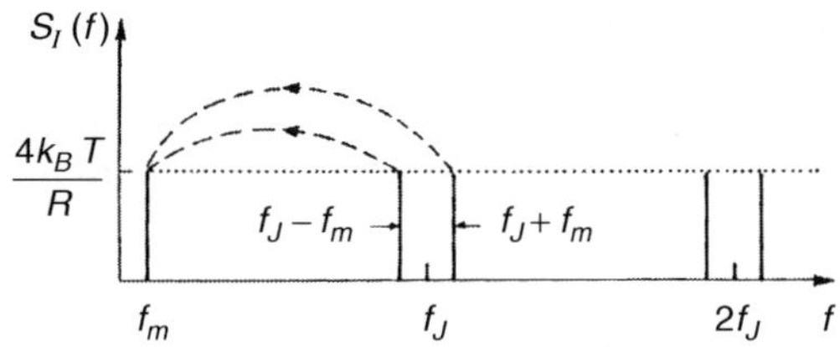

Fig. 12.2 Schematic representation of the noise terms in Eq. (12.13). The Nyquist noise generated in the resistor at frequency f_m contributes directly at f_m; noise generated at $f_J \pm f_m$ is mixed down to f_m.

the underdamped junction, this process produces a series of voltage pulses randomly spaced in time. Thus, the time average of the voltage is nonzero even though $I < I_0$, and the I–V characteristic is "noise-rounded" at low voltages [13]. Because this thermal activation process reduces the observed value of the critical current, there is a minimum value of I_0 for which the two sides of the junction remain coupled together. This condition is

$$\Gamma \equiv \frac{2\pi k_B T}{I_0 \Phi_0} = \frac{k_B T}{|U_0|} \lesssim 1. \tag{12.12}$$

For $T = 4.2\,\mathrm{K}$, we find $I_0 \lesssim 0.2\,\mu\mathrm{A}$.

The second consequence of thermal fluctuations is voltage noise. In the limit $\beta_C \ll 1$ and for $I > I_0$, the spectral density of this noise at a measurement frequency f_m that we assume to be much less than the Josephson frequency f_J is given by [14, 15]

$$S_V(f_m) = \left[1 + \frac{1}{2}\left(\frac{I_0}{I}\right)^2\right] \frac{4k_B T R_d^2}{R} \qquad \left\{ \begin{array}{c} \beta_C \ll 1, \\ I > I_0, \\ f_m \ll f_J \end{array} \right\}. \tag{12.13}$$

The first term on the right-hand side of Eq. (12.13) represents the Nyquist noise current generated at the measurement frequency f_m flowing through the dynamic resistance $R_d = dV/dI$ to produce a voltage noise (Fig. 12.2). The second term, $(1/2)(I_0/I)^2(4k_BT/R)R_d^2$, represents Nyquist noise generated at frequencies $f_J \pm f_m$ mixed down to the measurement frequency by the Josephson oscillations and the inherent nonlinearity of the junction.

12.2.2 The quantum Langevin equation

At sufficiently high bias current, the Josephson frequency f_J exceeds k_BT/h, and Eq. (12.10) becomes a quantum Langevin equation for which the spectral density of $I_N(t)$ is $(2hf/R)\coth(hf/2k_BT)$. The spectral density of the voltage noise across the junction is [14, 16]

$$S_V(f_m) = \left[\frac{4k_BT}{R} + \frac{2eV}{R}\left(\frac{I_0}{I}\right)^2 \coth\left(\frac{eV}{k_BT}\right)\right] R_d^2. \tag{12.14}$$

We have assumed that $hf_m/k_BT \ll 1$, so that the first term on the right-hand side of Eq. (12.14) remains in the thermal limit. Thus, quantum corrections [16] to the observed voltage noise become important in the limit $eI_0R/k_BT \gg 1$ provided the term $(1/2)(I_0/I)^2$ is not too small. In the limit $hf \gg 2k_BT$, the spectral density of the current noise in the resistor R reduces to the quantum value $2hf/R$. In this limit, the second term on the right-hand side of Eq. (12.14), $(2eV/R)(I_0/I)^2R_d^2$, represents noise mixed down from zero-point fluctuations near the Josephson frequency.

12.2.3 Observation of quantum fluctuations

Zero-point fluctuations were first observed in a current-biased, resistively shunted Josephson junction using the circuit shown in Fig. 12.3 [17]. The voltage noise was measured at three frequencies by means of two LC resonant circuits connected separately or in parallel to a low-noise, room-temperature amplifier. Measurements at the three frequencies allowed the subtraction of a small contribution of $1/f$ noise from the junction; the measured voltage noise and current noise of the preamplifier were also subtracted. In the low-frequency limit, the spectral density of the voltage noise across a given tank circuit with inductance L_t was $Q^2S_V(0)$, where the quality factor $Q = \omega L_t/R_d$. Thus, the quantity $S_V(0)/R_d^2$ was independent of Q, and could be compared directly with the prediction of Eq. (12.14) rewritten as

$$\frac{S_V(f_m)}{R_d^2} = \left[\frac{4k_BT}{R} + \frac{2eV}{R}\left(\frac{I_0}{I}\right)^2 \coth\left(\frac{eV}{k_BT}\right)\right] \tag{12.15}$$

using measured values of I_0, R, I, V, T, and L_t. We note that the term $(2eV/R)(I_0/I)^2 \times \coth(eV/k_BT)$ can also be written as $(4eV/R)(I_0/I)^2\{[\exp(2eV/k_BT)-1]^{-1}+1/2\}$, that is, essentially as the Planck distribution plus the zero-point energy.

The results are shown in Fig. 12.4. Figure 12.4(a) shows $S_V(0)/R_d^2$ versus voltage (proportional to frequency) for the junction at 4.2 K. The open circles

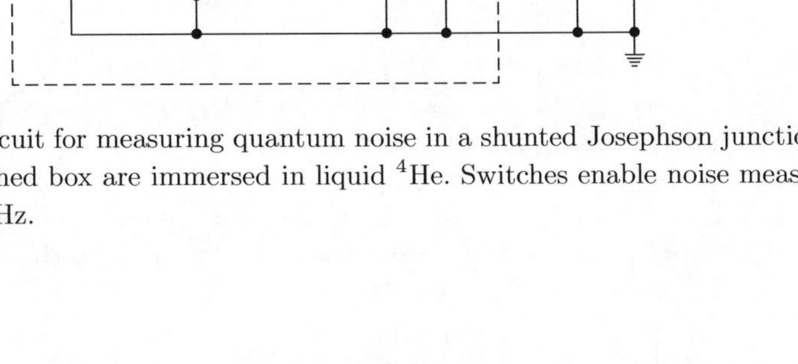

Fig. 12.3 Circuit for measuring quantum noise in a shunted Josephson junction. Components within the dashed box are immersed in liquid ^{4}He. Switches enable noise measurements at 70, 106, and 183 kHz.

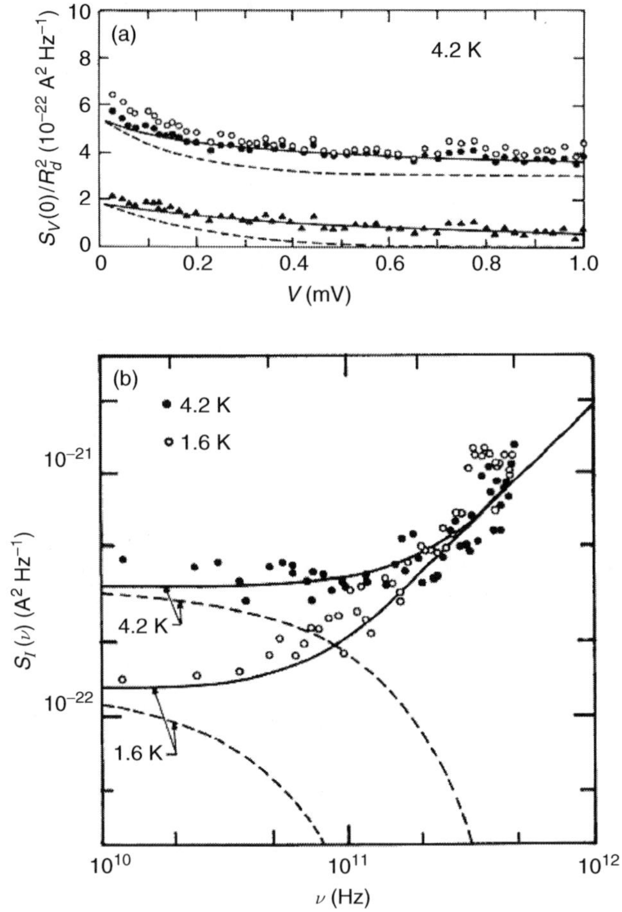

Fig. 12.4 Quantum noise in a resistively shunted Josephson junction with $I_0 = 0.514\,\mathrm{mA}$, $R = 0.70\,\Omega$, and $\beta_C \approx 0.2$. (a) Open circles show the total measured noise across the junction; solid circles below show the noise remaining after correction for amplifier and $1/f$ noise. The upper solid line is the prediction of Eq. (12.15), the dashed line is the prediction with the zero-point term removed. Solid triangles are measured mixed-down noise, the solid line is the prediction with the zero-point term included, and the dashed line is the prediction without the zero-point term. (b) Measured spectral density of the current noise $S_I(f)$ generated by the resistance R at 4.2 K (solid circles) and 1.6 K (open circles). Solid lines are the corresponding predictions of Eq. (12.15) with measured values of $\nu = 2eV/h$, R, and T. Dashed lines are predictions in the absence of the zero-point term.

show the total measured noise and the solid circles show the noise after corrections. The solid line in the upper plots is the prediction of Eq. (12.15), whereas the dashed line is the prediction with the zero-point term subtracted, that is, $(4eV/R)(I_0/I)^2\{[\exp(2eV/k_BT)-1]^{-1}\}$. The lower set of plots shows the mixed-down noise, obtained by subtracting $4k_BT/R$ from the solid circles, and the solid line is the predicted value $(2eV/R)(I_0/I)^2\times$ coth(eV/k_BT). The lower dashed line is the prediction with the zero-point term removed. The plots in Fig. 12.4(a) show very clearly that the zero-point term is required to fit the experimental data.

We can extract from the data the measured spectral density of the current noise $S_I(f)$ generated by the resistance R. We multiply each value of the mixed-down noise by $2(I/I_0)^2$, and set $2eV = h\nu$. The results are plotted in Fig. 12.4(b) for 4.2 K (solid circles) and 1.6 K (open circles). The solid lines are the corresponding predictions of Eq. (12.15) with measured values of $\nu = 2eV/h$, R, and T. The agreement between the predictions and the data is rather good, especially bearing in mind that there are no fitting parameters. The dashed lines represent the prediction in the absence of the zero-point term, and fall off dramatically at the higher frequencies.

These results demonstrate, first, the existence of a zero-point term in the spectral density of the current noise of a resistor in thermal equilibrium and, second, that these fluctuations give rise to the limiting voltage noise in a current-biased resistively shunted Josephson junction in the quantum limit for $I > I_0$. It should be emphasized that the observation of the zero-point term is entirely due to the nonlinearity of the Josephson junction, which mixes down high-frequency noise near the Josephson frequency. One can think of the zero-point fluctuations as randomly modulating the tilt of the washboard, a process that requires no energy but that modulates the rate at which the phase difference evolves with time. Furthermore, the good agreement between the data and the model predictions justifies the use of a quantum Langevin equation to calculate quantum noise in an overdamped, current-biased Josephson junction in the free-running regime $I > I_0$. This gives us some confidence in the use of a quantum Langevin approach to calculate the noise in a dc SQUID in the quantum limit.

12.3 The dc SQUID

12.3.1 Equations of motion: the classical Langevin equation

Figure 12.5 shows the model for the dc SQUID. Two Josephson junctions are connected in parallel on a superconducting loop of inductance L. Each junction is resistively shunted to eliminate hysteresis on the I–V characteristics. When we current-bias the SQUID into the voltage state and apply a monotonically increasing magnetic flux Φ, the critical current and I–V characteristic are modulated with period Φ_0. The SQUID is generally operated near the steepest region of the V–Φ curve, which occurs at about $(n + 1/2)\Phi_0/2$, where the flux-to-voltage transfer coefficient $V_\Phi \equiv (\partial V/\partial\Phi)_I$ is a maximum. Thus, the SQUID produces an output voltage $\delta V = V_\Phi\delta\Phi$ in response to a small applied flux $\delta\Phi$, and is effectively a flux-to-voltage transducer.

Our goals are to calculate V_Φ, the spectral densities of the voltage noise $S_V(\Phi)$, circulating current noise $S_J(f)$, and their cross-correlation spectrum $S_{VJ}(f)$. The SQUID inductance is L, and each junction has a critical current I_0 and a self-capacitance C and is shunted with a resistor R. The phase differences across the two junctions are δ_1 and δ_2, respectively, and the associated resistors have independent Nyquist noise currents I_{N1} and I_{N2}. The equations of motion are [18, 19]

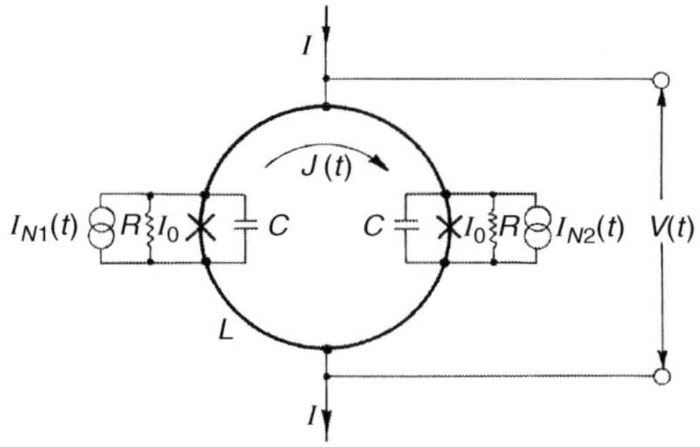

Fig. 12.5 Model for dc SQUID.

$$V = \frac{\hbar}{4e}(\dot{\delta}_1 + \dot{\delta}_2), \tag{12.16}$$

$$J = \frac{\Phi_0}{2\pi L}\left(\delta_1 - \delta_2 - \frac{2\pi\Phi}{\Phi_0}\right), \tag{12.17}$$

$$\frac{\hbar C}{2e}\ddot{\delta}_1 + \frac{\hbar}{2eR}\ddot{\delta}_1 = \frac{I}{2} - J - I_0 \sin\delta_1 + I_{N1}, \tag{12.18}$$

$$\frac{\hbar C}{2e}\ddot{\delta}_2 + \frac{\hbar}{2eR}\ddot{\delta}_2 = \frac{I}{2} + J - I_0 \sin\delta_2 + I_{N2}. \tag{12.19}$$

Equation (12.16) relates the voltage to the average rate of change of phase; Eq. (12.17) relates the current in the loop, J, to $\delta_1 - \delta_2$ and to Φ; and Eqs. (12.18) and (12.19) are Langevin equations coupled via J. These equations have been solved numerically for a limited range of values of the noise parameter, reduced inductance $\beta_L \equiv 2LI_0/\Phi_0$, and hysteresis parameter $\beta_C = 0$. For typical SQUIDs in the ^{4}He temperature range, $\Gamma = 0.05$. Full details of Claudia Tesche's simulations can be found in [18]. The first task is to compute the dependence of the critical current on applied flux, a purely static problem. A much more complicated calculation is to find the time-averaged values of the I–V characteristic as a function of applied flux, from which one can compute the time-averaged voltage V versus Φ, and hence find V_Φ. One also computes the current J circulating around the SQUID loop.

12.3.2 Current-voltage characteristic, flux-to-voltage transfer function, and noise

As an example of the results, Fig. 12.6 shows the time-averaged I–V characteristic of a SQUID for three values of magnetic flux. Noise rounding at low voltages is clearly visible. Figure 12.7 shows three key results: V_Φ, $S_V(0)$, and flux noise $S_\Phi^{1/2}(0)$ versus I/I_0. Figure 12.7(a) shows that V_Φ peaks as a function of bias current, at a value that depends on the applied flux. For each value of flux, the peak occurs at the maximum value of the dynamic resistance; the peak is highest when the flux is $\Phi_0/4$. The spectral density of the noise voltage was computed as a function of bias current at fixed flux,

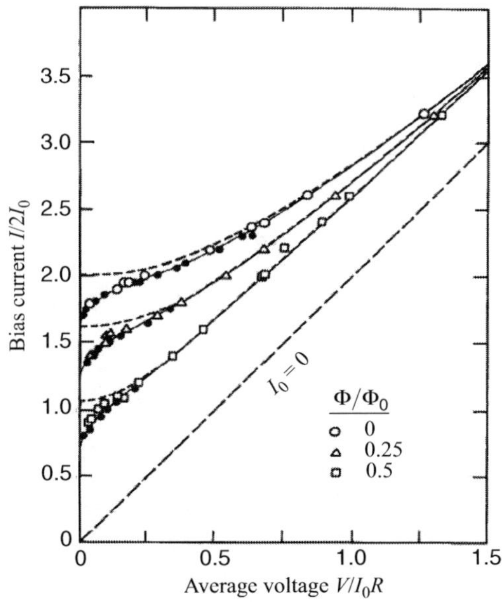

Fig. 12.6 Time-averaged current-voltage characteristics of a dc SQUID. Simulations were with independent Nyquist noise sources in the shunting resistors, $\Gamma = 0.05$, $\beta_L = 1.0$, and three values of flux. Dashed lines are noise-free characteristics. Open and solid symbols are for two different methods of implementing the noise.

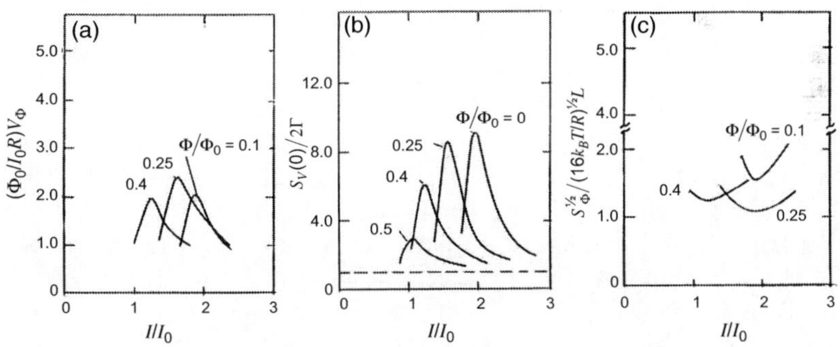

Fig. 12.7 Computed properties of a dc SQUID: (a) flux-to voltage transfer coefficient; (b) voltage noise spectral density; (c) flux noise. Parameters are $\Gamma = 0.05$ and $\beta_L = 1$.

and found to be white at frequencies well below the Josephson frequency. Figure 12.7(b) shows that, for each value of flux, the noise spectral density peaks smoothly at the value of I where V is a maximum. These results are combined in Fig. 12.7(c), in which we plot the ratio $S_V^{1/2}(0)/V_\Phi$ to yield the flux noise $S_\Phi^{1/2}(0)$ versus bias current for three values of flux. We note that the minimum in the flux noise is substantially broader in bias current that the peaks in the voltage noise and transfer function, and that the lowest flux noise occurs at $\Phi_0/4$.

A convenient way of comparing the flux noise in SQUIDs with different parameters is in terms of the noise energy per unit bandwidth

$$\varepsilon = \frac{S_\Phi}{2L}. \tag{12.20}$$

From a series of simulations, one finds that the noise energy has a minimum (not to be confused with the minimum noise temperature, discussed later) when $\beta_L = 1$. For $\Gamma = 0.05$, $\beta_C = 1$, and $\Phi = (n + 1/2)\Phi_0/2$ and at the value of I at which V is a maximum, the optimized results can be summarized as follows:

$$V_\Phi \approx \frac{R}{L}, \tag{12.21}$$

$$S_V(0) \approx 16k_BTR, \tag{12.22}$$

$$S_\Phi(0) \approx \frac{16k_BTL^2}{R}, \tag{12.23}$$

$$\varepsilon \approx \frac{9k_BTL}{R} \approx 16k_BT(LC)^{1/2}. \tag{12.24}$$

Equation (12.21) shows that V_Φ can be written as a characteristic frequency, and Eq. (12.22) shows that the voltage noise spectral density is about eight times the Nyquist noise in a resistance $R/2$ (the parallel resistance of the two shunt resistors). To obtain the last expression in Eq. (12.24), we set $R = (\Phi_0/2\pi I_0 C)^{1/2}$ ($\beta_C = 1$). The resulting expression shows that the noise energy scales with T, reflecting its origin in Nyquist noise, and inversely with the characteristic frequency $(LC)^{-1/2}$. Thus, in the classical limit, lowering the temperature reduces the noise energy. Furthermore, "smaller is better"—lowering the loop inductance and junction capacitance while keeping $\beta_L = \beta_C = 1$ reduces the noise energy. These results have been found to be excellent predictors for the performance of practical SQUIDs.

As we shall see in our discussion of amplifiers, however, the noise energy is not a complete specification of the SQUID, because it does not account fully for the circulating current noise. Claudia Tesche calculated the current noise, and its spectral density $S_J(0)$ is plotted in Fig. 12.8(a). For fixed flux, the current noise peaks as a function of bias current. As the flux is increased from zero, the peak current noise increases, diverging at $\Phi_0/2$. For a SQUID with $\beta_L = 1$, $\Gamma = 0.05$, and $\Phi = (n + 1/2)\Phi_0/2$, the peak spectral density of the current noise is [19]

$$S_J(f) = \frac{11k_BT}{R}. \tag{12.25}$$

Furthermore, the current noise is partially correlated with the voltage noise across the SQUID, as shown in Fig. 12.8(b). For the same parameter values, the peak cross-spectral density is [19]

$$S_{VJ}(f) = 12k_BT. \tag{12.26}$$

The correlation arises because the current noise generates a flux noise, which, in turn, contributes to the total voltage noise across the junction, provided $V_\Phi \neq 0$.

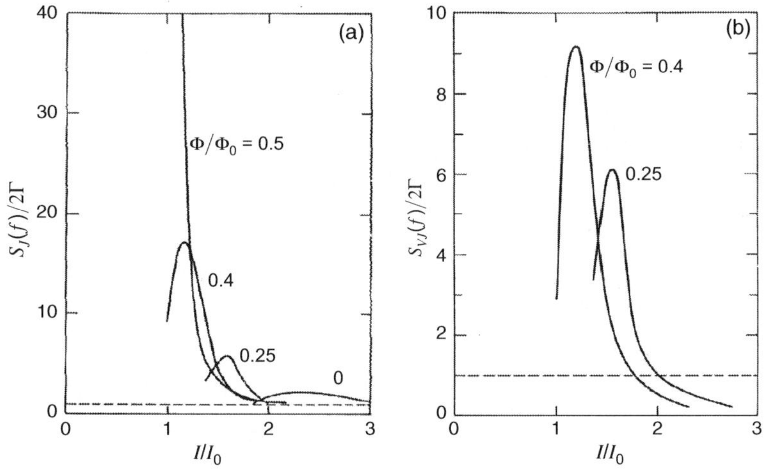

Fig. 12.8 Computed properties of a dc SQUID: (a) current noise spectral density; (b) cross spectral density. Parameters are $\Gamma = 0.05$ and $\beta_L = 1$.

12.3.3 Practical dc SQUIDs

Modern dc SQUIDs are made from thin films with the aid of either photolithography or electron-beam lithography, and come in a great variety of designs. A widely used design that is available commercially was introduced by Mark Ketchen and Jeffrey Jaycox [20] and is shown in Fig. 12.9. The SQUID body consists of a square washer with a spiral, superconducting input coil deposited on it with an intervening insulating layer. Such devices are typically fabricated in batches of several hundred on oxidized Si wafers. Except for the resistive shunts, the entire structure is made of Nb. The junctions are patterned from Nb/AlOx/Nb [21]. In this process, following the deposition of the Nb base electrode and a thin Al layer, the Al is oxidized in a reduced pressure of oxygen and the Nb counter-electrode is deposited. The entire trilayer is formed without removing the wafer from the controlled atmosphere of the deposition system. The junction areas are defined by anodizing a small ring of the counter-electrode, and

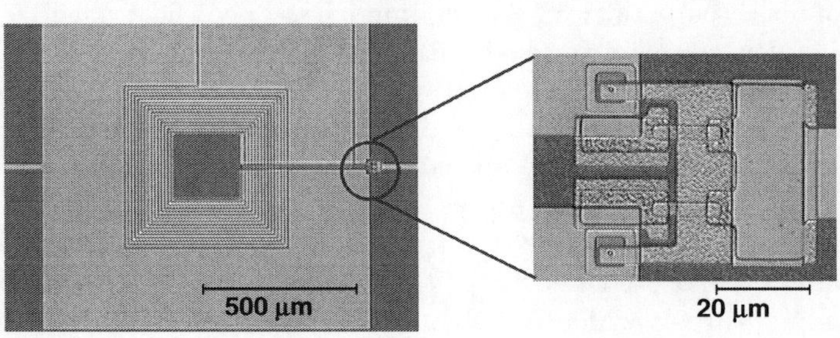

Fig. 12.9 Square-washer dc SQUID showing the spiral input coil (left) and an expanded view of the junctions (right).

the base electrode is etched to form the SQUID washer. Subsequent processes add the Nb layer that forms the input and flux modulation coils and makes the connection to the counter electrode, the shunt resistors (typically Mo or Pd), and the final Nb layer that connects the innermost turn of the input coil. The insulation between each layer is usually SiO_2, and patterning is performed with reactive ion etching. Typical loop inductances are 100–400 pH and the shunt resistances are a few ohms.

Design guidelines for the square-washer SQUID were given by Ketchen and Jaycox [20], who showed that a square washer (with no slit) with inner and outer edges d and w has an inductance $L(\text{loop}) = 1.25\mu_0 d$ in the limit $w \gg d$. They gave the following expressions for the inductances L and L_i of the SQUID and spiral coil, respectively, and for the mutual inductance M_i between them:

$$L = L(\text{loop}) + L_j, \tag{12.27}$$

$$L_i = n^2(L - L_j) + L_s, \tag{12.28}$$

$$M_i = n(L - L_j), \tag{12.29}$$

$$\alpha^2 = \left(1 - \frac{L_j}{L}\right)\left[1 + \frac{L_s}{n^2(L - L_j)}\right]. \tag{12.30}$$

Here, L_j is the parasitic inductance associated with the junctions and slit, n is the number of turns on the input coil, and L_s is the stripline inductance of this coil, which is generally much smaller than L_i for $n \geq 20$. Measured parameters are generally in good agreement with these predictions.

12.4 Low-frequency SQUID amplifiers

12.4.1 Noise temperature

Before delving into the theory of SQUID amplifiers, it is convenient first to introduce the concept of noise temperature T_N. For simplicity, we consider a field effect transistor (FET) amplifier that at low frequencies has a high input impedance and does not load the source. Its voltage gain is A. Referred to its input terminals, the FET has a virtual voltage noise e_N and an uncorrelated actual current noise i_N that develops a voltage noise $i_N R_i$ across a resistance R_i connected across its input. Our goal is to determine the optimum value of R_i that minimizes T_N.

The voltage noise at the amplifier output is $V_o = A(e_N + i_N R_i)$. Since e_N and i_N are uncorrelated, the spectral density of this noise is $S_{V_o} = A^2(S_e + S_i R^2)$. We define T_N as the temperature at which the Nyquist noise associated with R_i would produce the equivalent output noise spectral density $4k_B T R_i A^2$. Equating these two quantities gives $T_N = (S_e/R_i + S_i R_i)/4k_B$. Finally, differentiating with respect to R_i yields the optimized values

$$R_i^{\text{opt}} = \left(\frac{S_e}{S_i}\right)^{1/2} \tag{12.31}$$

and

$$T_N^{\text{opt}} = \frac{(S_e S_i)^{1/2}}{2k_B}. \tag{12.32}$$

12.4.2 SQUID amplifier theory, noise, and optimization

We now discuss the theory, operation and performance of SQUID amplifiers at frequencies up to, say, 100 MHz. In this frequency range, we can treat the amplifier in the lumped-circuit approximation or "op-amp" approximation—in fact, as we shall see, the SQUID amplifier is in many ways the dual of a semiconductor operational amplifier. We recognize at the beginning that the SQUID is a complicated, highly nonlinear device. In particular, the capacitance between the input coil and the SQUID washer may attenuate the coupling of the SQUID to the input circuit at the Josephson frequency. In the limits in which there is no attenuation or very high attenuation, the calculations of the influence of the SQUID on the input circuit and vice versa are straightforward. In practice, the coupling is likely to be somewhere between the two limits, so that the mutual influences are nontrivial calculations. Consequently, we consider a rather simplified model based on more detailed publications [22–24].

If we imagine "looking" into the input terminals of a coil to SQUID, we will "see" a dynamic impedance $\mathcal{Z}$ in the SQUID loop that can be written in the form [23]

$$\frac{1}{\mathcal{Z}} = \frac{1}{j\omega\mathcal{L}} + \frac{1}{\mathcal{R}}, \tag{12.33}$$

where $j = \sqrt{-1}$. The dynamic inductance $\mathcal{L}$ and dynamic resistance $\mathcal{R}$ are not simply related to L and R, but vary with bias current and flux; for example, $1/\mathcal{L}$ is zero for certain values of flux.

The configuration of a tuned amplifier is shown in Fig. 12.10(a). An input voltage source $V_i(t)$ with source resistance R_i is connected in series with a capacitor C_i, the input coil of a SQUID, and some stray inductance L_s. The output voltage from the SQUID is V_o. In general, the presence of the input circuit modifies all the SQUID parameters, including the noise terms [22, 23]. By the same token, the terms $\mathcal{L}$ and

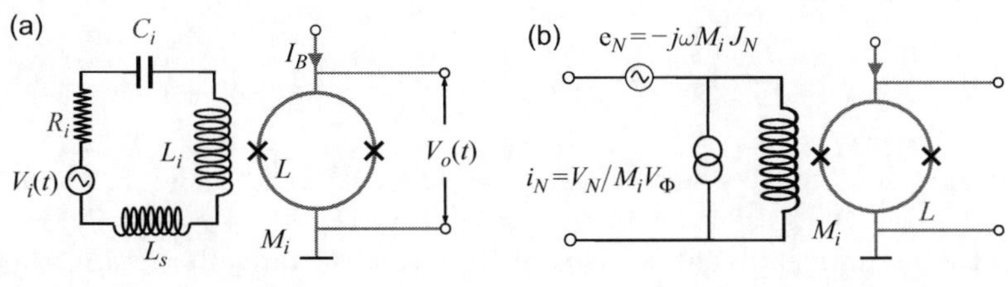

Fig. 12.10 SQUID tuned amplifier: (a) configuration of model; (b) representation of SQUID noise sources transferred to the terminals of the input coil.

$\mathcal{R}$ introduce additional inductances and resistances into the input circuit. Strictly speaking, we should take these corrections into account. If the effective coupling coefficient between the source and the SQUID is sufficiently weak, however, as we shall see, these mutual interactions are negligible. For the purpose of illustration, we assume a SQUID with given values of M_i, L_i, L, V_Φ, $S_V(f)$, $S_J(f)$, and $S_{VJ}(f)$, and find the values of C_i and R_i that optimize the noise temperature.

Tuned amplifier: on resonance In the weak-coupling limit, the SQUID noise voltage is represented as a virtual current i_N in the input circuit V_N/M_iV_Φ. The noise current J_N in the SQUID loop induces an actual voltage $e_N = j\omega M_i J_N$ into the input circuit (Fig. 12.10(b)); since, on resonance, the impedance of the input circuit is R_i, the current generated is $j\omega M_i J_N/R_i$. It is important to note that this current is in quadrature with J_N and thus with the contribution that J_N makes to the voltage noise across the SQUID. We note that the noise terms e_N and i_N are the duals of those for an FET, for which the current is actual and the voltage is virtual. Inserting the spectral densities of e_N and i_N into Eqs. (12.31) and (12.32) immediately yields

$$R_i^{\text{res}} = \omega M_i^2 V_\Phi \left(\frac{S_J}{S_V}\right)^{1/2} \approx \alpha^2 \omega L_i, \tag{12.34}$$

and, on resonance,

$$T_N^{\text{res}} = \omega \frac{(S_J S_V)^{1/2}}{2k_B V_\Phi} \approx \frac{42 f T}{V_\Phi} \approx \frac{5 f \varepsilon(f)}{k_B}, \tag{12.35}$$

where we have used Eqs. (12.21), (12.22), and (12.25). We note that the cross spectral term S_{VJ} drops out of T_N^{res}. This is because the current induced into the input circuit is in quadrature with J_N, and is thus uncorrelated with the contribution of J_N to the voltage noise across the SQUID. Furthermore, as noted earlier, although $\varepsilon(f)$ does not fully characterize a SQUID amplifier, within the framework of the model it does enable one to predict T_N.

We now introduce the quality factor of the tuned circuit,

$$Q = \omega \frac{L_i + L_s}{R_i} \approx \frac{L_i + L_s}{\alpha^2 L_i} = \frac{1}{\alpha_e^2}, \tag{12.36}$$

where we have used Eq. (12.34). Here,

$$\alpha_e^2 = \alpha^2 \frac{L_i}{L_i + L_s} = \frac{M_i^2}{L_i(L_i + L_s)} \tag{12.37}$$

is the *effective* coupling coefficient between the SQUID and the input circuit. Since $Q\alpha_e^2 \approx 1$, we see that even a moderately high-Q implies that α_e is small, thereby justifying the assumption that we can neglect the mutual interaction of the SQUID and input circuit.

One can readily calculate the gain on resonance. For $\alpha^2 \ll 1$, an input signal V_i produces an output voltage $V_o \approx (V_i/R_i^{\text{res}})M_i V_\Phi$. Thus, the power gain is given by

$G \approx (V_o^2/R_D)/(V_i^2/R_i)$, where $R_D \equiv \partial V/\partial I$ is the dynamic output resistance of the SQUID. Assuming $R_D \approx R$, we find

$$G \approx \frac{V_\Phi}{\omega}, \tag{12.38}$$

where we have used Eqs. (12.21) and (12.34). We see that G is the ratio of the characteristic frequency V_Φ to the signal frequency f, and in this sense is reminiscent of a parametric amplifier. The dc SQUID mixes up the signal to a high frequency, and down-converts it to the signal frequency with gain.

Tuned amplifier: optimized noise temperature Operating a SQUID tuned amplifier at the resonant frequency, however, does not give the lowest noise temperature, and we now consider the off-resonance case. In the weak-coupling limit, the noise current J_N induces a voltage $j\omega M_i J_N$ into the input circuit, and hence a current $j\omega M_i J_N/Z_i$, where

$$Z_i \approx R_i + j\omega(L_i + L_s) + \frac{1}{j\omega C_i} \tag{12.39}$$

is the impedance of the input circuit. In general, this current is not in quadrature with J_N, since Z_i is complex. This noise current, in turn, induces a flux in the SQUID loop and finally a voltage $-j\omega M_i^2 J_N V_\Phi/Z_i$ across the SQUID. Thus, the noise voltage across the SQUID in the presence of the input circuit is

$$V_N^{'} = V_N - j\omega M_i \frac{M_i V_\Phi}{Z_i} J_N, \tag{12.40}$$

where V_N is the noise voltage of the bare SQUID, which we assume to be unchanged by the input circuit in the limit of small α_e. The spectral density of $V_N^{'}$ is found to be

$$\begin{aligned} S_V^{'}(f) &= S_V(f) + \omega^2 M_i^2 \frac{M_i^2 V_\Phi^2}{|Z_i|^2} S_J(f) \\ &\quad - \left\{ \frac{2\omega M_i^2 V_\Phi}{|Z_i|^2} \left[\omega(L_i + L_s) - \frac{1}{\omega C_i} \right] \right\} S_{VJ}(f). \end{aligned} \tag{12.41}$$

We now apply a sinusoidal input signal frequency $\omega/2\pi$, with a mean-square amplitude $\langle V_i^2 \rangle$. The mean-square signal at the output of the SQUID is

$$\langle V_o^2 \rangle = \frac{M_i^2 V_\Phi^2}{|Z_i|^2} \langle V_i^2 \rangle. \tag{12.42}$$

The signal-to-noise ratio is

$$\frac{S}{N} = \frac{\langle V_o^2 \rangle}{S_V^{'}(f) B} \tag{12.43}$$

in a bandwidth B. We introduce the noise temperature T_N for the amplifier by setting $S/N = 1$ with $\langle V_i^2 \rangle = 4k_B T_N R_i B$. This procedure implies that the output noise power generated by the SQUID is equal to the output noise power generated by the resistor R_i when it is at a temperature T_N. We optimize T_N with respect to R_i and C_i for a given value of L_i to find

$$R_i^{\text{opt}} = \left[\alpha^2 \frac{\omega(L_i + L_s)LV_\Phi}{S_V}\right](S_V S_J - S_{VJ}^2)^{1/2}, \tag{12.44}$$

$$\frac{1}{\omega C_i^{\text{opt}}} = \omega(L_i + L_s)\left(1 + \frac{\alpha^2 S_{VJ} LV_\Phi}{S_V}\right), \tag{12.45}$$

and

$$T_N^{\text{opt}} = \frac{\pi f}{k_B V_\Phi}(S_V S_J - S_{VJ}^2)^{1/2}. \tag{12.46}$$

We note from Eq. (12.46) that the optimum noise temperature occurs off resonance. For the values of the spectral densities given in Eqs. (12.22), (12.25), and (12.26), $T_N^{\text{opt}}/T_N^{\text{res}} \approx 0.4$.

Finally, we remark that this theory is concerned only with the noise temperature of the amplifier itself. Nyquist noise from the input resistor may add a contribution that exceeds the amplifier noise. When the value of T_N is well below T, the optimization procedure outlined above does not necessarily give the lowest system noise.

12.4.3 Experimental results

Hilbert and Clarke [24] made several radiofrequency amplifiers with both tuned and untuned inputs, flux biasing the SQUID near $\Phi = (2n + 1)\Phi_0/4$. There was no flux-locked loop. The measured parameters were in good agreement with predictions. For example, for an amplifier with $R = 8\,\Omega$, $L = 0.4\,\text{nH}$, $L_i = 5.6\,\text{nH}$, $M_i = 1\,\text{nH}$, and $V_\Phi = 3 \times 10^{10}\,\text{s}^{-1}$ at 4.2 K, they found $G = 18.6 \pm 0.5\,\text{dB}$ and $T_N = 1.7 \pm 0.5\,\text{K}$ at 93 MHz. The predicted values were 17 dB and 1.1 K, respectively.

12.5 High-frequency SQUID amplifiers: the quantum limit

12.5.1 Noise and optimization in the quantum limit

For a linear, phase-preserving amplifier, the quantum-limited noise temperature is given by

$$T_N^{\text{opt}} = \frac{hf}{k_B}. \tag{12.47}$$

A convenient way of expressing the noise temperature T_N is in terms of Caves' added noise number $A \geq 1/2$ [25]:

$$T_N = \left(A + \frac{1}{2}\right)\frac{hf}{k_B}. \tag{12.48}$$

Clearly, $A = 1/2$ for a quantum-limited amplifier.

The noise temperature for an optimized SQUID tuned amplifier at $T = 0$ was computed by Roger Koch [26]. The approach was to replace the thermal noise current in Eqs. (12.18) and (12.19) with the quantum noise current with spectral density $2hf/R$, compute the quantities V_Φ, S_V, S_J, and S_{VJ} in the quantum regime, and optimize the parameters for lowest noise temperature. The results are summarized in Fig. 12.11, which shows seven computed and derived quantities versus flux at a constant bias current for $R = 40\,\Omega$, $\beta_L = 1$, and three values of β_C. The peak values of V_Φ, S_V, S_J, and S_{VJ} for $\beta_C = 1$ are substantially higher than those for $\beta_C = 0.25$ and 0.5, reflecting a higher dynamic resistance. The values of values of V_Φ and S_V were used to compute ε (Fig. 12.11(f)), which for the lower values of β_C has a minimum below $\hbar$. Finally, V_Φ and the three noise terms were used in Eqs. (12.46) and (12.48) to compute A, which had a minimum value of 0.5, off resonance. The estimated computational

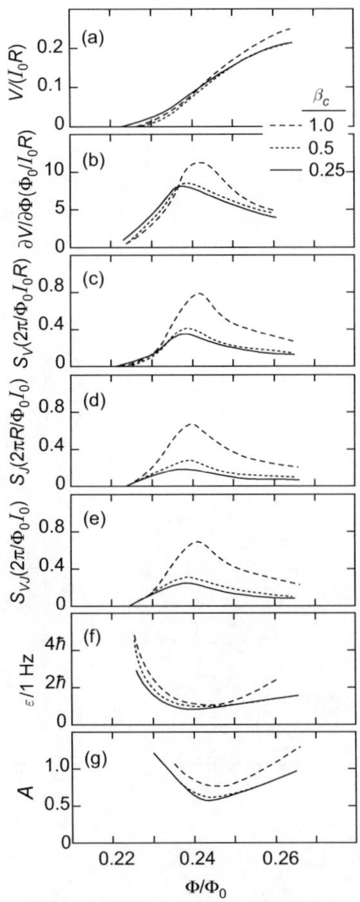

Fig. 12.11 Simulations for a SQUID tuned amplifier at $T = 0$: (a) V; (b) V_Φ; (c) S_V; (d) S_J; (e) S_{VJ}; (f) ε; (g) A. Parameters are $\beta_L = 1$ and $R = 40\,\Omega$.

accuracy was ±15%. Needless to say, the prediction that the SQUID with a tuned input circuit should ideally be a quantum-limited amplifier assumes the validity of both the quantum Langevin equation and the optimization procedure developed for the thermal limit.

To demonstrate quantum-limited amplification convincingly, one requires $T_N > T$. For a lowest practical operating temperature of 20 mK, this implies that the signal frequency should be greater than about 0.5 GHz. At the time the theory was developed (1981), SQUIDs were invariably used at much lower frequencies, and there was no motivation to develop gigahertz devices.

12.5.2 Harmonic balance treatment of the dc SQUID

In this subsection, we discuss an analytical treatment of the SQUID in the paradigm of the scattering theory of circuits. This formalism is especially useful for a high-frequency analysis of the SQUID and enables simultaneous evaluation of the properties such as power gain and noise temperature of the device by treating coherent input signals and noise on the same footing.

The circuit representation for this analysis is shown in Fig. 12.12. Using the input-output theory of circuits [27], we model the response of the circuit through a scattering matrix specifying the relationship between the incoming and outgoing waves for all modes of interest (temporal and spatial). For convenience, we employ a modified version of the SQUID equations (12.16)–(12.19) introduced in Section 12.3:

$$\frac{\hbar C}{2e}\ddot{\delta}^C + \frac{\hbar}{2eR}\dot{\delta}^C = \frac{I}{2} - I_0 \sin\delta^C \cos\delta^D + 2I_{RF}^{\text{in},C}(t), \tag{12.49}$$

$$\frac{\hbar C}{2e}\ddot{\delta}^D + \frac{\hbar}{2eR}\dot{\delta}^D = \frac{I_0}{\pi\beta_L}\left(2\pi\frac{\Phi}{\Phi_0} - 2\delta^D\right) - I_0 \cos\delta^C \sin\delta^D + 2I_{RF}^{\text{in},D}(t), \tag{12.50}$$

where we have introduced the common (C) and differential (D) combinations of the junction variables $\delta^C = (\delta_1 + \delta_2)/2$ and $\delta^D = (\delta_1 - \delta_2)/2$. Here, $I_{RF}^{\text{in},C}$ and $I_{RF}^{\text{in},D}$ represent the input driving signals for the common and differential modes. The symbol $\eta = I_0/I$, the inverse of the reduced bias current, is the expansion parameter of choice

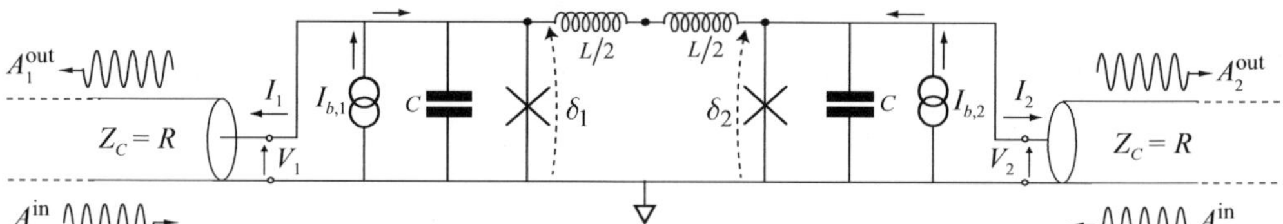

Fig. 12.12 Equivalent circuit of Fig. 12.5 for input-output analysis. The static bias current I is modeled as the even mode current $-(I_{b,1} + I_{b,2})$ imposed by the two external biases, while the external flux Φ is represented by the odd mode combination $L(I_{b,1} - I_{b,2})/2$. The signals processed by the SQUID are modeled as incoming and outgoing waves travelling on semi-infinite transmission lines, representing the shunt resistances across the two junctions.

since $\eta \lesssim 1.0$ for the SQUID to operate in the running state, which is the regime of interest for voltage amplifiers. We first calculate the steady-state response of the system, putting $I_{RF}^{\text{in},C,D}(t) = A^{\text{in},C,D}(t)/\sqrt{R} = 0$, to determine the static working point of the device. This step involves a self-consistent evaluation of the strength of the various Josephson harmonics that have the role of pumps, analogous to the pump employed in conventional parametric amplifiers [28]. The rf dynamics are calculated as a perturbation series in η around the working point imposed by the static bias. For this purpose, one needs to consider all the mixing processes between the low-frequency input signal and the Josephson harmonics generated by the nonlinearity of the junctions biased in the running state, using the harmonic balance of Eqs. (12.49) and (12.50). As the bias current I is reduced toward I_0, the higher-order Josephson harmonics become increasingly important as their strength grows with increasing $\eta = I_0/I$. This approach enables us to write a scattering matrix of the SQUID in the C and D basis, which, through a simple transformation, bears a direct connection to the measured scattering parameters for the microstrip SQUID discussed in Section 12.6.2. Details of this treatment can be found in [29].

Figure 12.13 shows the power gain of the device evaluated as the ratio of power delivered at the output to that dissipated at the input [28], taking into account the dressing of the input and output impedances of the device by matched loads. The fit to the curve in Fig. 12.13(b) is of the form $1/f^2$, where f is the input signal frequency. We note that this frequency dependence is different from that of Eq. (12.38), as the

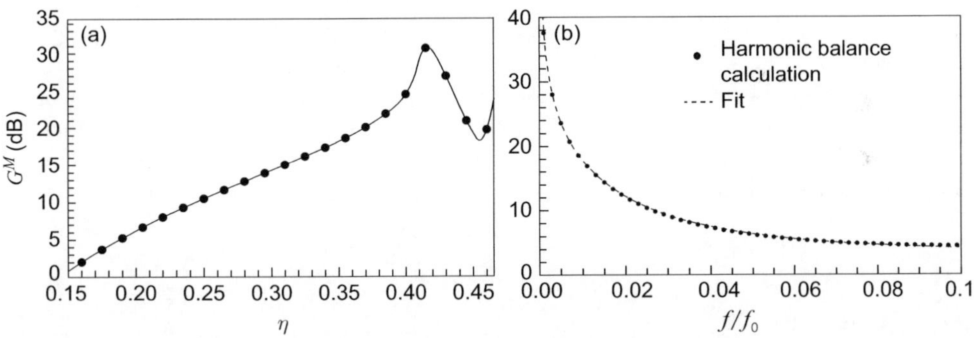

Fig. 12.13 Power gain of a dc SQUID calculated using the harmonic balance analysis including first ($2f_J$) and second ($3f_J$) Josephson harmonics. The parameters used for the calculation were $\Phi = \Phi_0/4$, $\beta_L \equiv 2LI_0/\Phi_0 = 1$, and $\beta_C' \equiv 2\pi IR^2C/\Phi_0 = 1$. (a) Power gain G^M versus normalized bias parameter $\eta = I_0/I$ calculated for a fixed input frequency $f = 0.01f_0$. Here, $f_0 = I_0R/\Phi_0$ is the Josephson plasma frequency for $\beta_C = 1$; for critical current densities readily achievable with Nb tunnel junctions, f_0 is of the order of 100 GHz. The solid curve is an interpolating polynomial of degree two. (b) Power gain G^M versus normalized input frequency f/f_0 calculated with bias parameter fixed at $\eta = 0.455$, the optimum value for attaining minimum noise temperature (see Fig. 12.14(a)) at low frequencies ($f \ll f_0$). The fit is of the form G^M(linear units) $= [0.006/(f/f_0)^2] + 2$.

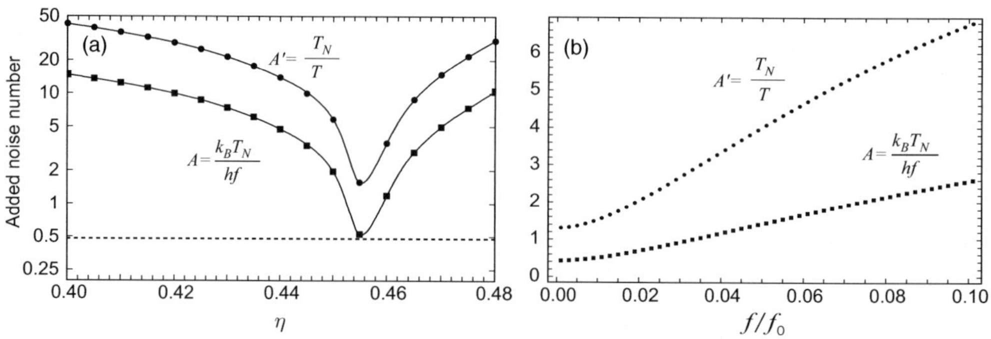

Fig. 12.14 Caves' added noise number for a dc SQUID, calculated using the harmonic balance analysis including the first ($2f_J$) and second ($3f_J$) Josephson harmonics, as a function of (a) $\eta = I_0/I$ for $f = 0.01f_0$ and (b) normalized input frequency f/f_0 with $\eta = 0.455$. In both plots, round markers show the noise number $A' = T_N/T$ obtained in the thermal regime $k_BT \gg hf$, while square markers show the noise number $A = k_BT_N/hf$ calculated in the quantum regime $k_BT \ll hf$. The solid curves represent interpolating polynomials. The quantum calculation gives a minimum value for $A = k_BT_N/hf \approx 0.5$, attained at $\eta = 0.455$, corresponding to one-half photon of added noise (shown by the dashed line). The optimal value of the bias current for minimum added noise ($I = 2.2\ I_0$) lies below the optimum bias current for maximum gain ($I = 2.4\ I_0$) (Fig. 12.13(a)), as noted previously in Section 12.4.2.

harmonic balance analysis, for simplicity, involves no input circuit to couple flux into the loop; rather, the signal is coupled directly through the differential mode current of the device.

Figure 12.14 shows plots of the noise number of the device as a function of bias current and input frequency, calculated by replacing the input current signal $I_{\mathrm{RF}}^{\mathrm{in}}$ carried by the transmission lines with a quantum noise current with a spectral density introduced earlier in Section 12.2.2,

$$S_I(f) = \frac{2hf}{R} \coth\left(\frac{hf}{2k_BT}\right). \tag{12.51}$$

The spectral density contribution of each mode is weighted by its respective scattering coefficient in the input channel to calculate the noise spectral densities S_V, S_J, and S_{VJ}, which are used in conjunction with Eq. (12.46) to calculate the added noise number. The result in Fig. 12.14(a) shows that, for the optimal bias condition, the dc SQUID can reach the quantum limit of amplification. The calculation yields an added noise number of one-half photon at sufficiently low input signal frequencies compared with the Josephson frequency, in accordance with the results of simulations of the quantum Langevin equation [26]. However, the added noise number increases with the input frequency as shown by both thermal and quantum calculations (Fig. 12.14(b)).

12.6 The microstrip SQUID amplifier

12.6.1 Principles, gain, and tuning

The original motivation to develop a SQUID amplifier with high gain and low noise at gigahertz frequencies was the need for such a device for the Axion Dark Matter Experiment (ADMX) [30]. The immediate challenge with the square washer SQUID—which has been so successful at lower frequencies—is the parasitic capacitance between the coil and the washer that rolls off the response at frequencies above typically 100 MHz. This problem was overcome by Michael Mück, who moved one wire so that the input signal, instead of being connected to the two ends of the input coil, was connected to one end of the input coil and the washer [31]. Thus, the signal is propagated along the microstrip formed by the coil inductance and its capacitance to the washer. When the length of the coil corresponds to $\lambda/2$, where λ is the signal wavelength on the microstrip, one expects to see a resonance that couples the signal strongly to the SQUID.

A microstrip consists of a superconducting strip of width w separated from an infinite superconducting sheet by an insulator with dielectric constant ϵ and thickness d. We assume that the thicknesses of the two superconductors are much greater than the superconducting penetration depth λ_s and that $w \gg d$. The capacitance and inductance per unit length of the microstrip are given by $C_s = \epsilon\epsilon_0 w/d$ ($\mathrm{F\,m^{-1}}$) and $L_s = (\mu_0 d/w)(1 + 2\lambda_s/d)$ ($\mathrm{H\,m^{-1}}$) [32]. Here, $\epsilon_0 = 8.85 \times 10^{-12}\,\mathrm{F\,m^{-1}}$ and $\mu_0 = 4\pi \times 10^{-7}\,\mathrm{H\,m^{-1}}$ are the permittivity and permeability of free space and $c = 1/(\epsilon_0\mu_0)^{1/2} = 3 \times 10^8\,\mathrm{m\,s^{-1}}$ is the velocity of light in vacuum. We have assumed the permeability $\mu = 1$. The factor $1+2\lambda_s/d$ accounts for the penetration of the magnetic field into the (identical) superconductors. The velocity of an electromagnetic wave on the microstrip is thus given by $\bar{c} = (L_s C_s)^{-1/2} = c/[\epsilon(1+2\lambda_s/d)]^{1/2}$ and its characteristic impedance by $Z_s = (L_s/C_s)^{1/2} = (d/w)[\mu_0(1+2\lambda_s/d)/\epsilon\epsilon_0]^{1/2}$. For a microstrip of length l with its two ends either open or terminated with resistances greater than Z_0, the fundamental frequency occurs when $l = \lambda/2$ [32]:

$$f_0(L_s) = \frac{c}{2l[\epsilon(1 + 2\lambda_s/d)]^{1/2}}. \tag{12.52}$$

The resonant frequency of the microstrip SQUID, however, is complicated by the inductive loading produced by the SQUID. Equation (12.28) implies that L_s should be replaced by n^2L/l, where we have assumed that $L \gg L_s$ and neglected L_j, resulting in the resonant frequency [33]

$$f_0(n^2L) = \frac{1}{2n(lLC_s)^{1/2}}. \tag{12.53}$$

Figure 12.15(a) shows the circuit of a microstrip SQUID amplifier (MSA). In the first experiments, a sweep oscillator was coupled to the MSA input via a room-temperature 100 dB attenuator and a cold 20 dB attenuator that prevented noise from the generator from saturating the SQUID. The cold attenuator also presented an impedance of 50 Ω to both the input coaxial line and the MSA. A second cold 4 dB attenuator coupled the output of the SQUID to a room-temperature postamplifier. The gain of the system excluding the MSA was calibrated by disconnecting the MSA and connecting together the input and output attenuators. All gain measurements were referred to the baseline thus obtained. Figure 12.15(b) shows gain versus frequency for four MSAs with progressively shorter coils. The peak gain was about 18 dB, and occurred at progressively higher frequencies as the coil length was reduced. These frequencies were given approximately by Eq. (12.53).

In many applications, it is desirable to tune the frequency at which the maximum gain occurs. Tuning is accomplished by connecting a GaAs varactor diode across the otherwise open end of the coil and the washer [34]. The capacitance of the diode can be varied by changing the value of the reverse bias voltage. Changing the capacitance modifies the phase shift of the electromagnetic wave when it is reflected, thereby increasing or decreasing the effective length of the microstrip and lowering or raising the peak frequency. In the experiments, the capacitance of two diodes in parallel (to increase the tuning range) could be varied from 1 to 10 pF by changing the bias

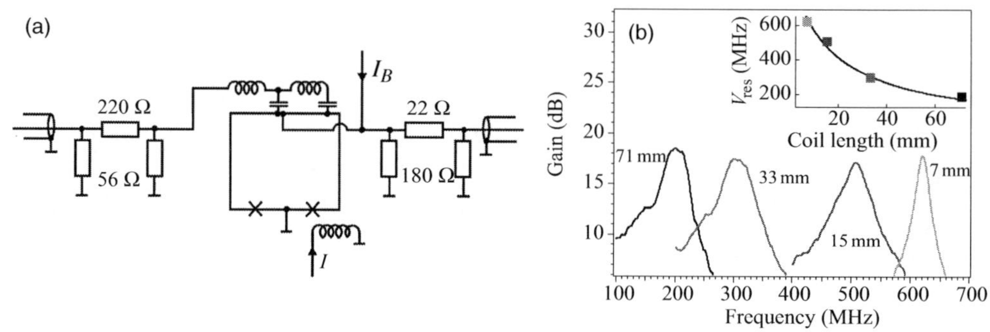

Fig. 12.15 Microstrip SQUID amplifier. (a) Circuit configuration; the input coil is shown as a distributed line. (b) Gain versus frequency for four coil lengths.

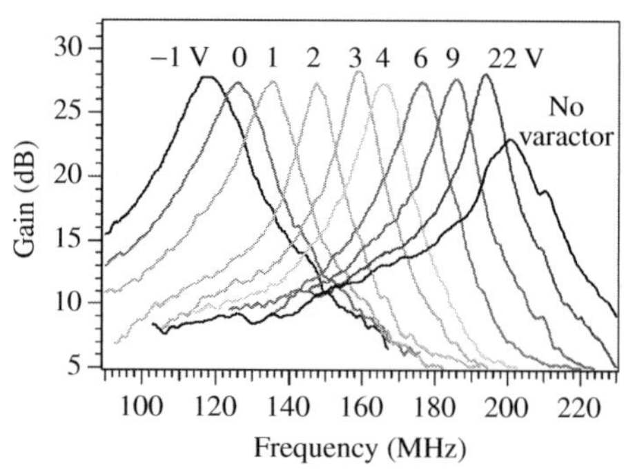

Fig. 12.16 Tuning the MSA. Gain versus frequency for a 31-turn SQUID at 4.2 K for nine values of reverse bias voltage applied to the varactor diodes connected between the open end of the coil and the washer.

voltage from 1 V to −22 V. The diodes were connected in series with a capacitor to avoid applying a static voltage to the microstrip. The gain for optimized current and flux biases for a SQUID with 31 turns is shown in Fig. 12.16 for nine values of the diode capacitance. We see that the peak frequency is progressively lowered, from 195 MHz to 117 MHz, as the capacitance is increased. The maximum gain is constant at about 28 dB over this range. In the absence of the varactor diodes, the peak frequency is about 200 MHz. The dependence of the peak frequency on the varactor capacitance is in reasonable agreement with a simple model [34]. The presence of the varactors increases the gain, most likely by increasing the degree of positive feedback from the output to the input.

12.6.2 Scattering parameters and input matching

To maximize the gain of the MSA, it is essential to know its input impedance, which is generally complex. A two-port network can be described by a scattering matrix that relates the voltage V^+ incident at one port with the voltage V^- reflected from a second port [35]. The scattering parameter is defined as $S_{ij} = V_i^-/V_j^+$, where S_{11} is the input reflection coefficient with the output port terminated by a matched load, and S_{21} is the forward gain. Figure 12.17(a) shows the configuration for a reflection measurement of S_{11} [36]. The vector network analyzer (VNA) and the various cables were calibrated by replacing the input of the MSA, in turn, with an open circuit, a short circuit, and a 50 Ω resistor.

Measurements were made on an MSA [36] with the following parameters. For the SQUID, $L = 450\,\text{pH}$, $I_0 \approx 2\,\mu\text{A}$, $C \approx 0.2\,\text{pF}$, $R \approx 20\,\Omega$, $\beta_L \approx 0.9$, and $\beta_C \approx 0.5$; for the coil, $n = 11$, $l \approx 15\,\text{mm}$, $w \approx 5\,\mu\text{m}$, $d \approx 400\,\text{nm}$, and $\epsilon \approx 5.5$. Figure 12.18

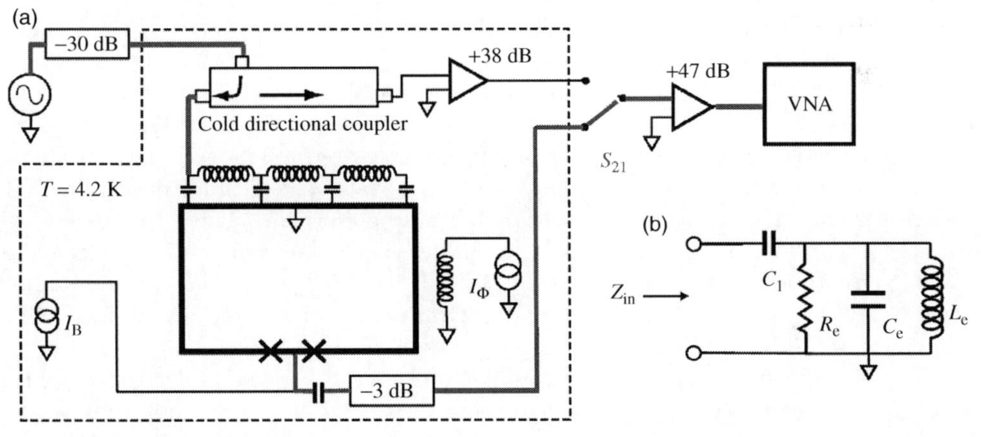

Fig. 12.17 (a) Experimental configuration for measurements of S_{11} and S_{21} at 4.2 K; the configuration for S_{11} is shown. The signal generator is incorporated in the VNA. The cold 3 dB attenuator reduces power re-reflected at the output of the MSA. (b) Equivalent circuit model of low-loss transmission line. The static capacitance is C_1, and the resonance circuit is represented by $R_eL_eC_e$.

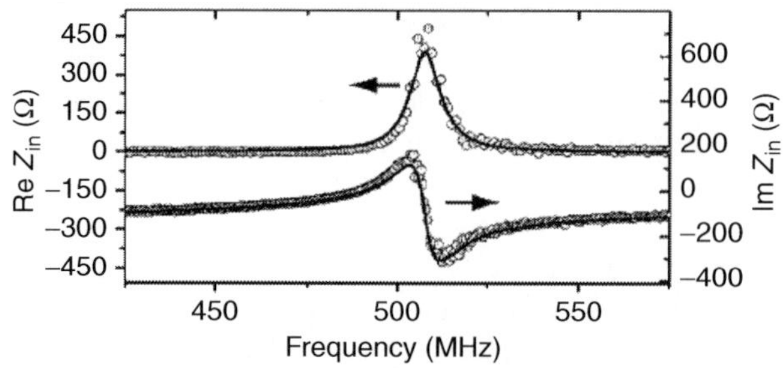

Fig. 12.18 Real and imaginary parts of the input impedance of the model shown in Fig. 12.17(b) measured at $\Phi_0/4$. Solid lines are fits to the measured S_{11} data converted to impedance (open circles).

shows S_{11}, converted to input impedance, versus measurement frequency with a flux bias $(n + 1/2)\Phi_0/2$. For a low-loss transmission line, these resonance curves can be described [35] by the input impedance

$$Z_{\text{in}} = Z_0 \frac{Z_L + Z_0 \tanh(\gamma l)}{Z_0 + Z_L \tanh(\gamma l)}. \tag{12.54}$$

Here, Z_0 is the characteristic impedance, Z_L is the terminating impedance, and $\gamma = \alpha + i\beta$ is the complex propagation constant. The data are an excellent fit to Eq. (12.54) with $Z_L = \infty$, $\alpha = 1.9\,\text{Np/m}$, $\beta = 209\,\text{m}^{-1}$, and the resonant frequency $f_r = 506.2\,\text{MHz}$. For the given microstrip geometry one finds $Z_0 \approx 16\ \Omega$, which is reasonably consistent with the measured value of $14\,\Omega$, and $\bar{c} \approx 0.33c$ [32], which predicts a resonant frequency of about 3 GHz. If, instead, we use Eq. (12.53), we find a resonant frequency of about 700 MHz, which is not too far removed from the measured value. Since the effect on f_r is much greater than on Z_0, we conclude that the loading is in the form of a lumped—rather than distributed—inductance. We note that changes in both the current and flux biases affect the S parameters.

With $Z_L = \infty$, the impedance can be equivalently described by the parallel $R_eL_eC_e$ circuit shown in Fig. 12.17(b); C_1 represents the static capacitance of the line. This model gives a direct measure of the circuit parameters [33]. Figure 12.19 shows the variations of f_r, V, Z_0, C_e, L_e, and R_e with bias flux. The variations are very mixed. The sign of V_Φ is positive for $\Phi < \Phi_0/2$ and negative for $\Phi > \Phi_0/2$. Both Z_0 and L_e are roughly asymmetric about $(n+1/2)\Phi_0$. On the other hand, f_r, C_e, and R_e exhibit no evident systematic behavior.

We determined the forward scattering parameter S_{21}—essentially the amplitude gain—by connecting the output signal from the SQUID to the room-temperature postamplifier and measuring it with the VNA (Fig. 12.17(a)). We calibrated the gain by replacing the MSA with a short circuit. The real and imaginary parts of S_{21} are shown in Fig. 12.20(a). From the circuit model in Fig. 12.20(b), we predict S_{21} to be $M_iV_\Phi i_L/V_i$, where M_i is the mutual inductance between the coil and the SQUID, i_L is the rf current in the microstrip, and V_i is the input voltage. This circuit model is as before, with the addition of a voltage source with an impedance R_s of $50\,\Omega$ and a

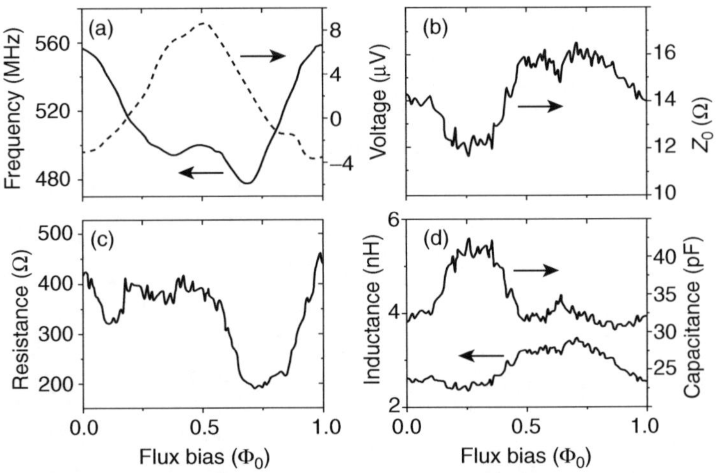

Fig. 12.19 Dependence of MSA parameters on flux bias: (a) resonant frequency f_r (solid) and static SQUID voltage V (dashed); (b) microstrip impedance Z_0; (c) resistance R_e; (d) capacitance C_e; inductance L_e.

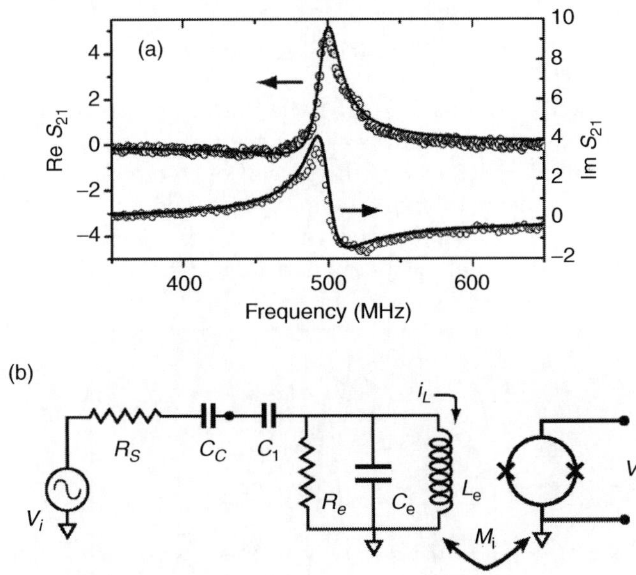

Fig. 12.20 Measurements of S_{21}. (a) Measured real and imaginary parts versus frequency (circles) at $(2n + 1/2)\Phi_0/2$. Solid curves are predictions of the equivalent circuit model shown in (b).

coupling capacitor C_c. The solid lines are predictions from the expression for S_{21} using the values $C_1 = 3.2\,\text{pF}$, $C_c = 222\,\text{pF}$, $L_e = 2.4\,\text{nH}$, $C_e = 40.9\,\text{pF}$, and $R_e = 400\,\Omega$ from the measurement of S_{11}, and fitting an overall scale factor corresponding to the transimpedance $M_i V_\Phi = 23.0\,\Omega$. The static values $V_\Phi = 40\,\mu V/\Phi_0$ and $M_i = 3.5\,\text{nH}$ yield a transimpedance of $68\,\Omega$, about a factor of 3 higher. This implies a reduction in M_i or V_Φ from its low-frequency value; possibly, both are reduced.

The central goal of these measurements is to determine the coupling of the input circuit for the MSA required to optimize the gain and noise temperature. The S_{11} results indicate that the intrinsic quality factor Q of the MSA at 4.2 K is typically 40–80. The Q values found from the S_{21} measurements with a source impedance of $50\,\Omega$, however, are typically 5–20, implying that the source impedance significantly damps the resonator. Decreasing the coupling between the source and the MSA would increase Q, but at the same time would reduce the signal coupled to the resonator. The problem of coupling a resonator to a real source impedance is solved using critical coupling, that is, matching the real impedance at resonance to the source impedance by means of a series capacitor. The circuit model resulting from the S_{11} measurement can be used to estimate the required series coupling capacitance [35] $C_c = (16\pi^2 f_r^3 R_s^2 R_e C_e)^{-1/2}$. For the circuit parameters of a different MSA with nine turns, $f_r = 0.816\,\text{GHz}$, $R_e = 714\,\Omega$, and $C_e = 4.4\,\text{pF}$, we find $C_c \approx 1.4\,\text{pF}$. The measured gain for three values of C_c is shown in Fig. 12.21. The maximum gain occurs for $C_c = 2.2\,\text{pF}$. The resonator is clearly overcoupled for $C_c = 10\,\text{pF}$ and undercoupled for $C_c = 0.5\,\text{pF}$. These results are in good agreement with the predictions from the equivalent circuit model.

These measurements demonstrate that the important properties of the MSA can be represented by a low-loss transmission line leading to an equivalent circuit model. By measuring the input impedance with the coil open-circuited, one can predict the maximum gain and frequency response. One can also design the input circuit to give maximum gain by critically coupling the source to the microstrip resonator. Needless to say, the model parameters are strongly dependent on the values of I_0, β_C, β_L, and

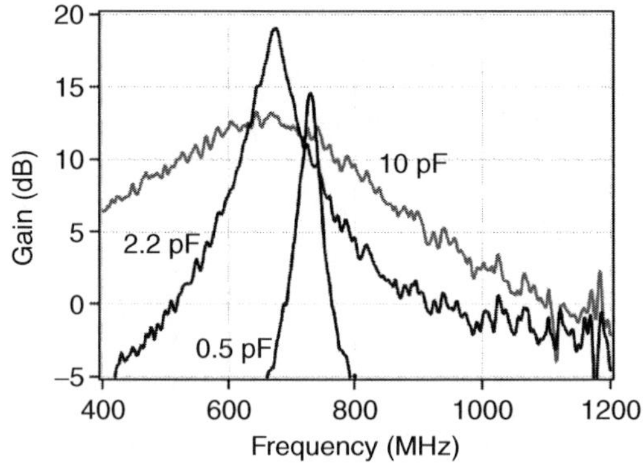

Fig. 12.21 Measured gain versus frequency for a nine-turn MSA for three different input coupling capacitors. The microstrip resonator is undercoupled (0.5 pF), approximately critically coupled (2.2 pF), and overcoupled (10 pF).

R; since I_0 may change with temperature, the S parameters should ultimately be measured at the desired operating temperature.

12.6.3 Noise temperature

Numerous measurements have been made of the noise temperature of the MSA. For brevity, we shall describe only the most recent measurements, carried out at millikelvin temperatures [37], with the circuit parameters based on the results of S-parameter measurements. The inner and outer dimensions of the Nb washer were 0.2 and 1 mm, corresponding to a loop inductance (including the slit) of about 400 pH. Each of the two Nb–AlO_x–Nb Josephson tunnel junctions, with dimensions of $2 \times 2\,\mu m^2$, had the following approximate parameters: $I_0 \approx 4\,\mu A$, $C \approx 0.2\,pF$, $R \approx 16\,\Omega$, $\beta_L \approx 1.6$, $\beta_C \approx 0.6$, and $V_\Phi \approx 100\,\mu V/\Phi_0$. The eight-turn coil, with a 5 µm width and 15 µm pitch, produced a resonance varying from 620 to 645 MHz, depending on the SQUID static current and flux biases. At low temperatures, dissipation in the resistive shunts typically raises the electron temperature to 120–150 mK, thereby increasing the Nyquist noise [38]. To reduce this temperature, we connected a $500 \times 500\,\mu m^2$ cooling fin [38], fabricated from 500 nm thick Au–Cu alloy, to a corner of each shunt resistor. Figure 12.22 shows a schematic diagram of the experiment. The MSA, together with its bias and coupling circuitry, was housed in a copper box, lead-plated on the inside, that shielded against radiofrequency interference and changes in ambient magnetic fields. The box, surrounded with a Cryoperm shield, was mounted on the mixing chamber of a dilution refrigerator, and cooled to temperatures ranging from 40 mK to 1 K. All bias lines were heavily filtered. A GaAs high-electron-mobility transistor (HEMT) postamplifier with a gain of 18 dB and a noise temperature T_p of about 1.4 K was installed in the helium bath. To reduce out-of-band noise from the HEMT

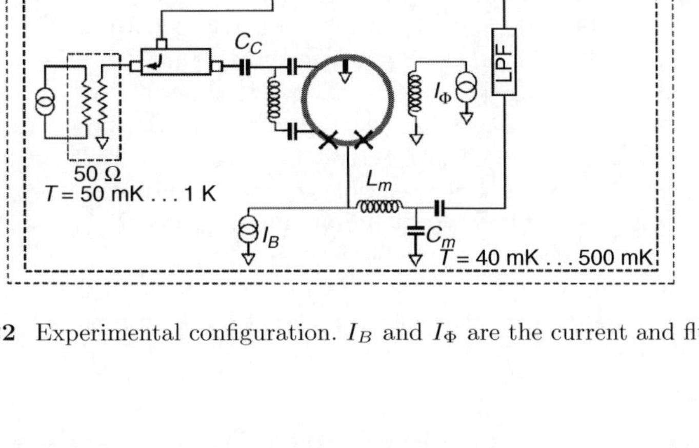

Fig. 12.22 Experimental configuration. I_B and I_Φ are the current and flux biases.

that could couple to the MSA, we inserted a coaxial low-pass filter, with a cutoff frequency of 1 GHz, between the MSA output and HEMT input. The MSA output was matched to the 50 Ω impedance of the filter with a lossless L match, consisting of a 12 pF off-chip capacitor (C_m in Fig. 12.22) connected across the series inductance (L_m, approximately 5 nH) of the on-chip output leads of the MSA. The gain of the MSA was optimized by means of a 2.2 pF capacitor (C_c) in series with the input resonator, critically coupling it to the 50 Ω source impedance. The gain of the MSA was determined by connecting a signal from a VNA to the input of the MSA via a cold directional coupler. The output power was amplified and measured with the VNA. A calibration run determined the loss and electrical length of the signal path with the MSA replaced with a short.

The value of T_N was measured over a wide frequency range using the hot-cold load technique. With the MSA input connected to a 50 Ω resistor, the output power consists of the noise power contributed by the amplifier $P_N = Gk_BT_NB$ and the amplified Nyquist noise power of the resistor $P = (1/2)GhfB\coth(hf/2k_BT) \approx Gk_BTB$. The hot-cold load technique involves two measurements of the output power, $P_1 = Gk_B(T_1 + T_N)B$ and $P_2 = Gk_B(T_2 + T_N)B$, with the 50 Ω source at different temperatures T_1 and T_2. The ratio $R \equiv P_2/P_1$ yields $T_N = (T_2 - RT_1)/(R - 1)$. The variable-temperature source consisted of a 50 Ω SMA termination, a wire-wound heater, and a RuOx thermometer embedded in a block of oxygen-free, high-conductivity copper. The temperature of the block was regulated with feedback. The optimal current and flux bias points were determined with automated scans.

Noise power spectra were acquired with the 50 Ω load resistor at 100 and 300 mK; typically, 5000 spectra were averaged together to produce the values of P_1 and P_2. At each frequency, the ratio P_2/P_1 was used to determine the system noise temperature T_S. Obtaining T_N from T_S requires two corrections. The first is to subtract T_p/G to correct for the HEMT noise, taking into account the 0.3 dB loss between the MSA and the HEMT. This reduction is about 12 mK at resonance. The second correction is for the measured insertion loss of the directional coupler and the cable loss between the 50 Ω resistor and the MSA, about 0.2 dB, corresponding to a 2 mK correction to T_N.

Figure 12.23(a) shows the measured gain and lowest measured noise temperature versus frequency at 45 mK. The minimum noise temperature was 48±5 mK, a factor of 1.6 above the value $T_Q = 29.4$ mK at 612 MHz, for a gain of 20.4 dB and a bandwidth of 21 MHz; the corresponding added noise number $A = (T_N/T_Q) - 1/2 = 1.1$. This value of T_N is a factor of about 30 lower than that of the best GaAs HEMT amplifiers. An interesting feature of Fig. 12.23(a) is the frequency dependence of T_N. The lowest value of T_N, 48±5 mK, occurs slightly below resonance at about 612 MHz, whereas the value on resonance is 66 ± 5 mK. This behavior was seen consistently in three separate runs of the experiment, and is in qualitative agreement with the predictions of Eq. (12.46).

Finally, the entire process was repeated as the temperature was progressively raised to 1000 mK. Figure 12.23(b) shows the minimum measured T_N versus T. We see that T_N scales linearly with decreasing temperature until saturating at a value above T_Q at $T \approx 100$ mK. In separate experiments at 100 kHz, it was found that the flux noise also flattened out at $T \approx 100$ mK, demonstrating that hot electrons limited the ultimate noise temperature [38].

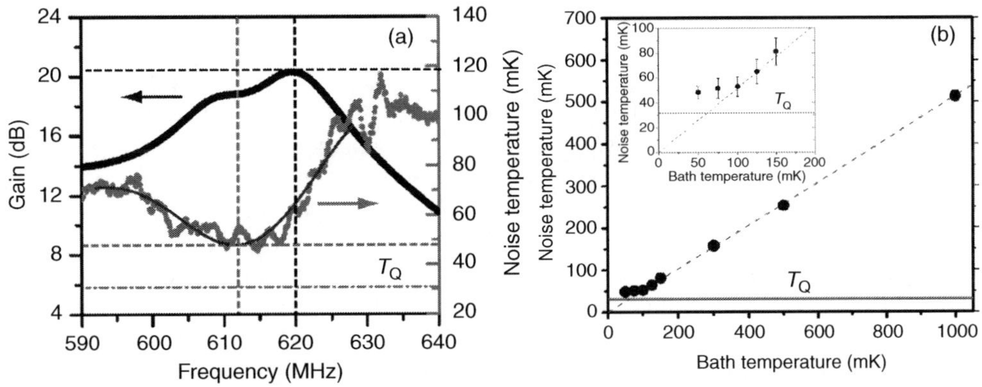

Fig. 12.23 (a) Gain and noise temperature of MSA at 45 mK. The narrow solid line is a fit to the noise temperature data. The dash-dot line represents $T_Q = hf/k_B$. (b) Minimum measured noise temperature versus bath temperature T. The dashed line is a fit showing the proportionality of the noise temperature to T. The solid line represents $T_Q = hf/k_B$. The inset shows the five lowest temperature points.

In concluding this section, we note that there are at least two other approaches to using the dc SQUID as an amplifier, both of which separate the role of the resonator from the structure of the SQUID [39, 40].

12.7 Concluding remarks

The dc SQUID configured as an amplifier has an essential role in a remarkably wide range of physics. Hundreds of thousands of SQUIDs are used to read out signals from transition edge sensors (TESs) on telescopes detecting the cosmic background radiation that started its journey some 14 billion years ago, just after the Big Bang. Optimally, these detectors are limited by photon shot noise. These far-infrared measurements are critically important to our understanding of the formation of the Universe. Similar sensors are used to search for weakly interacting massive particles (WIMPs). Energy-resolving TES calorimeters will be used in the upcoming International X-Ray Observatory and to analyze γ- and X-ray spectra in nuclear safeguards applications. The MSA is a critical component in ADMX, which will make a definitive search for the axion in the next few years. Without the near quantum-limited noise temperature of the MSA, ADMX would not exist. Yet another major application that has developed over the last few years is the use of near quantum-limited dc SQUID amplifiers to perform measurements on superconducting qubits (quantum bits).

Given this plethora of applications, one would like to have a quantitative understanding of the gain and noise temperature of SQUID amplifiers. In this chapter, we have attempted to lay out the basic physics of these devices, and to provide guidance for their design. At frequencies below (say) 100 MHz, these principles have been well understood for more than quarter of a century. Only in the last decade, however, with initial impetus from the stringent requirements of ADMX, has a thorough understanding of SQUID amplifiers in the microwave regime been realized. Although the essential idea of a quantum-limited SQUID amplifier—based on the quantum Langevin equation—goes back 30 years, the recent development of the scattering theory of circuits has deepened our understanding of the quantum limit. A crucial feature of both treatments is that the partial correlation of the voltage noise across the SQUID and the circulating noise current around the SQUID loop is essential for obtaining the quantum limit. Furthermore, the quantum-limited noise temperature must necessarily occur at a frequency below the frequency of maximum gain. The scattering-theory approach also makes clear that the quantum limit can be achieved only at signal frequencies far below the Josephson frequency.

We look forward to seeing new experimental approaches to dc SQUID amplifiers in the coming years—particularly at microwave frequencies—and hope that these notes will help to guide these developments.

Acknowledgments

One of us (J.C.) wishes to acknowledge the enormous contributions to the work described in this chapter by Marc-Olivier André, Jost Gail, Cristoph Heiden, Claude Hilbert, Mark Ketchen, Darin Kinion, Roger Koch, John Martinis, Michael Mück, Claudia Tesche, Cristian Urbina, Dale van Harlingen, and Fred Wellstood. The writing of this chapter was funded by the US Department of Energy, Office of High Energy Physics, under contract number DE-FG02-11ER41765 (J.C.); by the NSF under grant number DMR-1006060 (MD and AK); and by the Office of the Director of National Intelligence (ODNI), Intelligence Advanced Research Projects Activity (IARPA), through the Army Research Office. All statements of fact, opinion, or conclusions contained herein are those of the authors and should not be construed as representing the official views or policies of IARPA, the ODNI, or the US Government (J.C., M.D., and A.K.).

References

[1] London, F. (1950). *Superfluids*. Wiley, New York.
[2] Josephson, B. D. (1962). *Phys. Lett.*, **1**, 251–253.
[3] Josephson, B. D. (1965). *Adv. Phys.*, **14**, 419–451.
[4] Jaklevic, R. C., Lambe, J., Silver, A. H., and Mercereau, J. E. (1964). *Phys. Rev. Lett.*, **12**, 159–160.
[5] Zimmerman, J. E., Thiene, P., and Harding, J. T. (1970). *J. Appl. Phys.*, **41**, 1572–1580.
[6] Mercereau, J. E. (1970). *Rev. Phys. Appl.*, **5**, 13–20.
[7] Nisenoff, M. (1970). *Rev. Phys. Appl.*, **5**, 21–24.
[8] Koelle, D., Kleiner, R., Ludwig, F., Dantsker, E., and Clarke, J. (1999). *Rev. Mod. Phys.*, **71**, 631–686.
[9] Clarke, J. and Braginski, A. I. (eds.) (2004). *The SQUID Handbook.* Volume I: *Fundamentals and Technology of SQUIDs and SQUID Systems.* Wiley-VCH, Weinheim.

[10] Clarke, J. and Braginski, A. I. (eds.) (2006). *The SQUID Handbook.* Volume II: *Applications of SQUIDs and SQUID Systems.* Wiley-VCH, Weinheim.
[11] Stewart, W. C. (1968). *Appl. Phys. Lett.*, **39**, 277–280.
[12] McCumber, D. E. (1968). *J. Appl. Phys.*, **39**, 3113–3118.
[13] Ambegaokar, V. and Halperin, B. I. (1969). *Phys. Rev. Lett.*, **22**, 1364–1366.
[14] Likharev, K. K. and Semenov, V. K. (1972). *Pisḿa Zh. Eksp. Teor. Fiz.*, **15**, 625–629. [*JETP Lett.*, **15**, 442–445].
[15] Vystavkin, A. N., Gubankov, V. N., Kuzmin, L. S., Likharev, K. K., Migulin, V. V., and Semenov, V. K. (1974). *Phys. Rev. Appl.*, **9**, 79–109.
[16] Koch, R. H., Van Harlingen, D. J., and Clarke, J. (1980). *Phys. Rev. Lett.*, **45**, 2132–2135.
[17] Koch, R. H., Van Harlingen, D. J., and Clarke, J. (1981). *Phys. Rev. Lett.*, **47**, 1216–1219.
[18] Tesche, C. D. and Clarke, J. (1977). *J. Low Temp. Phys.*, **27**, 301–331.
[19] Tesche, C. D. and Clarke, J. (1979). *J. Low Temp. Phys.*, **37**, 397–403.
[20] Ketchen, M. B. and Jaycox, J. M. (1982). *Appl. Phys. Lett.*, **40**, 736–738.
[21] Rowell, J. M., Gurvitch, M., and Geerk, J. (1981). *Phys. Rev. B*, **24**, 2278–2281.
[22] Martinis, J. M. and Clarke, J. (1985). *J. Low Temp. Phys.*, **61**, 227–236.
[23] Hilbert, C. and Clarke, J. (1985). *J. Low Temp. Phys.*, **61**, 237–262.
[24] Hilbert, C. and Clarke, J. (1985). *J. Low Temp. Phys.*, **61**, 263–280.
[25] Caves, C. M. (1982). *Phys. Rev. D*, **26**, 1817–1839.
[26] Koch, R. H., Van Harlingen, D. J., and Clarke, J. (1980). *Appl. Phys. Lett.*, **38**, 380–382.
[27] Yurke, B. (2004). In *Quantum Squeezing* (ed. P. D. Drummond and Z. Ficek), pp. 53–98. Springer-Verlag, Berlin.
[28] Clerk, A. A., Devoret, M. H., Girvin, S. M., Marquardt, F., and Schoelkopf, R. J. (2010). *Rev. Mod. Phys.*, **82**, 1155–1208.
[29] Kamal, A., Clarke, J., and Devoret, M. H. (2012). *Phys. Rev. B*, **86**(14), 144510.
[30] Bradley, R., Clarke, J., Kinion, D., Rosenberg, L. J., van Bibber, K., Matsuki, S., Mück, M., and Sikivie, P. (2003). *Rev. Mod. Phys.*, **75**, 777–817.
[31] Mück, M., André, M.-O., Clarke, J., Gail, J., and Heiden, C. (1998). *Appl. Phys. Lett.*, **72**, 2885–2887.
[32] Ramo, S., Whinnery, J. R., and van Duzer, T. (1965). *Fields and Waves in Communication Electronics.* Wiley, New York.
[33] Mück, M. and Clarke, J. (2000). *J. Appl. Phys.*, **88**, 6910–6918.
[34] Mück, M., André, M.-O., Clarke, J., Gail, J., and Heiden, C. (1999). *Appl. Phys. Lett.*, **75**, 3545–3547.
[35] Collin, R. E. (2001). *Foundations for Microwave Engineering.* IEEE, New York.
[36] Kinion, D. and Clarke, J. (2008). *Appl. Phys. Lett.*, **92**, 172503.
[37] Kinion, D. and Clarke, J. (2011). *Appl. Phys. Lett.*, **98**, 202503.
[38] Wellstood, F. C., Urbina, C., and Clarke, J. (1994). *Phys. Rev. B*, **49**, 5942–5955.
[39] Spietz, L., Irwin, K., and Aumentado, J. (2009). *Appl. Phys. Lett.*, **95**, 092505.
[40] Ribeill, G. J., Hover, D., Chen, Y.-F., Zhu, S., and McDermott, R. (2011). *J. Appl. Phys.*, **110**, 103901.

13

Quantum information science: experimental implementation with trapped ions

T. MONZ*, P. SCHINDLER*, D. NIGG*, and R. BLATT*, †

*Institut für Experimentalphysik, Universität Innsbruck
Technikerstraße 25, 6020 Innsbruck, Austria
†Institut für Quantenoptik und Quanteninformation der Österreichischen Akademie der Wissenschaften, Technikerstraße 21a, 6020 Innsbruck, Austria

Quantum Machines: Measurement and Control of Engineered Quantum Systems.
Edited by M. Devoret, B. Huard, R. Schoelkopf, and L. F. Cugliandolo.

Chapter Contents

This chapter serve as an extension to previous Les Houches lecture notes on ion-trap-based quantum computation [1, 2].

13.1 Introduction

Classical computers are fast, reliable, and cheap, and can perform an amazing number of tasks. But they also have limitations and face challenges: more and more data needs to be handled, smaller and smaller chips struggle with increasing heat production, and the smaller electronics become more fragile. How would it be to store information on the smallest unit we currently feel confident to ever control reliably: a single atom? Every book ever written could be stored on the tip of a fingernail; shuttling around single atoms or electrons ought to produce little wasted energy and heat. But—is this feasible? Which tools and operations are mandatory? According to DiVincenzo [3], one has to be able to initialize a quantum register, implement a universal set of gate operations on it, and read it out—all this with high fidelity and sufficient speed such that decoherence processes do not destroy the information. Once such a set is successfully implemented in a system, it can be used to implement every conceivable quantum algorithm (while the actual details of the set of gates might differ a lot from setup to setup). A universal set of quantum operations consists of arbitrary single-qubit operations and an entangling two-qubit operation, typically a controlled-NOT (CNOT) gate. In the following, quantum information processing with trapped ions will be discussed. Step by step, we will follow DiVincenzo's criteria, from storing ions, initializing them, performing single-qubit operations on an ion, to finally explaining how a CNOT gate can be implemented between two ions in a linear Paul trap [4]. With these tools at hand, quantum states, their analysis, and basic quantum algorithms will be discussed. In the final part of this chapter, advanced topics such as quantum error correction will be mentioned. This overview will be concluded with a discussion of how ion traps, currently one of the most promising systems for quantum information processing, may be scaled up to operate with hundreds of quantum bits.

13.2 Ion-trap quantum computers: the concept

From a theoretical point of view, a qubit can be associated with a two-level system. It follows that any physical system that can be well described as a two-level system might be a suitable candidate as a qubit. In the following, we will focus on a single $^{40}Ca^+$ ion, trapped in a linear Paul trap [4]. ^{40}Ca can be found in the second column of the periodic table, meaning that it has two valence electrons. When it is singly ionized, it features only a single valence electron and thus its level scheme will be similar to that of a hydrogen atom (see Fig. 13.1(a)). In addition, ^{40}Ca has no nuclear spin, offering a particularly simple level structure without hyperfine splitting. Photoionization of ^{40}Ca is straightforward with off-the-shelf diode lasers at 422 nm and 375 nm [5]. While there are numerous different quantum states in a $^{40}Ca^+$ ion, its level scheme can be further simplified to a V-shaped system consisting of the ground state $S_{1/2}$ coupled to $P_{1/2}$ via the $S_{1/2} \leftrightarrow P_{1/2}$ dipole transition at a wavelength of 397 nm and a metastable

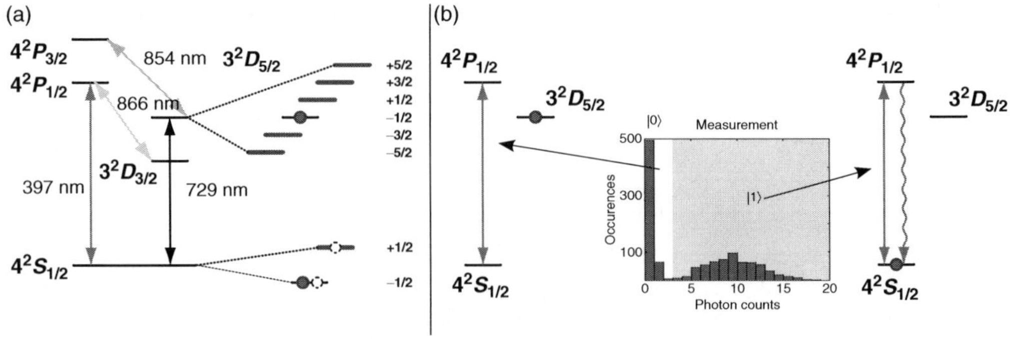

Fig. 13.1 $^{40}Ca^+$ for ion-trap quantum computation. (a) A simple level structure with all transition wavelengths available as lasers off-the-shelf facilitate the experimental setup. (b) Electron shelving [12] on the $S \leftrightarrow P$ transition yields a discrimination between $|S\rangle$ and $|D\rangle$ better than 99%.

$D_{5/2}$ state, accessible via the $S_{1/2} \leftrightarrow D_{5/2}$ quadrupole transition at a wavelength of 729 nm. Note that the lifetime of the D state is about 1 s [6, 7] (corresponding to a transition linewidth of about 1 Hz), which is considerably longer than all interactions described in the following. The qubit is encoded in $S_{1/2}$ and $D_{5/2}$, in particular the Zeeman levels $S_{1/2}(m = -1/2) = |1\rangle$ and $D_{5/2}(m = -1/2) = |0\rangle$. Using σ^- polarized light at the $S_{1/2} \leftrightarrow P_{1/2}$ transition, the qubit is initialized in $S_{1/2}(m = -1/2) = |S\rangle$ using an optical pumping process. The qubit is manipulated with laser light, resonant with the $S_{1/2}(m = -1/2) \leftrightarrow D_{5/2}(m = -1/2)$ transition. This laser directly affects the quality of the performed operations and has a linewidth of about 5 Hz [8]. The ions are stored in a linear Paul trap, which is characterized by trap frequencies of 1 MHz axially and 4 MHz radially. These axial and radial modes serve to describe the quantized motion of the ions and show up as sidebands in the excitation spectrum of a single ion, as presented in Fig. 13.2. These trap frequencies result in a ion localized to about 11 nm when it is cooled to the ground state. Cooling the motion of the ions close to their ground state is mandatory to perform high-fidelity quantum operations. We employ Doppler cooling [9, 10] to about 15 phonons in the axial sideband, followed by subsequent sideband cooling [11] on the axial sideband to provide the required cooling. Here, the sidebands can be spectroscopically resolved and laser light resonantly couples $|S\rangle \otimes |n\rangle \leftrightarrow |D\rangle \otimes |n-1\rangle$, with $|n\rangle$ representing the excitation of the harmonic oscillator used to describe the motion of the trapped ion. The lifetime of the $D_{5/2}$ is too long to achieve an acceptable cooling rate, and therefore the lifetime is artificially reduced via the $P_{3/2}$ level by applying quench light at 854 nm. With this procedure, the motional ground state is reached in roughly 8 ms and at a fidelity of better than 99%. State discrimination of the qubit between $|S\rangle$ and $|D\rangle$ is performed by scattering photons on the $S_{1/2} \leftrightarrow P_{1/2}$ transition at 397 nm: If the qubit is in state $|S\rangle$, photons are scattered, and detected by a photomultiplier tube (PMT) or a CCD camera. Should the qubit be in state $|D\rangle$, the ion remains dark. This state-discrimination technique is also referred to as electron shelving [12] (see also Fig. 13.1(b)). The CCD

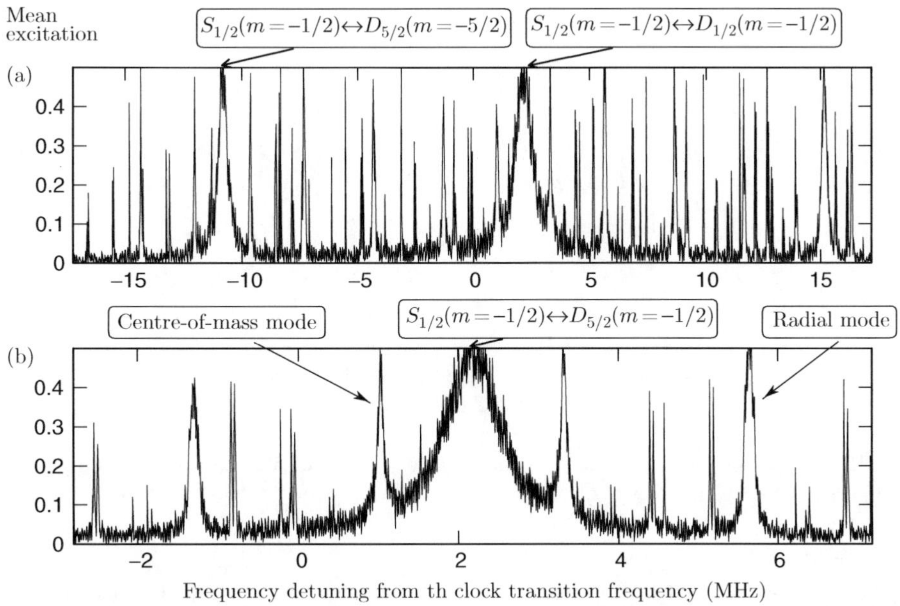

Fig. 13.2 Excitation spectrum of a $^{40}Ca^+$ ion in a linear Paul trap when illuminated with 729 nm light close to the quadrupole transition. (a) The applied magnetic field of about 3 G results in a Zeeman splitting. Here, transitions are accompanied by sideband transitions corresponding to the motion of the ion within the trap. (b) A close-up look at the transition $|S\rangle \leftrightarrow |D\rangle$ shows the axial mode at about 1 MHz. Other modes such as the radial modes at about 4 MHz are clearly recognized.

camera has the advantage of spatial resolution, which allows us to read out individual qubits, while the PMT can only provide information on how many qubits were in state $|S\rangle$ but not which one. The PMT features a higher detection efficiency and can provide time-resolved information about the scattered photons which can be used for achieving higher detection fidelities [13, 14]. Considering that this measurement process provides classical information about the excitation of the ion, whether it was bright (for the $|S\rangle$ or $|1\rangle$ state) or dark (for the $|D\rangle$ or $|0\rangle$ state), its behaviour can be interpreted as a measurement of the σ_z Pauli operator with the two projectors $|1\rangle\langle 1|$ and $|0\rangle\langle 0|$.

13.2.1 Coherent manipulation of ion-qubits

The unitary operation acting on the qubit can be described by

$$U = \exp\left(-i\frac{\theta}{2}\sigma_\phi\right) = \cos\left(\frac{\theta}{2}\right)\mathbb{1} + i\sin\left(\frac{\theta}{2}\right)[\sigma^+\exp(i\phi) + \sigma^-\exp(-i\phi)], \quad (13.1)$$

with $\sigma_\phi = \cos(\phi)\sigma_x + \sin(\phi)\sigma_y$, $\sigma_{(x,y)}$ being the corresponding Pauli operators, and $\theta = \Omega t$, with Ω the Rabi frequency and t the time for which the laser illuminates the

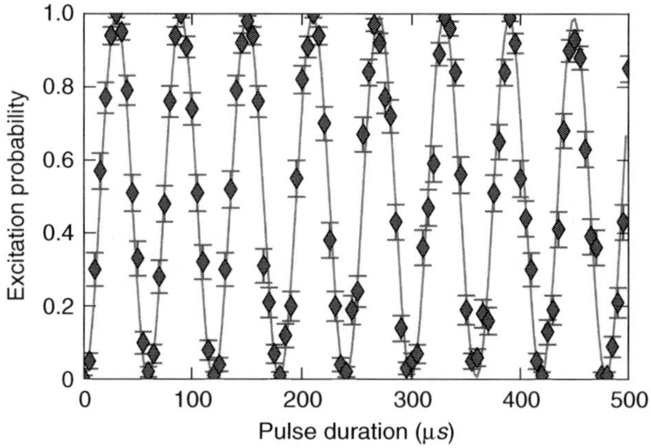

Fig. 13.3 Rabi oscillations of a single qubit: after a time of about 20 μs, the qubit is flipped from $|0\rangle$ to $|1\rangle$ and vice versa.

ion. The excitation of a qubit over time when subject to resonant excitation is shown in Fig. 13.3, depicting the measured Rabi oscillations. Using resonant interaction, any single-qubit operation can be implemented by choosing interaction time and phase. This opens up a wide range of experiments that can be implemented with only a single ion. In particular, Ramsey experiments [15] of various sorts allow the investigation of magnetic field fluctuations in the vicinity of the ion, and of laser frequency fluctuations, as well as the use of the ion as a reference oscillator for frequency standards [16]. In a similar way, composite pulse sequences such as "spin echo" pulses or dynamical decoupling [17, 18] can be used to render a qubit insensitive to a wide range of noise sources. In particular, coherence times of up to 100 ms [19] have been achieved in the presented setup (see Fig. 13.4), notably longer than the timescale of local operations, which is of the order of tens of microseconds.

For arbitrary quantum computation, in addition to full control of the individual qubits, an interaction between at least two qubits (usually resulting in a maximally entangled state) is needed. For now, let us focus on two qubits. Loading additional ions into an ion trap is straight forward. Given that the axial trap frequency is smaller than both radial frequencies, the ions will arrange themselves in a linear string. Owing to the Coulomb force, the ions find a stable equilibrium position with an inter-ion distance of about 3–4 μm. This is larger than the wavelength of the qubit transition at 729 nm, and therefore illuminating a single qubit with a well-focused beam is possible. The remaining question for single-qubit manipulation within a long ion string is how to subsequently address different ions on timescales of tens of microseconds. One possibility is to change the relative voltage on the tips of the Paul trap and thereby move ions of the string in and out of the focus of a stationary beam. The challenge here is to move ions rapidly without exciting the motional state of the string, which has been demonstrated recently [20, 21]. A notably simpler approach follows the idea of deflecting the beam onto a specific ion. This is performed with an electro-optical

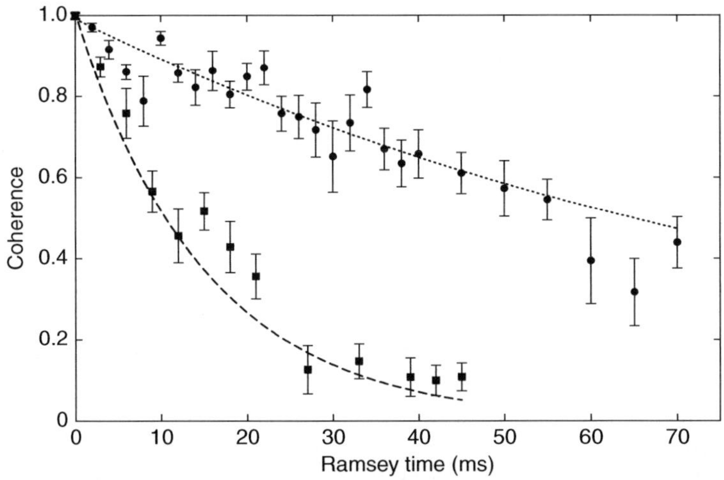

Fig. 13.4 Coherence versus time for a single ion qubit: Ramsey experiments on the $S(m = -1/2) \leftrightarrow D(m = -1/2)$ transition (circles) show an exponential decay with a time constant of about 100 ms. A similar investigation of the $S_{m=-1/2} \leftrightarrow D_{m=-5/2}$ transition (squares), which is about a factor 5 more sensitive to magnetic field fluctuations, yields a time constant of about 15 ms.

deflector (EOD) that acts like a prism made of a material with a voltage-dependent index of diffraction. Thus, the deflection angle changes with the voltage applied to the material. The main limitations here are the angle that can be covered with the EOD and the time it takes to apply the required voltage to the EOD. A third possibility to change the beam position with respect to the individual ions employing a micromechanical mirror system [22].

The next step toward quantum computation is to entangle two ions within a trap. However, employing laser light resonant at the $|S\rangle \leftrightarrow |D\rangle$ transition, on both ions simultaneously, can not entangle them. A wide beam illuminating the entire ion string induces collective operations on the entire string of the form $\exp[-i(\theta/2)S_\phi]$, with $S_\phi = \sum_k \sigma_\phi^{\{k\}}$ and $\sigma_\phi^{\{k\}}$ being the corresponding operation σ_ϕ on ion k. So how can we make the ions "talk" to each other? The motion of the individual qubits is strongly coupled via the Coulomb repulsion between the ions. Multiple ions in the trap can be treated as coupled harmonic oscillators, and thus, their motion can be described by normal modes similar to classical harmonic oscillator. Using the common mode where all ions oscillate in unison allows cooling an entire string of ions to the motional ground state of the axial mode by applying the laser light to a single ion [23]. In a similar way, the quantized motion of the ion string can be exploited as a data bus. Taking into account the motional state of the ion string, individual illumination of an ion at a frequency $\nu_l + \nu_{\mathrm{ax}}$, with ν_l the transition frequency and ν_{ax} the trap frequency, results in an interaction of the form

$$\cos\left(\frac{\theta}{2}\right)\, \mathbb{1} - i\sin\left(\frac{\theta}{2}\right)\, [\sigma^+ a^\dagger \exp(i\phi) + \sigma^- a \exp(-i\phi)],$$

where $\theta = \Omega_{\text{sb}} t$ is the rotation angle, $a^\dagger(a)$ is the raising (lowering) operator of the harmonic oscillator describing the quantized motion of the ion, Ω_{sb} is the Rabi frequency, and t is the illumination time. This transition is called the blue sideband transition and its Rabi frequency is $\Omega_{\text{sb}} = \sqrt{n+1}\ \eta\ \Omega_{\text{car}}$ (in the first-order approximation of the Lamb-Dicke regime [24]), with Ω_{car} the Rabi frequency on the carrier and the Lamb-Dicke parameter given by [25]

$$\eta = \sqrt{\frac{\hbar}{2\mu\nu_{\text{ax}}}}\mathbf{k}\cdot\mathbf{e}_{\text{ax}},$$

where m is the mass of the ion, ν_{ax} is the trap frequency for the chosen mode along the direction $\mathbf{e}_{\text{ax}}$ relative to the wavevector of the incident laser beam defined by $\mathbf{k}$. For the geometry and trap parameters in our laboratory in Innsbruck, the Lamb-Dicke parameter is usually about 3%.

In the following, a simple example will illustrate how interactions on the blue sideband can be employed to create entangled states and mediate state-dependent interactions along the ion string. Consider a register of two ions in the electronic and motional ground state $|SS, n=0\rangle$. A $\pi/2$ pulse on the blue sideband transition on qubit 1 (counted from the right, $|SS\rangle \equiv |S\rangle_2 |S\rangle_1$, with the indexes indicating the ion label), will result in the state $\frac{1}{\sqrt{2}}(|SS, n=0\rangle + |SD, n=1\rangle)$. A subsequent carrier π operation on qubit 2 maps into: $\frac{1}{\sqrt{2}}(|DS, n=0\rangle + |DD, n=1\rangle)$. A blue sideband operation does not couple to the state $|D, n=0\rangle$, and therefore a final π pulse on the blue sideband applied on ion 2 maps the quantum state onto $\frac{1}{\sqrt{2}}|DS + SD, n=0\rangle$, a maximally entangled Bell state. This state, as well as the locally equivalent other three Bell states, have been investigated by Roos *et al.* [26], with the density matrices describing the state depicted in Fig. 13.5. This technique can be generalized for long ion chains to efficiently create W states. To date, this technique has been employed to create genuine multipartite entanglement between up to eight ion-qubits [27].

This scheme for creating W states, however, has one drawback: it only works for the initial state $|S\ldots S, n=0\rangle$. If the first qubit, which generates the motional superposition, starts in state $|D\rangle$, the procedure fails. In that sense, the presented recipe works for state preparation, but does not correspond to a unitary operation that can form a universal set of operations. However, it is possible to achieve an entangling operation exactly with a few adaptations. Consider, again, two ion qubits, initialized in state $|SS, 0\rangle$. On ion 2, we implement two subsequent $\pi/2$ pulses of opposite phases, which effectively corresponds to an identity operation. The idea now is to induce a π phase shift on the second qubit between the two Ramsey pulses, if and only if the first qubit was in state $|S\rangle$. If that phase shift is applied, the second Ramsey pulse will effectively map the entire sequence into a bit flip on ion 2. If the phase shift is not applied, the identity operation is applied on ion 2. This protocol is equivalent to a controlled bit flip (or CNOT), with qubit 1 as control and qubit 2 as target. The main challenge is the state-dependent phase shift, which can be implemented using a scheme suggested by Cirac and Zoller [28] in which a π pulse on the blue sideband of the control qubit will create a phonon if the qubit has been in state $|S\rangle$. With the motion shared between the qubits, an effective 2π pulse on the blue sideband of the

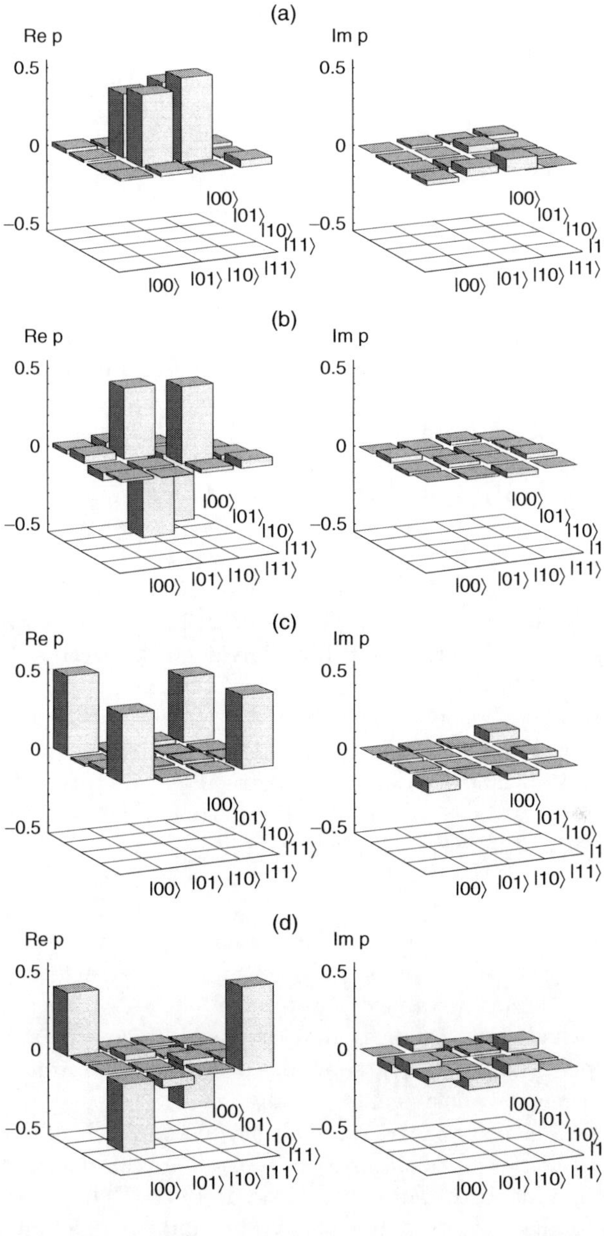

Fig. 13.5 Real and imaginary parts of the density matrices of the four Bell states: (a) $|\psi^+\rangle$; (b) $|\psi^-\rangle$; (c) $|\phi^+\rangle$; (d) $|\phi^-\rangle$. The fidelities achieved have been 91%, 90%, 91%, and 88%, respectively.

target qubit will result in a controlled phase gate. The coupling strength on the blue sideband depends on the phonon number, and thus, a 2π pulse on the blue sideband in the presence of zero phonons is not equivalent to the operation in the presence of one phonon. This difference in coupling strengths, however, can be exploited to perform the desired conditional phase shift with the aid of composite pulse sequences as described in [29, 30]. Finally, given that the controlled phase gate is enclosed by Ramsey pulses, the unitary operation is now equivalent to the desired controlled bit-flip operation.

13.2.2 State and process tomography

The brief introduction into the generation of entangled quantum states and implementation of arbitrary quantum computation raises the question how the generation of a particular state can be validated. Ideally, one would like to obtain information about the entire output state or, alternatively, the process that has been realised. Both can be done, via quantum state and process tomography [31], as will be explained in the following: Consider the state space of a single qubit. Here, the identity operation together with the three Pauli operators span an orthonormal basis. As such, any quantum state can be decomposed into that basis: $\rho = \sum_i \mathrm{tr}(\rho\, O_i)O_i$, with O_i the elements of the operator basis. As the identity operator has a trivial expectation value, any single-qubit density matrix is completely described by the expectation values of the three Pauli operators. Interpreting this equivalence between a density matrix and its set of expectation values for a chosen operator basis, state tomography boils down to measuring these expectation values. However, electron shelving on the qubit provides only information about σ_z. How is it possible to obtain information about the expectation values of σ_x and σ_y? Using the Heisenberg picture, adding an operation after the desired state has been prepared, but before it is measured, can be interpreted as a measurement in a different basis; for example, applying $\exp[i(\pi/4)\sigma_y]$ before the measurement in the basis $\{|0\rangle\langle.|\,,|1\rangle\langle.|\}$ can also be interpreted as a measurement in basis $\{|+\rangle\langle.|\,,|-\rangle\langle.|\}$. From this, a direct procedure to implement state tomography for a single qubit can be constructed. The remaining issue is of statistical nature: the above introduction to state tomography and its reconstruction relies on obtaining expectation values—the mean value of an infinite number of measurements. In a real experiment, the measurement time is limited, resulting in a statistical uncertainty in the derived expectation value. The acquired data may result in unphysical density matrixes with negative eigenvalues. Using constrained reconstructions that are restricted to physical density matrices, such as maximum-likelihood reconstruction, these statistical errors can be taken into account. The final question now is: how can this procedure be extended to more than a single qubit? The number of parameters defining a density matrix is $4^N - 1$, with N the number of qubits. Obtaining data on all possible observables for $\{\sigma_x, \sigma_y, \sigma_z\}^{\otimes N}$ would only yield 3^N data points and therefore be incomplete with respect to the $4^N - 1$ parameters. However, using the CCD camera, not only the expectation value of each Pauli is obtained—in fact, all 2^N projectors of the Pauli observable can be distinguished. As such, the 3^N Pauli settings for state tomography yield 6^N outcomes, which is overcomplete with respect to the $4^N - 1$ parameters. The main drawback here is the exponentially increasing number of measurement settings, as well as the numerically demanding reconstruction. The big advantage, on the other hand, is that one gains knowledge about the entire quantum state. Therefore, every question about the quantum state can be investigated and, in principle, answered, such as the fidelity achieved with respect to a desired quantum state, or whether the quantum state shows genuine multipartite entanglement. Using full state tomography and an extension of the scheme presented above to create W states on N qubits of the form $W_N = N^{-1/2}(|0\ldots01\rangle + |0\ldots010\rangle + \ldots + |10\ldots0\rangle)$, we have been able to perform full state tomography of up to eight qubits (as shown in Fig. 13.6), with fidelities for $3, 4, 5, 6, 7$, and 8 qubits of 82%, 85(1)%, 76(1)%, 79(1)%, 76.3(3)%, 72.2(1)%, respectively, and proof of genuine multipartite entanglement in all created states using an entanglement witness [27].

Quantum state tomography allow the characterization of the final state of a quantum algorithm, but is it also possible to analyze the algorithm itself? Any process acting on a quantum system can be described using the operator-sum representation. Here, a processes $\mathcal{E}$ is entirely described by the χ matrix with respect to a chosen operator basis A_i, usually the operators $\{\mathbb{1}, \sigma_x, \sigma_y, \sigma_z\}^{\otimes N}$. For an input state ρ_{in} and a process $\mathcal{E}$, the output state ρ_{out} is then given by

$$\rho_{\text{out}} = \mathcal{E}(\rho_{\text{in}}) = \sum_{i,j} \chi_{i,j} A_i \rho_{in} A_j^\dagger .$$

But how can this χ matrix be determined from an experiment? Considering that the χ matrix includes information about how any arbitrary input state is mapped onto the output state space, it is obvious that the entire input state space needs to be

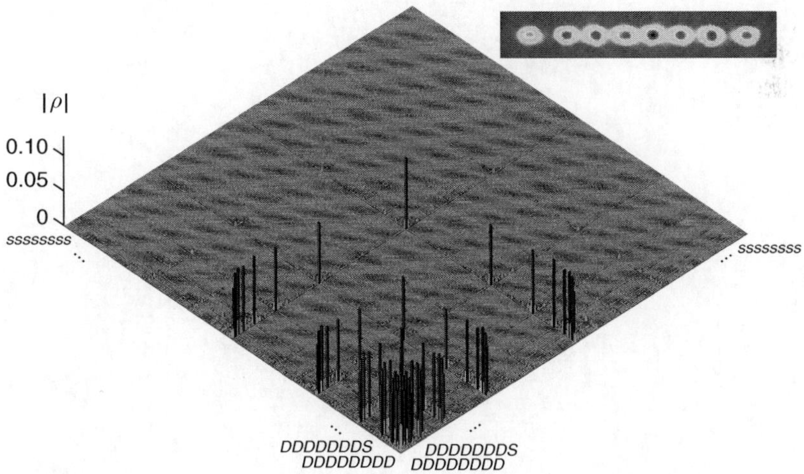

Fig. 13.6 Absolute values of the reconstructed density matrix. Distributing a single phonon over a string of eight ion qubits and subsequently applying suitable bit flips generates this multipartite entangled quantum state. With respect to the ideal state $\frac{1}{\sqrt{8}}(|0\ldots01\rangle + |0\ldots010\rangle + \ldots + |10\ldots0\rangle)$, a fidelity of 72.2(1)% is achieved.

sampled during such a process. Performing state tomography on the output state for each of those input states allows one to determine the χ matrix. However, as with the issue of negative eigenvalues for density matrices, a simple linear inversion to obtain the χ matrix results in nonphysical processes. Here again, a maximum-likelihood estimation is able to infer the most likely physical processes with respect to the obtained data. How is process tomography then performed in detail? In order to sample the input state space, the input states $\{|0\rangle, \frac{1}{\sqrt{2}}(|0\rangle + |1\rangle), \frac{1}{\sqrt{2}}(|0\rangle + i|1\rangle), |1\rangle\}$ are used for each qubit. On each of those 4^N input states, state tomography with 3^N settings is performed. In total, this corresponds to $(4 \cdot 3)^N$ measurement settings—a scaling that is very unfavorable and results in process tomographies being performed on at most three qubits [32]. With respect to processes acting on fewer qubits, quantum process tomography has been performed in Innsbruck on single-qubit gate operations with fidelities of about 99%, defined by $F_\chi = \mathrm{tr}(\chi_{\mathrm{exp}}\chi_{\mathrm{ideal}}\}$, with χ_{exp} describing the realized process and χ_{ideal} describing the desired operation. The implementation of a

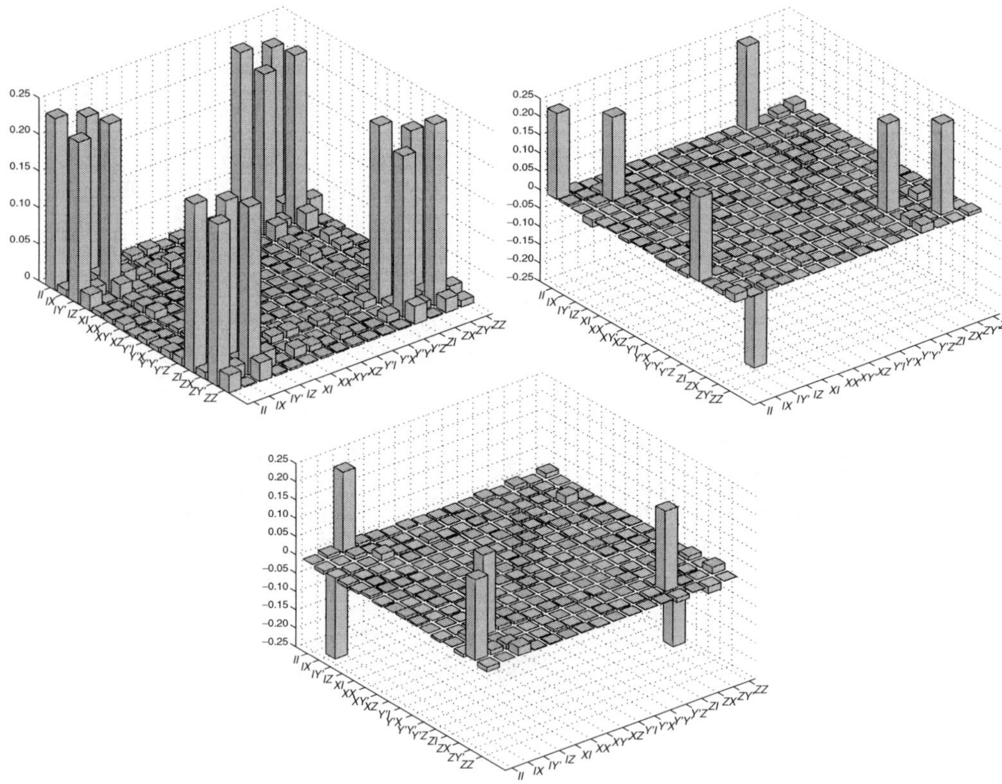

Fig. 13.7 Absolute, real, and imaginary parts of the reconstructed χ matrix for a CNOT operation as described in [30]. The two-qubit operator basis is described here in terms of the Pauli operator basis $\{\mathbb{1}, X, Y, Z\}^{\otimes 2}$.

CNOT gate as outlined above [30] has achieved fidelities as high as 90.8(6)%, with the corresponding χ matrix of the process shown in Fig. 13.7. A different, but related, question is whether the output states are close to the expected output states.

13.3 Quantum teleportation and entanglement swapping

In summary, single-qubit operations at fidelities of 99% and a CNOT implementation at about 93% fidelity have been discussed. These constitute a universal set of gates and, as such, allow for arbitrary quantum computation. In the following, two fundamental experiments of quantum information processing will be discussed based on the tools of quantum information processing: push-button teleportation and deterministic entanglement swapping. Quantum teleportation is a recipe that transfers quantum information between qubits that have never interacted directly with each other. At first glance, this does not seem to be difficult—in a classical computer, one can measure a bit, learn what state it is in, and then, using classical communication, realize a copy of the original bit anywhere else. In quantum mechanics, such an approach is not possible. A measurement projects a qubit onto one of the eigenstates of the measurement operator—but this is not necessarily the original state. With only a single copy of a quantum state, it is therefore not possible to infer information about the original qubit without destroying it, as stated by the noncloning theorem [33]. With this knowledge, one might then argue that if it is not possible to measure a single-copy qubit without destroying its information, then one cannot transfer its information to a distant location without physically moving the carrier of the information. This, however, turns out to be wrong: consider qubit 1 in an arbitrary pure quantum state $|\psi_{\text{in}}\rangle = \alpha\,|0\rangle_1 + \beta\,|1\rangle_1$, with the index indicating the qubit number, and two entangled qubits in the state $|\psi^+\rangle = \frac{1}{\sqrt{2}}(|01\rangle_{23} + |10\rangle_{23})$. The state of the entire three-qubit system (with the input initially in qubit 1, and the Bell state between qubits 2 and 3 as indicated by the indices) is now $|\psi_{\text{in}}\rangle_1 \otimes |\psi^+\rangle_{23}$. Using the Bell states $|\phi^\pm\rangle = \frac{1}{\sqrt{2}}(|00\rangle \pm |11\rangle), |\psi^\pm\rangle = \frac{1}{\sqrt{2}}(|01\rangle \pm |10\rangle)$, the original state can be rewritten as

$$
\begin{aligned}
|\psi_{\text{in}}\rangle_1 \otimes |\phi^+\rangle_{23} =& \frac{1}{2}(|\phi^+\rangle_{12} \otimes (\sigma_x\,|\psi_{\text{in}}\rangle) + |\phi^-\rangle_{12} \otimes (\sigma_z\sigma_x\,|\psi_{\text{in}}\rangle) \\
&+ |\psi^+\rangle_{12} \otimes |\psi_{\text{in}}\rangle + |\psi^-\rangle_{12} \otimes (-\sigma_z\,|\psi_{\text{in}}\rangle))\,.
\end{aligned} \tag{13.2}
$$

This means that measuring qubits 1 and 2 in the Bell basis projects the third qubit—regardless of its distance to qubit 1 and without ever interacting with it—into the original state (and a conditional operation based on the Bell state measurement result). The content of qubit 1 is transferred to qubit 3 without ever gaining information about the original qubit when qubits 1 and 2 are measured in the Bell basis, classical communication is used to transfer information about the measurement result to qubit 3, and finally a correcting operation is applied on qubit 3. We have already discussed how a universal set of operations allows us to perform arbitrary quantum computation. Now, which other tools are necessary to run the presented teleportation algorithm? The missing components are how one measure ion qubits in an entangled basis and whether it is possible to perform the classical feedforward from the measurement toward the remaining qubit? Let us first discuss the problem of measuring two qubits in the Bell basis. This measurement process ideally ought to project an input state onto one of the four Bell states. However, the measurement process can only yield classical information. If the classical information tells us uniquely which Bell state has been detected, then we have indeed measured in the Bell basis. Consider the four input states $\{|00\rangle, |01\rangle, |10\rangle, |11\rangle\}$. When a Hadamard operation, effectively a $\pi/2$ pulse, has been applied on the first qubit (counted from the right-hand side), one obtains the states $|0\rangle \otimes \frac{1}{\sqrt{2}}(|0\rangle + |1\rangle)$, $|0\rangle \otimes \frac{1}{\sqrt{2}}(|0\rangle - |1\rangle)$, $|1\rangle \otimes \frac{1}{\sqrt{2}}(|0\rangle + |1\rangle)$, $|1\rangle \otimes \frac{1}{\sqrt{2}}(|0\rangle - |1\rangle)$. Finally, consider a CNOT operation, controlled by the first qubit, to be applied onto these states. This maps the computational basis states onto $\frac{1}{\sqrt{2}}(|00\rangle + |11\rangle)$, $\frac{1}{\sqrt{2}}(|00\rangle - |11\rangle)$, $\frac{1}{\sqrt{2}}(|01\rangle + |10\rangle)$, $\frac{1}{\sqrt{2}}(|01\rangle - |10\rangle)$—the Bell states. This means that the combination of a Hadamard gate and a CNOT operation effectively maps between the computational basis and the Bell basis. As such, the application of Hadamard and CNOT, prior to the electron shelving, corresponds to a measurement in the Bell basis, although the measurement via electron shelving projects the qubits in the computational basis. In the particular example above, the first qubit (counted from the right) would convey the information whether $|\phi^{\pm}\rangle$ or $|\psi^{\pm}\rangle$ has been created, while the second qubit would convey the information about the phase of the superposition. Subsequently, depending on which of the four Bell states has been detected, the remaining qubit can be corrected: if qubit 1 was found in state $|1\rangle$, the remaining qubit needs to be manipulated by a Z bit flip and if qubit 2 was found in state $|1\rangle$, the remaining qubit needs to be rotated by an additional X flip, to deterministically teleport the information from qubit 1 to qubit 3. This raises two final questions: (i) How can a qubit be measured within an ion string without disturbing the quantum information encoded in the other ions? (ii) How does the classical feedforward work in an ion trap? Question (i)—how the state of a single ion can be measured without disturbing the remaining ions—is straightforward to answer. So far, the level structure of the $^{40}Ca^{+}$ ion has been simplified to a two-level system. However, the entire level scheme is considerably more elaborate. In particular, we can also drive transitions from $S_{1/2}(m = -1/2) \leftrightarrow D_{5/2}(m = -5/2)$. Using a π pulse on that transition, a qubit is now encoded in the two D states $D_{5/2}(m = -1/2)$ and $D_{5/2}(m = -5/2)$. This encoding renders the qubit insensitive to the electron-shelving technique that scatters photons on the $S_{1/2} \leftrightarrow P_{1/2}$ transition—the qubit always remains dark and no information about its state can be extracted. For the example of teleportation with three ions, this means that during the Bell state measurement process, the other two ions are "hidden" in the D-state manifold while the remaining ion is projected. The scattered photons are then measured with the PMT within the coherence time of the qubit. The missing building block to implement this recipe in an experiment is (ii) feedforward of the measurement result to perform the conditional X, Z operations. Considering that photons are scattered and detected by a PMT, a counter can be employed to determine whether more counts have been detected than a specific threshold, which, subsequently, sets a digital signal to either high or low. This digital signal is used to trigger conditionally the correcting laser pulse for the ion qubit, as described in [30, 34]. All these key features have been implemented in

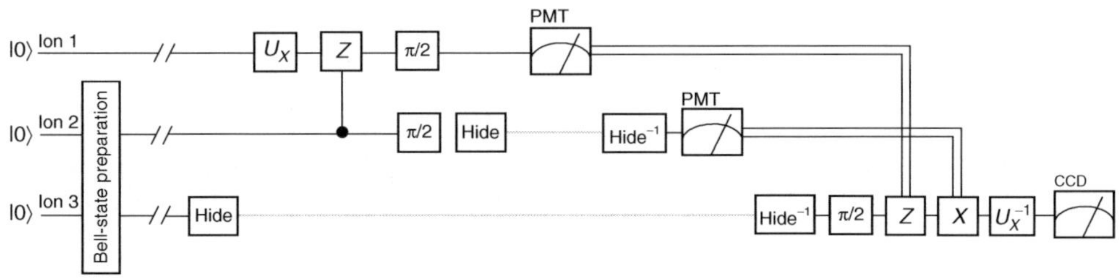

Fig. 13.8 Quantum circuit for deterministic quantum teleportation. Qubit 1 is initialized in an arbitrary quantum state while qubits 2 and 3 are prepared in a Bell state. A controlled phase gate following the ideas of Cirac and Zoller [28] is enclosed by two Ramsey pulses—effectively creating a controlled NOT gate. Combined with a subsequent Hadamard operation on the second qubit, this section of the quantum circuit efficiently maps from the Bell bases $\{|\phi^{\pm}\rangle, |\psi^{\pm}\rangle\}$ to the product basis $\{|00\rangle, |01\rangle, |10\rangle, |11\rangle\}$. Subsequent feedforward on qubit 3, based on the measurement results from qubits 1 and 2, teleports the input state from qubit 1 onto qubit 3.

the ion-trap experiment [35] with the realized quantum circuit shown in Fig. 13.8: arbitrary input states in qubit 1 are generated and a Bell pair in qubits 2 and 3 provides the resource for the teleportation protocol. A Bell-state measurement is performed on qubits 1 and 2, and its outcome is used to deterministically generate the input state at qubit 3—a qubit that never directly interacted with the original qubit. Using process tomography, this procedure was evaluated and found to work with mean fidelities of 83(1)% [36].

Careful analysis of the teleportation procedure shows that quantum information is destroyed at its original location and reappears at the target location. While the initial qubit and the target qubit never interacted with each other, both sites share an entangled qubit pair. At this point, one might argue that it is rather simple to generate entanglement between ions right next to each other but that distributing entanglement over entire networks might be unfeasible, given its fragile nature. In addition, it would be highly challenging to move qubits within their generally short coherence time from one place to another. Which raises the question: how can entanglement be distributed? Quantum teleportation shows that quantum information can be moved between qubits that never interacted directly. This protocol can be extended to directly create entanglement between remote qubits [37], which can even be done in the presence of errors [38]. Consider the following four-qubit state (with qubit number indicated by the index): $|\Psi\rangle_{1234} = |\psi^-\rangle_{12} \otimes |\psi^+\rangle_{34}$, which can be rewritten as

$$\begin{aligned} |\Psi\rangle_{1234} = -\frac{1}{2}(&|\phi^+\rangle_{14} \otimes |\phi^+\rangle_{23} + |\phi^-\rangle_{14} \otimes |\phi^-\rangle_{23} \\ &- |\psi^+\rangle_{14} \otimes |\psi^+\rangle_{23} + |\psi^-\rangle_{14} \otimes |\psi^-\rangle_{23}). \end{aligned} \tag{13.3}$$

It follows that a Bell-state measurement (and classical feedforward of correcting pulses) on qubits 2 and 3 allows deterministic creation of entanglement between qubits

1 and 4, although they never ever directly interacted with each other. This extension of quantum teleportation toward remote creation of entanglement is referred to as entanglement swapping [37], and has successfully been implemented in various photon experiments and with trapped ions [39]. Combined with entanglement purification [40], entanglement swapping can be used to realize quantum repeaters [38] that can create entanglement between distant quantum systems and thus enables quantum networks.

While the presented experiments have been important proofs of principle, the achieved fidelities in the 90% region have been low, compared with error thresholds for fault-tolerant quantum computation [41] at about 10^{-4}. The two main error sources are crosstalk between the qubits and error susceptibility of the employed CNOT gate. Crosstalk can be defined by the ratio of the Rabi frequency of an addressed qubit in comparison with the next-neighbor qubit. Here, the finite beamwaist of a focused beam of about 1 μm compared with an ion distance of about 3 μm results in crosstalk of about 3%—present during every gate operation. Beside crosstalk, motional state initialization is another limitation. Sideband cooling initializes the motional state better than 99% [29]. Considering that the motional state serves as a data bus, any subsequent computation after undesired creation of phonons due to crosstalk on the sideband operations will result in a faulty result. Besides the additional challenge of high-fidelity state initialisation of long and heavy ion strings, heating of the motional state [42] will also affect the motional data bus and limit the fidelities that can be achieved.

13.4 A new universal set of quantum gates for high-fidelity ion-trap quantum computers

13.4.1 The Mølmer and Sørensen quantum gate

Although technical improvements can reduce these errors, it seems more advantageous to use a different approach that is less sensitive to noise sources. A different interaction to create maximally entangled states follows the idea of Mølmer and Sørensen (MS)[43]. Illuminating the ions with a bichromatic light field detuned by $\Delta \approx 10\,\text{kHz}$ from the red and blue sideband transitions results in a time evolution of the harmonic oscillator equivalent to an off-resonantly driven harmonic oscillator. In phase space, the harmonic oscillator follows a circle and, after a time $t = 1/\Delta$, returns to its initial state—regardless of the starting point. Interestingly, owing to the coupling between the electronic and motional states, the Rabi frequency Ω_b on the blue sideband can be chosen such that the enclosed area of the circle is equal to $\pi/2$ [43]. Considering an initial state $|0\ldots0\rangle$ of an even number of qubits, this interaction results in the maximally entangled state $\frac{1}{\sqrt{2}}(|0\ldots0\rangle + |1\ldots1\rangle)$, or so-called Greenberg-Horne-Zeilinger (GHZ) states [44], regardless of the number of qubits (for more details see [45]). This entangling interaction is independent of the initial motional state in the first-order Lamb-Dicke approximation. GHZ states have been prepared starting from a thermal motional state with a mean phonon number $\langle n\rangle \approx 20(2)$ (initialized by Doppler cooling

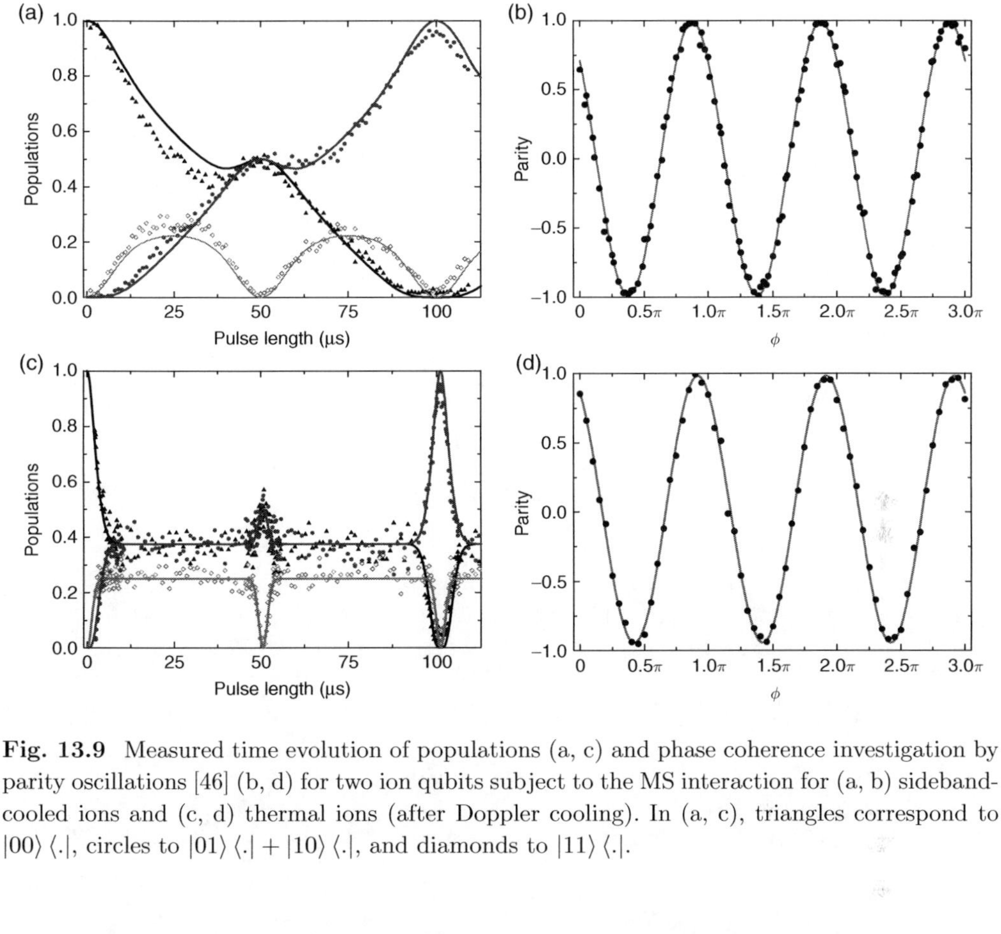

Fig. 13.9 Measured time evolution of populations (a, c) and phase coherence investigation by parity oscillations [46] (b, d) for two ion qubits subject to the MS interaction for (a, b) sideband-cooled ions and (c, d) thermal ions (after Doppler cooling). In (a, c), triangles correspond to $|00\rangle\langle.|$, circles to $|01\rangle\langle.| + |10\rangle\langle.|$, and diamonds to $|11\rangle\langle.|$.

only), reaching fidelities of up to 97.4(1)% [45], as shown in Fig. 13.9. The MS interaction is able to generate a GHZ-type state regardless of the number of ions. It has been successfully applied to generate genuine multiparticle entangled states for up to 14 ions. An investigation of the coherence properties of the created states is shown in Fig. 13.10. The fidelities for 2, 3, 4, 5, 6, 8, 10, 12, and 14 ions were measured to be 98.6(2)%, 97.0(3)%, 95.7(3)%, 94.4(5)%, 89.2(4)%, 81.7(4)%, 62.6(6)%, 47.4(7)%, and 50.8(9)%, respectively [47].

The MS gate is insensitive with respect to the initial motional state and returns high-fidelity quantum gate operations. However, the MS operation, acting on all qubits simultaneously, differs notably from the two-qubit CNOT gate operation, which is commonly used to decompose algorithms into circuits and, subsequently, pulse sequences. In the following, we will briefly explain a distinct universal set of gates based on the MS gate and subsequently how unitary operations can be decomposed into a sequence of operations from this set.

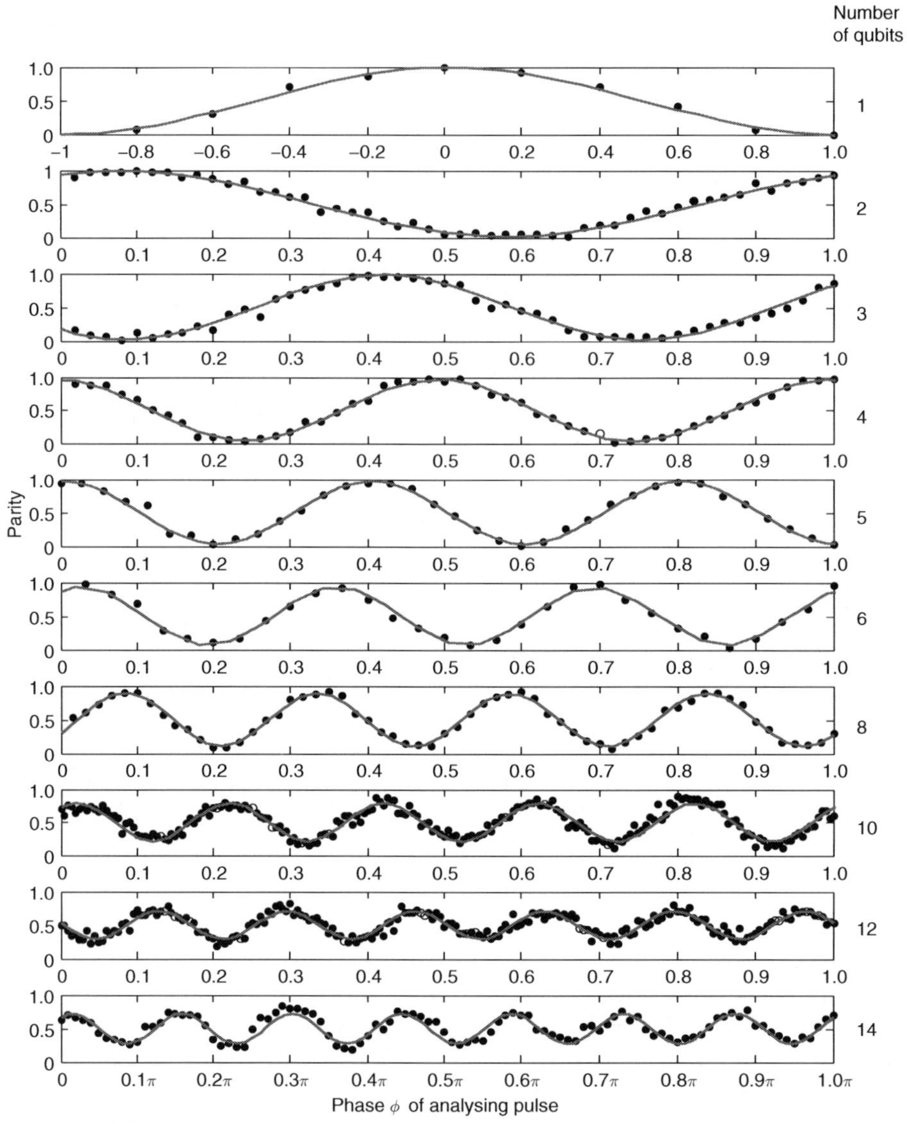

Fig. 13.10 Parity oscillations [46] observed on {2, 3, 4, 5, 6, 8, 10, 12, 14}-qubit GHZ states to investigate the coherence properties of the quantum state. The visibility of the signal, combined with measurements on the populations of the quantum state, allows us to calculate the fidelity of the realized state with respect to an ideal GHZ state. (Error bars are smaller than the marker size.)

13.4.2 ac Stark shifts for low cross-talk single-qubit operations

The maximally entangling MS interaction combined with arbitrary single-qubit operations forms a universal set of gates. Here, the global beam, which is for instance employed for the MS gate interaction and homogeneously illuminates the entire ion string, seemingly causes challenges. The phase of the MS gate is defined by the phase of the light field of the global beam. The tightly focused laser beam capable of addressing single qubits, however, follows a different beam path and its phase is not stable with respect to the global beam. So how can single-qubit operations be implemented with a fixed phase relation to the global beam? We follow a refocusing approach where any single-qubit operation of the kind $\exp(i\theta\sigma_x^{(1)})$ is decomposed into $\exp[-i(\pi/2)\sigma_z^{(1)}]\exp[-i(\theta/2)S_x]\exp[-i(\pi/2)\sigma_z^{(1)}]\exp[i(\theta/2)S_x]$. This sequence of operations will result in the identity operation on all qubits except the first, as the two global operations cancel each other out. The first qubit will, owing to the additional π phase shift, experience the desired operation. Here, the phase shift is induced by a far off-resonant beam causing an ac Stark shift. There are two advantages to this approach: (i) The ac Stark shift does not depend on the phase of the light field—therefore, the phase relation between the global and local beams is no longer important; and (ii) the ac Stark shift depends on the intensity of the light, not the electric field. It follows that crosstalk is substantially reduced for the ac Stark shift, resulting in a crosstalk of less than 1%. With this method, any local operation can be implemented on the qubit register [48], thus completing the alternative universal set of gates.

13.4.3 Implementing repetitive quantum error correction in an ion-trap quantum computer

The main problem now is to derive sequences of operations that efficiently realize a desired unitary operation. Using our universal set of gates (collective operations, single-qubit phase shifts, and the entangling MS gate), we employ an adapted gradient ascent pulse engineering (GRAPE) algorithm [48] to derive a sequence of operations for a key ingredient of large-scale quantum computation: quantum error correction. In our realization, the quantum information of a single qubit is encoded in one logical qubit consisting of three ion qubits. In particular, we intend to protect the information against single-qubit phase-flip errors with the help of two ancilla qubits. We seek to realize the three-qubit Calderbank-Shor-Steane code [31, 49, 50] and encode a single qubit in the following way:

$$\alpha\,|0\rangle + \beta\,|0\rangle \rightarrow \alpha\,|+++\rangle + \beta\,|---\rangle\ .$$

Reversing the encoding efficiently maps the information about which of the four possibilities (no error or phase flips on either ions $\{1,2,3\}$) occurred onto the two ancilla qubits. Considering that we only want to protect the information on qubit 1, it is sufficient to correct for only one out of four cases (when a phase-flip on qubit 1 has been detected). This controlled unitary effectively corresponds to a controlled-controlled NOT (CCNOT or Toffoli) operation. The circuit model for this encoding is depicted in Fig. 13.11, and its breakdown into a sequence of operations is shown below. Interestingly, the two CNOT operations required for the encoding can be realized by a single MS gate only. In a similar fashion, our GRAPE algorithm [48] derives a sequence for decoding and error correction consisting only of three MS gate operations and a few local operations. Using process tomography, the ideal case (no error) has been investigated, resulting in a fidelity of 90.1(2)% with respect to the identity operation (meaning the output is directly the input state). Further evaluation shows that the fidelity for any of the three single-qubit phase-flip errors remains 89.1(2)% [51]. This result can be interpreted in the following way: within the capabilities of the implementation, all allowed errors are perfectly corrected, as they yield the same fidelities as the implementation without errors. In addition, the behaviour of the implemented quantum error correction algorithm is investigated with respect to various noise sources (such as correlated and uncorrelated phase noise). The details of these measurements are discussed in [51].

A significantly more challenging aspect is to apply quantum error correction repetitively within an implementation. At the end of an error correction round, the ancilla

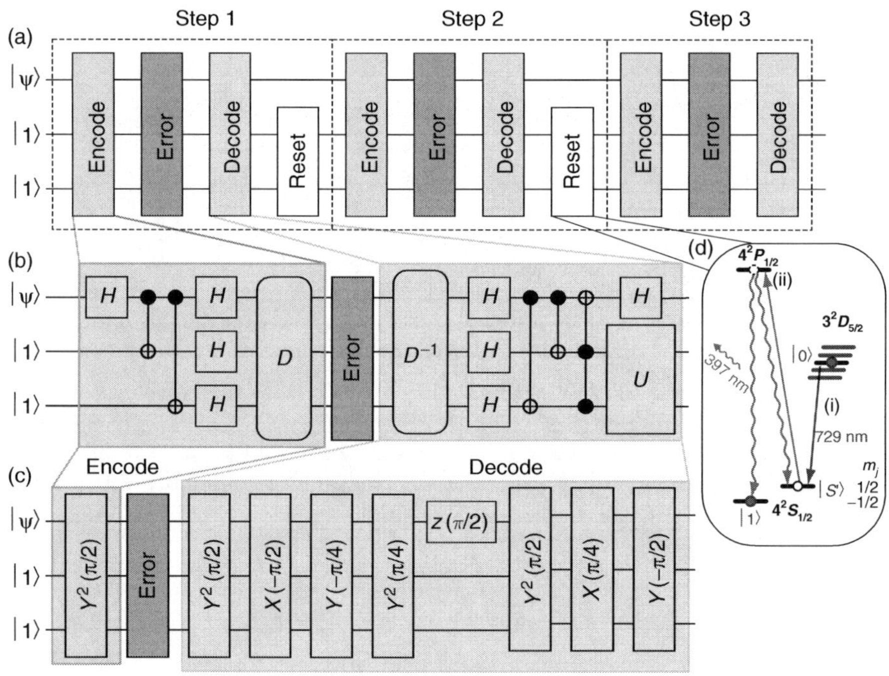

Fig. 13.11 Repetitive quantum error correction. (a) Schematic of three subsequent error correction steps. (b) Quantum circuit for quantum error correction of a single-qubit via encoding in a three-qubit system, including encoding, decoding, and correction. (c) Pulse sequence that implements the desired unitary operations. (d) Sketch of the reset process to reinitialize the ancilla qubits for subsequent applications in the iterative quantum error correction.

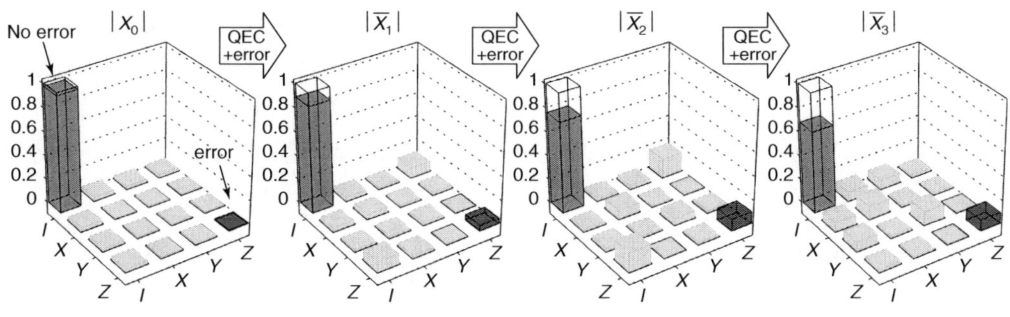

Fig. 13.12 Absolute values of the χ matrices describing the action of the implemented quantum error correction protocol on a single qubit. Ideally only the identity operation ought to be present. With more and more correction steps, the fidelity slowly degrades, following a Markovian behaviour.

qubits carry information about which error has occurred, and thus, the ancillas need to be reset before they can enter the next iteration of quantum error correction. Such a task is experimentally challenging to implement. Granted a quantum register of three qubits, two qubits need to be reset without destroying any quantum information on the third qubit. This is performed by transferring the population of the ancillas from the $|D\rangle$ to the $+1/2$ substate of the $S_{1/2}$ state, followed by an optical pumping step with σ^- light at the detection transition. The optical pumping light does not couple to the original qubit states [51]. Combining the reset with the previously discussed quantum error correction, up to three consecutive correction steps have been implemented in a linear Paul trap. After each step, full process tomography is performed, with the χ matrices shown in Fig. 13.12. The achieved fidelities versus the number of steps are 89.1(2)%, 76.3(2)%, and 68.3(3)%.

Using the same universal set of gate operations and tools such as the qubit reset or on-demand amplitude and phase damping, several other experiments have recently been realized, including a digital quantum simulator [52] and an open-system quantum simulator [53]. These experiments have proven the applicability of ion traps for small-scale quantum information processing, but the scalability of such systems has not yet been addressed.

13.5 Scalable ion-trap quantum computers

The ultimate goal of quantum information processing is to control an arbitrary number of qubits at a fault-tolerant level. The investigations described above have illustrated that while two qubits can be controlled almost at a fault-tolerant level [54], larger ion strings are notably harder to control. From that point of view, it is plausible to follow the approach to store several qubits, but work only on a few at a time. Investigations show that this approach might pave the way toward large-scale quantum computation [55]. However, this approach comes with high technological demands. Ion strings need to be stored and split, and these subunits of previously large ion strings need to be efficiently transferred between the different computational zones. This can be achieved by segmenting the dc electrodes of Paul traps into several parts, each at an individual voltage. Considering an inter-ion distance of few micrometers, it is evident that the size of the segments needs to be of the same order of magnitude. However, it turns out that such micrometer-sized traps are difficult to manufacture: For mechanical production, the required precision can barely be achieved; on the other hand, the structures are usually too large to be created with well-known lithographic techniques as used in semiconductor fabrication. Beside this technical problem of creating traps of appropriate size, such traps naturally have a shorter ion-surface distance than macroscopic ion traps. This proximity of the ions to surfaces makes them more sensitive to decohering effects. Here especially, motional heating is a prominent noise source that prevents high-fidelity quantum computation. The increased motional heating close to surfaces is presumably due to surface patch potentials. Currently, a promising technique to suppress undesired heating is cooling down the ion trap to cryogenic temperatures. Once these problems are overcome, one can expect significant progress in large-scale control of such quantum systems.

13.6 Summary and perspectives

Ion-trap systems are currently among the most prominent candidates for high-fidelity control of quantum systems. For two qubits, operations close to fault-tolerant levels have already been achieved. Single-qubit readout within a register and classical feedforward are straightforward to implement. A toolbox of high-fidelity quantum operations exists and is routinely employed for the implementation of both quantum computations and quantum simulations. Advanced schemes such as repetitive quantum error correction have been successfully implemented. Finally, genuine multipartite entanglement of up to 14 qubits has been proven. The knowledge gained in these experiments is currently being transferred to a novel design of segmented ion traps that promise scalable ion-trap-based quantum computation.

Acknowledgments

We thank Michael Chwalla and Esteban Martinez for their continuous support in the laboratory. We gratefully acknowledge support by the European Commission AQUTE, through the Foundations and Applications of Quantum Science (SFB FoQus), by the Austrian Science Fund (FWF), and by the Institut für Quanteninformation GmbH.

References

[1] Blatt, R., Häffner, H., Roos, C., Becher, C., and Schmidt-Kaler, F. (2004). In *Quantum Entanglement and Information Processing. École d'été de Physique des Houches Session LXXIX, 30 July – 25 August 2003* (ed. D. Estève, J. M. Raimond, and J. Dalibard), pp. 223–255. Elsevier, Amsterdam.

[2] Wineland, D. J. (2004). In *Quantum Entanglement and Information Processing. École d'été de Physique des Houches Session LXXIX, 30 July – 25 August 2003* (ed. D. Estève, J. M. Raimond, and J. Dalibard), pp. 261–293. Elsevier, Amsterdam.

[3] DiVincenzo, D. P. (2000). *Fortschr. Phys.*, **48**, 771.

[4] Paul, W. and Steinwedel, H. (1953). *Z. Naturforsch. A*, **8**, 448.

[5] Lucas, D. M., Ramos, A., Home, J. P., McDonnell, M. J., Nakayama, S., Stacey, J. P., Webster, S. C., Stacey, D. N., and Steane, A. M. (2004). *Phys. Rev. A*, **69**, 012711.

[6] Barton, P. A., Donald, C. J. S., Lucas, D. M., Stevens, D. A., Steane, A. M., and Stacey, D. N. (2000). *Phys. Rev. A*, **62**, 032503.

[7] Kreuter, A., Becher, C., Lancaster, G. P. T., Mundt, A. B., Russo, C., Häffner, H., Roos, C., Hänsel, W., Kaler, F. S., and Blatt, R. (2005). *Phys. Rev. A*, **71**, 032504.

[8] Chwalla, M., PhD thesis. (2009). University of Innsbruck, Austria.

[9] Kastler, A. (1950). *J. Phys. Radium*, **11**, 255.

[10] Zel'dovich, Y. (1974). *JETP Lett.*, **19**, 74.

[11] Wineland, D. J. and Dehmelt, H. (1975). *Bull. Am. Phys. Soc.*, **20**, 637.

[12] Dehmelt, H. (1975). *Bull. Am. Phys. Soc.*, **20**, 60.

[13] Myerson, A. H., Szwer, D. J., Webster, S. C., Allcock, D. T. C., Curtis, M. J., Imreh, G., Sherman, J. A., Stacey, D. N., Steane, A. M., and Lucas, D. M. (2008). *Phys. Rev. Lett.*, **100**, 200502.

[14] Burrell, A. H., Szwer, D. J., Webster, S. C., and Lucas, D. M. (2010). *Phys. Rev. A*, **81**, 040302.

[15] Ramsey, N. F. (1990). *Rev. Mod. Phys.*, **62**, 541.

[16] Chou, C. W., Hume, D. B., Koelemeij, J. C. J., Wineland, D. J., and Rosenband, T. (2010). *Phys. Rev. Lett.*, **104**, 070802.

[17] West, J. R., Fong, B. H., and Lidar, D. A. (2010). *Phys. Rev. Lett.*, **104**, 130501.

[18] Viola, L., Knill, E., and Lloyd, S. (1999). *Phys. Rev. Lett.*, **82**, 2417.

[19] Monz, T., PhD thesis. (2011). University of Innsbruck, Austria.

[20] Walther, A., Ziesel, F., Ruster, T., Dawkins, S. T., Ott, K., Hettrich, M., Singer, K., Kaler, F. S., and Poschinger, U. (2012). *Phys. Rev. Lett.*, **109**, 080501.

[21] Bowler, R., Gaebler, J., Lin, Y., Tan, T. R., Hanneke, D., Jost, J. D., Home, J. P., Leibfried, D., and Wineland, D. J. (2012). *Phys. Rev. Lett.*, **109**, 080502.

[22] Kim, C., Knoernschild, C., Liu, B., and Kim, J. (2005). In *IEEE/LEOS Optical MEMS 2005—International Conference on Optical MEMS and Their Applications*, pp. 96–97. IEEE, New York.

[23] Rohde, H., Gulde, S. T., Roos, C. F., Barton, P. A., Leibfried, D., Eschner, J., Schmidt-Kaler, F., and Blatt, R. (2001). *J. Opt. B: Quantum Semiclassical Opt.*, **3**, S34.

[24] Roos, C., PhD thesis. (2000). University of Innsbruck, Austria.

[25] Stenholm, S. (1986). *Rev. Mod. Phys.*, **58**, 699.

[26] Roos, C. F., Lancaster, G. P. T., Riebe, M., Häffner, H., Hänsel, W., Gulde, S., Becher, C., Eschner, J., Kaler, F. S., and Blatt, R. (2004). *Phys. Rev. Lett.*, **92**, 220402.

[27] Häffner, H., Hansel, W., Roos, C. F., Benhelm, J., Chek-al-kar, D., Chwalla, M., Korber, T., Rapol, U. D., Riebe, M., Schmidt, P. O., Becher, C., Gühne, O., Dür, W., and Blatt, R. (2005). *Nature*, **438**, 643.
[28] Cirac, J. I. and Zoller, P. (1995). *Phys. Rev. Lett.*, **74**, 4091.
[29] Schmidt-Kaler, F., Haffner, H., Riebe, M., Gulde, S., Lancaster, G. P. T., Deuschle, T., Becher, C., Roos, C. F., Eschner, J., and Blatt, R. (2003). *Nature*, **422**, 408.
[30] Riebe, M., Kim, K., Schindler, P., Monz, T., Schmidt, P. O., Körber, T. K., Hänsel, W., Häffner, H., Roos, C. F., and Blatt, R. (2006). *Phys. Rev. Lett.*, **97**.
[31] Nielsen, M. A. and Chuang, I. L. (2000). *Quantum Computation and Quantum Information.* Cambridge University Press, Cambridge.
[32] Monz, T., Kim, K., Hänsel, W., Riebe, M., Villar, A. S., Schindler, P., Chwalla, M., Hennrich, M., and Blatt, R. (2009). *Phys. Rev. Lett.*, **102**, 040501.
[33] Wootters, W. K. and Zurek, W. H. (1982). *Nature*, **299**, 802.
[34] Riebe, M., PhD thesis. (2005). University of Innsbruck, Austria.
[35] Schmidt-Kaler, F., Häffner, H., Gulde, S., Riebe, M., Lancaster, G. P. T., Deuschle, T., Becher, C., Hänsel, W., Eschner, J., Roos, C. F., and Blatt, R. (2003). *Appl. Phys. B: Lasers Opt.*, **77**, 789.
[36] Riebe, M., Chwalla, M., Benhelm, J., Häffner, H., Hänsel, W., Roos, C. F., and Blatt, R. (2007). *New J. Phys.* **9**, 211.
[37] Żukowski, M., Zeilinger, A., Horne, M. A., and Ekert, A. K. (1993). *Phys. Rev. Lett.*, **71**, 4287.
[38] Briegel, H. J., Dür, W., Cirac, J. I., and Zoller, P. (1998). *Phys. Rev. Lett.*, **81**, 5932.
[39] Riebe, M., Monz, T., Kim, K., Villar, A. S., Schindler, P., Chwalla, M., Hennrich, M., and Blatt, R. (2008). *Nature Phys.*, **4**, 839.
[40] Reichle, R., Leibfried, D., Knill, E., Britton, J., Blakestad, R. B., Jost, J. D., Langer, C., Ozeri, R., Seidelin, S., and Wineland, D. J. (2006). *Nature*, **443**, 838.
[41] Shor, P. W. (1996). In *Proceedings of 37th Annual Symposium on Foundations of Computer Science*, pp. 56–65. IEEE, Los Alamitos, CA.
[42] Wineland, D. J., Monroe, C., Itano, W. M., Leibfried, D., King, B. E., and Meekhof, D. M. (1998). *J. Res. Natl. Inst. Stand. Tech.*, **103**, 259.
[43] Sørensen, A. and Mølmer, K. (2000). *Phy. Rev. A*, **62**, 022311.
[44] Greenberger, D. M., Horne, M. A., and Zeilinger, A. (2007). Going beyond Bell's theorem. arXiv:0712.0921.
[45] Kirchmair, G., Benhelm, J., Zähringer, F., Gerritsma, R., Roos, C. F., and Blatt, R. (2009). *New J. Phys.*, **11**, 023002.
[46] Leibfried, D., Knill, E., Seidelin, S., Britton, J., Blakestad, R. B., Chiaverini, J., Hume, D. B., Itano, W. M., Jost, J. D., Langer, C., Ozeri, R., Reichle, R., and Wineland, D. J. (2005). *Nature*, **438**, 639.
[47] Monz, T., Schindler, P., Barreiro, J. T., Chwalla, M., Nigg, D., Coish, W. A., Harlander, M., Hänsel, W., Hennrich, M., and Blatt, R. (2011). *Phys. Rev. Lett.*, **106**, 130506.
[48] Nebendahl, V., Häffner, H., and Roos, C. F. (2009). *Phys. Rev. A*, **79**, 012312.
[49] Calderbank, A. R. and Shor, P. W. (1996). *Phys. Rev. A*, **54**, 1098.

[50] Steane, A. M. (1996). *Phys. Rev. Lett.*, **77**, 793.
[51] Schindler, P., Barreiro, J. T., Monz, T., Nebendahl, V., Nigg, D., Chwalla, M., Hennrich, M., and Blatt, R. (2011). *Science*, **332**, 1059.
[52] Lanyon, B. P., Hempel, C., Nigg, D., Müller, M., Gerritsma, R., Zähringer, F., Schindler, P., Barreiro, J. T., Rambach, M., Kirchmair, G., Hennrich, M., Zoller, P., Blatt, R., and Roos, C. F. (2011). *Science*, **334**, 57.
[53] Barreiro, J. T., Muller, M., Schindler, P., Nigg, D., Monz, T., Chwalla, M., Hennrich, M., Roos, C. F., Zoller, P., and Blatt, R. (2011). *Nature*, **470**, 486.
[54] Benhelm, J., Kirchmair, G., Roos, C. F., and Blatt, R. (2008). *Nature Phys.*, **4**, 463.
[55] Kielpinski, D., Monroe, C., and Wineland, D. J. (2002). *Nature*, **417**, 709.

Part II

Seminars

14

An introduction to laser cooling optomechanical systems

J. G. E. Harris

Departments of Physics and Applied Physics Yale University
New Haven, CT 06520, USA

Quantum Machines: Measurement and Control of Engineered Quantum Systems.
Edited by M. Devoret, B. Huard, R. Schoelkopf, and L. F. Cugliandolo.

Chapter Contents

14.1 Quantum machines and optomechanics

A quantum machine may be defined as a system that has a few degrees of freedom and where each of these degrees of freedom can be controlled and read out. Two-level systems (TLSs) and harmonic oscillators (HOs) are perhaps the most common choices for these degrees of freedom. In a quantum machine, TLSs (which may be realized, for example, by nuclear spins, electron spins, or nonlinear circuits) often serve as logical qubits, while HOs (which may be realized, for example, by the normal modes of electromagnetic or mechanical resonators) often serve as buses between TLSs [1]. In some implementations, HOs may also be used as memory, or even as logical elements themselves [2].

Part of the appeal of using TLSs and HOs to build a quantum machine is their simplicity. They are are familiar from introductory quantum mechanics classes, and their basic properties are well known—when they are in isolation. However, to achieve control and readout of the quantum machine's degrees of freedom, they must be coupled to many other degrees of freedom that are not part of the machine. These ancillary degrees of freedom may be associated with amplifiers, magnet coils, photodiodes, piezoelectric actuators, or room-temperature electronics. In general, these ancillary degrees of freedom will not be as well controlled as those that are part of the machine, and as a result they will represent a type of bath (though not necessarily a thermal equilibrium bath).

At the same time, the machine's degrees of freedom will also couple to other degrees of freedom that are external to both the machine and its control/readout. These external degrees of freedom may include phonons in the host material, nearby nuclear spins or other magnetic impurities, patch charges, broken chemical bonds, or modes of the electromagnetic field that are not part of the machine, its control, or readout. These external degrees of freedom will also constitute a bath, and, unlike the control readout apparatus, are likely to be in thermal equilibrium.

This thermal bath will impact the machine to a degree that depends upon the temperature of the bath T and the strength of its couplings to the machine's various degrees of freedom. T may range from 300 K in a room-temperature experiment down to as low as about 10 mK (the lower practical limit for most cryogenic experiments). The systems considered in this chapter will nearly always have level spacings that are considerably smaller than $k_B T$.

The particular quantum machine that will be the focus of this chapter is the cavity optomechanical system: an optical resonator coupled to a mechanical resonator. This might not sound like a promising concept for a quantum machine, since each of these elements is a harmonic oscillator, and it is often said that the harmonic oscillator is the most classical of all machines. It will be helpful to begin this lecture with a discussion of why this is the case.

14.2 Harmonic oscillators

14.2.1 Definitions

Let us start by briefly revisiting the basic description of a harmonic oscillator. A more detailed description is provided in Chapter 3 of this volume. If we consider a particle that has mass m, position x, and momentum p and is subject to a potential $V(x) = \frac{1}{2}kx^2$, then it is straightforward to show that the ground-state wavefunction of this system is (in the x basis) a Gaussian of width $x_{\mathrm{ZPF}} \equiv \sqrt{\hbar/2m\omega_m}$ where $\omega_m \equiv \sqrt{k/m}$ is the oscillator's natural frequency. In the p basis, the wavefunction is also a Gaussian, and has width $p_{\mathrm{ZPF}} \equiv m\omega_m x_{\mathrm{ZPF}}$. From these expressions, it can be seen that the ground state of the harmonic oscillator is a minimum-uncertainty wavefunction, since the product of these two widths is $x_{\mathrm{ZPF}}p_{\mathrm{ZPF}} = \hbar/2$.

In what follows, it will often be useful to replace the operators corresponding to the oscillator's position and momentum ($\hat{x}$ and $\hat{p}$) with two other operators known as the lowering and raising operators:

$$\begin{aligned} \hat{a} &\equiv x_{\mathrm{ZPF}}\left(\hat{x} + \frac{i}{m\omega_m}\hat{p}\right), \\ \hat{a}^\dagger &\equiv x_{\mathrm{ZPF}}\left(\hat{x} - \frac{i}{m\omega_m}\hat{p}\right). \end{aligned} \tag{14.1}$$

From these definitions, it is straightforward to show that $[\hat{a}, \hat{a}^\dagger] = 1$ and that the Hamiltonian of the oscillator can be written as

$$\hat{H} = \frac{1}{2}k\hat{x}^2 + \frac{1}{2m}\hat{p}^2 = \hbar\omega_{\mathrm{m}}\left(\hat{a}^\dagger\hat{a} + \frac{1}{2}\right). \tag{14.2}$$

14.2.2 Coherent states

In addition to the ground state, it is useful to consider a class of states (of which the ground state is a member) called *coherent states*. These states can be derived by starting with the ground-state wavefunction and displacing it (in x) by an amount $\alpha' x_{\mathrm{ZPF}}$ and then multiplying it by $e^{i\alpha'' p_{\mathrm{ZPF}}x/\hbar}$. This latter action is equivalent to displacing the wavefunction (in p) by an amount $\alpha'' p_{\mathrm{ZPF}}$. Each coherent state is thus characterized by the two dimensionless numbers α' and α'', which we can combine into a single complex number $\alpha \equiv \alpha' + i\alpha''$. Each coherent state is clearly still a minimum-uncertainty wavepacket, and has a mean position and momentum given by

$$\langle\alpha|\,\hat{x}\,|\alpha\rangle = \alpha' x_{\mathrm{ZPF}} \tag{14.3}$$

and

$$\langle\alpha|\,\hat{p}\,|\alpha\rangle = \alpha'' p_{\mathrm{ZPF}}. \tag{14.4}$$

It is also straightforward to show that these coherent states are eigenstates of the lowering operator with (complex) eigenvalues α; that is, $\hat{a}\,|\alpha\rangle = \alpha\,|\alpha\rangle$.

The dynamics of coherent states are most easily calculated in the Heisenberg picture, in which the time dependence of the operator $\hat{a}(t)$ is determined by

$$\dot{\hat{a}} = -\frac{i}{\hbar}[\hat{H}, \hat{a}] = -i\omega_m\hat{a}, \tag{14.5}$$

which has solution $\hat{a}(t) = \hat{a}(0)e^{-i\omega_m t}$. As a result, the action of this (time-dependent, Heisenberg-picture) operator on a coherent state is $\hat{a}(t)\,|\alpha\rangle = \hat{a}(0)e^{-i\omega_m t}\,|\alpha\rangle = e^{-i\omega_m t}\alpha\,|\alpha\rangle$. Thus, a state that is initially an eigenstate of the operator $\hat{a}$ with eigenvalue $\alpha(0)$ remains an eigenstate of $\hat{a}$ with eigenvalue $e^{-i\omega_m t}\alpha(0)$. This tells us that an oscillator initially in a coherent state $|\alpha(0)\rangle$ will evolve into the coherent state $|\alpha(0)e^{-i\omega_m t}\rangle$.

So an oscillator that starts in a coherent state will always be found in a coherent state, and its time evolution can be represented graphically as executing a circle in the complex α plane with frequency ω_m. Since the complex number α encodes the expectation values of the oscillator's position and momentum $\langle\hat{x}(t)\rangle$ and $\langle\hat{p}(t)\rangle$ (Eqs. (14.3) and (14.4)), this evolution in the complex α plane with frequency ω_m is reminiscent of the evolution of the corresponding classical oscillator in the classical phase-space plane. In fact, it requires only a few lines of algebra to show that $\langle\hat{x}(t)\rangle$ and $\langle\hat{p}(t)\rangle$ have exactly the same time dependence as the classical oscillator's position and momentum $x(t)$ and $p(t)$. We can make this correspondence stronger: since the full wavefunction (in the x basis) of the state $|\alpha(0)e^{-i\omega_m t}\rangle$ consists of a minimum-uncertainty wavepacket centered at $\langle\hat{x}(t)\rangle = x(t)$, or, in the p basis, a minimum-uncertainty wavepacket centered at $\langle\hat{p}(t)\rangle = p(t)$, the time evolution of a coherent state can be thought of as mimicking the classical behavior of the oscillator to the closest degree permitted by quantum mechanics.

From the preceding description, we can see that if we take an isolated harmonic oscillator in its ground state, displace its position and momentum by some amount, allow it to evolve for some time, and then ask what will be the average value of measurements of either position or momentum, then quantum mechanics and classical mechanics return the same value. Note that the thought experiment of the preceding sentence is well-posed for either quantum mechanics or classical mechanics: although one does not usually refer to a "ground state" in classical mechanics, or ask about the *average* value of the outcomes of measurements, both are meaningful. The former refers to the case $x = 0$, $p = 0$; the latter reflects the fact that classically we can write $\langle\hat{x}(t)\rangle = x(t)$ and $\langle\hat{p}(t)\rangle = p(t)$.

This is not to suggest that the isolated harmonic oscillator has no uniquely quantum features. For example, repeated measurements of the oscillator's position (after identical preparations) will not all return a single value $x(t)$, but will have a range of values that reflect the finite width of the wavefunction (i.e., roughly x_{ZPF}). Furthermore, we note that the similarity with the classical oscillator has been demonstrated only for coherent states. Other states, such as the energy eigenstates (also known as Fock states) exhibit nonclassical effects much more explicitly [3]. Thus, it might seem that oscillators prepared in Fock states (as opposed to coherent states) could serve as better building blocks for a quantum machine. Although this idea has merit, we will show below that the simplest control schemes tend to produce coherent states rather than Fock states.

14.2.3 Driven oscillators and the correspondence limit

To investigate what can be achieved with the exertion of some control, let us now consider an isolated harmonic oscillator to which we apply an arbitrary time-dependent force $F(t)$. This type of drive is referred to as a linear drive; the distinction is with nonlinear drives (i.e., those in which F would also depend upon x).

The Hamiltonian for this system is

$$\hat{H} = \hbar\omega_m \hat{a}^\dagger \hat{a} - f(t)(\hat{a}^\dagger + \hat{a}), \tag{14.6}$$

where we have defined the rescaled force $f(t) \equiv 2x_{\mathrm{ZPF}}F(t)$ and removed the constant term $\hbar\omega_m/2$ from Eq. (14.2). The Heisenberg equation of motion for the operator $\hat{a}$ is

$$\dot{\hat{a}} = -\frac{i}{\hbar}[\hat{H}, \hat{a}] = -i\omega_m \hat{a} - i\frac{f(t)}{\hbar}. \tag{14.7}$$

Since this is a first-order differential equation, it would be helpful to cast one side as a total derivative and then solve the problem by direct integration. We can do this by multiplying both sides of Eq. (14.7) by $e^{i\omega_m t}$,

$$e^{i\omega_m t}(\dot{\hat{a}} + i\omega_m \hat{a}) = -i\frac{f(t)}{\hbar}e^{i\omega_m t}, \tag{14.8}$$

and rewriting this in terms of the total derivative:

$$\frac{d}{dt}(\hat{a}e^{i\omega_m t}) = -i\frac{f(t)}{\hbar}e^{i\omega_m t}. \tag{14.9}$$

Multiplying by dt and integrating from an initial time $t = 0$ to a final time $t = T$ gives

$$\int_{\hat{a}(0)e^{i\omega_m 0}}^{\hat{a}(T)e^{i\omega_m T}} d(\hat{a}e^{i\omega_m t}) = -\frac{i}{\hbar}\int_0^T f(t)e^{i\omega_m T}\,dt,$$

$$\hat{a}(T)e^{i\omega_m T} - \hat{a}(0) = -\frac{i}{\hbar}\int_0^T f(t)e^{i\omega_m T}\,dt,$$

$$\hat{a}(T) = \hat{a}(0)e^{-i\omega_m T} - \frac{i}{\hbar}\int_0^T f(t)e^{i\omega_m (t-T)}\,dt. \tag{14.10}$$

Without a specific choice for the form of $f(t)$, we can't proceed any further. However, this formal solution (14.10) still provides valuable insight into the dynamics of the linearly driven harmonic oscillator. To see this, consider the action of the time-dependent operator $\hat{a}(t)$ on a coherent state:

$$\hat{a}(t)\,|\alpha\rangle = \left[\hat{a}(0)e^{-i\omega_m T} - \frac{i}{\hbar}\int_0^T f(t)e^{i\omega_m (t-T)}\,dt\right]|\alpha\rangle. \tag{14.11}$$

This expression is easy to evaluate, since there is only one operator inside the square brackets and we know its action on a coherent state:

$$\hat{a}(t)\,|\alpha\rangle = \left[\alpha e^{-i\omega_m T} - \frac{i}{\hbar}\int_0^T f(t)e^{i\omega_m (t-T)}\,dt\right]|\alpha\rangle \equiv \alpha(t)\,|\alpha\rangle. \tag{14.12}$$

Here, the number $\alpha(t)$ (identified with the quantity in the square brackets) is the eigenvalue of $\hat{a}(t)$.

From Eq. (14.12), we can see that an oscillator initially in a coherent state will remain in a coherent state regardless of the form of $f(t)$. This may seem like bad news if one wants to build a quantum machine from a (linearly driven) harmonic oscillator. In fact, the news gets worse: not only does the oscillator remain in a coherent state, it remains in exactly the coherent state that classical mechanics would predict, given the same time-dependent force $F(t)$.

To see this, consider the same system described by classical mechanics. The Hamiltonian function is

$$H(x,p) = \frac{1}{2}kx^2 + \frac{1}{2m}p^2 - F(t)x \tag{14.13}$$

and the equations of motion are

$$\dot{x} = \frac{\partial H}{\partial p}, \tag{14.14}$$

$$\dot{p} = -\frac{\partial H}{\partial x}. \tag{14.15}$$

Eq. (14.15) immediately gives $\dot{x} = p/m$, which can be combined with Eq. (14.15) to give

$$m\ddot{x} + kx = F(t). \tag{14.16}$$

To find a formal solution to Eq. (14.16) in analogy with Eq. (14.11), we first find the homogenous solution by setting $F(t) = 0$. This gives

$$\begin{aligned} x_{\text{hom}} &= x(0)\cos(\omega_m t) + \frac{p(0)}{m\omega_m}\sin(\omega_m t), \\ p_{\text{hom}} &= p(0)\cos(\omega_m t) - m\omega_m x(0)\sin(\omega_m t), \end{aligned} \tag{14.17}$$

where we have also made use of Eq. (14.15).

To get the inhomogeneous part of the solution, we first find the oscillator's response to a delta-function force arriving at time t': $F_G(t) = p_0\delta(t-t')$. Here, the constant p_0 has units of momentum and the delta function has units of 1/time. By integrating the relationship $\dot{p} = F$ over an infinitesimal period of time that includes t', we find that the action of F_G is to change the oscillator's momentum by an amount p_0. By assumption, F_G is zero for all subsequent times, so the subsequent evolution of the oscillator is given by Eq. (14.17) with $x(0) = 0$ and $p(0) = p_0$, giving us the Green's functions

$$\begin{aligned} x_G(t,t') &= \frac{p(0)}{m\omega_m}\sin[\omega_m(t-t')], \\ p_G(t,t') &= p(0)\cos[\omega_m(t-t')] \end{aligned} \tag{14.18}$$

and the corresponding inhomogeneous solutions

$$x_{\rm in}(t) = \frac{1}{m\omega_m^2}\int_0^t F(t')\sin[\omega_m(t-t')]\,dt',$$
$$p_{\rm in}(t) = \frac{1}{\omega_m}\int_0^t F(t')\cos[\omega_m(t-t')]\,dt'. \tag{14.19}$$

Putting this all together for the full solution, we have

$$x(t) = x(0)\cos(\omega_m t) + \frac{p(0)}{m\omega_m}\sin(\omega_m t) + \frac{1}{m\omega_m^2}\int_0^t F(t')\sin[\omega_m(t-t')]\,dt'$$
$$p(t) = p(0)\cos(\omega_m t) - m\omega_m x(0)\sin(\omega_m t) + \frac{1}{\omega_m}\int_0^t F(t')\cos[\omega_m(t-t')]\,dt'. \tag{14.20}$$

We can make a direct comparison with the result of the quantum calculation in Eq. (14.11) by recalling that the time dependence of the coherent state is described by

$$\alpha(t) = \alpha(0)e^{-i\omega_m T} - \frac{i}{\hbar}\int_0^T f(t)e^{i\omega_m(t-T)}\,dt.$$

If we recall the relations $\langle\hat{x}\rangle = x_{\rm ZPF}\,{\rm Re}\,\alpha$, $\langle\hat{p}\rangle = x_{\rm ZPF}m\omega_{\rm m}\,{\rm Im}\,\alpha$, $f(t) = 2x_{\rm ZPF}F(t)$, and $x_{\rm ZPF} = \sqrt{\hbar/2m\omega_m}$, then we can rewrite Eq. (14.20) as

$$\langle\hat{x}(t)\rangle = \langle\hat{x}(0)\rangle\cos(\omega_m t) + \frac{\langle\hat{p}(0)\rangle}{m\omega_m}\sin(\omega_m t) + \frac{1}{m\omega_m^2}\int_0^t F(t')\sin[\omega_m(t-t')]\,dt',$$
$$\langle\hat{p}(t)\rangle = \langle\hat{p}(0)\rangle\cos(\omega_m t) - m\omega_m\langle\hat{x}(0)\rangle\sin(\omega_m t) + \frac{1}{m\omega_m^2}\int_0^t F(t')\sin[\omega_m(t-t')]\,dt'. \tag{14.21}$$

Comparing Eq. (14.21) with Eq. (14.12), we can see that, as with the free oscillator, a linearly driven oscillator that is initialized in a coherent state remains in a coherent states, with a wavefunction consisting of a minimum-uncertainty wavepacket whose trajectory follows exactly that of the corresponding classical oscillator.[1]

Thus one cannot gain access to the "less-classical" Fock states by starting with an oscillator and applying a linear drive to it. This is in contrast with a two-level system, in which a transition from the ground state to an excited-energy eigenstate can be induced by applying a linear drive at the appropriate frequency for the appropriate time (i.e., realizing a π pulse). One way to picture the difference between the harmonic oscillator and the two-level system in this situation is to note that the harmonic oscillator's spectrum consists of a ladder of equally spaced energy levels. This equal

[1] The fact that these relationships hold for arbitrarily small $|\alpha|$ means that the harmonic oscillator is always in the correspondence limit (i.e., not just for large $|\alpha|$).

spacing means that a drive whose frequency is chosen to couple the ground state $|0\rangle$ to the next Fock state $|1\rangle$ will invariably also couple $|1\rangle$ to $|2\rangle$, $|1\rangle$ back down to $|0\rangle$, $|3\rangle$ to $|2\rangle$, $|2\rangle$ back down to $|1\rangle$, and so on. Thus, as the drive builds a population in $|1\rangle$, some of this population will also be transferred to $|2\rangle$, and so on, with the result that the oscillator will be in a superposition of many Fock states. Our full solution of the Heisenberg equation of motion indicates that this superposition corresponds to the coherent state in Eq. (14.12).

In a two-level system, the story is quite different. Whatever physical system is being used to implement a two-level system will also have a ladder of states, but if the spacings between the states are not even, then a drive that is resonant between some pair of states $|g\rangle$ and $|e\rangle$ will not, in general, be resonant with other states. As a result, the system evolves only within the Hilbert space spanned by $|g\rangle$ and $|e\rangle$.

These observations lead us to conclude that accessing interesting quantum effects in harmonic oscillators will typically require the introduction of some nonlinearity. If the oscillator is realized using a mode of an optical cavity, nonlinearity could be introduced by inserting atoms or a solid with nonlinear electrical susceptibility into the cavity. In some circumstances, it is also possible to apply a nonlinear drive to a harmonic oscillator (it is left as an exercise to show that a drive of the form $f(t)(\hat{a}^\dagger + \hat{a})^2$ inserted into Eq. (14.6) will take an oscillator to a noncoherent state). Even measurement can be used to provide nonlinearity to an otherwise-linear oscillator, since if one makes a strong measurement of the energy, the oscillator will be left in a Fock state regardless of its initial state.

Our motivation for having devoted this long introduction to the tendency of oscillators to behave classically is that in the rest of this chapter we will focus specifically on cavity optomechanical systems, in which we attempt to build quantum machines out of harmonic oscillators.

14.3 Optomechanics

14.3.1 Optomechanical coupling

The basic model of a cavity optomechanical system contains two components: an electromagnetic oscillator and a mechanical oscillator [4, 5]. Given the preceding discussion, this might not seem like a promising approach to building a quantum machine, since each of these components is a harmonic oscillator. Furthermore, linearly coupling harmonic oscillators to each other does not result in easier access to quantum effects. This is because any system of linearly coupled harmonic oscillators can be decomposed into normal modes, which are themselves *uncoupled* harmonic oscillators.

Optomechanical systems circumvent this problem by exploiting a nonlinear coupling between the electromagnetic and mechanical oscillators. This nonlinearity arises from the parametric nature of the optomechanical coupling, which is illustrated in Fig. 1. The device shown in Fig. 1 consists of an optical cavity whose length is changed by the displacement of a mechanical oscillator (which we will call "the mirror"). The Hamiltonian of such a system can be written as

$$\hat{H} = \hbar\omega_{\text{cav}}(\hat{x})\hat{a}^\dagger\hat{a} + \hbar\omega_{\text{m}}\hat{b}^\dagger\hat{b}, \tag{14.22}$$

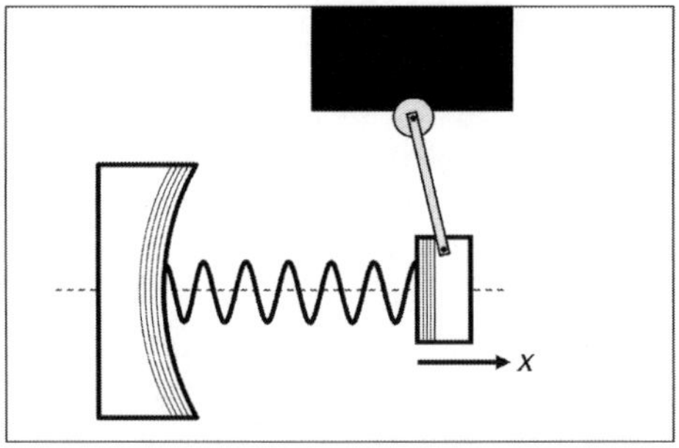

Fig. 14.1 Schematic illustration of an optomechanical system. A Fabry-Perot cavity is formed between two mirrors. The left-hand mirror is fixed in position, but the right-hand mirror is free to move within a harmonic potential, illustrated here as a pendulum. The optical mode inside the cavity (thick line) can be excited by a laser entering via the left-hand mirror's (very small) transmission. The cavity mode is detuned as the right-hand mirror moves and the optomechanical coupling arises from this detuning, as described in the text.

where $\hat{a}(\hat{b})$ is the photon (phonon) lowering operator of the cavity (mirror's motion), and $\omega_{\text{cav}}(\hat{x})$ is the cavity's resonant frequency, which depends on the (operator-valued) displacement $\hat{x}$ of the mirror. Treating $\omega_{\text{cav}}(\hat{x})$ as a parameter that instantaneously follows the mechanical oscillator's displacement is a Born-Oppenheimer approximation that is valid when the oscillator's motion is very slow compared with the round-trip time of light in the cavity. This is an excellent approximation in most optomechanics experiments.

For small displacements, we can Taylor-expand $\omega_{\text{cav}}(\hat{x})$ and rewrite Eq. (14.22) as

$$\hat{H} = \hbar\omega_{\text{cav}}^{(0)}\hat{a}^\dagger\hat{a} + \hbar\omega_{\text{m}}\hat{b}^\dagger\hat{b} + \hbar g(\hat{b}^\dagger + \hat{b})\hat{a}^\dagger\hat{a} + \ldots, \tag{14.23}$$

where $\omega_{\text{cav}}^{(0)} = \omega_{\text{cav}}(0)$, and $g = x_{\text{ZPF}}\partial\omega_{\text{cav}}/\partial x\big|_{x=0}$. The nonlinearity of this system is evident from the three-operator terms in Eq. (14.23), which describe the optomechanical coupling. These terms are equivalent to radiation pressure, as they describe a force that is proportional to the number of photons in the cavity.

Although the illustration in Fig. 14.1 is highly schematic, the Hamiltonian in Eq. (14.23) provides a basic description of nearly all optomechanical systems realized to date, including those based on whispering-gallery modes in toroids, discs, and spheres; Fabry-Perot cavities; and photonic crystals. However, Eq. (14.23) only describes an isolated system, whereas a quantum machine will require both readout and driving (e.g., a laser beam driving the cavity and/or a piezoelectric element driving the mechanical oscillator). In addition to the "intentional" driving that might be supplied by the experimenter, "unintentional" drives (e.g., from the device's thermal environment) will also be present, and will also lead to damping.

14.3.2 Coupling to a thermal bath

To begin a discussion of these topics, let us start by describing a simple harmonic oscillator coupled to a thermal bath at temperature T. When an oscillator is removed from pristine isolation and coupled to a thermal bath, the first dramatic consequence is that its oscillations are damped and can be described by the equation of motion

$$m\ddot{x} + \beta\dot{x} + kx = 0, \tag{14.24}$$

where β is a constant that reflects the strength of the oscillator's coupling to the bath's degrees of freedom.

Despite the fact that this equation of motion is presented in many undergraduate physics courses, it is clearly not the correct equation of motion for an oscillator coupled to a thermal bath. This can be seen by noting that for all solutions of this equation, $x \to 0$ and $\dot{x} \to 0$ as $t \to \infty$. This contradicts the equipartition theorem, which states that in thermal equilibrium with a bath, each degree of freedom must have an average energy of $k_B T/2$. If we interpret the "average" in this case as a time average, then equipartition requires that the oscillator's trajectory satisfy

$$\begin{aligned} \langle \mathrm{KE} \rangle &= \lim_{T\to\infty} \frac{1}{T} \int_0^T \frac{1}{2} k x^2(t)\, dt = \frac{k_B T}{2}, \\ \langle \mathrm{PE} \rangle &= \lim_{T\to\infty} \frac{1}{T} \int_0^T \frac{1}{2} m \dot{x}^2(t)\, dt = \frac{k_B T}{2}. \end{aligned} \tag{14.25}$$

These requirements are not satisfied by Eq. (14.24).

Nevertheless, the damping described by Eq. (14.24) is quite real. To resolve this contradiction, there must be a time-varying force $F_L(t)$ (known as the Langevin force) on the right-hand side of Eq. (14.24) to ensure that $\langle \mathrm{KE} \rangle = \langle \mathrm{PE} \rangle = k_B T/2$; that is,

$$m\ddot{x} + \beta\dot{x} + kx = F_L(t). \tag{14.26}$$

This requirement still leaves us considerable freedom regarding the exact form of $F_L(t)$. However, we can make some reasonable assumptions in order to proceed. For example, the physical origin of $F_L(t)$ is the thermal fluctuations of the many degrees of freedom constituting the bath, so we assume that it is uncorrelated from one instant to the next:

$$\langle F_L(t) F_L(t') \rangle = \delta(t - t') S_{FF}^T \tag{14.27}$$

where S_{FF}^T is an (as-yet-unspecified) constant that has units of $\mathrm{N}^2\,\mathrm{Hz}^{-1}$ and can be thought of as specifying the strength of $F_L(t)$. It follows from Eq. (14.27) (via the Wiener-Khinchin theorem) that $\left|\tilde{F}_L[\omega]\right|^2$, the power spectrum of $F_L(t)$, is constant (i.e., white) with magnitude $\left|\tilde{F}_L[\omega]\right|^2 = S_{FF}^T$.

This provides us with some information about $F_L(t)$, though certainly not enough to solve Eq. (14.26) for $x(t)$. However, we can still learn most of what we want by noting that the linearity of the harmonic oscillator means that it responds independently to each Fourier component of the drive. Specifically, we know (e.g., from solving

Eq. (14.24)) that, given a force of amplitude $F[\omega]$ applied at frequency ω, a harmonic oscillator's (complex) amplitude of motion $x[\omega]$ satisfies:

$$\frac{x[\omega]}{F[\omega]} = \frac{1/m}{\omega_m^2 - \omega^2 - i\omega\omega_m/Q} \equiv G[\omega], \tag{14.28}$$

where we have introduced the quality factor $Q \equiv m\omega_m/\beta$. If we are not interested in the phase relationship between the applied force and the resulting motion, but only want the magnitude of motion for a certain magnitude of the driving force, we use Eq. (14.28) to write

$$S_{xx}[\omega] = \left|G[\omega]\right|^2 S_{FF}[\omega] = S_{FF}[\omega]\frac{1}{m^2}\left[\frac{1}{(\omega_m^2 - \omega^2)^2 + (\omega\omega_m/Q)^2}\right]. \tag{14.29}$$

This is relevant to our discussion of the motion of an oscillator coupled to a thermal bath, since we have argued that $S_{FF}[\omega] = S_{FF}^T$ (i.e., a constant), giving

$$S_{xx}^T[\omega] = S_{FF}^T \frac{1}{m^2} \frac{1}{(\omega_m^2 - \omega^2)^2 + (\omega\omega_m/Q)^2}. \tag{14.30}$$

The actual value of S_{FF}^T is still unknown, but we can combine Eqs. (14.25) and (14.30) to find the value for S_{FF}^T that leads to an $x(t)$ that would be consistent with the equipartition theorem. We do this by using the relationship

$$\langle x^2 \rangle = \frac{1}{2\pi}\int_0^\infty S_{xx}^T[\omega]\, d\omega. \tag{14.31}$$

Rewriting the right-hand side of Eq. (14.31) using Eq. (14.30), and the left-hand side using Eq. (14.25) gives the condition that S_{FF}^T must satisfy in order to be consistent with the equipartition theorem:

$$\begin{aligned} \frac{k_B T}{k} &= \frac{1}{2\pi}\int_0^\infty \left|G[\omega]\right|^2 S_{FF}^T[\omega]\, d\omega \\ &= S_{FF}^T \frac{1}{2\pi m^2}\int_0^\infty \frac{1}{(\omega_m^2 - \omega^2)^2 + (\omega\omega_m/Q)^2}\, d\omega. \end{aligned} \tag{14.32}$$

This expression provides us with a useful insight: for a given oscillator (and a given damping), the force exerted by the bath (or rather the force's power spectral density) is proportional to the bath temperature.

The integral in Eq. (14.32) can be evaluated using the method of residues. The evenness of the integrand allows us to extend the lower limit of integration to $-\infty$, while the asymptotic (i.e., $\omega \to \infty$) behavior of S_{xx}^T allows the contour to be closed in either the upper or lower half of the complex ω plane. Either choice will enclose one of the two poles in S_{xx}^T, giving

$$S_{FF}^T = 4k_B T m\gamma_m, \tag{14.33}$$

where $\gamma_m \equiv \omega_m/Q$ is the oscillator's linewidth. This connection between the damping of an oscillator (i.e., γ_m) and the fluctuating force that drives its thermal motion (i.e., S_{FF}^T) is an example of the more general fluctuation-dissipation theorem.

From the point of view of the mechanical oscillator, Eq. (14.33) is significant because it tells us that one cannot reduce an oscillator's thermal motion simply by increasing its damping (so long as this damping comes from a thermal bath).

14.3.3 Coupling to two thermal baths

Just to check that we haven't missed any opportunities in this regard, let us briefly ask what would happen if we coupled the oscillator to two separate thermal baths. In particular, we would like to know to which temperature the oscillator will equilibrate. To answer this question, we assume that these two baths are not coupled to each other, that they are at temperatures T_1 and T_2, and that the oscillator's coupling to each bath is given by the coefficients β_1 and β_2. Following the reasoning leading to Eq. (14.26), the equation of motion for this oscillator will be

$$m\ddot{x} + \beta_1\dot{x} + \beta_2\dot{x} + kx = F_L^{(1)}(t) + F_L^{(2)}(t), \tag{14.34}$$

and the Langevin force associated with each bath will satisfy $S_{FF}^{(T,i)} = 4k_B T_i m\gamma_i$, where $\gamma_i = \beta_i/m$. We can rewrite Eq. (14.34) as

$$m\ddot{x} + \beta_{\text{tot}}\dot{x} + kx = F_L^{(\text{tot})}(t), \tag{14.35}$$

where $\beta_{\text{tot}} = \beta_1 + \beta_2$, and $F_L^{(\text{tot})}(t) = F_L^{(1)}(t) + F_L^{(2)}(t)$. If two random quantities are uncorrelated, their power spectral densities add, so the power spectral density of $S_{FF}^{(T,\text{tot})} = 4k_B m(T_1\gamma_1 + T_2\gamma_2)$. To see what temperature the oscillator will equilibrate to, we note that Eq. (14.35) is identical to the equation of motion for an oscillator coupled to a single bath, Eq. (14.26), that is, a damped oscillator driven by a white-noise force. As a result, its motion will be identical to that of an oscillator with damping $\beta_{\text{tot}} = \beta_1 + \beta_2$ at a temperature

$$T_{\text{eff}} = \frac{S_{FF}^{(T,\text{tot})}}{4k_B m\gamma_{\text{tot}}} = \frac{\gamma_1 T_1 + \gamma_2 T_2}{\gamma_1 + \gamma_2}, \tag{14.36}$$

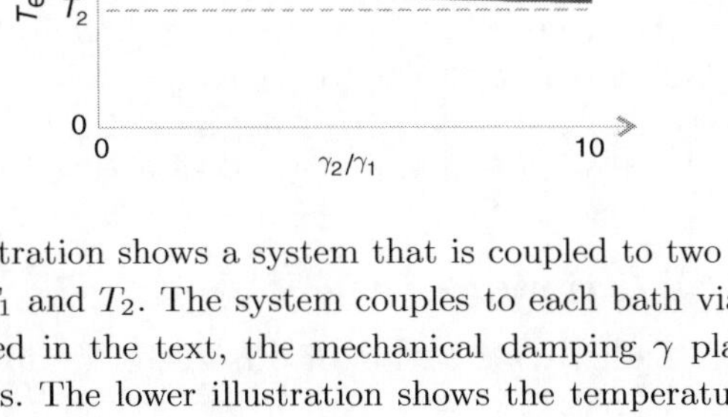

Fig. 14.2 The top illustration shows a system that is coupled to two separate baths, each at different temperatures T_1 and T_2. The system couples to each bath via thermal conductances G_1 and G_2. As described in the text, the mechanical damping γ plays the same role as G for mechanical oscillators. The lower illustration shows the temperature at which the system equilibrates as a function of the strength of the two couplings γ_1 and γ_2. In this plot, we have taken $T_1 = 2T_2$.

which is the weighted average of the two bath temperatures, where the weighting is determined by the strength of the coupling to each bath.

This may seem like a trivial result, but it helps to illustrate an important point, namely, that the equilibrium temperature of the oscillator is the same temperature one would find if one considered the simple thermodynamics problem illustrated in Fig. 14.2, that is, a system coupled to two baths at temperatures T_1 and T_2, each via thermal conductances $G_i \propto \gamma_i$. For the time being, it is clear that considering a pair of baths has not produced a new means for decreasing the thermal motion of an oscillator. So far, the only way seems to be to lower the temperature of the environment.

14.4 Laser cooling

14.4.1 Introduction to cold damping

So far, our discussion has only considered damping that arises from interactions between the oscillator and a thermal bath (i.e., a system that is in thermal equilibrium). If we could instead arrange for damping to occur via some means other than thermal equilibration, then the fluctuation-dissipation theorem would not apply and it might be possible to find a damping mechanism that would actually reduce the oscillator's thermal motion.

Since our basic definition of a "machine" in this chapter presupposes that the oscillator's motion can be monitored and controlled, one way to exert nonthermal damping would be to monitor $x(t)$, calculate from this record the instantaneous velocity $\dot{x}(t)$, and apply a force proportional to this quantity. This approach would certainly lead to damping as described in Eq. (14.24). It would also introduce some fluctuating force (resulting, for example, from errors in the measurement of $x(t)$) as in Eq. (14.26), but the relationship between the damping produced by the feedback and the additional force noise would not necessarily be given by Eq. (14.33). As a result, this type of feedback can be used to reduce the thermal motion of an oscillator. It is known as "feedback cooling," and has been used to cool mechanical oscillators since at least the 1950s [6].

However, in some systems, a qualitatively similar process occurs naturally, that is, without the application of an external control loop. The example that we will consider in the remainder of this chapter is a linearly driven oscillator that is parametrically coupled to a second oscillator. Cavity optomechanics realizes this scheme quite naturally, since a laser serves as a linear drive of the optical cavity, and the cavity is parametrically coupled to the mirror's motion, as described in Eq. (14.23) [7, 8].

14.4.2 Qualitative description of cold damping

Before resorting to an explicit calculation of this effect, we first provide a qualitative discussion. Figure 14.3 shows a plot of the optical power circulating in a cavity as a function of the position of the cavity's end-mirror. Strictly speaking, the quantity plotted is the equilibrium circulating power—the curve in Fig. 14.3 will only be realized

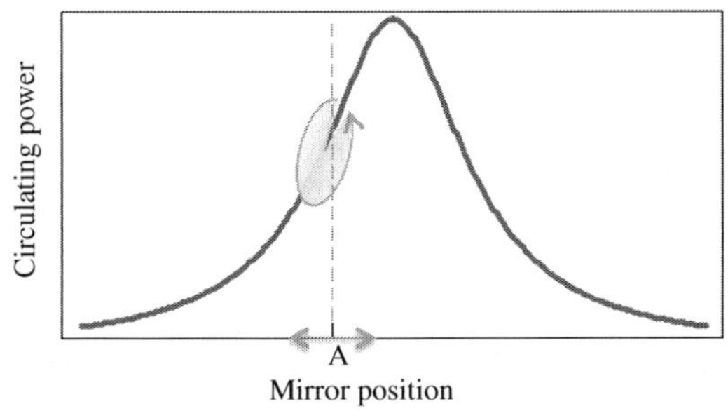

Fig. 14.3 Plot of the steady-state circulating power in a Fabry-Perot cavity as a function of the position of one of the cavity's end-mirrors. If the mirror is initially located at the point A and undergoes oscillations with amplitude indicated by the double arrow, then the instantaneous circulating power will lag behind the steady-state value, and will trace out the loop indicated in the figure. As described in the text, this loop results in laser cooling of the mirror.

in the adiabatic limit, that is, when the mirror's position is varied much more slowly than the cavity's ringdown time $1/\kappa$. Additionally, the plot in Fig. 14.3 assumes that the cavity is illuminated by a monochromatic beam (i.e., a noiseless laser).

Since the radiation pressure on a mirror is simply proportional to the optical intensity incident upon it, Fig. 14.3 also represents the force-versus-displacement curve $F(x)$ for the radiation pressure (again assuming that the mirror's motion is adiabatic with respect to the cavity field). As a result, if we assume that the end-mirror's position x is initially on the side of resonance (i.e., point A in Fig. 14.3) and undergoes very small oscillations, this will cause the amount of light in the cavity, and hence the force on the mirror, to oscillate. If the mirror motion is adiabatic with respect to the cavity field (i.e., $\omega_m \ll \kappa$), then the force on the mirror will follow the curve in Fig. 14.3, that is, its time-dependence can be written as $F(x(t))$. However, if we allow for the fact that the mirror motion is not infinitely slow compared with the cavity ringdown, then the amount of light in the cavity at any given instant will not be given by $F(x(t))$; rather, it will lag behind this value. As a result, the trajectory of $F(t)$ will trace out a loop, as illustrated in Fig. 14.3.

Since this is a loop in the force-displacement plane, work is done by (or on) the system, with the magnitude of the work being proportional to the area of the loop, and the sign of the work determined by the sense of the rotation about the loop. On the left-hand side of the resonance in Fig. 14.3 (i.e., if the laser driving the cavity is red-detuned from the cavity resonance), the sign of the loop is such that work is being done by the mirror's motion. This means that energy is lost from the mirror with each cycle of its motion, corresponding to damping. Note that there is no fluctuating force associated with this process, as we have assumed that the frequency and amplitude of the driving laser are perfectly constant.

This last point is quite important, as it is equivalent to stating that the damping is not being provided by a system in thermal equilibrium. This is because the spectrum of a noiseless laser (which would be a delta function centered at the laser frequency) is manifestly different from the spectrum associated with a thermal equilibrium state of the electromagnetic field (i.e., a blackbody spectrum). As a result, the fluctuation-dissipation theorem cannot be applied to this system, and damping may be able to reduce the oscillator's thermal motion.

14.4.3 Quantitative description of cold damping

We can provide a more quantitative description of this process by writing down the equations of motion for the oscillator and the cavity. The oscillator's equation of motion is

$$m\ddot{x} + \beta\dot{x} + kx = F_L(t) + F_{\text{opt}}(t) + F_{\text{ext}}(t), \tag{14.37}$$

where the damping coefficient β, and the thermal Langevin force $F_L(t)$ satisfy $S_{\text{FF}}^T = 4k_B T m\gamma_{\text{m}}$. $F_{\text{opt}}(t) = N(t)\hbar\omega_{\text{opt}}/L$ is the force exerted on the oscillator by the light in the cavity; here N is the number of photons in the cavity, ω_{opt} is their frequency, and L is the cavity length. The reduced Planck's constant $\hbar$ has appeared because we are writing the radiation pressure in terms of the number of photons in the cavity; however, the treatment so far is entirely classical ($\hbar$ would disappear if we wrote the radiation pressure in terms of the circulating power in the cavity rather than the photon number). Last, $F_{\text{ext}}(t)$ is an external force whose utility will become clear later.

For the optical cavity's equation of motion, we will make a gross oversimplification and assume that the *equilibrium* number of photons in the cavity, $N_{\text{eq}}(x)$, depends upon the oscillator's position x via the curve in Fig. 14.3, while the *actual* number of photons in the cavity, $N(t)$, simply decays toward the instantaneous $N_{\text{eq}}(x(t))$ with a relaxation rate κ:

$$\dot{N} = [N_{\text{eq}}(x) - N]\kappa.$$

Furthermore, we assume that the oscillator's motion will be small enough that we can replace the full form of $N_{\text{eq}}(x)$ with its Taylor expansion about the oscillator's equilibrium position:

$$\dot{N} = [N_{\text{eq}}(0) + xN'_{\text{eq}}(0) - N]\kappa. \tag{14.38}$$

This equation of motion is not capable of describing optomechanical systems that are out of the so-called "bad-cavity limit" (i.e., $\omega_m \ll \kappa$), but it will be seen that it still captures the basic physics of laser cooling and will provide considerable insight even into the quantum limits to this cooling. More complete treatments of the cavity dynamics are provided in Chapters 8 and 9.

The behavior of the system described by the coupled equations (14.37) and (14.38) can be solved by taking $F_{\text{ext}}(t) = F_1 e^{i\omega t}, x(t) = x_0 + x_1 e^{i\omega t}$, and $N(t) = N_0 + N_1 e^{i\omega t}$ (at present, we are only interested in determining the system's response to F_{ext}, so we ignore $F_L(t)$ for now). Here, F_1, x_0, and N_0 are real constants, with the latter two representing the displaced equilibrium position of the oscillator and the mean number of photons in the cavity. The complex constants x_1 and N_1 specify the system's dynamics. After some algebra, we have

$$x_1 = \frac{F_1}{m} \frac{1}{\left(\omega_m^2 - \frac{N'_{\text{eq}} \hbar \omega_{\text{opt}}}{Lm} - \omega^2\right) + i\left(\frac{\omega \omega_m}{Q} - \frac{N'_{\text{eq}} \hbar \omega_{\text{opt}} \omega}{\kappa Lm}\right)}. \tag{14.39}$$

This has the same form as a harmonic oscillator with a new resonant frequency

$$\omega_{m,\text{eff}} = \sqrt{\omega_m^2 - \frac{N'_{\text{eq}} \hbar \omega_{\text{opt}}}{Lm}} \tag{14.40}$$

and a new quality factor

$$Q_{m,\text{eff}} = \frac{1}{Q^{-1} - \frac{N'_{\text{eq}} \hbar \omega_{\text{opt}}}{\kappa Lm\omega_m}}, \tag{14.41}$$

or, equivalently, a new damping rate

$$\gamma_{m,\text{eff}} = \omega_{m,\text{eff}} \left(Q^{-1} - \frac{N'_{\text{eq}} \hbar \omega_{\text{opt}}}{\kappa Lm\omega_m}\right) \equiv \gamma_m + \gamma_{\text{opt}}. \tag{14.42}$$

The shift in the mechanical oscillator's resonance frequency described by Eq. (14.40) is known as the "optical spring," while the damping induced by the radiation pressure, Eq. (14.42), is known as "optical damping." Note that the sign of the optical damping is determined by the sign of N'_{eq}, that is, whether the driving laser is detuned to the red or blue with respect to the cavity resonance. In addition, N'_{eq} is proportional to the power of the driving laser P_{in}; as a result, $\gamma_{\text{opt}} \propto P_{\text{in}}$.

Since the mechanical element coupled to the driven optical cavity is still behaving as a harmonic oscillator (albeit with new resonant frequency and damping), its response to the white-noise force $F_L(t)$ exerted by the thermal bath will simply be characterized by a displacement power spectral density $S_{xx}[\omega] = \left|G_{\text{eff}}[\omega]\right|^2 S_{FF}$. Note that the mechanical susceptibility $G_{\text{eff}}[\omega] \equiv x_1/F_1$ is modified by the optomechanical coupling, while S_{FF} is not.

As a result, the mirror's motion will be equivalent to thermal motion at temperature

$$T_{\text{eff}} = \frac{S_{FF}^T}{4k_B m \gamma_{\text{eff}}} = T\frac{\gamma_m}{\gamma_{\text{eff}}}. \tag{14.43}$$

Thus, we can see that if the driving laser is detuned to the red of the cavity (to ensure the correct sign for N'_{eq}), then the mechanical oscillator's effective temperature decreases as $1/P_{\text{in}}$ (at least for large P_{in}), suggesting that T_{eff} could be decreased arbitrarily simply by increasing P_{in}. We have arrived at this conclusion because we have found a way to introduce damping (via the lag between the radiation pressure and the mirror's motion) without adding any fluctuating force. This type of damping is known as "cold damping."

14.4.4 Qualitative description of quantum-limited cooling

In reality, the laser that drives the cavity (and in doing so produces the cold damping) must also exert a fluctuating force on the oscillator. This is because any real laser produces an oscillating electromagnetic field whose phase and amplitude fluctuate. Classically, these fluctuations could arise from thermal fluctuations in the laser source. Even in the absence of thermal fluctuations, quantum mechanics dictates that fluctuations in the amplitude and phase of an oscillating electromagnetic field must satisfy a Heisenberg-like relation [3]. In the spirit of our heuristic description of cavity dynamics, we will for the time being describe the quantum fluctuations of the laser beam in terms of photon shot noise. Such a description will miss some important points, but it provides some useful intuition for the role of quantum noise in those optomechanical systems for which $\omega_m \ll \kappa$. In Section 14.4.5, we will refine this picture of quantum noise somewhat.

Shot noise can be thought of as arising from the random (Poisson-distributed) arrival times of photons in a beam of light. This randomness leads to apparent fluctuations in the beam's intensity with a white power spectral density $S_{II}^{(\mathrm{SN})}[\omega] \propto I$ where I is the laser intensity. These fluctuations in the laser intensity lead to fluctuations in the number of photons in the cavity, and hence to a fluctuating force on the mechanical oscillator. The Fourier components that will have the greatest impact on the mechanical oscillator are those near ω_m. Since we are assuming that $\omega_m \ll \kappa$, we can say that the intracavity fluctuations at frequencies near ω_m will closely track the fluctuations of the driving laser. As a result, the laser's shot noise leads to a fluctuating force on the mirror with power spectral density $S_{FF}^{(\mathrm{SN})}[\omega] \propto S_{II}^{(\mathrm{SN})}[\omega] \propto I \propto P_{\mathrm{in}}$ (again, assuming $\omega \approx \omega_m$ and $\omega_m \ll \kappa$).

As a result, the equation of motion, Eq. (14.37), must be modified to include the force noise arising from the laser's photon shot noise:

$$m\ddot{x} + \beta\dot{x} + kx = F_L(t) + F_{\mathrm{opt}}(t) + F_{\mathrm{ext}}(t) + F_{\mathrm{SN}}(t), \tag{14.44}$$

where $F_{\mathrm{SN}}(t)$ has a frequency-independent power spectral density $S_{FF}^{(\mathrm{SN})} \propto P_{\mathrm{in}}$. Since $F_{\mathrm{SN}}(t)$ is random, it doesn't change the oscillator's mechanical susceptibility; thus, the optical spring and optical damping will be as in Eqs. (14.40) and (14.42). However, when we calculate the effective temperature of the oscillator, we must include the total force noise acting on the oscillator (i.e., $F_L(t)$ and $F_{\mathrm{SN}}(t)$, which we take to be uncorrelated):

$$T_{\mathrm{eff}} = \frac{S_{FF}^{(\mathrm{tot})}}{4k_B m \gamma_{\mathrm{eff}}} = \frac{1}{4k_B m} \frac{S_{FF}^T + S_{FF}^{(\mathrm{SN})}}{\gamma_m + \gamma_{\mathrm{opt}}}. \tag{14.45}$$

We can gain some insight into Eq. (14.45) by first considering the limit $P_{\mathrm{in}} = 0$, that is, with no driving laser. Then, the force noise is just due to the thermal bath, that is, $S_{FF}^{(\mathrm{tot})} = S_{FF}^T$, and the mechanical damping is just the bare damping: $\gamma_{\mathrm{opt}} = 0$. In this case, Eq. (14.45) gives $T_{\mathrm{eff}} = T$, as would be expected.

In the opposite limit (when P_{in} is large enough that both $S_{FF}^{(\mathrm{tot})}$ and γ_{eff} are dominated by the driving laser) P_{in} drops out of the right-hand side of Eq. (14.45), which is to say that further increasing P_{in} will not lower T_{eff}. Physically, this is because the driving laser produces both a damping force and a fluctuating force, with both proportional to P_{in}. This situation is strongly reminiscent of our discussion of the fluctuation-dissipation result in Section 14.3.2; however, the present result is not due to the equipartition theorem of statistical mechanics, but rather to the quantum nature of the electromagnetic field.

An important question to ask at this point is: what is the lowest temperature that can be reached? If we denote this temperature as T_{opt}, we have

$$T_{\text{opt}} = \frac{S_{FF}^{(\text{SN})}}{4k_B m\gamma_{\text{opt}}} \tag{14.46}$$

(we are still assuming that both the fluctuating force and the damping are dominated by the optical field). So we would like to have the smallest possible value of $S_{FF}^{(\text{SN})}$ and the largest possible value of γ_{opt}. From this point of view, the worst arrangement would be to have the laser driving the cavity precisely at resonance, since then $\gamma_{\text{opt}} = 0$ (as can be seen from Fig. 14.3) while $S_{FF}^{(\text{SN})}$ will be large (the resonant drive will maximize the intensity of light in the cavity, and hence $S_{FF}^{(\text{SN})}$). On the other hand, at very large detunings, we again have $\gamma_{\text{opt}} = 0$. If we take the optimal detuning to be about $\kappa/2$ (to maximize N'_{eq} and ensure that a substantial fraction of the driving light enters the cavity), we have

$$\gamma_{\text{opt}} = \frac{P_{\text{in}} L F^3 \omega_{\text{opt}}}{\pi^3 c^3 m} \tag{14.47}$$

and

$$S_{FF}^{(\text{SN})} = \frac{8P_{\text{in}} F^2 \hbar\omega_{\text{opt}}}{\pi^2 c^2}, \tag{14.48}$$

giving

$$T_{\text{opt}} = \frac{\hbar}{k_B}\kappa \tag{14.49}$$

as the lowest temperature to which the mirror can be cooled. As indicated above, if the detuning Δ is not chose optimally, then the large-P_{in} limit of T_{eff} can be greater than T_{opt}:

$$\lim_{P_{\text{in}}\to\infty} T_{\text{eff}} = T_{\min}(\Delta) \geq T_{\text{opt}}.$$

Comparing Figs. 14.2 and 14.4, it appears that in an optomechanical system, the mechanical oscillator is coupled to two baths: the thermal bath at temperature T and an optical "bath" whose temperature $T_{\min}$ depends on Δ, with a minimum value T_{opt} for the appropriate choice of Δ. The strength of the oscillator's coupling to this optical "bath" is set by the intensity of the light field driving the cavity. Last, this optical "bath" is not thermal: its fluctuations arise from the quantum properties of the light field. Nevertheless, the mirror responds to it in the same way that it would respond to a thermal bath with $T = T_{\min}$. In Section 14.4.5, we consider these facts in light of our earlier discussion of coherent states, as well as the discussions of quantum noise in Chapters 2 and 3.

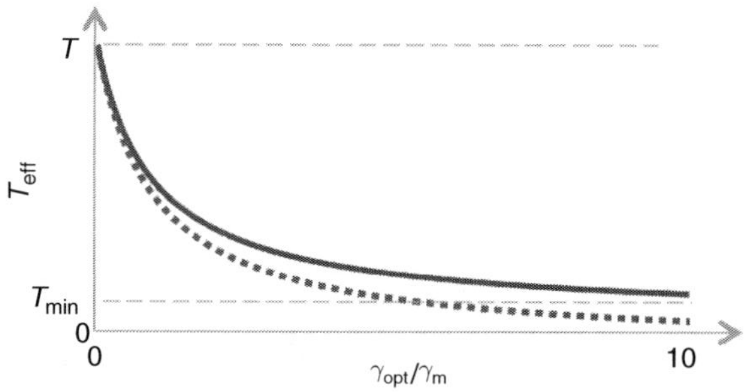

Fig. 14.4 A plot of the laser-cooled temperature of a mirror as a function of the laser power incident on the cavity. Here, the bath temperature is T and the laser power is proportional to γ_{opt}, the amount of damping added to the mirror by the laser. The dashed line shows the calculation assuming a perfectly noiseless laser, Eq. (14.43), in which the temperature is reduced arbitrarily as the laser power is increased. The solid line shows the calculation including a simple model for photon shot noise, Eq. (14.45), which leads to a minimum temperature for the laser cooling denoted by T_{min}, which here is taken to be $T/10$.

14.4.5 Quantitative description of quantum-limited cooling

As shown in Chapter 3, each normal mode of the electromagnetic field can be described as a harmonic oscillator. Sources of electromagnetic radiation such as lasers can be thought of as applying a linear, harmonic drive to one or more modes of the electromagnetic field (e.g., by moving electrical charges in an oscillatory manner).[2] As described in Section 14.2.3, the application of such a linear drive creates a coherent state.

As described in Section 14.2.2, a coherent state can be considered to be a classical oscillation (i.e., a sinusoidal wave with well-defined amplitude and phase) plus an uncertainty (in each quadrature of oscillation) equal in magnitude to the oscillator's zero-point motion. For our purposes, we can treat this uncertainty as stochastic classical fluctuations that are superposed on the classical oscillation. Within this picture, the power spectrum $S_{EE}^{(\text{ext})}[\omega]$ of the electromagnetic field produced by the laser is as illustrated in Fig. 14.5(a): a narrow peak corresponding to the classical oscillations and a white-noise background corresponding to the quantum fluctuations. This is the spectrum of the field that is incident upon the cavity.

For a cavity with perfectly reflecting mirrors, each normal mode of the electromagnetic field inside the cavity is also a harmonic oscillator. If the cavity mirrors are slightly transmissive, then the normal mode inside the cavity is weakly coupled to the modes outside the cavity; as a result, the electromagnetic field outside the cavity can be treated as a linear drive applied to the cavity mode. The linearity of the optical cavity ensures that each Fourier component of this drive is filtered by the cavity's susceptibility, with the result that $S_{EE}^{(\text{cav})}[\omega]$, the spectrum inside the cavity, is as shown

[2] This is a gross oversimplification of a real laser, but will suffice for the present discussion.

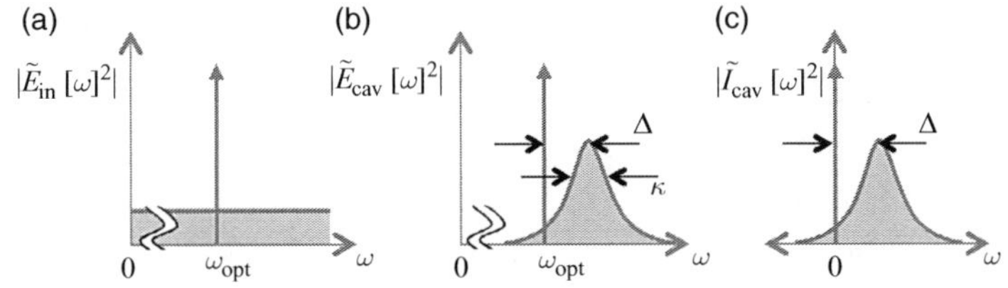

Fig. 14.5 Schematic illustrations of the electromagnetic spectra. (a) Spectrum of the electric field incident on the cavity, including the laser line at frequency ω_{opt} and the white background associated with the zero-point fluctuations of the electric field. (b) The same spectrum after passage into the cavity, which applies a Lorentzian filter to the spectrum in (a). The detuning Δ between the laser and the cavity is indicated, as is the cavity linewidth κ. (c) Spectrum of intensity fluctuations associated with the field fluctuations plotted in (b). The force acting on the mirror is proportional to intensity, and so also has this spectrum.

in Fig. 14.5(b): the laser line remains a narrow peak (albeit with a different amplitude and phase, depending upon its detuning Δ from the cavity resonance) and the quantum noise is filtered to have the Lorentzian shape of the cavity's susceptibility.

Figures 14.5(a, b) show the spectrum of the field oscillations; however the mirror responds to radiation pressure, which is proportional to the square modulus of the field. Figure 14.5(c) shows $S_{\mathrm{II}}[\omega]$, the spectrum of the intensity fluctuations in the cavity. The classical laser produces a constant intensity within the cavity, as expected. However, the beating between the laser and each Fourier component of the electric field noise in Fig. 14.5(b) results in intensity fluctuations that have the Lorentzian profile of the cavity's susceptibility and are centered at $\omega = \Delta$.

Since the radiation pressure on the mirror is proportional to the intensity inside the cavity, the force noise resulting from the quantum fluctuations of the electromagnetic field, $S_{\mathrm{FF}}^{(\mathrm{cav})}[\omega]$, will also have the form shown in Fig. 14.5(c). Note that the form of $S_{FF}^{(\mathrm{cav})}[\omega]$ is not that of a thermal bath, and so it is not immediately obvious how to assign either an effective temperature or a damping strength to this "bath." However, the mirror's high Q ensures that its motion is mostly determined by the components of $S_{FF}^{(\mathrm{cav})}[\omega]$ near $\omega = \pm\omega_m$. As shown in Chapter 2, this means that the effective temperature of this "bath" is given by

$$\frac{S_{FF}^{(\mathrm{cav})}[+\omega_m]}{S_{FF}^{(\mathrm{cav})}[-\omega_m]} = e^{\hbar\omega_m/k_B T_{\mathrm{eff}}} \tag{14.50}$$

and the damping is given by

$$\gamma_{\mathrm{opt}} \propto S_{FF}^{(\mathrm{cav})}[+\omega_m] - S_{FF}^{(\mathrm{cav})}[-\omega_m] \tag{14.51}$$

Examination of Fig. 14.5(c) shows that, for $\Delta \to 0$, Eq. (14.50) gives $T_{\mathrm{eff}} \to \infty$ and Eq. (14.51) gives $\gamma_{\mathrm{opt}} \to 0$, which is to say that the optical "bath" appears quite hot, but is only weakly coupled to the mirror's motion. $T_{\rm eff}$ is also large in the limit of $\Delta \to \infty$. The lowest value of $T_{\rm eff}$ is found for $\Delta \sim \kappa/2$, where

$$T_{\rm opt} = \frac{\hbar}{k_B} \frac{\kappa}{\sqrt{3}}, \tag{14.52}$$

in rough agreement with our (even more) heuristic estimate in Section 14.4.4.

As a final point, we note that if one wishes to cool the mirror to its ground state (i.e., $T_{\rm eff} \ll \hbar\omega_m/k_B$), then Eqs. (14.49) and (14.52) both indicate that the mechanical frequency must be greater than the cavity linewidth, that is, $\omega_{\rm m} \gg \kappa$, also known as the "good-cavity limit" or the "resolved-sideband limit." This is precisely the opposite limit from what we have assumed in all the preceding analysis, indicating that the tools developed in this chapter are not adequate for describing ground-state cooling. A more thorough treatment of cooling in optomechanical systems (which replaces the simplified cavity equation of motion, Eq. (14.38), with a more realistic description) that can describe ground-state cooling is presented in Chapters 8 and 9. The results derived here remain valid in the appropriate limits, and $\omega_m \gg \kappa$ remains a prerequisite for ground-state cooling.

14.5 Conclusion and summary

In this chapter, we have presented a heuristic description of laser cooling in optomechanical systems. The main results are:

- A linearly driven quantum harmonic oscillator behaves identically to the corresponding classical oscillator, albeit with an uncertainty equivalent to the (undriven) oscillator's zero-point motion.
- Coupling an oscillator to a thermal bath produces damping and fluctuations; the two are linked so as to satisfy the equipartition theorem.
- The connection between an oscillator and its thermal bath is analogous to a thermal conductivity proportional to $1/Q$.
- Cooling can be achieved by applying a linear drive to an oscillator that is parametrically coupled to a second oscillator (a situation that is naturally realized in optomechanical systems).
- The quantum fluctuations in this drive set a limit to the lowest temperature that can be realized. In the "bad-cavity limit," this lowest temperature is set by the cavity linewidth. To reach the ground state of an oscillator, the optomechanical system must be in the resolved sideband limit.

Our hope is that deriving these results in an intuitive manner can serve as a foundation for more sophisticated treatments, particularly those in Chapters 8 and 9.

Acknowledgments

It is a pleasure to acknowledge assistance from S. M. Girvin (particularly for the derivation leading to Eq. (14.21)), F. Marquardt, A. A. Clerk, and K. W. Lehnert, as well as from S. W. Hoch with the preparation of this manuscript.

References

[1] Haroche, S. and Raimond, J. M. (2006). *Exploring the Quantum: Atoms, Cavities, and Photons*. Oxford University Press, Oxford.

[2] Knill, E., Laflamme, R., and Milburn, G. J. (2001). A scheme for efficient quantum computation with linear optics. *Nature*, **409**, 46.

[3] Scheich, W. P. (2001). *Quantum Optics in Phase Space*. Wiley-VCH, Weinheim.

[4] Marquardt, F. and Girvin, S. M. (2009). Optomechanics, an introduction. *Physics*, **2**, 40.

[5] Kippenberg, T. J. and Vahala, K. J. (2007). Cavity Opto-Mechanics. *Opt. Express*, **15**, 17172.

[6] Milatz, J. M. W. and Van Zolingen, J. J. (1953). The Brownian motion of electrometers. *Physica*, **19**, 181.

[7] Braginsky, V. B. and Manukin, A. B. (1967). Ponderomotive effects of electromagnetic radiation. *Sov. Phys. JETP*, **52**, 986.

[8] Braginsky, V. B., Manukin, A. B., and Tikhonov, M. Y. (1970). Investigation of dissipative ponderomotive effects of electromagnetic radiation. *Sov. Phys. JETP*, **31**, 829.

15
Tomography schemes for characterizing itinerant microwave photon fields

C. Eichler,* D. Bozyigit,* C. Lang,*
L. Steffen,* J. Fink,* and A. Wallraff*

*Department of Physics, ETH Zürich, CH-8093, Zürich, Switzerland

Quantum Machines: Measurement and Control of Engineered Quantum Systems.
Edited by M. Devoret, B. Huard, R. Schoelkopf, and L. F. Cugliandolo.

Chapter Contents

A wide range of experiments studying microwave photons localized in superconducting cavities have made important contributions to our understanding of the quantum properties of radiation. Typically, experiments are performed in the context of cavity quantum electrodynamics (QED) with Rydberg atoms [1–3] or superconducting circuits [4–6].

More recently, propagating (itinerant) microwave fields on a single-photon level have been generated [7], but their quantum properties have not yet been studied with the same intensity. This is partly due to the fact that efficient single-photon counters in this frequency domain are still under development [8, 9]. However, it has been shown that characteristic quantum properties of propagating microwave photons, such as antibunching, can be observed in correlation measurements using linear amplifiers and quadrature amplitude detectors [10, 11]. In the context of circuit QED, propagating microwaves are also used to control [12] and read out the quantum state of artificial atoms [13, 14] and to observe phenomena such as resonance fluorescence [15, 16].

Here, we discuss techniques for the reconstruction of the quantum state of itinerant microwave photon fields [17–20]. Efficient methods, applicable to linear amplification and quadrature detection schemes, to separate the detected microwave signal from the noise added by the amplifier are presented. We show how to reconstruct the corresponding density matrix or Wigner function from the measurement data. As an explicit example, we discuss measurements of single-photon Fock states and their superposition with the vacuum reconstructed by analyzing the moments of the measured amplitude distribution up to fourth order.

15.1 Optical and microwave frequency field detection

In many instances, a radiation field of interest can be described by a single propagating mode a that reaches the detector within a specific window in time. The single mode a can be isolated from the continuum of modes by integrating the continuous signal over a chosen temporal profile as discussed in [20, 36]. For the full reconstruction of the quantum state of the field both the photon number statistics and all coherences between the different contributing Fock states have to be experimentally determined. This is typically achieved by measuring generalized field quadrature components $\hat{X}_\phi \equiv \frac{1}{2}(ae^{-i\phi}+a^\dagger e^{i\phi})$ instead of the photon number $a^\dagger a$, which naturally allows for exploring the full phase space, that is, the off-diagonal elements of the density matrix in the number-state basis.

In optical systems, where number statistics are naturally obtained using photon counters, such a field quadrature measurement can be realized using homodyne detection schemes. In this approach, the field of interest is combined on a beamsplitter with a strong coherent field of a local oscillator, such that the difference of the photocurrents at the two beamsplitter outputs is proportional to a specific field quadrature $\hat{X}_\phi$ of the input field (see Fig. 15.1(a)). The quadrature phase ϕ can be tuned by changing the local oscillator phase. Instead, microwave field quadratures are usually measured by mixing the field with a local oscillator and sampling the electric field directly using

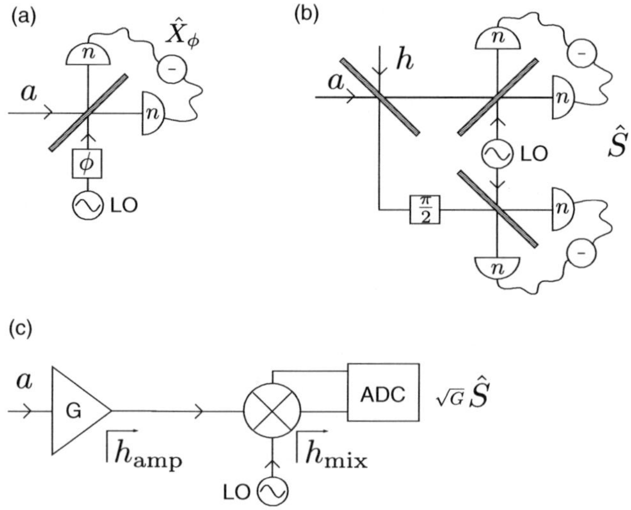

Fig. 15.1 Typical field quadrature detection schemes for optical and microwave fields. (a) Optical homodyne detection is based on photocounters and a strong local oscillator field. (b) Measurement of two conjugate quadratures (i.e., the complex amplitude $\hat{S}$) can be achieved by adding a beamsplitter and a second optical homodyne detector. Alternatively, one can use a heterodyne detection scheme. (c) At microwave frequencies, direct sampling of the electrical field is possible. The amplification stage required for detection adds a minimal amount of noise, but allows for the simultaneous detection of two conjugate quadratures naturally.

analog-to-digital converters (ADCs). However, these ADCs are only sensitive enough to detect large-amplitude fields that contain a macroscopic number of photons per sampling time, so that a linear amplification stage is required in the process of detection, as shown in Fig. 15.1(c). The noise added during this amplification process is typically the main limitation on the detection efficiency of microwave fields, as discussed below.

Instead of measuring a single field quadrature for different phases ϕ, one can measure two conjugate field quadratures simultaneously to get all the information required for a complete quantum state reconstruction [22–26]. Such a measurement can be realized by using a beamsplitter and two quadrature detectors at each output (see Fig. 15.1(b)). The beamsplitter necessarily introduces an additional mode h through its open port. This mode adds (at least) the vacuum noise to the signal with which the simultaneous detection of conjugate variables preserves Heisenberg's uncertainty principle. Taking the beamsplitter transformations $a \to (a+h)/\sqrt{2}$ and $h \to (a-h)/\sqrt{2}$ into account, the two detected field quadratures at the beamsplitter outputs correspond to the real, $\hat{X} \equiv \hat{X}_0$, and imaginary, $\hat{P} \equiv \hat{X}_{\pi/2}$, parts of the complex amplitude $a + h^\dagger$. This holds for both the optical and the microwave cases. However, for microwaves, we still have to consider the transformation of the signal mode due to the linear amplification stage. A generic phase-preserving linear amplifier transformation can be modeled as [27–29]

$$a \rightarrow \sqrt{G}a + \sqrt{G-1}h^{\dagger}_{\text{amp}}, \tag{15.1}$$

where h_{amp} is an additional bosonic mode accounting for the noise added by the amplifier. Again, in the ideal (i.e., quantum-limited) case, h_{amp} is in the vacuum state, and for a more realistic scenario in a thermal state. Combining the amplification transformation with the beamsplitting at the mixing stage (cf. Fig. 15.1(c)) and dividing by $\sqrt{G/2}$, we find the relation

$$\hat{S} \equiv a + h^{\dagger} = \hat{X} + i\hat{P}, \tag{15.2}$$

with the total noise mode $h = \sqrt{(G-1)/G}\,h_{\text{amp}} + \sqrt{1/G}\,h_{\text{mix}}$. Here, we have defined the complex amplitude operator $\hat{S}$ representing the two conjugate quadratures as a single complex number. In the limit of large gain $G \gg 1$, the total noise is dominated by the amplifier noise $h \approx h_{\text{amp}}$ and the following noise contributions can be neglected. Furthermore, we notice that, once we amplify the field phase-insensitively, at least the vacuum noise is added independently of whether we detect only one quadrature or two conjugate quadrature components. Once the signal is amplified, it is thus natural to detect two conjugate quadratures, since the signal-to-noise ratio is unaffected by the necessary splitting of the signal. Commercial IQ mixers therefore have a built-in beamsplitter and two output ports for the two conjugate quadratures $\hat{X}$ and $\hat{P}$ (or I and Q in microwave circuit language).

It is important to mention that there is a detection scheme using linear amplifiers that is ideally noiseless for one quadrature component. This is achieved by replacing the phase-insensitive amplifier by a phase-sensitive one, which can, in the quantum limit, be modeled by the squeezing transformation [28–30]

$$a \rightarrow \sqrt{G}e^{-i\phi}a + \sqrt{G-1}e^{i\phi}a^{\dagger}, \tag{15.3}$$

with tunable phase ϕ. Amplifiers have recently been built that are described by this transformation and are working close to the quantum limit [31–33]. The quadrature $\hat{X}_{\phi}$ is noiselessly amplified, while its conjugate quadrature is deamplified. The detection scheme is thus equivalent to an optical homodyne detection [18, 24, 34].

As a concluding remark, we note that while for optical fields the simultaneous detection of two conjugate quadratures requires a more complicated setup than for number detection, it is the natural measurement observable for microwave fields, and we will therefore focus on it in this work in the context of quantum state reconstruction.

15.2 Quantum state reconstruction based on single-channel field quadrature detection

The goal of quantum state reconstruction is the estimation of the density matrix ρ_a that characterizes the state of the field mode a. This can be achieved experimentally by preparing the state many times and performing a certain set of measurements on these states. Depending on the set of measurements, there are different representations of the density matrix that are most closely related to the direct measurement results. In the case of field quadrature detection, these representations are the phase-space distributions such as the Husimi Q function or the Wigner function. In this section, we introduce these phase-space distributions and the concept of statistical moments. We also discuss their connection to the Fock basis representation of the density matrix.

15.2.1 Introduction to phase-space distributions

First we introduce the Glauber-Sudarshan P distribution, from which all other quasi-probability distributions can be derived. Owing to the non-orthogonality of coherent states, $\langle\alpha|\beta\rangle = e^{-\frac{1}{2}|\alpha|^2-\frac{1}{2}|\beta|^2}e^{\alpha^*\beta}$, an arbitrary density matrix ρ_a can be expanded as a linear combination of projectors $|\alpha\rangle\langle\alpha|$ onto coherent states [21, 35]:

$$\rho_a = \int_\alpha P_a(\alpha)|\alpha\rangle\langle\alpha|. \tag{15.4}$$

Here we have introduced the notation $\int_\alpha \equiv \int_{\mathbb{C}} d^2\alpha$ for integrals over the complex plane and the Glauber-Sudarshan P distribution $P_a(\alpha)$. The density matrix is thus uniquely defined by $P_a(\alpha)$, which is always real-valued but can be negative and can contain singularities proportional to derivatives of the Dirac δ distribution to all orders [36]. As can be seen from its definition, Eq. (15.4), the P function reduces to a two-dimensional Dirac distribution $P_a(\alpha) = \delta^2(\alpha-\beta)$ for coherent states $|\beta\rangle$. Coherent states thus appear as single points in phase space, with no statistical spread, similar to their classical counterparts. The vacuum fluctuations of the field are not present in this quasi-probability distribution. For this reason, and owing to its possible negativities, the P function is not directly measurable. However, it is very useful as a theoretical tool, since its statistical moments directly correspond to the normally ordered moments of the field operator,

$$\langle(a^\dagger)^m a^n\rangle = \int_\alpha (\alpha^*)^m\alpha^n \; P_a(\alpha), \tag{15.5}$$

and because of its analogy to the probability distributions of classical fields. In addition to the P distribution, we introduce the Husimi Q function as

$$Q_a(\alpha) = \frac{1}{\pi}\langle\alpha|\rho_a|\alpha\rangle, \tag{15.6}$$

which generates the antinormally ordered moments

$$\langle a^n(a^\dagger)^m\rangle = \int_\alpha (\alpha^*)^m\alpha^n \; Q_a(\alpha)\;. \tag{15.7}$$

Substituting Eq. (15.4) into the definition of the Q function, we note that it is related to the P function by a Gaussian convolution. For coherent states, it becomes a two-dimensional Gaussian distribution with variance 1 centered around the coherent-state amplitude. Half of these fluctuations can be attributed to the intrinsic vacuum fluctuations of the quantum field, the other half being due to the minimum added noise when directly measuring a Q function. The fact that there are finite fluctuations embedded in the Q-function representation makes it a completely positive and measurable function, as discussed below.

Both distributions are special cases of the *s-parametrized* quasi-probability distribution (QPD) $W(\alpha, s)$,

$$Q(\alpha) = W(\alpha, -1), \tag{15.8}$$

$$P(\alpha) = W(\alpha, 1), \tag{15.9}$$

which was introduced by Cahill and Glauber [37] as a generalized phase-space representation of the density matrix where the parameter $s \in (-\infty, +1]$. For different values of s, the QPDs are related by a Gaussian convolution [37]

$$W(\alpha, s) = \frac{2\pi^{-1}}{t-s} \int_\beta \exp\left(-\frac{2|\alpha-\beta|^2}{t-s}\right) W(\beta, t) \tag{15.10}$$

for all $t > s$. An intuitive interpretation of the parameter $1 - s$ relates to the amount of noise that is contained in the distribution in units of half photons. For $s = 0$, we obtain the Wigner function $W(\alpha) \equiv W(\alpha, 0)$, which contains the intrinsic vacuum fluctuations but no additional noise due to measurement. For $s = 1$, we get the P function, where even the intrinsic vacuum fluctuations are removed. On the other hand, for $s = -1$, we get the Q function, where both the vacuum fluctuations and the minimal added detection noise are included. As discussed below, additional classical detection noise leads to a further decrease in the parameter s when identifying measured distributions with a generalized QPD.

15.2.2 Measurement of phase-space distributions

In order to understand the connection between generalized QPDs and measured distributions in different experimental situations, let us assume that the complex amplitude $\hat{S} = a + h^\dagger$ as introduced above is the measured observable and that the results S of repeated measurements are stored in a two-dimensional histogram $D^{[\rho_a]}(S)$, where the two axes represent the real and imaginary part of S. From this measured histogram, all statistical moments can be evaluated numerically as

$$\langle (\hat{S}^\dagger)^n \hat{S}^m \rangle_{\rho_a} = \int_S (S^*)^n S^m \; D^{[\rho_a]}(S) \; . \tag{15.11}$$

Under the reasonable assumption that the signal mode a and the noise mode h are uncorrelated, $\rho = \rho_a \otimes \rho_h$, one can decompose these moments into products of signal and noise moments.

$$\langle (\hat{S}^\dagger)^n \hat{S}^m \rangle_{\rho_a} = \sum_{i,j=0}^{m,n} \binom{n}{j} \binom{m}{i} \langle (h^\dagger)^i h^j \rangle \langle a^{m-i} (a^\dagger)^{n-j} \rangle. \tag{15.12}$$

Here, we have chosen an operator ordering where the signal moments $\langle a^m (a^\dagger)^n \rangle$ appear antinormally ordered and the noise moments $\langle (h^\dagger)^m h^n \rangle$ normally ordered. Note that since $\hat{S}$ is a normal operator $[\hat{S}, \hat{S}^\dagger] = 0$, one can express Eq. (15.12) also with opposite ordering as shown below.

From statistics, it is well known that the probability distribution for the sum of two independent random variables $a + h^\dagger$ is given by the convolution of the individual distributions for a and $h^\dagger$. As a result, the probability distribution $D^{[\rho_a]}(S)$ is related to the convolution [38]

$$D^{[\rho_a]}(S) = \int_\alpha P_h(S^* - \alpha^*) Q_a(\alpha), \tag{15.13}$$

where the specific combination of P and Q functions is related to the normal and antinormal ordering of the moments in Eq. (15.12).

Let us discuss special cases of Eq. (15.13) that are of particular experimental relevance. In optical experiments, where the measurement of S can be realized using a double homodyne or heterodyne detection, the noise mode h is nearly in the vacuum state and we are close to the ideal situation where $P_h(\beta) = \delta^{(2)}(\beta)$ is a two-dimensional Dirac δ-function [36], resulting in

$$D^{[\rho_a]}(S) = Q_a(S). \tag{15.14}$$

In contrast, for microwave fields, the noise mode h is typically in a thermal state with mean photon number N_0 ranging from 0.5 to 10 if parametric or SQUID amplifiers are used [18, 39, 40] or between 30 and 150 if the first amplification is performed by a high-electron-mobility transistor (HEMT) amplifier [11, 16, 17, 20]. In this case, $P_h(\alpha) = e^{-|\alpha|^2/N_0}/\pi N_0$ acts as a Gaussian filter and we measure the smoothed QPD

$$D^{[\rho]}(S) = W(S, -1 - 2N_0). \tag{15.15}$$

Note that finite thermal noise in h is equivalent to optical homodyne detection with finite detection efficiency η, for which the measured distribution is given by $D^{[\rho]}(S) = W(S, 1 - 2\eta^{-1})$ [41]. This leads to the intuitive picture that added noise can be understood as a reduced detection efficiency $\eta = 1/(1 + N_0)$.

We conclude that, under the reasonable assumption of h being in a thermal state not correlated with a, the histogram measurement of $\hat{S}$ is a direct measurement of the generalized QPD and therefore contains all information about the density matrix. In contrast to other reconstruction schemes, only a single observable $\hat{S}$ has to be measured.

However, in many experiments, the system noise number is larger than the mean photon number of the detected field, $N_0 > \langle a^\dagger a \rangle$, and consequently, different measured probability distributions may appear very close to a Gaussian distribution for very different quantum states. One therefore has to find methods to transform the measured QPD into other representations from which a direct estimation of the properties of the state, such as the fidelity with respect to some expected density matrix, is possible.

15.2.3 Statistical moments and the Fock-space density matrix

One way of quantifying the properties of a quantum state is to analyze the statistical moments $\langle (a^\dagger)^n a^m \rangle$ of the field operator, since quantities such as the mean amplitude, the mean photon number and the variance in the photon number can be extracted immediately. In this section, we will discuss the approach developed in [20] to extract these moments from the measured distributions in the presence of significant amplifier noise N_0. The basic idea is to deconvolve the QPDs for the field operators a and h order by order.

Let us rewrite Eq. (15.12) with a different choice of ordering:

$$\langle (\hat{S}^\dagger)^n \hat{S}^m \rangle_{\rho_a} = \sum_{i,j=0}^{n,m} \binom{m}{j} \binom{n}{i} \langle (a^\dagger)^i a^j \rangle \langle h^{n-i} (h^\dagger)^{m-j} \rangle, \tag{15.16}$$

where the signal moments $\langle (a^\dagger)^n a^m \rangle$ are normally ordered. In this form, we see that once the antinormally ordered moments of the noise mode $\langle h^n (h^\dagger)^m \rangle$ are known, we can solve this set of linear equations for $\langle (a^\dagger)^n a^m \rangle$. Making use of Eq. (15.11), we find that a reference measurement $D^{[|0\rangle\langle 0|]}(S)$, where a is prepared in the vacuum, gives direct access to the moments $\langle h^n (h^\dagger)^m \rangle$, since all normally ordered moments in a with $n, m \neq 0$ are then $\langle (a^\dagger)^n a^m \rangle = 0$ and Eq. (15.16) reduces to

$$\langle (\hat{S}^\dagger)^n \hat{S}^m \rangle_{|0\rangle\langle 0|} = \langle h^n (h^\dagger)^m \rangle \ . \tag{15.17}$$

In many experiments, such a reference measurement with a in the vacuum is available since the source of radiation can be cooled into the ground state or very close to it [42]. The identity in Eq. (15.17) can be understood in a very simple way. Since the roles of a and h in our detection model can be exchanged, the situation with a in the vacuum state can be interpreted as an ideal Q-function measurement for the noise mode h and the moments generated by this distribution are exactly the antinormally ordered ones. Finally, we invert Eq. (15.16) to extract the desired moments $\langle (a^\dagger)^n a^m \rangle$.

In principle, one can evaluate the moments of the measured histograms to arbitrary order. However, there are limitations in the accuracy with which the moments $\langle (a^\dagger)^n a^m \rangle$ can be determined, depending on the integration time and the detection efficiency. As investigated theoretically by da Silva *et al.* [10], the statistical error on the moments in fact increases with increasing order. The result, which is based on Chebyshev's equation [43], shows that the number of measurements that are necessary to extract a moment of order M with a given confidence scales with $(1 + N_0)^M$. The measurement time necessary to determine higher-order moments with some fixed precision thus scales exponentially with increasing order.

The state of a single mode of the radiation field has an infinite number of degrees of freedom (i.e., an infinite-dimensional Hilbert space). This makes it in principle impossible to exactly reconstruct a state, because an infinite amount of information is to be acquired [44]. However, in many cases, we can settle for an approximation that requires only a finite amount of information from the measurements [44]. For our purposes, we present two important classes of states that allow a significant reduction of the required information.

The class of *Gaussian states*, which comprises coherent, thermal, and squeezed states, are completely determined by their statistical moments up to second order,

$$\{\langle a\rangle, \langle a^\dagger a\rangle, \langle a^2\rangle\}. \tag{15.18}$$

All higher-order moments can be directly calculated from these three independent moments. Therefore, the experimental effort for a reconstruction approximated by a Gaussian state reduces to the extraction of these three moments. In order to analyze how close to a Gaussian the reconstructed state really is, one can determine the third-order cumulants and check how well they approach zero.

A second class of states that can be reconstructed using finite measurement time are the states with finite photon number N, which satisfy $\langle n|\rho_a|m\rangle = 0$ for $m, n > N$ in the Fock basis $\{|n\rangle\}$. For these states, the normally ordered moments

$$\langle (a^\dagger)^m a^n\rangle = 0 \qquad \text{for } m \text{ or } n > N \tag{15.19}$$

vanish and the state is completely determined by the finite set of moments

$$\{\langle (a^\dagger)^n a^m\rangle\} \qquad \text{with } m \text{ and } n \leq N. \tag{15.20}$$

It is important to mention that it necessarily follows from $\langle (a^\dagger)^N a^N\rangle = 0$ that there are no Fock states $|n\rangle$ with $n \geq N$ contributing to the density matrix. If $\langle (a^\dagger)^N a^N\rangle < \epsilon$ can be verified experimentally, one knows an upper bound

$$\epsilon > \langle (a^\dagger)^N a^N\rangle = \sum_{n\geq N} \langle n|\rho_a|n\rangle \frac{n!}{(n-N)!} \geq \sum_{n\geq N} \langle n|\rho_a|n\rangle \tag{15.21}$$

for the sum of higher-order Fock-state populations. The approximation made when truncating the Hilbert space is thus well controlled.

If such a truncation is possible, the moments can be mapped to a density matrix in Fock representation by evaluating the sum [45]

$$\langle m|\rho_a|n\rangle = \frac{1}{\sqrt{n!m!}} \sum_{j=0}^{\infty} \frac{(-1)^l}{l!} \langle (a^\dagger)^{n+l} a^{m+l}\rangle \tag{15.22}$$

up to terms of order $2N$.

For the reconstruction of specific quantum states (Gaussian states and small-photon-number states), the described procedure is very efficient, since the evaluation of moments from the measured distributions as well as finding the solution of Eq. (15.16) requires only small computational effort. Furthermore, the moment representation provides a very intuitive picture to extract fundamental properties of the quantum state, as discussed below.

15.3 Experimental reconstruction of single-photon Fock states

We have experimentally implemented the state reconstruction procedure described above for single-photon Fock states and their superposition with the vacuum [20]. The photon source in these experiments is realized as a single-sided superconducting resonator coupled to a transmon qubit. By swapping single-qubit excitations coherently into the resonator, one can be prepare Fock states in the resonator, which then decay exponentially into the transmission line. We observe the field in the transmission line using a detection scheme as shown in Fig. 15.1(c) with a commercial HEMT amplifier.

We store the results S of repeated measurements of $\hat{S}$ in a two-dimensional histogram with 1024×1024 entries which corresponds to a discretized version of the probability distribution $D^{[\rho_a]}(S)$. To extract the properties of mode a alone, we perform two measurements, one in which mode a is left in the vacuum serving as a reference measurement for the noise and a second in which the state of interest $|\psi\rangle$, such as a Fock state $|1\rangle$, is prepared. In practice, both histograms are accumulated in an interleaved fashion, changing between the two cases every 25 μs to avoid systematic errors due to drifts. The measured histograms for both the vacuum $D^{[|0\rangle\langle 0|]}$ and for a Fock state $D^{[|1\rangle\langle 1|]}$ are dominated by the noise added by the amplifier (see Fig. 15.2(a, b)). However, when calculating the numerical difference of both histograms, we already clearly observe the circular symmetric character of the single-photon phase space distribution (see Fig. 15.2(c)). The small deviation from an ideal circular symmetry is explained by a slight coherent admixture of the vacuum $|0\rangle$ to the single-photon Fock

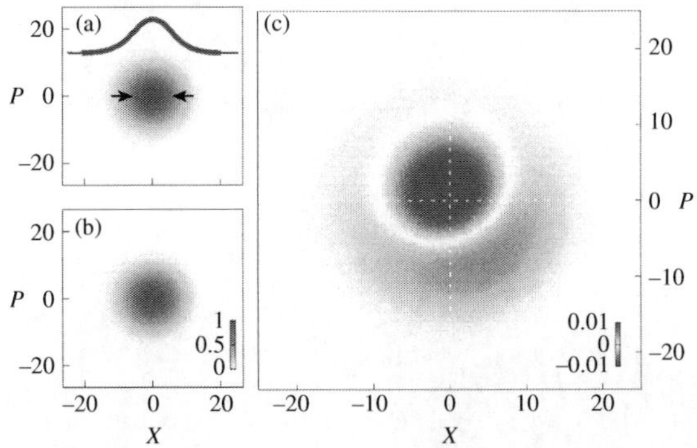

Fig. 15.2 (a) Measured quadrature histogram $D^{[|0\rangle\langle 0|]}(S)$ for a in the vacuum where $S = X + iP$. The inset shows a horizontal cut through the histogram (thick curve). The distribution is well described by a normal distribution (thin curve) with width $\sigma = 5.7$ (indicated by the arrows) corresponding to a system noise temperature of $T_{\text{noise}} \approx 21\,\text{K}$. (b) Quadrature histogram $D^{[|1\rangle\langle 1|]}(S)$ for preparation of single-photon Fock states. (c) Difference of $D^{[|1\rangle\langle 1|]}(S)$ and $D^{[|0\rangle\langle 0|]}(S)$. Note the two different shading scales, both given in units of $D^{[|1\rangle\langle 1|]}(0)$, indicating the small difference between the two histograms in the central part, the values go from 0 to 0.01 as the shading goes from light to dark, while in the surrounding "ring," the values go from 0 to −0.01 as the shading darkens.

state $|1\rangle$ due to small errors in the state preparation. In addition, we observe the expected increase in the probability of measuring larger amplitude values when the single-photon source is turned on compared with the vacuum case.

To further analyze the data, we calculate the moments $\langle (\hat{S}^\dagger)^n \hat{S}^m \rangle_{\rho_a}$ of both measured histograms up to order $n + m = 4$ using Eq. (15.11), from which we can determine the normally ordered moments $\langle (a^\dagger)^n a^m \rangle$ for the signal using Eqs. (15.16) and (15.17). The result for the preparation of a single-photon Fock state is shown in Fig. 15.3(a).

We note that the quadrature histograms are normalized such that the zeroth-order moments are always unity for all prepared states, which is equivalent to the normalization of the histogram to unity. The off-diagonal elements in the moment matrix express coherences between different photon number states. They vanish for states with circular symmetric phase-space distributions such as pure Fock states or thermal states. For the Fock state $|1\rangle$ (Fig. 15.3(a)), we observe that all off-diagonal moments are close to zero. In addition, we note that the fourth-order moment $\langle (a^\dagger)^2 a^2 \rangle$ is also close to zero, indicating antibunching of the prepared single-photon states [11]. In

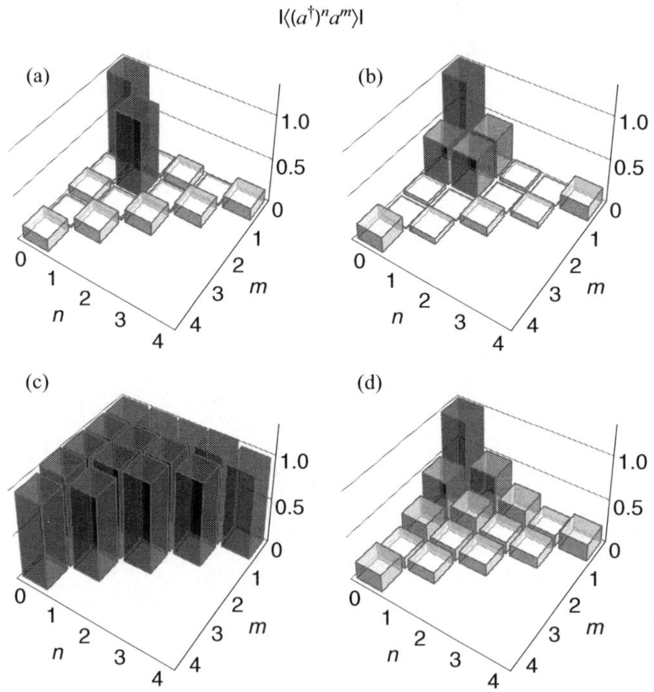

Fig. 15.3 Absolute value of the normally ordered moments $|\langle (a^\dagger)^n a^m \rangle|$ up to fourth order for a single-photon Fock state (a), a superposition state $(|0\rangle - |1\rangle)/\sqrt{2}$ (b), and two coherent states with amplitude $\alpha = 1$ (c), and $\alpha = 0.5$ (d).

contrast, a thermal state with the same mean photon number would display vanishing off-diagonal moments but finite diagonal fourth-order moments. Experimentally, for the single-photon Fock state, the aforementioned residual coherent admixture of the vacuum state leads to a nonvanishing small mean amplitude $|\langle a \rangle| = 0.044$ and a slightly reduced mean photon number $\langle a^\dagger a \rangle = 0.91$. For an integration time of 12 hours for each state, which is equivalent to the generation of 5×10^{10} single photons, we find the errors of the fourth-order moments to be approximately ± 0.1, where the statistical error in the moments is known to increase exponentially with increasing order [10]. In comparison, the estimated statistical errors for the first-, second-, and third-order moments are approximately 1.5×10^{-3}, 4.5×10^{-3}, and 1.5×10^{-2}, respectively. The errors have been estimated from the standard deviation of the moments acquired in repeated measurements of the distributions.

We have also prepared and analyzed superposition states of the type $(|0\rangle + e^{i\phi}|1\rangle)/\sqrt{2}$ (see Fig. 15.3(b)). The relative phase ϕ is controlled by the phase of the corresponding qubit state that is mapped into the resonator. For this class of states, the mean amplitude ideally equals the mean photon number $|\langle a \rangle| = \langle a^\dagger a \rangle = 0.5$, which is a signature of the nonclassicality of this state. The first equality remains approximately valid even if the state is slightly mixed with the vacuum. We have been able to use this property to determine the effective gain G of our amplifier chain, because first- and second-order moments have different characteristic scalings with G. This allowed us to scale the X and P axes of the histograms (Fig. 15.2) such that they correspond to the real and imaginary parts of $a + h^\dagger$. From our measurement data, we extract $|\langle a \rangle| = 0.466$, which is close to the expected value.

To further confirm the validity of our scheme, we have generated coherent states $|\alpha\rangle$ with amplitude $\alpha = 1$ and $\alpha = 0.5$ by applying 10 ns square coherent pulses with controlled amplitude to the weakly coupled input port of the resonator. The moments of the coherent states are given by $\langle (a^\dagger)^n a^m \rangle = (\alpha^*)^n \alpha^m$. For $\alpha = 1$, all moments are observed to be close to 1 (Fig. 15.3(c)), as expected. This also demonstrates that systematic errors in the detection chain, such as small nonlinearities, are negligible, since all moments take their expected values. For $\alpha = 0.5$ (Fig. 15.3(d)), the measured moments are observed to decay exponentially with $\langle (a^\dagger)^n a^m \rangle = 0.5^{n+m}$, as expected. The fourth-order moments appear larger than the third-order ones, owing to their larger statistical error.

As discussed in Section 15.2.3 (see Eq. (15.22)), one can also use the measured moments to reconstruct the quantum state of mode a in terms of a Fock-space density matrix or a Wigner function, as demonstrated in [20].

In conclusion, we have discussed how the joint measurement of two conjugate quadratures, expressed as a single observable $\hat{S}$, can be used for an efficient characterization of the quantum state of traveling bosonic fields. Our method has been shown to work even in the presence of significant system noise $N_0 \gg 1$. Increasing the total detection efficiency by making use of low-noise parametric amplifiers [32, 39, 40, 46] may allow the reconstruction of more complicated states in much shorter times as well. The availability of efficient methods for both the generation and characterization of microwave quantum fields offers the possibility to use those fields as carriers of quantum information over short distances.

Acknowledgments

We acknowledge fruitful discussions with A. Blais, M. P. da Silva, M. Devoret, K. Lehnert, and G. Milburn. This work was financially supported by the European Research Council (ERC) through a Starting Grant and by ETHZ.

References

[1] Haroche, S. and Raimond, J.-M. (2006). *Exploring the Quantum: Atoms, Cavities, and Photons*. Oxford University Press, Oxford.

[2] Raimond, J. M., Brune, M., and Haroche, S. (2001, Aug). Manipulating quantum entanglement with atoms and photons in a cavity. *Rev. Mod. Phys.*, **73**, 565–582.

[3] Sayrin, C., Dotsenko, I., Zhou, X., Peaudecerf, B., Rybarczyk, T., Gleyzes, S., Rouchon, P., Mirrahimi, M., Amini, H., Brune, M., Raimond, J.-M., and Haroche, S. (2011). Real-time quantum feedback prepares and stabilizes photon number states. *Nature*, **477**, 73–77.

[4] Blais, A., Huang, R.-S., Wallraff, A., Girvin, S. M., and Schoelkopf, R. J. (2004). Cavity quantum electrodynamics for superconducting electrical circuits: an architecture for quantum computation. *Phys. Rev. A*, **69**, 062320.

[5] Wallraff, A., Schuster, D. I., Blais, A., Frunzio, L., Huang, R.-S., Majer, J., Kumar, S., Girvin, S. M., and Schoelkopf, R. J. (2004). Strong coupling of a single photon to a superconducting qubit using circuit quantum electrodynamics. *Nature*, **431**, 162–167.

[6] Hofheinz, M., Wang, H., Ansmann, M., Bialczak, Radoslaw C., Lucero, E., Neeley, M., O'Connell, A. D., Sank, D., Wenner, J., Martinis, J. M., and Cleland, A. N. (2009). Synthesizing arbitrary quantum states in a superconducting resonator. *Nature*, **459**, 546–549.

[7] Houck, A. A., Schuster, D. I., Gambetta, J. M., Schreier, J. A., Johnson, B. R., Chow, J. M., Frunzio, L., Majer, J., Devoret, M. H., Girvin, S. M., and Schoelkopf, R. J. (2007). Generating single microwave photons in a circuit. *Nature*, **449**, 328–331.

[8] Chen, Y.-F., Hover, D., Sendelbach, S., Maurer, L., Merkel, S. T., Pritchett, E. J., Wilhelm, F. K., and McDermott, R. (2011). Microwave photon counter based on Josephson junctions. *Phys. Rev. Lett.*, **107**, 217401.

[9] Romero, G., García-Ripoll, J. J., and Solano, E. (2009). Microwave photon detector in circuit QED. *Phys. Rev. Lett.*, **102**, 173602.

[10] da Silva, M. P., Bozyigit, D., Wallraff, A., and Blais, A. (2010). Schemes for the observation of photon correlation functions in circuit qed with linear detectors. *Phys. Rev. A*, **82**, 043804.

[11] Bozyigit, D., Lang, C., Steffen, L., Fink, J. M., Eichler, C., Baur, M., Bianchetti, R., Leek, P. J., Filipp, S., da Silva, M. P., Blais, A., and Wallraff, A. (2011). Antibunching of microwave-frequency photons observed in correlation measurements using linear detectors. *Nature Phys.*, **7**, 154–158.

[12] Chow, J. M., DiCarlo, L., Gambetta, J. M., Motzoi, F., Frunzio, L., Girvin, S. M., and Schoelkopf, R. J. (2010). Optimized driving of superconducting artificial atoms for improved single-qubit gates. *Phys. Rev. A*, **82**, 040305.

[13] Palacios-Laloy, A., Mallet, F., Nguyen, F., Bertet, P., Vion, D., Esteve, D., and Korotkov, A. N. (2010). Experimental violation of a Bell's inequality in time with weak measurement. *Nature Phys.*, **6**, 442–447.

[14] Schuster, D. I., Wallraff, A., Blais, A., Frunzio, L., Huang, R.-S., Majer, J., Girvin, S. M., and Schoelkopf, R. J. (2005). AC Stark shift and dephasing of a superconducting qubit strongly coupled to a cavity field. *Phys. Rev. Lett.*, **94**, 123602.

[15] Astafiev, O., Zagoskin, A. M., Abdumalikov, A. A. Jr., Pashkin, Yu. A., Yamamoto, T., Inomata, K., Nakamura, Y., and Tsai, J. S. (2010). Resonance fluorescence of a single artificial atom. *Science*, **327**, 840–843.

[16] Lang, C., Bozyigit, D., Eichler, C., Steffen, L., Fink, J. M., Abdumalikov, A. A., Baur, M., Filipp, S., da Silva, M. P., Blais, A., and Wallraff, A. (2011). Observation of resonant photon blockade at microwave frequencies using correlation function measurements. *Phys. Rev. Lett.*, **106**, 243601.

[17] Eichler, C., Bozyigit, D., Lang, C., Baur, M., Steffen, L., Fink, J. M., Filipp, S., and Wallraff, A. (2011). Observation of two-mode squeezing in the microwave frequency domain. *Phys. Rev. Lett.*, **107**, 113601.

[18] Mallet, F., Castellanos-Beltran, M. A., Ku, H. S., Glancy, S., Knill, E., Irwin, K. D., Hilton, G. C., Vale, L. R., and Lehnert, K. W. (2011). Quantum state tomography of an itinerant squeezed microwave field. *Phys. Rev. Lett.*, **106**, 220502.

[19] Menzel, E. P., Deppe, F., Mariantoni, M., Araque Caballero, M. Á., Baust, A., Niemczyk, T., Hoffmann, E., Marx, A., Solano, E., and Gross, R. (2010). Dual-path state reconstruction scheme for propagating quantum microwaves and detector noise tomography. *Phys. Rev. Lett.*, **105**, 100401.

[20] Eichler, C., Bozyigit, D., Lang, C., Steffen, L., Fink, J., and Wallraff, A. (2011). Experimental state tomography of itinerant single microwave photons. *Phys. Rev. Lett.*, **106**, 220503.

[21] Carmichael, H. J. (1999). *Statistical Methods in Quantum Optics 1: Master Equations and Fokker-Planck Equations.* Springer-Verlag, Berlin.

[22] Arthurs, E. and Kelly, J. L. Jr., (1965). On the simultaneous measurement of a pair of conjugate observables. *Bell. Syst. Tech. J.*, **44**, 725–729.

[23] Braunstein, S. L., Caves, C. M., and Milburn, G. J. (1991). Interpretation for a positive P representation. *Phys. Rev. A*, **43**, 1153–1159.

[24] Caves, C. M. and Drummond, P. D. (1994). Quantum limits on bosonic communication rates. *Rev. Mod. Phys.*, **66**, 481–537.

[25] Welsch, D.-G., Vogel, W., and Opatrny, T. (2009). Homodyne detection and Quantum state reconstruction. arXiv:0907.1353.

[26] Yuen, H. and Shapiro, J. (1980). Optical communication with two-photon coherent states—Part III: Quantum measurements realizable with photoemissive detectors. *IEEE Trans. Inf. Theory*, **26**, 78–92.

[27] Haus, H. A. and Mullen, J. A. (1962). Quantum noise in linear amplifiers. *Phys. Rev.*, **128**, 2407–2413.

[28] Caves, C. M. (1982). Quantum limits on noise in linear amplifiers. *Phys. Rev. D*, **26**, 1817–1839.

[29] Clerk, A. A., Devoret, M. H., Girvin, S. M., Marquardt, Florian, and Schoelkopf, R. J. (2010). Introduction to quantum noise, measurement, and amplification. *Rev. Mod. Phys.*, **82**(2), 1155–1208.

[30] Yurke, B. and Buks, E. (2006). Performance of cavity-parametric amplifiers, employing Kerr nonlinearites, in the presence of two-photon loss. *J. Lightwave Technol.*, **24**, 5054–5066.

[31] Yurke, B. (1987). Squeezed-state generation using a Josephson parametric amplifier. *J. Opt. Soc. Am. B*, **4**, 1551–1557.

[32] Castellanos-Beltran, M. A., Irwin, K. D., Hilton, G. C., Vale, L. R., and Lehnert, K. W. (2008). Amplification and squeezing of quantum noise with a tunable Josephson metamaterial. *Nature Phys.*, **4**, 929–931.

[33] Hatridge, M., Vijay, R., Slichter, D. H., Clarke, J., and Siddiqi, I. (2011). Dispersive magnetometry with a quantum limited squid parametric amplifier. *Phys. Rev. B*, **83**, 134501.

[34] Lvovsky, A. I. and Raymer, M. G. (2009). Continuous-variable optical quantum-state tomography. *Rev. Mod. Phys.*, **81**, 299–332.

[35] Glauber, R. J. (1963). Coherent and incoherent states of the radiation field. *Phys. Rev.*, **131**, 2766–2788.

[36] Carmichael, H. J. (2008). *Statistical Methods in Quantum Optics 2: Non-Classical Fields.* Springer-Verlag, Berlin.

[37] Cahill, K. E. and Glauber, R. J. (1969). Density operators and quasiprobability distributions. *Phys. Rev.*, **177**, 1882–1902.

[38] Kim, M. S. (1997). Quasiprobability functions measured by photon statistics of amplified signal fields. *Phys. Rev. A*, **56**, 3175–3179.

[39] Vijay, R., Slichter, D. H., and Siddiqi, I. (2011). Observation of quantum jumps in a superconducting artificial atom. *Phys. Rev. Lett.*, **106**, 110502.

[40] Kinion, D. and Clarke, J. (2008). Microstrip superconducting quantum interference device radio-frequency amplifier: Scattering parameters and input coupling. *Appl. Phys. Lett.*, **92**, 172503.

[41] Leonhardt, U. and Paul, H. (1993). Realistic optical homodyne measurements and quasiprobability distributions. *Phys. Rev. A*, **48**, 4598–4604.

[42] Fink, J. M., Steffen, L., Studer, P., Bishop, Lev S., Baur, M., Bianchetti, R., Bozyigit, D., Lang, C., Filipp, S., Leek, P. J., and Wallraff, A. (2010). Quantum-to-classical transition in cavity quantum electrodynamics. *Phys. Rev. Lett.*, **105**, 163601.

[43] Rosenthal, J. S. (2009). *A First Look at Rigorous Probability Theory.* World Scientific, Singapore.

[44] Bužek, V., Adam, G., and Drobný, G. (1996). Quantum state reconstruction and detection of quantum coherences on different observation levels. *Phys. Rev. A*, **54**, 804–820.

[45] Herzog, U. (1996). Generating-function approach to the moment problem for the density matrix of a single mode. *Phys. Rev. A*, **53**, 2889–2892.

[46] Bergeal, N., Schackert, F., Metcalfe, M., Vijay, R., Manucharyan, V. E., Frunzio, L., Prober, D. E., Schoelkopf, R. J., Girvin, S. M., and Devoret, M. H. (2010). Phase-preserving amplification near the quantum limit with a Josephson ring modulator. *Nature*, **465**, 64–68.

16
Using a "frictionless" pendulum for quantum measurement

I. SIDDIQI

Quantum Nanoelectronics Laboratory
Department of Physics, University of California, Berkeley, CA 94720, USA

Quantum Machines: Measurement and Control of Engineered Quantum Systems.
Edited by M. Devoret, B. Huard, R. Schoelkopf, and L. F. Cugliandolo.

Chapter Contents

Electrical amplifiers have a key role in experimental physics. They raise the energy of a signal of interest to a level sufficient to overwhelm the noise of recording and processing electronics. At the heart of an amplifier is a nonlinear element, such as a continuous medium with a Kerr nonlinearity, or a discrete circuit element such as a transistor. The Josephson tunnel junction [1], two superconducting electrodes separated by a thin tunneling barrier, is a unique two-terminal electrical element that can be operated as either a nonlinear resistor or an inductor depending on whether the magnitude of the current $I(t)$ flowing across it is respectively greater or less than a critical value I_0. The dynamics of the junction obey the equations of a damped, driven pendulum. This nonlinearity can be harnessed in a number of different circuit topologies to realize sensitive amplifiers with near-quantum-limited noise performance [2–5]. In such devices, an input circuit, such as a superconducting island or loop, acts as a third terminal to couple an external signal, such as an electric charge or magnetic flux, respectively, to the junction critical current I_0. Small variations in I_0 can significantly change the transport properties of the junction under appropriate bias conditions, thus giving rise to amplification.

These resonant circuits can be driven with a variety of different time-varying currents to realize microwave-frequency amplifiers. In the case of a harmonic drive, the steady state response of the oscillator can be single-valued or exhibit bistability, depending on the detuning from the small-signal resonance frequency and intensity of the excitation. When biased in a region where only a single stable driven solution exists and the resonant frequency varies significantly with excitation intensity, it is possible to observe parametric amplification. Here, the oscillator is continuously irradiated with a fixed-frequency drive tone that biases it in a region of nonlinearity, and an additional weak input signal modulates I_0, or more generally the oscillator frequency, and is amplified. Both phase-sensitive and insensitive modes of amplification have been achieved in superconducting parametric amplifiers [6–8], with either zero or half a photon of added noise, respectively.

In regions where the oscillator can occupy one of two dynamical states [9], it is possible to realize a threshold or digital amplifier [10, 11]. Rather than simply applying a continuous driving field to bias the oscillator, here measurement sensitivity is achieved by entering the bistable region using a pulsed excitation and subsequently recording which dynamical state has been populated. Small changes in the oscillator frequency can significantly alter the probability to occupy one state or the other. There are two basic types of pulse sequences that can be used to arm this type of amplifier: an amplitude-modulated signal of fixed frequency or a frequency-modulated signal (a so-called "chirp") of fixed amplitude. The oscillator dynamics and the role of fluctuations associated with each of these protocols are very different. With amplitude modulation, the system can be modeled as initially occupying one attractor and then switching (or not) to the other when appropriate bias conditions have been reached [12]. Under a frequency-chirped drive, the oscillator phase locks (or not) to the driving field, and tracks one attractor or the other as they separate in phase space with no switching events between them. This phenomenon is known as autoresonance [13].

In both cases, the ultimate resolution of these digital amplifiers is set by the sharpness of the threshold to populate one dynamical state or the other with respect to an input parameter, which in this case is typically I_0. A finite width to this threshold is observed since these electrical oscillators are subject to quantum fluctuations [14] associated with the finite impedance of the embedding circuit. This zero-point motion acts as a noise source that broadens the dynamical transition between metastable states and therefore sets a quantum limit to the measurement sensitivity. In the case of amplitude modulation, quantum noise enhances the escape rate from one attractor to the other, much in the same way that thermal fluctuations affect an activated process in equilibrium with a noise bath [5, 15]. Under the influence of a rapidly chirped drive, the system in not in equilibrium with the noise bath, and quantum fluctuations only set the initial conditions of the oscillator and thus randomize the precise value of the dynamical threshold [16, 17].

References

[1] Josephson, B. D. (1964). *Rev. Mod. Phys.*, **36**, 216.
[2] Zorin, A. B. (1996). *Phys. Rev. Lett.*, **76**, 4408–4411.
[3] Schoelkopf, R. J., Wahlgren, P., Kozhevnikov, A. A., Delsing, P., and Prober, D. E. (1998). *Science*, **280**, 1238–1242.
[4] Muck, M., Welzel, C., and Clarke, J. (2003). *Appl. Phys. Lett.*, **82**, 3266–3268.
[5] Vijay, R., Devoret, M. H., and Siddiqi, I. (2009). *Rev. Sci. Instrum.*, **80**, 111101.
[6] Castellanos-Beltran, M. A. and Lehnert, K. W. (2007). *Appl. Phys. Lett.*, **91**, 083509.
[7] Bergeal, N., Schackert, F., Metcalfe, M., Vijay, R., Manucharyan, V. E., Frunzio, L., Prober, D. E., Schoelkopf, R. J., Girvin, S. M., and Devoret, M. H. (2010). *Nature*, **465**, 64–68.
[8] Hatridge, M., Vijay, R., Slichter, D. H., Clarke, J., and Siddiqi, I. (2011). *Phys. Rev. B*, **83**, 134501.
[9] Siddiqi, I., Vijay, R., Pierre, F., Wilson, C. M., Frunzio, L., Metcalfe, M., Rigetti, C., Schoelkopf, R. J., Devoret, M. H., Vion, D., and Esteve, D. (2005). *Phys. Rev. Lett.*, **94**, 027005.
[10] Siddiqi, I., Vijay, R., Pierre, F., Wilson, C. M., Metcalfe, M., Rigetti, C., Frunzio, L., and Devoret, M. H. (2004). *Phys. Rev. Lett.*, **93**, 207002.
[11] Naaman, O., Aumentado, J. Friedland, L., Wurtele, J. S., and Siddiqi, I. (2008). *Phys. Rev. Lett.*, **101**, 117005.
[12] Landau, L. D. and Lifshitz, E. M. (1960). *Mechanics*, Chap. 5. Pergamon Press, Oxford.
[13] Fajans, J. and Friedland, L. (2001). *Am. J. Phys.*, **69**, 1096–1102.
[14] Dykman, M. (2012). *Fluctuating Nonlinear Oscillators.* Oxford University Press, Oxford.
[15] Dykman, M. I. (2007). *Phys. Rev. E*, **75**, 011101.
[16] Barth, I., Friedland, L., Sarid, E., and Shagalov, A. G. (2009). *Phys. Rev. Lett.*, **103**, 155001.
[17] Murch, K. W., Vijay, R., Barth, I., Naaman, O., Aumentado, J., Friedland, L., and Siddiqi, I. (2011). *Nature Phys.*, **7**, 105.

17 Quantum Bayesian approach to circuit QED measurement

A. N. Korotkov

Department of Electrical Engineering
University of California Riverside, CA 92521-0204, USA

Quantum Machines: Measurement and Control of Engineered Quantum Systems.
Edited by M. Devoret, B. Huard, R. Schoelkopf, and L. F. Cugliandolo.

Chapter Contents

We present a simple formalism describing the evolution of a qubit in the process of its measurement in a circuit QED setup. When a phase-sensitive amplifier is used, the evolution depends on only one output quadrature, and the formalism is the same as for a broadband setup. When a phase-preserving amplifier is used, the qubit evolution depends on two output quadratures. In both cases, a perfect monitoring of the qubit state and therefore a perfect quantum feedback is possible.

17.1 Introduction and qualitative discussion

The goal of this chapter is to present a physical picture of the process of continuous quantum measurement of a qubit in the circuit quantum electrodynamics (QED) setup [1–5] (Fig. 17.1), extending or reformulating the previous theoretical descriptions [6–9]. Understanding qubit evolution in the process of measurement is important for developing intuition, which is useful in many cases, in particular in designing various schemes for quantum feedback [10–15]. When a quantum measurement is discussed [16], there are usually two different types of questions to answer: we can either focus on obtaining information on the initial state (before measurement) or focus on the quantum state after the measurement (i.e., evolution in the process of measurement). Let us emphasize that we consider the latter problem here and essentially extend the collapse postulate by describing continuous evolution "inside" the collapse timescale.

In the circuit QED setup (Fig. 17.1), a qubit interacts with a gigahertz-range microwave resonator, whose frequency changes slightly depending on whether the qubit is in the state $|0\rangle$ or $|1\rangle$ [1–9]. In turn, this frequency shift affects the phase (and, in general, the amplitude) of a probing microwave, which is transmitted through the resonator (in another setup, the microwave is reflected from the resonator, but the difference is not important). The outgoing microwave is amplified, following which, the radiofrequency (rf) signal is downconverted by mixing it with the original microwave tone, so that the low-frequency (<100 MHz) output of the IQ mixer provides information on the qubit state. The output noise is mainly determined by the first amplifying stage, the preamplifier. With the recent development of near-quantum-limited superconducting parametric amplifiers [5, 17, 18], it is natural to use them

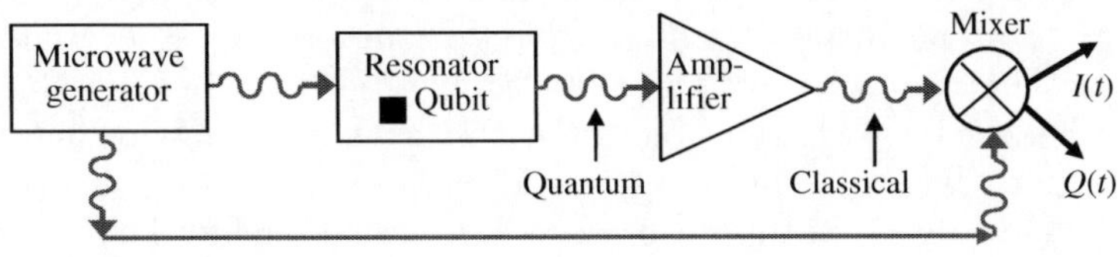

Fig. 17.1 Schematic of the circuit QED setup. A microwave field of frequency ω_m is transmitted through (or reflected from) the resonator of frequency ω_r, which changes slightly, $\omega_r \pm \chi$, depending on the qubit state. After amplification, the microwave is sent to the IQ mixer, which produces two quadrature signals: $I(t)$ and $Q(t)$. For a phase-preserving amplifier, we define $I(t)$ as the quadrature carrying information on the qubit state, while for a phase-sensitive amplifier, we define $I(t)$ as corresponding to the amplified quadrature.

as preamplifiers [5], instead of cryogenic high-electron-mobility transistors (HEMTs) [1–4], which usually have noise temperatures above 3 K.

Continuous quantum measurement in the circuit QED setup in some sense falls in between a qubit measurement by a quantum point contact (QPC) or a single-electron transistor (SET), the theory of which was developed over a decade ago [19–22] and continuous quantum measurement in optics, for which the theory of quantum trajectories was developed even earlier [23–25]. Nevertheless, the circuit QED setup differs from both these cases, and this is probably the reason why there is still a confusion about the proper physical description of the measurement process. The measurement by the QPC or SET is of the broadband type, meaning that the monitored frequency band starts from zero. In contrast, the circuit QED setup is of the narrowband type: we deal only with a relatively narrow band around the probing microwave frequency ω_m. This necessarily involves two orthogonal quadratures:[1] we work with rf signals of the type $A(t)\cos(\omega_m t) + B(t)\sin(\omega_m t)$ and there are essentially two signals $A(t)$ and $B(t)$ instead of only one in the broadband case. In this sense, the circuit QED setup is similar to the optical (especially cavity QED) setup [23–25]; however, there is an important difference—in the circuit QED case, the outgoing microwave is amplified (Fig. 17.1) before being mixed with the original microwave, while there is no amplification stage in the standard optical setup. The operation will obviously depend on whether a phase-sensitive or a phase-preserving amplifier is used, since a phase-preserving amplifier necessarily adds the half quantum of noise into any quadrature [7, 26–29]. Notice that the quantum trajectory theory for the circuit QED setup was developed in [9]; however, the amplifier stage was essentially missing in the analyzed model.

In this chapter, we consider the simplest circuit QED case, assuming a dispersive regime [6], exactly resonant microwave frequency, absence of the Rabi drive, and sufficiently wide resonator and amplifier bandwidths for the Markov approximation to be valid. Some generalizations are rather straightforward; however, our goal is a simple picture in a simple case.

A description of continuous qubit measurement is essentially a description of the quantum backaction. Following the same quantum Bayesian framework as for the measurement by QPC/SET [19, 20] (see [30] for a review), we will discuss two kinds of measurement backaction onto the qubit, which we call here "spooky" and "realistic." *The "spooky" (or "quantum," "informational," "nonunitary") backaction does not have a physical mechanism* and therefore cannot be described by the Schrödinger equation (in contrast to what people often think when trying to find a mechanism for the quantum collapse); however, *it is a commonsense consequence of acquiring information* on the qubit state in the process of measurement. This is essentially the same backaction that is discussed in the EPR paradox [31] and in Bell inequality violation [32]; the only difference is that in our case the information is incomplete and therefore we have to use the quantum Bayes rule [30, 33, 34] instead of the projective collapse rule. In contrast, the "realistic" (or "classical," "unitary") backaction has a

[1] We use the term "quadrature" in a somewhat sloppy way.

physical mechanism: in the circuit QED case, it is a fluctuation of the number of photons in the resonator, which affects the phase of the qubit state. The "realistic" backaction is usually discussed in the standard theories of circuit QED measurement [6–8]. Actually, there is a certain spookiness even in the "realistic" backaction (it may be affected by a delayed choice, as discussed in Section 17.5); however, we do not want to emphasize it here, to keep the picture simple. When we measure the z coordinate of the qubit state on the Bloch sphere (the basis states $|1\rangle$ and $|0\rangle$ correspond to the north and south poles), then the "spooky" backaction changes the z coordinate and leads to the state evolution along the meridian lines, while the "realistic" backaction leads to the evolution around the z axis (i.e., along the parallels).

It is important to notice that when the probing microwave leaves the resonator after interaction with the qubit, one quadrature of the microwave carries information about the qubit state, while the orthogonal quadrature carries information on the fluctuating number of photons in the resonator [6–9]. Therefore, if a phase-preserving amplifier is used, then the "spooky" and "realistic" backactions are fully separated and correspond to two orthogonal quadratures $I(t)$ and $Q(t)$ measured after the mixer (it is trivial to choose the proper linear combinations of the IQ mixer outputs). The signals $I(t)$ and $Q(t)$ are necessarily noisy, and the measurement backactions are stochastic; however, there is a correlation (full correlation in the ideal case) between the output noise and the backaction noise in both channels. As a result (derived later), for a quantum-limited phase-preserving amplifier and in the absence of extra decoherence, the measured quadratures $I(t)$ and $Q(t)$ give us *full information about the backaction*, so that a *random evolution of the qubit wavefunction can be monitored precisely* (a useful analogy is with a Brownian particle under a microscope: we cannot predict its motion, but we can monitor it). This is what is needed, in particular, for arranging perfect quantum feedback control of the qubit state. It is interesting to notice that for an ensemble-averaged evolution (in which the random but monitorable qubit evolution is replaced by dephasing), exactly one half of the ensemble dephasing Γ comes from the "spooky" backaction and the other half comes from the "realistic" backaction.

In the case of a phase-sensitive amplifier, it is sufficient to measure after the mixer only the quadrature that was amplified; let us still denote it by $I(t)$, though now its phase is determined by the amplifier instead of the microwave-qubit interaction. In this case, the "spooky" and "realistic" backactions are in general mixed (not separated), because there is only one output signal $I(t)$. This situation corresponds exactly to the broadband measurement by the QPC/SET with a correlation between the output and "realistic" backaction noises [30]. The situation simplifies when the amplified quadrature is the one that carries information about the qubit state (z coordinate). Then, in the quantum-limited case, the "realistic" backaction is completely absent: we cannot measure the photon number fluctuation and correspondingly it does not fluctuate (in the imperfect case, the effect of the remaining "realistic" backaction can be described by an extra dephasing). So we are left with only the "spooky" backaction, and the quantum measurement description coincides with the simpler theory of measurement by a symmetric QPC [30], which does not produce the "realistic" backaction. In contrast, in the case when the photon-number quadrature is amplified, we do not obtain any information on the qubit z coordinate, and therefore there is no "spooky" backaction, but only the "realistic" one. In the general case, when the amplified quadrature makes an arbitrary angle φ with the qubit-information quadrature, both types of backaction are present, and their strength depends on φ. It is important to mention that the ensemble dephasing rate Γ does not depend on φ, as required by causality. In particular, in the quantum-limited case, the contribution $\Gamma \cos^2 \varphi$ comes from the "spooky" backaction, while $\Gamma \sin^2 \varphi$ comes from the "realistic" backaction.

Let us emphasize that both the phase-sensitive and phase-preserving amplifiers permit exact monitoring of the qubit state and therefore perfect quantum feedback. The necessary condition in both cases is that the detection system be quantum-limited.

In the following sections, a formal description of these results is presented. We start by reviewing the Bayesian approach for the broadband qubit measurement, then briefly discuss the difference between phase-preserving and phase-sensitive amplifiers, and then present the formalism of the narrowband continuous measurement of a qubit in the circuit QED setup. In Conclusion we briefly discuss generalizations of the formalism, quantum feedback, and the causality principle. We note that our approach can be converted into the formal language of positive-operator-valued-measure (POVM)-type generalized quantum measurement [35, 36] (then separation of the "spooky" and "realistic" backactions corresponds to the decomposition of the measurement operator into diagonal and unitary parts—see later), and our results for the case of a phase-sensitive amplifier are very similar to the results of Gambetta *et al.* [9].

17.2 Broadband measurement

In this section, we review the Bayesian formalism [19, 20, 30] for the broadband measurement of a qubit, considering only the simple case without additional evolution, and thus emphasizing the main physical idea of the formalism. We start with the broadband formalism because it is simpler than for the narrowband (circuit QED) measurement and it can be used as a natural step in understanding the circuit QED setup.

For definiteness, let us assume that the qubit is a double quantum dot populated with one electron (Fig. 17.2), and the states $|0\rangle$ and $|1\rangle$ correspond to the electron localized in one or the other dot. The qubit is measured by a small-transparency tunnel junction (a model of a QPC), whose barrier height depends on the electron location, so that the two qubit states correspond to different average currents I_0 and I_1 through the QPC. The voltage across the QPC is sufficiently large to make the detector output classical (Markov approximation), and $|\Delta I| \ll |I_c|$, where $\Delta I = I_1 - I_0$ is the response and $I_c = (I_0 + I_1)/2$ is the mean value; this weak response assumption allows us to

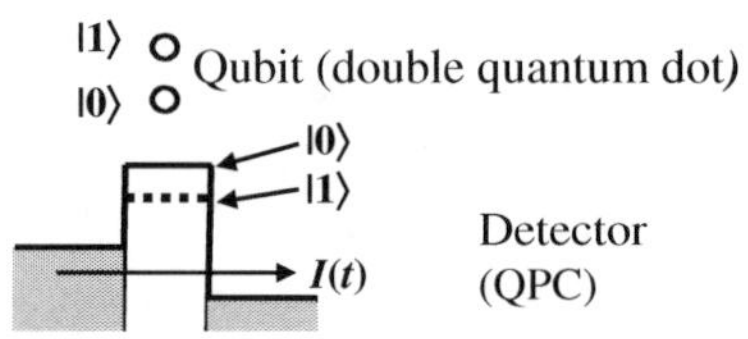

Fig. 17.2 Schematic of a broadband measurement setup: a double-quantum-dot qubit is measured by a QPC (tunnel junction). The output signal $I(t)$ is the QPC current.

consider the QPC current $I(t)$ as a quasicontinuous noisy signal (see [30] for a detailed discussion of the required assumptions; the formalism needs only a minor change if $\Delta I \sim I_c$). Then the output signal of the detector is

$$I(t) = I_c + (\Delta I/2)\, z(t) + \xi(t), \quad S_\xi(\omega) = S, \tag{17.1}$$

where $z = \rho_{11} - \rho_{00}$ is the z component of the Bloch-sphere representation of the qubit density matrix $\rho(t)$, and $\xi(t)$ is the white shot noise with spectral density $S = 2eI_c$ (we use the single-sided definition for the spectral density, in which the signal variance ("power") corresponds to $\int_0^\infty S(\omega)\, d\omega/2\pi$; the definition of S is smaller by a factor of 1/2 in [7] and a factor of $1/4\pi$ in [8] and [34]). We emphasize that the detector signal $I(t)$ is classical, and the qubit state $\rho(t)$ is practically unentangled from the detector, but obviously depends on $I(t)$.

The detector Hamiltonian and the qubit-detector interaction Hamiltonian are given in [19, 20, 30]; they are not really important for our discussion here. For simplicity, let us assume that the qubit Hamiltonian is zero, $H_{qb} = 0$, so that the qubit evolution is due to the measurement only. In this case, the qubit evolution during time t happens to be determined only by the time-averaged value of the measured detector output

$$\overline{I}_m(t) = \frac{1}{t} \int_0^t I(t')\, dt', \tag{17.2}$$

which would contain full information for a classical measurement.

Because of the correspondence principle, the evolution of the diagonal elements of the qubit density matrix ρ ($\mathrm{Tr}\, \rho = 1$) should correspond to the classical evolution of probabilities, which are given by the classical Bayes rule. The Bayes rule says that an updated (a posteriori) probability of a system state is proportional to the initial (a priori) probability and the probability (likelihood) of the obtained measurement result assuming this particular state. In our case, $\overline{I}_m(t)$ is the measurement result, and its probability for the qubit in the basis state $|j\rangle$ has the Gaussian distribution

$$P_{|j\rangle}(\overline{I}_m) = \frac{1}{\sqrt{2\pi D}} \exp\left[-\frac{(\overline{I}_m - I_j)^2}{2D}\right], \quad D = \frac{S}{2t}, \tag{17.3}$$

where D is the variance, which decreases with the measurement time t. Therefore, the correspondence principle demands the Bayesian evolution

$$\frac{\rho_{11}(t)}{\rho_{00}(t)} = \frac{\rho_{11}(0)}{\rho_{00}(0)} \frac{\exp[-(\overline{I}_m(t) - I_1)^2/2D]}{\exp[-(\overline{I}_m(t) - I_0)^2/2D]}, \tag{17.4}$$

which in our terminology is due to the "spooky" backaction; it cannot be described by the Schrödinger equation, but follows from common sense.

If the phase of the qubit state is not affected by the measurement process (no "realistic" backaction), then an arbitrary initial wavefunction $|\psi(0)\rangle = \sqrt{\rho_{00}(0)}\, |0\rangle + e^{i\phi}\sqrt{\rho_{11}(0)}\, |1\rangle$ becomes $|\psi(t)\rangle = \sqrt{\rho_{00}(t)}\, |0\rangle + e^{i\phi}\sqrt{\rho_{11}(t)}\, |1\rangle$ with the same phase ϕ; therefore, for an arbitrary mixed state, we get

$$\rho_{01}(t) = \rho_{01}(0) \frac{\sqrt{\rho_{00}(t)\, \rho_{11}(t)}}{\sqrt{\rho_{00}(0)\, \rho_{11}(0)}}. \tag{17.5}$$

Equations (17.4) and (17.5) describe the "spooky" backaction.

Now assume that due to the qubit-detector interaction (e.g., Coulomb interaction), each electron passing through the detector rotates the qubit phase ϕ by a small amount $\Delta\phi$. From the measured result $\overline{I}_m(t)$, we know exactly how many electrons passed through ($n_e = \overline{I}_m(t)t/q$, with q being the electron charge), and can easily introduce the corresponding phase factor into Eq. (17.5):

$$\rho_{01}(t) = \rho_{01}(0)\,\frac{\sqrt{\rho_{00}(t)\,\rho_{11}(t)}}{\sqrt{\rho_{00}(0)\,\rho_{11}(0)}}\,\exp[iK\overline{I}_m(t)\,t], \tag{17.6}$$

where $K = \Delta\phi/q$. The nonstochastic factor $\exp(iKI_c t)$ can obviously be ascribed to the qubit Hamiltonian; however, this is not important here. The factor $\exp[iK\overline{I}_m(t)\,t]$ in Eq. (17.6) is the effect of the "realistic" backaction. It may or may not be present in a particular physical situation; for example, $K = 0$ for measurement by a symmetric QPC, while $K \neq 0$ in an asymmetric QPC or SET case.

Finally, if there is an extra pure dephasing of a qubit with rate γ, then Eq. (17.6) becomes

$$\rho_{01}(t) = \rho_{01}(0)\,\frac{\sqrt{\rho_{00}(t)\,\rho_{11}(t)}}{\sqrt{\rho_{00}(0)\,\rho_{11}(0)}}\,\exp[iK\overline{I}_m(t)\,t]\,e^{-\gamma t}. \tag{17.7}$$

Eqs. (17.4) and (17.7) are *the main starting point of the Bayesian formalism* [30]. It is then easy to include a nonzero qubit Hamiltonian H_{qb} by differentiating Eqs. (17.4) and (17.7) with respect to time (paying attention to whether the Stratonovich or Itô definition of the derivative is used) and adding terms due to H_{qb}. Energy relaxation and other mechanisms of the qubit state evolution can be included in the same way. Actually, there are many ways to derive the Bayesian equations (17.4) and (17.7) [19–22, 30, 37, 38], but we focus here only on their meaning, not on their derivation.

Notice that averaging of Eqs. (17.4) and (17.7) over the measurement result $\overline{I}_m$ (i.e. ensemble averaging) with the probability distribution

$$P(\overline{I}_m) = \rho_{00}(0)P_{|0\rangle}(\overline{I}_m) + \rho_{11}(0)P_{|1\rangle}(\overline{I}_m) \tag{17.8}$$

gives the same evolution as for a pure dephasing: the diagonal matrix elements of ρ do not evolve, while the off-diagonal element ρ_{01} decays as $\rho_{01}(0)\,e^{-\Gamma t}$ (neglecting the non-stochastic phase evolution) with the ensemble dephasing rate [30]

$$\Gamma = \frac{(\Delta I)^2}{4S} + \frac{K^2 S}{4} + \gamma, \tag{17.9}$$

which has clear contributions from the "spooky" backaction, "realistic" backaction, and additional dephasing.

In the case $\gamma = 0$, an initially pure qubit state remains pure; in other words, we can monitor evolution of a qubit wavefunction. This property can be used as the definition of a quantum-limited detector [19, 20, 30]. The quantum efficiency η can then be naturally defined as $\eta = 1 - \gamma/\Gamma$. If, for some reason, the "realistic" backaction is considered as dephasing (i.e., only in the averaged way), then the quantum efficiency can be defined as $\tilde{\eta} = 1 - \gamma/\Gamma - K^2 S/4\Gamma$ (here the definitions of $\tilde{\eta}$ and η are exchanged compared with the definitions in [30]). In other words, $\tilde{\eta} = (\Delta I)^2/4S\Gamma$ is the relative contribution of only the "spooky" backaction in the ensemble dephasing Γ. In particular, this definition is relevant to the peak-to-pedestal ratio of the Rabi spectral peak [39, 40], which is equal to $4\tilde{\eta}$. As an example, if $\gamma = 0$ and the contributions in Eq. (17.9) from the "spooky" and "realistic" backactions are equal to each other (as in the circuit QED setup with a phase-preserving amplifier), then $\eta = 1$ but $\tilde{\eta} = 1/2$.

A nonideal detector ($\eta < 1$) can be modeled in two equivalent ways [41]: we either add an extra dephasing γ to the qubit or add an extra noise to the output of the ideal detector. Only the total dephasing Γ, response ΔI, total output noise S, and correlation factor $K = \delta\langle\phi\rangle/\delta(\overline{I}_m t)$ are physical (i.e., experimentally measurable) parameters, while the distribution of the nonideality between the extra dephasing and additional output noise is a matter of convenience (here $\phi = \arg(\rho_{01})$, and the notation $\langle\phi\rangle$ reminds us about averaging over additional classical noise at the output). We emphasize that the Bayesian formalism deals only with the experimentally measurable parameters ΔI, S, K, Γ, and the output signal $I(t)$.

In the ideal case ($\eta = 1$), the evolution equations (17.4) and (17.6) can be translated into the language of POVM-type generalized measurement. In this approach, the effect of measurement is described as [35, 36]

$$|\psi(t)\rangle = \frac{M_R|\psi(0)\rangle}{||M_R|\psi(0)\rangle||}, \quad \rho(t) = \frac{M_R\rho(0)M_R^\dagger}{\mathrm{Tr}[M_R^\dagger M_R\rho(0)]}, \tag{17.10}$$

where M_R is the so-called measurement (Kraus) operator, corresponding to the result R. The probability of the result R is $P_R = ||M_R|\psi(0)\rangle||^2$ using wavefunctions or $P_R = \mathrm{Tr}[M_R^\dagger M_R\,\rho(0)]$ using density matrices; therefore, the POVM elements $M_R^\dagger M_R$ should satisfy the completeness condition $\sum_R M_R^\dagger M_R = \mathbb{1}$. The relation between this approach and the quantum Bayesian approach can be understood via the operator decomposition

$$M_R = U_R\sqrt{M_R^\dagger M_R}\,, \tag{17.11}$$

where U_R is unitary and the square root of the positive operator $M_R^\dagger M_R$ is defined in the natural way in the diagonalizing basis. It is easy to see that $\sqrt{M_R^\dagger M_R}$ is essentially the quantum Bayes rule (in the diagonalizing basis); in our terminology, it corresponds to the "spooky" backaction, while U_R corresponds to the "realistic" backaction. For the discussed setup, the result R is $\overline{I}_m(t)$, the "spooky" backaction $\sqrt{M_R^\dagger M_R}$ should be determined by the probabilities $P_{|j\rangle}(\overline{I}_m)$ given by Eq. (17.3), and the "realistic" backaction U_R is given by the phase factor in Eq. (17.6). Therefore, the corresponding measurement operator is

$$\begin{aligned} M(\overline{I}_m) &= \exp(-iK\overline{I}_m t\,\sigma_z/2) \\ &\times \left[\sqrt{P_{|0\rangle}(\overline{I}_m)}\,|0\rangle\langle 0| + \sqrt{P_{|1\rangle}(\overline{I}_m)}\,|1\rangle\langle 1|\right], \end{aligned} \tag{17.12}$$

where σ_z is the Pauli matrix.

17.3 Phase-preserving versus phase-sensitive amplifiers

Before discussing microwave amplifiers, let us consider a measurement of an oscillator, for example, a mechanical resonator with frequency ω_r and mass m. This is a very well-studied problem [7, 16], so we will only discuss a way to understand the results. A classical resonator position x oscillates as $x_c(t) = A\cos(\omega_r t) + B\sin(\omega_r t)$ (in this section, x stands for the usual spatial coordinate, not for the Bloch-sphere coordinate). The corresponding quantum state is called the "coherent state" in optical language; it is represented by the wavefunction $\psi(x,t) = \psi_{\rm gr}[x - x_c(t)]\exp(ip_c x/\hbar)$, where $\psi_{\rm gr}(x)$ is the ground state and $p_c = m\dot{x}_c(t)$ is the classical momentum. So the coherent state is essentially the ground state with oscillating center position. Notice that continuous quantum measurement of a resonator position can be described in the same Bayesian way [42] as in Section 17.2; for example the "spooky" backaction gives the evolution $\psi(x,t) = \psi(x,0)\exp\{-[\overline{I}_m(t) - I(x)]^2/4D\}/\text{Norm}$, where $I(x)$ is the average detector signal for the resonator position x, and Norm is a normalization constant (see Eqs. (17.4) and (17.5)). The time step t in this case should be chosen much shorter than ω_r^{-1} so that the unitary evolution and evolution due to measurement may be simply added.

Let us consider the following game. Charlie prepares an oscillator in a coherent state with quadratures A and B, and gives it to David, and David's goal is to find A as accurately as possible. An optimal strategy is rather obvious: David should make a projective measurement of x at time $t = 2\pi n/\omega_r$ with any integer n (to avoid contribution from the B term), and the measurement result is the best estimate of A (if the measurement is done at $t = (2\pi n + \pi)/\omega_r$, then the result should be multiplied by -1). Even though the strategy is optimal, the inaccuracy of David's result is obviously the width (standard deviation) $\sigma_{\rm gr} = \sqrt{\hbar/2m\omega_r}$ of the ground-state shape $|\psi_{\rm gr}(x)|^2$; in energy units, this inaccuracy corresponds to one half of the energy quantum.

Now assume that David cannot make projective measurements, but only "finite-strength" (i.e., imprecise) measurements. The best accuracy $\sigma_{\rm gr}$ can still be achieved if the measurement is done in the simple but very clever "quantum nondemolition" (QND) way: many finite-strength measurements are made at times $t = 2\pi n/\omega_r$; this is called "stroboscopic" measurement [16]. Since the oscillator returns to the same state after a period $2\pi/\omega_r$, the unitary evolution is not important, and many finite-strength measurements (described by the Bayesian equation above) are "stacked" to produce a strong, essentially projective, measurement. More generally, the necessary condition to have the best accuracy $\sigma_{\rm gr}$ for A is that the measurement is not sensitive to the quadrature B.

Now assume that David is only allowed to make a continuous measurement with unmodulated weak strength (so that the inaccuracy achieved after ω_r^{-1} is much larger than $\sigma_{\rm gr}$). Then the "spooky" backaction gets mixed with the unitary evolution, essentially adding noise to the monitored evolution, so that, after a while, the resonator state becomes mostly determined by the backaction and almost independent of the initial state. As the result, the best accuracy for measurement of A becomes $\sqrt{2}\,\sigma_{\rm gr}$, which in energy units corresponds to two half-quanta [16], which is twice as bad as for the projective or stroboscopic measurement. However, the continuous monitoring gives us information about B in the same way as for A, so the accuracy of B measurement is also $\sqrt{2}\,\sigma_{\rm gr}$. Therefore, in some sense, continuous phase-insensitive measurement brings the same total information as phase-sensitive (e.g., stroboscopic) measurement; however, in our game, only half of this information is useful for David's goal.

After discussing measurement of the resonator state, it is easy to understand the quantum limits for high-gain microwave amplifiers. Now suppose Charlie prepares a coherent state of a microwave resonator with quadratures A and B, and gives it to David to find A, and David uses an amplifier for amplification of the microwave field, which slowly leaks from the resonator until it is empty. There is only classical signal processing after the amplifier, so amplification is essentially the quantum measurement. The results are the same as above [7, 16, 28, 29]: a phase-sensitive amplifier, which amplifies only the A quadrature and "de-amplifies" (attenuates) the B quadrature, can measure A with accuracy $\sigma_{\rm gr}$, while a phase-preserving amplifier can measure A only with accuracy $\sqrt{2}\,\sigma_{\rm gr}$ (and also measures B with the same accuracy $\sqrt{2}\,\sigma_{\rm gr}$). Technically, the accuracy is limited by the noise at the amplifier output, so this noise should forbid measuring A with accuracy better than $\sigma_{\rm gr}$ by a phase-sensitive amplifier, and better than $\sqrt{2}\,\sigma_{\rm gr}$ by a phase-preserving amplifier. Therefore, in the quantum-limited case, the output noise power of a phase-preserving amplifier (per quadrature) is twice as large as for a phase-sensitive amplifier with the same gain; this is often called an "additional noise," corresponding to one half of the energy quantum (one more half-quantum is present in both cases) [7, 16, 28, 29]. It may seem somewhat confusing that this result does not depend on the rate with which the microwave leaks from the resonator. So let us check the scaling: for k times slower leakage (k times larger Q factor), the microwave amplitude is $\sqrt{k}$ times smaller, but the accumulation time is k times longer, and therefore the measured signal for the quadrature A is $\sqrt{k}$ times larger, which is the same factor as for the noise accumulation. Therefore, the signal-to-noise ratio that determines the A accuracy does not depend on the leakage rate (resonator bandwidth).

17.4 Narrowband (circuit QED) measurement

Using the discussion of microwave amplifiers in Section 17.3, it is easy to extend the Bayesian approach for a broadband quantum measurement to the narrowband circuit QED setup.

We consider the standard circuit QED setup [1–9], in which a qubit interacts with a microwave resonator, and assume the dispersive regime with the Hamiltonian

$$H = (\hbar\omega_{\rm qb}/2)\,\sigma_z + \hbar\omega_r a^\dagger a + \hbar\chi a^\dagger a\sigma_z, \tag{17.13}$$

where $\omega_{\rm qb} = \omega_{\rm qb,\,bare} + \chi$ is the Lamb-shifted qubit frequency with no photons in the resonator, $\chi = g^2/(\omega_{\rm qb,bare} - \omega_r)$ is the effective coupling (with g being the Jaynes-Cummings coupling), ω_r is the bare resonator frequency, the Pauli operator σ_z acts on the qubit state in the energy basis $\{|0\rangle, |1\rangle\}$, and the resonator creation and annihilation operators are $a^\dagger$ and a. Notice that the resonator frequency increases by 2χ when the qubit state changes from $|0\rangle$ to $|1\rangle$; conversely, the qubit frequency increases by 2χ for each additional photon in the resonator. To measure the qubit state, a microwave field with frequency ω_m is either transmitted through or reflected from the resonator, then amplified and sent to the IQ mixer, which measures both quadratures relative to the original microwave tone (Fig. 17.1). The qubit state affects the resonator frequency and therefore affects the phase (and in general amplitude) of the outgoing microwave.

An elementary Fabry-Pérot analysis gives the classical (complex) microwave field F_r inside the resonator:

$$F_r = \frac{2F_{\rm in}t_{\rm in}/\kappa\tau_{\rm rt}}{1 - 2i(\omega_m - \omega_r)/\kappa}, \tag{17.14}$$

where $F_{\rm in}$ is the applied incident field, $t_{\rm in}$ is the transmission amplitude of the barrier from the incident side, κ is the resonator bandwidth due to the microwave leakage from both sides (the Q factor is ω_r/κ), and the round-trip time is $\tau_{\rm rt} = 2\pi/\omega_r$ for a half-wavelength resonator and $\tau_{\rm rt} = \pi/\omega_r$ for a quarter-wavelength resonator. A similar formula with the same denominator describes a lumped resonator. In the presence of the qubit, the resonator frequency ω_r in this formula is substituted by $\omega_r \pm \chi$, depending on the qubit state. Notice that for the quantum measurement analysis, *there is no difference between the cases of transmission and reflection* for the same F_r and κ, because the field leaking from the resonator is determined only by F_r and κ. (The reflection case has a technical advantage of dealing with an outgoing microwave field of half the strength for the same measured signal.) However, an important parameter is the collected fraction $\eta_{\rm col} = \kappa_{\rm col}/\kappa$ of the leaking microwave power; we will often assume the ideal case $\eta_{\rm col} = 1$ (for the transmission setup, this requires strongly asymmetric coupling, $|t_{\rm in}| \ll |t_{\rm out}|$).

For simplicity, we assume the resonant case, $\omega_m = \omega_r$; then the ensemble qubit dephasing due to measurement is [6, 7]

$$\Gamma = 8\chi^2\bar{n}/\kappa, \tag{17.15}$$

where $\bar{n}$ is the average number of photons in the resonator. It is easy to include Rabi oscillations in the model; however, we do not do so for simplicity and also for a more transparent analogy with Section 17.2, in which we considered a qubit with zero Hamiltonian, evolving only due to measurement; this case exactly corresponds to the circuit QED Hamiltonian in Eq. (17.13) in the rotating frame.

We will need several assumptions to describe the qubit state evolution in the process of measurement. First, for the validity of the dispersive approximation in Eq. (17.13), we need sufficiently large qubit-resonator detuning, $|\omega_{\rm qb} - \omega_r| \gg |g|$, and not too many photons in the resonator, $\bar{n} \ll (\omega_{\rm qb} - \omega_r)^2/g^2$ (we do not consider the recently discovered nonlinear regime [43]). Second, to use the Markov approximation for the evolution, we need the so-called "bad-cavity" assumption: $\Gamma \ll \kappa \ll \omega_r$ (if the qubit evolves due to Rabi oscillations with frequency Ω_R, we also need $\Omega_R \ll \kappa$). This assumption means that the photons leave the resonator much faster than evolution of the qubit state, and therefore there is practically no entanglement between the qubit and the unmeasured microwave field. This assumption also implies that the two resonator states for the qubit states $|0\rangle$ and $|1\rangle$ are almost indistinguishable, $\bar{n}(\chi/\kappa)^2 \ll 1$. Third, we use the "weak-response" assumption, which requires a small

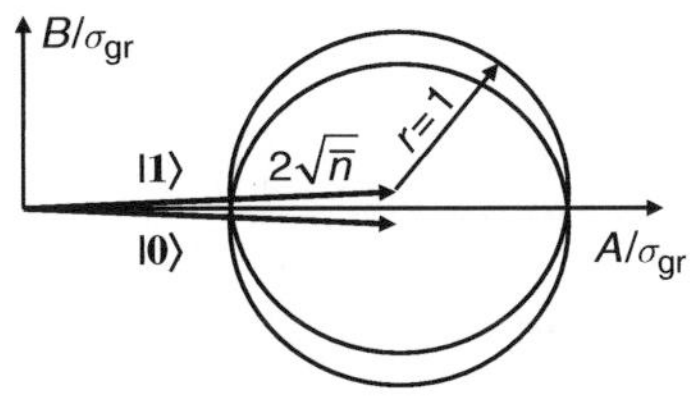

Fig. 17.3 Phase-space representation: for each qubit state, the coherent state with $\langle F_r \rangle = A\cos(\omega_r t) + B\sin(\omega_r t)$ in the resonator is shown [44] as an "error circle" with radius 1, shifted by $2\sqrt{\bar{n}}$ from the origin. Axes are normalized by the standard deviation $\sigma_{\rm gr}$ of the ground state. The B quadrature carries information about the qubit state, while the A-quadrature corresponds to the number of photons in the resonator.

phase difference between the two resonator states, $|\chi|/\kappa \ll 1$. This means that each outgoing photon carries only a little information about the qubit state. Notice that for $\bar{n} \gtrsim 1$ the previous assumption $\kappa \gg \Gamma$ automatically implies the weak-response, and even for $\bar{n} \ll 1$ the weak-response assumption is not always needed. Fourth, we will neglect the qubit energy relaxation due to measurement [6, 7], which can be added later.

A coherent state in the resonator with average $\bar{n}$ photons and zero average phase corresponds to the oscillation of the field expectation value $\langle F_r(t)\rangle = 2\sqrt{\bar{n}}\,\sigma_{\rm gr}\cos(\omega_m t)$, where $\sigma_{\rm gr}$ is the ground-state width (root-mean-square (rms) uncertainty) and we assume $\omega_m = \omega_r$. (Notice that the amplitude $\sigma_{\rm gr}$ corresponds to one quarter of a photon.) Interaction with the qubit slightly changes the phase, $\cos(\omega_m t \mp 2\chi/\kappa)$, depending on the qubit state, so that

$$\langle F_r(t)\rangle = A\cos(\omega_m t) + B\sin(\omega_m t), \;\; A = 2\sqrt{\bar{n}}\,\sigma_{\rm gr}, \tag{17.16}$$
$$B = \pm\frac{4\chi}{\kappa}\sqrt{\bar{n}}\,\sigma_{\rm gr} = \frac{4\chi}{\kappa}\sqrt{\bar{n}}\,\sigma_{\rm gr}\, z,$$

where z is the qubit Bloch coordinate. Thus, the small B quadrature carries information about the qubit state, while the larger A quadrature may give us information on the fluctuations of the photon number in the resonator. In the optical representation (Fig. 17.3) with axes $A/\sigma_{\rm gr}$ and $B/\sigma_{\rm gr}$, the two resonator states for the qubit states $|0\rangle$ and $|1\rangle$ are shown as two "error circles" [44] with rms uncertainty 1 along any direction and distance $2\sqrt{\bar{n}}$ between the origin and circle centers (if axes $A/2\sigma_{\rm gr}$ and $B/2\sigma_{\rm gr}$ are used, then the distance is $\sqrt{\bar{n}}$, while the uncertainty is 1/2).

17.4.1 Phase-sensitive amplifier

Let us start with the case when a phase-sensitive amplifier is used in the circuit QED setup (Fig. 17.4a). Also, we first assume the most ideal case: the amplifier is quantum-limited, it amplifies the optimal B quadrature, there is no microwave collection loss ($\kappa_{\rm col}/\kappa = 1$), and there is no extra noise or dephasing. Then, as discussed in Section 17.3, measuring the microwave contents of the resonator once (by fully emptying it), we can measure the B quadrature with imprecision $\sigma_{\rm gr}$. Therefore, in the

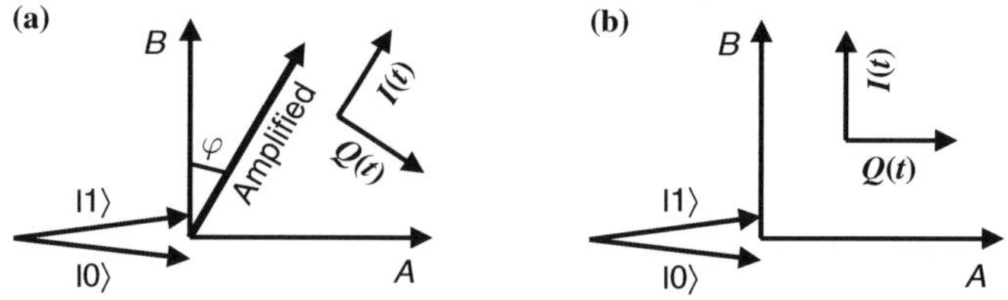

Fig. 17.4 Relations between relevant quadratures. The quadratures A and B are for the microwave field in the resonator. (a) For a phase-sensitive amplifier, the amplified quadrature makes an angle φ with the informational B quadrature. The corresponding quadrature at the mixer output is defined as $I(t)$ (the output $Q(t)$ is then useless). (b) For a phase-preserving amplifier, we define output quadratures $I(t)$ and $Q(t)$ as corresponding to the resonator quadratures B and A.

continuous measurement for time t, the B quadrature is measured with imprecision $\sigma_{\rm gr}/\sqrt{\kappa t}$, which converts into the imprecision $\sqrt{\kappa/t}/(4\chi\sqrt{\bar{n}})$ of the qubit z coordinate. Following the language of Section 17.2, let us discuss the signal and noise at the output of the setup. There are two outputs of the IQ mixer; however, only the amplified quadrature carries any information, so let us denote the corresponding output of the mixer (or their linear combination) by $I(t)$. Then the response $\Delta I = I_1 - I_0$ corresponds to $\Delta z = 2$ and $\Delta B = 8(\chi/\kappa)\sqrt{\bar{n}}\,\sigma_{\rm gr}$. For measurement during time t, the variance $(\kappa/t)/(4\chi\sqrt{\bar{n}})^2$ of the z coordinate converts into the variance $(\kappa/t)(\Delta I/8\chi\sqrt{\bar{n}})^2$ of the measured output $\bar{I}_m = (1/t)\int_0^t I(t')\,dt'$. Equating it with $D = S/2t$, we find the (single-sided) spectral density of the $I(t)$ noise:

$$S_{\rm min} = \frac{(\Delta I_{\rm max})^2\kappa}{32\chi^2\bar{n}}, \tag{17.17}$$

where we have replaced S with $S_{\rm min}$ and ΔI with $\Delta I_{\rm max}$ to remind ourselves that we are considering the quantum-limited case, and the response is maximized by amplifying the optimal quadrature. Notice that, since $\Delta I_{\rm max} \propto \chi\sqrt{\bar{n}/\kappa}$, the noise $S_{\rm min}$ does not depend on the qubit or resonator properties; it is essentially the amplified vacuum noise and depends only on the amplifier gain.

Obtaining information on the qubit z coordinate via the signal $I(t)$ with response $\Delta I_{\rm max}$ and noise $S_{\rm min}$, we necessarily cause the "spooky" backaction described by Eqs. (17.4) and (17.5). As discussed in Section 17.2, this is a consequence of the corresponding principle or just common sense. Now averaging the ρ_{01} evolution in Eq. (17.4) over the measurement result $\bar{I}_m$ with its probability distribution, Eqs. (17.8) and (17.3), we see that the "spooky" backaction dephases an ensemble of qubits with a rate (see Eq. (17.9)) $(\Delta I_{\rm max})^2/4S_{\rm min} = 8\chi^2\bar{n}/\kappa$. This rate coincides with the total ensemble dephasing in Eq. (17.15), and therefore the qubit state cannot additionally fluctuate for any other reason. Thus, we have derived an important result: *in the ideal case with phase-sensitive amplifier, there is only the "spooky" backaction and no "realistic" backaction.* This means that the *number of photons in the resonator does not fluctuate* (otherwise there would be an additional dephasing), which makes sense since we cannot measure the A quadrature, carrying information on the photon number. Notice that it is also easy to prove this result when $\omega_m \neq \omega_r$. Then, from Eq. (17.14), we obtain that the informational quadrature amplitude is multiplied by a factor $[1 + 4(\omega_m - \omega_r)^2/\kappa^2]^{-1/2}$ compared with Eq. (17.16). The response $\Delta I_{\max}$ is multiplied by same factor, while the noise $S_{\min}$ does not change. Therefore, the "spooky" backaction contribution to the ensemble dephasing is multiplied by a factor $[1 + 4(\omega_m - \omega_r)^2/\kappa^2]^{-1}$, which again coincides with the result [6, 7] for the total ensemble dephasing Γ. This proves the absence of the "realistic" backaction for the nonresonant case $\omega_m \neq \omega_r$ as well.

Now let us consider the case when an ideal phase-sensitive amplifier amplifies the A quadrature (we again assume $\omega_m = \omega_r$ for simplicity). Then we do not get any information on the qubit z coordinate, and therefore there is no "spooky" backaction, but there is the "realistic" backaction due to the fluctuating number of photons. The description of evolution in this case is essentially the standard one [6, 7]. Let us still denote by $I(t)$ the output signal from the mixer, corresponding to the amplified quadrature. For measurement during time t, we measure the A quadrature with imprecision $\sigma_{\rm gr}/\sqrt{\kappa t}$. This is consistent with the fluctuation of the number N of emitted photons: $\text{var}(N) = \bar{N}$, $\bar{N} = \bar{n}\kappa t$. The correlation function of the photon number in the resonator depends on time as $\exp(-\kappa t/2)$ [6, 7], which means that each extra photon inferred from the $I(t)$ fluctuation spends (on average) a time $2/\kappa$ in the resonator and therefore changes the qubit phase ϕ by $4\chi/\kappa$ (the correlation time $2/\kappa$ is essentially the lifetime of the field, not power [6, 7]). Then the ϕ variance is $\text{var}(\phi) = (4\chi/\kappa)^2\bar{n}\kappa t$ and the corresponding ensemble dephasing is $\text{var}(\phi)/2t = 8\chi^2\bar{n}/\kappa$. As expected, this reproduces the standard result from Eq. (17.15) for the ensemble dephasing, while for individual qubit evolution we have the correlation discussed above: each additional photon inferred from the $I(t)$ fluctuation changes ϕ by $4\chi/\kappa$. For the same amplifier gain and noise as for measuring the B quadrature, we get $\delta\sqrt{\bar{n}} = (4\chi/\kappa)\sqrt{\bar{n}}\,(\delta I_m)/\Delta I_{\max}$, and therefore the correlation is $K = \delta\langle\phi\rangle/\delta(\bar{I}_m t) = 32(\chi^2/\kappa)\,\bar{n}/\Delta I_{\max} = \Delta I_{\max}/S_{\min}$. It is easy to check that $K^2 S_{\min}/4$ (see Eq. (17.9)) coincides with the ensemble dephasing Γ from Eq. (17.15), as expected for the presence of only the "realistic" backaction.

Finally, assume that the phase-sensitive amplifier amplifies the quadrature, which makes an angle φ with the optimal B quadrature and an angle $\pi/2 - \varphi$ with the A quadrature. The measured signal $I(t)$ still denotes the output of the IQ mixer, corresponding to the amplified quadrature; now it gives information about both the B and A quadratures, with factors $\cos\varphi$ and $\sin\varphi$, respectively. Combining the "spooky" and "realistic" backactions, we get the same formulas as for the broadband detection of Section 17.2:

$$\frac{\rho_{11}(t)}{\rho_{00}(t)} = \frac{\rho_{11}(0)}{\rho_{00}(0)} \exp\left[\frac{\tilde{I}_m(t)\Delta I}{D}\right], \quad D = \frac{S}{2t}, \tag{17.18}$$

$$\frac{\rho_{01}(t)}{\rho_{01}(0)} = \frac{\sqrt{\rho_{00}(t)\,\rho_{11}(t)}}{\sqrt{\rho_{00}(0)\,\rho_{11}(0)}} \exp[iK\tilde{I}_m(t)\,t], \tag{17.19}$$

$$\Delta I = I_1 - I_0 = \Delta I_{\max}\cos\varphi, \quad K = K_{\max}\sin\varphi, \tag{17.20}$$

$$K_{\max} = \Delta I_{\max}/S, \tag{17.21}$$

$$\tilde{I}_m(t) = \frac{1}{t}\int_0^\infty I(t')\,dt' - \frac{I_0 + I_1}{2}, \tag{17.22}$$

$$I(t) = \frac{I_0 + I_1}{2} + \frac{\Delta I}{2} z(t) + \xi(t), \;\; S_\xi(\omega) = S. \tag{17.23}$$

Here we have introduced $\tilde{I}_m$ by subtracting the constant $(I_0 + I_1)/2$ from $\overline{I}_m$ and have performed some simple algebra to convert Eq. (17.4) into Eq. (17.18); the qubit rotating frame corresponds to $\bar{n}$ photons, $K_{\rm max}$ is the correlation for A-quadrature amplification discussed above, $\Delta I_{\rm max}$ is the response for B-quadrature amplification, and $S = S_{\rm min}$. Notice that the total ensemble dephasing in Eq. (17.9) does not depend on φ:

$$(\Delta I_{\rm max}\cos\varphi)^2/4S_{\rm min} + (K_{\rm max}\sin\varphi)^2 S_{\rm min}/4 = \Gamma. \tag{17.24}$$

So far, we have discussed only the ideal case. There are several mechanisms for nonideality. First, the qubit may have an additional environmental dephasing $\gamma_{\rm env}$. This will lead to an extra factor $e^{-\gamma_{\rm env}t}$ in Eq. (17.19) and increase the ensemble dephasing Γ by $\gamma_{\rm env}$. Following the definitions in Section 17.2, the corresponding quantum efficiency is $\eta_{\rm env} = (1 + \gamma_{\rm env}\kappa/8\chi^2\bar{n})^{-1}$. Second, not all microwave power leaking from the resonator may be collected and amplified. This can be characterized by the collection efficiency $\eta_{\rm col} = \kappa_{\rm col}/\kappa$ and multiplies the response ΔI and correlation K by the factor $\sqrt{\eta_{\rm col}}$, while not affecting the output noise S. Third, if the phase-preserving amplifier is not quantum-limited, it introduces additional noise $S_{\rm add}$ compared with the quantum limit $S_{\rm min}$ (given by Eq. (17.17) when $\eta_{\rm col} = 1$). The corresponding amplifier efficiency is $\eta_{\rm amp} = S_{\rm min}/(S_{\rm min} + S_{\rm add})$. This does not affect ΔI, but multiplies K by $\eta_{\rm amp}$ (because for uncorrelated Gaussian-distributed random numbers x_1 and x_2, averaging of x_1 for a fixed sum $x_1 + x_2$ gives a correlation $\langle x_1\rangle/(x_1 + x_2) = {\rm var}(x_1)/[{\rm var}(x_1) + {\rm var}(x_2)]$).

If all three mechanisms of nonideality are present, then the evolution can still be described [41] by Eqs. (17.18)–(17.23), but S is now the total (experimental) output noise, $\Delta I_{\rm max}$ is the experimental response for $\varphi = 0$, so that $S = (\Delta I_{\rm max})^2\kappa/(32\chi^2\bar{n}\,\eta_{\rm col}\,\eta_{\rm amp})$, the correlation $K = \delta\langle\phi\rangle/\delta(\tilde{I}_m t)$ is still given by Eqs. (17.20) and (17.21), and the only change is the extra factor in Eq. (17.19):

$$\frac{\rho_{01}(t)}{\rho_{01}(0)} = \frac{\sqrt{\rho_{00}(t)\,\rho_{11}(t)}}{\sqrt{\rho_{00}(0)\,\rho_{11}(0)}}\exp[iK\tilde{I}_m(t)\,t]\,e^{-\gamma t}, \tag{17.25}$$

$$\gamma = \Gamma - (\Delta I_{\rm max})^2/4S, \tag{17.26}$$

where the ensemble dephasing is now $\Gamma = 8\chi^2\bar{\eta}/\kappa + \gamma_{\rm env}$. We emphasize that the qubit evolution depends only on the experimentally measurable parameters $\Delta I_{\rm max}$, S, Γ, φ, and the output signal $I(t)$.

The quantum efficiencies can be expressed as

$$\eta = \frac{(\Delta I_{\rm max})^2}{4S\Gamma} = \eta_{\rm amp}\,\eta_{\rm col}\,\eta_{\rm env}, \;\; \tilde{\eta} = \eta\cos^2\varphi, \tag{17.27}$$

where, as in Section 17.2, η is the relative contribution to Γ from both the "spooky" and "realistic" backactions, while $\tilde{\eta}$ is the relative contribution from only the "spooky" backaction (see also [45]). The definition of $\tilde{\eta}$ corresponds to replacing the "realistic" backaction factor $\exp(iK\tilde{I}_m t)$ in Eq. (17.25) with the corresponding ensemble dephasing $\exp(-K^2St/4)$. As mentioned in Section 17.2, the peak-to-pedestal ratio of the spectral peak of continuous Rabi oscillations is $4\tilde{\eta} = 4\eta\cos^2\varphi$.

17.4.2 Phase-preserving amplifier

Now assume that a phase-preserving amplifier (Fig. 17.4b) is used (this includes parametric amplifiers, HEMTs, etc.). Now both the A and B quadratures of Eq. (17.16) are amplified independently with the same gain. Correspondingly, both quadratures at the IQ mixer output carry physical information, instead of only one quadrature in the case of a phase-sensitive amplifier. Let us denote by $I(t)$ the output of the IQ mixer, corresponding to the B quadrature; thus, $I(t)$ provides information on the qubit z coordinate. The output signal for the orthogonal quadrature is denoted by $Q(t)$; it corresponds to the A quadrature in the resonator and provides information on the fluctuating number of photons. The main difference from the case of a phase-sensitive amplifier is that now the "spooky" and "realistic" backactions are related to two different output signals: $I(t)$ and $Q(t)$.

Let us start with the quantum-limited case and assume an amplifier with the same gain as in the phase-sensitive case, so that the $I(t)$-channel response is the same as the optimal phase-sensitive response: $\Delta I = \Delta I_{\max}$. The "spooky" backaction is always described by the quantum Bayes formulas (17.4)–(17.5), but now the noise S of the output $I(t)$ is twice as large as the value in Eq. (17.17) for the phase-sensitive amplifier, $S = 2S_{\min}$ (see the discussion in Section 17.3). Therefore, the "spooky" evolution is twice as slow as in the phase-sensitive case with $\varphi = 0$. The signal $Q(t)$ has the same noise $S = 2S_{\min}$. It is again twice as large as for the phase-sensitive case with $\varphi = \pi/2$, and therefore, the correlation factor $K = \delta\langle\phi\rangle/\delta[\int_0^t Q(t')\,dt']$ for the "realistic" backaction is twice as small: $K = K_{\max}/2$ (this reduction is similar to the effect of a nonideal amplifier discussed above). We see that $K = \Delta I/S$, and the ensemble dephasing is at least $(\Delta I)^2/4S + K^2S/4 = (\Delta I)^2/2S = (\Delta I_{\max})/4S_{\min}$. This again coincides with $\Gamma = 8\chi^2\bar{n}/\kappa$, and therefore there can be no additional evolution of the qubit besides these "spooky" and "realistic" backactions.

Thus, in the ideal case, the qubit evolution is

$$\frac{\rho_{11}(t)}{\rho_{00}(t)} = \frac{\rho_{11}(0)}{\rho_{00}(0)} \exp[\tilde{I}_m(t)\Delta I/D], \quad D = \frac{S}{2t}, \tag{17.28}$$

$$\frac{\rho_{01}(t)}{\rho_{01}(0)} = \frac{\sqrt{\rho_{00}(t)\,\rho_{11}(t)}}{\sqrt{\rho_{00}(0)\,\rho_{11}(0)}} \exp[iK\tilde{Q}_m(t)\,t], \tag{17.29}$$

$$\Delta I = I_1 - I_0, \quad K = \Delta I/S, \tag{17.30}$$

$$\tilde{Q}_m(t) = \frac{1}{t}\int_0^\infty Q(t')\,dt' - \langle Q\rangle, \quad S_Q = S_I = S, \tag{17.31}$$

where $\langle Q\rangle$ is the average value of $Q(t)$ (which depends on $\bar{n}$), $\tilde{I}_m(t)$ is defined by Eq. (17.22), and the channels $I(t)$ and $Q(t)$ both have the same (uncorrelated) noise $S = (\Delta I)^2\kappa/(16\chi^2\bar{n})$. Notice that $(\Delta I)^2/4S = K^2S/4 = 4\chi^2\bar{n}/\kappa$, and therefore, *in the phase-preserving case, the ensemble dephasing* Γ *contains equal contributions* $\Gamma/2$ *from the "spooky" and "realistic" backactions.* We emphasize that Eqs. (17.28) and (17.29) still *allow us to monitor a qubit wavefunction* if the initial qubit state is pure.

A nonideal case can be analyzed in the same way as for the phase-sensitive amplifier. An extra dephasing $\gamma_{\rm env}$ of the qubit is described by $\eta_{\rm env} = (1+\gamma_{\rm env}\kappa/8\chi^2\bar{n})^{-1}$, imperfect collection efficiency is described by $\eta_{\rm col} = \kappa_{\rm col}/\kappa$, and the amplifier efficiency is $\eta_{\rm amp}$. We define $\eta_{\rm amp} = S_{\rm ql}/S$ for a phase-preserving amplifier by comparing its output noise S (per quadrature) with the quantum limit for a phase-preserving amplifier, $S_{\rm ql} = 2S_{\rm min}$, so that $\eta_{\rm amp} = 1$ in the quantum-limited case. We also define $\tilde{\eta} = S_{\rm min}/S$ by comparison with a phase-sensitive amplifier having the same gain, so that $\tilde{\eta}_{\rm amp} = \eta_{\rm amp}/2$ and obviously $\tilde{\eta}_{\rm amp} \leq 1/2$. Similarly to the phase-preserving case, incomplete microwave collection multiplies the response ΔI and correlation K by a factor $\sqrt{\eta_{\rm col}}$ but does not change the noise S; the extra noise in the amplifier multiplies K by $\eta_{\rm amp}$ but does not change ΔI.

The qubit evolution can still be described by Eqs. (17.28)–(17.31), with the only change being in Eq. (17.29):

$$\frac{\rho_{01}(t)}{\rho_{01}(0)} = \frac{\sqrt{\rho_{00}(t)\,\rho_{11}(t)}}{\sqrt{\rho_{00}(0)\,\rho_{11}(0)}}\,\exp[iK\tilde{Q}_m(t)\,t]\,e^{-\gamma t}, \tag{17.32}$$

$$\gamma = \Gamma - 2(\Delta I)^2/4S, \tag{17.33}$$

where now S is the total (experimental) noise per quadrature, ΔI is the experimental response, and $\Gamma = 8\chi^2\bar{\eta}/\kappa + \gamma_{\rm env}$ is the total ensemble dephasing. The qubit evolution is determined by the parameters ΔI, S, Γ, and output signals $I(t)$ and $Q(t)$.

The quantum efficiencies are

$$\eta = \eta_{\rm amp}\eta_{\rm col}\eta_{\rm env} = (\Delta I)^2/2S\Gamma, \quad \tilde{\eta} = \eta/2. \tag{17.34}$$

Here, η describes the fraction of Γ due to the contribution from both the "spooky" and "realistic" backactions. The efficiency $\tilde{\eta} = \tilde{\eta}_{\rm amp}\eta_{\rm col}\eta_{\rm env}$ describes the fraction from only the "spooky" contribution; it corresponds to replacing the term $\exp(iK\tilde{Q}_m\,t)$ in Eq. (17.32) with the dephasing term $\exp(-K^2t/4S)$. In particular, the peak-to-pedestal ratio of the Rabi spectral peak for the signal $I(t)$ is $4\tilde{\eta} = 2\eta$.

Let us mention that Eqs. (17.28)–(17.30) for the ideal phase-preserving case can also be obtained from Eqs. (17.18)–(17.21) for the phase-sensitive case in the following way. Let us think about a phase-preserving amplifier as a phase-sensitive amplifier, in which the angle φ changes rapidly with time, and we have to average over φ. When the coefficients $\cos\varphi$ and $\sin\varphi$ in Eq. (17.20) are substituted into Eqs. (17.18) and (17.19), we see a natural formation of the quadratures $\tilde{I}_m$ and $\tilde{Q}_m$ of the phase-preserving setup. Then the exponential factor in Eq. (17.18) becomes $\exp(\tilde{I}_m\Delta I_{\rm max}/D)$, and the exponential factor in Eq. (17.19) becomes $\exp(iK_{\rm max}\tilde{Q}_m t)$. Now let us take into account that the average response is $\Delta I = \overline{\cos^2\varphi}\,\Delta I_{\rm max} = \Delta I_{\rm max}/2$, and the phase-sensitive amplifier noise S splits equally between the $I(t)$ and $Q(t)$ quadratures (the orthogonal, de-amplified quadrature is noiseless). The mutual cancellation of these two factors of 2 leads to the same form of Eq. (17.28) as in Eq. (17.18) and the relation $K = \Delta I/S$ in Eq. (17.30).

One more way to understand the relation between the ideal phase-sensitive and phase-preserving cases is the following. Instead of using a phase-preserving amplifier, let us split the outgoing microwave into two equal parts and use phase-sensitive amplifiers with $\varphi = 0$ and $\varphi = \pi/2$ in the two channels. To keep the same noise S per channel, we increase the gain by a factor $\sqrt{2}$, which also compensates the signal loss at the splitter. Then the channel $\varphi = 0$ produces the "spooky" backaction in Eq. (17.28), while the channel $\varphi = \pi/2$ informs us of the "realistic" backaction in Eq. (17.29), and the relation (17.30) between K and ΔI is the same as between K_{max} and ΔI_{max} in Eq. (17.21).

In the ideal case ($\eta = 1$), the qubit evolution description can be translated into the language of the POVM-type measurement. In the same way as in Section 17.2, Eqs. (17.28) and (17.29) can be converted into the measurement operator

$$M(\overline{I}_m, \tilde{Q}_m) = \exp(-iK\tilde{Q}_m t\, \sigma_z/2) \times \left[\sqrt{P_{|0\rangle}(\overline{I}_m)}\, |0\rangle\langle 0| + \sqrt{P_{|1\rangle}(\overline{I}_m)}\, |1\rangle\langle 1|\right], \quad (17.35)$$

where the probabilities $P_{|j\rangle}$ are given by Eq. (17.3). Similarly, Eqs. (17.18) and (17.19) for the case of a phase-sensitive amplifier can be converted into the same measurement operator (17.35), in which $\tilde{Q}_m$ is replaced with $\tilde{I}_m$.

17.5 Conclusion

We have presented a simple physical picture of qubit evolution due to its measurement in the circuit QED setup. The "spooky" backaction is universal, it is caused by gradual extraction of information about the qubit state. The "realistic" backaction is due to a specific mechanism: fluctuation of the photon number in the resonator. For a phase-sensitive amplifier, the qubit evolution is described by Eqs. (17.18) and (17.25); it is determined by the output signal $I(t)$, which corresponds to the amplified quadrature. For a phase-preserving amplifier, the evolution is described by Eqs. (17.28) and (17.32); it is determined by two output signals $I(t)$ and $Q(t)$, where $I(t)$ now corresponds to the quadrature, which provides information about the qubit state (B quadrature in the resonator) and $Q(t)$ corresponds to the orthogonal A quadrature, which gives us a record of the photon number fluctuations in the resonator.

While the circuit QED setup differs significantly from both the broadband quantum measurement setup [30] and the standard optical setup [10] we see that the description of qubit evolution is exactly the same as in both these cases if a phase-sensitive amplifier is used. The description is only slightly different when a phase-preserving amplifier is used: we should assign the "spooky" and "realistic" backactions to the separate output signals $I(t)$ and $Q(t)$ instead of only one signal. It is also useful to think about the phase-preserving case via the model in which we split the outgoing microwave (the quantum signal) into two equal parts and then use 90°-shifted phase-sensitive amplifiers for these two channels.

We have intentionally considered only the simplest case, because most of the further steps and generalizations are quite straightforward [30]. In particular, it it very simple to include Rabi oscillations and energy relaxation of the qubit state. For that, we have to take the time derivatives of the evolution equations and add the terms due to Rabi oscillations and energy relaxation. If the Stratonovich definition of the derivative is used, we get the equations of the Bayesian formalism [30]; if the Itô derivative is used, we get the equations of the quantum trajectory formalism [9, 10]. Generalization to measurement of several entangled qubits is also straightforward [30]. We have considered only the resonant case, $\omega_m = \omega_r$; however, generalization to the case $\omega_m \neq \omega_r$ is quite simple [9]: we just need a different definition of the informational B quadrature and photon-fluctuation A quadrature. In our formalism, we have implicitly assumed sufficiently wide bandwidth of the amplifier (much larger than the ensemble dephasing Γ and Rabi frequency Ω_R). If this is not the case, the formalism should change significantly. However, we believe that, in most practical cases, we can take this effect into account by adding a classical narrowband filter to the classical signal at the amplifier output; this will correspond to passing the signals $I(t)$ and $Q(t)$ through low-pass filters. A much more serious change in the theory is required when the resonator bandwidth κ is comparable to Γ or Ω_R; this still has to be done.

Understanding the difference between the "spooky" and "realistic" backactions is important for designing quantum feedback control of Rabi oscillations [13–15]. The simplest case is when a phase-sensitive amplifier amplifies the informational B quadrature. Then there is no "realistic" backaction, and the feedback loop should only modulate the amplitude of the Rabi drive (i.e., the Rabi frequency Ω_R); this case has been well studied for the broadband setup [13–15]. The situation is different for a phase-preserving amplifier. Then, we need *two feedback channels*: the first (usual) channel should modulate the Rabi frequency Ω_R to compensate the "spooky" backaction, while the second channel should compensate the "realistic" backaction by modulating the qubit frequency $\omega_{\rm qb}$ or the frequency of the Rabi drive, ω_R. The controller for the second feedback channel is quite simple: it should compensate the contribution $iK\tilde{Q}(t)$ to the qubit phase derivative $\dot{\phi}(t)$ due to the K term in Eq. (17.32). Therefore, the controller is

$$\Delta(\omega_{\rm qb} - \omega_R) = -K[Q(t) - \langle Q \rangle]; \tag{17.36}$$

that is, we should apply the signal $Q(t)$ directly to modulate $\omega_{\rm qb}$ or ω_R. The second feedback channel essentially eliminates the K term in Eq. (17.32) and decreases the ensemble dephasing Γ by $K^2S/4 = (\Delta I)^2/4S$. Correspondingly, in the absence of the first (main) feedback channel, the peak-to-pedestal ratio of the Rabi peak increases from 2 to 4 in the quantum-limited case. The first feedback channel should be the same as for the broadband setup; it depends on the signal $I(t)$ and can be realized using various ideas for the controller ("direct," Bayesian, "simple," etc. [13–15]). Note that *without the second channel the feedback performance is determined by the quantum efficiency* $\tilde{\eta}$, *while with the second channel it is determined by* $\tilde{\tilde{\eta}} = \tilde{\eta}/(1-\eta+\tilde{\eta})$ (this is one more combination of the terms in Eq. (17.9), which can be used for the definition [30] of quantum efficiency). The case of a phase-sensitive amplifier, which amplifies a non-optimal quadrature ($\varphi \neq 0$, $\varphi \neq \pi/2$) is similar to the case of a phase-preserving amplifier, but both feedback channels should start with the same signal $I(t)$. *In both the phase-sensitive and phase-preserving setups, perfect feedback control is possible in the quantum-limited case* $\eta = 1$.

Discussion of the "spooky" and "realistic" backactions in the circuit QED setup necessarily raises the question of causality. When the microwave leaves the resonator, it does not yet "know" in which way it will be measured (phase-preserving or phase-sensitive, which angle φ, etc.). Moreover, when a circulator is used for the outgoing microwave, the field in the resonator and the qubit can never "know" in a realistic way which method of measurement is used. Nevertheless, the qubit evolution strongly depends on the measurement method. As we have discussed, the "spooky" evolution moves the qubit state along the meridians of the Bloch sphere, the "realistic" backaction moves the state along the parallels, and the measurement method determines whether the qubit experiences the "spooky" or the "realistic" backaction (or their combination). In this sense, the "realistic" backaction is not fully realistic: it has a physical mechanism, but whether this mechanism works or not is determined in a spooky way. Causality requires that we cannot pass "useful" information to the qubit by choosing the measurement method. This means that *the ensemble-averaged evolution of the qubit cannot depend on the measurement method (this is the general requirement of causality in quantum mechanics).* It is surely satisfied in our circuit QED setup.

Acknowledgments

The author thanks Michel Devoret, Konstantin Likharev, Patrice Bertet, and Farid Khalili for useful discussions. This work was supported by ARO MURI grant W911NF-11-1-0268 and by NSA/IARPA/ARO grant W911NF-10-1-0334.

References

[1] Wallraff, A., Schuster, D. I., Blais, A., Frunzio, L., Huang, P. S., Majer, J., Kumar, S., Girvin, S. M., and Schoelkopf, R. J. (2004). *Nature*, **431**, 162.

[2] DiCarlo, L., Reed, M. D., Sun, L., Johnson, B. R., Chow, J. M., Gambetta, J. M., Frunzio, L., Girvin, S. M., Devoret, M. H., and Schoelkopf, R. J. (2010). *Nature*, **467**, 574.

[3] Fragner, A., Goppl, M., Fink, J. M., Baur, M., Bianchetti, R., Leek, P. J., Blais, A., and Wallraff, A. (2008). *Science*, **322**, 1357.

[4] Palacios-Laloy, A., Mallet, F., Nguyen, F., Bertet, P., Vion, D., Esteve, D., and Korotkov, A. N. (2010). *Nature Phys.*, **6**, 442.

[5] Vijay, R., Slichter, D. H., and Siddiqi, I. (2011). *Phys. Rev. Lett.*, **106**, 110502.

[6] Blais, A., Huang, R. S., Wallraff, A., Girvin, S. M., and Schoelkopf, R. J. (2004). *Phys. Rev. A*, **69**, 062320.

[7] Clerk, A. A., Devoret, M. H., Girvin, S. M., Marquardt, F., and Schoelkopf, R. J. (2010). *Rev. Mod. Phys.*, **82**, 1155.

[8] Gambetta, J., Blais, A., Schuster, D. I., Wallraff, A., Frunzio, L., Majer, J., Devoret, M. H., Girvin, S. M., and Schoelkopf, R. J. (2006). *Phys. Rev. A*, **74**, 042318.
[9] Gambetta, J., Blais, A., Boissonneault, M., Houck, A. A., Schuster, D. I., and Girvin, S. M. (2008). *Phys. Rev. A*, **77**, 012112.
[10] Wiseman, H. M. and Milburn, G. J. (1993). *Phys. Rev. Lett.*, **70**, 548.
[11] Hofmann, H. F., Mahler, G., and Hess, O. (1998). *Phys. Rev. A*, **57**, 4877.
[12] Wang, J. and Wiseman, H. M. (2001). *Phys. Rev. A*, **64**, 063810.
[13] Ruskov, R. and Korotkov, A. N. (2002). *Phys. Rev. B*, **66**, 041401(R).
[14] Zhang, Q., Ruskov, R., and Korotkov, A.N. (2005). *Phys. Rev. B*, **72**, 245322.
[15] Korotkov, A. N. (2005). *Phys. Rev. B*, **71**, 201305(R).
[16] Braginsky, V. B. and Khalili, F. Ya. (1992). *Quantum Measurement.* Cambridge University Press.
[17] Bergeal, A., Schackert, F., Metcalfe, M., Vijay, R., Manucharyan, V. E., Frunzio, L., Prober, D. E., Schoelkopf, R. J., Girvin, S. M. and Devoret, M. H. (2010). *Nature*, **465**, 64.
[18] Castellanos-Beltran, M. A., Irwin, K. D., Hilton, G. C., Vale, L. R., and Lehnert, K. W. (2008). *Nature Phys.*, **4**, 929.
[19] Korotkov, A. N. (1999). *Phys. Rev. B* **60**, 5737.
[20] Korotkov, A. N. (2001). *Phys. Rev. B*, **63**, 115403.
[21] Goan, H. S., Milburn, G. J., Wiseman, H. M., and Sun, H. B. (2001). *Phys. Rev. B*, **63**, 125326.
[22] Goan, H. S. and Milburn, G. J. (2001). *Phys. Rev. B*, **64**, 235307.
[23] Wiseman, H. M. and Milburn, G. J. (1993). *Phys. Rev. A*, **47**, 642.
[24] Carmichael, H. J. (1993). *An Open System Approach to Quantum Optics.* Springer, Berlin.
[25] Wiseman, H. M. and Milburn, G. J. (2009). *Quantum Measurement and Control.* Cambridge University Press.
[26] Haus, H. A. and Mullen, J. A. (1962). *Phys. Rev.*, **128**, 2407.
[27] Giffard, R. P. (1976). *Phys. Rev. D*, **14**, 2478.
[28] Caves, C. M. (1982). *Phys. Rev. D*, **26**, 1817.
[29] Devyatov, I. A., Kuzmin, L. S., Likharev, K. K., Migulin, V. V., and Zorin, A. B. (1986). *J. Appl. Phys.*, **60**, 1808.
[30] Korotkov, A. N. (2003). In *Quantum Noise in Mesoscopic Physics* (ed. Yu. Nazarov), p. 205. Kluwer, Dordrecht.
[31] Einstein, A., Podolsky, B., and Rosen, N. (1935). *Phys. Rev.*, **47**, 0777.
[32] Aspect, A., Dalibard, J., and Roger, G. (1982). *Phys. Rev. Lett.*, **49**, 1804.
[33] Caves, C. M. (1986). *Phys. Rev. D*, **33**, 1643.
[34] Gardiner, C. W. and Zoller, P. (2004). *Quantum Noise.* Springer-Verlag, Berlin.
[35] Nielsen, M. A. and Chuang, I. L. (2000). *Quantum Computation and Quantum Information.* Cambridge University Press, Cambridge.
[36] Kraus, K. (1983). *States, Effects, and Operations.* Springer-Verlag, Berlin.
[37] Jordan, A. N. and Korotkov, A. N. (2006). *Phys. Rev. B*, **74**, 085307.
[38] Wei, H. and Nazarov, Yu. V. (2008). *Phys. Rev. B*, **78**, 045308.
[39] Korotkov, A. N. (2001). *Phys. Rev. B*, **63**, 085312.

[40] Korotkov, A. N. and Averin, D. V. (2001). *Phys. Rev. B*, **64**, 165310.
[41] Korotkov, A. N. (2003). *Phys. Rev. B*, **67**, 235408.
[42] Ruskov, R., Schwab, K., and Korotkov, A. N. (2005). *Phys. Rev. B*, **71**, 235407.
[43] Reed, M. D., DiCarlo, L., Johnson, B. R., Sun, L., Schuster, D. I., Frunzio, L., and Schoelkopf, R. J. (2010). *Phys. Rev. Lett.*, **105**, 17360.
[44] Walls, D. F. and Milburn, G. J. (2008). *Quantum Optics*. Springer-Verlag, Berlin.
[45] Averin, D. V. (2003). In *Quantum Noise in Mesoscopic Physics* (ed. Yu. Nazarov), p. 229. Kluwer, Dordrecht.

18
Superconducting quantum circuits: artificial atoms coupled to 1D modes

Y. NAKAMURA

Research Center for Advanced Science and Technology (RCAST)
The University of Tokyo, 4-6-1 Komaba, Meguro-ku,
Tokyo 153-8904, Japan

Quantum Machines: Measurement and Control of Engineered Quantum Systems.
Edited by M. Devoret, B. Huard, R. Schoelkopf, and L. F. Cugliandolo.

Chapter Contents

18.1 Introduction

Although we are usually not aware of it, according to quantum field theory, we are living in a deep sea of quantum fluctuations [1]. We do not recognize it partly because it is hidden by the dominant thermal fluctuations and, more fundamentally, because we cannot perceive vacuum fluctuations (or can we?).

Even if we forget about parallel universes or hidden higher dimensions, the omnipresent quantum modes in our three-dimensional (3D) world (plus an additional dimension of time) could potentially be a plentiful resource for quantum information technologies such as quantum computing and communication [2]. However, there is an issue here in that too many degrees of freedom are always present, far beyond our practical controllability. Therefore, quantum engineering typically starts with looking at how to sculpture the vast quantum vacuum into a tractable subspace in which we can master all the relevant modes. The rest of the universe then provides tools for controlling these.

Atoms have historically been leading examples of such quantum subspaces. The discretized eigenmodes of atomic excitations are well localized in space and isolated from the continuum of excitations in the surrounding electromagnetic field (despite finite corrections due to vacuum fluctuations, i.e., the Lamb shift). On the other hand, one can also separate discrete electromagnetic modes from the continuum of the vacuum. For example, a Fabry-Pérot cavity supports standing-wave modes between mirrors. Although those modes have a finite volume, here we consider them zero-dimensional (0D) as they are confined on a cavity. One of the main research topics on cavity quantum electrodynamics (QED) and circuit QED has been coherent interaction between an atom and a 0D cavity mode in the strong-coupling regime with little damping and decoherence [3, 4].

18.2 0D and 1D, not more

Indeed, we are not necessarily forced to use such 0D electromagnetic modes for coherent interaction between atoms and photons. For example, an atom in the excited state spontaneously emits a photon in a 3D space, into the continuum of modes: this is also a coherent process as a whole. However, it is obvious that keeping track of the photon wavepacket emitted into the 4π solid angle is not an easy task. No practical device can manipulate and detect such a photon efficiently. The situation is similar and is not greatly simplified in the 2D case. Therefore, the only possibility left is in 1D.

Naturally in optics, flying photons in 1D modes are commonly used for quantum information technologies. Those modes are defined both by lenses/mirrors in free space and by optical fibers/waveguides. They travel over long distances with small loss and carry quantum information either in their path, polarization, or phase degrees of freedom.

It is the demand for greater functionality in such networks of 1D modes, beyond those realized by linear optical components, that has attracted attention to combinations of atoms and 1D modes. Various physical systems in the optical domain have been investigated toward that goal. For example, single emitters such as an atom, a molecule, or a quantum dot are coupled with a focused light mode. However, it remains a big challenge to achieve strong enough coupling in those systems. Replacing the focused laser beam with more tightly confined 1D modes such as in optical nanofibers and waveguides or surface plasmon modes along metallic nanowires may improve the coupling.

On the other hand, combinations of superconducting qubits as artificial atoms and superconducting microwave transmission lines as 1D channels offer unique opportunities. Although they work in the microwave domain, four orders of magnitude below the optical one in energy scale, they are perfectly suitable for realizing concepts in quantum optics.

The setup has the following advantages.

(i) The superconducting artificial atoms are lithographically fabricated on chip and fixed at the desired locations. Their energy levels can be flexibly designed and dynamically tuned. One can often utilize more than two levels and selection rules in the transitions among them.
(ii) The size of such artificial atoms ranges from 1 μm to 1 mm, several orders of magnitude larger than natural atoms. As a result, the associated dipole moment, either electric or magnetic, is typically much larger than those of atoms. This readily enables very strong coupling to the electromagnetic modes.
(iii) The microwave modes in the superconducting transmission line are well confined in 1D and have little loss. Thanks to the one-dimensionality, the spatial mode matching between transmitting waves is simply achieved.

Even in 1D, the fact that an atom coupled to a transmission line emits photons in both directions, forward and backward, still makes it complicated to keep track of the flow of quantum information. This can be avoided by putting the atom in front of a mirror; for a microwave transmission line, it is either a short or an open end. The atom is now coupled to the end of a semi-infinite transmission line. Then, the output photon wavepacket only goes backward because of reflection. Finally, it is necessary to spatially separate the output and the input modes, which can be realized by using nonreciprocal devices such as circulators. At present, only bulky off-chip ferrite circulators are available. However, there are proposals for integrating nonreciprocal elements in on-chip Josephson-junction circuits, either passive [5] or active [6].

18.3 Strong coupling in 1D

A frequently quoted condition for the strong coupling in cavity QED is $g \gg \kappa, \gamma, \gamma_\varphi$, where g is the coupling strength between the atom and the 0D cavity mode, κ is the cavity linewidth, γ is the qubit decay rate into other channels, and γ_φ is the pure dephasing rate of the atom. What is the corresponding criterion that characterizes the coupling strength in the case of a transmission line? The condition to be fulfilled is $\Gamma_1 \gg \gamma, \gamma_\varphi$, where Γ_1 is the spontaneous emission rate of the atom into the 1D modes. This basically means that the atom has to emit the photon into the 1D modes, not elsewhere, and before it is dephased. In the bad-cavity regime in cavity QED, the atom decay rate into the cavity output modes is given by $\Gamma_1 = 4g^2/\kappa$, and the cooperativity is defined as $C = 4g^2/\kappa\gamma = \Gamma_1/\gamma$. Similarly, in the 1D case, the condition can be represented as $C = \Gamma_1/\gamma \gg 1$.

Such a strong-coupling effect between the artificial atom and the transmission line is clearly evidenced in experiments on Rayleigh scattering, where a two-level artificial atom is coupled to a transmission line and driven by a weak continuous microwave [7–9]. Nearly perfect (∼99%) extinction of the transmission has been observed at the resonance frequency. This is due to destructive interference between the transmitted field and the absorbed and re-emitted field. As a result, all the input power is reflected back by the single atom in the transmission line.

18.4 More functionalities with three-level atoms

Artificial atoms in superconducting circuits are not genuine two-level systems, but are effective ones that additionally have higher energy levels. One can use those levels for introducing more functionalities.

For example, a third level of a superconducting artificial atom coupled to a transmission line can be used as a single-atom microwave switch [9, 10]. The atom has a ladder (or Ξ) system, in which only transitions between neighboring energy levels are allowed. The direct dipole transition from the ground state to the second excited state is prohibited by the selection rule resulting from the symmetry of the eigenfunctions. When the upper transition is driven strongly, the transmission coefficient at the resonance frequency of the lower transition recovers from zero to unity, while the reflection coefficient decreases accordingly. This apparent electromagnetically induced transparency (EIT) can be interpreted as the result of Autler-Townes splitting of the first excited state. Because of the relatively large relaxation rate in the second excited state, the narrow transparency window due to the interference effect in EIT is not seen [11].

Another example is microwave amplification by stimulated emission of radiation with a single atom coupled to a transmission line (a maser, but without a cavity) [12], where a Δ system that allows all the transitions among the three lowest energy levels is used. The operation principle is very similar to that of the Er-doped fiber amplifiers commonly used in opto-communication networks. The second excited state is excited by the pump frequency and then relaxes to the first excited state, where the population inversion is generated. Stimulated emission and relaxation to the ground state are then induced by the input signal. Note that in this case there is only one three-level atom coupled to the transmission line.

A Λ system is yet another type of three-level system and has two stable ground states in which qubit information can be stored. An interesting property appears when the excited state has equal relaxation rates to each ground state and when the two relaxation paths are distinguishable, either by polarization selectivity or by frequency selectivity. For such an atom placed in front of a mirror, upon the input of a photon that couples to one of the transitions for which the ground state is prepared, the atom state is swapped and a photon in the other transition is emitted back. This is accounted for by the same interference argument as for the extinction of transmission by the Rayleigh scattering in 1D modes.

By using this property, we can implement a SWAP gate between the photon and the atom, and similarly a $\sqrt{\text{SWAP}}$ gate when the input photon frequency is detuned by a certain amount. Also, by combining a series of these gates in a train of three pulses reflected by the Λ-type atom, we can realize a photon-photon entangling gate just by the fully passive atomic node [13]. A similar idea can be applied to a Δ system whose top level has equal relaxation rates to the other two states. In this case, the system works as a unit-efficiency down-converter for the input photons exciting the ground state to the second excited state [14].

Unfortunately, such a symmetric Λ system that has stable ground states has been elusive in superconducting artificial atoms. Nevertheless, a "pseudo" Λ system realized in a current-biased Josephson junction is now used as the first single-microwave-photon detector [15–17]. The system consists of two quantized states in a well of the tilted cosine potential in a current-biased Josephson junction and a continuum of excitations to which the upper state in the well can escape across the bump of the cosine. By adjusting the tilt of the cosine potential with the bias current, the intra-well relaxation rate and the escape rate can be equalized. To obtain 100% efficiency of single-photon detection, the junction should be placed in front of a mirror and be prepared in the ground state in the well. Then, as discussed above for the quantum gate or the down-converter, the input photon on resonance with the intra-well transition is completely absorbed and the tunneling escape takes place. This can be detected as a switching event of the Josephson junction from the superconducting state to the voltage state, generating a click of the detector.

Again, interference is key for achieving the high-efficiency detection of a single microwave photon. This is also the case for all other phenomena discussed here and the key for the microwave quantum photonics in 1D-mode networks.

18.5 Conclusions

Cavity QED and circuit QED have been predominantly developed in atom(s)-in-a-cavity setups. We have discussed here how, in superconducting circuits, artificial atoms in a continuum of 1D modes also offer interesting opportunities for investigating their quantum optical effects. Thanks to their large dipole moments and the strong field confinement in the 1D modes, artificial atoms achieve, without the aid of a cavity or an ensemble, unprecedentedly strong coupling (large cooperativity) to the 1D modes in microwave transmission lines. This ensures that, upon the interaction between the atoms and the flying photons, quantum information does not leak out from the system.

For such superconducting circuits, the number of components available in the microwave quantum optics toolbox are growing rapidly. Although we could not discuss them here, there are developing technologies for on-demand single-photon sources [18, 19] and squeezed vacuum sources [20], as well as two-mode squeezing generators using a Josephson ring oscillator [21] and the dynamical Casimir effect [22]. On the measurement side, Josephson parametric amplifiers have become a standard tool for low-noise measurements of microwave signals [23], and homodyne approaches for analyze itinerant microwave field have been developed [24, 25]. We expect much progress in quantum engineering for handling itinerant microwave photons in the coming years.

Acknowledgments

The author would like to thank Kazuki Koshino, Oleg Astafiev, Abdufarrukh Abdumalikov, and the members of the RIKEN Macroscopic Quantum Coherence team for valuable discussions.

References

[1] Milonni, P. W. (1994). *The Quantum Vacuum*. Academic Press, San Diego.

[2] Lloyd, S. (2000). Ultimate physical limits to computation. *Science*, **406**, 1047.

[3] Haroche, S. and Raimond, J. M. (2006). *Exploring the Quantum: Atoms, Cavities, and Photons*. Oxford University Press, Oxford.

[4] Schoelkopf, R. J. and Girvin, S. M. (2008). Wiring up quantum systems. *Nature*, **451**, 664.

[5] Koch, J., Houck, A. A., Le Hur, K., and Girvin, S. M. (2010). Time-reversal-symmetry breaking in circuit-QED-based photon lattices. *Phys. Rev. A*, **82**, 043811.

[6] Kamal, A., Clarke, J., and Devoret, M. H. (2011). Noiseless non-reciprocity in a parametric active device. *Nature Phys.*, **7**, 311.

[7] Shen, J.-T. and Fan, S. (2005). Coherent single photon transport in a one-dimensional waveguide coupled with superconducting quantum bits. *Phys. Rev. Lett.*, **95**, 213001.

[8] Astafiev, O., Zagoskin, A. M., Abdumalikov, A. A. Jr., Pashkin, Yu. A., Yamamoto, T., Inomata, K., Nakamura, Y., Tsai, J. S. (2010). Resonance fluorescence of a single artificial atom. *Science*, **327**, 840.

[9] Hoi, I.-C., Wilson, C. M., Johansson, G., Palomaki, T., Peropadre, B., and Delsing, P. (2011). Demonstration of a single-photon router in the microwave regime. *Phys. Rev. Lett.*, **107**, 073601.

[10] Abdumalikov, A. A., Jr., Astafiev, O., Zagoskin, A. M., Pashkin, Yu. A., Nakamura, Y., and Tsai, J. S. (2010). Electromagnetically induced transparency on a single artificial atom. *Phys. Rev. Lett.*, **104**, 193601.

[11] Anisimov, P. M., Dowling, J. P., and Sanders, B. C. (2011). Objectively discerning Autler-Townes splitting from electromagnetically induced transparency. *Phys. Rev. Lett.*, **107**, 163604.

[12] Astafiev, O. V., Abdumalikov, A. A., Jr., Zagoskin, A. M., Pashkin, Yu. A., Nakamura, Y., and Tsai, J. S. (2010). Ultimate on-chip quantum amplifier. *Phys. Rev. Lett.*, **104**, 183603.

[13] Koshino, K., Ishizaka, S., and Nakamura, Y. (2010). Deterministic photon-photon $\sqrt{\mathrm{SWAP}}$ gate using a Λ-system. *Phys. Rev. A*, **82**, 010301(R).

[14] Koshino, K. (2009). Down-conversion of a single photon with unit efficiency. *Phys. Rev. A*, **79**, 013804.

[15] Chen, Y.-F., Hover, D., Sendelbach, S., Maurer, L., Merkel, S. T., Pritchett, E. J., Wilhelm, F. K., and McDermott, R. (2011). Microwave photon counter based on Josephson junctions. *Phys. Rev. Lett.*, **107**, 217401.

[16] Romero, G., García-Ripoll, J. J., and Solano, E. (2009). Microwave photon detector in circuit QED *Phys. Rev. Lett.*, **102**, 173602.

[17] Peropadre, B., Romero, G., Johansson, G., Wilson, C. M., Solano, E., and García-Ripoll, J. J. (2011). Approaching perfect microwave photodetection in circuit QED. *Phys. Rev. A*, **84**, 063834.

[18] Houck, A. A., Schuster, D. I., Gambetta, J. M., Schreier, J. A., Johnson, B. R., Chow, J. M., Frunzio, L., Majer, J., Devoret, M. H., Girvin, S. M., and Schoelkopf, R. J. (2007). Generating single microwave photons in a circuit. *Nature*, **449**, 328.

[19] Bozyigit, D., Lang, C., Steffen, L., Fink, J. M., Eichler, C., Baur, M., Bianchetti, R., Leek, P. J., Filipp, S., da Silva, M. P., Blais, A., and Wallraff, A. (2011). Antibunching of microwave frequency photons observed in correlation measurements using linear detectors. *Nature Phys.*, **7**, 154.

[20] Mallet, F., Castellanos-Beltran, M. A., Ku, H. S., Glancy, S., Knill, E., Irwin, K. D., Hilton, G. C., Vale, L. R., and Lehnert, K. W. (2011). Quantum state tomography of an itinerant squeezed microwave field. *Phys. Rev. Lett.*, **106**, 220502.

[21] Bergeal, N., Schackert, F., Frunzio, L., and Devoret, M. H. (2012). Two-mode correlation of microwave quantum noise generated by parametric down-conversion. *Phys. Rev. Lett.*, **108**, 123902.

[22] Wilson, C. M., Johansson, G., Pourkabirian, A., Simoen, M., Johansson, J. R., Duty, T., Nori, F., and Delsing, P. (2011). Observation of the dynamical Casimir effect in a superconducting circuit. *Nature*, **479**, 376.

[23] Castellanos-Beltran, M. A., Irwin, K. D., Hilton, G. C., Vale, L. R., and Lehnert, K. W. (2008). Amplification and squeezing of quantum noise with a tunable Josephson metamaterial. *Nature Phys.*, **4**, 928.

[24] da Silva, M. P., Bozyigit, D., Wallraff, A., and Blais, A. (2010). Schemes for the observation of photon correlation functions in circuit QED with linear detectors. *Phys. Rev. A* , **82**, 043804.

[25] Eichler, C., Bozyigit, D., Lang, C., Steffen, L., Fink, J. M., and Wallraff, A. (2011). Experimental tomographic state reconstruction of itinerant microwave photons. *Phys. Rev Lett.*, **106**, 220503.

19
A superconducting artificial atom with two internal degrees of freedom

O. Buisson

Institut Néel
CNRS, Université Joseph Fourier, INP-Grenoble, Grenoble, France

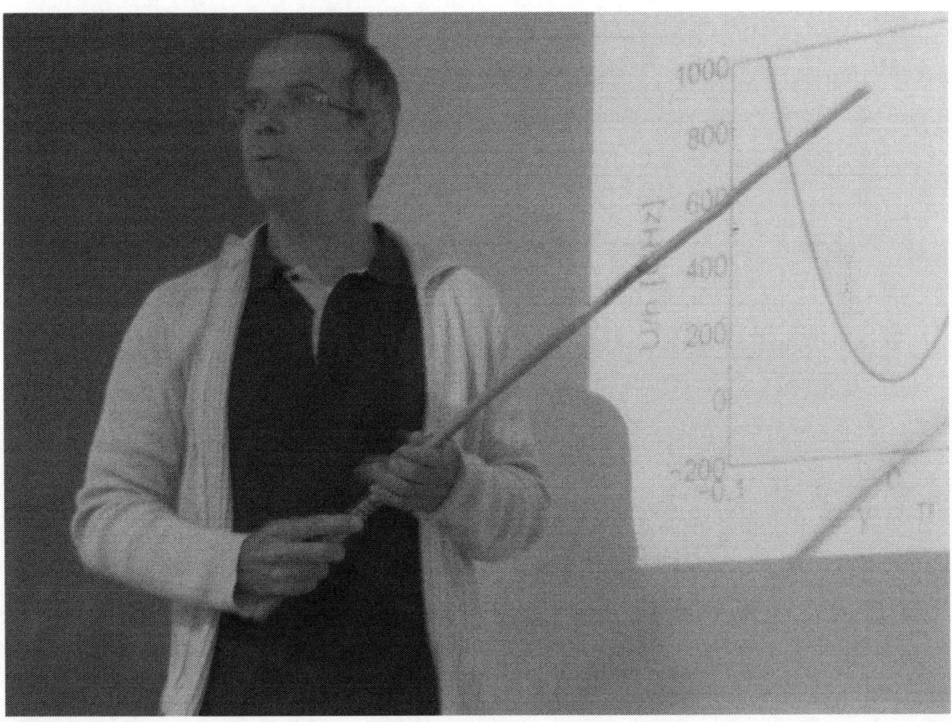

Quantum Machines: Measurement and Control of Engineered Quantum Systems.
Edited by M. Devoret, B. Huard, R. Schoelkopf, and L. F. Cugliandolo.

Chapter Contents

The richness of the energy-level structure and the diversity of the associated transitions are at the heart of atomic and quantum optics experiments. They result from the multiple degrees of freedom existing in these systems. In contrast, studies on superconducting artificial atoms are focused on devices with only one degree of freedom. In our recent work, by adding a large inductance to a dc-SQUID phase qubit loop, we decouple the junctions' dynamics and create a superconducting artificial atom with two internal degrees of freedom. This device can be described by two anharmonic oscillators labeled s (symmetric) and a (antisymmetric) because the current oscillations at each junction can be either in phase or in phase opposition, respectively. Spectroscopic measurements (Fig. 19.1(a)) show an anticrossing of two levels, which demonstrates a nonlinear coupling between the two modes s and a [1, 2]. We show independent coherent control of the two degrees of freedom and we use here one of the coupling terms to demonstrate a time-resolved up and down frequency conversion in the microwave domain (Fig. 19.1(b, c)). This device can be a new building block for superconducting electronics, providing substantial benefits with regard to nondestructive transmon readout [3] and parametric effects.

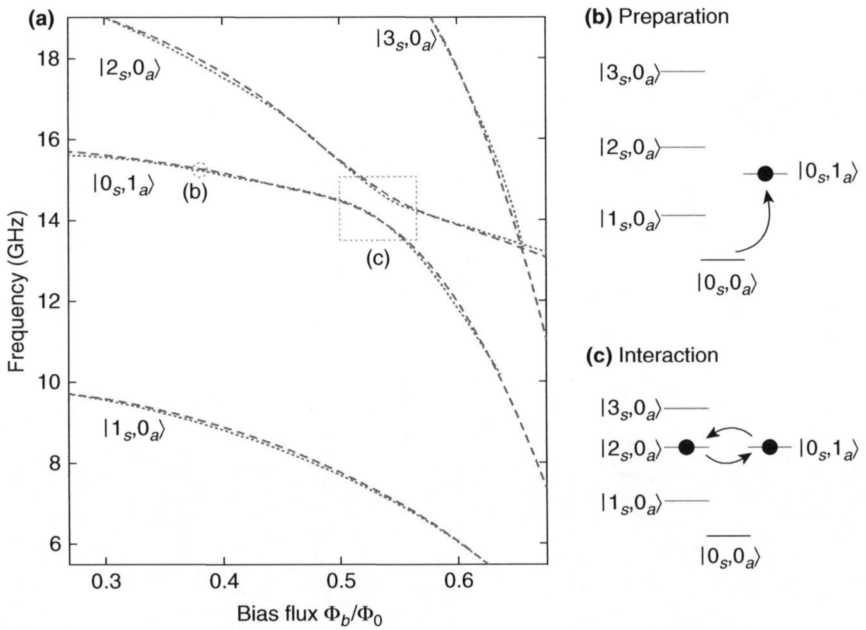

Fig. 19.1 (a) Energy spectrum versus applied magnetic flux (in units of the flux quantum Φ_0) measured by microwave spectroscopy of an artificial atom with two internal degrees of freedom. At $\Phi = 0.54\Phi_0$, the two characteristic frequencies of modes s and a are about 7.2 and 14.4 GHz, respectively, and we observe a level anticrossing. (b, c) Quantized energy spectra of the two anharmonic oscillators at $\Phi = 0.37\Phi_0$ and $\Phi = 0.54\Phi_0$. In (b), the system is prepared in the state $|0_s, 1_a\rangle$. In (c), the nonlinear coupling between the two oscillators leads to coherent oscillations between states $|0_s, 1_a\rangle$ and $|2_s, 0_a\rangle$, producing up and down frequency conversion in time domain.

References

[1] Lecocq, F., Pop, I. M., Matei, I., Dumur, E., Feofanov, A., Naud, C., Guichard, W., and Buisson, O. (2012). *Phys. Rev. Lett.*, **108**, 107001.
[2] Lecocq, F., Claudon, J., Buisson, O., and Milman, P. (2011). *Phys. Rev. Lett.*, **107**, 197002.
[3] Diniz, I., Dumur, E., Buisson, O., and Auffèves, A. (2013). *Phys. Rev. A,* **87**, 033837.

Printed and bound by CPI Group (UK) Ltd, Croydon, CR0 4YY